Lecture Notes in Electrical Engineering

Volume 723

The book series *Lecture Notes in Electrical Engineering* (LNEE) publishes the latest developments in Electrical Engineering - quickly, informally and in high quality. While original research reported in proceedings and monographs has traditionally formed the core of LNEE, we also encourage authors to submit books devoted to supporting student education and professional training in the various fields and applications areas of electrical engineering. The series cover classical and emerging topics concerning:

- Communication Engineering, Information Theory and Networks
- Electronics Engineering and Microelectronics
- Signal, Image and Speech Processing
- Wireless and Mobile Communication
- Circuits and Systems
- Energy Systems, Power Electronics and Electrical Machines
- Electro-optical Engineering
- Instrumentation Engineering
- Avionics Engineering
- Control Systems
- Internet-of-Things and Cybersecurity
- Biomedical Devices, MEMS and NEMS

For general information about this book series, comments or suggestions, please contact leontina.dicecco@springer.com.

To submit a proposal or request further information, please contact the Publishing Editor in your country:

China

Jasmine Dou, Editor (jasmine.dou@springer.com)

India, Japan, Rest of Asia

Swati Meherishi, Editorial Director (Swati.Meherishi@springer.com)

Southeast Asia, Australia, New Zealand

Ramesh Nath Premnath, Editor (ramesh.premnath@springernature.com)

USA, Canada:

Michael Luby, Senior Editor (michael.luby@springer.com)

All other Countries:

Leontina Di Cecco, Senior Editor (leontina.dicecco@springer.com)

**** This series is indexed by EI Compendex and Scopus databases. ****

More information about this series at http://www.springer.com/series/7818

Atif Iqbal · Hasmat Malik · Ahmed Riyaz ·
Kouzou Abdellah · Sertac Bayhan
Editors

Renewable Power for Sustainable Growth

Proceedings of International Conference on Renewal Power (ICRP 2020)

Springer

Editors
Atif Iqbal
Department of Electrical Engineering
Qatar University
Doha, Qatar

Ahmed Riyaz
Department of Electrical Engineering
BGSB University
Rajouri, Jammu and Kashmir, India

Sertac Bayhan
Energy Centre
Qatar Environment and Energy
Research Institute
Doha, Qatar

Hasmat Malik
ICE Division
NSUT Delhi
New Delhi, Delhi, India

BEARS
NUS Campus
University Town, Singapore

Kouzou Abdellah
Faculty of Science and Technology
University of Djelfa
Djelfa, Algeria

ISSN 1876-1100 ISSN 1876-1119 (electronic)
Lecture Notes in Electrical Engineering
ISBN 978-981-33-4079-4 ISBN 978-981-33-4080-0 (eBook)
https://doi.org/10.1007/978-981-33-4080-0

This Springer imprint is published by the registered company Springer Nature Singapore Pte Ltd.
The registered company address is: 152 Beach Road, #21-01/04 Gateway East, Singapore 189721, Singapore

Preface

Renewable power is booming, as innovation brings down costs and starts to deliver on the promise of a clean energy future. The solar and wind generation are breaking records and being integrated into the Electricity grids without compromising reliability. This means that renewables are increasingly displacing the fossil fuels in the power sector, offering the benefit of lower emissions of carbon and other types of pollution. But not all sources of energy marketed as "renewable" are beneficial to the environment. Biomass and large hydroelectric dams create difficult tradeoffs when considering the impact on wildlife, climate change, and other issues. In this book the discussion is about the different types of renewable energy sources—and how you can use these emerging technologies at your own home. The huge growth of renewable energy sources in the modern power systems is based on many years of committed research and technology development. New challenges and possibilities are arising with increasing renewable shares, but cost-effectiveness, reliability, robustness and security of the system must always be maintained. This requires new energy paradigm and new advanced solutions. Renewable energy, often referred to as clean energy, comes from natural sources or processes that are constantly replenished. For example, sunlight or wind keep shining and blowing, even if their availability depends on time and weather. While renewable energy is often thought of as a new technology, harnessing nature's power has long been used for heating, transportation, lighting, and more. Wind has powered boats to sail the seas and windmills to grind grain. The sun has provided warmth during the day and helped kindle fires to last into the evening. But over the past 500 years or so, humans increasingly turned to cheaper, dirtier energy sources such as coal and gas.

First of all we are thankful to the contributors of this edited book. We are indebted to authors experts in the field of Renewable Energy Sources and Technology, Power Generation, Transmission and Distribution, Smart Grid Technologies and Applications, Transportation Electrification and Automotive Technologies and Energy Management and Control System and substance who have authored and co-authored their original research findings. We have combined the state-of-art and advanced techniques of Renewable Energy Sources and Technology that will be highly beneficial to the researchers, students and practicing engineers. There are 77 chapters included in this volume have undergone strict quality check by peer review process in several rounds. We thankfully acknowledge the help, cooperation and support in each stage of production of the book by all authors. This is a unique single volume that have bring all the relevant topics ranging from Power Generation, Transmission and Distribution, Distributed Energy Generations and Smart Buildings.

Now that we have increasingly innovative and less-expensive ways to capture and retain wind and solar energy, renewables are becoming a more important power source, accounting for more than one-eighth of U.S. generation. The expansion in renewables is also happening at scales large and small, from rooftop solar panels on homes that can sell power back to the grid to giant offshore wind farms. Even some entire rural communities rely on renewable energy for heating and lighting.

Non-renewable energy sources are also typically found in specific parts of the world, making them more plentiful in some nations than others. By contrast, every country has access to sunshine and wind. Prioritizing nonrenewable energy can also improve national security by reducing a country's reliance on exports from fossil fuel–rich nations.

Many nonrenewable energy sources can endanger the environment or human health. For example, oil drilling might require strip-mining Canada's boreal forest, the technology associated with fracking can cause earthquakes and water pollution, and coal power plants foul the air. To top it off, all these activities contribute to global warming.

Humans have been harnessing solar energy for thousands of years to grow crops, stay warm, and dry foods. According to the National Renewable Energy Laboratory, US "more energy from the sun falls on the earth in one hour than is used by everyone in the world in one year." Today, we use the sun's rays in many ways—to heat homes and businesses, to warm water, or power devices.

Chapter "Parameter Extraction of PV Cell: A Review" of the book presents a review of Extraction of parameters of photovoltaic (PV) cell, which is an important part of studying the performance of PV module/array/cell. Generally manufacturer of PV module does not specify these parameters in datasheet provided by them. The parameters which are to be extracted are: Shunt resistance, Illumination current, Diode reverse saturation current, series resistance and Diode ideality factor. These parameters vary with changing atmospheric condition and do not remain constant. The IV and PV characteristics are dominated by variation of these parameters in the photovoltaic module. As the manufacturing cost of PV cells are lowered during last two decades and the improvement of efficiency leads to increase in demand of photovoltaic energy in recent times and the continuous reduction in conventional energy resources and pollution problem encourages the global electrical energy utilization in the form of solar energy. The present work leads to centralize the work on parameters extraction of PV cell in the literature which can be extremely useful for the researchers and as well as for the global PV module manufacturer.

Chapter "Power Enhancement by Transmitting AC-DC Power Simultaneously—An Experimental Implementation" describes the Power transfer enhancement as an issue because of thermal and stability limitations associated with AC transmission.

Preface

Renewable power is booming, as innovation brings down costs and starts to deliver on the promise of a clean energy future. The solar and wind generation are breaking records and being integrated into the Electricity grids without compromising reliability. This means that renewables are increasingly displacing the fossil fuels in the power sector, offering the benefit of lower emissions of carbon and other types of pollution. But not all sources of energy marketed as "renewable" are beneficial to the environment. Biomass and large hydroelectric dams create difficult tradeoffs when considering the impact on wildlife, climate change, and other issues. In this book the discussion is about the different types of renewable energy sources—and how you can use these emerging technologies at your own home. The huge growth of renewable energy sources in the modern power systems is based on many years of committed research and technology development. New challenges and possibilities are arising with increasing renewable shares, but cost-effectiveness, reliability, robustness and security of the system must always be maintained. This requires new energy paradigm and new advanced solutions. Renewable energy, often referred to as clean energy, comes from natural sources or processes that are constantly replenished. For example, sunlight or wind keep shining and blowing, even if their availability depends on time and weather. While renewable energy is often thought of as a new technology, harnessing nature's power has long been used for heating, transportation, lighting, and more. Wind has powered boats to sail the seas and windmills to grind grain. The sun has provided warmth during the day and helped kindle fires to last into the evening. But over the past 500 years or so, humans increasingly turned to cheaper, dirtier energy sources such as coal and gas.

First of all we are thankful to the contributors of this edited book. We are indebted to authors experts in the field of Renewable Energy Sources and Technology, Power Generation, Transmission and Distribution, Smart Grid Technologies and Applications, Transportation Electrification and Automotive Technologies and Energy Management and Control System and substance who have authored and co-authored their original research findings. We have combined the state-of-art and advanced techniques of Renewable Energy Sources and Technology that will be highly beneficial to the researchers, students and practicing engineers. There are 77 chapters included in this volume have undergone strict quality check by peer review process in several rounds. We thankfully acknowledge the help, cooperation and support in each stage of production of the book by all authors. This is a unique single volume that have bring all the relevant topics ranging from Power Generation, Transmission and Distribution, Distributed Energy Generations and Smart Buildings.

Now that we have increasingly innovative and less-expensive ways to capture and retain wind and solar energy, renewables are becoming a more important power source, accounting for more than one-eighth of U.S. generation. The expansion in renewables is also happening at scales large and small, from rooftop solar panels on homes that can sell power back to the grid to giant offshore wind farms. Even some entire rural communities rely on renewable energy for heating and lighting.

Non-renewable energy sources are also typically found in specific parts of the world, making them more plentiful in some nations than others. By contrast, every country has access to sunshine and wind. Prioritizing nonrenewable energy can also improve national security by reducing a country's reliance on exports from fossil fuel–rich nations.

Many nonrenewable energy sources can endanger the environment or human health. For example, oil drilling might require strip-mining Canada's boreal forest, the technology associated with fracking can cause earthquakes and water pollution, and coal power plants foul the air. To top it off, all these activities contribute to global warming.

Humans have been harnessing solar energy for thousands of years to grow crops, stay warm, and dry foods. According to the National Renewable Energy Laboratory, US "more energy from the sun falls on the earth in one hour than is used by everyone in the world in one year." Today, we use the sun's rays in many ways—to heat homes and businesses, to warm water, or power devices.

Chapter "Parameter Extraction of PV Cell: A Review" of the book presents a review of Extraction of parameters of photovoltaic (PV) cell, which is an important part of studying the performance of PV module/array/cell. Generally manufacturer of PV module does not specify these parameters in datasheet provided by them. The parameters which are to be extracted are: Shunt resistance, Illumination current, Diode reverse saturation current, series resistance and Diode ideality factor. These parameters vary with changing atmospheric condition and do not remain constant. The IV and PV characteristics are dominated by variation of these parameters in the photovoltaic module. As the manufacturing cost of PV cells are lowered during last two decades and the improvement of efficiency leads to increase in demand of photovoltaic energy in recent times and the continuous reduction in conventional energy resources and pollution problem encourages the global electrical energy utilization in the form of solar energy. The present work leads to centralize the work on parameters extraction of PV cell in the literature which can be extremely useful for the researchers and as well as for the global PV module manufacturer.

Chapter "Power Enhancement by Transmitting AC-DC Power Simultaneously—An Experimental Implementation" describes the Power transfer enhancement as an issue because of thermal and stability limitations associated with AC transmission.

way to uplift the voltage gain significantly. The converter behavior and suitability for high voltage is analyzed in continuous conduction mode and later the performances are verified through simulation results at a frequency of 10 kHz.

Chapter "Techno-Economic Analysis of Hybrid Rural Microgrid for Isolated Hilly Area in Indian Scenario" describes the techno-economic analysis of Hybrid Rural Microgrid for Isolated Hilly Area in Indian Scenario. Hybrid renewable resources extensively used to electrify the rural areas. To provide the reliable and economic supply of power in the rural region where grid extension is infeasible integration of the locally available energy resource is the best option. In the present work, economical and optimized design for electrification of isolated rural community proposed with the combination of locally available energy resources. The model comprises of a PV/battery/wind and diesel generator. The microgrid realized in the cluster of the village of Uttrakhand state of India. The residential load of the community investigated using the HOMER energy for the least cost and optimal design. The solar, wind potential and diesel generator used to perform the economic and feasibility study of the suggested model. The net present cost (NPC) and cost of energy (COE) analyzed for the hybrid system. Further model investigated utilizing the sensitivity of different parameters such as variation in diesel price intermittency in solar and wind potential. The study compares the performance of the system and suggests a viable configuration for the rural community.

Chapter "Solar Inverter with Integrated Monitoring and Control System" describes a Solar Inverter with Integrated Monitoring and Control System. In a dynamic world where a constant effort has to be made in order to keep up with the energy requirements of rapidly growing population, we need remedies to monitor energy wastage until the ways to derive sufficient amount of energy from renewable energy resources are invented. This paper proposes a system wherein power is obtained from solar energy and has a monitoring system put in place using an Arduino board and NodeMCU. NodeMCU has an integrated ESP8266 Wifi module that will act as a node for various loads to be monitored. IoT is used in this system to create an environment that will keep in check the energy used by different loads.

Chapters "Analysis of Underground Renewable Energy Storage Tunnels Subjected to Capricious Superstructures", "A Survey on Solar Power for Present and Future Perspective in Indian Market", "Trajectory Tracking of Quadrotor Using LQR Controller", "Performance Analysis of PV Array Connection Schemes Under Mismatch Scenarios", "Techno-Economic Analysis of Diesel/Wind/PV/Battery Hybrid Energy System for Androth Island", "A Scientific Study on Effect of Polarization in Calculation of Rain Attenuation Using ITU-R Model", "MATLAB-Based Modelling and Simulations for the Low and High-Temperature Module Power Generation of PV Panels in Kuala Lumpur and Genting Highlands, Malaysia ", "Systematic Study of Maximum Power Point Tracking Methods Used In Photovoltaic Based Systems", "Hybrid Video Watermarking Based on LWT, SVD and SWT Using Fused Images for Data Security", "Modeling and Simulation of Solar PV Based Grid-Tied Multilevel Inverter", "Optimization Algorithm Based Maximum Power Point Tracking Techniques for Solar PV Systems", "Smart Street Light System for Smart Cities Using IoT", "Energy Audit in an Indian University—

This paper presents a model at which transmission lines are loaded near to their maximum ampacity unlike existing AC power systems. Proposed concept is known as simultaneous ac-dc transmission where the existing AC lines can carry AC current with DC super-imposed on it without major structural modifications. Power transfer capability of 70 km transmission line is enhanced by converting a fraction of AC power from the existing system to DC power which will be put in the sending end transformer through the neutral. Transformer zigzag connection eradicates the issue of core saturation that results from superimposed DC component. Trans-mission line data is provided by Qatar general electricity and water corporation (Kahramaa). Theoretical and real-time simulation results show that power transfer capability has been enhanced by 89%. Transmission angle is increased beyond 30°. MATLAB\Simulink and Real time digital simulator (RTDS) are two different software used to model and simulate the ac-dc transmission sys-tem offline and in real-time respectively. Laboratory prototype shows that power transfer capability has been enhanced by 33%. Practical results and waveforms had been recorded using the power analyzer PA400.

In Chapter "Step by Step Design and Simulation of Boost Controller for L and LCL Filters for EV Fast Charging Systems" discusses the adoption of Electric Vehicles (EV) as an important measure for greenhouse gas reduction in many countries worldwide. The increased production of EV in the market will be associated with continuous demand of installing and maintaining Charging Stations (CS) in public areas. The charging station consist of different power converters for processing the power. Filter is an important component of the EV charging station. Correct design of filter is important for the overall satisfactory operation of the EV chargers. This paper aims to provide designing steps of a closed loop controller of the filter in the EV battery charging system. The contributions of the paper are (1) Step by step design procedure for closed loop input power controller for DC-DC converter, (2)Simulations model development using systematic approach for closed loop controller for LCL and L filter to validate the design procedure, (3)General overview of EV charging system for researchers working on EV CS. The authors believe that this paper is useful for interest-ed pupils linked to EV CS research and working to make its adoption quicker and more convenient.

Chapter "A Novel High Gain Boost Converter for Interfacing Low Voltage PV in a DC Microgird" describes a high gain boost converter for interfacing low voltage PV in a DC microgird. Low terminal voltage at the output of a solar photovoltaic needs efficient boost converters for establishing a proper communication and energy transfer to the DC microgrids. In order to serve this purpose, the traditional boost converter is required to operate at some large duty ratio which is perilous to the converter performance and would create undesirable effects such as more losses, spikes of voltages and poor transient response. A boost converter with significantly high gain at low duty ratio has been presented in this article. The proposed converter is well suited for integration of a solar PV with the DC bus of higher voltage levels in a DC microgrid. This particular converter is simple in structure and easily controlled with a single train of pulses. The advantages of switched inductor circuit embedded with the capacitor is availed in an intelligent

A Case Study", "Study of Effectiveness of Autonomous Solar Energy Systems", "Analysis of Combined Effect of Temperature and Wind on Solar Power Production", "State flow Realization and Performance Evaluation of Selected MPPT techniques", "Cost Analysis of 18 kW Solar Photovoltaic System for Smart Cities Growth in India", "The Role of Power Electronics in the Field of Photovoltaic System: A Study" and "A Review for Energy Generation Analysis and Comparison in China, Indonesia and Ireland with India" discusses various technological advancement on renewable energy sources related to present and future Perspective.

Further, Chapters "Comparison of Direct Torque Control and Indirect Field-Oriented Control for Three-Phase Induction Machine", "Rotor Flux Oriented Control of Three Phase Induction Motor Using Sliding Mode Controller and Rotor Flux Estimator.", "Performance Analysis of Induction Motor Utilizing SPRS Based on MOSFET/IGBT Cuk Converter Topologies, "Design Kinematics and Control for a Differential Drive Mobile Robot", "Initial sizing and Sensitivity Analysis of a Personal Air Vehicle", "Internet of Things Applications in Electric Vehicles–A Review", "An Overview of Using Hydrogen in Transportation Sector as Fuel" and "Performance Assessment of Variable Speed Induction Motor by Advanced Modulation Techniques" describes various control techniques related to electrical vehicle and drives for efficient energy solutions.

Chapters "Study of Different Topology for Multilevel Converter and Inverter Modulating in Cascaded H Bridge Solid State Transformer", "Triple Switch DC to DC Converter for High Voltage Boost Application- Revista", "Modelling and Analysis of Bridgeless PFC Boost Convertor", "Performance Analysis of Different Sliding Mode Controller on Single Link Inverted Pendulum", "A Resilient Hybrid Output Converter with Inherent Cross Regulation Avoidance Feature", "Review of Various Load Frequency Controllers", "Performance Assessment of Unipolar Control Technique Based Cascaded H-bridge Multilevel Inverter", "Realization of a Flyback DC-DC Converter for Experimentation Assisted Teaching in Power Electronics, Nine Step Multilevel Inverter Output Analysis Using the EP Approach ", Case Study of Synchronization of Solar Power Converter", "High Gain DC-DC Converter for Modular Multilevel Converter Applications", "Level Shifted Carrier Based Pulse Width Modulation for Modular Multilevel", "An Improved 15-Level Asymmetrical Multilevel Inverter with Reduced Switch Count", "THD Analysis of 5 Level, 7 Level and 9 Level CHB—Multi level Inverters Using SPWM Switching Approach", "Comparative Evaluation of Bipolar and Unipolar Control Technique Based Cascaded H-bridge Multilevel Inverter", "Performance Comparison of Dual Stage Photovoltaics Based Water Pumping Systems" and "Buck Boost Converter with No Dead or Overlap-Times" details various converters for efficient conversion of electrical power.

Chapters "Design and Analysis of Grounding Grid Performance Using ETAP 16.0.0 Software", "Grid-Interactive Microgrid Charging Infrastructural Network for Electric Vehicles with International Standards", "Transformer Based DC-to-AC Grid Connected Multi-level Inverter Topology", "Deregulation and Its Effect on Indian Power Industry", "Performance Analysis of Five Level Cascaded H-Bridge

Inverter", "Improvement of Voltage Stability Margin in a Radial Distribution System", "Generation of HVDC from Voltage Multiplier Using Opto-Isolator and Marx Generator", "VP-ZA-LMS Control Algorithm for Double Stage Grid Integration of Solar PV System", "Planning and Economical optimization of Grid-Connected Renewable Energy Resource-Based Microgrid", "Comparative Analysis of TPS and EPS of IBDC for Power Management", "A Review Paper on Analysis, Planning of Electricity Generation in Turkey, United Arab Emirates and Germany and Comparison with India", "A Transformer-Less Ultra-Gain Switched Inductor Boost Converter for DC Microgrid Applications", "Performance Analysis of Grid Connected Solar Energy Conversion System Under Varying Conditions" and "A Review Paper on Comparison, Analysis and Planning of Electricity Generation in Australia, Argentina, New Zealand, Mexico with India" discusses various technological advancements for electrical power grid.

Chapters "A Comparative Study of Costs Function in Multivariate Stratified Double Sampling Design", "Fabricated CdO-ZnO films: A Potential Candidate for Optoelectronics Device Applications", "Efficient Object Removal and Region Filling Image", "A Stratified Optional Mathematical Model", "Space Heating for Kashmir Valley: Issues, Challenges and Remedies", "A Review of Power Factor Correction and Reduction in Total Harmonic Distortion for LED Drivers", "Fuzzy Discrete Event System (FDES): A SurveyFuzzy Discrete Event System (FDES): A Survey", "Mathematical Modelling of Water Heater and HVAC for Demand Response Analysis", "A Mathematical Approach to Speech Enhancement for Speech Recognition and Speaker Identification Systems", "A Survey of Recent Trends in Two-Stage Object Detection Methods" and "A Comparative Analysis of Small- and Medium-Scale Industrial Development in Amhara Region, Ethiopia" discusses various mathematical modelling techniques that are very much helpful particularly for young researchers to formulate the modern-day mathematical models of the system.

Similarly, authors of other chapters have discussed various important aspects and their research in the relevant areas. We thanks all the contributors of this book for their valuable effort in producing high class literature for research community. We are sincere thankful to the Intelligent Prognostic Private Limited India to provide the all type of technical and non-technical facilities, cooperation, and support in each stage to make this book in reality.

We wish to thank our colleagues and friends for their insight and helpful discussion during the production of this edited book. We are sincere thankful to the Intelligent Prognostic Private Limited India to provide the technical and non-technical help. We would like to highlight the contribution of Prof. Haitham Abu-Rub, Texas A&M University at Qatar, Prof. Sukumar Mishra, IIT Delhi, Prof. Imtiaz Ashraf, Aligarh Muslim University, India, Prof. Majid Jamil, Jamia Millia Islamia Delhi, Prof. Mairaj ud din Mufti, NIT Srinagar, Prof. Y. R. Sood and Prof. R. K. Jarial, NIT Hamirpur, Prof. Narender Kumar, DTU Delhi, Prof. Anand Parey, IIT Indore, India.

We would like to express our love and affection to our family members, Shadma (wife of Prof. Atif Iqbal), Abuzar, Abubaker (Sons of Prof. Atif Iqbal), Noorin (daughter of Prof. Atif Iqbal) whose passions, care and love have made this book a reality. We would like to express our gratitude to Dr. N. Fatema (wife of Dr. Hasmat Malik), Zainub and Ayesha (daughter of Dr. Hasmat Malik) for their intense feeling of deep affection. We further would like to express our love and affection to our family members, Asiya Naaz (wife of Mr. Ahmed Riyaz), Ashhab, Azhar (Sons of Mr. Ahmed Riyaz), Masira (daughter of Mr. Ahmed Riyaz). Furthermore, we would like to express our deep gratitude to all family members of Prof. Kouzou Abdellah and Prof. Sertac Bayhan for their intense feeling of deep affection.

Doha, Qatar	Atif Iqbal
Woodlands, Singapore	Hasmat Malik
Rajouri, Jammu and Kashmir, India	Ahmed Riyaz
Djelfa, Algeria	Kouzou Abdellah
Doha, Qatar	Sertac Bayhan

Contents

About the Editors

Atif Iqbal, Fellow IET (UK), Fellow IE (India) and Senior Member IEEE, Vice-Chair, IEEE Qatar section, D.Sc. (Poland), Ph.D. (UK) Associate Editor, IEEE Trans. On Industrial Electronics, IEEE ACCESS, Editor-in-Chief, I'manager journal of Electrical Engineering, Former Associate Editor IEEE Trans. On Industry Application, Former Guest Associate Editor IEEE Trans. On Power Electronics. Full Professor at the Dept. of Electrical Engineering, Qatar University and Former Full Professor at the Department of Electrical Engineering, Aligarh Muslim University (AMU), Aligarh, India. Recipient of Outstanding Faculty Merit Award academic year 2014–2015 and Research excellence awards 2015 and 2019 at Qatar University, Doha, Qatar. He received his B.Sc. (Gold Medal) and M.Sc. Engineering (Power System and Drives) degrees in 1991 and 1996, respectively, from the Aligarh Muslim University (AMU), Aligarh, India and Ph.D. in 2006 from Liverpool John Moores University, Liverpool, UK. He obtained D.Sc. (Habilitation) from the Gdansk University of Technology in Control, Informatics and Electrical Engineering in 2019. He has been employed as a Lecturer in the Department of Electrical Engineering, AMU, Aligarh since 1991 where he served as Full Professor until August 2016. He is a recipient of Maulana Tufail Ahmad Gold Medal for standing first at B.Sc. Enginering (Electrical) Exams in 1991 from AMU. He has received several best research papers awards e.g. at IEEE ICIT-2013, IET-SEISCON-2013, SIGMA 2018, IEEE CENCON 2019, IEEE ICIOT 2020, and Springer ICRP 2020. He has published widely in International Journals and Conferences his research findings related to Power Electronics, Variable Speed Drives, and Renewable Energy Sources. Dr. Iqbal has authored/co-authored more than 420 research papers and four books and several chapters in edited books. He has supervised several large R&D projects worth more than multi-million USD. He has supervised and co-supervised several Ph.D. students. His principal area of research interest is Smart Grid, Complex Energy Transition, Active Distribution Network, Electric Vehicles drivetrain, Sustainable Development, and Energy Security, Distributed Energy Generation, and multiphase motor drive system.

Hasmat Malik (M'16) received Diploma in Electrical Engineering from Aryabhatt Government Polytechnic Delhi, B.Tech. degree in Electrical and Electronics Engineering from the GGSIP University, Delhi, M.Tech. degree in Electrical Engineering from National Institute of Technology (NIT) Hamirpur, Himachal Pradesh, and Ph.D. in Power System from Electrical Engineering Department, Indian Institute of Technology (IIT) Delhi, India. He is currently Postdoctoral Scholar at BEARS, University Town, NUS Campus, Singapore, and Assistant Professor (on-Leave) at Division of Instrumentation and Control Engineering, Netaji Subhas University of Technology Delhi, India. He is a life member of Indian Society for Technical Education (ISTE), Institution of Electronics and Telecommunication Engineering (IETE), International Association of Engineers, Hong Kong (IAENG), and International Society for Research and Development, London (ISRD), and a member of the Institute of Electrical and Electronics Engineers (IEEE), USA, and Mir Labs, Asia. He has published more than 100 research articles, including papers in international journals, conferences and book chapters. He is Guest Editor of Special Issue of Journal of Intelligent and Fuzzy Systems, 2018, (SCI, Impact Factor 2019:1.637), (IOS Press). He received the POSOCO Power System Award (PPSA 2017) for his Ph.D. work for research and innovation in the area of power system. He has received best research papers awards at IEEE INDICON-2015 and full registration fee at IEEE SSD-2012 (Germany). He has supervised 23 PG students. He is involving in several large R&D projects. His principle area of research interests is artificial intelligence, machine learning and big data analytics for renewable energy, smart building and automation, condition monitoring and online fault detection and diagnosis (FDD).

Kouzou Abdellah (IEEE senior member) is Full Professor at the University of Djelfa, Algeria. He was a researcher with Munich Technical University in Germany in 2010, 2011 and 2012. He is a collaborator researcher at Texas A&M University at Qatar. He has participated and led several research projects. He is Founder of the Power Electronics and Power Quality research group at the University of Djelfa. He is a member of many editorial boards in several scientific journals and a member of the steering committees in several national and international conferences. He was an expert in several national and international scientific activities and project evaluations. He has published more than 320 papers. His main research interests include power quality issues, power electronics devices, application of power electronics in renewable energies, sensorless and predictive control, fault diagnosis in power electronics converters, smart grid and smart buildings, reliability and diagnostics in power electronics converters.

Ahmed Riyaz received his M.Tech. (Power System & Drives) and B.Tech. (Electrical) Engineering from the Department of Electrical Engineering, Aligarh Muslim University, Aligarh (AMU), India, in 2009 and 2007, respectively, and he is pursuing Ph.D degree with the Electrical Engineering Department, Indian Institute of Technology, Dhanbad, India. He has worked as Senior Research Fellow in an MHRD sponsored research project at IIT Roorkee (India) developing online open source simulation tools. He has research and teaching experience of ten years. He is currently working as a faculty member at BGSB University, Government of J&K (India), and developed multi-phase drives lab simultaneously handling many key responsibilities in administration. He is also Nodal Officer (Academics) for a World Bank sponsored project worth one hundred ten million Indian Rupees at BGSB University. His area of research includes multi-phase (more than three-phase) electric drives, power electronic converters and renewable energy.

Sertac Bayhan (SMIEEE) received the M.Sc. and Ph.D. degrees in Electrical Engineering from Gazi University, Ankara, Turkey, in 2008 and 2012, respectively. His undergraduate studies are also at the same university, and he graduated as valedictorian. In 2008, he joined the Electronics and Automation Department, Gazi University, as a Lecturer, where he was promoted to Associate Professor in 2017. From 2014 to 2018, he also worked at Texas A&M University at Qatar as Associate Research Scientist. Dr. Bayhan is currently working in Qatar Environment and Energy Research Institute (QEERI) as Scientist. His research interests lie in the areas of advanced control of PV and wind systems, microgrids and smart grid applications. He has led multi-PI projects with collaborators from all over the world. He has authored 120 high-impact journal and conference papers. He is the co-author of two books and three book chapters. Dr. Bayhan has been an active senior member of IEEE. Because of the visibility of his research, he has been recently elected as Chair of IES Power Electronics Technical Committee and selected as Co-Chair of IEEE-IES Student and Young Professional Activity Program. He currently serves as Associate Editor for the IEEE Journal of Emerging and Selected Topics in Industrial Electronics and IEEE TRANSACTIONS on INDUSTRIAL ELECTRONICS and Guest Editor for the IEEE TRANSACTIONS on INDUSTRIAL INFORMATICS. Dr. Bayhan was the recipient of many prestigious international awards, such as the Research Fellow Excellence Award in recognition of his research achievements and exceptional contributions to the Texas A&M University at Qatar in 2018, the Best Paper and Presentation Recognition at the 41st and 42nd Annual Conference of the IEEE Industrial Electronics Society in 2015 and 2016, Research Excellence Travel Awards in 2014 and 2015 (Texas A&M University at Qatar) and Researcher Support Awards from the Scientific and Technological Research Council of Turkey (ten times).

Parameter Extraction of PV Cell: A Review

Mohammad Shahabuddin, Mohammed Asim, and Adil Sarwar

Abstract Extraction of parameters of photovoltaic (PV) cell is an important part of studying the performance of PV module/array/cell. Generally, manufacturer of PV module does not specify these parameters in datasheet provided by them. The parameters which are to be extracted are: Shunt resistance, R_{sh}, illumination current I_{ph}, diode reverse saturation current I_{o}, series resistance R_{se}, and diode ideality factor. These parameters vary with changing atmospheric condition and do not remain constant. The IV and PV characteristics are dominated by variation of these parameters in the photovoltaic module. As the manufacturing cost of PV cells is lowered during last two decades and the improvement of efficiency lead to increase in demand of photovoltaic energy in recent times. Also the continuous reduction in conventional energy resources and pollution problem encourages the global electrical energy utilization in the form of solar energy. The present work leads to centralize the work on parameters extraction of PV cell in the literature which can be extremely useful for the researchers and as well as for the global PV module manufacturer.

Keywords Parameter extraction photovoltaic (PV) cell · MATLAB/SIMULINK · MPPT · Metaheuristic

1 Introduction

Finding out the accurate values of PV parameters has been a preliminary work for the researchers to obtain a complete solution of nonlinear behavior of solar PV module under varying atmospheric conditions. Various types of techniques are involved in formulating parameter extractions of solar PV panel, but due to the involvement of complexity and nonlinear behavior of solar PV panel, it is not an easy task to sort out exact values of parameters. It requires making certain assumptions to dig out the

M. Shahabuddin (✉) · M. Asim
Department of EE, Integral University, Lucknow, India
e-mail: asimamu@gmail.com; mshahab@iul.ac.in

A. Sarwar
Department of EE, AMU, Aligarh, India

A. Iqbal et al. (eds.), *Renewable Power for Sustainable Growth*, Lecture Notes in Electrical Engineering 723, https://doi.org/10.1007/978-981-33-4080-0_1

all five parameters. Various techniques are presented in literature. However, there is a failure of focused efforts to integrate all of this work into a single platform that provides an intermediate point for researchers to perform a specific design of the module frameworks. [1, 2]. (There is a gap to sort out the work of different researchers and contribution in the area of performance of solar PV module under varying environment conditions).

2 Parameter Extraction

Continuous, constant cut in available limited conventional energy resources leads the researchers to think about future problem of generation of electrical energy [3]. The results come out in the shape of renewable energy resources. The sun is the base of generation of all types of energy universally. Among different kinds of non-conventional resources of energy, the photovoltaic energy is preferable due to its negligible maintenance cost, free availability, and continuous reduction of production cost. But because of variations in solar irradiation and operating temperature of PV module, the continuous availability of maximum power cannot be assured, so some means are adopted to track the maximum power of photovoltaic module. The output of the PV parameters will be installed to predict the behavior of the PV module. These frames should be removed as they can be provided by the manufacturer. One diode model used to extract a parameter due to its simplicity and clarity. In one diode model, there is a current source, a diode, series resistance, and the equivalent shunt resistance as shown in Fig. 1. Information such as open circuit voltage (V_{oc}), short circuit current (I_{sc}), voltage at maximum power point (V_{mpp}), current at maximum power point (I_{mpp}), and the constants (K_v) and (K_i) are generally given in the data sheet provided by the manufacturer. The parameters to be extracted are: illumination current (I_{ph}), diode reverse saturation current (I_o), series resistance (R_s), parallel resistance (R_{sh}), and diode ideality factor (a).

In Fig. 1, the current at output I is given by the equation:

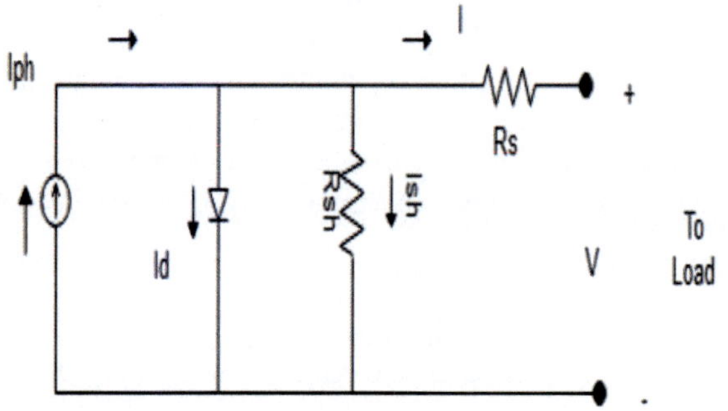

Fig. 1 Single diode model of PV cell

$$I = I_{ph} - I_d - I_{sh} \tag{1}$$

$$I = I_{ph} - I_o\left(e^{\frac{q(V+IR_S)}{akt}-1}\right) - \left[\frac{V + IR_s}{R_{sh}}\right] \tag{2}$$

where

- I current at output
- I_{ph} illumination current
- I_d current in diode
- I_{sh} shunt current
- I_o reverse saturation current of diode
- q charge of electron
- V terminal voltage
- R_s series resistance
- a diode ideality factor
- k Boltzman constant
- t cell operating temperature.

3 Types of Techniques of Parameter Extraction

In order to enhance the use of photovoltaic energy, solar PV parameters calculation is an important task. The different techniques available in literature are discussed in the following five categories which are as follows.

3.1 Using Impp and V_{mpp}

With the help of only two known values of current at maximum power point, i.e., I_{mpp} and voltage at maximum power point, i.e., V_{mpp}, the parameters values are estimated after knowing the temperature at operating point and extra costs are saved because it does not employ any additional hardware [4]. The simulation work is done using MATLAB/SIMULINK software and the experimental work is done using model REC-AE 220 solar module to verify the validity of the conducted simulation. One diode model is convenient in terms of simplicity and accuracy.

A simplified procedure is presented in Fig. 2 for the evaluation of PV parameters. Generally, it is always not been possible to measure the operating temperature of PV cell directly, so PV model is formulated for the estimation of operating temperature of PV cell. The various complicated problem related to the environmental issues are analyzed.

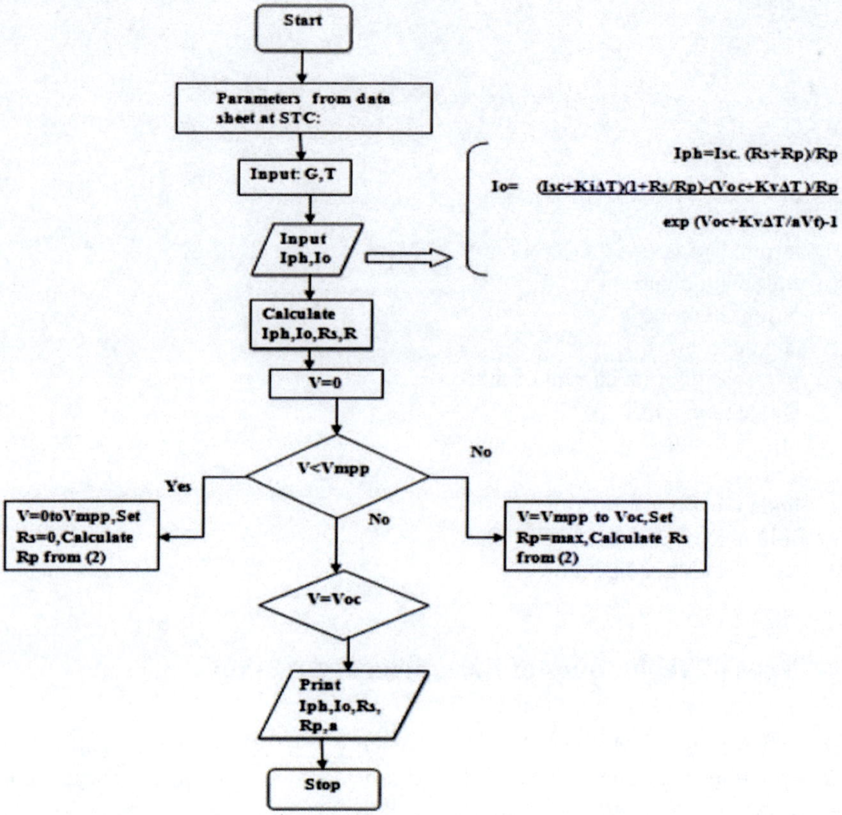

Fig. 2 Flow chart using V_{mpp} and I_{mpp}

3.2 *Numerical and Analytical Extraction*

For the first time, parameters extraction technique is applied for the desensitized cell using newer techniques of metaheuristic algorithm. With use of two types of techniques, one is analytical and other is numerical, the five parameters are extracted [4]. With the help of numerical method, more accurate extraction can be done as compare to analytical method.

As compare to other two, i.e., EAs and GA, the PSO is the best method for the most accurate extraction of these parameters.

By considering variations in solar irradiance and operating PV cell temperature, the proposed model can be compared with the data provided by the manufacturer. It is found that the results very closely matched with the manufacturer's data sheet. The methods adopted are capable of giving offline calculations as shown in Fig. 3 about the five unknown parameters with the explicit single diode mode.

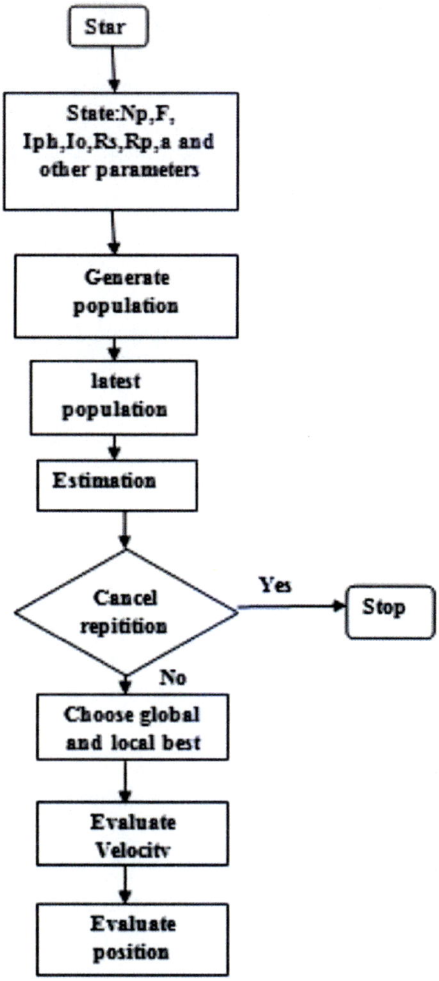

Fig. 3 Flow chart for numerical and analytical extraction

3.3 *By Module Integrated Converter (MIC)*

In most of the cases, in PV module, a switched converter is employed for enhancing the power of the module; for this, an MPPT tracker is incorporated with the module. Another method for the parameter extraction and MPPT the Module Integrated Converter (MIC) technique is introduced which overcomes all other previously adopted techniques for parameters extraction and for MPPT. It does not employ expensive pyranometer except for a preliminary tuning of the pyranometer [5]. Due to the absence of any pyranometer, the system reduces the cost as compare to the other techniques. The current at short circuit ($\grave{I}_{sc}$) is supposed to be almost equivalent

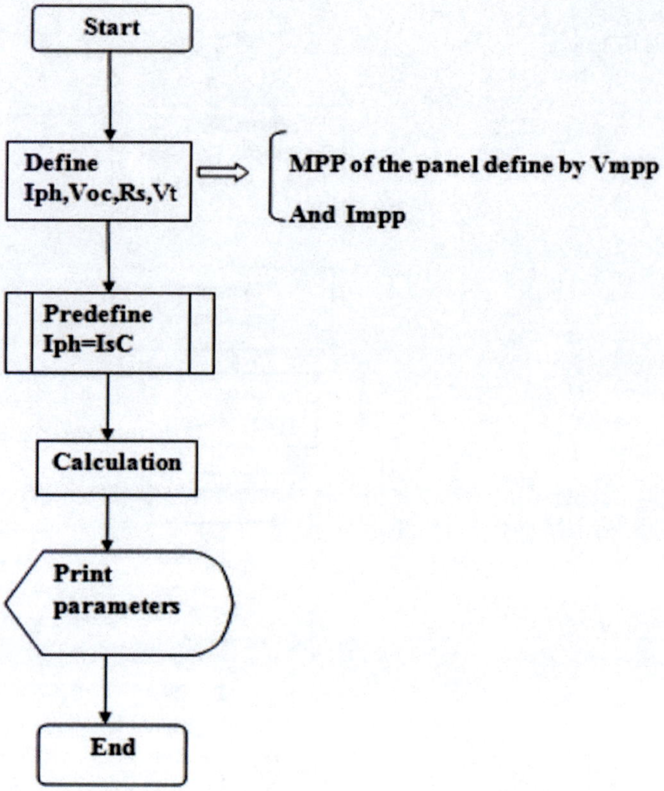

Fig. 4 Flow chart for MIC

to illumination current (I_{ph}) and other four parameters are extracted by forming the equations with I-V characteristics (Fig. 4).

3.4 *By Different Models*

There are various modeling techniques which are described in the literature which are taken into considerations for computing the values obtained from the model and from the actual modules [5]. Different authors have used various iteration methods. Some of the authors approximated the values from five unknown parameters and calculated the remaining four parameters. Equations are formed using different stages of modules like voltage at open circuit conditions, current at short circuit condition plus MPPT conditions. Looking at the formulated equations, these parameters are calculated and compared with the data sheets. Values are obtained from single diode model and it is compared with two diode model. In order to express the behavior

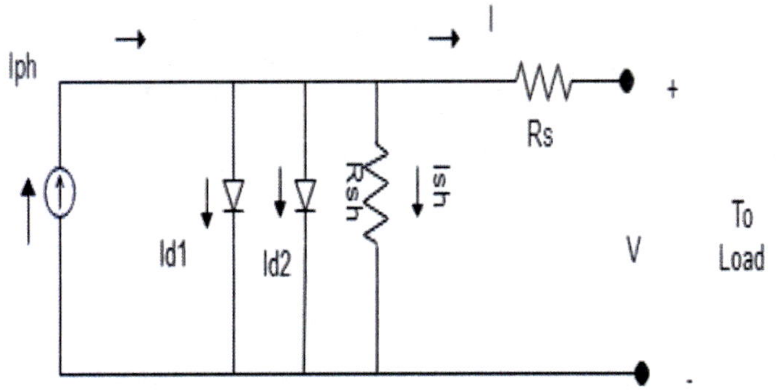

Fig. 5 Model of PV cell with double diodes

of PV module in terms of mathematical expressions like PV and IV curves, the modeling of PV cell/module is done under varying environmental conditions, i.e., varying insolation and temperature (Fig. 5).

In order to find out the five unknown parameters of PV model, double or triple modules are taken into consideration for comparing the values obtained from the model and from the actual modules (Fig. 6).

3.5 ANN Parameter Extraction

Artificial neuronetwork-based parameter extraction method is one which minimizes error, i.e., error under 1.0% for every parameter leaving reverse current Io plus current at diffusion that rise up 11.0%. The nonlinear behavior of I-V and P-V characteristics makes it lengthy process for the extraction of five parameters but this technique which utilizes ANN which is most accurate and simple method for this work. ANN-based method is one of the most widely used methods for parameter extraction of a one diode model [6].

Because of the limited functionality of the proposed ANN model which only can work in the particular range of the parameters, a predefined ranges of parameters are chosen which represents lowest, highest, and moderate values of series resistance R_{se}, parallel resistance R_{sh}, diode reverse current I_o, illumination current I_{ph}, and factor of diode ideality a.

Using evolutionary algorithm, parameters extraction is done for the desensitized solar cells (DSCs) for one diode equivalent circuit model [7]. The DSCs are photo-electrochemical devices and its operation principle is different from conventional solar cells. For the dysenscitised solar PV cell, it is find out that the particle swarm optimization metaheuristic optimization is the best one for parameter extraction.

With the use of flower pollination algorithm (FPA), solar PV parameter extraction is done and not a long back the thing is find out that evolutionary algorithm gives very accurate output in getting the solution of nonlinear problems. FPA is a nature-inspired

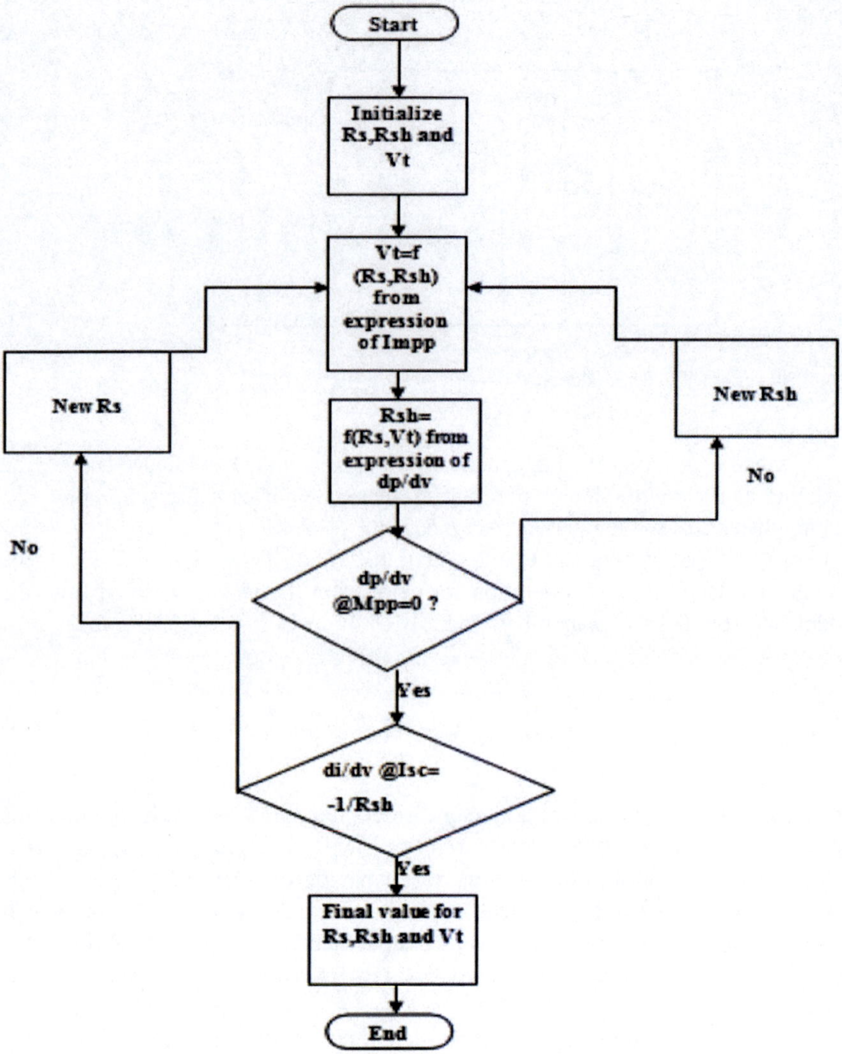

Fig. 6 Flow chart for using different models

algorithm, an alternative of solar PV parameter extraction having various advantages over other methods like: highly feasible in tracking, avoid parameter convergence, faster convergence, and superior computational speed.

A wind-driven optimization algorithm (WDA) is proposed which minimizes different types of error in power of different types of module and in different environmental conditions. The proposed algorithm is verified in terms of accuracy and flexibility using different environmental conditions. And the algorithm is presented as a strong method for recognizing the parameter of a twelve parameters DDM of the solar PV model [8–11].

For the future research, some recommendations are provided based on the conducted review. The whole research work for the solar PV parameter extractions is classified into three categories for the ease and these categories are reviewed.

4 Comparative Analysis

Various sorts of parameter extraction techniques have been invented earlier changing from the very basic one diode model to the recent ANN-based model [8–11]. A detailed analytical analysis with stress on the subsequent criteria has been performed as shown in Table 1.

The comparative analysis done for the parameters extraction can be beneficial for the researchers and academicians to assist in selection of parameter extraction technique as per their requirements depending upon the environmental conditions and financial conditions. A thorough differentiation data taking into consideration as the above-mentioned affecting different parameter extraction techniques of PV model has been presented.

The performance technique is gauged in terms of accuracy, simplicity, reliability, suitability, and complexity. The complexity in the hardware has been eliminated due to parameter extraction from the data provided by the manufacturer in his manual in case of single diode model.

As it is observed from the comparative analysis of all the techniques mentioned above, as for accuracy is concerned, the parameter extraction through module integrated converter (MIC) and artificial neuronetwork (ANN) is more accurate as compare to others leaving different model technique moderate while parameter extraction through V_{mpp}, I_{mpp} and numerical and analytical methods are having low accuracy. In terms of simplicity parameter extraction by V_{mpp}, I_{mpp} and by different

Table 1 Comparison of techniques

Parameters	Parameter extraction through I_{mpp} and V_{mpp}	Parameter extraction through numerical and analytical methods	Parameter extraction through module integrated converter (MIC)	Parameter extraction through different models	Parameter extraction through ANN
Accuracy	Bellow	Bellow	Upper	Middle	Upper
Simplicity	Upper	Middle	Bellow	Upper	Middle
Reliability	Middle	Bellow	Middle	Middle	Upper
Suitability	Upper	Middle	Middle	Bellow	Middle
Complexity	Bellow	Upper	Middle	Middle	Bellow

models are simpler methods of parameter extraction as compare to other methods. Artificial neuronetwork (ANN) method of parameter extraction is more reliable, with moderate suitability and low complexity. In terms of suitability and complexity V_{mpp}, I_{mpp} is preferred while due to high complexity of numerical and analytical method is not preferred.

5 Conclusion

A review of different parameters extraction techniques adopted has been done in this paper. The parameter extraction by I_{mpp} and V_{mpp} method, numerical and analytical extraction, Module Integrated Converter (MIC), different models, ANN parameter extraction, has been discussed and a comparative analysis has been done.

From the comparative analysis of parameters extraction of various techniques discussed above, it is concluded that: as for accuracy is concerned, the parameter extraction through module integrated converter (MIC) and artificial neuronetwork (ANN) is more accurate as compare to others leaving different model technique moderate while parameter extraction through V_{mpp}, I_{mpp} and numerical and analytical methods are having low accuracy. In terms of simplicity, parameters extraction by V_{mpp}, I_{mpp} method and by different models are simpler methods of parameter extraction as compare to other methods. Artificial neural network (ANN) method of parameter extraction is more reliable, with moderate suitability and low complexity. In terms of suitability and complexity V_{mpp}, I_{mpp} is preferred while due to high complexity of numerical and analytical method is not preferred.

Acknowledgements The author would like to acknowledge Integral University for providing the MCN number "IU/R&D/2020-MCN000794."

References

1. Asim M, Tariq M, Mallick MA, Ashraf I, Kumari S, Bhoi AK (2018) Critical evaluation of offline MPPT techniques of solar PV for stand-alone applications. In: Advances in smart grid and renewable energy. Springer, Singapore, pp 13–21
2. Tariq M, Bakhsh FI, Khan SA, Asim M (2010) Solar based space cooling system in building-a case study of Maulana Azad library, AMU. Int J Recent Trends Eng Technol 4(4)
3. Asim M, Mallick MA, Malik A, Saaqib M (2015) Modelling and simulation of 5 parameter model of solar cell. Int J Electron Electr Comput Syst 4
4. Stojanović I, Brajević I, Stanimirović PS, Kazakovtsev LA, Zdravev Z (2017) Application of Heuristic and Metaheuristic Algorithms in Solving Constrained Weber Problem with Feasible Region Bounded by Arcs, Mathematical Problems in Engineering, vol 2017, Article ID 8306732, p 13 https://doi.org/10.1155/2017/8306732
5. Murtaza AF, Munir U, Chiaberge M, Di Leo P, Spertino F (2018) Variable parameters for a single exponential model of photovoltaic modules in crystalline-silicon. Energies 11:2138. https://doi.org/10.3390/en11082138www.mdpi.com/journal/energies

6. Gastli A, Ben-Brahim L, Rhouma MBH (2015) ANN-based extraction approach of PV cell equivalent circuit parameters. In: 2015 17th European conference on power electronics and applications (EPE'15 ECCE-Europe), Geneva, pp 1–10
7. Shahabuddin M, Riyaz A, Asim M, Shadab MM, Sarwar A, Anees A (2018) Performance based analysis of solar PV emulators: a review. IEEE. 978-1-5386-4254-2/18/$31.00©2018
8. Malik H et al (eds) (2019) Applications of artificial intelligence techniques in engineering, vol 2. In: Advances in intelligent systems and computing 697. Springer Nature Singapore Pte Ltd. https://doi.org/https://doi.org/10.1007/978-981-13-1822-1
9. Malik H et al (eds) (2019) Applications of artificial intelligence techniques in engineering, vol 1. In: Advances in intelligent systems and computing 698. Springer Nature Singapore Pte Ltd. https://doi.org/https://doi.org/10.1007/978-981-13-1819-1
10. Iqbal A et al (eds) (2020) Soft computing in condition monitoring and diagnostics of electrical and mechanical systems, vol 1096. In: Advances in intelligent systems and computing. Springer, Singapore. https://doi.org/https://doi.org/10.1007/978-981-15-1532-3
11. Iqbal A et al (eds) (2020) Meta heuristic and evolutionary computation: algorithms and applications, vol 1096. In: Studies in computational intelligence. Springer, Singapore. https://www.springer.com/gp/book/9789811575709

A Comparative Study of Cost Function in Multivariate Stratified Double Sampling Design

Ziaul Hassan Bakhshi

Abstract This paper deals with sample size in stratified double sampling in objective function where costs are taken unknown (with and without), respectively. An equivalent deterministic form of objective function has been obtained by using modified E-model in the case of random cost function. Numerical illustrations have been presented.

Keywords Optimum allocation · Modified E-model · Stratified double sampling

1 Introduction

In the study of sampling, double sampling is used, if strata weights are unknown in survey problem. In the beginning, a sample with SRS is chosen, and for finding the unknown strata weights, from each stratum, sampled units are recorded. We obtain stratified random sample after taking subsamples of simple random from the selected units of the strata.

To determine sample numbers which require in each stage is the complicated issue which gives the desired accuracy for the maximum economy. There are two factors on which efficiency of double sampling depends: (i) the mathematical relationship's precision and (ii) direct measurement's cost compared to indirect estimates [1–9].

Neyman [5] firstly introduced double sampling design in survey problem. In multivariate surveys, Kokan [3] suggested NLPP solution but did not confer its uses about double sampling. Kokan and Khan [4] discussed a summarized approach of an allocation problem for multivariate type of problem and conferred uses of double sampling also. An optimum double sampling is expressed as a mathematical programming problem (MPP) [8, 9].

Z. H. Bakhshi (✉)
School of Basic Sciences and Technology, IIMT University, Meerut, UP, India
e-mail: bakhshistat@gmail.com

A. Iqbal et al. (eds.), *Renewable Power for Sustainable Growth*, Lecture Notes in Electrical Engineering 723, https://doi.org/10.1007/978-981-33-4080-0_2

Prekopa [6] and Diaz-Garcia [1] discussed the constraint case where sampling variances are random. For finding this type of problem, where objective function has unknown costs, modified E-model is applied which had been discussed by Javed and Bakhshi [2].

In this paper, we have considered costs as (with and without) random variables, and this type of problem can be expressed as stochastic nonlinear programming problem. By the use of modified E-model, we transformed this as an equivalent deterministic objective function. By LINGO software, the solution has been achieved.

2 Formulation of Stratified Double Sampling Problem

Suppose that population of finite size N is divided into k groups (called strata), with N_i ($i = 1, 2, \ldots, k$) units belong to the ith strata, i.e.,$\sum_{i=1}^{k} N_i = N$, which are widely different in means and which are homogeneous within themselves. For obtaining unbiasedly, average or the total study character, strata weights are used. In the case of unknown weights, we choose n' sample to obtain strata weights and again do the process such that $\sum_{i=1}^{k} n_i = n$.

Suppose W_i: $\frac{N_i}{N}$ is the proportion of units which comes under the ith stratum and w_i: $\frac{n_{i'}}{n'}$ is the proportion of 1st sample units which comes under the ith stratum. $\overline{y}_{\text{std}}$ is an unbiased estimator of $\overline{Y}$ i.e.,

$$\overline{y}_{\text{std}} = \sum_{i=1}^{k} w_i \overline{y}_i$$

where in the ith stratum, sample mean $\overline{y}_i$ is the variable under study.

$$V\left(\overline{y}_{\text{std}}\right) = \sum_{i=1}^{k} \left[w_i^2 + \frac{w_i(1 - w_i)}{n'} \right] \frac{S_i^2}{n_i} + \sum_{i=1}^{k} \frac{w_i \left(\overline{Y}_{hj} - \overline{Y}\right)^2}{n'} \tag{2.1}$$

By neglecting fpc. On the population in the specific strata and the expression $\frac{w_i(1-w_i)}{n'}$ becomes very small with comparability to w_i^2 [Cochran (1970)], we get

$$V\left(\overline{y}_{\text{std}}\right) = \frac{\sum_{i=1}^{k} w_i^2 S_i^2}{n} + \frac{\sum_{i=1}^{k} w_i \left(\overline{Y}_{hj} - \overline{Y}\right)^2}{n'}$$

Let it $\sum_{i=1}^{k} w_i^2 S_i^2$ by V_n and $\sum_{i=1}^{k} w_i \left(\overline{Y}_{hj} - \overline{Y}\right)^2$ by $V_{n'}$,
We get

$$V\left(\overline{y}_{\text{std}}\right) = \frac{V_n}{n} + \frac{V_{n'}}{n'} \tag{2.2}$$

The cost function is

$$C = C_1 n' + C_2 n + C_0$$

where C_1 is lesser than C_2 and C_0 is the fixed cost.

To select n and n' is the issue in such a style that costs are minimized subject to the variance (the resistance ranges) of different characters.

Therefore, this problem can be written as:

$$\begin{aligned} &\text{Min.}\, C = nC_n + n'C_{n'} + C_0 \quad C_0 \text{ is the fixed cost} \\ &\text{Subject to } \frac{V_n}{n} + \frac{V_{n'}}{n'} = V_j \end{aligned} \tag{2.3}$$

where V_j is the fixed variance.

Here, the above equation describes the matter of minimization the cost subject to prearranged fixed variance. Here, we will describe the cost based on two cases.

i. when costs are without random;
ii. when costs are with random unknown.

Case (I) If Costs are Known, (Without Random Cost)
The formulation of the problem is:

$$\left.\begin{aligned} &\underbrace{\text{Min.}}_{n,n'}\, C = \sum_{i=1}^{m} C_i n_i + C'n' + C_0 \\ &\text{Subject to } \tfrac{V}{n} + \tfrac{V'}{n'} = V_j \\ &\text{where } V = \sum_{i=1}^{m} W_i^2 S_i^2 \quad \text{and } V' = \sum_{i=1}^{m} W_i \left(\overline{Y}_j - \overline{Y}\right)^2 \\ &\qquad\qquad n_i \geq 2, n' \geq 5 \end{aligned}\right\} \tag{2.4}$$

Case (II) When Cf3.osts are Unknown, (With Random Cost)
For this, we firstly describe it by the use of modified E-model approach.

3 Modified E-Model Approach

We write the given the problem as:

$$\left.\begin{array}{l}\underbrace{\text{Min.}}_{n,n'} C = \sum_{i=1}^{m} C_i n_i + C'n' + C_0 \\ \text{Subject to } \frac{V}{n} + \frac{V'}{n'} = V_0 \\ \text{where } V = \sum_{i=1}^{m} W_i^2 S_i^2 \text{ and } V' = \sum_{i=1}^{m} W_i \left(\overline{Y_j} - \overline{Y}\right)^2 \\ \qquad n_i \geq 2, n' \geq 5\end{array}\right\} \quad (3.1)$$

Here, costs c_i are random variables which are considered as independently and normally distributed. Thus, the objective cost function will be distributed as normally with average $E\left(\sum_{i=1}^{m} n_i + c'n' + C_0\right)$ and variance $V\left(\sum_{i=1}^{m} c_i n_i + c'n' + C_0\right)$.

If $c_i \sim N(\mu_i, \sigma_i^2)$, then p.d.f. can be written as

$$f(c_i) = \frac{1}{\sigma_i\sqrt{2\pi}} e^{-\frac{1}{2\sigma_i^2}(c_i-\mu_i)^2}, i = 1, \ldots, m.$$

The joint distribution of costs $(c_1, \ldots, c_m)$ can be as follows:

$$f(\underline{c'}) = \frac{1}{\prod_{i=1}^{m} \sigma_i (2\pi)^{m/2}} e^{-\frac{1}{2\sigma_i^2}\sum_{i=1}^{m}(c_i-\mu_i)^2}$$

$$\begin{aligned} E\left(\sum_{i=1}^{m} c_i n_i + C_0\right) &= E\left(\sum_{i=1}^{m} c_i n_i\right) + C_0 \\ &= \sum_{i=1}^{m} n_i E(c_i) + n'c'_\mu + C_0 \\ &= \sum_{i=1}^{m} n_i \mu_i + n'c'_\mu + C_0, \text{ (say)} \end{aligned} \quad (3.2)$$

$$\begin{aligned} V\left(\sum_{i=1}^{m} c_i n_i + C_0\right) &= V\left(\sum_{i=1}^{m} c_i n_i\right) + V\left(c'n'\right) \\ &= \sum_{i=1}^{m} n_i^2 \operatorname{Var}(c_i) + n'^2 \sigma_{c'}^2 \\ &= \sum_{i=1}^{m} n_i^2 \sigma_{c_i}^2 + n'^2 \sigma_{c'}^2. \text{ (Say)} \end{aligned} \quad (3.3)$$

By the use of above values, this NLPP becomes

$$\left.\begin{aligned}
&\min_n .C = K_1 E\left(\sum_{i=1}^{m}(c_i n_i) + c'n' + C_0\right) + K_2\sqrt{V\left(\sum_{i=1}^{m}(c_i n_i) + c'n' + C_0\right)}\\
&\text{Subject to}\\
&\frac{\mathrm{V}}{\mathrm{n}} + \frac{V'}{n'} = V_0 \quad i = 1, 2, \ldots, m.\\
&\text{and} \quad V = \sum_{i=1}^{m} W_i\left(\overline{y}_i - \overline{y}_{\text{std}}\right)^2 \quad \text{and} \quad V' = \sum_{i=1}^{m} W_i^2\ S_i^2\\
&\qquad n_i \geq 2, n' \geq 5
\end{aligned}\right\} \tag{3.4}$$

In Eq. (3.4), objective function is deterministic, and $K_1 + K_2 = 1$ where K_1 and K_2 tell about the relative importance of mean and variance of $\left(\sum_{i=1}^{m} c_i n_i + c'n' + C_0\right)$, [see, pp. 599, Rao (1979)]. Therefore, Eq. (3.4) can be written as:

$$\left.\begin{aligned}
&\min_n\ C = K_1\left(\sum_{i=1}^{m} n_i \mu_{c_i} + n'c'_\mu + C_0\right) + K_2\left(\sum_{i=1}^{m} n_i^2\sigma_{c_i}^2 + n'^2\sigma_{c'}^2\right)^{1/2}\\
&\text{Subject to}\\
&\text{and} \quad V = \sum_{i=1}^{m} W_i\left(\overline{y}_i - \overline{y}_{\text{std}}\right)^2 \quad \text{and} \quad V' = \sum_{i=1}^{m} W_i^2\ S_i^2 \quad i = 1, 2, \ldots, m.\\
&n_i \geq 2, n' \geq 5 \quad i = 1, 2, \ldots, m\\
&\frac{\mathrm{V}}{n} + \frac{V'}{n'} = V_0 \quad i = 1, 2, \ldots, m.
\end{aligned}\right\} \tag{3.5}$$

Here, in Eq. (3.5), objective function is given in terms of expected cost values and variance of $\left(\sum_{i=1}^{m} c_i n_i + c'n' + C_0\right)$; then, estimators of $E\left(\sum_{i=1}^{m} c_i n_i + c'n' + C_0\right)$ and variance $V\left(\sum_{i=1}^{m} c_i n_i + c'n' + C_0\right)$ will be used.

Therefore, an equivalent deterministic problem can be written as:

$$\left.\begin{aligned}
&\min_n C = K_1 \hat{E}\left(\sum_{i=1}^{m} n_i \mu_{c_i} + c'n' + C_0\right) + K_2\sqrt{\hat{V}\left(\sum_{i=1}^{m} n_i^2\sigma_{c_i}^2 + n'^2\sigma_{c'}^2 + C_0\right)}\\
&\text{Subject to}\\
&\frac{\mathrm{V}}{n} + \frac{V'}{n'} = V_0 \quad i = 1, 2, \ldots, m.\\
&\text{and} \quad V' = \sum_{i=1}^{m} W_i\left(\overline{y}_i - \overline{y}_{\text{std}}\right)^2 \quad \text{and} \quad V = \sum_{i=1}^{m} W_i^2\ S_i^2 \quad i = 1, 2, \ldots, m.\\
&\qquad n_i \ \geq 2, n' \geq 5 \quad i = 1, 2, \ldots, m.
\end{aligned}\right\} \tag{3.6}$$

$$\begin{aligned}
\text{Estimator of } E\left(\sum_{i=1}^{m} c_i n_i + c'n' + C_0\right) &= E\left(\sum_{i=1}^{m} c_i n_i\right) + E\left(c'n'\right) + C_0\\
&= \sum_{i=1}^{m} n_i E(c_i) + n'c'_\mu + C_0\\
&= \sum_{i=1}^{m} n_i \bar{c}_i + n'c'_\mu + C_0
\end{aligned} \tag{3.7}$$

$$\text{Estimator of } V\left(\sum_{i=1}^{m} c_i n_i + c'n' + C_0\right) = \hat{V}\left(\sum_{i=1}^{m} c_i n_i + c'n' + C_0\right)$$
$$= \hat{V}\left(\sum_{i=1}^{m} c_i n_i\right) + V(c'n')$$
$$= \sum_{i=1}^{m} n_i^2 \hat{V}(c_i) + n'^2 \sigma_{c'}^2$$
$$= \sum_{i=1}^{m} n_i^2 \sigma_{c_i}^2 + n'^2 \sigma_{c'}^2 \tag{3.8}$$

Therefore, an equivalent objective function with given accuracy can be written as

$$\left.\begin{array}{l}
\underset{n}{\text{Min.}}\, C = K_1\left(\sum_{i=1}^{m} n_i\, \overline{c}_i + n'c'_\mu + C_0\right) + K_2\sqrt{\left(\sum_{i=1}^{m} n_i^2 \sigma_{c_i}^2\right) + n'^2 \sigma_{c'}^2 + C_0} \\
\text{Subject to} \\
\frac{V}{n} + \frac{V'}{n'} = V_0 \quad i = 1, 2, \ldots, m. \\
\text{and } \; V' = \sum_{i=1}^{m} W_i \left(\overline{y}_i - \overline{y}_{\text{std}}\right)^2 \; \text{ and } \; V = \sum_{i=1}^{m} W_i^2\, S_i^2 \quad i = 1, 2, \ldots, m. \\
\qquad n_i \geq 2, n' \geq 5 \quad i = 1, 2, \ldots, m
\end{array}\right\} \tag{3.9}$$

4 Numerical Illustration

Consider the problem with the following data information where population is divided into two groups (Table 1).

From the mentioned data, NLPP can be written as:

a. **when the cost function is known**

Table 1 Data based on two strata

ith stratum	n_i	n'	W_i	S_i^2	c_i
1	40	154	22	5.55	15
2	46	189	21.45	3.4	22

$$\left.\begin{aligned}&\underset{n,n'}{\text{Min.}}\, C = 15n_1 + 22n_2 + 28n' + 25\\&\text{Subject to}\\&\frac{1.11}{n_1} + \frac{1.032}{n_2} + \frac{1.28}{n'} = 0.211\\&n_1, n_2 \geq 2 \quad \text{and} \quad n' \geq 5\end{aligned}\right\}$$

Using LINGO software, this NLPP is determined. Since sample sizes are n_1, n_2 and n', respectively, so it should be an integer. We obtain the optimal value of the problem after 46,700 iterations, as follows:
$C = 1283, n_1 = 14, n_2 = 12\ n' = 28.$

b. **When the cost function is random (unknown)**

$$\left.\begin{aligned}&\underset{n,n'}{\text{Min.}}\, C = 0.65\left(19n_1 + 24n_2 + 33n' + 25\right) + 0.35\left(25\,n_1^2 + 16\,n_2^2 + 36n'^2\right)^{1/2}\\&\text{Subject to}\\&\frac{1.11}{n_1} + \frac{1.032}{n_2} + \frac{1.28}{n'} = 0.211\\&n_1, n_2 \geq 2 \quad \text{and} \quad n' \geq 5\end{aligned}\right\}$$

Using LINGO software, this NLPP is determined. Since sample sizes are n_1, n_2 and n', respectively, so it should be an integer. We obtain the optimal value of the problem after 75,213 iterations given as:

$$C = 7241.85, n_1 = 14, n_2 = 12\ n' = 28$$

5 Conclusion

The throughout study of cost function in the objective function in stratified double sampling has been presented where cost parameters are considered as normal random variable which is converted into an allocation problem. To obtain optimum allocation, many successful attempts have been made. By applying LINGO computer program, optimum solution can be obtained very easily.

References

1. Diaz Garcia JA, Garay Tapia MM (2007) Optimum allocation in stratified surveys: stochastic programming. Comput Stat Data Anal 51:3016–3026
2. Javaid S, Bakhshi ZH, Khalid MM (2009) Optimum allocation in stratified sampling with random costs. Int Rev Pure Appl Math 5(2):363–370

3. Kokan AR (1963) Optimum allocation in multivariate surveys. J Roy Statist Sot Ser A 126:557–567
4. Kokan AR, Khan S (1967) Optimum allocation in multivariate surveys: an analytical solution. J Roy Stat Soc Ser B 2:115–125
5. Neyman J (1938) Contribution to the theory of sampling human populations. J Am Stat Assoc 33:101–116
6. Prekopa A (1995) Stochastic programming. In: Series mathematics and its applications. Kluwer Academic Publishers
7. Rao JNK (1973) On double sampling for stratification and analytical surveys. Biometrika 60:125–133
8. Iqbal A et al (eds) (2020) Soft computing in condition monitoring and diagnostics of electrical and mechanical systems, vol 1096. In: Advances in intelligent systems and computing. Springer, Singapore. https://doi.org/10.1007/978-981-15-1532-3
9. Iqbal A et al (eds) (2020) Meta heuristic and evolutionary computation: algorithms and applications, vol 1096. In: Studies in computational intelligence. Springer, Singapore. https://www.springer.com/gp/book/9789811575709

Power Enhancement by Transmitting AC–DC Power Simultaneously—An Experimental Implementation

Marwa Alfouly, Safa Ahmed, Atif Iqbal, Hassan Mehrjerdi, and Faiz Ahmad

Abstract Power transfer enhancement has become an issue because of thermal and stability limitations associated with AC transmission. This paper presents a model at which transmission lines are loaded near to their maximum ampacity unlike the existing AC power systems. The proposed concept is known as simultaneous AC–DC transmission where the existing AC lines can carry AC current with DC superimposed on it without major structural modifications. Power transfer capability of 70-km transmission line is enhanced by converting a fraction of AC power from the existing system to DC power which will be put in the sending end transformer through the neutral. Transformer zigzag connection eradicates the issue of core saturation that results from superimposed DC component. Transmission line data is provided by Qatar general electricity and water corporation (Kahramaa). Theoretical and real-time simulation results show that power transfer capability has been enhanced by 89%. Transmission angle is increased beyond 30°. MATLAB\Simulink and real-time digital simulator (RTDS) are two different softwares used to model and simulate the AC–DC transmission system offline and in real time, respectively. Laboratory prototype shows that power transfer capability has been enhanced by 33%. Practical results and waveforms had been recorded using the power analyzer PA400.

M. Alfouly · S. Ahmed · A. Iqbal (✉) · H. Mehrjerdi
Department of Electrical Engineering, Qatar University, Doha, Qatar
e-mail: atif.iqbal@qu.edu.qa

M. Alfouly
e-mail: me1401768@qu.edu.qa

S. Ahmed
e-mail: sa1303637@qu.edu.qa

H. Mehrjerdi
e-mail: hasan.mehrjerdi@qu.edu.qa

F. Ahmad
Department of Science and Technology, Saharsa Engineering College, Bihar, India
e-mail: Faizahmad831@gmail.com

A. Iqbal et al. (eds.), *Renewable Power for Sustainable Growth*, Lecture Notes in Electrical Engineering 723, https://doi.org/10.1007/978-981-33-4080-0_3

Keywords Power transmission · AC/DC simultaneous power transmission · Harmonic injection · Zigzag transformer

1 Introduction

It is desired to investigate approaches that achieve a significant improvement in power transfer capability. One approach is through transmitting AC–DC power simultaneously which allows power lines to operate near to their maximum ampacity [1]. DC is injected into transmission line and retrieved back using zigzag transformer. Line conductor carries AC current along with one-third of DC current. Figure 1 shows the structure for the developed laboratory prototype. It is aimed to carry out the load flow analysis for the developed model and evaluate its feasibility [2–4].

2 Design of AC–DC Line

2.1 Design Specifications

Figure 2a, b illustrates the corresponding circuit of the model [5].

The presented scheme is an AC single circuit where a six-pulse rectifier bridge is used to obtain DC power. The secondary winding of the zigzag transformer is connected to the bridge through the neutral point at the sending side. DC power is converted back to AC through six-pulse inverter at the receiving side. Transformer zigzag connection is necessary to eliminate the problem of core saturation because of DC current, and it combines the advantages of both Y and Δ connections [1].

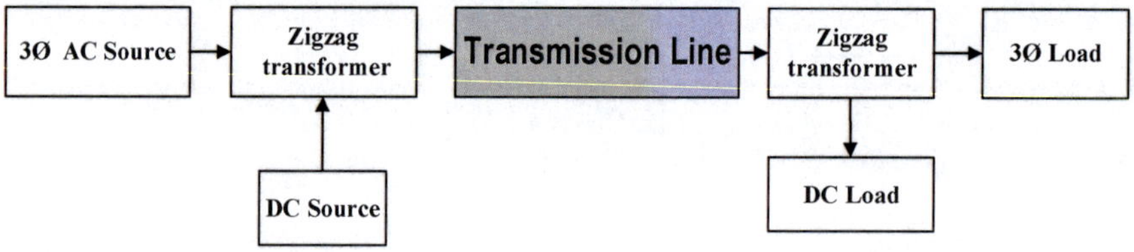

Fig. 1 Block diagram for a laboratory prototype model

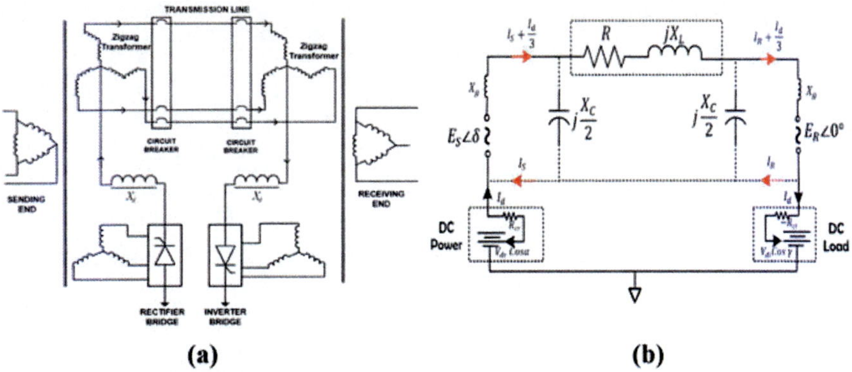

Fig. 2 Combined AC–DC power scheme **a** schematic, **b** equivalent circuit

3 Model Specifications

3.1 Mathematical Modeling for the Combined AC–DC Power System

Design constraints imply that transmission line is operating close to its maximum capacity. Hence,

$$I_{\text{Thermal}} = I = \sqrt{I_{\text{a}}^2 + \left(\frac{1}{3}I_{\text{d}}\right)^2} \quad (1)$$

I The total current (both ac and dc) following in the line conductors.
I_{a} AC current in rms flowing in the line conductors.
$1/3I_{\text{d}}$ DC current flowing in the line conductors.

Using (1), line losses can be expressed as follows:

$$P_{\text{L}} \cong 3I^2R \quad (2)$$

Conductor current as a factor of the thermal limit, *i*th is:

$$I_{\text{a}} = xI_{\text{th}} \text{ and } I_{\text{dc}} = \sqrt{1 - x^2}I_{\text{th}} \quad (3)$$

Two zero crossings are acquired if $\left({}^{I_{\text{d}}}/_{3I_{\text{a}}}\right) < \sqrt{2}$. Peak voltage of the compound AC/DC line is:

$$V_{\text{ac/dc(max)}} = \sqrt{2}V_{\text{ac}} + V_{\text{dc}} \quad (4)$$

Table 1 Transmission line data

Parameters	Values
Frequency	50 Hz
Voltage rating	220 kV
Transmission capacity	150 MVA
Thermal limit	437 MVA
Route length	70.0 km
AC resistance	0.1068 Ω/km
DC resistance	0.0885 Ω/km
Capacitive reactance	2.34580052 Ω/km
Inductive reactance	03693 Ω/km

$V_{\frac{ac}{dc(max)}}$ Peak voltage of the compound AC/DC line.
V_{ac} RMS AC voltage of the enhanced mode.
V_{dc} DC voltage in enhanced model.

Based on (4), two zero crossings in the LN voltage are achieved if $\left({}^{V_d}/_{V_{ac}}\right) < \sqrt{2}$. However, line-line voltage has no DC component, and its maximum value is $\sqrt{6}V_{ac}$. Assuming $\left({}^{V_d}/_{V_{ac}}\right) < k$,

$$\frac{P_{dc}}{P_{ac}} = \frac{V_{dc} I_{dc}}{3 V_{ac} I_a \cos\theta} = \frac{k\sqrt{1 - x^2}}{x \cos\theta} \tag{5}$$

Using (5), total power P_t is

$$P_t = P_{ac} + P_{dc} = \left(1 + \frac{k\sqrt{1 - x^2}}{x \cos\theta}\right) P_{ac} \tag{6}$$

3.2 *Simultaneous AC–DC Model*

The proposed concept is applied to a transmission line of 70 km length in Qatar. Table 1 summarizes the rating of the line.

3.3 *Transmission Network*

According to Table 1, voltage of the existing AC line per phase in rms (V_{ph}) equals to 127 kV. Based on (4), the maximum DC voltage and the minimum AC voltage are:

$$V_{\mathrm{dc}} = \frac{V_{\mathrm{ph}}}{\sqrt{2}} = 89.815\,\mathrm{kV} \text{ and } V_{\mathrm{ac}} = \frac{V_{\mathrm{ph}}}{2} = 63.509\,\mathrm{kV} \tag{7}$$

$$I_{\mathrm{th}} = \frac{437 \times 10^6}{\sqrt{3}\left(220 \times 10^3\right)} = 1.147\,\mathrm{kA} \tag{8}$$

$$I_{\mathrm{ac}} = \frac{150 \times 10^6}{\sqrt{3} \times 122 \times 10^3} = 709.86\,\mathrm{A} \tag{9}$$

For safety, set the current to 100 A below its thermal limit. The DC current injected to the line is calculated by applying (3)

$$x = 0.6761 \text{ and } I_{\mathrm{dc}} = 2.324\,\mathrm{kA} \tag{10}$$

In this case, two zero crossings are ensured that can be mathematically verified by (11)

$$\frac{I_{\mathrm{dc}}}{3I_{\mathrm{ac}}} = \frac{2324}{3 \times 709.86} = 1.0914 \leq \sqrt{2} \tag{11}$$

Assuming unity power factor, $k = 1.134$ and applying (6), power enhancement is such that $P_{\mathrm{t}} = 2.237\ P_{\mathrm{ac}}$

Figure 3 shows that maximum power transfer occurs at transmission of 43° where the power transfer is 279 MW which implies power upgradation of 89%.

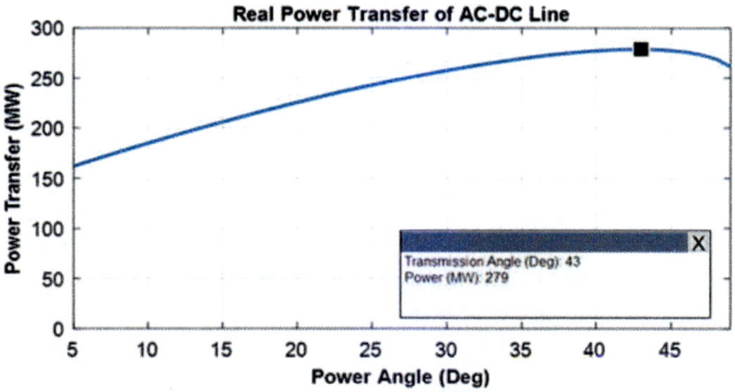

Fig. 3 Real power transfer of AC–DC line

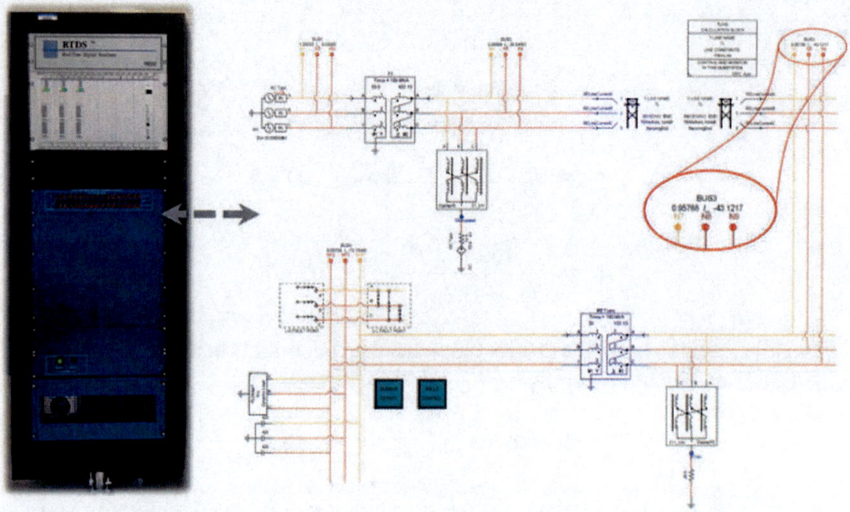

Fig. 4 Developed RSCAD model of the AC–DC system

4 Real-Time Simulation

4.1 The Developed Model for Transmitting AC–DC Power Simultaneously on RSCAD

Figure 4 displays RSCAD model of the system.

4.2 Real-Time Simulation Results

Operating voltage is 220 kV for the AC–DC line.

Figure 5 shows the operating phase voltages of the combined AC–DC power line.

Thermal current of the line is 1.147 kA. It is designed to operate about 100 A below the thermal current (1.050 kA) for safety. Figure 6 shows the RMS current at the sending and receiving sides.

DC current injected is successfully retrieved as shown in Fig. 7. For the AC circuit breaker to function probably during faults by maintaining zero crossings as shown in Fig. 8

Figure 9 shows the power recorded during real-time simulation at the transmission line ends.

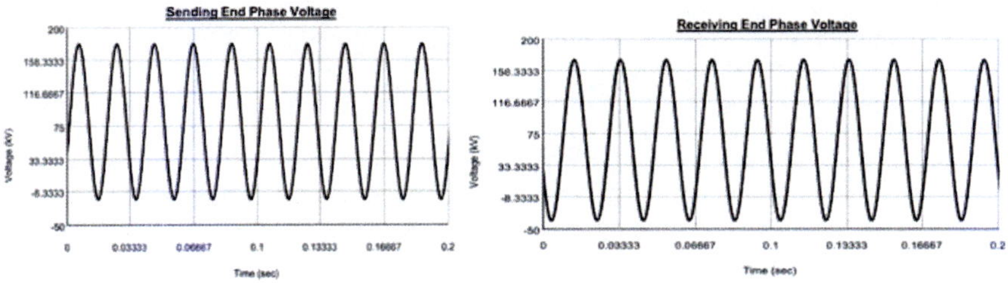

(a) (b)

Fig. 5 **a** Phase A sending side voltage, **b** phase A receiving side voltage

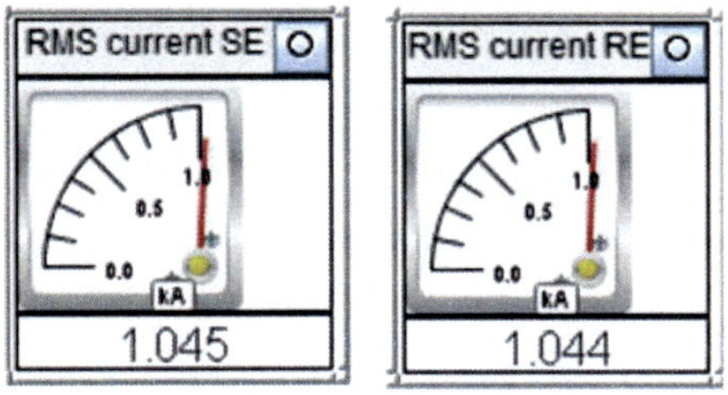

(a) (b)

Fig. 6 Current of AC–DC line **a** sending side, **b** receiving side

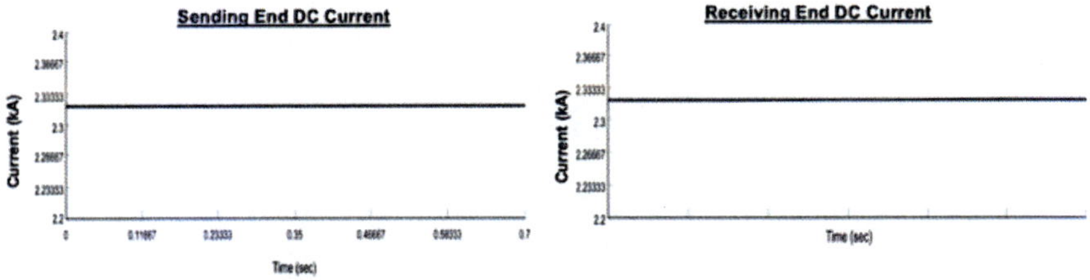

(a) (b)

Fig. 7 **a** Sending side DC current, **b** receiving side DC current

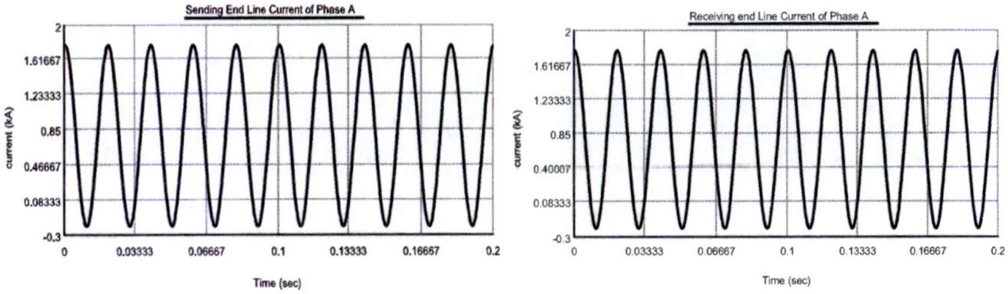

(a) (b)

Fig. 8 **a** Sending side line current, **b** receiving side line current

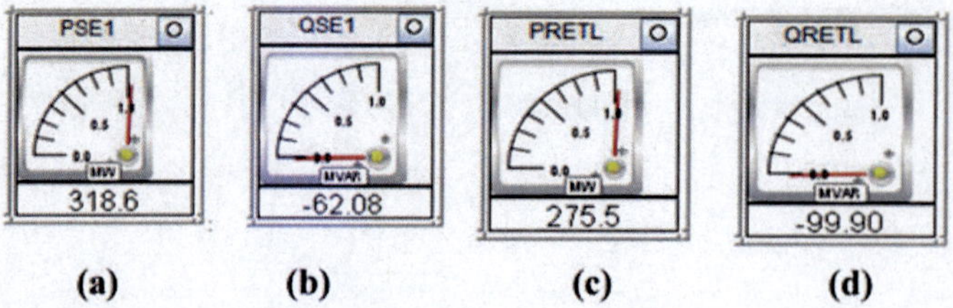

Fig. 9 Power flow on 70-km transmission line **a** real power at the sending side, **b** M_{var} at the sending side, **c** real power at the receiving side, **d** M_{var} at the receiving side

5 Hardware Implementation

Experimental setup is shown in Fig. 10. Line model (MV1420) is a lumped π circuit packaged in 410 × 245 × 160 mm box.

Figure 11 shows DC current at both sides of the line. Figure 12 shows the combined AC–DC waveforms at both terminals of the line.

Figure 13 shows the receiving side operating conditions.

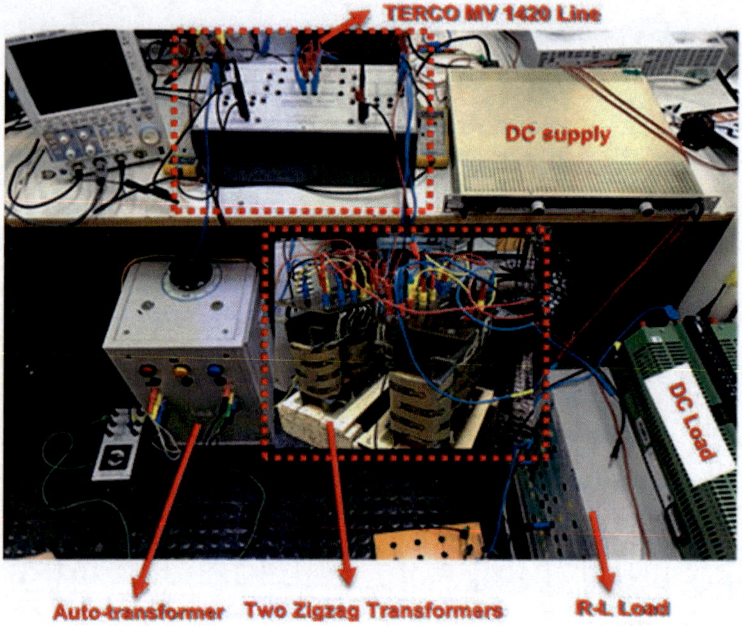

Fig. 10 Experimental setup

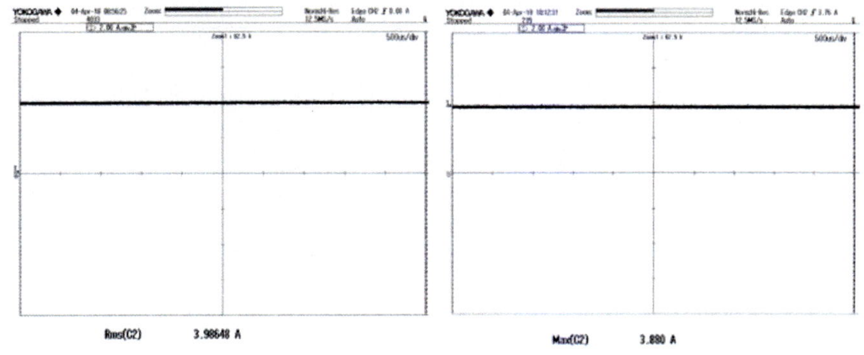

(a) (b)

Fig. 11 DC current **a** sending side, **b** receiving side

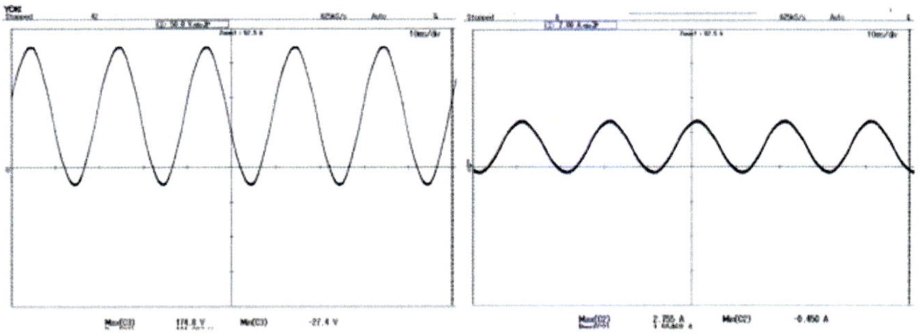

(a) (b)

Fig. 12 Operating conditions at the sending side **a** sending side phase voltage, **b** sending side line current

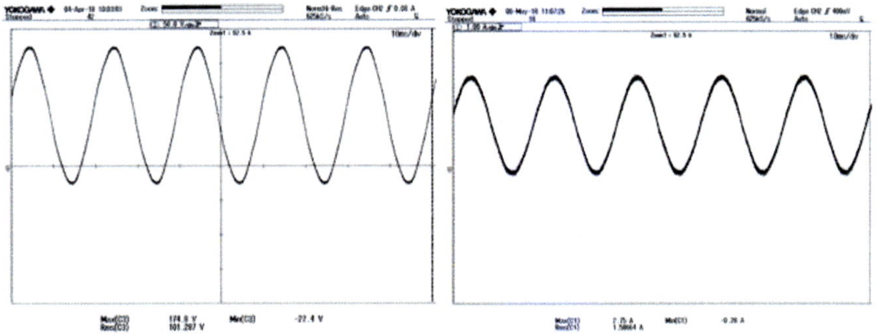

(a) (b)

Fig. 13 Receiving side operating conditions **a** phase voltage at the receiving side, **b** line current at the receiving side

6 Conclusion

Theoretical design and real-time simulation have been performed. Results indicate enhancement in power transfer capability and increase in transmission angle; about 87% upgradation in transferred power and corresponding transmission angle is 43°. A scaled-down model for transmitting AC–DC power simultaneously is designed. The feasibility of transmitting AC–DC power simultaneously is verified. Power transfer capability has been enhanced by 33%. Further work will include working on longer transmission line for higher power transfer capability.

References

1. Rahman H, Khan BH (2008) Stability improvement of power system by simultaneous AC–DC power transmission. Electr Power Syst Res 78(4):756–764. ISSN 0378-7796. https://doi.org/10.1016/j.epsr.2007.05.020. https://www.sciencedirect.com/science/article/abs/pii/S0378779607001344
2. Lantero A (2017) The war of the currents: AC versus DC power. Energy Gov [Online]. Available https://energy.gov/articles/war-currents-ac-vs-dc-power. Accessed 17 Oct 2017
3. Iqbal A et al (eds) (2020) Soft computing in condition monitoring and diagnostics of electrical and mechanical systems, vol 1096. In: Advances in intelligent systems and computing. Springer, Singapore. https://doi.org/10.1007/978-981-15-1532-3
4. Iqbal A et al (eds) Meta heuristic and evolutionary computation: algorithms and applications, vol 1096. In: Studies in computational intelligence. Springer, Singapore. https://www.springer.com/gp/book/9789811575709
5. Kondiparthi S et al Power transfer enhancement in transmission line by combining AC–DC transmission. IJERA 1(2):194–201

Step-by-Step Design and Simulation of Boost Controller for L- and LCL-Filters for EV Fast Charging Systems

Anas Ashraf Berbar, Amith Khandakar, Annaufal Rizqullah, Syed Rahman, Damyan Kraev, Atif Iqbal, and Mohammad Rafi Ahmad

Abstract Adoption of electric vehicles (EVs) is an important measure for greenhouse gas reduction in many countries worldwide. The increased production of EV in the market will be associated with continuous demand of installing and maintaining charging stations (CSs) in public areas. The charging station consists of different power converters for processing the power. Filter is an important component of the EV charging station. Correct design of filter is important for the overall satisfactory operation of the EV chargers. This paper aims to provide designing steps of a closed-loop controller of the filter in the EV battery charging system. The contributions of the paper are (1) step-by-step design procedure for closed-loop input power controller for DC–DC converter, (2) simulation model development using systematic approach for closed-loop controller for LCL- and L-filter to validate the design procedure, (3) general overview of EV charging system for researchers working on EV CS. The authors believe that this paper is useful for interested pupils linked to EV CS research and working to make its adoption quicker and more convenient.

Keywords EV fast charging · LCL- and L-filter · Simulation · Controller design

---

A. A. Berbar (✉) · A. Khandakar · A. Rizqullah · S. Rahman · D. Kraev · A. Iqbal · M. R. Ahmad
Department of Electrical Engineering, Qatar University, Doha, Qatar
e-mail: aa1403703@qu.edu.qa

A. Khandakar
e-mail: amitk@qu.edu.qa

A. Rizqullah
e-mail: ar1605250@qu.edu.qa

S. Rahman
e-mail: syed.rahman.rahman@gmail.com

D. Kraev
e-mail: dk1505452@qu.edu.qa

A. Iqbal
e-mail: atif.iqbal@qu.edu.qa

M. R. Ahmad
e-mail: dk1505452@qu.edu.qa

A. Iqbal et al. (eds.), *Renewable Power for Sustainable Growth*, Lecture Notes in Electrical Engineering 723, https://doi.org/10.1007/978-981-33-4080-0_4

1 Introduction

There is a significant increase in demand for energy resources, especially in the transportation sector. However, the present energy supplies are mostly from fossil fuels, which is impractical to depend on due to its quantity limitations. There are other drawbacks of such energy sources such as global warming, carbon emissions and greenhouse gases (GHGs) [1–5]. On the other hand, electric vehicles (EVs) are becoming a popular alternative in the mobility sector due to its comparative advantages. The increased production of EV in the market will be associated with the continuous demand for charging stations (CSs) in public areas.

Charging systems have been established in various countries and categorized into three levels depending on the speed/time of the charging system. Standards for each country is different, but in general, the levels of charging are Level-I charging which requires 8–12 h for a full charge, Level-II charging taking 4–6 h for a full charge, Level-III charging which provides 80% charge in around 30 min, but this type of charging is not compatible with all EVs [6–11]. Table 1 shows generally the levels of EV charging.

Fast chargers are produced to keep up with the EV revolution and consumers demands and to make EV adoption in the community faster [12]. In this paper, an important component of fast EV charger is proposed which will be described in several stages as shown in Fig. 1. The first stage of the charging system is the filter, and choosing a suitable filter type is crucial for the line harmonics, especially when

Table 1 EV charging levels

Charging levels	Voltage	Current (A)	Power level (kW)	Charger type	Charging time
AC level-I	120 V_{ac} (US) 230 V_{ac} (EU)	12	1.4	On-board	14–18 h
AC level-II	240 V_{ac} (US) 400 V_{ac} (EU)	17	4	On-board	4–6 h
DC level-I	120–240 V_{dc}	80	36	Off-board	1 h
DC level-II	240–400 V_{dc}	200	90	Off-board	20–30 min
DC level-III	400–600 V_{dc}	400	240	Off-board	<10 min

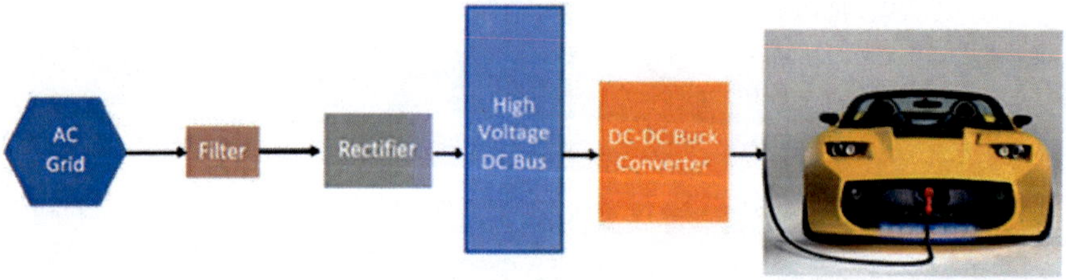

Fig. 1 Open-loop system block diagram

operating at high switching frequency and bidirectional power flow. Generally, L- and LCL-filters are the most commonly used in the rectifier side. LCL-filter is chosen due to its advantages over the L-filter such as smaller inductance values and less total harmonic distortion (THD) [13]. However, the order of the filter is increased. For this paper, the LCL-filter is designed according to [14]. LCL-filter is shown in Fig. 4.

The second stage is the AC/DC converter with an active power factor correction (PFC). A controlled rectifier is controlled by power devices such as IGBTs and MOSFETs and with the help of control pulses, which are generated by comparing a sawtooth waveform with input phase voltages, which determines the input power quantity and quality. It is important to design a PFC controller properly because a substandard controller might damage the power supply source due to its high THD. The third stage is the DC/DC synchronous buck converter to supply required voltage to the battery linked to an open-loop control for constant current (CC) mode and then switching to constant voltage (CV) mode. In CC mode, the objective is to supply the maximum current until the state of charge (SoC) is 80% in the minimum time possible and then switch to CV mode. In CV mode, the aim is to maintain the output voltage at certain DC level, and the current will flow through based on the potential difference until the SoC reaches to 100%. Figure 2 shows the closed-loop block diagram, Fig. 3 shows the CC-CV charging modes, and Fig. 5 shows the second and third stages of the CS (Fig. 4).

Considering the importance of EV adoption, in the community, a lot of research is being conducted in all the blocks of CS. These researches are done to begin the revolution of fast charging station of EV. In [15], the authors have discussed the importance of filter design and harmonic disturbances in the frequency range between 2 and 150 kHz, which are often referred to by the term supraharmonics. They developed a modeling procedure of an EV battery charger emitting supraharmonics in order to quantify them. In [16], the authors have presented the current control design procedure using LCL-filter. None of the papers in the literature has demonstrated the full design procedure of the filters used in the EV chargers. Thus, the objective of this paper is focused on a systematic design procedure for the second stage of

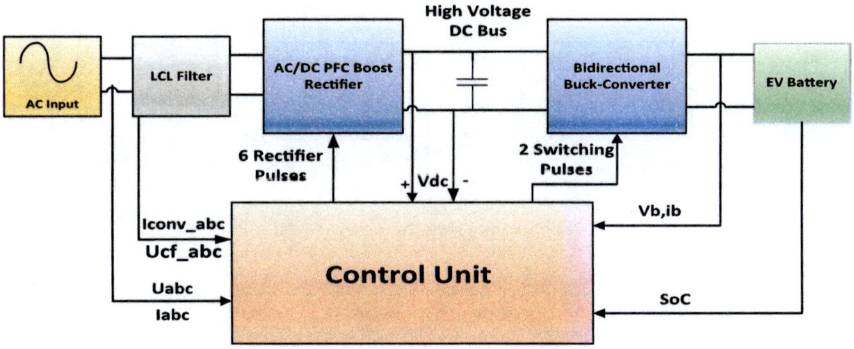

Fig. 2 Closed-loop system block diagram

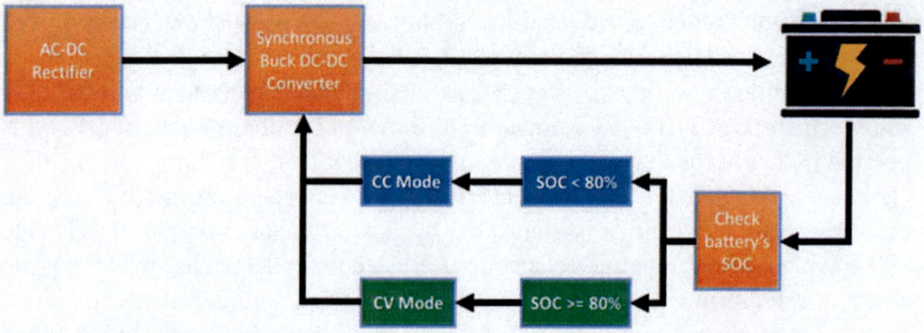

Fig. 3 CC-CV charging mode

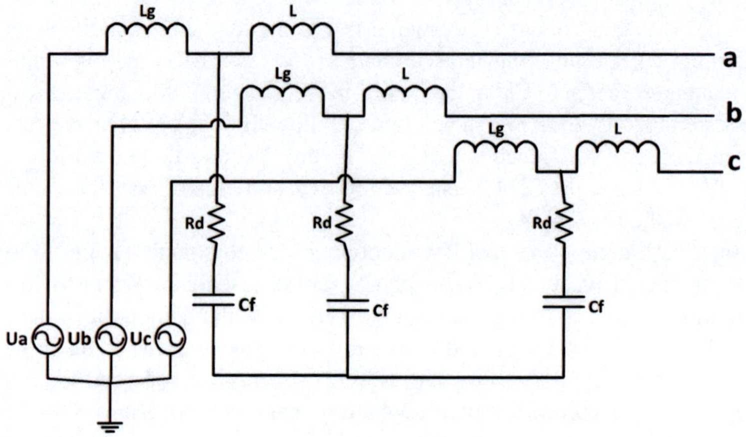

Fig. 4 LCL-filter

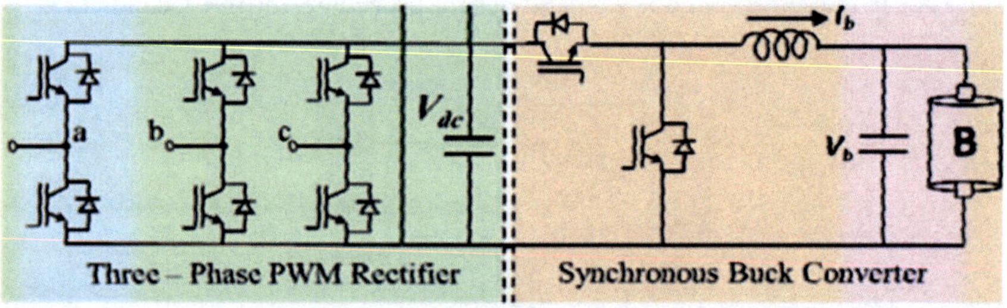

Fig. 5 Preliminary stages of the EV charging stations

EV charger, i.e., a closed-loop boost controller for AC/DC converter. This paper has demonstrated the simulation results of LCL- and L-filter used in conjunction with the fast EV chargers. The rest of the paper is divided into the sections below. The upcoming section presents a detailed discussion about the methodology adopted in this study for the LCL- and L-filter. Section 3 discusses about the simulated results, and finally, Sect. 4 articulates the future direction of the project.

2 Design Methodology

One of the ways to control the supplied reactive power and output DC voltage is to control the pulses of the AC/DC rectifier. This designed controller totally depends on the type of filter used in the front-end converter. However, there are four main inputs which are not affected by the type of filter used. These are U_{abc}, I_{abc}, V_{dc} and V_{dc*} as shown in Fig. 6.

(1) Draw the single-phase equivalent model of the system. This can help in finding the number of controllers needed and the parameters to be controlled.
(2) Decide the parameters needed in the controller from the previous step.
(3) Convert all parameters found in step (2) from the stationary phase coordinate system (abc) to the rotating coordinate system ($dq0$) using park transformation. (In the $dq0$ system, d-axis controls the output DC voltage, q-axis controls the reactive power supplied from the source. In other words, it controls the power factor [17]).
(4) Design a loop controller for the output DC voltage from the d-axis current values generated in step (3), taking into consideration that the output DC voltage is controlled by output DC current. In certain filters, the output current does not depend directly to the input current, and it will be clear from the equivalent model drawn in step (1). The d-axis controller should start from the output DC voltage.
(5) Design a loop controller for the power factor from the q-axis current values generated in step (3). The q-axis controller should start from the input current. For a unity power factor, $Iq^* = 0$.

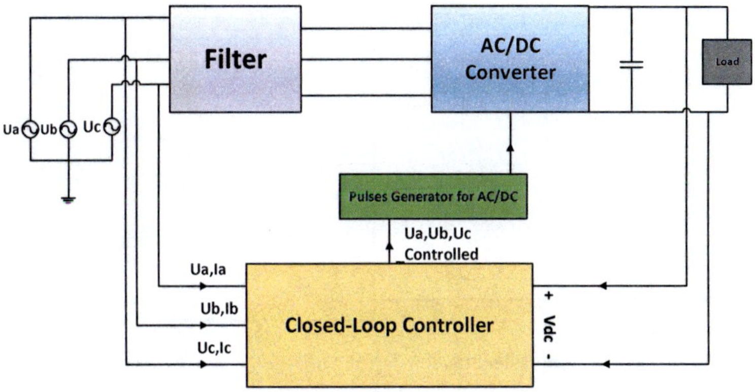

Fig. 6 Basic controller block diagram

(6) Choosing reasonable values of PI controllers, i.e., integral and proportional terms for PI controllers are crucial and they are highly affected by the filter component values. In general, PI controller terms for a well-designed filter should be [18]:

$0 < K_p\_\text{voltage} < 1.$
$K_i\_voltage > 1$
$K_p\_current > 1$
$K_i\_current > 1$

For applications other than EV charging, PI values may be out of range as well as special cases of filter designs [18].

The design steps were used for the closed-loop controllers for the two types of filters: L-filter and LCL-filter.

2.1 Controller Design for L-Filter

The step-by-step procedure of designing a controller for an L-filter converter is explained below in detail. It has to be noted that a constant three-phase balanced power supply is used.

1. The single-phase equivalent model of an L-Filter is shown in Fig. 7
2. From step (1), it can be seen that the output voltage depends on the output current which is directly related to the input current. Therefore the parameters needed in the controller are V_{dc}, $V_{\text{dc}*}$, U_{abc}, I_{abc}.
3. The three vector parameters (abc) found in step (2) are to be converted to ($dq0$) as shown in Fig. 8.
4. As mentioned in step (2), the Vdc depends on Iabc, so in d-axis, the controlled output DC voltage should be a reference for the I_{d}; afterward, the output of the controller should be subtracted from U_{d}, which can be seen in Fig. 9a.

In all figures, $\text{PI} = Kp + \frac{K_i}{s}$ and is inclusive of limiter block as shown in Fig. 9b.

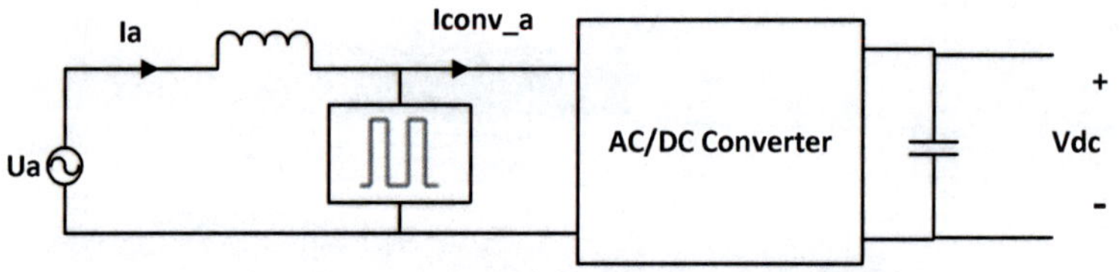

Fig. 7 Single-phase L-filter equivalent model

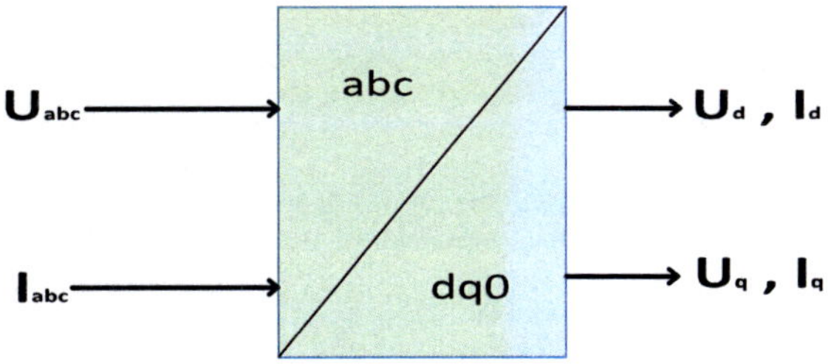

Fig. 8 L-filter abc to $dq0$ conversion

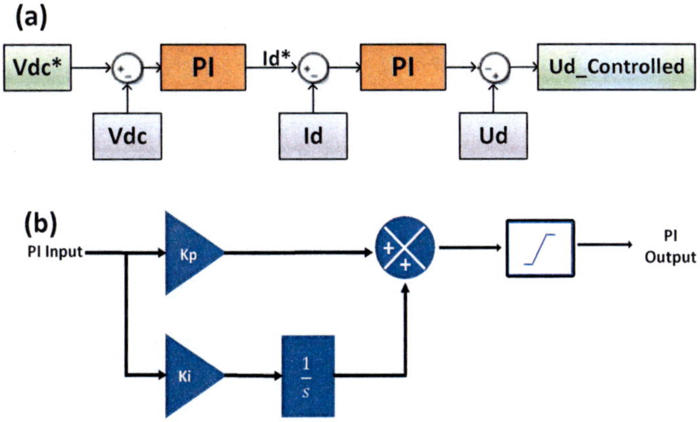

Fig. 9 L-filter direct-axis controller and PI block

5. In the q-axis, I_q can be controlled by having a reference of 0, and then, the output should be subtracted from U_q to generate the controlled U_q as can be shown in Fig. 10.

6. The chosen values of PI controllers are:

$$K_{p}\_\text{voltage} = 0.1$$
$$K_{i}\_\text{voltage} = 100$$
$$K_{p}\_\text{current} = 5$$
$$K_{i}\_\text{current} = 20.$$

The closed-loop block diagram is similar to the basic one shown in Fig. 6.

Fig. 10 L-filter quadrature-axis controller

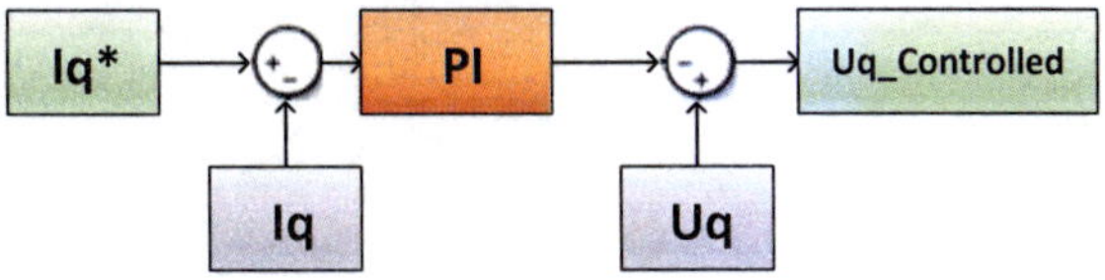

2.2 Controller Design for LCL-Filter

The step-by-step procedure to design a controller for an LCL-filter converter will be explained in detail. A constant three-phase balanced power supply is used.

(1) The single-phase equivalent model of an LCL-filter is shown in Fig. 11.
(2) From step (1), it can be noted that controlling the output voltage is equivalent to controlling the output current which is related to the converter current and the converter current depends on the voltage across C_f which depends on the input current. As a result, the parameters needed are V_{dc}, V_{dc*}, U_{abc}, I_{abc}, I_{Conv_abc}, U_{Cf_abc}.
(3) The three vector parameters (abc) found in step (2) are to be converted to ($dq0$) before starting the control as shown in Fig. 12.
(4) To control U_d from step (2), the controller should start by controlling V_{dc} and gives a reference to I_{d_conv}; then, the controlled output is a reference for U_{d_cf},

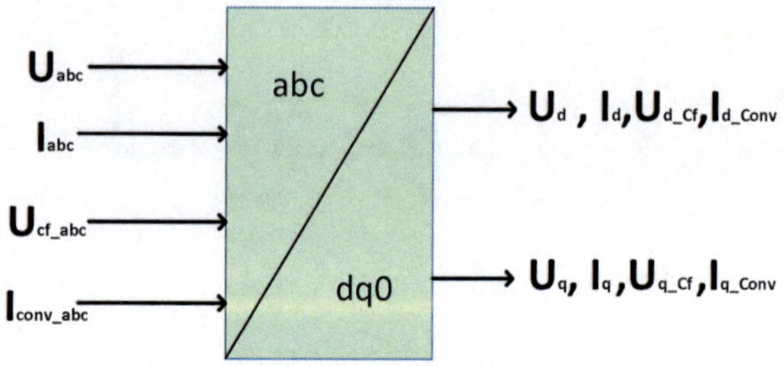

Fig. 11 LCL-filter abc parameters conversion to $dq0$

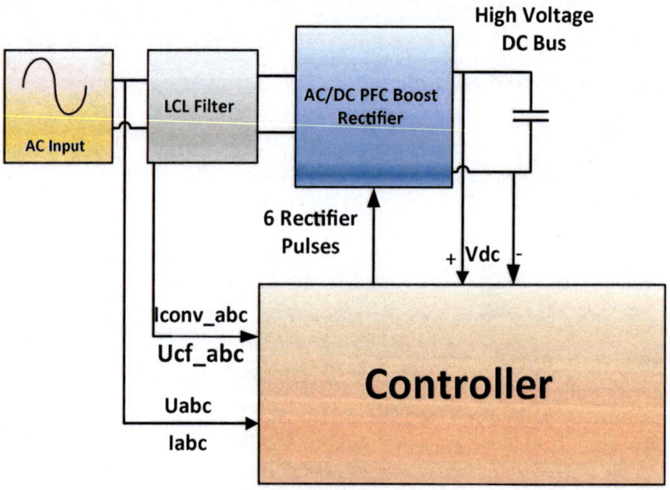

Fig. 12 LCL-filter closed-loop controller block diagram

and the output is the reference for Id; finally, the output is to be subtracted from U_d to generate the controlled V_d as shown in Figs. 13 and 14.

(5) In the *q*-axis, the controller starts from Iq to control the power factor; however, the controlled output from Iq should be a reference for U_{q_cf}, and the output is a reference for I_{q_conv}, finally subtracted from U_q to generate the controlled U_q, which can be seen in Fig. 15

(6) The chosen values of PI controllers are:

$$K_p\_\text{voltage} = 0.8$$
$$K_i\_\text{voltage} = 50$$
$$K_p\_\text{current} = 5$$
$$K_i\_\text{current} = 20.$$

LCL-filter closed-loop controller block diagram is shown in Fig. 13.

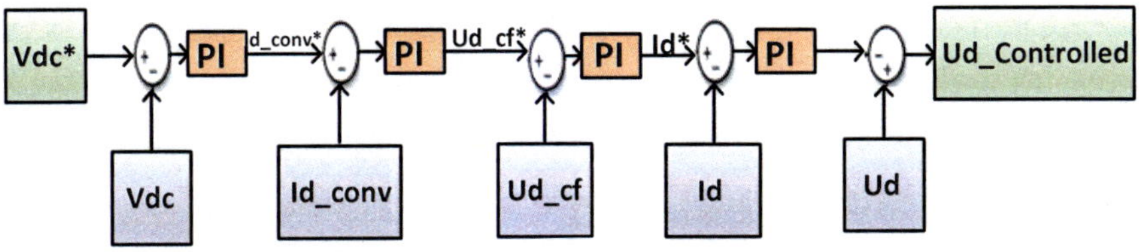

Fig. 13 LCL-filter *d*-axis control

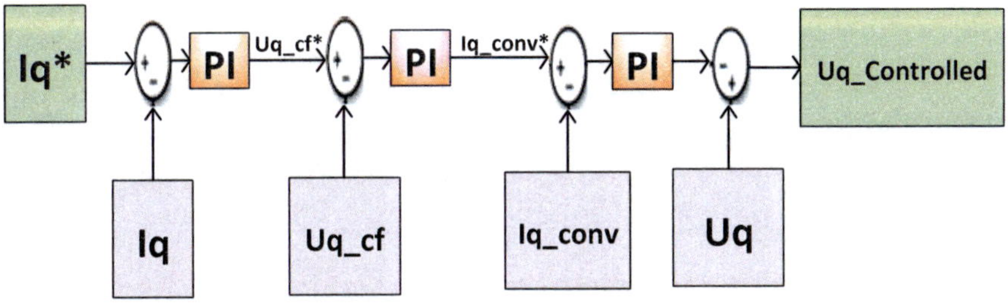

Fig. 14 LCL-filter *q*-axis control

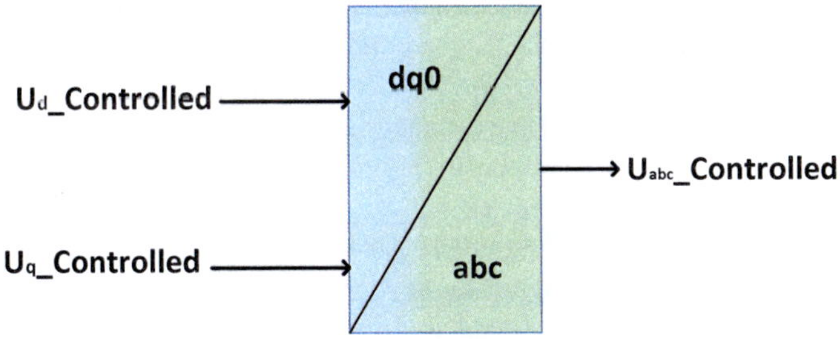

Fig. 15 Conversion of dq0 to abc

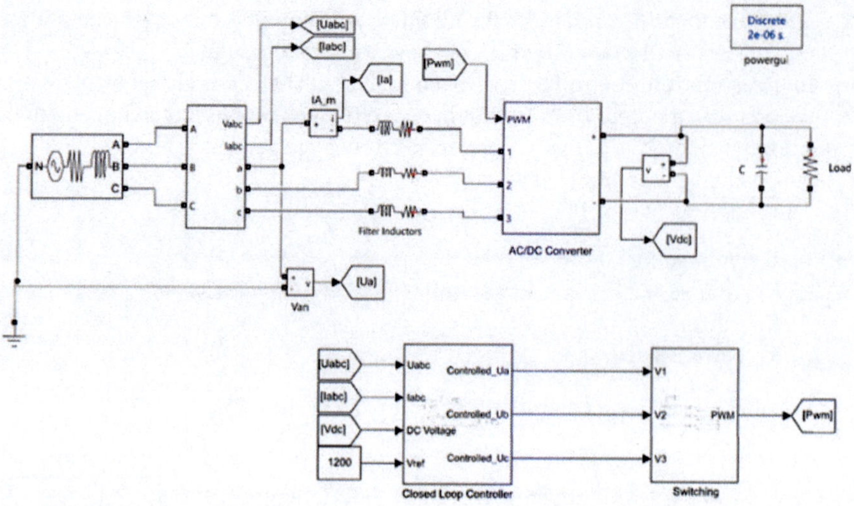

Fig. 16 Filter simulation system

The last step to generate the pulses of the AC/DC converter and to achieve the controlled pulses for all cases is to convert the $U_{\text{d\_Controlled}}$ and $U_{\text{q\_Controlled}}$ to abc domain as shown in Fig. 16.

U_{abc} controlled each phase is then compared with the sawtooth of a 100 kHz frequency to generate the pulses for the AC/DC Converter.

3 Simulation Results

3.1 L-Filter Simulation Results

The L-filter design has been simulated on MATLAB/Simulink. The overall system is shown in Fig. 16. The AC three-phase source used is a balanced three-phase voltage source with a frequency of 50 Hz and 415 V line to line. The line inductance value is 0.001 mH, and line resistance value is 0.1 mΩ. The reference DC voltage is 1200 V and the switching frequency of AC/DC converter is 100 kHz. Power devices used in the circuitry are MOSFETs (Fig. 17).

The result of the filter design is a regulated DC voltage with unity power factor. Figure 18 shows the output DC voltage, and Fig. 19 shows U_{a} and I_{a} for unity power factor.

It can be noticed that in the DC output voltage transient state, there is an overshoot, and this can be reduced by changing the integral term of controlling the voltage. However, it will affect the transient response time to reach the steady state, and this depends on the application as well as components used (Fig. 20).

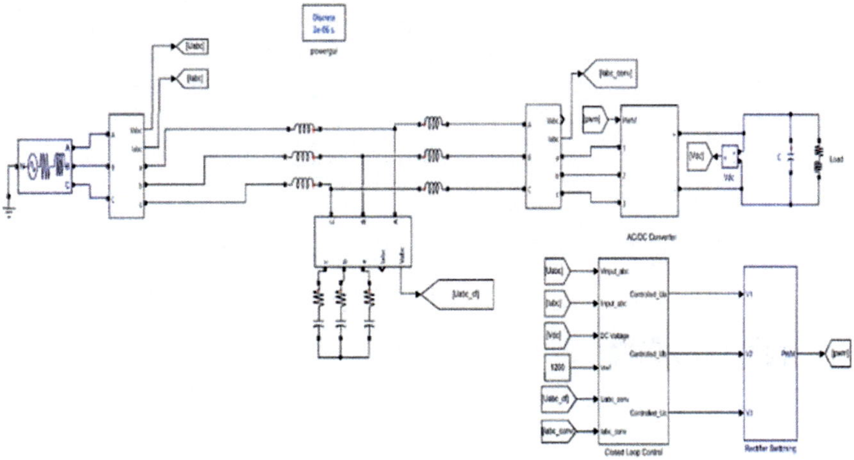

Fig. 17 Simulation system

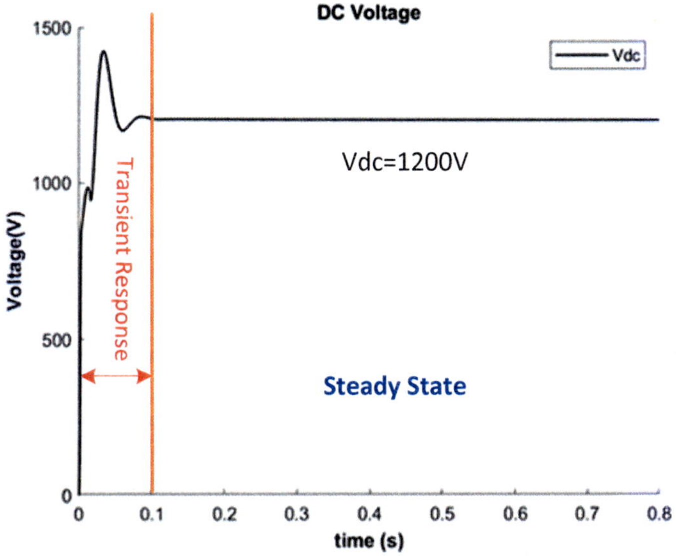

Fig. 18 L-filter controller DC output

3.2 LCL-Filter Simulation

Using the MATLAB/Simulink platform with the same assumptions as in the L-filter simulation. Figure 18 shows the overall system of the LCL-filter closed-loop rectifier (Fig. 21).

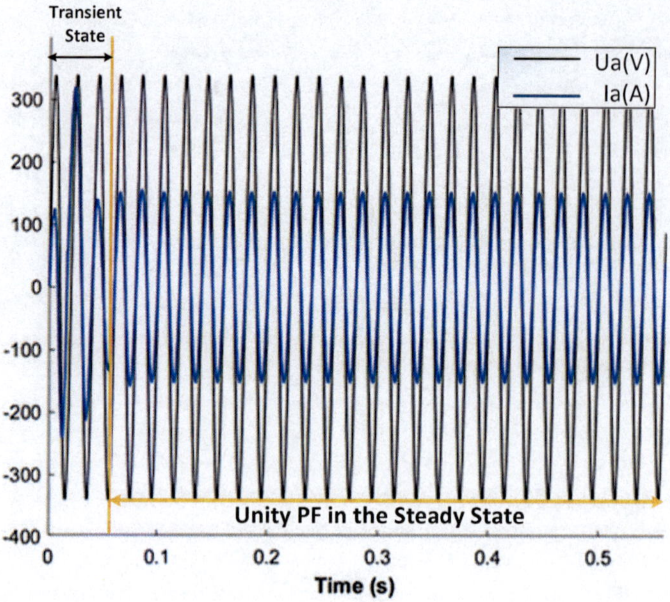

Fig. 19 U_a, I_a in the L-filter controller

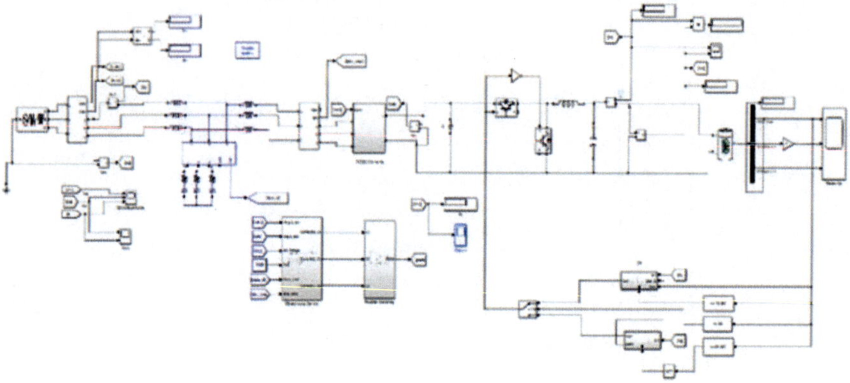

Fig. 20 Full charging system simulation

Figure 22 shows the output DC voltage and Fig. 23 shows U_a and I_a for unity power factor.

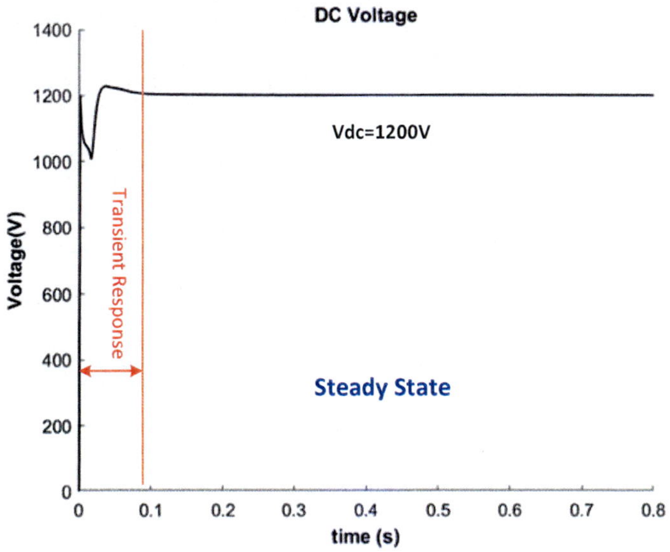

Fig. 21 LCL-filter controller output DC voltage

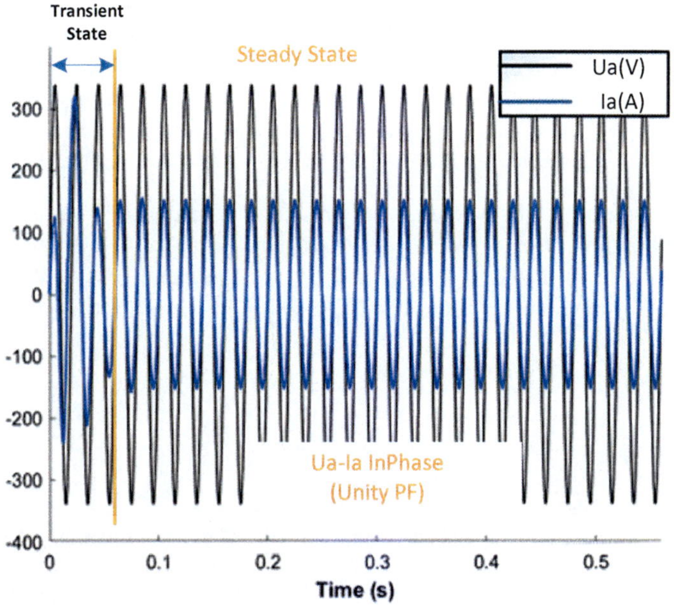

Fig. 22 LCL-filter controller U_a, I_a

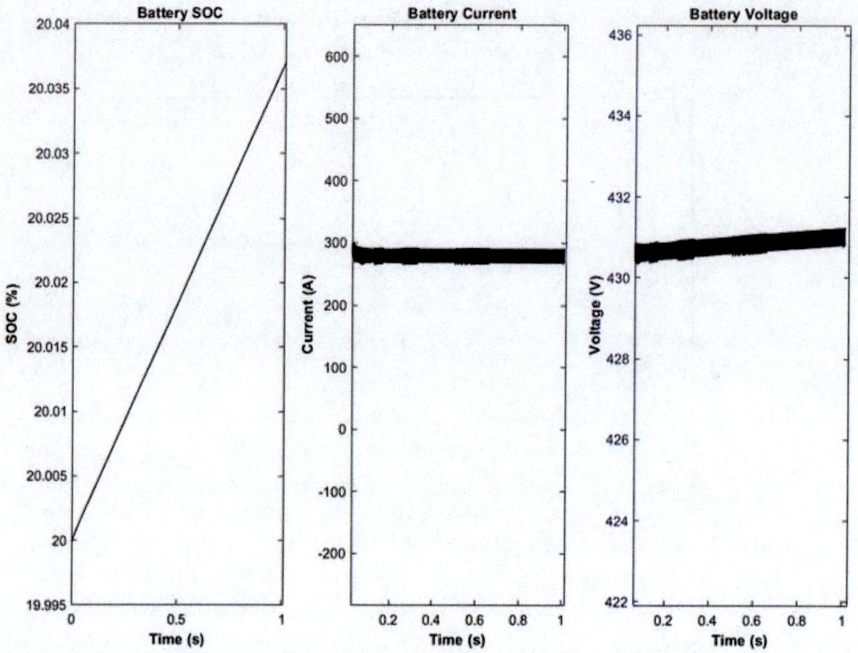

Fig. 23 CC-charging mode

3.3 CC-CV Full System Charging Results

For design verification, the whole system presented in this paper is designed on the MATLAB/Simulink platform and is shown in Fig. 21.

The simulation is done with the same LCL voltage parameters in the previous section and the same input source. The CC charging mode is shown in Fig. 24 and CV charging mode is shown in Fig. 24.

4 Conclusion

This paper presented a comprehensive yet simple procedure for design of LCL-filter for integration of EV charger to the utility grid. Firstly, different charging levels existing in the commercial space are discussed which signifies the need for reducing the charging time in order to promote EV culture and minimize range anxiety. Then, to achieve fast EV charging, topology consisting of PWM rectifier and buck converter is discussed. However, integration of this system to the grid requires design of LCL-filter and its associated control loop. For achieving this, decoupled vector control (based on $d-q$ axes control) is presented here. Finally, detailed simulation results are added for validation of the design procedure.

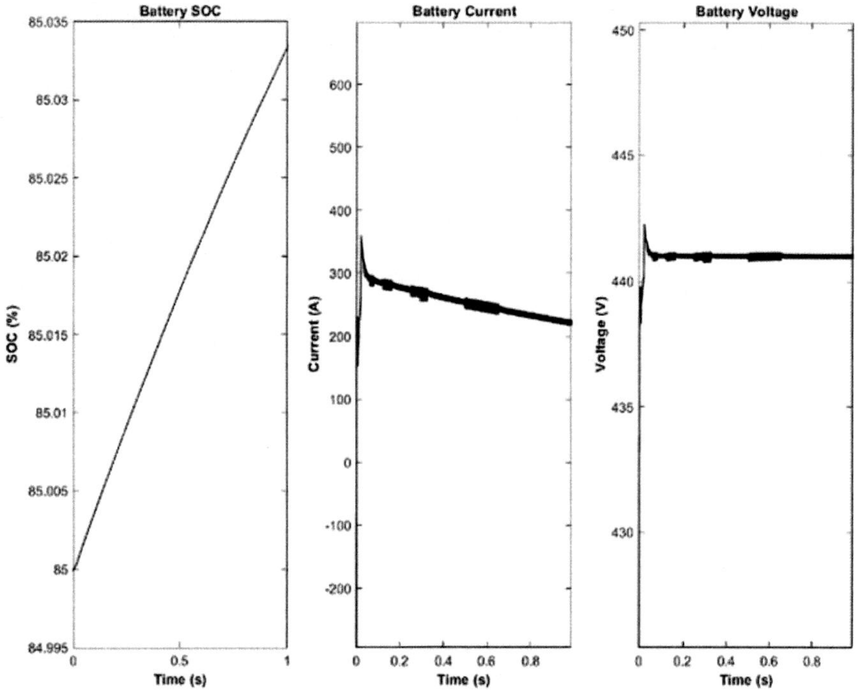

Fig. 24 CV charging mode

Acknowledgements This publication was made possible by UREP grant # [24-091-2-018] from the Qatar National Research Fund (a member of Qatar Foundation). The statements made herein are solely the responsibility of the authors.

References

1. Owusu PA, Asumadu-Sarkodie S (2016) A review of renewable energy sources, sustainability issues and climate change mitigation. Cogent Eng. 3(1):1–14
2. Harighi T, Bayindir R, Padmanaban S, Mihet-Popa L, Hossain E (2018) An overview of energy scenarios, storage systems and the infrastructure for Vehicle-to-Grid technology. Energies 11(8):1–18
3. Marshall BM, Kelly JC, Lee T-K, Keoleian GA, Filipi Z (2013) Environmental assessment of plug-in hybrid electric vehicles using naturalistic drive cycles and vehicle travel patterns: a Michigan case study. Energy Policy 58:358–370
4. Axsen J et al (2015) Electrifying vehicles: insights from the Canadian plug-in electric vehicle study. Simon Fraser University, Vancouver, Canada
5. Lopez-Behar D, Tran M, Froese T, Mayaud JR, Herrera OE, Merida W (2019) Charging infrastructure for electric vehicles in multi-unit residential buildings: mapping feedbacks and policy recommendations. Energy Policy 126:444–451

6. Serradilla J, Wardle J, Blythe P, Gibbon J (2017) An evidence-based approach for investment in rapid-charging infrastructure. Energy Policy 106:514–524
7. Domínguez-Navarro JA, Dufo-López R, Yusta-Loyo JM, Artal-Sevil JS, Bernal-Agustín JL (2019) Design of an electric vehicle fast-charging station with integration of renewable energy and storage systems. Int J Electr Power Energy Syst 105:46–58
8. Lam AYS, Leung KC, Li VOK (2017) Vehicular energy network. IEEE Trans Transp Electrif 3(2):392–404
9. Sujitha N, Krithiga S (2017) RES based EV battery charging system: a review. Renew Sustain Energy Rev 75:978–988
10. Iqbal A et al (eds) Soft computing in condition monitoring and diagnostics of electrical and mechanical systems, vol 1096. In: Advances in intelligent systems and computing. Springer, Singapore. https://doi.org/10.1007/978-981-15-1532-3
11. Iqbal A et al (eds) Meta heuristic and evolutionary computation: algorithms and applications, vol 1096. In: Studies in computational intelligence. Springer, Singapore. https://www.springer.com/gp/book/9789811575709
12. Gjelaj M, Arias NB, Traeholt C, Hashemi S (2019) Multifunctional applications of batteries within fast-charging stations based on EV demand-prediction of the users' behaviour. J Eng 2019(18):4869–4873
13. Hui L, Chao X, Chen C, Yibing W (2014) Simulation of three-phase voltage-source PWM rectifier with LCL filter. In: Advances in computer science and its applications. Springer, pp 901–909
14. Liserre M, Blaabjerg F, Hansen S (2005) Design and control of an LCL-filter-based three-phase active rectifier. IEEE Trans Ind Appl 41(5):1281–1291
15. Leroi C, De Jaeger E, Cassano S (2019) Modeling of harmonic propagation of fast DC EV charging station in a low voltage network. IEEE PowerTech
16. Soares R, Djekanovic N, Wallmark O, Loh PC (2019) Integration of magnified alternating current in battery fast chargers based on DC-DC Converters using Transformerless resonant filter design. IEEE Trans Transp Electrif
17. Tang L, Chen W (2008) Deduction of coordinate transform for instantaneous reactive power theory and analysis on the principle of harmonic current detection method. Power Syst Technol 5
18. Dannehl J, Wessels C, Fuchs FW (2008) Limitations of voltage-oriented PI current control of grid-connected PWM rectifiers with $ LCL $ filters. IEEE Trans Ind Electron 56(2):380–388

A Novel High Gain Boost Converter for Interfacing Low Voltage PV in a DC Microgird

Md. Samiullah, Imtiaz Ashraf, Atif Iqbal, and Khaliqur Rahman

Abstract Low terminal voltage at the output of a solar photovoltaic needs efficient boost converters for establishing a proper communication and energy transfer to the DC microgrids. In order to serve this purpose, the traditional boost converter is required to operate at some large duty ratio which is perilous to the converter performance and would create undesirable effects such as more losses, spikes of voltages and poor transient response. A boost converter with significantly high gain at low duty ratio has been presented in this paper. The proposed converter is well suited for integration of a solar PV with the DC bus of higher voltage levels in a DC microgrid. This particular converter is simple in structure and easily controlled with a single train of pulses. The advantages of switched inductor circuit embedded with the capacitor is availed in an intelligent way to uplift the voltage gain significantly. The converter behavior and suitability for high voltage is analyzed in continuous conduction mode, and later, the performances are verified through simulation results at a frequency of 10 kHz.

Keywords DC–DC converter · High gain · Switched inductor · DC microgrid

1 Introduction

With the worsening of the environmental condition, renewable sources of energy have gained a deep attention owing to the obvious reason of being environmental friendly and availability in abundance. Moreover, the demand for electrical energy is growing continuously leading to a huge burden on the conventional utility structure. Solar photovoltaics (PVs) are increasingly being integrated into power distribution in order to trim down the burden on the conventional power system [1]. Power electronic converters act as a mainstay for the appropriate utilization of these solar PVs owing to

Md. Samiullah (✉) · I. Ashraf · K. Rahman
Aligarh Muslim University, Aligarh, India
e-mail: md_samiullah@zhcet.ac.in

A. Iqbal
Qatar University, Doha, Qatar

A. Iqbal et al. (eds.), *Renewable Power for Sustainable Growth*, Lecture Notes in Electrical Engineering 723, https://doi.org/10.1007/978-981-33-4080-0_5

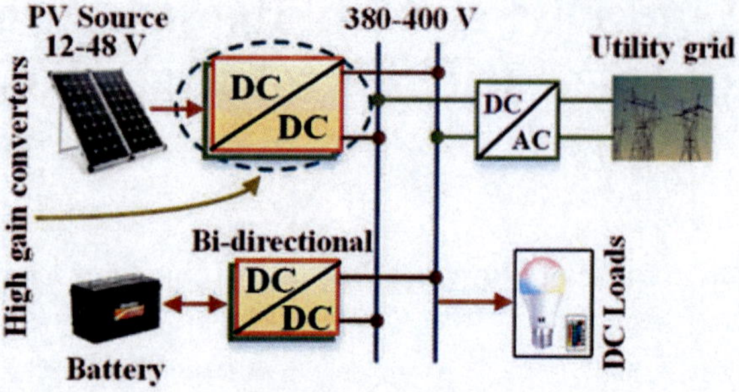

Fig. 1 A typical DC microgrid

their intermittency and uncertainty behavior. Distributed generation and microgrids are providing a source for interfacing solar photovoltaic in to the traditional structure of the power system [2, 3]. A DC-DC converter with suitable attributes is needed to properly link up the PV system to the microgrid. The output voltage at the terminals of a PV panel is low (12–48 V), and a high gain converter is the first need as can be observed from Fig. 1 [4].

For getting a high voltage to properly interface the photovoltaic unit with the microgrid, a conventional boost converter is needed to operate at around 90% of duty cycles, leading to many unwanted effects on the health of the converter and may rise EMI level, voltage spikes and ruinous transients [5].

In the literature, various topologies to realize high boost up DC-DC conversion are available. A broad classification of them is with or without transformers. A boost converter making use of transformer for isolation is the isolated one, which mainly moves around fly-back, forward, push–pull, and coupled inductor type [6]. Major limitations with these converters are their bulky size, core saturation, and leakage inductance. On the other hand, non-isolated converters do not use any transformers and therefore comparatively compact in size. They mostly use any boosting network such as capacitor-based voltage lifting circuits [7], switched inductor [8], and capacitor [9], interleaving and cascading of one another [10]. Each of the boosting circuit has its own pros and cons different from others. Nevertheless, there does not exist a single converter conveying all the desired benefits and fits at all the applications [11–13]. Therefore, the research in this area is still on progress and recent topologies with improved qualities are always on demand.

In this paper, a novel topology of a DC to DC high gain converter with many exceptional features has been introduced. Few among them includes its high gain at moderate duty ratio without using coupled inductor or transformer in the topology. Low voltage $R_{\mathrm{DS}}(\mathrm{ON})$ MOSFET can be used owing to the lower voltage stress across them, which decreases the conduction loss in the circuit and makes a high efficient converter. Switched inductor network in the circuit allows parallel magnetizing and discharging in series of the inductors which helps the converter to easily attain a significant high voltage. Furthermore, the switches being controlled simultaneously reduce the control complexity. Therefore, the proposed converter is a compact package of transformer-less circuit involving switched inductor to yield high voltage at the output with an improved efficiency.

2 The Proposed Converter

In Fig. 2, the layout of the novel converter proposed in the paper is shown. The converter design includes four inductors, six diodes along with three capacitors and two switches. The duo switches are controlled by the same gate pulses and do not involve a complex control strategy. Meanwhile, the identical condition is assumed for the devices being used in the circuit to evade the complexity and difficulty in analysis.

2.1 Steady-State Analysis During CCM

The proposed converter is analyzed in the continuous conduction mode in order to study the behavior and to find out the expression for voltage gain. The characteristics waveform during CCM has been plotted and shown in Fig. 5. The plot shows the voltages across and through the elements of the converter circuit. The analysis during CCM is done in the following two modes.

Mode I: (Duty period) from 0 to kT: In this period, the two switches of the circuit are switched ON simultaneously with the same gate signal and the equivalent model of the circuit during this interval is shown in Fig. 3. Only diodes D_1 and D_2 are forward biased, and the output voltage is maintained by the capacitor C_o. Inductors are magnetized in parallel through switches. The voltages available across the inductors in the circuit during this mode are presented as follows:

$$\left.\begin{aligned} v_{L1} &= v_{L2} = V_g \\ v_{L3} &= V_{C1} - V_{C2} \\ v_{L4} &= V_g - V_{C3} \end{aligned}\right\} \tag{1}$$

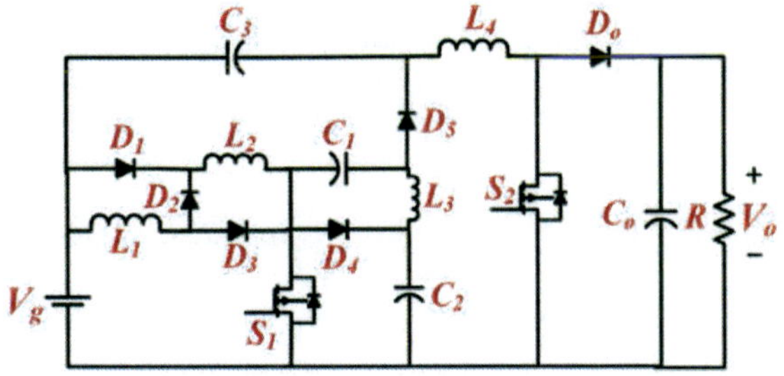

Fig. 2 Proposed converter's layout

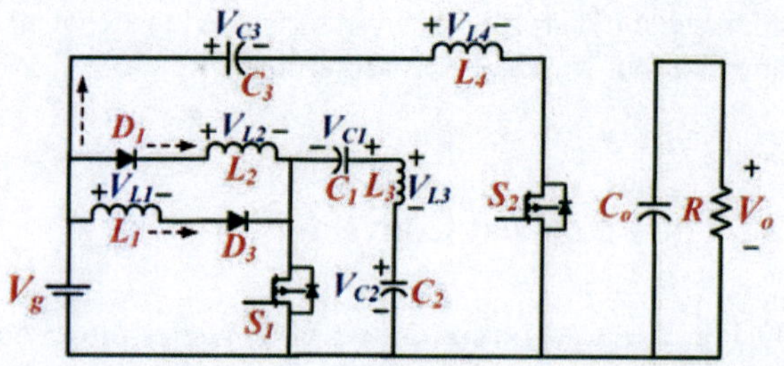

Fig. 3 Circuit structure during duty period

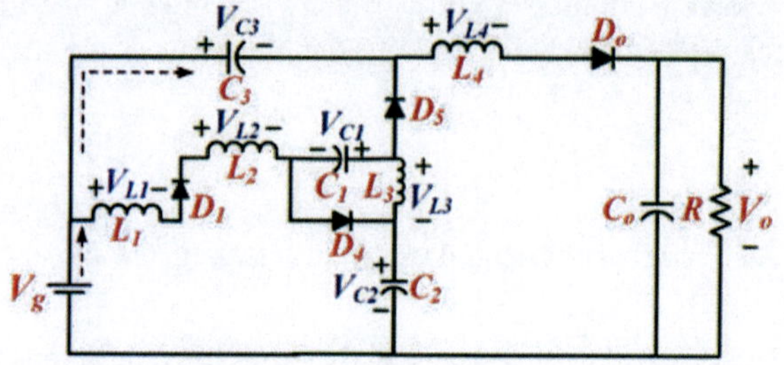

Fig. 4 Circuit structure during OFF-duty period

Mode II: (OFF-Duty period) from kT to T: This is the period when both switches are turned to their OFF period. Figure 4 shows the conducting model of the circuit. In the equivalent circuit, it is observed that inductors are discharging in series through diode D_1. The diode D_o being forward biased provides a conduction path for transferring energy from the input to the load and the capacitor C_o. The voltage across inductors can be evaluated as:

$$\left.\begin{aligned} v_{L1} &= (V_g - V_{C2})/2 \\ v_{L2} &= (V_{C1} + V_{C3})/2 \\ v_{L3} &= V_{C1} \\ v_{L4} &= V_g - V_{C3} - V_o \end{aligned}\right\} \tag{2}$$

Using principle of volt-second balance on inductor L_1, we come about with the following equations:

$$\int_0^{kT} V_g \mathrm{d}t + \int_{kT}^{T} \frac{(V_g - V_{C2})}{2} \mathrm{d}t \tag{3}$$

where k represents the duty ratio of the circuit and T is the time period of the gate pulses.

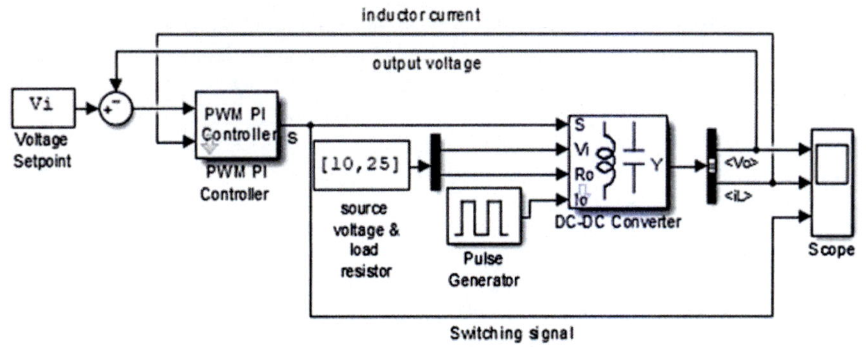

Fig. 5 Simulink model of the proposed converter with control blocks

After solving (3), the expression for V_{C2} is obtained as:

$$V_{C2} = \frac{(1+k)}{(1-k)} V_g \tag{4}$$

Using volt-second balance principle on inductor L_3, the following relation is obtained.

$$\int_0^{kT} (V_{C1} - V_{C2})\mathrm{d}t + \int_{kT}^{T} V_{C1}\mathrm{d}t \tag{5}$$

Once Eq. (5) is solved, we obtain the following relation for V_{C1}:

$$V_{C1} = kV_{C2} \tag{6}$$

Again, using principle of volt-second balance on inductor L_4, we come up with the following relation:

$$\int_0^{kT} V_g\mathrm{d}t + \int_{kT}^{T} \frac{(V_{C1} + V_{C3})}{2}\mathrm{d}t \tag{7}$$

After solving (7), and using the relations from Eqs. (4) and (6), the following relation is obtained.

$$G_{\mathrm{CCM}} = \frac{V_{\mathrm{o}}}{V_{\mathrm{in}}} = \frac{(1+k)^2}{(1-k)^2} \tag{8}$$

where G_{CCM} is the voltage boosting factor of the converter during continuous conduction mode and k represents the duty ratio. The voltage gain obtained in Eq. (8) is very high and an output voltage of significantly large magnitude can easily be acquired for low input supply from any renewable sources of energy such as fuel cells and photovoltaic.

3 Results and Discussion

The converter performance is finally examined through simulation results carried out in Simulink of MATLAB. The obtained results verified the analytical studies, and the converter performs well for the application it is designed for. The parameters chosen for the simulation are summarized in Table 1, and the simulation model is presented in Fig. 5. The capacitors and inductors values are chosen critically to make the converter operate in CCM. A low voltage of only 10 V is applied at the input and the duty ratio is maintained at 0.8 at a frequency of 10 kHz. Figure 6 presents the input voltage and the corresponding voltage at the output. The output voltage for an input of 10 V comes out to be around 790 V. This justifies the higher boosting factor of the analytical studies. Moreover, the low ripples in the output voltage also low justify the proper selection of the capacitance. Figures 7 and 8 show the voltages across switches S_1 and S_2, respectively. It can be observed that voltage across these semiconductor switches is not very high and low rating MOSFETs can be used.

Figure 10 depicts the capacitor C_1 voltage, where it is clearly seen that the voltage on capacitor C_1 is almost constant and this voltage is attained by the capacitor once the transient period is over. The ripples can further be minimized by increasing the value of the capacitance. The voltage profile of the capacitor C_2 has been presented in Fig. 11. Again, this capacitor maintains a constant voltage across it which is slightly higher than V_{C1}. The capacitor C_3 accumulates a negative constant potential across it of around −150 V. Moreover, inductor L_1 current profile is shown in Fig. 9 which possess the natural pattern of charging during switch ON period and discharging in switch OFF period. Figures 13 and 14 show the potential curve across diodes D_4 and D_5, respectively (Fig. 12).

Table 1 Simulation parameters

Parameters	Values
Input voltage	10 V
Duty ratio	0.8
Switching frequency	10 kHz
Inductances	$L_1 = L_2 = 20$ mH, $L_3 = L_4 = 10$ mH
Capacitances	$C_1 = 100$ µF, $C_2 = C_3 = 220$ µF, $C_o =$ 400 µF

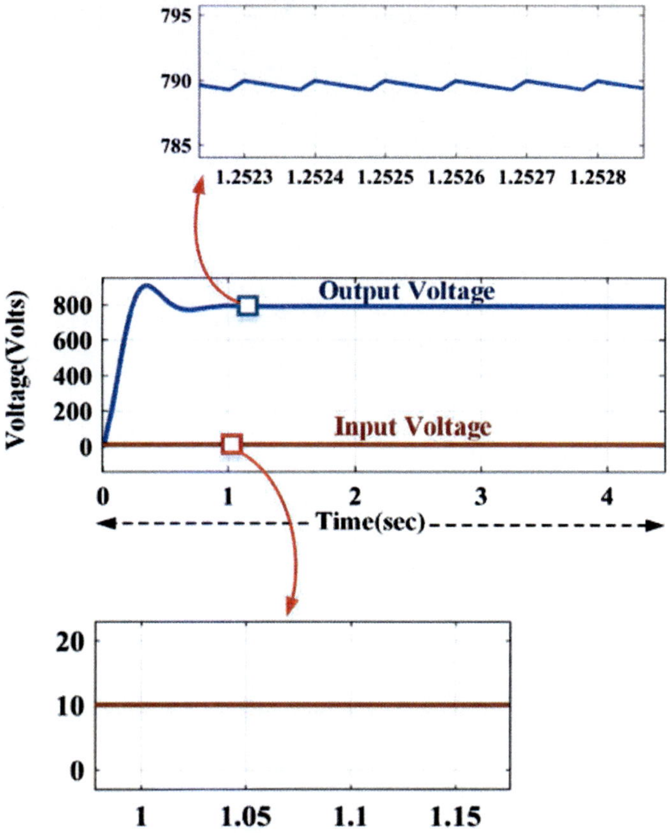

Fig. 6 Output voltage at duty cycle of 0.8

Fig. 7 Voltage across switch S_1

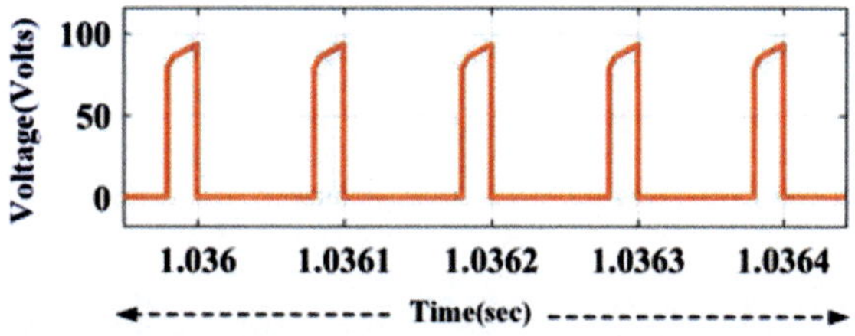

Fig. 8 Voltage across switch S_2

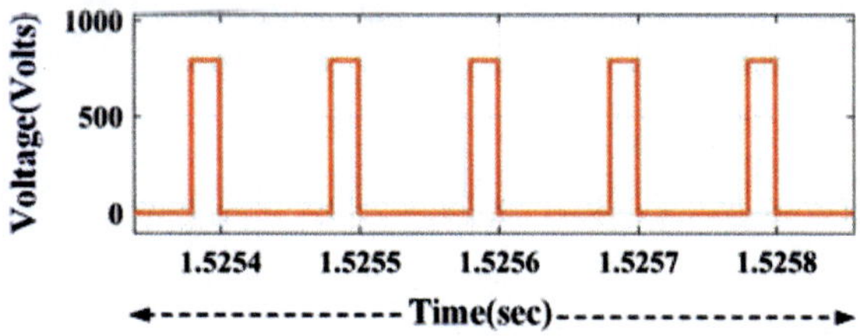

Fig. 9 Inductor L_1 current waveform

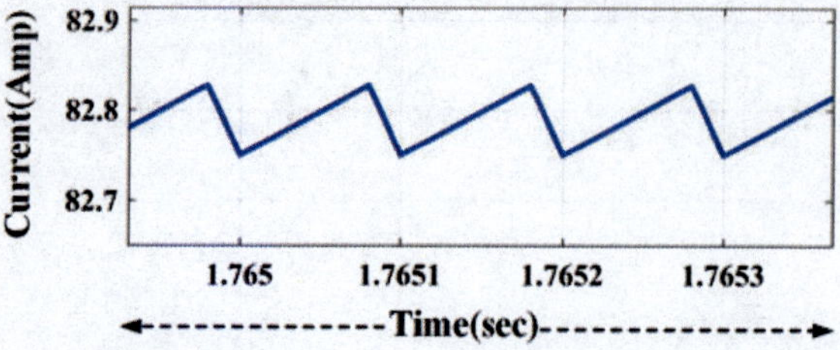

Fig. 10 Voltage across C_1

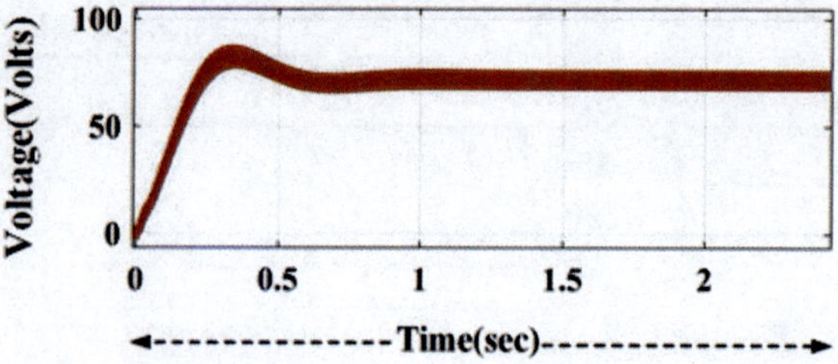

Fig. 11 Voltage across C_2

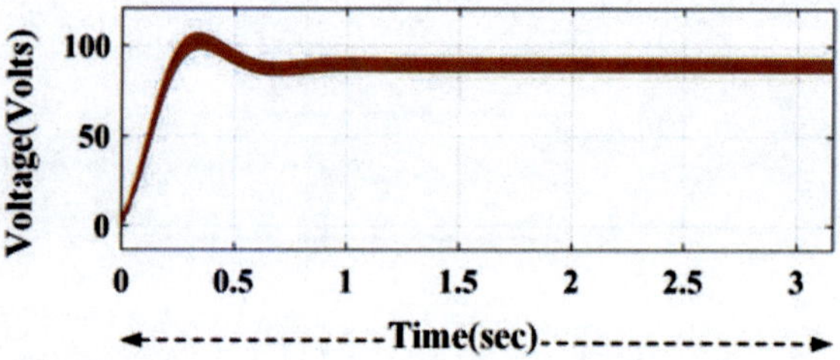

Fig. 12 Voltage across C_3

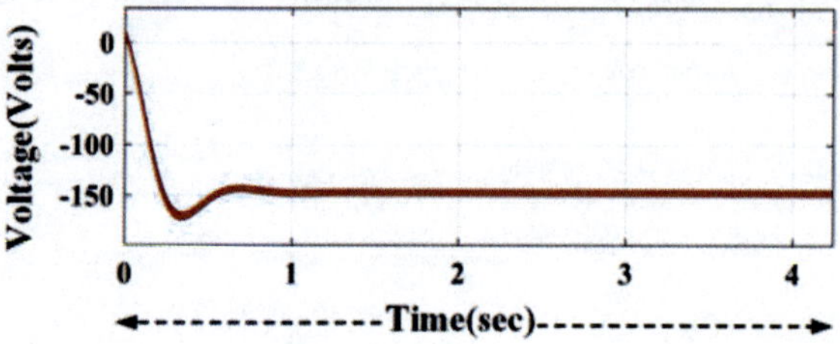

Fig. 13 Voltage across D_4

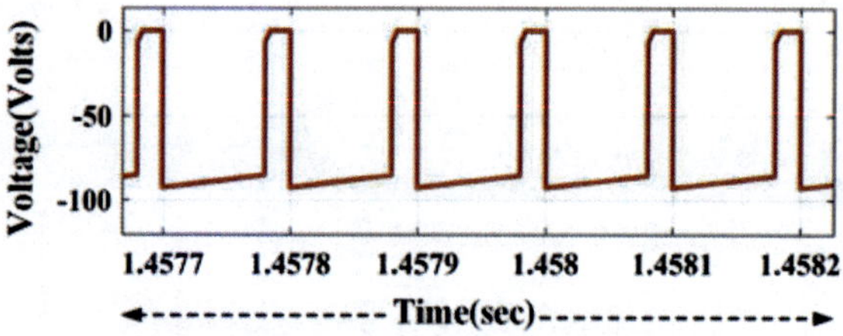

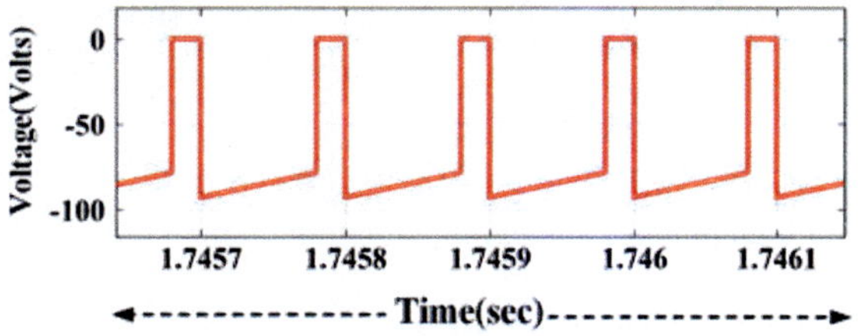

Fig. 14 Voltage across D_5

4 Conclusion

In this paper, a novel high gain boost converter topology has been introduced. The converter is well designed for DC microgrid applications where a low voltage generating system such as photovoltaic is to be integrated with the DC bus of high voltage. The structure of the converter is simple having two switches working simultaneously with the same gate pulses. The converter avails the advantages of switched inductor for stepping up the voltage by magnetizing in parallel and demagnetization in series. Moreover, the semiconductor devices being used in the circuit do not hold high potential across them, and therefore, low rating devices can be used for designing the prototype. The converter is analyzed during steady state of CCM, and later simulation of the same has been convened to verify the analytical findings. Overall, the converter performs well and justifies its suitability for the DC microgrid applications.

References

1. Walker GR, Sernia PC (2004) Cascaded DC-DC converter connection of photovoltaic modules. IEEE Trans Power Electron 19:1130–1139. https://doi.org/10.1109/TPEL.2004.830090
2. Gu Y, Chen Y, Zhang B et al (2019) High step-up DC-DC converter with active switched LC-network for photovoltaic systems. IEEE Trans Energy Convers 34:321–329. https://doi.org/10.1109/TEC.2018.2876725
3. Lakshmi M, Hemamalini S (2017) Nonisolated high gain DC-DC converter for DC microgrids. IEEE Trans Ind Electron 65:1205–1212. https://doi.org/10.1109/TIE.2017.2733463
4. Babaei E, Jalilzadeh T, Sabahi M et al (2019) High step-up DC-DC converter with reduced voltage stress on devices. Int Trans Electr Energy Syst 29:1–24. https://doi.org/10.1002/etep.2789
5. Yang LS, Liang TJ, Chen JF (2009) Transformerless DC-DC converters with high step-up voltage gain. IEEE Trans Ind Electron 56:3144–3152. https://doi.org/10.1109/TIE.2009.2022512
6. He L, Liao Y (2016) An advanced current-autobalance high step-up converter with a multi-coupled inductor and voltage multiplier for a renewable power generation system. IEEE Trans Power Electron 31:6992–7005. https://doi.org/10.1109/TPEL.2015.2509159
7. Shahir FM, Babaei E, Farsadi M (2018) Analysis and design of voltage-lift technique based non-isolated boost dc-dc converter. IET Power Electron 11:1083–1091. https://doi.org/10.1049/iet-pel.2017.0259
8. Salvador MA, Lazzarin TB, Coelho RF (2018) High step-up DC-DC converter with active switched-inductor and passive switched-capacitor networks. IEEE Trans Ind Electron 65:5644–5654. https://doi.org/10.1109/TIE.2017.2782239

9. Makowski MS, Maksimovic D (1995) Performance limits of switched-capacitor DC-DC converters. PESC Rec IEEE Annu Power Electron Spec Conf 2:1215–1221. https://doi.org/10.1109/pesc.1995.474969
10. Tang Y, Wang T, Fu D (2015) Multicell switched-inductor/switched-capacitor combined active-network converters. IEEE Trans Power Electron 30:2063–2072. https://doi.org/10.1109/TPEL.2014.2325052
11. Li W, Zhao Y, Deng Y, He X (2010) Interleaved converter with voltage multiplier cell for high step-up and high-efficiency conversion. IEEE Trans Power Electron 25:2397–2408. https://doi.org/10.1109/TPEL.2010.2048340
12. Iqbal A et al (eds) (2020) Soft computing in condition monitoring and diagnostics of electrical and mechanical systems, vol 1096. In: Advances in intelligent systems and computing. Springer, Singapore. https://doi.org/10.1007/978-981-15-1532-3
13. Iqbal A et al (eds) (2020) Meta heuristic and evolutionary computation: algorithms and applications, vol 1096. In: Studies in computational intelligence. Springer, Singapore. https://www.springer.com/gp/book/9789811575709

Techno-Economic Analysis of Hybrid Rural Microgrid for Isolated Hilly Area in Indian Scenario

Md. Mustafa Kamal, Arshad Mohammad, and Imtiaz Ashraf

Abstract Hybrid renewable resources extensively used to electrify the rural areas. To provide the reliable and economic supply of power in the rural region where grid extension is infeasible integration of the locally available energy resource is the best option. In the present work, economical and optimized design for electrification of isolated rural community is proposed with the combination of locally available energy resources. The model comprises of a PV/battery/wind and diesel generator. The microgrid realized in the cluster of the village of Uttrakhand state of India. The residential load of the community investigated using the HOMER energy for the least cost and optimal design. The solar, wind potential, and diesel generator used to perform the economic and feasibility study of the suggested model. The net present cost (NPC) and cost of energy (COE) analyzed for the hybrid system. Further model investigated utilizing the sensitivity of different parameters such as variation in diesel price intermittency in solar and wind potential. The study compares the performance of the system and suggests a viable configuration for the rural community.

Keywords Microgrid · Cost of energy · Solar · Wind · Renewable resource · Electrification

1 Introduction

For the overall growth of a country, energy plays a vital role due to population growth, urbanization, and industrial development. As per the International Energy Agency (IEA), more than 1.2 billion persons do not have a reliable supply of electricity worldwide. Further 95% of people are residents of Asian and African nations, in which 80% of the people residing in rural regions. Additional 2.4 billion get an irregular supply of power due to the runtishly and unreliable rural grid infrastructure. The current electrification program in rural regions focuses mostly on an extension

Md. M. Kamal (✉) · A. Mohammad · I. Ashraf
Aligarh Muslim University, Aligarh 202002, India
e-mail: mustafakamalece@gmail.com

A. Iqbal et al. (eds.), *Renewable Power for Sustainable Growth*, Lecture Notes in Electrical Engineering 723, https://doi.org/10.1007/978-981-33-4080-0_6

of the centralized grid. As per 2011 censuses in developing countries like India, still, more than 77 million depend on kerosene for lighting [1]. The remote rural population has either partially electrified or no access to electricity. In Indian more, the 44% population has no grid connectivity. Rural electrification for the development of remote areas is necessary. Even in the electrified village, available power is either low quality or irregular. More than 9000 villages in India still never connected to the grid because of remoteness and geographical constraints. Electrification using the national grid not viable for these remote areas due to higher costs for developing infrastructure. Electrification of these areas by islanded microgrids proliferates due to low basic cost compared to the central grid, successful model for energy micro-financing business, and technological advancement [2]. The infrastructure for the electrification in rural areas is either unfeasible or costlier for remote regions. Renewable energy-based resources perceptively used to ensure the electrification of the rural community. For the electrification of the rural population, the integrated energy model is more trustworthy and economical [3]. The structurer used for the electrification of the rural region is different from urban areas. Urban grid power supply infrastructure used new techniques such as demand-side management and power scheduling to increase the effectiveness in electrification [4]. Available energy resources in the area can be combined to form framework, which works on both grid-connected and standalone mode for electrification of the rural community. Hybrid energy system popularly is also known as integrated renewable energy system (IRES) model. The objective of the development of IRES model is to provide cost-effective energy supply to the rural community. To integrate renewable energy resources, careful and useful planning is necessary to maximize the benefits and efficiency for customers [5]. IRES models are widely used by the researcher to suggest the electrification of rural, remote areas. Worldwide authors reported power electrification using a modified integration of resources. Indian authors reported a few literature on designing of the hybrid model [6] [7, 8]. Diverse configuration of hybrid energy systems with DG and storage proven a reliable and economical solution for the energy need of rural areas. Kamal et al. [9–11] modeled the integrated energy system for the remote rural village for the hilly region and minimize the energy cost with different optimization algorithms. Das et al. [12] examined the possible application of an integrated energy model using HOMER for least cost of operation in the rural region of Bangladesh.

2 Study Area

Uttarakhand formed as the 27th state of the Republic of India, with its capital is in Dehradun. It has thirteen districts in which for the case study Tehri Garhwa. Its headquartered is in Theri. According to the 2011 census, the population of the district is 618,913. The region bounded by Dehradun in the west, Ruderparyag in the east, Pauri Garhwal in the south, and Uttarkashi in the north. The locality spread in an area of 36,242 square kilometers. Still, a large number of villages are the un-electrified or intermittent power supply. Total, 202 villages electrified, also 57 un-electrified out of 250 hamlets. The government is promoting renewable-based electrification of remote areas. Uttaranchal renewable development agency (UREDA) electrified twenty-four villages using renewable resources in the region.

3 Load Assessment

The region contains of a group of twelve un-electrified settlements. Hence, electricity consumption is low due to its low population density. Mainly energy consumption increases in the evening because most of the light and apparatus switched on, but in the daytime, only agriculture load derives the load profile. For load estimation, the extensive survey conducted and asked a few questions regarding which types of load they used and how much time they used. To supply the energy resource, the expected load of the area explored. The electrical load and energy consumed for cooking purposes in the study area have surveyed. Domestic loads of each house require a water pump, fan, television, radio, compact fluorescent lamp (CFL), press-iron, etc. Commercial loads consist of small shops of the village, mini diary, flourmills, commercial water pump. Agriculture load includes different loads, which used for agriculture purposes such as crop threshing machine, water pump, thrashing machine, agriculture apparatus, and fodder cutting machine. Community loads include hospital, school, community hall, streetlight, post office, market, etc. The energy demand, which shown in Fig. 1, peak energy demand is 148.67 kW respectively. In the study area, the household survey performed to get which types are load they used. The load profile consists of different types of loads, which used in commercial, domestic, and community loads.

The solar photovoltaic system (SPV) consists of several PV modules. Solar radiation data taken for the geography located between 28.47 latitude to 78.14 longitudes.

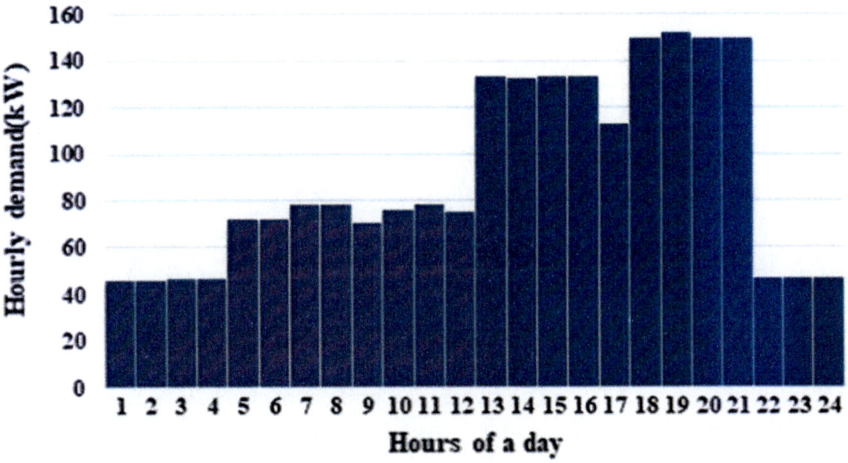

Fig. 1 Load of the area

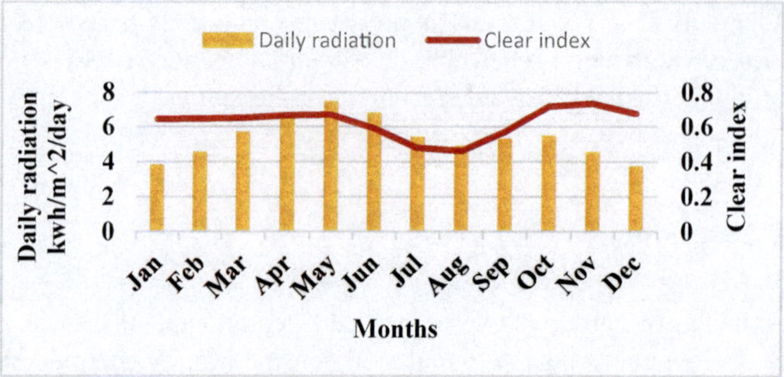

Fig. 2 Monthly average insolation and clear index of the region

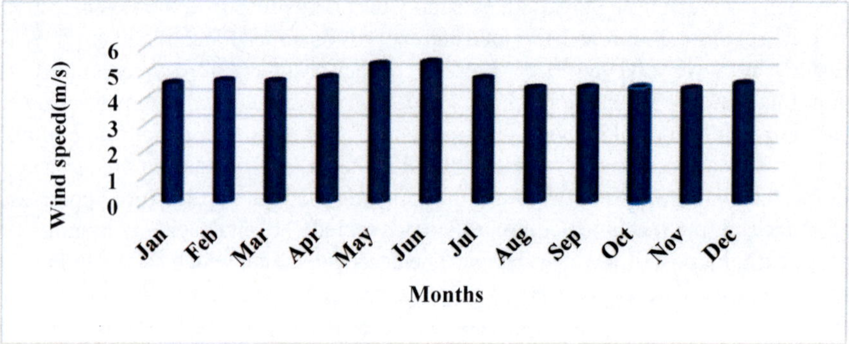

Fig. 3 Average monthly wind speed of region

The area has annual average solar insolation is 5.32 kWh/m^2, and in May, maximum insolation is 7.420 kWh/m^2.

The region has enormous solar potential; the total annual solar energy potential recorded as 22,369 kWh/m^2/yr. Figure 2 displays the monthly insolation and clearness index of the investigated area. Figure 3 suggests the monthly average wind speed of the region.

4 Problem Formulation

The microgrid is an alternative to come across the energy requirement for the un-electrified communities. Microgrid integrates the locally available energy resource to form a hybrid energy model and can work with grid-connected as well as standalone mode. To electrify the local community by using solar/wind/diesel/storage device formed a standalone microgrid. The schematic of standalone microgrid iss shown in Fig. 4. The proposed model is to provide reliable and economical electrification of

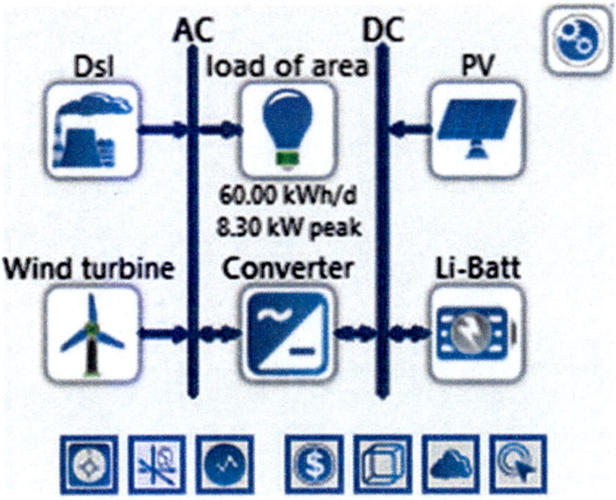

Fig. 4 Schematics of the hybrid model

Table 1 Specifications of component used in modeling of microgrid

Parameters	Wind turbine	PV	Battery	Converter	Diesel
Rating	8.1 kW	1 kW	6 V	1 kW	10 kW
Capital cost ($)	29,000.00	3300.00	500.00	900.00	$6500.00
Replacement cost ($)	25,000.00	2300.00	500.00	900.00	5500.00
O&M cost ($)	400.00/yr	8.00/yr	10.00/yr	00.00	0.20/h
Lifetime	20 yrs	25 yrs	500 cycles	20 yrs	12,000 h

rural region. Storage devices used to store energy when the production of energy in surplus amount and supply the energy in peak hours.

For electrification of the rural community, a standalone energy system with battery is proposed for the above-stated load profile. The specification of components used listed in Table 1

4.1 *Energy Management Strategy and Optimization Method*

A battery bank used in the project to counter the intermittent nature of renewable energy resources. The energy resources depends climate conditions, and the temperature of the area is not always available. When generated, energy is more than required load battery bank store the energy, and it will be used when no power is available. For the uninterrupted supply of energy, diesel sets also provided.

The proposed model is to electrify the rural villages where grid extension is unfeasible. The model optimized using hybrid optimization mode for electric renewables (HOMER) and developed by National Renewble Energy Laboratory (NREL), the USA. The software used to simulate, optimized, economic analysis, and sensitivity analysis of the proposed model. For the optimization HOMER software, simulate the all-possible combination of components based on the requirement of energy every year. For the sensitivity analysis, the parameters of component varied and the effect studied on the model. For the feasible solution of the model, all constraints must be satisfied. To analyses, the integrated model following criteria considered.

The total net present cost (NPC) of the integrated model comprises all costs and expenses in the project period. It consists of initial capital costs, emission penalties, O&M cost and replacement costs, fuel costs, and cost included in purchasing cost from the grid. Total annualized cost consists of the combination of capital costs, O&M costs and replacement cost for all the components. The NPC can modeled by Eq. (1)–(3).

$$C_{\mathrm{NPC}} = \frac{C_{\mathrm{anu}}}{\mathrm{CF}(i, m)} \tag{1}$$

$$\mathrm{CF}(j, m) = \frac{j(1 + j)^m}{(1 + j)^m - 1} \tag{2}$$

$$j = \frac{r - f}{1 + f} \tag{3}$$

where

CF is the capital recovery factor,
C_{anu} is total annualized cost,
r is the rate of interest,
f is the yearly inflation frequency, and
m is the lifespan of the plan.

Levelized cost of energy is a significant factor for the economic calculation of the energy model. It is measured in terms of the average cost to produce one kWh of the energy by the system. LCOE is the percentage of the yearly cost of energy produced to the total beneficial energy created by the system.

5 Results and Discussion

The proposed system is to deliver a consistent supply of electricity of the community wherever grid extension is not possible. The results of the optimized case and sensitivity analysis were simulated for the different combinations of the components in the integrated model. For the PV capacity of 6.88 kW, Li-ion battery has 43 numbers of strings, and diesel capacity has 10 kW; the cost of energy is $0.6122 and the net present cost is $173,329. Summary of simulation result is shown in Fig. 5. Figures 6 and 7 show the results of different simulation parameters.

Figure 5 suggests the total optimal cost of the project.

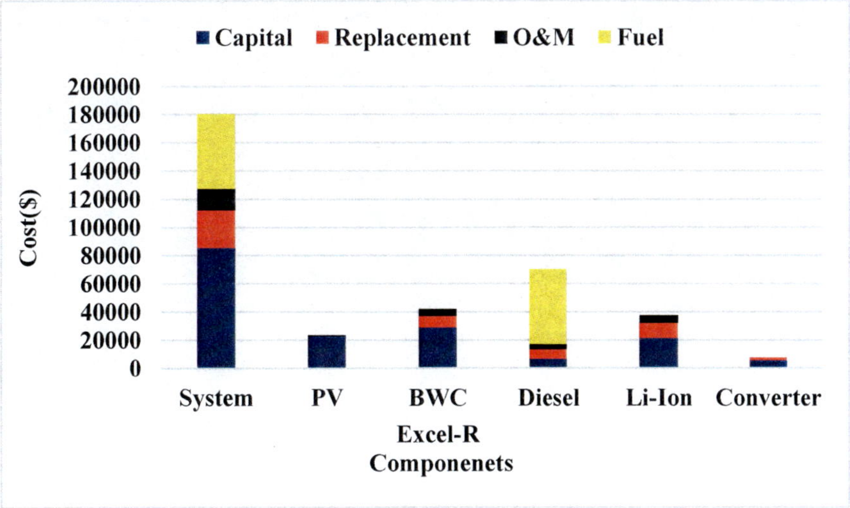

Fig. 5 Total optimized cost

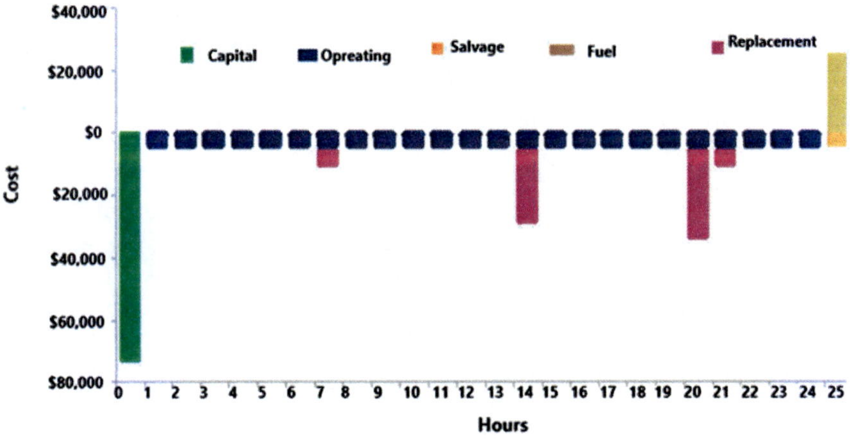

Fig. 6 Cashflow of the system

Figure 7 displays the optimized monthly energy production for the least cost of operation.

For the analysis and reliable solution of electrification of the rural community, the different combinations of available energy resources are tested for lest the cost of energy as well as net present cost. Table 2 shows the unusual combination of renewable energy resources.

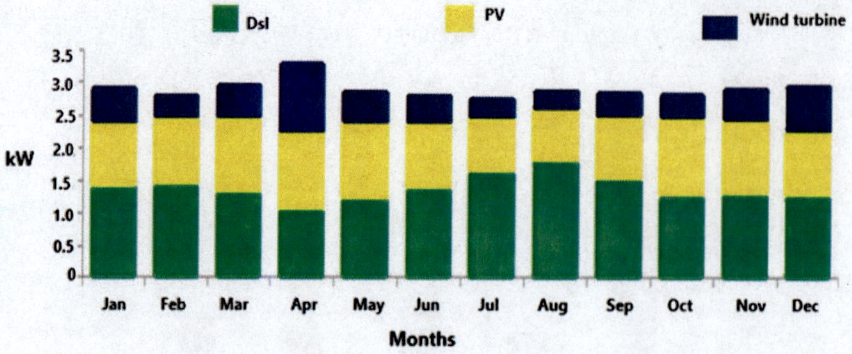

Fig. 7 Monthly electricity production by diesel and PV system

Table 2 Optimization result

Combination	LCOE ($)	NPC ($)
PV//battery/wind/diesel/converter	0.612	173,329.00
PV/diesel/battery/converter	0.524	148,230.00
battery/converter/diesel	0.622	176,030.00
wind/battery/diesel/PV	0.653	184,935.00
PV/converter/battery	0.777	218,260.00
PV/converter/battery/wind	0.824	231,432.00
PV/battery/diesel	0.839	237,487.00

For the sensitivity analysis of the intermittent condition of renewable energy resources, consider the availability of renewable resources depends on the topographical position and climatic condition of the area. Diesel price subjected to market variation for the sensitivity analysis three different diesel prices considered. Table 3 summarizes the LCOE and total NPC for the variation of parameters such as radiation wind speed, insolation, and diesel price variation.

6 Conclusion

In the existing work, a hybrid model of the energy system, which comprises of solar/wind/diesel/battery, proposed to electrify the rural community where grid extension is not possible. The proposed microgrid system is worked in standalone mode. The study of the energy system carried out for the twelve un-electrified villages of Uttarakhand (India). The simulation results suggest that the lowest cost of energy for the integrated model is $0.6122, LCOE, and total NPC is $173,329. For the feasibility study, different combinations of energy sources were analyzed. It found that to electrify the remote community in the proposed area PV/wind/battery/diesel provide

Table 3 Diesel price sensitivity results

Diesel price ($/L)	Solar isolation	Wind speed	LCOE ($)	Total NPC ($)
0.980	4.00	4.00	0.524	148,230.00
0.980	4.50	4.50	0.508	148,230.00
0.980	4.80	4.66	0.490	143,939.00
0.980	5.32	5.00	0.500	141,676.00
1.00	4.00	4.00	0.492	145,098.00
1.00	4.50	4.50	0.513	142,835.00
1.00	4.80	4.66	0.528	141,889.00
1.00	5.32	5.00	0.505	142,855.00
1.20	4.00	4.00	0.552	156,399.00
1.20	4.50	4.50	0.541	153,074.00
1.20	4.80	4.66	0.543	151,078.00
1.20	5.32	5.00	0.546	151,199.00

least cost compared to other combination. The sensitivity analysis for the variation of diesel price is carried out for the suggested model. This study suggested that the survey carried out for the rural community is useful for electrification planning.

References

1. Aklin M, Bayer P, Harish SP, Urpelainen J (2018) Economics of household technology adoption in developing countries: evidence from solar technology adoption in rural India. Energy Econ 72:35–46
2. Gandini D, de Almeida AT (2017) Direct current microgrids based on solar power systems and storage optimization, as a tool for cost-effective rural electrification. Renew Energy 111:275–283
3. Satsangi KP, Das DB, Babu GSS, Saxena AK (2019) Real time performance of solar photovoltaic microgrid in India focusing on self-consumption in institutional buildings. Energy Sustain Dev 52:40–51
4. Odou ODT, Bhandari R, Adamou R (2019) Techno-economic data for decentralised energy system sizing for rural areas in Benin. A case study of the village of Fouay. Data Br 26:104501
5. Ramirez Camargo L, Valdes J, Masip Macia Y, Dorner W (2019) Assessment of on-site steady electricity generation from hybrid renewable energy systems in Chile. Appl Energy 250:1548–1558
6. Junaid M, Kumar A, Mathew L (2017) Techno economic feasibility analysis of different combinations of PV-wind-diesel-battery hybrid system for telecommunication applications in different cities of Punjab, India. Renew Sustain Energy Rev 76:577–607
7. Iqbal A et al (eds) (2020) Soft computing in condition monitoring and diagnostics of electrical and mechanical systems, vol 1096. In: Advances in intelligent systems and computing. Springer, Singapore. https://doi.org/10.1007/978-981-15-1532-3
8. Iqbal A et al (eds) Meta heuristic and evolutionary computation: algorithms and applications, vol 1096. In: Studies in computational intelligence. Springer, Singapore. https://www.springer.com/gp/book/9789811575709

9. Kamal MM, Ashraf I, Fernandez E (2020) Energy resource allocation for distributed energy resource based power generation in a rural microgrid. IEEE Int Conf ICEECOT:571–575
10. Kamal MM, Ashraf I, Fernandez E, Alam A (2020) Resource allocation, utilization and feasibility study of a rural Microgrid. IEEE Int Conf UPCON:1–6
11. Kamal M, Ashraf I, Fernandez E (2019) Energy resource planning for a rural microgrid: comparison of results using different optimization algorithms. In: 2019 2nd international conference on advance computing communication paradigm. ICACCP 2019, pp 1–6
12. Mandal S, Das BK, Hoque N (2018) Optimum sizing of a stand-alone hybrid energy system for rural electrification in Bangladesh. J Clean Prod 200:18–27

Fabricated CdO-ZnO Films: A Potential Candidate for Optoelectronics Device Applications

Rayees Ahmad Zargar, Malik Mubasher Hassan, and Muzafer Ahmad Sheer Gogre

Abstract Fabrication of transparent conductive oxides (TCO)-based material has gained much importance because of its exceptional industrial applications. This Chapter reports preparation and characterizations of pure CdO and Zn-doped $Cd_{0.75}Zn0._{25}O$ screen printing films on glass substrates. For structural characteristics, X-ray diffraction (XRD) shows nature of polycrystalline for pure CdO film, whereas $Cd_{0.75}Zn0._{25}O$ film shows mixed phase of CdO and ZnO with maximum diffraction at (111) plane. For optical property, UV–visible spectra show optical band gap energies of 2.43 eV for CdO and 2.87 eV for $Cd_{0.75}Zn0._{25}O$ film followed by direct transition. The DC conduction mechanism from two-probe method confirms the semiconductor nature of the films.

Keywords TCO · XRD · UV–visible · And conductivity screen printing

1 Introduction

Studies from the past few years transparent conductive oxides (TCO) films have shown a broad scope in the research as they show an essential diversity in their response to the light and hence in the field of optoelectronic devices [1–3]. Cadmium oxide (CdO) and zinc oxide (ZnO) represent II–VI classes of prominent wide band gap semiconductors and more importantly their absorption edge found within the ultraviolet region; this could play a key role for generation of electron–hole pair (e–h) [4]. CdO one of the famous n-type semiconductors having energy band gap of 2.48 eV

R. A. Zargar · M. A. S. Gogre
Department of Physics, Baba Ghulam Shah Badshah University, Rajouri, Jammu and Kashmir 185234, India
e-mail: rayeesphy12@gmail.com

M. M. Hassan (✉)
Department of ITE, Baba Ghulam Shah Badshah University, Rajouri, Jammu and Kashmir 185234, India
e-mail: mubasher2003@gmail.com

A. Iqbal et al. (eds.), *Renewable Power for Sustainable Growth*, Lecture Notes in Electrical Engineering 723, https://doi.org/10.1007/978-981-33-4080-0_7

has widely used in optoelectronics due to its exceptional electrical conductivity and good transparency in the visible spectrum region and is consider to be the best window material in CIS and CdTe hetero-junction [5]. Similarly, ZnO is also being n-type semiconductor material having energy band gap of 3.37 eV, very cheap, huge exciton binding energy of 60 meV as opposed to 25 meV of GaN, high thermal stability and environmental friendly. Due to these characteristics, ZnO has been consider as a promising dopant in the of field of optoelectronic devices [6]. Therefore, alloying of these two metal oxides not only modulates the optical band gap but also conductivity of the material and thus makes the composition more fruitful in the deposition of hetero-junctions-based photovoltaic devices [7].

This chapter highlights the easy fabrication and basic characterization of CdO and CdZnO films developed by simple screen printing method. Screen printing method has many distinct advantages as compared to the mentioned methods [1, 4–7]. The distinct features of screen printing coating such as very simple, low cost, eco-friendly, no material wastage, high porosity and cover large area and hence prove a truly easy and suitable deposition method [8]. The motive of this work describes the fabrication of large area CdO and CdZnO films via economical screen printing technique for the purpose of optoelectronics applications.

2 Experimental Measurement

For the fabrication of photovoltaic films, the following materials were used: CdO, ZnO, $CdCl_2$ and ethylene glycol, and these materials were identified as (AR)-grade and purchased from Aldrich company. Pure CdO and $Cd_{0.75}Zn_{0.25}O$ pastes were prepared separately with 10% of anhydrous $CdCl_2$, and ethylene glycole was added as dropwise (binder). The prepared paste of CdO and $Cd_{0.75}Zn_{0.25}O$ was used on cleaned glass substrates by means of screen printing technology. The full procedure of film casting is already reported in reference [9]. Further, the coated films were annealed in a heat controlled furnace at 550 °C for 10-min in order to evaporate the organic materials completely. The film thickness was found in the order of 5 μm by using a profilometer (Surftest SJ—301).

3 Results and Discussions

3.1 Analysis of X-Ray Diffraction (XRD)

Diffraction spectras of the CdO and $Cd_{0.75}Zn_{0.25}O$ films are depicted in Fig. 1. It has been observed that (111), (200) and (220) are confirmed CdO peaks as per JCPDS (05-0640) file. Whereas extra peaks of (100), (002), (101) and (110) are identified in $Cd_{0.75}Zn_{0.25}O$ film and these peaks are of ZnO confirmed from JCPDS

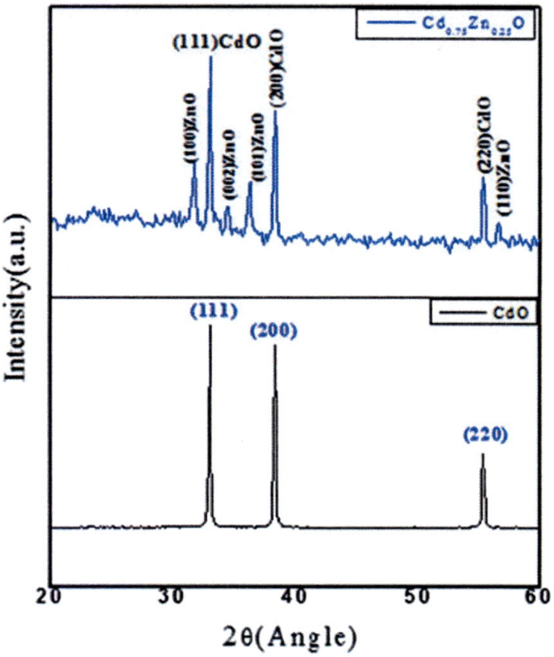

Fig. 1 XRD pattern of CdO-ZnO (prepared films)

(36-1451) file, respectively. In $Cd_{0.75}Zn_{0.25}O$ film the peaks shifted towards lower angle, intensities fall down and also, full width half maximum (FWHM) becomes broader upon Zn content. These changes may be because of unsymmetrical change in crystallites size (due to 25% ZnO doping). The various structural parameters via particle size,the dislocation density and strain can be estimated for the most intense diffraction peak (111) plane by using Eqs. (1), (2) and (3), respectively [10]. The calculated parameters are displayed in Table1. The well-known Debye–Scherrer's Eq. (1) has been applied for evaluating the particle size (P_s) of films.

$$P_s = \frac{0.94\lambda}{\beta \cos\theta} \tag{1}$$

where

λ source wavelength (1.54 Å) by using Cu-Kα,

Table 1 Analysis of various XRD parameters is given below

Sample	Particle size (nm) (P_s)	Dislocation × (10^{15} lines/m^2) (δ)	Strain × (10^{-3} lines/m^4) (ε)
CdO	24.51	1.67	2.27
CdO-ZnO	20.23	2.24	2.54

B full width at half maximum (FWHM) in terms of radian and
θ Braggs diffraction angle.

The dislocation density (δ) has been calculated by using Eq. (2).

$$\delta = \frac{1}{P_s^2} \tag{2}$$

Meanwhile, the strain (ε) possessed by the films was obtained by using Eq. (3). The variation in the parameters is due to decrease in particle size upon Zn doping and confirms the phase segregation in CdO-ZnO lattice [9].

$$\varepsilon = \frac{\beta \cos \theta}{4} \tag{3}$$

4 Optical Analysis

In modern technology, optical properties such as band gap tailoring play a significant role for designing efficient devices. Defused reflectance spectroscopy is used where light is scattered especially in powdered samples. The present paper discusses the nature and determination of optical energy band gap for pure and doped samples by using defused reflectance spectroscopy in the range of 400–1200 nm. Figure 2 shows the variation of wavelength dependence of diffused reflectance spectra of pure CdO and doped $Cd_{0.75}Zn_{0.25}O$ films. This spectrum shows blue shift, i.e. the shift towards longer wavelength side which indicates the increase in optical energy band gap, when zinc content is incorporated.

The energy band gap (E_g) can be obtained from the famous Tauc's law (4).

$$\alpha = \frac{B}{hv}(hv - E_g)^n \tag{4}$$

where E_g represents optical band gap, B is a constant and n represents the optical transitions (for direct case transition $n = 1/2$) [11]. The Kulbel–Munk function can be expressed from the linear extrapolation of the curve between $[F(R)hv]^2$ verses photon energy (hv), which gives the actual optical band gap of material [12]. The variation between photon energy and diffused reflectance is displayed in Fig. 3, and from this plot, the direct transition is clearly visible and optical band gap energies were estimated as 2.43 eV for pure CdO film and 2.83 eV for $Cd_{0.75}Zn0_{.25}O$ film. The optical band gap was enhanced upon doping of Zn from CdO to $Cd_{0.75}Zn_{0.25}O$ and is consistent with reported value of Ref. [13].

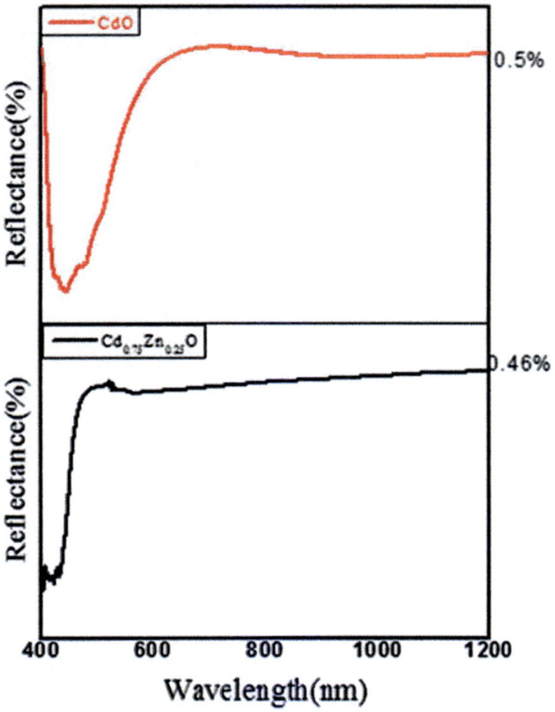

Fig. 2 Reflectance spectra of CdO-ZnO films

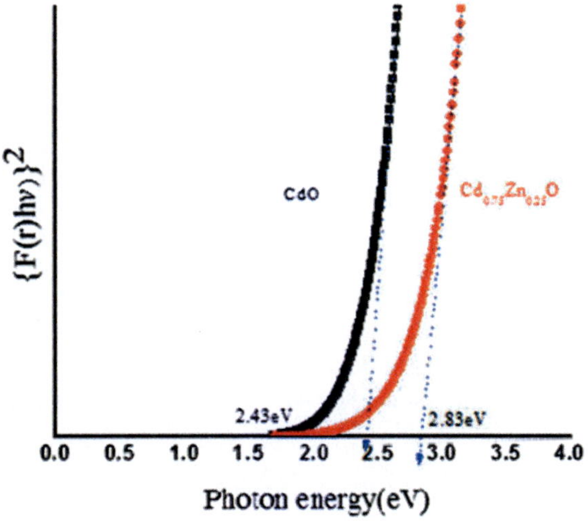

Fig. 3 Tauc's plot of of CdO-ZnO films

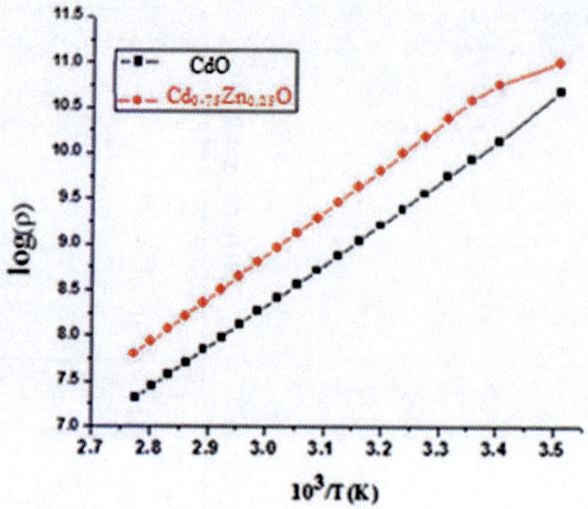

Fig. 4 DC resistivity measurement of CdO-ZnO films

5 Electrical Conduction Mechanism:

The electrical characterization plays a main role for the identification of photovoltaic material; in this report, the DC resistivity measurement of CdO and $Cd_{0.75}Zn_{0.25}O$ fabricated films was carried out by using Keithley 4200 two-probe system. Figure 4 depicts that $Cd_{0.75}Zn_{0.25}O$ film possess a high resistivity due to effects of grain boundary and larger band gap of the ZnO, and this could create a potential barrier hence causes reduction in conductivity [11]. The temperature dependence of dark electrical resistivity (DC) variation can be depicted by using a simple Arrhenius Eq. (5).

$$\rho = \rho_0 . \exp\left(\frac{E_a}{kT}\right) \tag{5}$$

where ρ_0 is the factor (pre exponential), E_a is the electron activation energy, T is the constant of Boltzmann's and k is the temperature (in Kelvin) respectively. Both the films show semiconducting type behaviour w.r.to temperature. The electron activation energy for CdO and $Cd_{0.75}Zn_{0.25}O$ fabricated films were calculated from the slope of graph as shown in Fig. 4. Activation energy was found to increase from 0.17 to 0.23 eV upon Zn doping. This increase in activation energy is a consequence of increase in resistivity that may be as a result of increase in band gap.

6 Conclusion

In summary, CdO and doped $Cd_{0.75}Zn0._{25}O$ films were successfully fabricated on glass substrates via low-cost coating method (screen printing). The analyses of various parameters were studied in view point of their photovoltaic applications. From the XRD, it was observed that decrease in crystalline nature of the fabricated film shows modification in crystal size and other parameters with the addition of ZnO into CdO. From the optical analysis, band gap of $Cd_{0.75}Zn0_{.25}O$ film increased upon Zn concentration. Also from electrical study shows that DC resistivity measurement of the $Cd_{0.75}Zn0_{.25}O$ film increased. This study appeals that CdO-ZnO is a desired wide band gap material and strongly recommend the use of these films in optoelectronic device applications.

Acknowledgements We express our gratitude and acknowledge to the BGSB University for providing the financial support under TEQIP-III research grant. Ref. No. BGSBU/TEQIP-III/RGS/006.

References

1. Dakhel A (2011) Int J Mod Phys B 25:277–282
2. Iqbal A et al (eds) (2020) Soft computing in condition monitoring and diagnostics of electrical and mechanical systems, vol 1096. In: Advances in intelligent systems and computing. Springer, Singapore. https://doi.org/10.1007/978-981-15-1532-3
3. Iqbal A et al (eds) (2020) Meta heuristic and evolutionary computation: algorithms and applications, vol 1096. In: Studies in computational intelligence. Springer, Singapore. https://www.springer.com/gp/book/9789811575709
4. Gupta RK, Ghosh K, Patel R, Mishra SR, Kahol PK (2008) Mater Lett 62(25):4103–4105
5. Kumar BR, Rao TS (2013) Appl Surf Sci 265:169–175
6. Zargar RA, Shah AH, Arora M, Mir FA (2019) Arab J Sci Eng 44:6631–6636
7. Chackrabarti S, Zargar RA, Joseph S, Aziz A, Hafiz AK (2016) Int J Optik 127:9966–9973
8. Zargar RA (2019) Mater Resh Exp 6:095909
9. Zargar RA, Bhat MA, Reshi HA, Khan SD (2018) Res Phys 9:1673–1676
10. Kumari R, Singh P, Kumar VJ (2018) Mater Sci Mater Electron 29:8989–8994
11. Zargar RA, Khan SD, Khan MS, Arora M, Hafiz AK (2014) Phys Res Int 2014:464809
12. Yakuphanoglu F, Ilican S, Caglar M, Caglar Y (2010) Superlattices Microstruct 47(6):732–743
13. Joseph B, Gopchandran KG, Manoj PK, Koshy P, Vaidyan VK (1999) Bull Mater Sci 22:921–926

Study of Different Topology for Multilevel Converter and Inverter Modulating in Cascaded H-Bridge Solid-State Transformer

Vimal Singh Bisht, Rakesh Thapliyal, Navneet Joshi, and Manoj Kumar Singh

Abstract This paper describes the utmost significant topologies and governor systems for multilevel converters and inverters along with their applications, enabling technology for modulating and implementation of power electronic-based distribution transformer. The motive of the work is to recognize a technique which is proficient of helping other features of a transformed used in distribution. SST has AC–DC and DC–AC stages received as they are capable of delivering good quality at the output. The desired essential voltage with minimum THD at the output is also achieved. The different modulation scheme for a solid-state transformer (SST) and overall model represents input rectifier AC–DC stage, isolated DC–DC converter and DC–AC output inverter stage. The model output of some DC–DC converter is connected to separate battery bank, which has been acting as distributed energy storage device (DESD) for solid-state transformer (SST). Mathematical modeling of controller is formulated and the dynamic characteristics of solid-state-based distribution transformer are verified by MATLAB.

Keywords Isolation bridge · Dual active bridge (DAB) multilevel cascaded converter · Multilevel inverter · Solid-state distribution transformer (SSDT) · Distributed energy storage device (DESD)

V. S. Bisht (✉)
Electronics Engineering Department, GEHU, Bhimtal, India
e-mail: bishtvimal@yahoo.com

R. Thapliyal
NIT Uttarakhand, Srinagar, Uttarakhand, India

N. Joshi
Mathematics Department, GEHU, Bhimtal, India

M. K. Singh
School of Computing, GEHU, Bhimtal, India

A. Iqbal et al. (eds.), *Renewable Power for Sustainable Growth*, Lecture Notes in Electrical Engineering 723, https://doi.org/10.1007/978-981-33-4080-0_8

1 Introduction

The transformer is extensively cast-off in power system engineering for performing some important functions like transformation of voltage from one circuit to another and isolation of two circuits. Meanwhile, the dimension of a transformer and traditional cast-off Cu-based iron is contrariwise proportional with the operating frequency, thus shooting the frequency helps us in using the core in a better way and also helps us in reducing the size of a transformer. Compared with the traditional transformer PE transformers are not only small in size but it also saves the use of oil, which is not at all required in PE transformer [1, 2].

In this paper, a different converter and inverter topologies are compared for integration with SST model. Organization of the paper is as: Section 3 presents basics of different topologies of converter suggesting the desired topology under different condition. In Sect. 4, the different DC–DC link connection methods has been used for SST. Section 5 deals with the operating modes of different inverter topologies for output. The simulation results and the proposed topology are discussed in Sect. 6 followed by the conclusion in Sect. 7.

2 AC–DC Converter Topologies

Rectifier composed of diodes and thyristor bridge converters. IGBT converters have been traditionally used to obtain voltage direct current from one side of the utility. The disadvantages of this AC–DC rectifier are they who allow the utility to break with the lower order harmonics, which are difficult to filter. In a power system, a load having lower value of PF takes more current for doing same work. As we all are aware of that higher amplitude values of current increases the loss in distribution wing of PS, thus, we need to have great power cable connected to the system. The easiest way to control the harmonic current is by using a passive [3–7] filter.

However, the use of modulation techniques such as SPWM and PWM AC–DC converter to control the switching speed for purely DC output waveform resulting reach near power single and also using the inductor is connected with the supply side of sinusoidal input current [8].

Distortion sine wave normal normally created by harmonics with nonlinear loads such as variable speed drives, arc devices, electronic ballast and power supplies. They can cause the equipment close to failure, voltage distortion and generate resonance with the utility. The THD levels of the signal are defined below in Eq. (1) as:

$$\mathrm{THD} = \frac{\sqrt{I_2^2 + I_3^2 + \cdots + I_N^2}}{I_1} = \frac{\sqrt{\sum_{N=2}^{N} I_N^2}}{I_1} \tag{1}$$

Relation between distortion power factor and total harmonic distortion is given by Eq. (2):

$$\text{Distoration}_{\text{pf}} = \sqrt{\frac{1}{1 + \text{THD}^2}} \tag{2}$$

2.1 Three-Phase Diode Rectifiers

Three-stage rectifier diodes are mostly used in engineering domain for providing the input voltage to the motor drives and DC–DC converters [8–10]. These rectifiers are of low in cost but they attract non-sinusoidal currents or reactive power deteriorating the superiority of the power system.

These aspects of non-sinusoidal nature of currents have a bad effect together on PF and PQ. The THD in current for diode rectifiers is generally up to 30% and the PF is less than 0.954. It is seen that even though the displacement power factor is unity, the distortion PF is very high due to the presence of huge harmonic content which results in low PF.

2.2 AC–DC Converter Topology

PWM converters are used for shifting frequency harmonics resulting in there higher value where they can be easily filtered. A six three-phase AC converter consists of external control voltage loop and current control internal loop. The current controller detects the input current and compares it to a reference sinusoidal current [2]. To acquire this, information up to date reference phase voltages utilities or current is required.

2.3 Conventional Cascaded Multilevel Converters

Primary stage of SST connected multilevel inverter H-bridge. The modular structure of this stage is input converts the AC to square wave with high frequency directly. Main challenges in control rectifier CHB keep DC link voltages to the desired reference value. When the implementation of high voltage converters, unequal loading lead it to disturb the DC voltage level. This unequal loading by maintaining control of the controller and PI control and DQ methods. The control device is adapted to the DC level by minimizing the error signal obtained at each level, and controlling the rectifier CHB rectifier output voltage level. Each value of input voltage and current is converted to DQ frame reference and compared with the desired DC level voltage and current which formed inner and outer current and voltage loops, respectively [1, 2].

3 Topologies for DC–DC Converter

3.1 Three-Phase DAB DC–DC Converter

The output for each phase is at 120° from each other. The rated output current is increased by inserting three phases in parallel and increasing the driving frequency to three times of the supply frequency. This reduces the ripples in current by great amount due to cancelation effects. The fed rate of DAB is organized by phase displacement between the low voltage sides of active bridges.

When three-phase DAB start operating in the boost mode the energy start flowing form LV side of transformer to the HV side. And now when the energy of high side DC flows to the circuit the LV side the DAB start working in the BUCK mode.

3.2 DC–DC Converter with Dual Active Bridge

All the switches present in the circuit of DAB are to be operated under constant frequency. The duty cycle choosen for the operation at constant frequency is taken as 50%.

Each rectifier CHB connected with level high DC-DC converter voltage across the DC capacitor 'C_i'. The DC voltage from step input is moderated at high level of frequency where square wave couples with high frequency transformer and then transformed to lower voltage DC. Here, the high frequency module connected to each DC–DC converter input and output square wave transformer controlled by the PI controller. High frequency link transformer maintains isolation between primary winding and secondary winding sides. In the isolation step, various types of DC–DC converters are used. Among the full-bridge topologies, the voltage zero voltage has better performance than other topologies.

4 Inverter Clamped with the Diode

From the figure given below, we can understand the working of a three-level diode clamp inverter. In the circuit, we can see capacitor's C_1 and C_2 are connected in series and three levels are allocated for defining the bus voltage of the circuit. The point where these two capacitors are meeting is classified as neutral point of the circuit. The voltage of the output is having three stages: (a) $V_{cc}/2$, (b) 0 and (c) $-V_{dc}/2$. For the voltages having level $V_{dc}/2$, the S_1 and S_2 need to be turned on and for level 0 S_1 and S_2 and need to be turned off. See Figs. 1 and 2.

Presence of $D1$ and $D^{'}1$ in the circuit makes if different form traditional one.

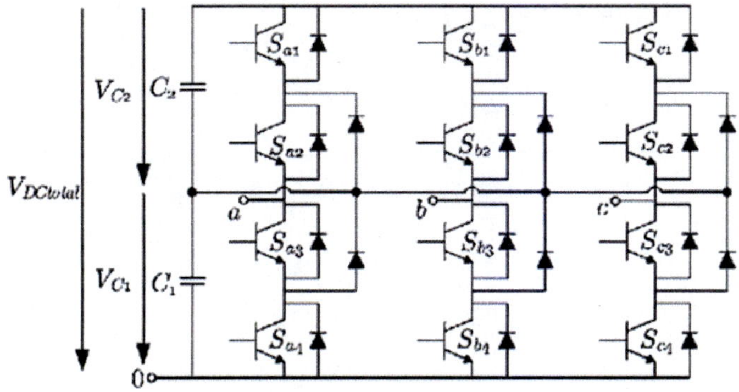

Fig. 1 Three-level diode clamp converter

Fig. 2 Five-level diode clamp converter

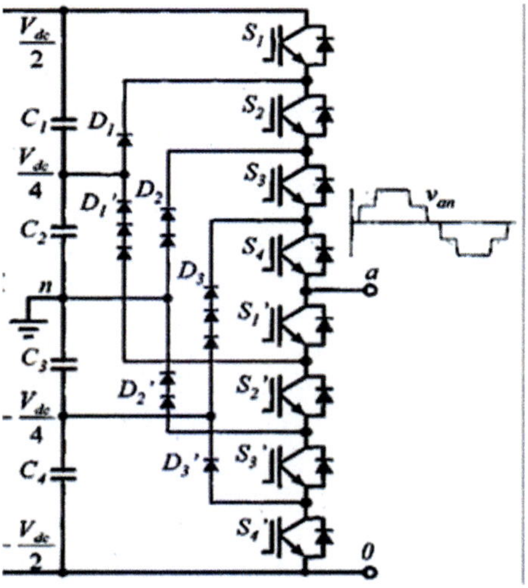

5 Cascaded Multicell Inverters (CMI)

CMI is another type of converter that is based on series DC converts having their separate sources. Figure 2 shows the power supply of the circuit having nine (9)-level inverter leg along having four (4) cells associated with each phase. Voltages generated by the individual cell is added and processed to obtain the resulting voltages. The resulting waveforms are almost sinusoidal in nature; this is achieved even without applying any filter.

6 Results and Discussion

6.1 Performance of CHB Rectifier

The first simulation investigates the behavior of cascaded high level voltage rectifier with DQ control strategy at four-level rectifier stages. All parameters calculation of CHB rectifier to obtained desired DC level voltage is shown in Table 1.

CHB four-level rectifier controller controls the gate pulses at switching frequency of $f_s = 4$ kHz where DC voltage across CHB rectifier is 4 kV, high voltage capacitor $C_i = 0.047$ F and input inductance $L_s = 2$ mH. Waveforms of the voltage can be clearly seen from Figs. 3 and 4.

Table 1 Value of input stage parameters

Parameter	Symbol	Value
Input voltage (rms)	V_{in}	11 kV
Input inductance	L_s	2 mH
DC bus capacitors	C_i	0.047 F
DC bus voltage	V_{dc}	4 kV
Number of series cells	N	4
Proportional gain (voltage)	K_{pv}	2
Integral gain (voltage)	K_{iv}	10
Proportional gain (current)	K_{pi}	80
Integral gain (current)	K_{ii}	1000
Frequency of the line	f_1	50 Hz
Frequency used for switching	f_s	4 kHz

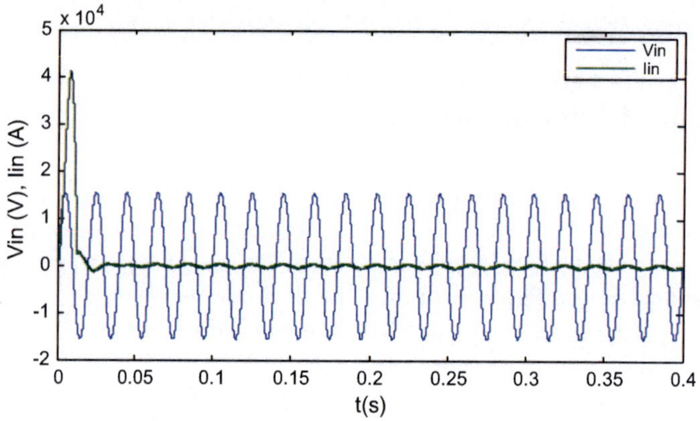

Fig. 3 I/P stage I and V waveform analysis

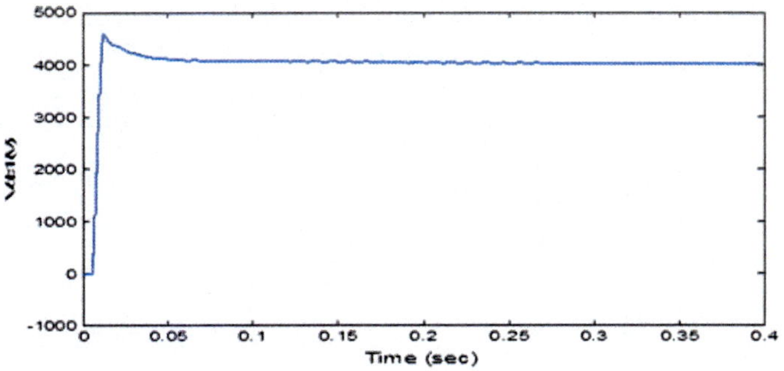

Fig. 4 Rectifier output DC voltage

Table 2 DC–DC converter parameters values

Parameter	Symbol	Value
Input DC bus voltage	V_{dci}	4 kV
Number of parallel cells	M	4
DC bus capacitors	C_i	0.047 F
Output DC voltage	V_{dco}	400 V
High frequency transformer rating	P_{hft}	50 kVA
Turns ratio	k	10:1
DC bus capacitor	C_o	200 μF
(PG): proportional gain	K_p	10
(IG): integral gain	K_i	1000
Switching frequency	f_s	25 kHz

6.2 Performance of the DC–DC Converter

DC–DC conversion has been carried out through high frequency link transformer at $f_s = 25$ kHz with proposed controller has elaborated in this isolation stage of SST. The main parameters used simulation is shown in Table 2 and the performance of the same can be seen in Fig. 5.

6.3 Inverter Stage Performance

For three-phase inverter, which is connected at other ends of SST characteristics of waveforms are as follows. Figures 6 and 7 shows the inverter performance which shows the three-phase voltages across output.

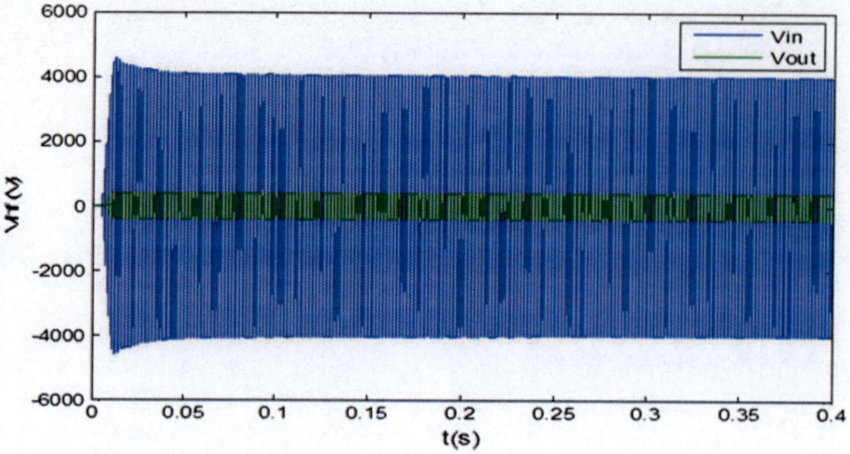

Fig. 5 DC–DC converter input and output DC voltage

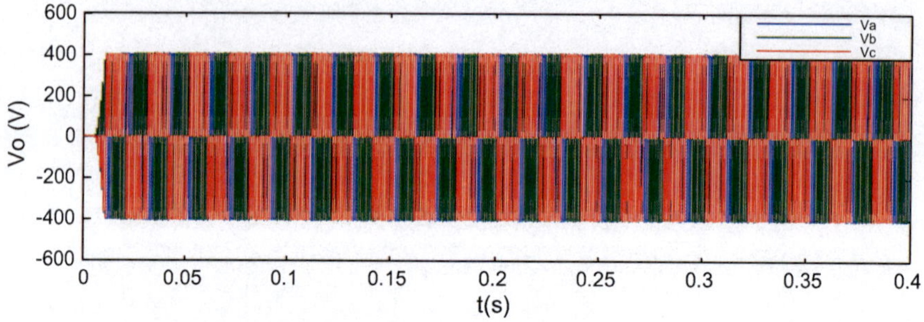

Fig. 6 Inverter output three-phase voltage

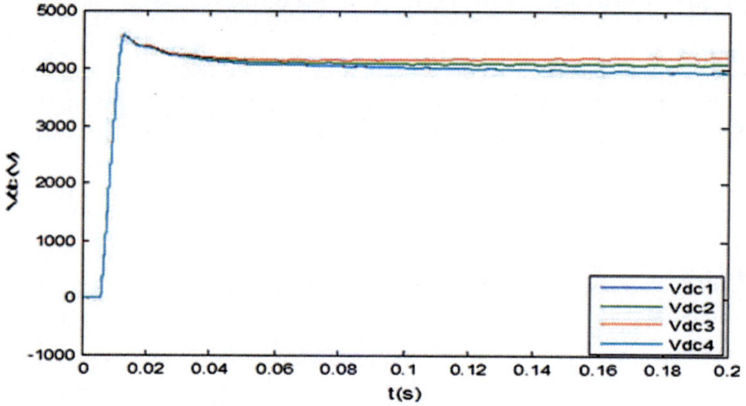

Fig. 7 Unequal load variation in DC voltage

6.4 *Effect of Unequal Loading in Performance of SST*

In the CHB rectifier, unequal loading by increasing 25% to 50% to investigate the performance of isolation stage output and overall SST output value of voltage and the current can be seen in the following figures. Figure 15 shows the DC link voltages across four levels CHB rectifier for loading as $R_1 = 20\ \Omega$, $R_1 = 25\ \Omega$, $R_1 = 30\ \Omega$ and $R_1 = 20\ \Omega$.

It is to be found that due to unbalance loading DC voltage across each level of CHB calculated obtained as $V_{\text{dc1}} = V_{\text{dc4}} = 3880$ V, $V_{\text{dc2}} = 4090$ V and $V_{\text{dc3}} = 4253$ V for simulation time $t = 0.3$ s. This effect of unbalance loading will not disturb the output performance.

6.5 *Temperature Rise Analysis for IGBT*

The semiconductor switches integrated with a diode such as diode IGBT model to allow both the losses in the semiconductor switch and the diode individually specify using only one set of lookup table. Due to the reverse polarity of the diode, the diode losses are added to the tables of the loss in the semiconductor switch by the extension of the table of the negative voltage and the current of the negative direction for losses diode in conduction, and the positive voltage and the direction of the negative current to the diode switching losses can be seen from Figs. 8, 9 and 10.

7 Conclusion

In this paper, a topology for various rectifiers, DC–DC converter and the inverter output have been discussed and proposed outcome of the multilevel topology suited to compound the series connection of multiple inverters sublevels. The complete proposed topology has been optimized for different purposes. This topology has been compared with the topologies presented in symmetrical CHB inverter multilevel and multilevel inverters asymmetric CHB. Tables 1 and 2 show the number of parameters SST topology and algorithms suggested topologies presented in using rectifier stage, and the average DC voltage is achieved. Rectifier controls the input power and high voltage reduce stress levels. DC–DC isolated ensure proper regulation of sides of the input and output. The simulated result shows good SST performance of the balance with the control strategy of the proposed state.

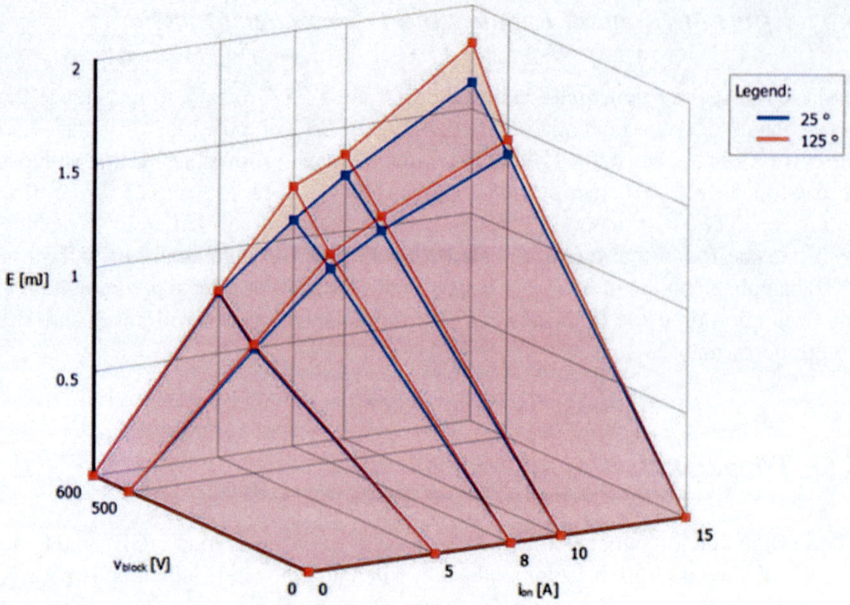

Fig. 8 Turn on loss by temperature rise

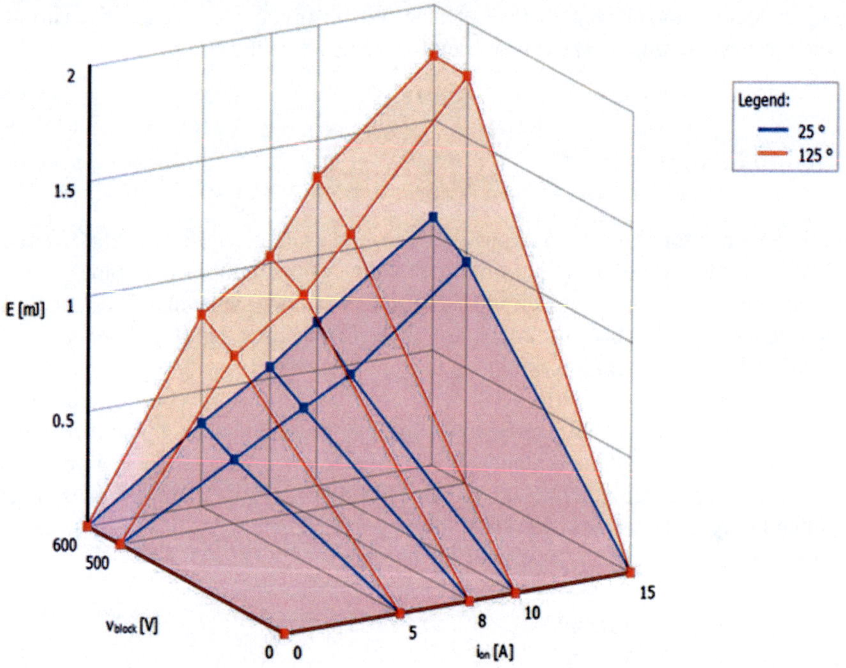

Fig. 9 Turn off loss by temperature rise

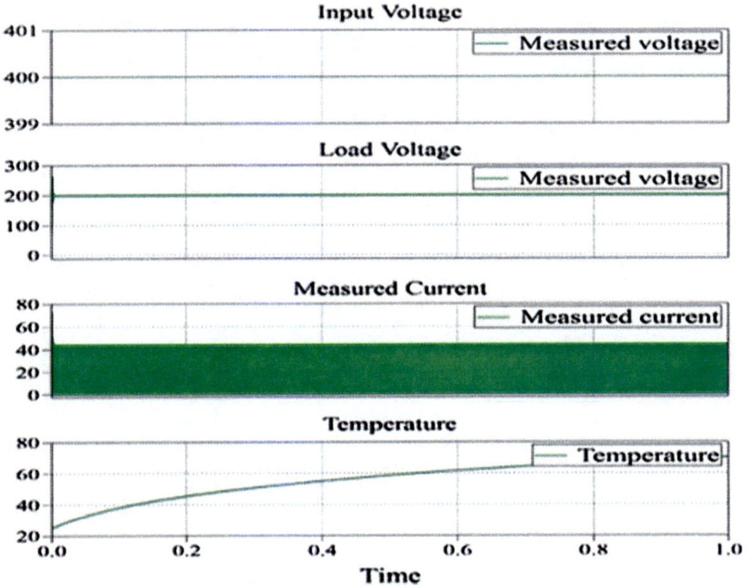

Fig. 10 Temperature rise in IGBT during switching

References

1. Iqbal A et al (eds) (2020) Soft computing in condition monitoring and diagnostics of electrical and mechanical systems, vol 1096. In: Advances in intelligent systems and computing. Springer, Singapore. https://doi.org/10.1007/978-981-15-1532-3
2. Iqbal A et al (eds) (2020) Meta heuristic and evolutionary computation: algorithms and applications, vol 1096. In: Studies in computational intelligence. Springer, Singapore. https://www.springer.com/gp/book/9789811575709
3. Gascoigne MNKW, Divan D, Baumann E (1992) Performance characterization of a high-power DCB DC-to-DC converter. Indust Appl IEEE Trans 28(6):1294–1301
4. Bhattacharya S, Zhao T, Wang G, Dutta S, Baek S, Du Y, Parkhideh B, Zhou X, Huang AQ (2010) Design and development of generation-I silicon based SST. In: Applied power electronics conference and exposition (APEC), 2010 twenty-fifth annual IEEE, pp 1666–1673, 21–25 Feb 2010
5. Zhao T, Yang L, Wang J, Huang AQ (2007) 270 kVA solid state transformer based on 10 kV SiC power devices. In: Electric ship technologies symposium. ESTS'07. IEEE, pp 145–149, 21–23 May 2007
6. Mao X, Ayyanar R, Falcones S (2009) A modular, interleaved AC-AC flyback topology for solid state transformer. In: Proceedings of FREEDM annual conference 2009, North Carry State University, Raleigh, NC, pp 221–224, 18–19 May 2009
7. Mao X, Ayyanar R (2009) Average and phasor models of single phase photo voltaic generators for analysis and simulation of large power distribution systems. In: Applied PE conference and exposition. APEC 2009. Twenty-fourth annual IEEE, pp 1964–1970, 15–19 Feb 2009
8. Vangen K, Melaa T, Adnanes AK (1992) Soft-switched high-frequency, high power DC/AC converter with IGBT. In: Power electronics specialists conference 1992. PESC'92 record, 23rd annual IEEE, vol 1, pp 26–33, 3 Jul 1992
9. De Doncker R, Divan D, Kheraluwala MH (1991) A three-phase soft-switched HP-density DC/DC converter for high-power applications. Indust Appl IEEE Trans 27(1):63–73

10. Yang L, Zhao T, Wang J, Huang A (2007) Design and analysis of a 270 kW five-level DC/DC converter for solid state transformer using 10 kV SiC power devices. In: PE specialists conference. PESC 2007. IEEE, pp 245–251, 17–21 June 2007
11. Gangyao, Huang X, Wang J, Zhao T, Bhattacharya S, Huang AQ (2010)Comparisons of 6.5 kV 2.5 A Si IGBT and 10-kV SiC MOSFET in SST application. In: Energy conversion congress and exposition (ECCE). IEEE, pp 100–104, 12–16 Sept 2010

Efficient Object Removal and Region Filling Image Refurbishing Approach

Mahroosh Banday

Abstract Inpainting is a technique to restore damaged portions in images such that it is imperceivable to the naked eye. The challenge is to fill the damaged regions of images such that it looks more discernible and visually plausible while restoring the unity of the picture. Inpainting has various applications such as restoration of damaged photographic images, paintings and removal of particular objects from photographs and images. In this paper, a novel inpainting approach has been introduced that removes objects from images in an indistinguishable manner. Although the exemplar-based traditional object removal approach has certain plus points such as ability to preserve textural, structural details of an image and also preserves lucidity in images, it does not preserve the visual quality of inpainted images. This paper proposes a robust "image refurbishing" algorithm for both synthetic and real unshaded textured images wherein the local information from synthesized patch is used to reconstruct the target area in the image in pixel-by-pixel manner, thus propagating structural as well as textural information simultaneously using a single efficient algorithm. Furthermore, the state-of-art inpainting methods available till date use air-brush masking method; however, in the proposed approach, masking is automatically done which helps in reducing the inpainting time. The inpainted images obtained using the proposed approach are visually appealing which depict the efficiency of the proposed method in removing desired large objects from photographs and other images. The experimental results of the proposed method compare favorably to the state-of-the-art object removal approaches, showing supremacy of the proposed technique over others in terms of computational efficiency, visual quality of inpainted images and inpainting time.

Keywords Inpainting · Image refurbishing · Exemplar inpainting · Object removal · Algorithm

M. Banday (✉)
Department of Electronics and Communication Engineering, NIT Srinagar, Srinagar, India
e-mail: bandaymahroosh@yahoo.com

A. Iqbal et al. (eds.), *Renewable Power for Sustainable Growth*, Lecture Notes in Electrical Engineering 723, https://doi.org/10.1007/978-981-33-4080-0_9

1 Introduction

Inpainting is the term borrowed from medieval art where the restoration artists are tasked to bring the old photographs and damaged paintings close to the original state by filling in the missing gaps in order to make the destroyed artwork plausible while restoring its uniqueness. Image inpainting is a new and active research area [1] during the present decade, and its objective is to revert deterioration (scratches and cracks in photographs) and to remove undesired writings and objects from the image [2]. The inpainting problem first appeared in telecommunication sector where it was required to fill the image blocks lost during the transmission.

Digital image inpainting refers to the technique that is used to reconstruct corrupted and missing portions in an image utilizing information from neighboring areas of that particular image in such a way that the alterations made in restored workpiece are apparently valid [3]. The purpose of inpainting is enhancing the overall quality of the image by reverting deterioration to make it more legible while conserving the unity of the image. In recent years, an important part of artistic and scientific work is stored in the form of image and film archive. Thus, digital image inpainting has become a hotspot in digital photography and film industry. Image inpainting has numerous applications such as photo and film restoration, disocclusion, text removal, special effects in movies, red eye removal. Digital image inpainting is actually a linear transformation [4, 5]. A large number of inpainting algorithms continuously come forth due to extensive technology advancements and research in this field and the goal of all is to enhance the quality of image. The general steps that determine the underlying methodology (Fig. 1) [6] of any inpainting algorithm are:

1. The figure determines the way in which the target region is to be filled, the purpose being to conserve the unity of the image. Thus, the first step is to identify/mask the target region (Ω) to be inpainted.
2. Structure of hole ($\partial\Omega$) is extended into the gap.

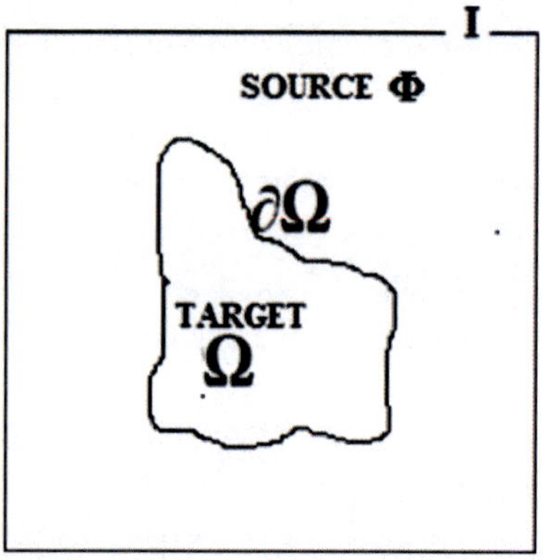

Fig. 1 Entire image I, the target area Ω to be restored and its contour $\partial\Omega$

3. The target region/gap is then filled with pixels $(x, y) \in \Omega$ that match with those at the inpainting domain ($\partial\Omega$) boundary.
4. Minute details are then inpainted, or in other gates, the texture is added.

In order to retain the image data and conserve the unity of the picture, various techniques of inpainting have been proposed by researchers. Generally, digital inpainting approaches can be categorized into two main classes—diffusion method and exemplar method.

The most basic diffusion techniques are pixel interpolation-based [7–10]. Bertalmio et al. [11] was the first to present the digital inpainting approach that has been inspired from the partial differential (PD) heat flow equations. The partial differential equation-based approach is modeled to reconstruct the corrupted areas by diffusal of information from the known image area into the target area along direction of the isophote [12]. The main problem of this method is that it cannot restore textured images where it gives blurry result.

Another class of inpainting methods is the exemplar-based technique. Each state-of-the-art exemplar algorithm is based on the technique of synthesizing a texture sample using the information from neighboring region. Criminisi et al. [13] presented a patch-based exemplar approach that works on standard texture images where the accurate match texels from the original source region are recognized and copied directly into the missing region. Exemplar-based method combines features of both diffusion methods and texture synthesis [14] to determine the fill-in process. On comparing with the PDE method, exemplar-based method gives better results for large missing regions, but the overall cost of exemplar-based method is huge because of its iterative nature. Moreover, in most of exemplar-based methods, accurate results cannot be achieved because large missing area is filled with small synthesized artificial patches producing observable artifacts [15]. Zhang et al. [16] presented an inpainting method for filling in missing regions using the superpixel segmentation approach and considering a minimal user input. Ding et al. [17] proposed an algorithm for inpainting images using geometric structures and texture having large missing regions. The method reproduces the textural information using a nonlocal Gaussian texture measure and also uses trimmed mean filter for enhancing pixel intensity so as to achieve inpainted images. In [18], the authors have proposed a pixel-based algorithm for removal of objects from the images, and algorithm repairs the affected region by using the similar contour pixels that are around the affected areas.

In this paper, we introduce a new and efficient pixel-based "image refurbishing" approach for removal of bigger objects from images. This method utilizes local pixel information from the synthesized sample texture for reconstructing the lost data in the target region, propagating both structural and textural information simultaneously, thereby providing improved visual quality. Unlike all inpainting algorithms, our algorithm masks the target region automatically which adds to the efficacy of our algorithm. Compared to the exemplar-based algorithms, our algorithm gives plausible inpainting results.

Paper Outline: Section 2 presents the idea of traditional exemplar-based inpainting approach. Section 3 of this paper presents the proposed refurbishing algorithm for object removal. The experimental results of the proposed algorithm and traditional exemplar technique are compared in Section 4 of this paper. Section 5 discusses the conclusion and future scope of the work.

2 Exemplar Inpainting Approach

The idea behind the exemplar inpainting technique refers to the use of texels or image blocks extracted from either the source region in same image or from another image belonging to a set of representative images. The aim is to fill the hole with best candidate patch from the same image source region [6, 19]. One of the most important algorithms that is given as a reference while discussing exemplar inpainting approach is the one put forward by Criminisi et al. [13]. This algorithm focuses on removal of large objects from the image while filling in hole that is left behind with the information from the neighboring area. The very essence of exemplar-based inpainting method is extending the linear structure of the image and finding the patches from the complementary area of the hole which contains the desired information that needs to be filled in the gap. A description of the underlying concept of exemplar inpainting method is presented here for better understanding.

Figure 2 depicts the procedure of exemplar-based inpainting approach [20, 21]. Let us denote the input image as I, target area and its boundary by Ω and $\partial\Omega = I - \Omega$, respectively, $\Phi = I - \Omega$ is known area, Ψ_{p} denotes the target patch, and the source patch/candidate patch is denoted by Ψq. This method starts the inpainting process by filling the missing regions first near the boundary of the target area (Ω) using texels blocks from the source area (Φ)[22]. The priority value $P(p)$ for all the boundary points is computed, and the patch centered at the point (p) of the boundary ($\partial\Omega$) having maximum priority is filled first. The priority of patch is computed as multiplication of data term $D(p)$ and confidence term $C(p)$. The confidence term $C(p)$ gives the count of pixels known in target texel, and data term $D(p)$ is a function that determines isophote at point (p) [23, 24].

Thus,

$$P(p) = C(p) \cdot D(p) \tag{1}$$

Here, $C(p)$ is the confidence term defined as:

$$C(p) = \frac{\sum_{q \in \Psi p \cap \Phi} C(q)}{|\Psi_{\mathrm{p}}|} \tag{2}$$

here, $|\Psi_{\mathrm{p}}|$ is the size of patch Ψ_{p}. And $D(p)$ is the data term given as:

$$D(p) = \frac{|\nabla^{\perp} I_{np} . \overrightarrow{np}|}{\alpha} \tag{3}$$

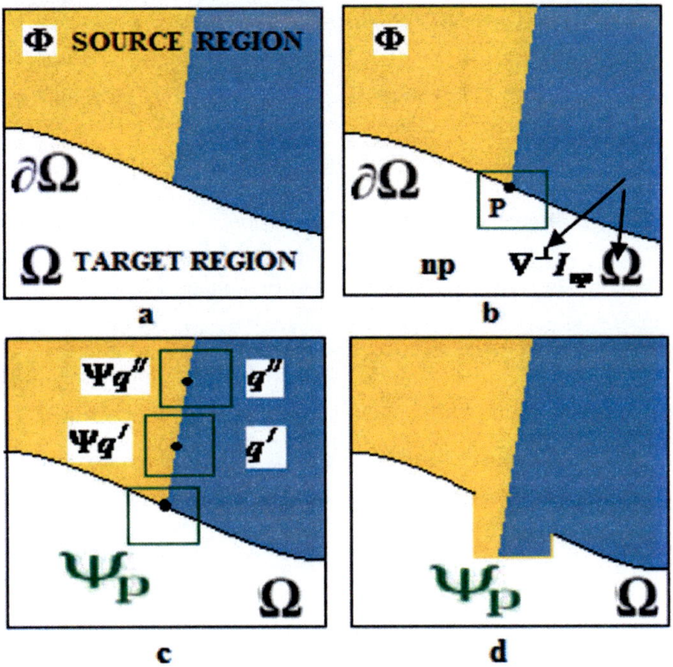

Fig. 2 Texture synthesis using exemplar-based method **a** image with ϕ as source area, Ω as target and ∂Ω as boundary. **b** restoration region delimited by the ψ_p patch having center at *p*. **c** $\left(\Psi'_q, \Psi''_q\right)$ as candidate patches in the source area. **d** best-match candidate patch extended to the area delimited by patch ψ_p, getting 1 fill-in of Ω partially

here, α is the factor of normalization (for gray-level image, $\alpha = 255$), $\overrightarrow{np}$ is the orthogonal unit vector to the contour (∂Ω) at point *p*, and $\nabla^{\perp} I_{np}$ is isophote which is located at point *p*. Taking initial value of $C(p) = 0; \forall P \varepsilon \Omega$ and $C(p) = 1; \forall P \varepsilon \Phi$, the below given steps are iterated till all the target region (Ω) is fully filled [13]:

1. Identify the region to be filled ∂Ω from target region (Ω).
2. Compute the priorities of patches for all patches centered at the boundary of Ω.
3. Select Ψ_p patch with highest patch priority $P(p)$.
4. Within the source Φ, the best-match Ψ_p patch selected that has the least sum-of-squared difference (SSD) with the Ψ_p patch with center at $p \forall p \varepsilon (\Psi p \cap \Phi)$.
5. Fill the region (Ω) in target area patch with the pixel data from the best candidate patch Ψ_q, $\forall q \varepsilon \Phi$.
6. Update the pixel confidence term at point p in unknown/target patch Ψ_p.

3 Proposed Image Refurbishing Algorithm

The main issue with traditional exemplar inpainting approach is that mostly, the target patch has not sufficient required source pixels due to which they do not completely make use of content redundancy [25] in image for a meaningful representation. This thing in turn has adverse effects on the final result leading to blurring in some portions of the target region. Image refurbishing, a more robust but easy method to select the target patch in such a way that it contains enough of pixels from the known region to produce qualitative results, is introduced in this paper. Moreover, our algorithm uses automatic masking unlike other exemplar-based approaches that use manual masking using some airbrush software. Our concept is to synthesize a bigger target patch and use that information in a pixel-by-pixel manner on the target region which has proved effective for removing large objects from unshaded textured images. We now proceed with the illustration of our algorithm as follows.

3.1 *Masking*

Masking of the target region is the foremost step in any inpainting technique. All the inpainting techniques till date use manual masking using some airbrush softwares [26] like Photoshop which of course is very time consuming. So, in order to overcome this cumbersome masking method, we have introduced an efficient, automatic yet easy masking method. First of all, the user points out the target region Ω that is to be removed by making freehand boundary $\partial\Omega$ around the target region Ω. Once the marking is done, our algorithm automatically masks the marked region. After that, we perform calculations on the masked image in order to get the coordinates of the bounding box around the mask. Knowing the coordinates, we crop that portion from the image and store its dimensions for use in patch synthesis.

3.2 *Color-Pixel Refurbishing*

Once the masking is complete, we utilize the dimensions of the cropped masked region in order to synthesize a patch of similar dimensions from the source area (Φ) that is defined as whole image I subtracted the target area, (Ω) i.e., $\Phi = I - \Omega$. Thus, the bigger best-matching patch Ψ_{Bq} from source area (Φ) is selected, and pixel information $P_{\Phi i}$ from this patch Ψ_{Bq} is propagated into the target region Ω till all pixels (P_{Ω}) in target area (Ω) are completely filled in.

As illustrated in Figs. 3 and 4, the target area (Ω) is filled with information $P_{\Phi i}$ from the source region (Φ) in an efficient manner. For clear understanding, the masked values (P_{Ω}) in the target area (Ω) were filled with most appropriate values of pixels ($P_{\Phi 6}$, $P_{\Phi 7}$, $P_{\Phi 10}$, $P_{\Phi 11}$) from the source region (Φ). The color-pixel refurbishing

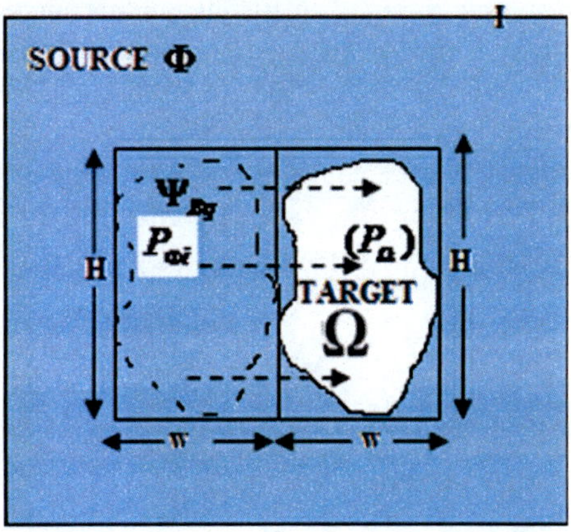

Fig. 3 Structure and texture propagation by image refurbishing algorithm

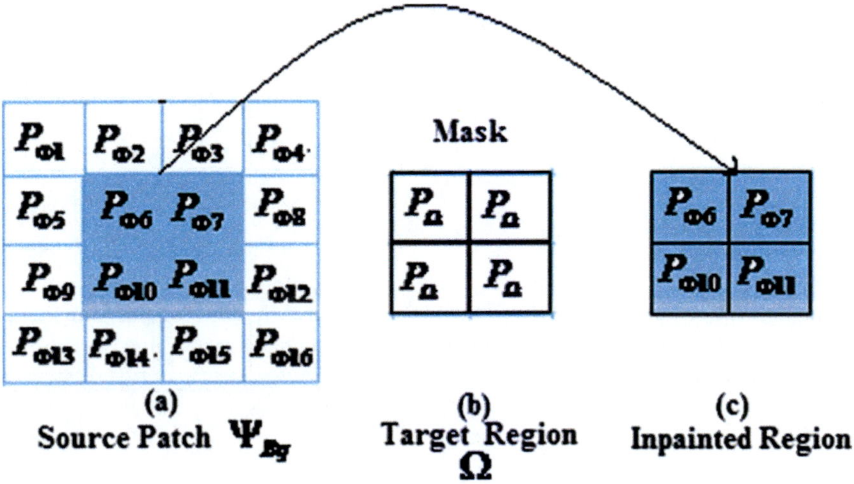

Fig. 4 Image refurbishing **a** synthesized source patch Ψ_{Bq}. **b** Masked target region Ω. **c** Inpainted/restored target area

process is governed by the following mathematical equation:

$$\sum_{i=x1}^{x2}\sum_{j=y1}^{y2}\sum_{k=1}^{3}\text{Im}(i,j,k) = P_{\Phi}$$
$$\forall \text{Im}(i,j,k)\varepsilon\Omega \quad \text{and} \quad P_{\Phi}\varepsilon\Phi \tag{4}$$

where P_{Φ} is the value of pixel outside masked area from source patch Ψ_{Bq}. The changes made in the inpainted region thus seem undetectable to the casual observer resulting in visually plausible outputs that look reasonable to the human eye. Formulating, a precise reckoning approach for depicting the success of our algorithm is an important yet useful task. To evaluate accurately, both the qualitative and quantitative approaches are used. The qualitative assessment of the results is done by human visual analysis (HVM), and the quantitative evaluation is done by analyzing and calculating mean square error (MSE) and peak signal-to-noise ratio (PSNR) of images. These correlation methods are mathematically expressed as:

$$\mathrm{MSE} = (1/mn)\sum_{i=0}^{m-1}\sum_{j=0}^{n-1}\|I(i,j) - U(i,j)\|^2 \tag{5}$$

Here, I is reference image and U denotes its inpainted version, i and j are the coordinates of the pixels and m and n represent the number of rows and columns of original image, respectively. Lesser the MSE, the better is quality of inpainted image. PSNR gives ratio of maximum signal power noise power that corrupts signal. The PSNR values are usually represented in dB (decibels). Greater the value of the PSNR, better is quality of inpainted image. PSNR is mathematically expressed as:

$$\text{Peak SNR} = 10\log_{10}\left(\frac{(2^n - 1)^2}{\mathrm{MSE}}\right) \tag{6}$$

where n is the number of bits.

Values of PSNR and MSE obtained from our algorithm are compared with the values obtained from traditional exemplar-based approach showing that our technique gives both qualitatively and quantitatively much better results than those obtained by exemplar-based algorithm.

4 Experimental Results and Discussion

For evaluating the performance of proposed approach, the experiments were conducted on MATLAB 13 b, and the results obtained were compared with traditional exemplar approach results. A set of 20 images has been utilized as test images so as to analyze performance efficiency of proposed approach. The quantitative analysis and evaluation of the proposed and traditional methods is done using peak signal-to-noise ratio (PSNR) and MSE. The result summary of MSE and peak SNR values is given in Table 1. The proposed "image refurbishing" approach shows improved MSE and peak SNR results than those obtained from traditional exemplar methods for all test images.

We tested our proposed technique on high-resolution football field image as depicted in Fig. 5a. Masked target area (the football) is shown in Fig. 5b. In the result obtained from exemplar method shown in Fig. 5c, the inpainted/restored grasses show

Table 1 Comparison of MSE and peak SNR of traditional exemplar method and proposed image refurbishing

Image	Resolution (dpi)	Exemplar-based method		Image refurbishing method	
		MS	PSNR (dB)	MSE	PSNR (dB)
Horse	220	3.0622	79.4275	0.3604	88.7200
		2.6787	80.0087	0.4284	87.9695
		3.8012	78.4887	0.4970	87.3239
Field	115	6.2913	76.3005	1.2965	83.1602
		4.7103	77.5574	1.1812	83.5645
		5.6131	76.7959	1.5573	82.3643
Man	130	4.3404	77.9126	0.9877	84.3415
		3.8935	78.3845	0.9994	84.2903
		4.1685	78.0881	1.0055	84.2643

Fig. 5 Inpainting football field image

blocky effect and some blur. However, our method overcomes this blocky effect and provides better visual quality without image blur as depicted in Fig. 5d.

A comparison of exemplar method (E) and the proposed "image refurbishing" method (P) is given in Table 1. Also, performance evaluation graphs with the comparison of peak SNR and MSE values for three different images are depicted in Fig. 6a,

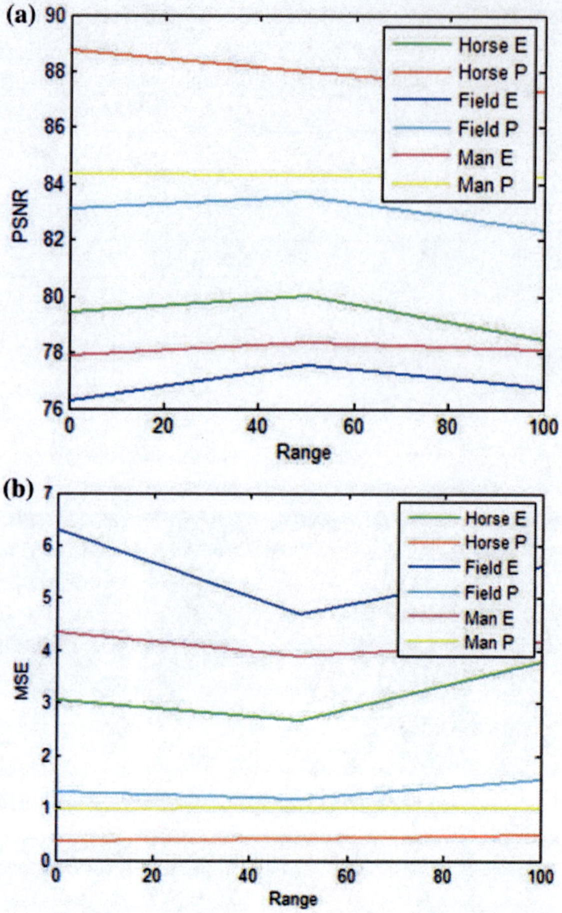

Fig. 6 Comparison of peak SNR and MSE using three different images

b, respectively. It shows that proposed approach has improved not only the quality of inpainted image but also the inpainting efficiency.

5 Conclusion

In this paper, a new and efficient "image refurbishing" approach has been presented for removing larger objects from images. The proposed approach marks target area using automatic masking unlike traditional inpainting methods, wherein masking is done manually, thereby reducing time required for inpainting an image. In our method, information from a synthesized sample patch from the same image is propagated into the target area in a pixel-by-pixel manner until all unknown pixels in target area are completely filled. The proposed method propagates both structure and texture in target area with single potent scheme. Automatic masking of the proposed method not only reduces the inpainting time manifold but also uses pixel information from a bigger patch. The proposed approach significantly outperforms work done in traditional methods both by increasing computational efficiency and visual quality at the same time. The results obtained depict superiority of the proposed algorithm over traditional exemplar approach, since its peak SNR value is high and MSE has low value. However, it is applicable to unshaded textured images and quality is reduced with complicated/irregular textures. The future work will concentrate on investigating extensions to modify the current algorithm further for processing complex textures in photographs as well as for removing objects from videos.

References

1. Wong A, Orchard J (2008) A nonlocal-means approach to exemplar-based inpainting. In: International conference on image processing. IEEE
2. Tuptewar DJ, Pinjarkar A (2017) Robust exemplar based image and video inpainting for object removal and region filling. In: International conference on intelligent computing and control (I2C2). Coimbatore, pp 1–4
3. Jaidilert S, Farooque G (2018) Crack detection and images Inpainting method for Thai mural painting images. In: 3rd international conference on image, vision and computing (ICIVC). IEEE, Chongqing, pp 143–148
4. Glaly YNE (2007) A thesis on development of PDE-based digital Inpainting algorithm applied to missing data in digital images
5. Gonzales RC, Woods RE (2002) Digital image processing, 2nd edn. Prentice Hall. ISBN: 0-201-18075-8
6. Alexandra IOF (2012) Digital Inpainting for artwork restoration: algorithms and evaluation. In: Master thesis report
7. Bertalmio M, Sapiro G, Caselles V, Ballester C (2000) Image Inpainting. In: Proceedings of SIGGRAPH. New Orleans, USA, pp 417–424
8. Chan T, Shen J (2001) Mathematical models for local non texture inpaintings. SIAM J Appl Math 62(3):1019–1043
9. Iqbal A et al (eds) (2020) Soft computing in condition monitoring and diagnostics of electrical and mechanical systems, vol 1096. In: Advances in intelligent systems and computing. Springer, Singapore. https://doi.org/10.1007/978-981-15-1532-3
10. Iqbal A et al (eds) (2020) Meta heuristic and evolutionary computation: algorithms and applications, vol 1096. In: Studies in computational intelligence. Springer, Singapore. https://www.springer.com/gp/book/9789811575709
11. Bertalmio M, Bertozzi AL, Sapiro G, Navier S (2001) Fluid dynamics, and image and video inpainting. In: Proceedings of conference on computer vision and pattern recognition. Hawai, pp 355–362, Dec 2001
12. Bertalmio M, Vese L, Sapiro G, Osher S (2003) Simultaneous structure and texture image inpainting. In: Proceedings of conference on computer vision and pattern recognition. IEEE
13. Criminisi A, Perez P, Toyama K (2004) Region filling and object removal by exemplar based image in-painting. IEEE Trans Image Process 13(9):1200–1212
14. Choi JH, Cheul HH (2013) An exemplar-based image inpainting method with search region prior. In: 2nd global conference on consumer electronics (GCCE). IEEE, pp 68–71, 978-1-4799-0892-9
15. Mahajan K, Vaidya MB (2012) Image in painting techniques: a survey. IOSR J Comput Eng 5(4):45–49

16. Zhang X, Hamann B, Pan X, Zhang C (2017) Superpixel-based image inpainting with simple user guidance. In: International conference on image processing (ICIP). IEEE, Beijing, pp 3785–3789
17. Ding D, Ram S, Rodríguez J (2019) Image inpainting using nonlocal texture matching and nonlinear filtering. IEEE Trans Image Process 28(4):1705–1719
18. Erkan U, Enginoğlu S, Thanh DNH (2019) An iterative image inpainting method based on similarity of pixels values. In: 6th international conference on electrical and electronics engineering (ICEEE). Istanbul, Turkey, pp 107–111
19. Christine G, Olivier LM (2014) Image inpainting-overview and recent advances. IEEE Signal Process Mag:127–144. ISSN: 1053-5888
20. Lixin Y, Chen C (2012) An effective exemplar-based image inpainting method. IEEE, 978-1-4673-2101-3
21. Venkatesh MV (2010) Digital inpainting algorithms and evaluation. In: Doctoral dissertations, University of Kentucky, Paper 55. https://uknowledge.uky.edu/gradschool_diss/55
22. Pandya N, Limbasiya B (2013) A survey on image inpainting techniques. Int J Curr Eng Technol 3(5):1828–1831
23. Tae-o-sot S, Nishihara A (2013) Exemplar-based image inpainting with patch shifting scheme. In: 17th international conference on digital signal processing (DSP), pp 1–5, Dec 2013
24. Qin Y, Wang F (2010) A curvature constraint Exemplar-based image inpainting. In: International conference on image analysis and signal processing (IASP), pp 263–267
25. Taschler ME (2006) A comparative analysis of image inpainting techniques. The University of York, pp 01–120
26. Neelima N, Arulvan M (2013) Object removal by region based filling inpainting. IEEE, 978-1-4673-5301-4

Analysis of Underground Renewable Energy Storage Tunnels Subjected to Capricious Superstructures

Md. Rehan Sadique, Mohammad Zaid, M. Wasif Naqvi, and M. F. Akhtar

Abstract The construction and development of different modes of energy storage has been increased with urbanization and development of smart cities. The advent of modern underground excavation techniques in developing nations like India has put on swing the construction of underground storage. Explicit study of underground structures for different field scenario has been mandatory due to changing geology, influence of local infrastructure, variation in groundwater table, etc. The development of infrastructure facilities on the surface above or near tunnels is an unavoidable situation. Hence, it must be dealt with intense care due to its ability of significantly affecting the stability of underground structures. Further, it may also stimulus the design-parameters of underground facilities. The load intensity of the superstructure is an important parameter for stability of these structures. The present study describes a three-dimensional mathematical modelling of an underground tunnel. Simulation has been performed by using Abaqus, which is based on a continuum finite element. The elastoplastic Mohr–Coulomb constitutive model of the soil has been incorporated in analysis. The number of storey of a superstructure has been varied to assess the consequence of the superstructure on the behaviour of an underground energy storage tunnel. The numerical model has been modelled, and the load changes in a stepwise manner, no load to 50-storey building. The results are compared in terms of displacement and stresses generated in soil, and it has been observed that the construction of a structure has significant influence on an existing underground structure.

Md. Rehan Sadique · M. Zaid (✉)
ZHCET, Aligarh Muslim University, Aligarh 202002, India
e-mail: mohammadzaid1@zhcet.ac.in

Md. Rehan Sadique
e-mail: rehan.sadique@gmail.com

M. W. Naqvi
Department of Civil and Environmental Engineering, University of Toledo, Toledo, OH, USA
e-mail: MohammadWasif.Naqvi@rockets.utoledo.edu

M. F. Akhtar
Department of Civil Engineering, University of Cincinnati, Cincinnati, OH, USA
e-mail: akhtarmn@mail.uc.edu

A. Iqbal et al. (eds.), *Renewable Power for Sustainable Growth*, Lecture Notes in Electrical Engineering 723, https://doi.org/10.1007/978-981-33-4080-0_10

Keywords Energy storage · Underground tunnel · Finite element · Mohr–Coulomb

1 Introduction

Urbanization has increased globally, with huge population drifting towards metropolitan cities and its suburb. The movement for work or elevate living standards towards cities has amplified the population density of these areas. For illustration, in case of Delhi (India), the change in population density is from 9340 persons/km^2 in 2001 to 11,297 persons/km^2 in 2011 [1] and hence forcing the planners and developers to go underground to accommodate rapid transportation system, infrastructure and storage. Geotechnical structures have been increasingly engaged as energy geostructures worldwide [2]. More and more underground caverns and tunnels have been constructed at an exponential rate. The increasing need for renewable energy sources has led to the expansion of shallow geothermal applications for heating and/or cooling of buildings [3]. Tunnels move through different geological condition and different overburden pressure [4–12]. Compressed air energy storage (CAES) systems are a relatively new technology for storing huge amount of energy [3]. These airtight tunnelling technologies can be applied to a variety of energy underground storage facilities, e.g., compressed natural gas (CNG), liquefied petroleum gas (LPG), demethyl ether (DME), etc. [13]. In urban areas, both types of construction have been in progress, i.e., construction of buildings over the existing tunnels and vice-versa.

In the open literature source, a lot of research paper is existing about different ground effect like settlement, sink hole, ground movement, soil subsidence, etc., due to the excavation of tunnelling under a constructed structure [14]. Khan et al. [14] have also summarized the different factors effecting the response of tunnel liners. Baruah and Bharadwaz [15] had analysed the surface subsidence in the term of ground trough of subsidence. The settlement trough had been reported in the route of the tunnel and also in the transversely direction of the tunnel. Ground deformation along with the surface settlement occurred due to tunnel boring machine (T.B.M.) or different tunnel coring equipment has also been illustrated [16]. Surface subsidence investigation due to the construction of tunnel has been reported by Maraš-Dragojević [17]. Nevertheless, the influence of additional loading due to the building erection on the existing underground structures has not been studied satisfactorily. Few researchers have considered the influence of deep foundation excavation and its loading on existing underground excavation and tunnel [18]. Kumar and Ayothiraman had considered the result of the vertically loaded pile on an urban tunnel in constructed in clay [19]. Calabrese and Monaco had investigated the stresses developed in an existing deep tunnel because of driving a pile [20]. Furthermore, an underground transportation tunnel in Bangalore (India) had been studied for performance under different static loads. The hydrostatic pressure of water table, self-weight due to soil and load of the traffic movement on the ground surface had been considered [212223].

Therefore, it has been concluded that there is still wide scope for further study related to effect of superstructure constructed above an existing underground structure. The numerical analysis carried out in the present paper deals with the effect of different storey buildings on the tunnel being used as an energy storage structure. The parametric study consists of variation of number of storey a building has, i.e., five to fifty. In addition, the insitu stresses are also included in the present paper for better simulation of field conditions.

2 Problem Definition

As a result of the shortage of land, gigantic structure were being constructed above the existing underground structures. These big infrastructures may bring a high risk to the stability of underground facilities. Further, the stability of structure itself will be endangered, if built on the external ground in the neighbourhood of tunnels. The present work has been carried out to observe the influence of construction of superstructure on the strength of the built tunnel. The tunnel has been assumed to be already built and having a concrete lining support system. It is found to be safe when tested under gravity load only. The weight due to a varying storey number of structure extending from five to fifty is enforced on the soil above the underground tunnel. The developed stresses in soil and liner have been observed. Further, ground movement due to load of the construction on the critical locations of underground structure has been analysed. The deviation of the parameters in the direction of tunnel axis has been observed. The validation of analysis has been performed using Terzaghi and Richart [24] solutions. The analysis provides significant outcomes on the values of important design parameters that recommends the tunnel permanency. Moreover, it extracts the anodyne construction perimeter of the structure on the green field above the underground tunnels.

3 Numerical Modelling

A three-dimensional finite element geometric has been constructed using finite element package. Figure 1a shows the typical meshed models of soil mass, and Fig. 1b shows the concrete lining.

3.1 *Tunnel Modelling*

The finite element model of an underground soil tunnel, constructed in an Indian city situated in the Indo-Gangetic delta, has been chosen. Tunnel has the diameter of 6.35 m with an inside concrete lining having thickness as 28 cm. The tunnel lining

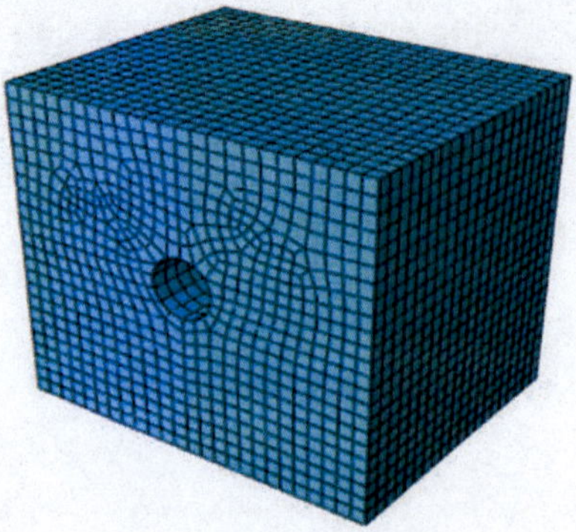

(i) Soil Mass model

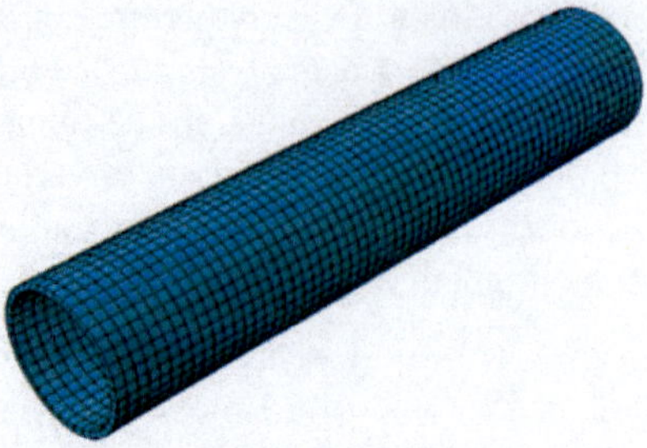

(ii) element characterization of Concrete lining

Fig. 1 3D model with mesh

Table 1 Properties of the model used

Sand of Yamuna	
Unit weight of sand	15 kN/m^3
Young's modulus of sand	25 MPa
Poisson ratio of sand	0.3
Friction angle of sand	37.5
Cohesion of sand	500 kPa
Concrete lining	
Density of lining	25 kN/m^3
Young's modulus of lining	23.48 GPa
Poisson ration of lining	0.2

has been meshed using C3D20R (quadratic hexahedron) elements available in finite element package Abaqus/CAE. Twenty-four hundred elements are present in tunnel lining, which has a mean element size about 0.5 m. The material properties opted for the lining have been stated in Table 1.

3.2 Modelling

The quadratic hexahedron reducing elements (C3D20R) of Abaqus have been applied to characterize the soil mass. The 12,260 number of C3D20R elements with size 1.5 m and aspect ratio as unity were present in the model. The Mohr–Coulomb material model has been utilized for providing the elastoplastic properties of the soil. Table 1 illustrates the properties of the soil employed in the model. The dimension of the soil model base/height/depth is 38 × 35 × 30 m^3. All the proportions, excluding the top, were kept such that all the aspects were at a distance three times the diameter of tunnels, measuring from the centre of the underground tunnel. Hence, gaining the satisfactory boundary condition, i.e., the stress contour became non-significant at this distance, and the effect of stress on the boundary is trivial beyond it. The underground tunnel is fixed at the midpoint of model, from the top surface at a depth of 12.5 m. It makes the simulation further representative.

3.3 Loading and Boundary Conditions

The base of the tunnel model has fixed support, as the soil extends to a semi infinite depth, which is several times higher than the depth of numerical model. Moreover, the vertical sides of the soil mass are roller supported, and therefore, the lateral displacement has been neglected in the present study. The vertical displacement has major contribution in causing damage in tunnel. The footing of 10 m × 10 m has been placed at the middle length of tunnel, and the load has been applied in the form of pressure. The magnitude of pressure corresponds to the weight of the total number of floors in a multi-storey building. There are seven different storey of a building have been varied, i.e., 5, 10, 15, 20, 25, 35 and 50-storey.

4 Result Validation

The tunnel model of the present paper has been modelled with the results presented by Terzaghi and Richart [24] through analytical equations. The two stresses around the tunnel, i.e., tangential and radial stresses are considered, which show close resemblance with the solution of Terzaghi and Richart [24], and plotted in Fig. 2. The radial distance has been considered along the hoop direction at the middle of tunnel length.

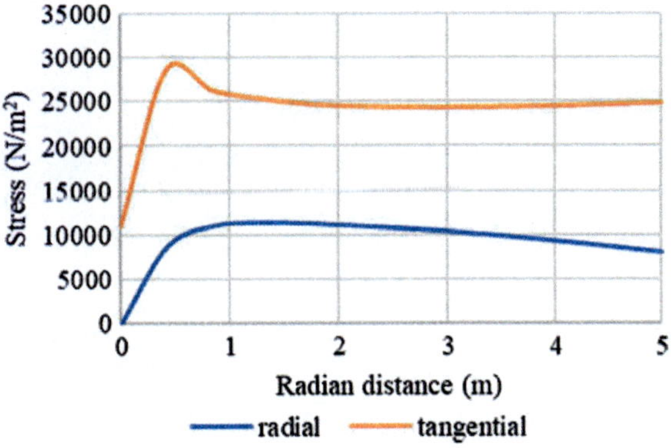

Fig. 2 Stress distribution around tunnel (validation of numerical results by Terzaghi and Richart [24] solution)

5 Results and Discussion

In the present paper, a 3D nonlinear finite element has been carried out to understand the influence of superstructure on an underlying energy storage system. The finite element modelling has been carried out by using Abaqus, and behaviour of concrete lining and soil has been observed.

5.1 *Stresses Plot of Concrete Lining*

The number of storey of the superstructure has been varied to incorporate the effect of surcharge loading. Figure 3 has been plotted to illustration to variation of radial and tangential stresses in the tunnel lining. The data points along the lining correspond

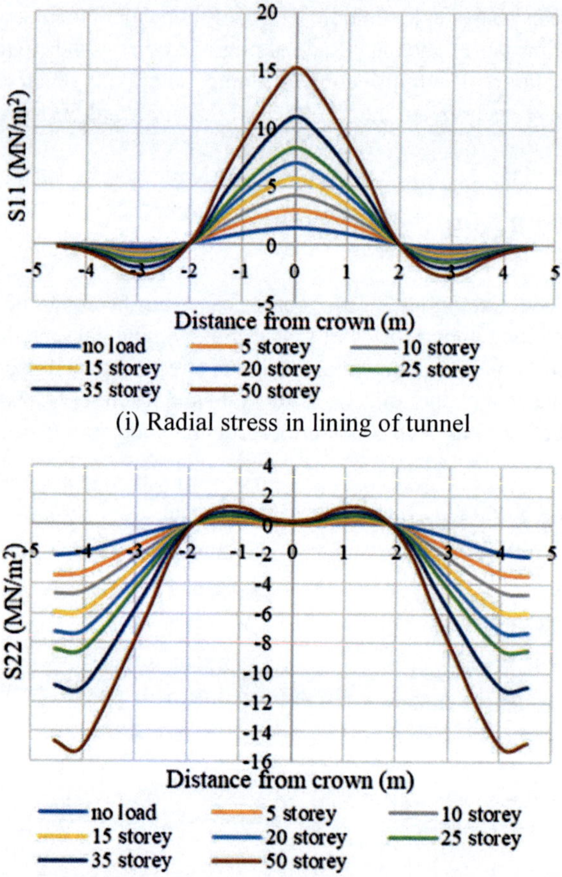

(i) Radial stress in lining of tunnel

(ii) Tangential stress in lining of tunnel

Fig. 3 Principal stresses variation at tunnel lining

to the boundary supporting the soil around the tunnel at the central section of the model. It has been observed that the stresses along the springline increase with increasing number of storey. Moreover, tensile nature of radial stresses has been observed at the crown position as shown Fig. 3a. However, the maximum tensile stresses that have been observed at crown show a decrease along the springline, and its value becomes negligible at 3 m from the crown. Afterwards, an increase in the compressive stresses has been observed after 3 m from the crown, magnitude of which initially decreases and becomes zero. The tensile stresses show higher deviation in magnitude as compared to compressive stresses in the radial direction. In case of no load, the tensile stress on the crown is 1.48 MN/m^2, while it increases by approximately by 10-*times* to a maximum of 15.27 MN/m^2 for the superstructure having 50th storey. Moreover, the maximum compressive stress at 3 m from the crown has magnitude of 0.24 MN/m^2 and 2.44 MN/m^2 in case of no load and 50th storey, respectively. However, minor deviation has been observed in the magnitude of compressive stresses in comparison with tensile stress.

In addition to radial stresses, the stress in the tangential direction also has significant effect on the tunnel stability as observed in Fig. 3b. In case of tangential stresses, tensile behaviour has been observed up to 2 m in the proximity of crown for different cases of superstructural load. In addition, severe increase in the magnitude of compressive stresses has been observed up to springer. The compressive stress in the tangential direction has a magnitude of 2 MN/m^2 and 15 MN/m^2 in case of no load and 50th storey load, respectively. Therefore, 7.5-*times* higher magnitude of compressive stress has been observed due to 50th storey load.

Figure 4 shows the variation of radial and tangential stresses, and it has been observed that both have maximum magnitude at the crown, i.e., below the point of loading. Moreover, S11 has higher magnitude as compared to S22; however, magnitude of both the stresses decreases with the distance farther from the position of loading. In addition, difference in the magnitude of stresses at the crown and at a point 15 m away from loading location increases with the number of storey of superstructure. In case of 50th storey load, S11 and S22 have a magnitude of 15.27 MN/m^2 and 0.2 MN/m^2, respectively, 1.5 MN/m^2 and 0.02 MN/m^2, respectively, in case of no load. Also, the response of tunnel due to load from superstructure has close relation with results reported by Ramamurthy [25].

5.2 Stresses Generated in Soil Surrounding the Tunnel

The stresses in the soil resulted due to superstructural load have significant effect on the stability of underground structures, and therefore, contours of stresses under varying loading condition have been shown in Fig. 5. The extent of stress magnitude and area of influence increases with the number of storey of superstructure. The load due to superstructure results in higher concentration of stresses up to a 5 m depth. The maximum stresses are observed in 50 storey loading case, i.e., 0.49 MN/m^2, and minimum stress has been found in case of 5 storey superstructure, i.e., 0.0.088

Fig. 4 Graph of principal stresses at the crown point along the tunnel or excavation

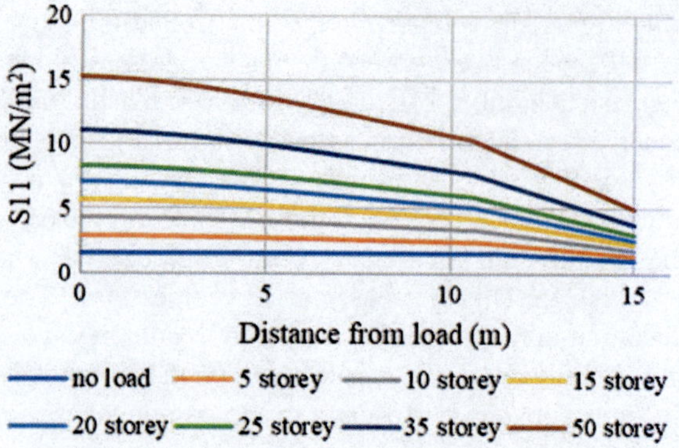

(i) Radial Stresses at the crown point along tunnel

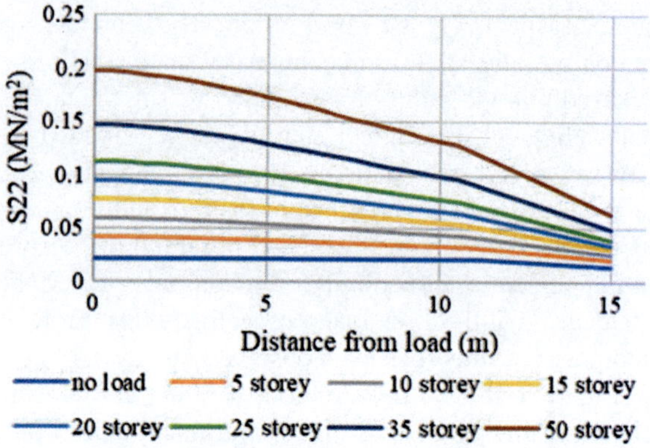

(ii) Tangential Stresses at crown point along tunnel-length

MN/m^2. The stress has higher concentration along the springline of tunnel. It has been concluded that significant monitoring system for tunnel is required while constructing a high-rise building above it by considering the bearing capacity and stress dissipation in soil.

5.3 Displacement at Critical Location

From the Fig. 6, it has been observed that displacement in the different cases of tunnel follows a similar pattern. In addition, magnitude of displacement decreases with increase in the distance between the location of loading and point of observation, i.e., tunnel crown. When the 50 storey superstructure has been constructed above the tunnel, it resulted in 80 mm of displacement, while 23 mm of displacement has been observed in no load case due to tunnelling itself. The maximum variation of 10 mm of displacement has been observed at crucial points, namely crown, invert and springer for the case of 50 storey. The vertical displacement in case of 50 storey

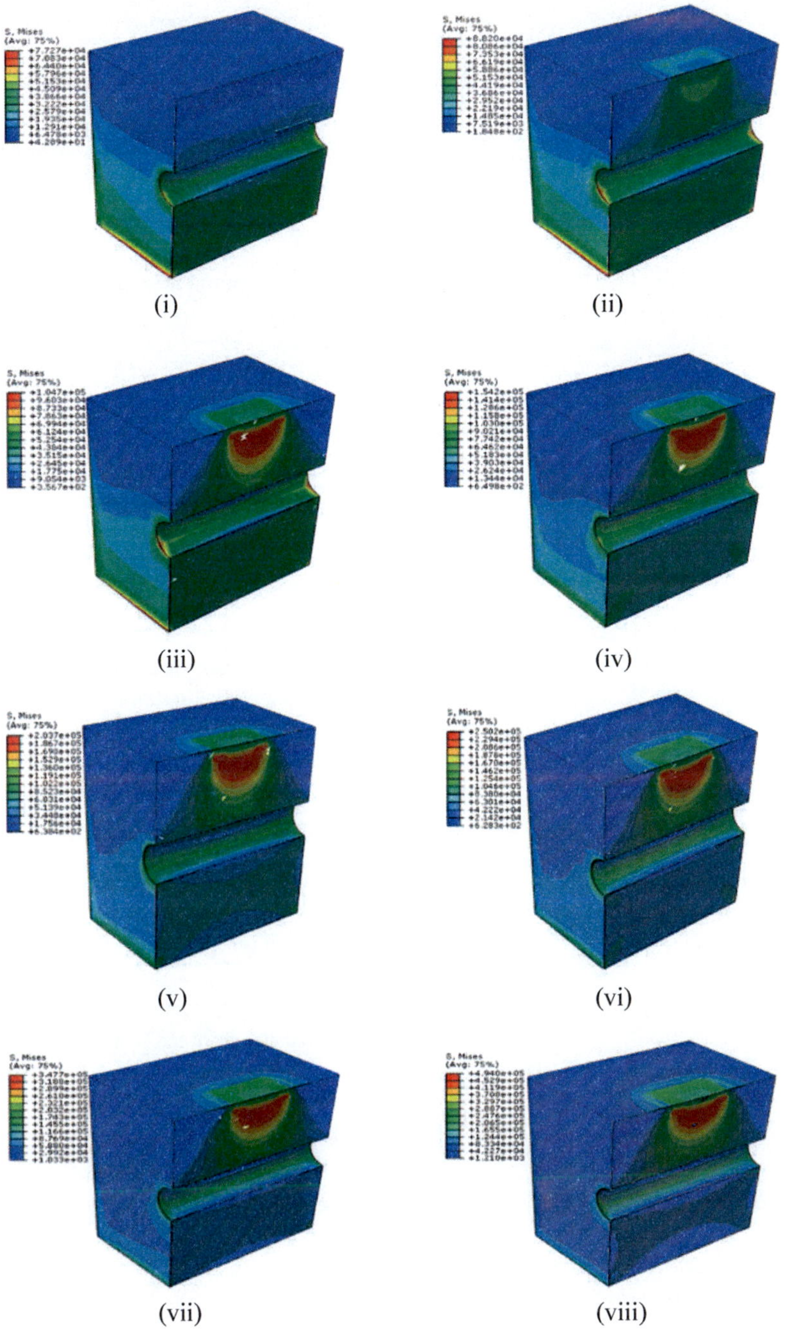

Fig. 5 Contours of stresses generated in soil due to applied load in case of **a** zero surcharge weight, **b** weight of 5-storey building, **c** weight of 10-storey building, **d** weight of 15-storey buildings, **e** 20-storey, **f** 25-storey, **g** 35-storey and **h** **50**-storey

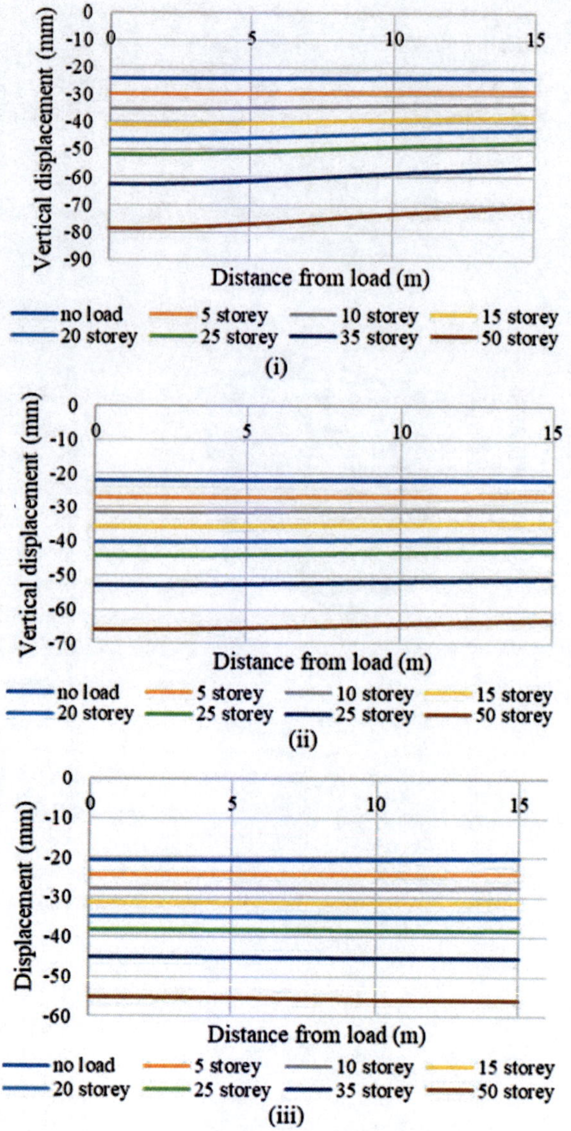

Fig. 6 Variation of displacement along critical underground tunnel points moving from centre to exit of tunnel **a** crown, **b** springer, **c** invert

is approximately 4-times higher than no load case at crown and springer locations, while in case of an invert, this variation is 3 times. Therefore, number of storey has substantial influence on the crucial points which causes instability of underground energy storage tunnels.

6 Conclusion

This study has been carried out to understand the influence of superstructure construction above the underground energy storage tunnel. The finite element technique has been adopted for the present analyses by using Abaqus. It has been concluded that the radial stresses on tunnel liner near crown are higher; however, tangential stresses are higher at the springline of tunnel. The nature of stresses numerically observed in the present study has a close agreement with that predicted analytically in the available literature. A high-rise building just above the energy storage structure may lead to failure of soil, which in turn deforms the underneath tunnel significantly. Further, it has been concluded that the construction of superstructure even with raft foundation should be strictly checked to avoid any underground disaster. The results may vary significantly with consideration of heterogeneous profile of the soil above the tunnel.

References

1. Government of NCT of Delhi (2012) Statistical abstract of Delhi 2012, 194
2. Barla M, Di A (2018) Donna energy tunnels: concept and design aspects. Undergr Space 3(4):268–276
3. Perazzelli P, Anagnostou G (2016) Design issues for compressed air energy storage in sealed underground cavities. J Rock Mech Geotech Eng 8(3):314–328
4. Zaid M, Sadique MR, Samanta M (2020) Effect of unconfined compressive strength of rock on dynamic response of shallow unlined tunnel. SN Appl Sci (2), 21–31. https://doi.org/10.1007/s42452-020-03876-8
5. Zaid M, Sadique MR, Alam MM, Samanta M (2020) Effect of shear zone on dynamic behavior of rock tunnel constructed in highly weathered granite. Geomech Eng Technopress 23(3):245–259. https://doi.org/10.12989/GAE.2020.23.3.245
6. Zaid M, Sadique MR (2020) Numerical modelling of internal blast loading on a rock tunnel. Adv Comput Design 5(4):417–443. https://doi.org/10.12989/ACD.2020.5.4.417
7. Zaid M, Sadique MR (2020) The response of rock tunnel when subjected to blast loading: Finite element analysis. Eng R e12293. https://doi.org/10.1002/eng2.12293
8. Zaid M, Sadique MR. (2020) Blast resistant behaviour of tunnels in sedimentary rocks. Int J Protective Struct. https://doi.org/10.1177/2041419620951211
9. Zaid M, Sadique MR (2021) Dynamic analysis of tunnels in western ghats of indian peninsula: effect of shape and weathering. In: Pathak KK, Bandara JMSJ, Agrawal R (eds) Recent Trends in Civl Engineering. Lecture Notes in Civil Engineering, vol 77. Springer, Singapore. https://doi.org/10.1007/978-981-15-5195-6_57
10. Zaid M, Mishra S, Rao KS (2020) Finite element analysis of static loading on urban tunnels. In: Latha Gali M, PRR (eds) Geotechnical Characterization and Modelling. Lecture Notes in Civil Engineering, vol 85. Springer, Singapore. https://doi.org/10.1007/978-981-15-6086-6_64
11. Naqvi MW, Akhtar MF, Zaid M et al. (2020) Effect of superstructure on the stability of underground tunnels. Transp Infrastruct Geotech. https://doi.org/10.1007/s40515-020-00119-6
12. Ali Khan M, Sadique MR, Zaid M (2019) Effect of stratification on underground opening: a numerical approach. In: Pulugurtha S, Ghosh I, Biswas S (eds) Advances in Transportation Engineering. Lecture Notes in Civil Engineering, vol 34. Springer, Singapore. https://doi.org/10.1007/978-981-13-7162-2_11
13. Kim H-M, Ryu D-W, Chung S-K, Song W-K (2009) Flexible and lined segment tunnel for underground compressed air energy storage (CAES). J Korean Soc Rock Mech 19(2):77–85

14. Khan MA, Sadique MR, Alam MM Stress analysis of tunnel in soft soil: a state of art report. Int J Adv Res Sci 6(03):187–195
15. Nabanita Baruah MPB (2014) Prediction of ground subsidence due to shallow tunnelling in soft soil by using finite element analysis. J Civ Eng Environ Technol 1:45–48
16. Hosseini S, Mohammad Nejad M (2012) Numerical and analytical investigation of ground surface settlement due to subway excavation. Geosci (2):185–191
17. Maras-Dragojevic S (2012) Analysis of ground settlement caused by tunnel construction. Casopis-Gradjevinar 64:573–581
18. Ayothiraman R, Arun Kumar S (2010) Effect of vertically loaded pile on existing urban tunnel in clay. In: Indian geotechnical conference, Bombay, pp 751–754
19. Ayothiraman R, Arun Kumar S (2011) Influence of vertical pile loading on existing tunnel lining in soft clay. In: 14th Australasian tunnelling conference, Auckland, New Zealand, pp 1–13
20. Calabrese M, Monaco P (2001) Analysis of stresses induced in an old deep tunnel by pile driving from the surface, pp 199–204
21. Stanley Jose A, Vivek S (2014) Static analysis of tunnels in urban areas for metro construction. Int J Eng Res Technol (IJERT) 3:264–269
22. Iqbal A et al (eds) (2020) Soft computing in condition monitoring and diagnostics of electrical and mechanical systems, vol 1096. In: Advances in intelligent systems and computing. Springer, Singapore. https://doi.org/10.1007/978-981-15-1532-3
23. Iqbal A et al (eds) Meta heuristic and evolutionary computation: algorithms and applications, vol 1096. In: Studies in computational intelligence. Springer, Singapore. https://www.springer.com/gp/book/9789811575709
24. Terzaghi K, Richart FE Jr (1952) Stresses in rock about cavities. Geotechnique 3(2):57–90
25. Ramamurthy T (2007) Closed form solutions for underground openings. Engineering in rocks for slopes, foundations and tunnels. PHI Publications, p 409

Solar Inverter with Integrated Monitoring and Control System

Gauri Goenka, Mazhar Ahmad, Prakriti Ghosh, Abhishek Kumar, and Mohd. Saquib Faraz

Abstract In a dynamic world where a constant effort has to be made in order to keep up with the energy requirements of rapidly growing population, we need remedies to monitor energy wastage until the ways to derive sufficient amount of energy from renewable energy resources are invented. This paper proposes a system wherein power is obtained from solar energy and has a monitoring system put in place using an Arduino board and NodeMCU. NodeMCU has an integrated ESP8266 Wi-Fi module that will act as a node for various loads to be monitored. IoT is used in this system to create an environment that will keep in check the energy used by different loads.

Keywords Solar-based power conversion system · Arduino · IoT · ESP8266 Wi-Fi module

1 Introduction

In this paper, we propose an inverter system which is powered by solar energy, thus acting as a hybrid systems. System aims at minimising the energy demand. To make the system more interactive and user friendly, we collect real-time data of the instantaneous current used by the loads and is processed to calculate power consumption and plotting it along with time on an IoT platform.

G. Goenka (✉) · M. Ahmad · P. Ghosh · A. Kumar · Mohd. S. Faraz
ADGITM, Guru Gobind Singh Indraprasth University, New Delhi, Delhi, India
e-mail: gaurigoenka1998@gmail.com

M. Ahmad
e-mail: mazharahmad20298@gmail.com

P. Ghosh
e-mail: prakritighosg5@gmail.com

A. Kumar
e-mail: 27abhi.ky@gmail.com

Mohd. S. Faraz
e-mail: mdsaquibfaraz@gmail.com

A. Iqbal et al. (eds.), *Renewable Power for Sustainable Growth*, Lecture Notes in Electrical Engineering 723, https://doi.org/10.1007/978-981-33-4080-0_11

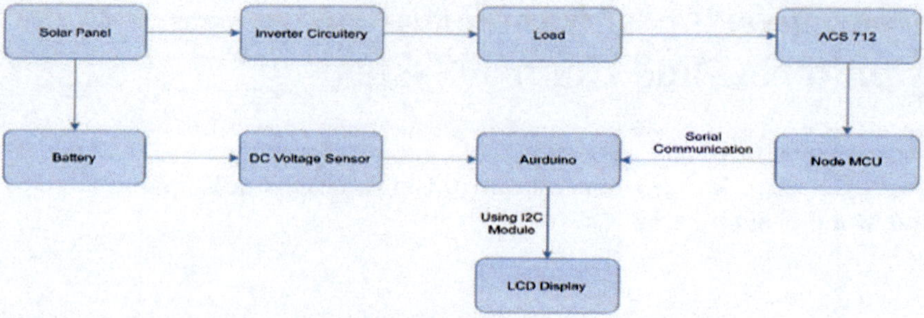

Fig. 1 Block diagram of entire system

In order to achieve this, we have divided our project into three parts.

(a) Power Source and Conversion System—Solar energy is promoted
(b) Power Monitoring System—Power consumed recorded to make smart decision
(c) Battery Management System—Put in place to increase battery life (Fig. 1).

2 Literature Review

IoT-based smart inverter using Raspberry Pi [1] In this paper, the authors have used Raspberry Pi as a controlling board. A solar panel is being used to power the inverter battery. An Internet of things (IoT) environment is developed which is used to monitor energy consumption by the loads.

Remote monitoring of solar inverter (An application of IoT) [2] Various sensors are used to devise a system which collects and feeds data to an Arduino board. A Wi-Fi module is used by the authors to feed data to the IoT platform which helps in representing the data collected.

An IoT-enabled smart inverter [3] The paper presented focuses on the wireless control of home appliances along with monitoring battery health. A voltage divider circuit is used to step down voltage of the battery, and the resulting voltage value is fed to the microcontroller board where an algorithm runs to map the voltage and give the exact voltage value. An LCD display is used to represent runtime of various loads. The data is represented on a mobile application, and in order to send the data to the mobile application, a Wi-Fi module is used.

Energy monitoring and control using the Internet of things [4] The authors have used a current transformer to monitor the current reading. An analog to digital converter is used to monitor and convert analog values of the current to the digital values which are fed to the Raspberry Pi board which is used as a processing board in this case.

Research on smart home energy monitoring system [5] The system proposed is a combination of smart metering system, smart grid system and home automation.

IoT-enabled smart inverter [6] A lead acid battery is used as an energy bank. The Wi-Fi module is used to send data to a Wi-Fi-enabled PC from the controller board. LCD displays runtime of loads.

3 Solar-Based Power Conversion System

A solar panel is used to charge a 12 V, 7Ah lead acid battery thus creating a hybrid environment for a battery to be charged. The 12 V DC battery is connected to the inverter circuit which produces a square wave AC voltage. In order to create the inverter circuit, we have a 555 timer IC which is used in the form of an astable multivibrator. It has a square wave as output. Two IR540 MOSFETs are used in push pull configuration to further amplify the square wave. A 12-0-12 V centre tapped transformer is used in reverse form in order to step up the voltage level. The circuit designed gave an AC square voltage of 52 Hz and has a root main square value of 180 V. The above system is designed in order to harness solar energy, thus reducing dependency on the traditional power sources.

4 Power Monitoring System

The power monitoring system is realised using ACS712 current sensor which can operate a 5 V and calculate instantaneous current up to 30 A, based upon the Hall effect. Voltage (Vrms) is 180 V, output of the inverter. In order to calculate the Irms, samples of instantaneous current are collected using ACS712 sensor. The loads chosen for this project have a power factor of 0.9. The NodeMCU board further calculates the energy utilised after every one hour with the help of real power calculated. The data is further sent to the ThingSpeak API using the TCP/IP protocol which further helps in visualising the data in real time with the help of plotting graphs. The loads in a house are scattered and cannot be connected to a single controlling board as it will require a large amount of wires which will unnecessarily increase the complexity of the project and add on to its cost. In order to avoid such a havoc, NodeMCU is used as it helps in creating more number of nodes and enables better communication. The power monitoring system helps the user to make a better decision by identifying loads which consume a greater amount of energy and switching them off if there is no necessity to use them at the moment. The above-explained system helps in collecting and logging data of power consumption which can be used to smartly monitor loads and make billing predictions as well (Fig. 2).

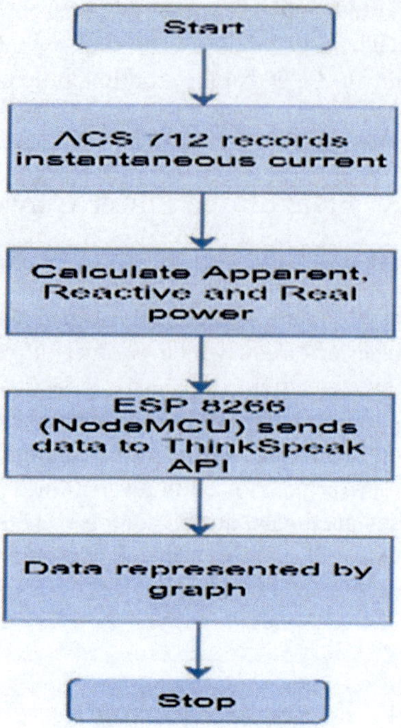

Fig. 2 Power monitoring system flow chart

5 Battery Management System

A battery management system is also provided along with the inverter system which helps the user to determine the battery life of the battery and runtime of loads. It also calculates the voltage of the battery which helps the user to keep a check on the battery, and if it drains out, it will help the user to know that it needs charging. In this system, we have used the DC voltage sensor based on Hall effect. The sensor used calculates the voltage of the battery using following formula [7, 8] (Figs. 3 and 4).

$$\text{Battery Life} = \frac{V_t \times C_b}{\text{Wattage}} \tag{1}$$

6 Result

Our project records real-time data of power consumed and is logged (via MQTT) in ThingSpeak API which makes it interactive for users as they can interpret graphs easily and make decision based on it. The data collected can also be used to devise energy-saving algorithms which can help in actuating loads according to their usefulness.

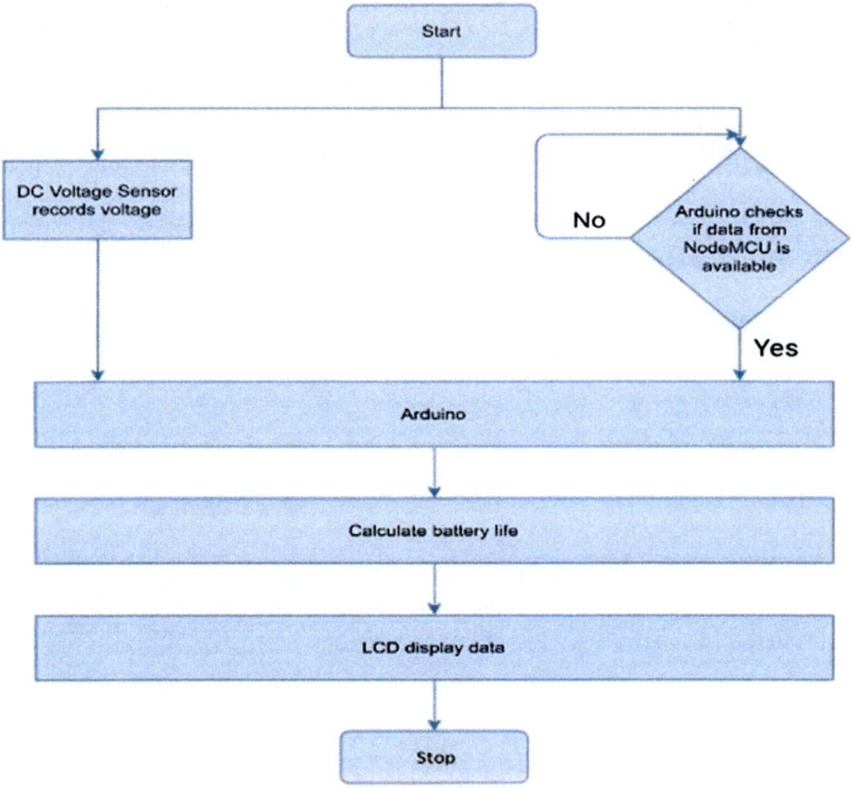

Fig. 3 Algorithm of battery management system

Fig. 4 Battery management system LCD

Table 1 Table showing runtime of different use cases for bulbs

S. No.	Operating bulb	Runtime (h)
1	9 W	7.4
2	12 W	5.6
3	9 + 12 W (Parallel connection)	3.2

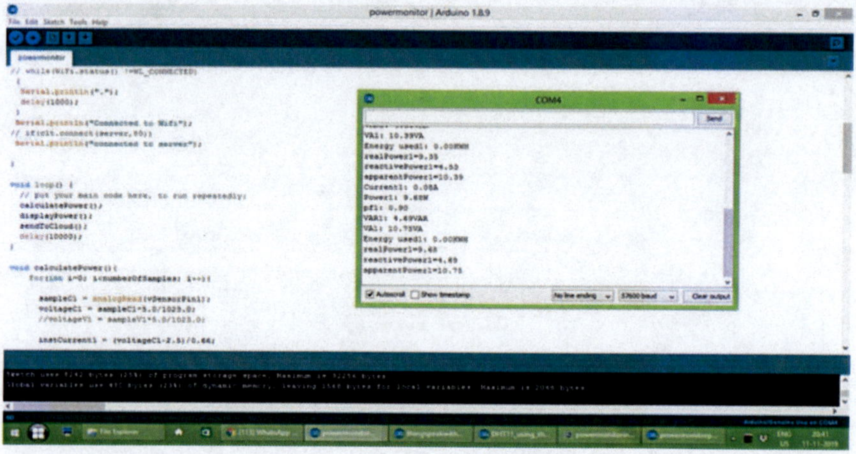

Fig. 5 Obtained result

The battery monitoring system helps in looking after battery health and gives data about battery voltage (undervoltage and overvoltage can be detected) and battery life (duration till which the battery in use will last) (Table 1; Figs. 5 and 6).

7 Conclusion

In fast developing world, we need such innovative solutions to the basic problems such as energy wastage as renewable resources as of now are difficult to harness, and thus to tackle growing energy demands and saving available resources, avoiding wastage is the only solution. The current need is to make such solutions more affordable and user friendly through the use of automation and if possible, AI. Such measures will ensure that these solutions are available to all, and it will make a contribution to the fast development of the world.

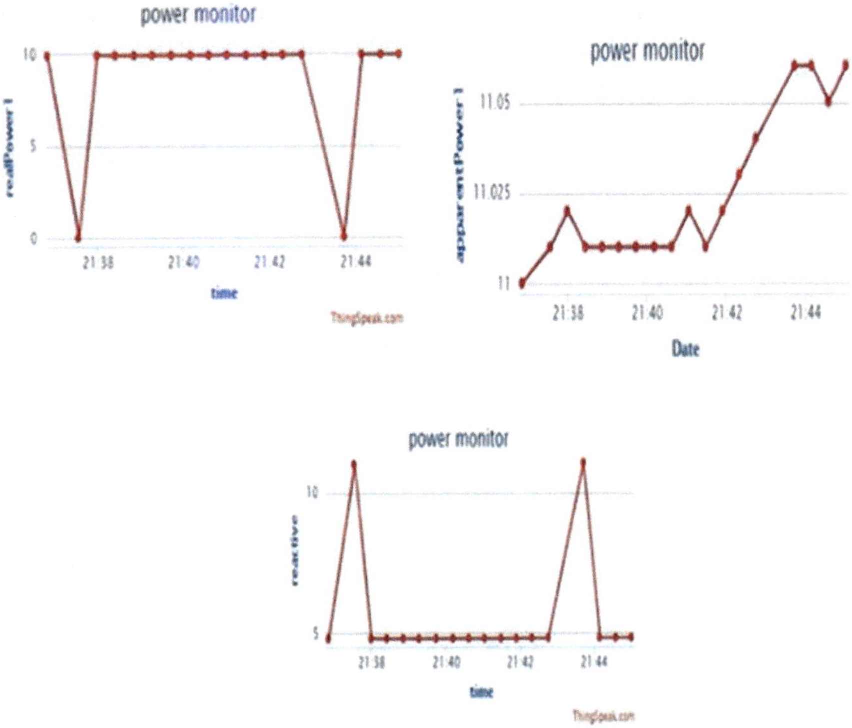

Fig. 6 Display of real, active, apparent power to the user using ThingSpeak API

References

1. Karundkar S, Kanthe S, More S, Marathe P, Gadade Y (2017) Remote monitoring of solar inverter: an application of IoT. Am J Eng Res (AJER) 6(7):70–74. Available: http://www.ajer.org/papers/v6(07)/K06077074.pdf
2. Joshi MA, Kavya S (2017) An IoT based smart inverter using raspberry pi. Int J Latest Technol Eng Manag Appl Sci (IJLTEMAS) 6(VS). Available: https://www.ijltemas.in/DigitalLibrary/Vol.6Issue5S/50-53.pdf
3. Baridar N, Gadgay B, Pujari V (2017) IoT enable smart inverter. Int J Res Appl Sci Eng Technol (IJRASET) 5(6) (2017). Available: https://www.ijraset.com/fileserve.php?FID=8298
4. Hartman WT, Hansen A, Vasquez E, El-Tawab S, Altaii K (2018) Energy monitoring and control using IoT. In: 2018 IEEE system and information engineering design symposium (SIEDS). IEEE, Charlottesville, VA, USA. Available: https://ieeexplore.ieee.org/document/8374723/authors#authors
5. Kiska M, Ivanov M, Volcko V, Janiga P (2015) Research on smart home energy management system. In: 2015 IEEE 16 th international scientific conference on electric power engineering (EPE). IEEE, Kouty nad Desnou, Czech Republic. Available: https://ieeexplore.ieee.org/document/7161102/authors#authors
6. Kumar A, Bijapur AA, Charitha B, Kulkarni KR, Narajan K (2016) An IoT based smart inverter. In: 2016 IEEE international conference on recent trends in electronics, information and communication technology (RTEICT). IEEE, Banglore, India. Available: https://ieeexplore.ieee.org/document/7808182/authors#authors

7. Iqbal A et al. (eds) (2020) Soft computing in condition monitoring and diagnostics of electrical and mechanical systems, vol. 1096. In: Advances in intelligent systems and computing. Springer, Singapore. https://doi.org/10.1007/978-981-15-1532-3
8. Iqbal A et al. (eds) (2020) Meta heuristic and evolutionary computation: algorithms and applications, vol 1096. In: Studies in computational intelligence. Springer, Singapore. https://www.springer.com/gp/book/9789811575709

A Survey on Solar Power for Present and Future Perspective in Indian Market

Abhinav K. Gautam, R. K. Chauhan, J. P. Pandey, and K. S. Verma

Abstract In the ancient world, there was no fossil fuel, mineral oil and coal available for utilization. Renewable energy sources especially solar have very good potential in India and the world as well. The small (KW) plants at the micro grid/local grid level and the large (MW) plants at the national/state grid level can produce the electricity. If the mineral oil and coal were not found as a fuel, the solar/renewable technology would have developed with its utmost potential. But this is the starting of the renewable technologies. The future is only for the renewable technologies where a lot of scope seems to be there with the points of the good business, good environment, good technologies, good research and developments, overall good livelihood. At present, we see the adverse effects of the polluted environment as a result of the removal of the breeds of birds, animals, flies—mostly butterflies, etc. Removal of the EM Waves from the environment is essential to save the ecological system on the earth. This could be achieved by sustainable and economic technological development for optimum utilization of the natural resources. Innovations in low voltage electricity generation and proper channelization of the mini grid/micro grid/local grid interactive power system concept are desirable for the future aspects. Optimum utilization of renewable resources like the hybrid solar photovoltaic, biomass/biomass gas, biogas and wind power plants does not mean to be far away. Limited utilization of the high-frequency-related equipment's and development of the low-frequency utilization technologies is essentially required for the medical fitness of the humanity and to save the life expectancy for today. Emphasis in this paper is to achieve the highest growth

A. K. Gautam (✉) · J. P. Pandey · K. S. Verma
Department of Electrical Engineering, Kamla Nehru Institute of Technology, Sultanpur, India
e-mail: abhi.knit2012@gmail.com

J. P. Pandey
e-mail: tojppandey@rediffmail.com

K. S. Verma
e-mail: ksv211@rediffmail.com

R. K. Chauhan
Department of Electrical and Electronics Engineering, IMS Engineering College, Ghaziabad, India
e-mail: mmmec.rkc@gmail.com

A. Iqbal et al. (eds.), *Renewable Power for Sustainable Growth*, Lecture Notes in Electrical Engineering 723, https://doi.org/10.1007/978-981-33-4080-0_12

and development in the solar technological innovations and suggestions. Commercial and deployment activities for the solar projects implementation may be the areas for developing EPC Solar NET metering/Gross metering and rooftops/ground mounted for better business solutions. The government should take keen interest and initiatives to collect and provide funds for these projects. A transparent fit in tariff (FIT) should grow to build up the mass scale solar business.

Keywords Renewable energy · Solar power · Government solar polices

1 Introduction

Despite the general economic crisis, India's energy demand is growing and continuously rising with the global warming and the disastrous consequences. The expected economic and population growth, together with urbanization and industrialization, point towards continued growth in energy demand. The energy demand is expected to increases 4.5% annually by 2035 from 3.5% which are previously projected demand during 2000–2017. The total use of electricity in India is 1010 kWh, compared with a world average of 3200 kWh [1], and several literatures have been cited in [2, 3]. There are only one solution for this problem is to focus on renewable energy. India concentrates on renewable energy and has been projected to generate up to 84,000 MW of hydroelectric power at 60% load factors. The government mainly concentrated in the country's southern part and also preparing to accelerate a hydro-development programme to install 50,000 MW of new capacity by 2026 (around the end of the 14th Five-Year Programme) [4]. Also revealed at the 2015 Paris Climate Summit was a "Solar Alliance," which seeks to increase its solar generating capacity to 175 GW by 2022. India is increasing its solar energy as the name implies, with aims to have 100 GW capacity by 2022 [5].

India is the nation of villages and towns where the integrated approach to empower the people of the community is essential. Electricity is required to improve the healthcare, water and agriculture, education and enterprises facilities in the villages, towns and cities in the country. Lighting system in the villages as well as in the towns is very poor. People community especially in the villages and small towns of India are living in energy poverty. Newly announced 19th EPS projects with a demand for electricity of 1743 TWh (6.59% CAGR) from 2017 and a peak load of 299 GW (6.32% CAGR) by 2027.

2 Government Perspectives

The energy sector of India is exciting to transform and build a stable, scalable, productive and supportable network by 2027 to deliver people with reliable and good energy through the use of creative technology and guidelines to meet the desires and expectations of all, through active shareholder sharing [6]. Technological transition makes it difficult to forecast of future mix energy of India, but even considering increased domestic supply energy and its higher energy efficiency, India's energy import dependence could grow from 36 to 55% by 2040. Our Indian government should take the interest to improve the condition of the existing villages and towns and also should take keen interest in developing the smart and new cities for economic and industrial growth of India. Though the concept of developing smart cities is good enough for a long period progress of the nation but at the time the development in the villages and towns systems should be the first priority. Integrated approach of using solar energy solutions with local grid concept at the sites is technically feasible to power the community lighting system. Such type of development model should be adopted where the people of India get their enterprises, their production, their machineries and equipment's powered by sufficient electricity, generate their economies, and achieve and avail all the facilities of a smart city. They get the qualitative educational facilities in the villages and small towns with new efficient computers, laptops, highly equipped laboratories, etc. They get the water and agriculture, powered water wells, clean drinking water, drainage and roads, pumps and year round crop irrigation systems powered by the solar photovoltaic systems. A good quality of transport and communication powered by the solar photo voltaic, medical and health facility with good quality of laboratories, diagnostic equipment's and vaccine refrigerators will definitely improve the living standard of the gentry of our India and the country as well. They will definitely fully utilize and digitalize their skill with their strengths. The standard of the people in this way definitely will improve and the villages and towns will get smartness—in living, growth and development.

Villages should be selected and electrified under the remote rural development programs. The use of solar power must be intensified and independent off grid solar power projects should be launched on the basis of local/micro grid technology. Grid tied/hybrid solar power projects should be sanctioned as the mega solar projects to feed the Indian or national grid. For that the national/Indian/private integration, installation and commissioning companies/institutions should be funded. Good initiatives for the investors are also required to solve the monetary and funding problems. National/Indian/private banks and institutions may also be involved for the direct investment in building the solar park. A good qualitative and transparent funding scheme with clear bidirectional tariff policies should be introduced.

2.1 Government Policies and Regulations to Promote Solar Energy in India

1. 1982: Commission of Alternatives sources of energy was created
2. 2003: Electricity Act, 2003
3. 2005: National Electricity Policy, 2005

4. 2006: National Tariff Policy, 2006
5. 2006: National Rural Electrification Policies, NREP, 2006
6. 2007: Semiconductor Policy, 2007
7. 2008: National Action Plan on Climate Change, NAPCC, 2008
8. 2009: Generation-based initiatives to encourage solar PV energy in India
9. 2010: Jawaharlal Nehru National Mission, JNNSM, 2010
10. 2011: Renewable Energy Certificates, RECs.

2.2 The NITI Aayog's National Energy Policy Aims to Improve Its Energy Sector Are as Follows

Smart grid has a very wide ranging vision of the forthcoming and is working enthusiastically toward achieving the targets and objectives propagated in the 13th and 14th Five-Year Plans [6]. Presently, the government is working on 13th Five-Year Plan from 2017 to 2022 and focused the following targets:

1. Reduction of transmission and distribution losses to up to 10% in all services
2. End of load shedding
3. Increase in power quality
4. Efficient estimation and shipment of renewable energy
5. Substructure and requirements for EV
6. 1200 kV AC network in operation
7. Distribute of smart grid goods to the abroad market.

2.3 The Future Plan of the Government in the Next Fourteenth Five Years Plan from 2022 to 2027 Will Investigate

1. Economically possible services
2. Constant 24 × 7 electricity supply to all customer.
3. 33% or extra renewable in power system.
4. Electric vehicle setup control as VPP
5. Transfer of smart grid goods, technologies and facility to abroad.
6. Clean cooking access through efficient and affordable gasoline.
7. Information technology network and CRM system for electric utilities delivered to further service providers, for example, water and gas delivery, land revenue collection.

3 Solar Energy

Solar energy has long been recognized as the green energy, ensures the energy security and the economic growth, renewable and carbon targets and is environment friendly.

Wind power interpretation for the maximum at 46% (around 36 GW), after that solar with a stake of 36% (30 GW) share. Biomass captured the residual market at 12% (9 GW) and small hydro ventures at 6% catered for 5 GW. In India, wind energy capacity has risen 1.7 times in the last four years. In addition to this, last year record 100 bn+ renewable electricity generation units. Solar power has risen from 2.6–28.18 GW in March 2019 by more than 11 times in the last five years [7]. As of 29 February 2020, solar installed capacity in the country exceeded 34.404 GW. India has the lowest cost of capital per MW for building solar power plants globally [8].

3.1 Benefits of Solar Energy Are Following

1. Power cut management
2. Reduction in fuel and electricity consumption
3. No smoke, no noise, clean and green energy
4. Increased production and minimum cost of energy
5. High societal impact and potential to commercialize
6. Increased battery life and ensures energy efficiency and energy security concerns.

1. **On Grid Connected**

   See Table 1.

2. **Solar Off-Grid Basis**

Scheme	Objectives	Funding assistance	Time duration
Off grid and dispersed solar photovoltaic application scheme—under phase II	Connection of an added off-grid solar capability of 118 MW power by 2020 over ensuing application: • 3,00,000 solar street lights • 25,00,000 solar study lamps • 100 MW power of off-grid solar power plants	• CFA of 30% of the cost for solar street lights and solar power plants in developed states and CFA of 90% of the system in northeastern states including Sikkim, J & K, HP, UK, Lakshadweep and A&N • CFA provides 85% cost for solar lamps	Till 31.03.2020

(continued)

(continued)

Scheme	Objectives	Funding assistance	Time duration
PM KUSUM	To achieve 25,750 MW through renewable sources up to 2022 with total CF support of Rs. 344,220,000,000/-	DISCOMs will purchase the produced renewable power at a feed-in tariff mandated by the respective state electricity regulatory commission, SERC	Till 31.12.2022
Atal Jyoti Yojna (Ajay) phase II	To install 304,500 Solar Street Lights through phase II	SSLs with a 12 W LED capacity will be installed in compliance with the MNRE description and 75% of the SSL device expense will be covered by the MNRE budget and the lasting 25% from the MPLADS fund	Till 31.03.2021
Scaling up access to renewable energy for efficient use in rural areas	Improving the use of safe and affordable renewable energy for efficient rural usages/life and the usage of fossil fuels in underserved and under-served areas only for Assam, MP and Odisha	MNRE release will release up to 30% of CFA as advance and balance 70% based on progress	June, 2020
Off-grid and open solar thermal technologies (CST)	Promoting off-grid solar thermal systems applications to meet the goals set by the JNN solar program	CFA will only be issued by each of the regional test centers (RTCs) for reimbursement-related inspection	2017–18 to 2019–20

4 Future Perspectives in Solar Power in India

For the better uses of solar power will contribute grow in GDP. So it is necessary to improve the solar power efficiency government should work on as following areas:

Table 1 Perspective of solar schemes

Scheme	Objectives	Funding assistance	Time duration
Solar-based park creation and ultra-mega-solar power projects	To improve the power from 20 to 40 kMW	Financial assistance up to Rs. 25 lakh/solar park DPR and Rs. 20 lakh per MW or 30% of the entire project price	Up to 2021–22
Founding of more than 5 kMW grid-associated SPV power projects beneath JNNSM PHASE-II	To boost grid power and distribute renewable energy projects nationally, reducing transmission costs and losses	Up to Rs. 10,000,000 MW for open category and Rs. 12,500,000 MW for projects in domestic category project	Financial year 2015-16 to 2018-19
Establishment of scattered grid-connected solar PV energy projects at A & N and Lakshadweep Island	Developing pollution-free islands by phasing out diesel usage for electricity production and contributing to the Iceland's NAPCC&G along with reducing power production costs	CFA of up to 40% of the project Cost	2016–17 to 2019–20
CPSU scheme phase-II for setting up 12,000 MW grid-connected solar PV power	To establish solar PV projects by government to facilitate national energy security and environment sustainability for government purpose	VGF of up to Rs. 70 lakhs/MW and 50% shareholding of Central/State Govt	2019–20 to 2022–23
Grid-connected solar PV rooftop scheme	For attaining collective capacity of 40 kMW from RTSS up to year 2022	CFA to residential sector up to 4 GW	Till 31.12.2022

4.1 Seamless Availabilities

The government should ensure the seamless solar projects implementation and should commit the challenges to be a forefront of the solar PV revolution underway in the Indian ensuring to achieve high energy efficiency and high energy yield. Our Indian government aim should be to take the initiatives to provide and deliver the cost competitive and affordable solar solutions for clean energy technologies with the following capabilities:

1. Financial modeling and economic analysis
2. System design and engineering
3. Integration, installation, construction and commissioning

4. Complete project management, operation and maintenance
5. EPC solutions for grid tied/hybrid/off-grid power plants
6. Real-time monitoring and controlling facilities.

Indian government should take the interest to improve the economics of manufacturing units of solar panels, inverters, batteries, and other utility equipment being used in solar energy systems to stop the imports of the cheap China solar products.

4.2 *Initiatives*

The Indian government should take interest and initiatives in developing solar technologies and establish the laboratories/virtual laboratories to perform research and development in three major branches of solar energy

A. Photovoltaic (also called solar electric systems)
B. Solar thermal systems (solar heating systems)
C. Solar cooling systems.

The sustainability and the use of solar systems in aviation and medical fields persist high-grade and efficient solar technological development. In this way, the solar R&D cell might cover the full spectrum from fundamental studies to commercialization, systems development and integration to improve performance and reliability in the following research areas:

1. Research and development in solar cells and modules technologies.
2. Research and development in solar fields and radiations
3. Research and development in solar system installation and integration
4. Research and development in solar thermal
5. Research and development in solar heating
6. Research and development in solar cooling
7. Research and development in silicon materials and devices
8. Research and development in polycrystalline thin-film materials and devices
9. Research and development in III-V multifunctional materials and devices
10. Research and development in solar power plants structures
11. Research and development in solar economics.

4.3 *Role of Industries and Institutions*

The institutions should go for R&D with planning with pilot project activities to work with local, Indian federal and state governments and private industries/institutions and organizations to deploy renewable energy technologies that are commercially available with energy efficiencies. The experts should plan to prepare the market for emerging technologies with the aim of project development, technical assistances and disaster resiliency and recovery by removing barriers to adoption and deployment resources to move the projects in forward directions. The researches with data preparation uniquely and continuously utilize the ideas along with academic institutions, industries, agencies, and laboratories. The data might be used for renewable energy, climate change studies, atmospheric research, conversion systems, carbon capture, technological development and testing.

4.4 Methodologies

Government should go for a firm planning, monitoring and developing the departments to offer the unique and the distinguished advantages in the fields of solar photovoltaic, thermal and electronic projects mutually that may broadly be classified as:

A. Solar photovoltaic projects
   (1) Solar photovoltaic power generating systems
   (2) Grid-interfaced SPV system:
      (a) Single-phase GISPV system with
      (b) Three-phase GISPV system with single stage or double stage
      (c) Stand-alone SPV system with single stage or double stage: The capacity range is from 1 Kw residential projects to Mw commercial projects.
      (d) Solar surface/Submersible pumping systems
      (e) Solar street lighting:
B. Solar thermal projects
   (1) Solar thermal power projects with flat plate collectors
   (2) Solar thermal power projects with latest evacuated glass tube technology
   (3) Solar chimney power plants
C. Solar electronic system
   (1) LED lighting arrangement
   (2) Solar street lighting system
D. Solar LED luminary.

4.5 Best-Fit Solutions

As it is seen from the data, the rate of CO_2 emissions is increasing every year with a rapid rate. Improving energy efficiencies is the key tool to reduce CO_2 emissions. This requires the optimum utilization of clean and renewable energy to secure future generation. The Indian government can work/look/distribute the projects to the customers to fulfill the commercially viable requirements:

1. Energy-efficient solutions
2. Energy savings and securities
3. Cleaner and reasonable alternative to diesel
4. Rooftop solutions for sustainability
5. Bilateral industry open access contracts
6. Off-grid energy access opportunities
7. Grid interface SPV power generation opportunities
8. Achievement of green ratings for companies
9. Power transmission and distribution execution opportunities
10. Solar lighting opportunities
11. Solar water heating opportunities.

4.6 Guidelines

For the projects, the Indian government should issue the proper guidelines and the ratings for the commercial projects such as:

1. Quality and engineering products
2. Challenges—always focus on cost effectiveness
3. Raise capability to utilize solar power
4. Upgradation and innovation
5. Right amount reliability and optimum design considerations
6. Following IEC/NEC/REC/MNRE standards
7. Meticulous spirit with strong commitment to meeting with all the requirements of esteemed customers
8. Design and development of high-level technology with high-precision and sustainability.

4.7 Activities and Development

Commercial and deployment activities for the solar projects implementation may be the areas of the following with the efforts of developing EPC solar, NET metering/Gross metering, and rooftops/ground mounted for better business solutions:

1. Land in the villages/towns/cities
2. Grounds/Roofs of the houses, industries, institutes, hospitals, schools, colleges, etc.
3. Government buildings
4. Petrol pumps
5. Sides of the streets, rivers, canals, etc.

6. Shadow free area in the hotels, orchids, and resort centers petrol pumps.
7. Housing societies/townships.

4.8 Interest

The government should take keen interest and initiatives to collect and provide funds for these projects. A transparent fit in tariff (FIT) should grow to build up the mass scale solar business. It can accelerate the investors to invest the money in such projects and form the public enterprises to provide the institutional support for the industries in setting up the commercial solar energy projects. Policies should be clear and supportive with financial subsidies and market strategies.

5 Conclusion

Solar power system is only the alternative to produce electricity. The government should inculcate and develop the transparent policies for solar business in such a way that it should become the business of the masses. Most of the people should participate and grow this business as electricity is required for the growth and all round development of the country since development of any nation depends on per capita generation and utilization of the electricity. Optimization of hybrid renewable electricity model consisting of photovoltaic arrays, Battery bank and combined heat and power systems along with micro grid tools to deploy the off-grid residential solutions can be coupled to fulfill the load demand in Indian rural. Fossil fuels, biomass, biomass gas, biogas-based small-scale CHP systems can be combined with the solar PV systems to improve the benefits of co-generation in getting electricity, space heating and cooling, water heating, etc. Power management systems are to be proposed to cope up the micro grid to mitigate and suppress the disturbances due to the integration of renewable electricity into such grids. Software algorithms should be designed and suggested to filter out the adverse effects of the power events and to improve the power smoothing index of the complex hybrid electrical network (grid) system.

References

1. Jeffrey R et al (2015) Solar resource-reserve classification and flow-based economic analysis. Solar Energy Sci Direct Elsevier 116:45–55
2. Iqbal A et al (eds) (2020) Soft computing in condition monitoring and diagnostics of electrical and mechanical systems, vol 1096. In: Advances in intelligent systems and computing. Springer, Singapore. https://doi.org/10.1007/978-981-15-1532-3

3. Iqbal A et al (eds) (2020) Meta heuristic and evolutionary computation: algorithms and applications, vol 1096. In: Studies in computational intelligence. Springer, Singapore. https://www.springer.com/gp/book/9789811575709
4. Bijarniya JP, Sudhakar K, Baredar P (2016) Concentrated solar power technology in India: a review. Renew Sustain Energy Rev 63:593–603
5. Bauner C, Crago CL (2015) Adoption of residential solar power under uncertainty: I implications for renewable energy initiatives. Energy Policy Elsevier 86:27–35
6. https://mnre.gov.in/solar/schemes/
7. Ulsrud K, Winther T, Palit D, Rohracher H (2015) Village level solar power in Africa: accelerating access to electricity sources through a socio-technical design in Kenya. Energy Res Soc Sci 5:34–44
8. Som T, Chakrobarty N (2014) Studies on economic feasibility of an autonomous power delivery system utilizing alternative hybrid distributed energy resources. IEEE Trans Power Syst 27(1)
9. IEA (2020) India's energy policy: key findings of the IEA's In-Depth Review
10. https://www.dfat.gov.au/geo/india/ies/refernces.html
11. Zhang J et al (2015) A suit of matrices for assessing the performance of solar power forecasting. Solar Energy Sci Direct Elsevier 111:157–175
12. https://www.investindia.gov.in/sector/renewable-energy
13. Aichmayer L et al (2015) Thermoeconomic analysis of a solar dish micro gas-turbine combined-cycle power plant. Energy Procedia Sci Direct Elsevier 69
14. Mileva A, Johnston J, Nelsom JH, Kammen DM (2016) Power system balancing for deep decarburization of the electricity sector. Appl Energy Elsevier 162:1001–1009
15. Nickerson TA, Hathaway BJ, Smith TM, Daridson JH (2015) Economic assessment of solar and conventional biomass gasification technologies: financial and policy implications under feedstock and product gas price uncertainty. Biomass Bioenergy Sci Direct Elsevier 74:47–57
16. Askari B, Sadogh MO, Ameri M (2015) Energy management and economics of a regeneration system considering the effect of solar PV, solar collector and fuel prices. Energy Sustain Dev Elsevier 26:43–55
17. Shouman ER, Khattab NM (2015) Future economics of concentrating solar power (CSP) for electricity generation in Egypt. Renew Sustain Energy Rev Elsevier 41:1119–1127
18. Chang CM et al (2015) SEEDS: a solar based energy efficient distributed server farm. IEEE Trans Syst Man Cybern Syst 45(1)
19. Ondraczek J (2014) Are we there yet? Improving solar PV economics and power planning in developing countries: the case of Kenya. Renew Sustain Energy Rev Elsevier 30:604–615
20. Hoppmann J, Volland J, Schmidt TS, Hoffmann VH (2014) The economic viability of battery storage for residential solar photovoltaic systems: a review and a simulation model. Renew Sustain Energy Rev Elsevier 39:1101–1118
21. Shah KK, Mundada AS, Pearce JM (2015) Performance of U.S. hybrid distributed energy systems: solar photovoltaic, battery and combined heat and power. Energy Conserv Manag Sci Direct 105:71–80
22. Verma AK, Singh B, Shahani DT, Jain C (2017) Grid-interfaced solar photovoltaic smart building with bidirectional power flow between grid and electric vehicle with improved power quality. Electr Power Compon Syst 44:480–494
23. https://solarify.in/blog/government-policies-regulations-solar-energy-india/
24. Kumar S, Bhattacharyya B, Gupta B, Present and future energy scenario in India. J Inst Eng India Ser B. https://doi.org/10.1007/s40031-014-0099-7
25. Sharma NK, Tiwari PK, Sood YR (2012) Solar energy in India: strategies, policies, perspectives and future potential. Renew Sustain Energy Rev 16:933–941
26. Kapoor K, Pandey KK, Jain AK, Nandan A (2014) Evolution of solar energy in India: a review. Renew Sustain Energy Rev 40:475–487
27. Tiewsoh LS et al (2019) Electricity generation in India: present state, future outlook and policy implications. Energies 12:1361

Comparison of Direct Torque Control and Indirect Field-Oriented Control for Three-Phase Induction Machine

Sawsan Sayed, Mena Elmenshawy, Mariam Elmenshawy, Qamar Bader, and Atif Iqbal

Abstract Induction motors turn out to be the economical workhorses in the industry for the AC motor drive application. To satisfy the steady-state and transient performance specification of AC drives, control techniques are required for the induction motors. This paper presents the different control methods of the induction machine such as the scalar control, direct torque control (DTC), and the indirect field-oriented control (IFOC). Also, the principle of operation with the required mathematical equations and the Simulink models for the DTC and IFOC methods are described in detail. Moreover, these methods are implemented using MATLAB/Simulink, and their results are verified. A comparison between these methods is presented.

Keywords Induction motor · MATLAB/Simulink · Motor drives

1 Introduction

In the 1980s, the DC motor drive was commonly used in the variable speed drives owing to the simplicity of its control and due to having decoupled armature and field current. However, there are several drawbacks of the DC motor such as regular maintenance requirement for the commutator, brushes, and brush holders, in addition, limited current carrying capability for high-speed applications [1–16]. Accordingly, the DC motor has been replaced by the squirrel cage induction motor that is

S. Sayed (✉) · M. Elmenshawy · M. Elmenshawy · Q. Bader · A. Iqbal
Department of Electrical Engineering, Qatar University, Doha, Qatar
e-mail: ssayed@qu.edu.qa

M. Elmenshawy
e-mail: malmenshawy@qu.edu.qa

M. Elmenshawy
e-mail: me1105930@student.qu.edu.qa

Q. Bader
e-mail: qb1204339@student.qu.edu.qa

A. Iqbal
e-mail: atif.iqbal@qu.edu.qa

A. Iqbal et al. (eds.), *Renewable Power for Sustainable Growth*, Lecture Notes in Electrical Engineering 723, https://doi.org/10.1007/978-981-33-4080-0_13

commonly employed because of the following advantages: high reliability, robustness, less regular maintenance, low cost and weight, and wide power range for the variable speed applications [1, 2]. Merging the merits of both the DC motor and the induction motor together into a variable speed brushless motor drives is the main reason for substituting the DC motor with an induction motor [1].

Generally, the AC drives control is more complex than DC drives control [15, 16]. There are different control methods for the induction motor, which are the scalar control or constant Volts/Hz control, and the vector control or the field-oriented control (FOC) [1]. The scalar control is the simplest and cheapest speed regulation approach for induction motor drives. The scalar control scheme can improve the steady-state operation, while the transient response remains poor. The vector control can solve the aforementioned problems [1].

The vector control or the FOC technique has been proposed in the 1970s, where both the magnitude and phase of the control variables are controlled. The principle of FOC depends on attaining decoupled torque and flux control variables through nonlinear coordinate transformation and controlling the direct-quadrature (dq) components. The FOC includes two methods which are the direct torque control (DTC) and the indirect FOC (IFOC) [1, 3].

The DTC method directly measures the rotor flux for the determination of the transformation angle. The flux is measured by mounting a sensor in the air gap that senses the flux. The flux estimated value may be reliable at the high-speed ranges; however, this is subjected to errors from harmonics at lower speeds. Accordingly, this method is not widely used due to the presence of a sensor that results in a complex construction of the motor, and its unreliability at low speeds. Therefore, the IFOC method is generally used instead of the DTC method in order to estimate the rotor flux by using the FOC model [1, 2]. By the IFOC technique, the flux estimated value is computed through either the voltage model estimator, such that by utilizing the stator voltages and currents, or the current model estimator, such that by using the stator current and rotor speed. The IFOC scheme disadvantages are dependence on the rotor resistance, in addition to being calculation-based model and inherent time consumption model [1, 3].

This paper presents the operation principle and Simulink models of the DTC and IFOC methods for an induction motor. In addition, a comparison between the two methods is presented with simulation results.

2 Direct Torque Control (DTC) Strategy

2.1 DTC Strategy Operation Principle

The DTC uses the induction motor model to evaluate and predict the required voltage that drives to the demanded torque and flux values [4]. This control scheme results in independent control of the motor flux and torque through the optimum selection of the inverter switches mode [4–7]. The calculations of the stator flux and electromagnetic torque are acquired by the stator voltages and currents. The inverter switches operate according to a voltage vector selection, that depends on restricting the flux error and torque error to some specific band, which is the hysteresis bands [4, 5]. The main building blocks for the implementation of the DTC control with the induction machine are the three-phase inverter, speed controller that generates the electromagnetic torque and the DTC controller [6]. The DTC controller is made of a flux and torque estimator block, hysteresis controllers, and sector selection and switching table blocks, such that the switching table output is the gating signals for the inverter switches [6].

Stator Flux Control. The stator flux vector in dq frame, $\boldsymbol{\lambda}_s = \left[\lambda_{ds}\ \lambda_{qs}\right]^{\mathrm{T}}$, can be represented as shown in Eq. 1, where $\boldsymbol{v}_s$ and $\boldsymbol{i}_s$ are the stator voltage and current vectors in the stationary reference frame, respectively [5]. This is referred to as the flux estimation based on the stator voltage model [5].

$$\boldsymbol{\lambda}_s = \int (\boldsymbol{v}_s - \boldsymbol{i}_s R_s)dt \text{ and } |\boldsymbol{\lambda}_s| = \sqrt{\lambda_{ds}^2 + \lambda_{ds}^2} \tag{1}$$

where R_s is the stator resistance.

Torque Control. The interaction among the stator flux and rotor flux produces the electromagnetic torque, T_e, that is represented by Eq. 2, where γ is the position between the stator flux and rotor flux (i.e., $\boldsymbol{\lambda}_s$ and $\boldsymbol{\lambda}_r$, correspondingly) [7].

$$T_e = \frac{3}{2}\frac{P}{2}\frac{L_m}{L_s L_r}\boldsymbol{\lambda}_s\ \boldsymbol{\lambda}_r \sin\gamma \tag{2}$$

where P represents the total poles. L_m is the magnetizing inductance for the induction motor. L_s is the motor stator self-inductance. L_r is the motor rotor self-inductance.

Hysteresis Controller. A torque and stator flux hysteresis controllers are required for DTC technique. The torque hysteresis controller is composed of three-level control that takes as input the torque error, ΔT_e, while gives output the torque error status, H_{Te}, as three digital outputs -1, 0, or 1 to the voltage vector selection. The torque error status decides the requirement of the torque to be increased, decreased, or remain the same with the conditions shown in Eq. 3, where $\Delta T_e = T_{e,\text{ref}} - T_e$ [5, 7]. $T_{e,\text{ref}}$ is the torque reference signal.

$$\begin{aligned}
&|H_{Te}| = 1 \text{ if } |T_e| < \left|T_{e,\text{ref}}\right| - |\Delta T_e| \therefore \text{ Torque Increase}\\
&|H_{Te}| = -1 \text{ if } |T_e| > \left|T_{e,\text{ref}}\right| + |\Delta T_e| \therefore \text{ Torque Reduce}\\
&|H_{Te}| = 0 \text{ if } \left|T_{e,\text{ref}}\right| - |\Delta T_e| \le |T_e| \le \left|T_{e,\text{ref}}\right| + |\Delta T_e| \therefore\\
&\quad \text{Torque Remain Unchanged}
\end{aligned} \tag{3}$$

A flux hysteresis control consists of a two-level control that takes as input the stator flux error, $\Delta\boldsymbol{\lambda}_s$, while its output is the stator flux status error, H_{λ_s}, that can be two values 0 or 1. The status H_{λ_s} is given to the switching table to decide the flux requirement of increasing/or decreasing with the conditions shown in Eq. 4 [5, 7]. This hysteresis controller keeps the stator flux constant to achieve control over the torque. $\boldsymbol{\lambda}_{s,ref}$ is the stator flux reference signal.

$$
\begin{aligned}
|H_{\lambda_s}| &= 1 \text{ if } |\boldsymbol{\lambda}_s| \le \left|\boldsymbol{\lambda}_{s,ref}\right| - |\Delta\boldsymbol{\lambda}_s| \therefore \text{ Flux Reduce} \\
|H_{\lambda_s}| &= 0 \text{ if } |\boldsymbol{\lambda}_s| \ge \left|\boldsymbol{\lambda}_{s,ref}\right| + |\Delta\boldsymbol{\lambda}_s| \therefore \text{ Flux Reduce}
\end{aligned}
\tag{4}
$$

Sector Selection and Switching Table for Flux and Torque Control. The output produced from the hysteresis controller of the stator flux and the torque, in addition to the sector selection table, provides the information for selecting the suitable switching voltage vector as shown in Table 1. Where the sector numbering $S(n)$ refers to the angle range shown in Eq. 5, for $n = 1$ to 6 [6].

$$
\begin{aligned}
&-30^\circ \le \theta < 30^\circ \rightarrow \text{Sector 1} \\
&30^\circ \le \theta < 90^\circ \rightarrow \text{Sector 2} \\
&90^\circ \le \theta < 150^\circ \rightarrow \text{Sector 3} \\
&150^\circ \le \theta < 180^\circ \rightarrow \text{Sector 4} \\
&-180^\circ \le \theta < -150^\circ \rightarrow \text{Sector 4} \\
&-150^\circ \le \theta < -90 \rightarrow \text{Sector 5} \\
&-90^\circ \le \theta < -30^\circ \rightarrow \text{Sector 6}
\end{aligned}
\tag{5}
$$

Three-Phase Inverter. A three-phase inverter of six switches is taken into consideration, thus, eight possible combinations of the voltage vectors are possible [5, 8].

Table 1 A switching table considering three-phase inverter of six switches by space vector

$H_{\lambda s}$	H_{Te}	S(1)	S(2)	S(3)	S(4)	S(5)	S(6)
1	1	V_2	V_3	V_4	V_5	V_6	V_1
	0	V_7	V_0	V_7	V_0	V_7	V_0
	– 1	V_6	V_1	V_2	V_3	V_4	V_5
0	1	V_3	V_4	V_5	V_6	V_1	V_2
	0	V_0	V_7	V_0	V_7	V_0	V_7
	–1	V_5	V_6	V_1	V_2	V_3	V_4

2.2 Simulink Modeling for DTC

The MATLAB/Simulink implementation of the DTC method is elaborated in this section. Figure 1 shows the complete Simulink blocks for the DTC integrated with the induction motor and the three-phase inverter.

The induction motor Simulink sub-block model consists of the motor stator flux, rotor flux, stator current, and rotor current equations in the stationary reference frame, as further elaborated in [8–10].

The calculation of the electromagnetic torque for the induction motor is based on (6), where $\boldsymbol{\lambda}_s$ and $\boldsymbol{i}_s$ vectors are both in the stationary reference frame [6].

$$T_e = \frac{3}{2}\frac{P}{2}(\boldsymbol{\lambda}_s \times \boldsymbol{i}_s) \tag{6}$$

The sub-block for the calculation of the rotor speed, ω_r, follows the equation presented in Eq. 7, where T_L is the load torque, B_m is the mechanical damping coefficient, and J is the rotor moment of inertia.

$$\omega_r = \frac{1}{J}\int (T_e - T_L) - \omega_r B_m \tag{7}$$

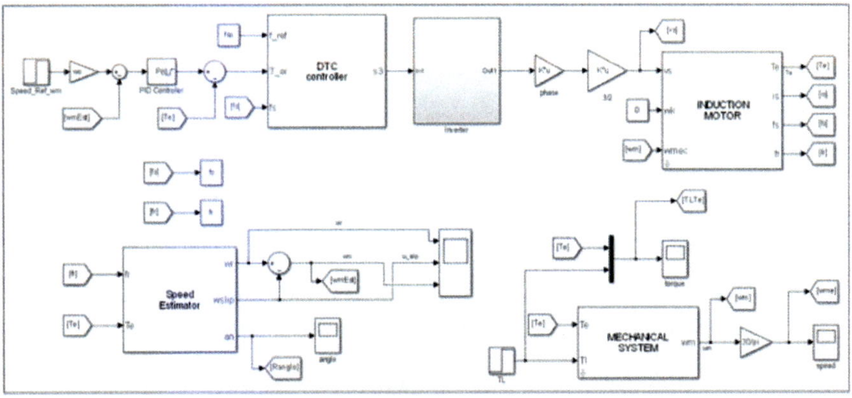

Fig. 1 Complete matlab/simulink implementation of the DTC for an induction motor

3 Indirect Field-Oriented Control (IFOC) Strategy

3.1 Indirect FOC Strategy Operation Principle

In the vector control, it is possible to acquire an independent control of the torque and flux, or it can be said that the IFOC relies on decoupling the torque and the flux to achieve the DC machine performance [11, 12].

For IFOC application with an induction machine, the inverter switching commands are generated by the control scheme to achieve the required electromagnetic torque at the motor shaft. The IFOC algorithm can be summarized as follows [1].

- The induction machine is connected with a current-controlled inverter of variable frequency and pulse width modulation (PWM). The error between the motor speed, ω_r, and the reference speed, ω_r^*, is given as an input to the speed controller. The output of the speed controller is the reference electromagnetic torque T_e^*.
- The reference q-axis stator current, i_{qs}^*, is obtained using the electromagnetic torque, as shown in the following equation.

$$i_{qs}^* = \frac{2}{3}\frac{2}{P}\frac{L_r}{L_m}\frac{T_e^*}{\hat{\lambda}_r} \tag{8}$$

where $\hat{\lambda}_r$ is the rotor flux linkage estimated value.

$$\hat{\lambda}_r = \frac{L_m i_{ds}}{1 + \tau_r s} \tag{9}$$

where i_{ds} is the d-axis stator current. The time constant of the rotor is described as $\tau_r = \frac{L_r}{R_r}$. R_r is the rotor resistance.

- The reference stator current in the d-axis, i_{ds}^*, is obtained by using the reference rotor flux linkage, $|\boldsymbol{\lambda}_r|^*$, as shown in the following equation.

$$i_{ds}^* = \frac{|\boldsymbol{\lambda}_r|^*}{L_m} \tag{10}$$

- The rotor flux position, θ_e, needed for the transformation blocks is attained by integrating the synchronous speed, ω_e, which is the summation of the rotor speed and the rotor slip speed, ω_{sl}.

$$\theta_e = \int \omega_e dt = \int (\omega_r + \omega_{sl})dt = \theta_r + \theta_{sl} \tag{11}$$

- The slip speed is obtained from the motor parameters and the q-axis stator current, i_{qs}, as shown in the following equation.

$$\omega_{sl} = \frac{L_m R_r}{|\boldsymbol{\lambda}_r| L_r} i_{qs} \tag{12}$$

3.2 *Simulink Modeling for IFOC*

The MATLAB/Simulink implementation for an induction motor with the IFOC scheme is presented in this section. Figure 2 shows the complete Simulink blocks integrated with the induction motor and the three-phase inverter.

By the IFOC method, the rotor flux is oriented along the motor's d-axis, assuming a synchronously rotating reference frame, as follows [13, 14].

$$\lambda_{dr} = \lambda_r^* = L_m i_{ds}^* \tag{13}$$

$$\lambda_{qr} = \frac{d\lambda_r^*}{dt} = 0 \tag{14}$$

where λ_{dr} is the d-axis rotor flux linkage. λ_{qr} is the q-axis rotor flux linkage.

In this case, the slip speed of the rotor is expressed as follows.

$$\omega_{sl} = \omega_{sl}^* = \frac{R_r}{L_r}\frac{i_{qs}^*}{i_{ds}^*} = \frac{1}{T_e}\frac{i_{qs}^*}{i_{ds}^*} \tag{15}$$

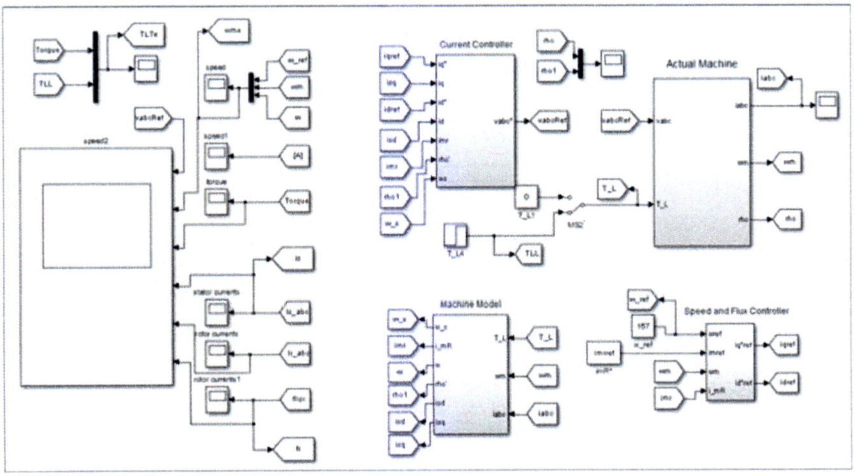

Fig. 2 Complete matlab/simulink implementation of IFOC for an induction motor

While the torque can be expressed as follows.

$$T_e = \frac{3P}{2}\frac{L_m^2}{L_r} i_{ds} i_{qs} \tag{16}$$

4 Simulation and Results

This section shows the MATLAB/Simulink modeling results for a three-phase induction machine with the DTC and IFOC techniques. The induction motor parameters and the overall system design conditions are presented in Table 2.

4.1 Direct Torque Control Simulation

The simulation conditions no-load torque 30 N.m and rotor speed reference 1200 rpm with step time 0.05 s are considered for the DTC method. While the load torque is set to 12 N.m with step time 0.3 s, The induction machine mechanical rotor speed

Table 2 Induction motor and system parameters for the DTC and IFOC schemes

Parameter	DTC scheme	IFOC scheme
R_s (Ω)	1.77	0.1
R_r (Ω)	1.34	0.1
L_{sl} (mH)	13.93	21
L_{rl} (mH)	12.12	21
L_m (mH)	369	19.9
Poles	4	4
Speed reference	1200 rpm with Step Time 0.05 s	157 rad/s
Flux reference (Wb)	0.996	44.321
T_L (N.m)	12 with step time 0.3 s	12 with step time 3 s
No-load torque (N.m)	30	17.36
J (Kg.m^2)	0.025	0.1
Additional parameters	DC-Link voltage = 700 V	Rater power = 7500 W
	B_m = 1e − 5 N.m.s	

Where L_{sl} is the stator leakage inductance. L_{rl} is the rotor leakage inductance

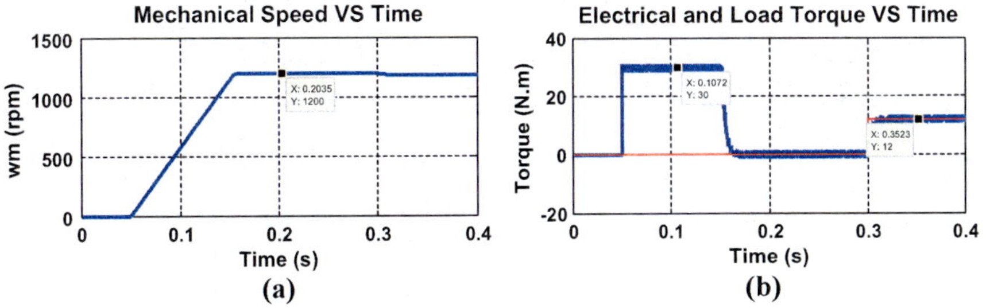

Fig. 3 Response of the (**a**) mechanical speed and (**b**) torque, with the DTC method

response is shown in Fig. 3a, and it can be seen that the machine takes some time until it reaches the expected speed value with a small deviation when the load torque is enabled. The electrical torque and load torque responses are presented in Fig. 3b. When the machine starts running at 0.05 s, then the electrical torque reaches the no-load condition that is 30 N.m. While during steady-state for the machine speed, the electrical torque drops to zero. When a load torque is applied at 0.3 s, the electrical torque follows it and in the steady state it reaches 12 N.m. The electrical torque response follows the load torque behavior, however, with some ripples.

4.2 Indirect Field-Oriented Control Simulation

The simulation of the IFOC method is based on no-load torque 17.36 N.m and rotor speed reference 157 rad/s. The load torque is set to 12 N.m with step time 3 s. The induction machine mechanical rotor speed response is displayed in Fig. 4a. Based on the results, the machine speed rises gradually till it reaches the expected speed value of 157 rad/s. The electrical torque and the load torque responses are presented in Fig. 4b. After the machine starts running, the electrical torque reaches the no-load condition of 17.36 N.m. While after the machine speed starts converging to the reference speed, the electrical torque drops to zero. When the load torque is applied

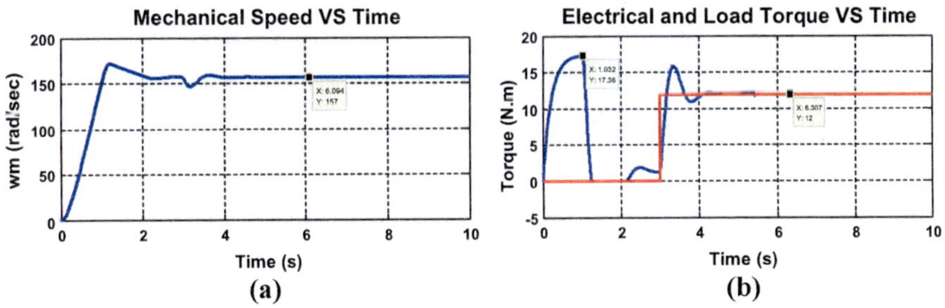

Fig. 4 Response of the (**a**) mechanical speed and (**b**) torque, with the IFOC method

Table 3 Comparison between the DTC and IFOC schemes

Point of comparison	DTC	IFOC
Reference frame transformation	Not required	Required
Control scheme	Simple	Complex
Stator current control	Not Required	Required
Motor parameter	R_s	R_s, R_r, L_{ls}, L_{lr}, L_m
Sensitivity to parameter variation	Less sensitive	Sensitive
PWM	Hysteresis	Carrier based, space vector PWM, hysteresis
Switching frequency	Variable	Fixed

at 3 s, the electrical torque follows it until reaching 12 N.m. As presented in Fig. 4b, the electrical torque response is following the load torque behavior.

A summary for the comparison between the control techniques is presented in Table 3 [11].

5 Conclusion

Conclusively, induction machine control is a necessity since it is a common machine that is always used in control systems of industrial motors. Accordingly, an induction machine drive with simple construction, low cost, rugged, and low maintenance is essential. In this paper, two control schemes were applied to an induction machine, which are the DTC and the IFOC. These schemes were studied and simulated using the MATLAB/Simulink. The key dissimilarity among the two vector control techniques is that field sensors are used for the DTC approach to acquire the rotor flux magnitude and position. However, in IFOC, the rotor flux position is acquired by the slip speed. The electrical torque response and the speed response for both control schemes were presented.

References

1. Lodhi RS, Thakur P (2013) Performance and comparison analysis of indirect vector control of three phase induction motor. Int J Emerg Technol Adv Eng 3(10):716–724
2. Vachhani UV, Patel AR (2017) Analysis of field oriented control strategy for induction motor. In: International conference on research and innovations in science, engineering and technology, pp 214–219. Kalpa Publications in Engineering
3. Singh G, Singh G (2014) Modeling and simulation of indirect field oriented control of three phase induction motor using fuzzy logic controller. Int J Eng Res Technol (IJERT) 3(8):1126–1130

4. Ozkop E, Okumus HI (2008) Direct torque control of induction motor using space vector modulation (SVM-DTC). In: 2008 12th International middle-east power system conference, pp 368–372. Aswan
5. Patra N (2013) Study of induction motor drive with direct torque control scheme and indirect field oriented control scheme using space vector modulation. Master Thesis
6. Tazerart F, Mokrani Z, Rekioua D, Rekioua T (2015) Direct torque control implementation with losses minimization of induction motor for electric vehicle applications with high operating life of the battery. Int J Hydrogen Energy 40(39):13827–13838
7. Allirani S, Jagannathan V (2012) Direct torque control technique for voltage source inverter fed induction motor drive. Int J Electr Eng 5(5):629–640
8. Leedy AW (2013) Simulink/Matlab dynamic induction motor model for use in undergraduate electric machines and power electronics courses. In: 2013 Proceedings of IEEE southeastcon, pp 1–6. FL
9. Douiri M, Nasserm T, Essadki A, Cherkaoui M (2010) Direct torque control of induction motor based on artificial neural networks with estimate and regulation speed using the MARS and neural PI controller. J Theor Appl Inf Technol 20:15–21
10. Fattahi SJ, Khayyat AA (2010) Direct torque control of brushless doubly fed induction machine. In: 2013 Proceedings of IEEE southeastcon, pp 1744–1747. Pisa
11. Gopal V (2017) Comparison between direct and indirect field oriented control of induction motor. Int J Eng Trends Technol (IJETT) 43(6):364–369
12. Guoa Z, Zhanga J, Suna Z, Zheng C (2017) Indirect field oriented control of three-phase induction motor based on current-source inverter. In: Procedia engineering 13th global congress on manufacturing and management, vol 174, pp 588–594
13. Masiala M, Vafakhah B, Salmon J, Knight AM (2008) Fuzzy self-tuning speed control of an indirect field-oriented control induction motor drive. IEEE Trans Ind Appl 44(6):1732–1740
14. Fayath RS, Ibrahim MM, Alwan MA (2005) Simulation of indirect field-oriented induction motor drive system using matlab/simulink software package. Sciences 31(1):83–94
15. Iqbal A et al (eds) (2020) Soft computing in condition monitoring and diagnostics of electrical and mechanical systems, vol 1096. In: Advances in intelligent systems and computing. Springer, Singapore. https://doi.org/10.1007/978-981-15-1532-3
16. Iqbal A et al (eds) (2020) Meta heuristic and evolutionary computation: algorithms and applications, vol 1096. In: Studies in computational intelligence. Springer, Singapore. https://www.springer.com/gp/book/9789811575709

Rotor Flux-Oriented Control of Three-Phase Induction Motor Using Sliding Mode Controller and Rotor Flux Estimator

H. M. Sabbir Ahmad, Mohammad Meraj, Atif Iqbal, and Nader Meskin

Abstract The work focuses on implementing rotor flux-oriented control (RFOC) for controlling the rotational speed and torque of a three-phase squirrel cage induction motor (IM). The speed controller has been implemented based on sliding mode controller (SMC). SMC is a robust control scheme which provides a means of achieving the control objective in the presence of model uncertainties and disturbances arising externally. Two different sliding mode controllers have been proposed in this paper. The first scheme is based on a discontinuous control input law which poses some performance limitations which have been illustrated though the simulation results. The second controller is designed using continuous control input law which overcomes the performance constraints of the first controller. A rotor flux estimation technique has been proposed to determine rotor flux position required for implementing RFOC. The performance of the sliding mode controllers and rotor flux estimator has been verified using computer simulation in Simulink using the mathematical model of a three-phase IM.

Keywords Rotor flux-oriented control · Sliding mode controller · Rotor flux estimation

1 Introduction

Owing to the necessity of IMs in several performance critical applications, there has been various control methods designed amongst which one of the widely known control methods is the field-oriented control (FOC) proposed in [1–3]. Conceptually, FOC is implemented by decoupling the torque and flux control loops to achieve the control requirement of the IM. There are multiple source of uncertainties that arises in the system including model parameter variations, for example, disturbances in external connected load, un-modelled dynamics and linearization also raises uncertainty. Thus, studies have been conducted by applying various different

H. M. Sabbir Ahmad (✉) · M. Meraj · A. Iqbal · N. Meskin
Department of Electrical Engineering, Qatar University, Doha, Qatar
e-mail: ha1607441@student.qu.edu.qa

A. Iqbal et al. (eds.), *Renewable Power for Sustainable Growth*, Lecture Notes in Electrical Engineering 723, https://doi.org/10.1007/978-981-33-4080-0_14

established techniques from control theory to address the performance degradation due to uncertainties. In [4, 5], authors have implemented predictive control for IMs. Adaptive control of IM has been proposed on [6]. Finally in [7, 8], the controllers have been designed using robust control and fuzzy control, respectively. One of the nonlinear control techniques that take model and parameter uncertainties and external disturbances is the SMC control and hence appears attractive for IMs [9].

Rotor flux-oriented control is a type of FOC/vector control which is implemented by aligning the rotor flux along the d-axis, thus requiring the position of the rotor flux at all times which can be measured using Hall effect sensors. Addition of sensors poses several cons including added costs, additional source of uncertainty, reduced overall reliability and installation difficulty. In [10, 11], the authors proposes an estimator design for an IM. The technique in [10] measures the stator current of the IM and uses it as a reference to compare it against the estimated stator current based on stator voltage–current model. In [11], authors propose estimation using unscented Kalman filter. Although UKF takes into consideration measurement noise, it is computationally expensive and thus poses an implementation challenge in embedded controllers.

This paper introduces two robust sliding mode controllers. Sliding mode controller is deemed appropriate since it considers uncertainties in the measurement of the mechanical parameters; for example, the load torque is a parameter which can vary over a wide range. This paper considers the moment of inertia of the machine as a source of measurement error. Besides that, the load torque is considered as a source of disturbances to the IM. The first controller is based on a discontinuous input control law, whereas the second input control law is continuous with a boundary layer. The rotor flux is estimated using the equations of the reduced order model of the IM which is included in the later sections of the paper. This paper is organized into five sections. The second section includes the reduced order model of the IM along with the concept of RFOC and rotor flux estimation for calculating the rotor flux position. Results from MATLAB and Simulink simulations are included in the third section along with interpretation and conclusions deduced from these results. The paper concludes with highlighting results and advantages of the proposed controllers in the fourth section.

2 Methodology

2.1 Reduced Order Model of Induction Machine

The reduced order model represents variables associated to three phases of the IM with two variables, thus reducing the overall order of the system. The transformation from three-phase '*abc*' to two-phase '*d-q*' frame as shown in Fig. 2 is achieved using the following matrix.

$$\begin{bmatrix} U_d \\ U_q \end{bmatrix} = \frac{2}{3}\begin{bmatrix} \cos(\omega t) & \cos\left(\omega t - \frac{2\pi}{3}\right) & \cos\left(\omega t + \frac{2\pi}{3}\right) \\ -\sin(\omega t) & -\sin\left(\omega t - \frac{2\pi}{3}\right) & -\sin\left(\omega t + \frac{2\pi}{3}\right) \end{bmatrix}\begin{bmatrix} U_a \\ U_b \\ U_c \end{bmatrix} \tag{1}$$

In (1), the term ωt represents the position of the d-axis in reference to phase a and ω is the angular velocity of the d-q reference frame. Assuming the d-q frame is rotating at arbitrary speed ω, the two-phase model of the can be represented as follows.

$$V_{ds} = r_s i_{ds} + p\psi_{ds} - \omega\psi_{qs} \tag{2}$$

$$V_{qs} = r_s i_{qs} + p\psi_{ds} + \omega\psi_{ds} \tag{3}$$

$$V_{dr} = r_r i_{dr} + p\psi_{dr} - (\omega - \omega_r)\psi_{qr} \tag{4}$$

$$V_{qr} = r_r i_{qr} + p\psi_{qr} - (\omega - \omega_r)\psi_{dr} \tag{5}$$

$$\psi_{ds} = L_s i_{ds} + L_m i_{dr} \tag{6}$$

$$\psi_{qs} = L_s i_{qs} + L_m i_{qr} \tag{7}$$

$$\psi_{dr} = L_r i_{dr} + L_m i_{ds} \tag{8}$$

$$\psi_{qr} = L_r i_{qr} + L_m i_{qs} \tag{9}$$

In case of stationary reference frame $\omega = 0$, for synchronously rotating reference frame $\omega = \omega_e$. In (2)–(9), $[V_{ds}, V_{qs}, V_{dr}, V_{qr}]$ and $[i_{ds}, i_{qs}, i_{dr}, i_{qr}]$ correspond to the stator and rotor voltages and currents along the d- and q-axis, respectively. $[\Psi_{ds}, \Psi_{qs}, \Psi_{dr}, \Psi_{qr}]$ represents the flux linkages along the d- and q-axis.

2.2 *Rotor Flux-Oriented Control*

RFOC is implemented using the reduced order of the model in synchronously rotating d-q frame. The control scheme aligns the rotor flux along the d axis. Thus, the rotor flux along synchronously rotating d-q frame becomes the following:

$$\psi_{qr}^{e} = 0;\ \psi_{dr}^{e} = \psi_{r}^{m};$$

Combining the rotor flux values, (4) and $V_{dr} = V_{qr} = 0$ gives the following

$$i_{dr}^{e} = 0$$

Substituting the value of i_{dr}^{e} in (8) results in the equation below.

$$i_{ds}^{e} = \frac{\psi_{dr}^{e}}{L_m} \tag{10}$$

Using the values of rotor flux along synchronously rotating d-q frame, the following is obtained from (9)

$$i_{qr}^{e} = \frac{-L_m}{L_r} i_{qs}^{e} \tag{11}$$

Thus, the torque equation for the reduced order IM under RFOC is as follows:

$$T_e = \frac{3}{4} p \frac{L_m}{L_r} \psi_r^m . i_{qs}^{e} = K_T . i_{qs}^{e} \tag{12}$$

As can be seen the electromechanical torque becomes a function of the rotor d-axis flux and the q-axis stator current in synchronously rotating *d-q* frame. The IM speed is controlled using the electrochemical torque. The torque is manipulated by keeping the rotor flux along *d*-axis fixed and changing the *q*-axis stator current. The rotor flux along *d*-axis is controlled using *d*-axis stator current. This paper implements RFOC in voltage control mode implying the currents along *d-q* axis are controlled using the voltage along *d-q* axis, respectively. Combining Eqs. (2) with (6), (7), (8) and (9) gives the following equation for calculating voltage reference along *d*-axis.

$$V_{ds}^{e*} = r_s i_{ds}^{e} + \tau_s' r_s \frac{di_{ds}^{e}}{dt} + \frac{L_m}{L_r} \frac{d\psi_{dr}^{e}}{dt} - \omega_e \tau_s' r_s i_{qs}^{e}$$

$$V_{ds}^{e*} = V_{ds}^{e'} + e_d ; \; V_{ds}^{e'} = r_s (i_{ds} + T_s' \frac{di_{ds}^{e}}{dt});$$

$$e_d = \frac{L_m}{L_r} \frac{d\psi_{dr}^{e}}{dt} - \omega_e T_s' r_s i_{qs}^{e}$$

$$V_{qs}^{e*} = r_s i_{qs}^{e} + T_s' r_s \frac{di_{qs}^{e}}{dt} - \frac{\omega_e L_m}{r_s L_r} \frac{d\psi_{dr}^{e}}{dt} - \omega_e T_s' i_{ds}^{e} \tag{14}$$

$$V_{qs}^{e*} = V_{qs}^{e'} + e_q ; \; V_{qs}^{e'} = r_s (i_{qs} + T_s' \frac{di_{qs}^{e}}{dt})$$

$$e_q = -\frac{\omega_e L_m}{L_r} \psi_{qr}^{e} - \omega_e \tau_s' r_s i_{ds}^{e}$$

$$\sigma = 1 - \frac{L_m^2}{L_s L_r} ; \; \tau_s' = \sigma \frac{L_s}{r_s} \tag{15}$$

The expression (14) and (15) gives the equation for calculating the d- and q-axis voltage value for any given d- and q-axis current reference. As can be seen, each equation contains two terms. The term e_d and e_q are considered as a feedforward terms which are fed to the voltage value required for controlling the d- and q-axis current, respectively.

2.3 Rotor Flux Angle Estimation

The rotor flux angle required for transforming three-phase variables to synchronously rotating d-q reference frame. It can be calculated from the rotor flux in stationary d-q reference frame. Substituting $\omega = 0$, $V_{dr} = V_{qr} = 0$ (as the rotor is made of squirrel cage winding) in (4) and (5) and combining with (8) and (9) renders the following two equations used for rotor flux estimation in stationary d-q reference frame.

$$P\hat{\psi}_{dr} = \frac{L_m}{\tau_r} i_{ds} - \frac{1}{\tau_r}\hat{\psi}_{dr} - \omega_r \psi_{qr} \tag{16}$$

$$P\hat{\psi}_{qr} = \frac{L_m}{\tau_r} i_{qs} - \frac{1}{\tau_r}\hat{\psi}_{qr} + \omega_r \psi_{dr}$$

$$\tau_r = \frac{L_r}{r_r} \tag{17}$$

Combining (19) and (20) the angular position of the synchronously rotating reference frame can be estimated as follows.

$$\hat{\theta}_e = \tan^{-1}\left(\frac{\hat{\psi}_{qr}}{\hat{\psi}_{dr}}\right)$$

2.4 Speed Controller Design

The following mechanical equation describes the dynamics of the rotational speed of the three-phase IM.

$$J\dot{\omega}_m + B\omega_m + T_L = T_e \tag{18}$$

In (19), J, B, T_L and T_e correspond to inertia, friction coefficient, load torque and electromechanical torque (given by (12)), respectively. Combining (12) and (18) the mechanical equation can represent as follows.

$$J\dot{\omega}_m + B\omega_m + T_L = K_T . i^e_{qs} \tag{19}$$

2.5 Robust Sliding Mode Controller Design

In practical circumstances, any modelling technique includes inaccuracies due to parametric uncertainties (unknown plant parameters, tolerance of the measuring device, etc.) and un-modelled dynamics (considering friction as linear). SMC is based on controlling an *n*th order system using a first-order system dynamics. With this notion, the mechanical equation in (19) can be written as below.

$$\omega_m = (a + \Delta a)\omega_m + (b + \Delta b)i_{qs}^e + (f + \Delta f) \tag{20}$$

where

$$a = \frac{-B}{J}; b = \frac{K_T}{J}; f = \frac{-T_L}{J}; \Delta a = \pm\frac{\Delta B}{J}; \Delta b = \pm\frac{\Delta K_T}{J}; \Delta f = \pm\frac{\Delta T_L}{J}$$

Δa, Δb and Δf represent the uncertainties in friction coefficient, uncertainties in torque constant, and disturbances and uncertainties in load torque. The sliding surface is $S(t)$ can be defined with integral error component as follows.

$$S(t) = e(t) - \int (a + bK)e(\tau)d\tau \tag{21}$$

where K is a linear feedback gain. Sliding mode occurs on the sliding surface when $S(t) = \dot{S}(t) = 0$ and therefore (21) can be represented using the equation below.

$$\dot{e}(t) = (a + bK)e(t) \tag{22}$$

In (22), $(a + bK)$ is the pole of the error dynamics when sliding mode occurs. For the error to converge to zero upon reaching $S = 0$, the following condition has to be obeyed.

$$(a + bK) \le 0; K \le 0 \tag{23}$$

The SMC is designed to ensure that sliding mode occurs by converging the error to zero. The tracking speed error is defined as:

$$e(t) = \omega_m(t) - \omega_m^*(t)$$

Taking the derivative of $e(t)$ yields the following:

$$\begin{aligned}
\dot{e}(t) &= (a + \Delta a)\omega_m(t) + (b + \Delta b)i_{qs}^e + (f + \Delta f) - \dot{\omega}_m^*(t) \\
\dot{e}(t) &= ae(t) + \left(a\omega_m^*(t) + bi_{qs}^e + f - \dot{\omega}_m^*(t)\right) \\
&\quad + \left(\Delta a\omega_m(t) + \Delta bi_{qs}^e + \Delta f\right)\dot{e}(t) = ae(t) + u_{eq}(t) + d(t)
\end{aligned} \tag{24}$$

where

$$u_{eq}(t) = a\omega_m^*(t) + bi_{qs}^e + f - \dot{\omega}_m^*(t)$$
$$d(t) = (\Delta a\omega_m(t) + \Delta bi_{qs}^e + \Delta f)$$

The controller is designed using two parts: the first part of which called the equivalent input control law is used to make $u_{eq}(t) = 0$ and the other part called the switching input control law is used to counteract the disturbances, control the convergence rate and ensure once sliding mode occurs $S(t)$ stays on the surface $S = 0$. The first part can be expressed as follows by setting $u_{eq}(t) = 0$.

$$i_{qseq}^e = \frac{1}{b}\left[-a\omega_m^*(t) - f + \dot{\omega}_m^*(t)\right] \tag{25}$$

The two switching input control law designed is discussed in following subsections.

Discontinuous switching law: The switching input control law is expressed as following.

$$i_{qssw}^e = Ke(t) - \beta sgn(S(t))$$
$$sgn(S(t)) = \begin{cases} 1 & \text{if } S(t) > 0 \\ -1 & \text{if } S(t) < 0 \end{cases} \tag{26}$$

The $sgn(S(t))$ term introduces extremely high-frequency switching of the input signal causing a phenomenon called chattering. Chattering has the following drawbacks.

1. It can excite un-modelled high-frequency dynamics of the system.
2. The bandwidth of the PWM inverter limits the maximum switching frequency possible that may degrade tracking performance.

Combining (25) and (26) gives the complete input control law given by the equation below.

$$i_{qs}^{e*} = Ke(t) - \beta sgn(S(t)) + \frac{1}{b}\left[-a\omega_m^*(t) - f + \dot{\omega}_m^*(t)\right] \tag{27}$$

The problem of chattering is overcome in the next choice of switching control input law which is described below. For the choice of input control law in (26), the error dynamics in state space form until sliding mode occurs can be found by substituting (27) in (24).

$$\dot{e}(t) = (a + K)e(t) - \beta sgn(S(t)) + d(t)$$
$$\dot{e}(t) = (a + K)e(t) + u(t)$$
$$u(t) = -\beta sgn(S(t)) + d(t) \tag{28}$$

Continuous switching law with boundary layer: To eliminate chattering, the $sgn(S(t))$ term in (26) is replaced, thus giving the following switching input control law.

$$i^{e}_{qssw} = Ke(t) - \beta \text{ sat}\left(\frac{S(t)}{\varphi}\right)$$

$$\text{sat}\left(\frac{S(t)}{\varphi}\right) = \begin{cases} \frac{S(t)}{\varphi} \text{ when} |S(t)| \leq \varphi \\ sgn\left(\frac{S(t)}{\varphi}\right) \text{ otherwise} \end{cases} \tag{29}$$

In (29), φ represents the width of the boundary layer. The $\text{sat}\left(\frac{S(t)}{\varphi}\right)$ terms provide a continuous switching law and eliminate the problem of chattering. The complete input control law is gained by combining (29) and (25).

$$i^{e*}_{qs} = Ke(t) - \beta \text{ sat}\left(\frac{S(t)}{\varphi}\right) + \frac{1}{b}\left[-a\omega^{*}_{m}(t) - f + \dot{\omega}^{*}_{m}(t)\right] \tag{30}$$

Similarly, the error dynamics in state space form for the control law given by (29) can be computed by substituting (30) into (24) which gives the following equation.

$$\dot{e}(t) = (a + K)e(t) - \beta \text{ sat}\left(\frac{S(t)}{\varphi}\right) + d(t)$$

$$(a + K)e(t) + u(t)$$

$$u(t) = -\beta \text{ sat}\left(\frac{S(t)}{\varphi}\right) + d(t) \tag{31}$$

For both choices of control inputs given by (26) and (29), the error dynamics contains a pole at $(a + K)$ which decides the rate of converge of the rotor speed tracking error. The sliding mode condition in (23) implies K has to be strictly negative. As a is constant the value of K is chosen such that speed tracking is achieved within reasonable time. The value of β in (27) and (31) is chosen to make $u(t)$ zero as per the following condition.

$$\beta \geq d(t)$$

3 Discussion and Results

The motor parameters used for simulation are tabulated below.

Table 1 Motor parameters

Parameter name	Parameter value	Parameter name	Parameter value
f (Hz)	50	ΔK_T (kg m^2 s^{-2} A^{-1})	0.01
P	4	ΔT_L (N.m)	8
P_{rated} (kW)	7.5	Rs (Ω)	0.1
I_{rated} (A)	26.8	Rr (Ω)	0.1
V_{rated} (V)	240	Lm (H)	0.001986
J (kgm^2)	0.1	L_s (H)	0.021
B (Nm (rad/s))	0.01	L_r (H)	0.021
K_T (kg m^2 s^{-2} A^{-1})	2.5	ΔK_T (kg m^2 s^{-2} A^{-1})	0.01
ΔB (kgm^2)	0.0005		

3.1 Response of Discontinuous and Continuous Controller

See Table 1.

3.2 Closed-Loop Response of Three-Phase IM Using Discontinuous Sliding Mode Controller

The simulation of three-phase IM using the discontinuous sliding mode controller for $\beta = 34$ and $K = -50$. The higher value of K was chosen for speed control of the IM because the response time of the controller depends on rotor speed dynamics and the dynamics of the q-axis current along synchronously rotating *d-q* frame while controlling the speed of the IM. The higher value of K was chosen for speed control of the IM because the response time of the controller depends on rotor speed dynamics and the dynamics of the q-axis current along synchronously rotating *d-q* frame. The simulation was done using the uncertainties in inertia and electromechanical torque constant along with disturbances in load torque using the values included in Table 1. The load torque was changed from 10 to 18 Nm at 30 s during simulation.

As can be observed from Fig. 1, the IM is able to track the reference rotational speed within an acceptable amount of time with $\pm 0.01\%$ error. Lastly, as anticipated the use of discontinuous input control law introduces chattering the q axis current and voltage reference as shown in Fig. 1b, c which may not be achievable in hardware due to the limitation in switching frequency thus will degrade the tracking performance.

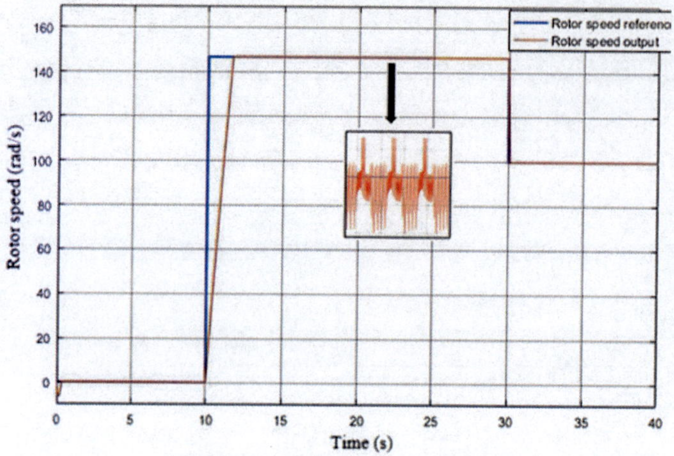

(a) Rotor speed tracking against time

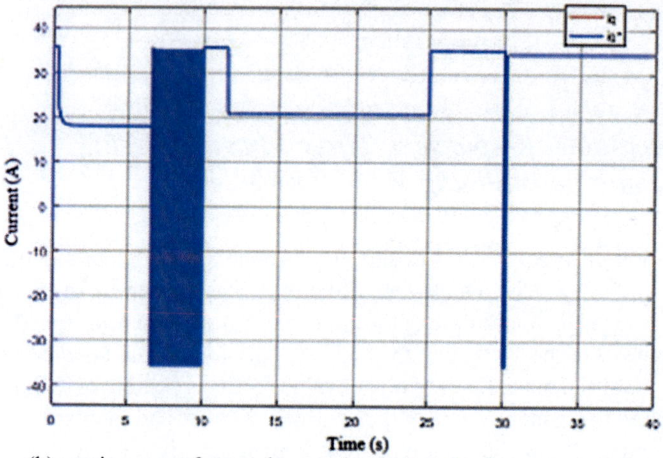

(b) q axis current for synchronously rotating d-q frame against time

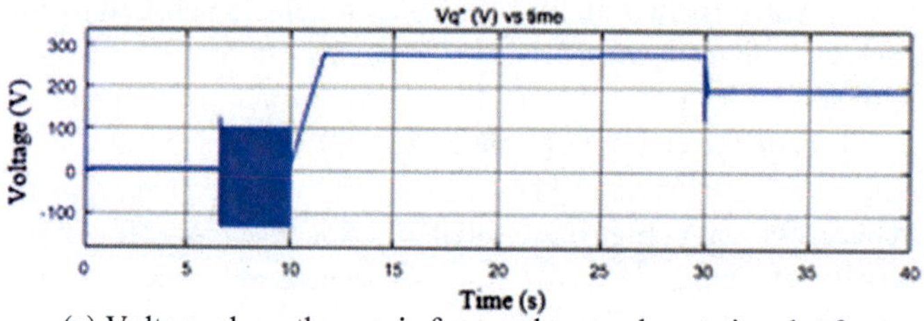

(c) Voltage along the q axis for synchronously rotating d-q frame

Fig. 1 Closed-loop parameter response for the three-phase IM using discontinuous sliding mode controller

3.3 *Closed-Loop Response of Three-Phase IM Using Continuous Sliding Mode Controller*

With $\beta = 34$ and $K = -50$, the response of three-phase IM using the continuous sliding mode controller implemented as shown in Fig. 2. The uncertainties and disturbances tabulated in Table 1 were included during simulation. The load torque was changes from 10 to 18 Nm at 25 s. Figure 2a shows the controller achieves good tracking with accuracy of ±0001% in less than 5 s. The chattering phenomenon is eliminated with the use of continuous SMC as can be seen from Fig. 2a, b, respectively.

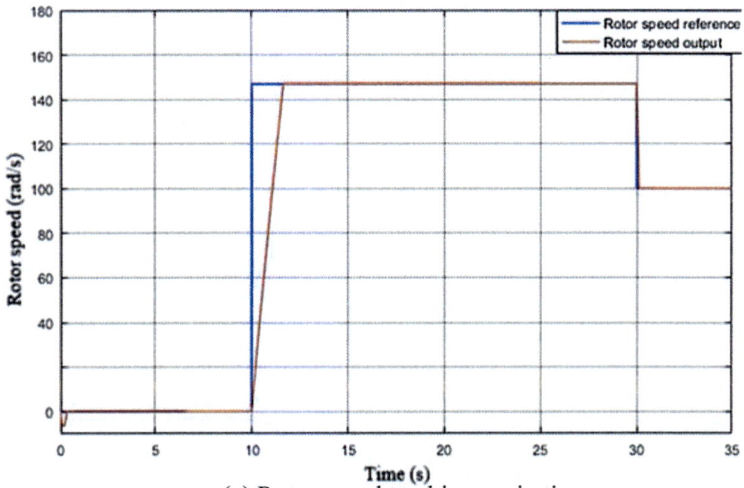

(a) Rotor speed tracking again time

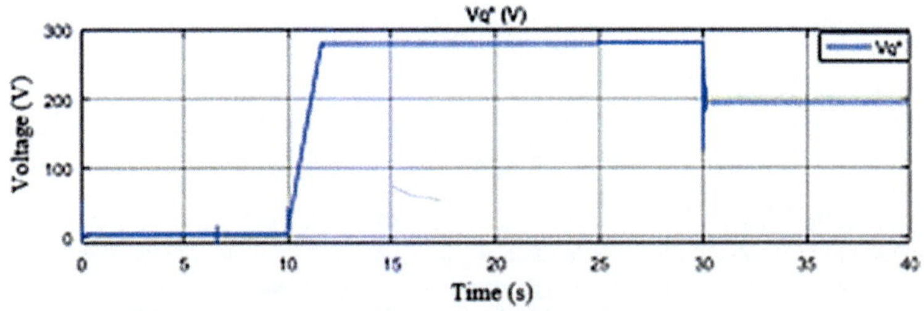

(b) Voltage along q axis for synchronously rotating d-q frame

Fig. 2 Closed-loop response of three-phase IM for the proposed continuous sliding mode controller

4 Conclusion

To conclude the presented results shows that both the proposed controllers are able to achieve good rotor speed tracking with very little error within an acceptable amount of time. The use of sliding mode controller provides reference tracking in the presence of parameter uncertainties and external disturbances. Finally, the discontinuous controller introduces chattering, which is overcome using the continuous input control law.

Acknowledgements This publication was made possible by the Graduate Sponsorship Research Award (GSRA) award (GSRA4-2-0518-17083) from the Qatar National Research Fund (QNRF), a member of the Qatar Foundation. Its contents are solely the responsibility of the authors and do not necessarily represent the official views of QNRF.

References

1. Holmes DG, McGrath BP, Parker SG (2012) Current regulation strategies for vector-controlled induction motor drives. IEEE Trans Ind Electron 59(10):3680–3689
2. Iqbal A et al (eds) (2020) Soft computing in condition monitoring and diagnostics of electrical and mechanical systems, vol 1096. In: Advances in intelligent systems and computing. Springer, Singapore. https://doi.org/10.1007/978-981-15-1532-3
3. Iqbal A et al. (eds) (2020) Meta heuristic and evolutionary computation: algorithms and applications, vol 1096. In: Studies in Computational Intelligence. Springer, Singapore. https://www.springer.com/gp/book/9789811575709
4. Guzinski J, Abu-Rub H (2013) Speed sensorless induction motor drive with predictive current controller. IEEE Trans Ind Electron 60(2):669–709
5. Alkorta P, Barambones O, Cortajarena JA, Zubizarrreta A (2014) Efficient multivariable generalized predictive control for sensorless induction motor drives. IEEE Trans Ind Electron 61(9):5126–5134
6. Orłowska-Kowalska T, Dybkowski M, Szabat K (2010) Adaptive sliding-mode neuro-fuzzy control of the two-mass induction motor drive without mechanical sensors. IEEE Trans Ind Electron 57(2):553–564
7. Sira-Ramrez H, Gonzlez-Montaez F, Corts-Romero JA, Luviano-Jurez A (2012) A robust linear field-oriented voltage control for the induction motor: experimental results. IEEE Trans Ind Electron 60(8):3025–3033
8. Suetake M, da Silva IN, Goedtel A (2011) Embedded DSP-Based compact fuzzy system and its application for induction-motor V/f speed control. IEEE Trans Ind Electron 58(3):750–760
9. Zhang LX (2013) Sensorless induction motor drive using indirect vector controller and sliding-mode observer for electric vehicles. IEEE Trans Ind Electron 62(7):3010–3018
10. Orłowska-Kowalska T, Dybkowski TM (2010) Stator-Current-Based MRAS estimator for a wide range speed-sensorless induction-motor drive. IEEE Trans Ind Electron 57(4):1296–1308
11. Jafarzadeh S, Lascu C, Fadali MS (2012) State estimation of induction motor drives using the unscented kalman filter. IEEE Trans Ind Electron 59(11):4207–4216

Performance Analysis of Induction Motor Utilizing SPRS Based on MOSFET/IGBT Ćuk Converter Topologies

Gourav Vardhan, O. P. Rahi, and Arun Kumar Bugaliya

Abstract In slip power recovery scheme (SPRS) to avoid the wastage of slip power in rotor resistance, it is used as feedback to main power supply. Several types of topologies are used for SPRS, i.e., boost, buck, buck–boost and Ćuk converter. Other than this, various switching devices like SCR, MOSFET, GTO, IGBT, etc., are used. Analysis of SPRS based on induction motor drive (IMD) utilizing MOSFET/IGBT Ćuk converter and voltage source inverter based on IGBT with pulse width modulated scheme is carried out in the paper. Various performance analyses of IMD have been executed by utilizing the IGBT and MOSFET in converters. Wound rotor induction motor (WRIM) model based on SPRS utilizing pulse width modulated (PWM) scheme and voltage source inverter (VSI) has been developed using Simulink/MATLAB toolbox. The effect of duty ratio on speed, DC link current, supply current, source reactive power, supply power factor and total harmonic distortion (THD) has been investigated. IGBT produces better results as compared to MOSFET shown by simulation results.

Keywords Chopper · Induction motor · IGBT · Slip power recovery scheme · Efficiency · MOSFET

1 Introduction

SPRS is among the best methods of WRIM speed control [1–14]. SPRS is widely used in industries where intermittent operation of drive is required in applications like cranes, hoists, lifts and conveyers because it is easy to control slip power by slip rings. To control speed of motor, slip power is supplied back to main power source. The features of the inverter can be controlled in the rotor circuit, and this controlling of features makes WRIM suitable for variable speed applications [1]. This drive operates in limited range of speed, and slip power is just a fractional part of motor

G. Vardhan (✉) · O. P. Rahi · A. K. Bugaliya
Department of Electrical Engineering, NIT Hamirpur, Hamirpur, HP, India
e-mail: vardhangourav10@gmail.com

A. Iqbal et al. (eds.), *Renewable Power for Sustainable Growth*, Lecture Notes in Electrical Engineering 723, https://doi.org/10.1007/978-981-33-4080-0_15

rating, so rating of converter is quite less and hence it results in less cost of converter [2]. Major drawbacks of SPRS are lesser supply power factor, high reactive power requirement and high content of harmonics [3, 4, 13, 14]. The presence of harmonics produces distortion in supply. Interest of researchers is increasing in the renewable energy area as a result of new upcoming technologies being introduced in converters [5].

Pulse width modulation (PWM) inverters and capacitive compensation approaches have been used for the improvement of power factor [6, 7]. A 12-pulse SCR-based inverter and IGBT-based boost converter have been used to get better power factor of supply, SPRD using VSI and boost converter decreases the THD of source, and hence there is improvement in the power factor [8].

Inverter along with chopper enhances power factor and efficiency of the SPRD and also the power quality [9]. Chopper control methodologies along with PWM voltage source inverter allows decoupled control of speed and nowadays the simultaneous speed control is attractive for researchers in this area [10].

2 Proposed Methodology

2.1 Conventional SPRS Model

The rotor windings are connected to slip rings through 3-phase full-wave rectifier. By varying the emf of inverter (V_2), we can control the power sent back. The emf of inverter can be changed easily just by changing value of firing angle of inverter.

To reduce DC ripples (I_{dc}), DC link inductor is there. Also, a recovery transformer is used to match the rectifier and inverter voltages [11]. By neglecting losses of stator and rotor drops, we can write

$$V_1 = \frac{3\sqrt{2}}{\pi} \times \frac{sV}{a} = 1.35 \times \frac{sV}{a} \tag{1}$$

$$V_2 = \frac{3\sqrt{2}}{\pi} \times \frac{V}{b} \cos\beta = 1.35 \times \frac{V}{b} \cos\beta \tag{2}$$

where V_1 = rectifier's output voltage, V_2 = inverter's output voltage, s = slip, β = inverter's firing angle, a = stator to rotor turn's ratio, b = transformer's turn ratio, V = supply voltage. For the safe commutation of the thyristor, maximum value of firing angle is kept at most 165°. By choosing firing angle appropriately required speed can be achieved [12].

Neglecting drop across the inductor, we can write

$$V_1 + V_2 = 0 \tag{3}$$

From Eqs. (1)–(3), we will get

$$s = \frac{a}{b}|\cos\beta| \tag{4}$$

On referring equivalent motor circuit to rotor side, approximating DC equivalent circuit and neglecting magnetizing branch, the equations for air gap power, DC link current and developed torque are given by

$$I_{dc} = \frac{V_1 + V_2}{2\left(sR_s' + R_r\right) + R_d} \tag{5}$$

Neglecting copper loss, we can write

$$sP_g = |V_2|I_d = 1.35 \times \frac{V}{b}\cos\beta \times I_{dc} \tag{6}$$

From Eqs. (4) and (6),

$$P_g = 1.35 \times \frac{V}{a} \times I_{dc} \tag{7}$$

$$P_m = T_d \times w_m(1 - s)\frac{2}{p} \tag{8}$$

$$P_m = (1 - s)P_g = T_d \times w_r \tag{9}$$

$$T_d = \frac{p}{2} \times \frac{P_g}{w_m} \tag{10}$$

From Eqs. (7) and (10),

$$T_d = \frac{p}{2} \times \frac{1.35V}{a \times w_m} \times I_{dc} \tag{11}$$

where I_{dc} = current at DC link, P_g = air gap power, T_d = torque developed, P_m = mechanical power.

w_m = synchronous speed in rad/s, w_r = rotational speed of rotor in rad/s, R_s' = rotor side referred stator resistance, R_r = resistance of rotor, p = number of poles.

2.2 SPRS Using MOSFET/IGBT Ćuk Converter

In conventional slip power scheme, there are many drawbacks but the main drawback is that more reactive power is drawn from source on increasing β above 90°.

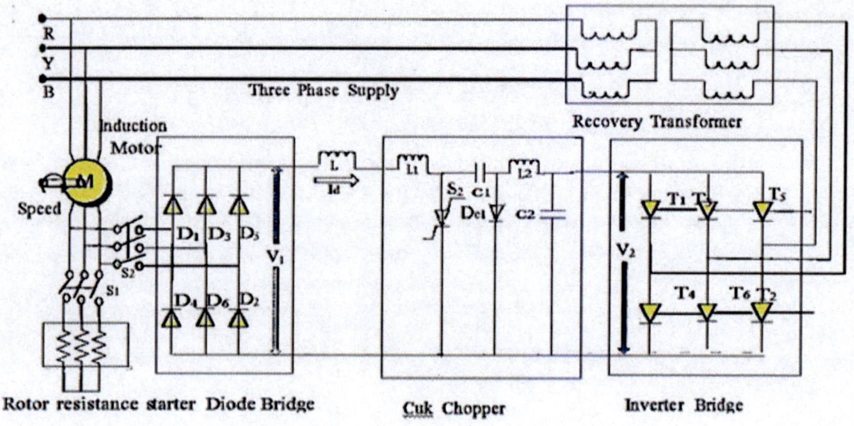

Fig. 1 SPRS simulink model using MOSFET/IGBT

Due to this, THD is increased and power factor is decreased. To overcome this problem depicted above, a scheme has been introduced in which MOSFET/IGBT Ćuk converter is utilized as depicted in Fig. 1. By changing the duty ratio (D) and keeping β of inverter constant, the speed control operation of SPRD is carried out. Voltage drop in inductor, voltage at inverter output, slip (s) and current at DC link are given by [8]:

$$V_2 = 1.35 \times \frac{V}{b}\left(\frac{-D}{1-D}\right)\cos\beta \tag{12}$$

$$s = \frac{a}{b}\left(\frac{-D}{1-D}\right)|\cos\beta| \tag{13}$$

$$I_{dc} = \frac{V_1 - V_2}{R_{dc}} \tag{14}$$

$$R_{dc} = s\left\{2R_s' + 3\left(\frac{X_s' + X_r}{\pi}\right)\right\} + R_r + R_d \tag{15}$$

where $D =$ duty ratio and $R_{dc} =$ DC link circuit resistance
where $X_s^{'} =$ rotor side referred stator reactance and $X_r =$ rotor reactance

$$P_g = \frac{|V_2|I_{dc}}{s} \tag{16}$$

$$\mathrm{T} = \frac{P_g}{w_m} = \frac{|V_{d2}|I_{dc}}{sw_m} \tag{17}$$

$$\cos\phi = \frac{P_s - P_r}{\sqrt{(P_s - P_r)^2 + (Q_s + Q_r)^2}} \tag{18}$$

where P_s = active power drawn from source, P_r = active power recovered from inverter, Q_s = reactive power drawn from source, Q_r = reactive power recovered from inverter

$$P_r = 1.35 \times V \times I_{dc}\left(\frac{-D}{1-D}\right)\cos\beta \tag{19}$$

$$Q_r = 1.35 \times V \times I_{dc}\left(\frac{-D}{1-D}\right)\cos\beta \tag{20}$$

From Eq. (20), it can be seen clearly that by decrement of D with constant β reactive power drawn will decrease. Hence, source power factor will increase as seen from Eq. (18).

3 Simulation Model of SPRS

In this paper, IGBT and MOSFET are used in Ćuk converter and then the performance of WRIM is analyzed by using the two switching devices. For this scheme, simulation model is developed using Simulink as depicted in Fig. 2. For this model, a 2.1 hp, 415 V, 50 Hz, 4-pole motor is used. The duty ratio (D) for the speed control is varied from 80 to 30%, and firing angle of inverter is at 91°. Various results are shown in Figs. 3, 4, 5, 6, 7 and 8.

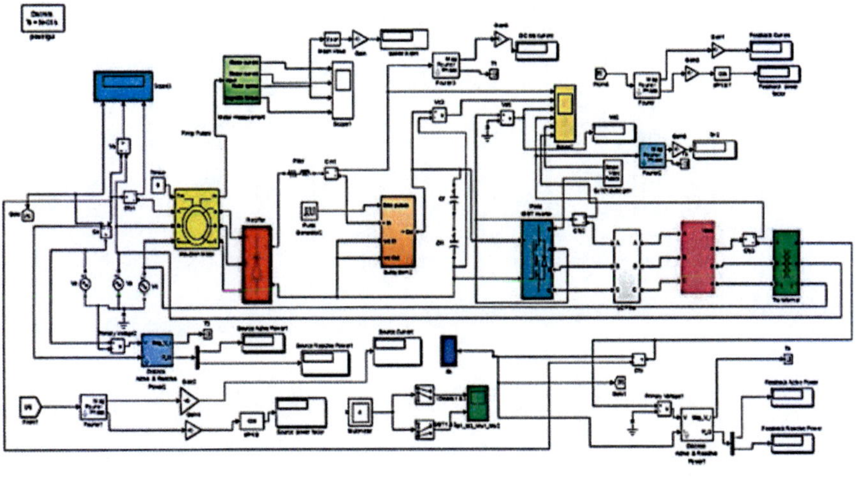

Fig. 2 SPRS simulink model using MOSFET/IGBT

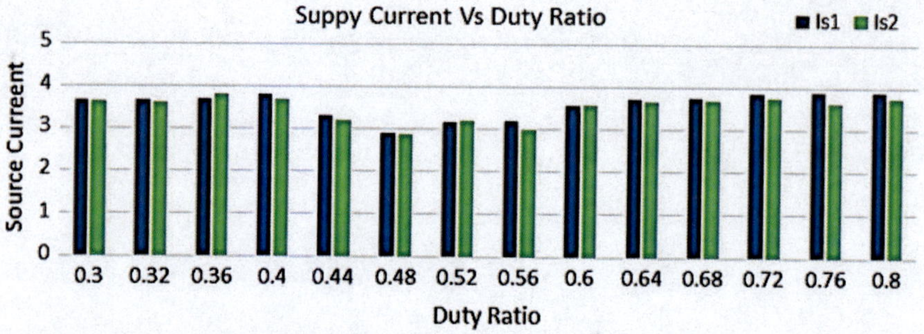

Fig. 3 Graph between supply current and duty ratio

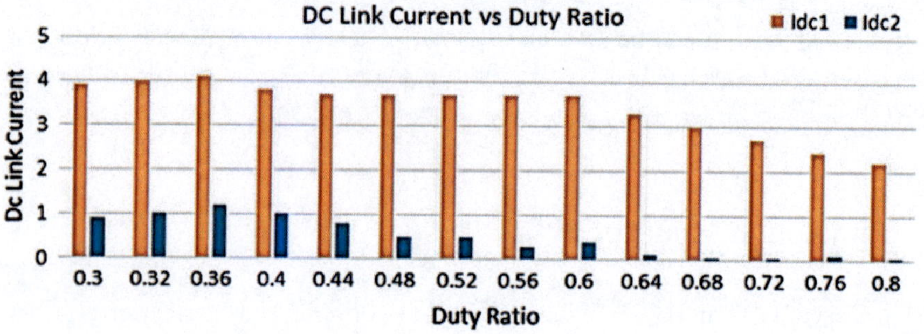

Fig. 4 Graph between DC link current and duty ratio

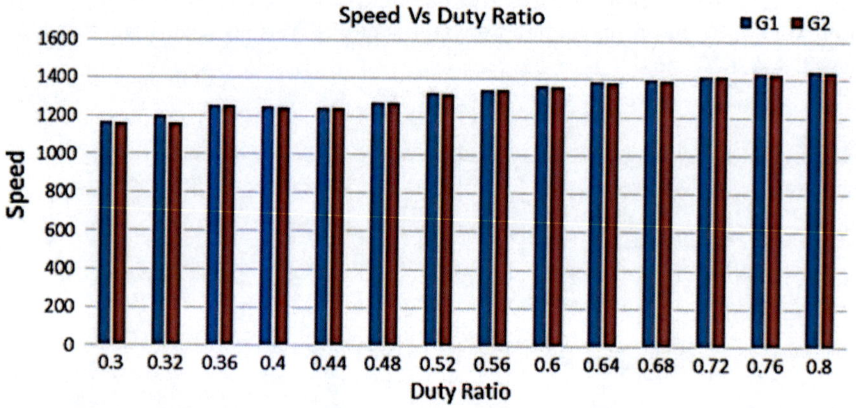

Fig. 5 Graph between speed (r.p.m) and duty ratio

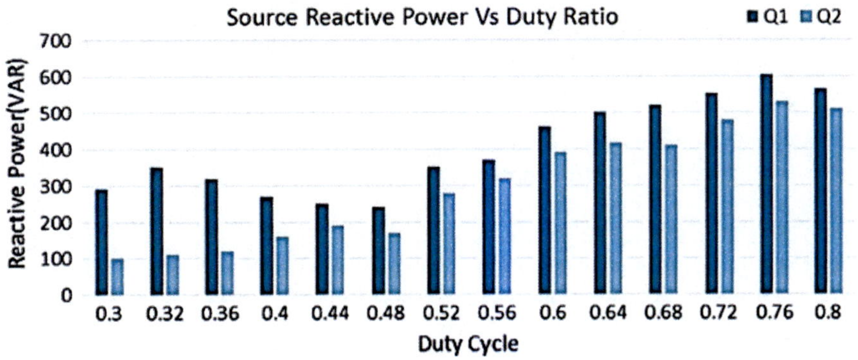

Fig. 6 Graph between reactive power (VAR) and duty ratio

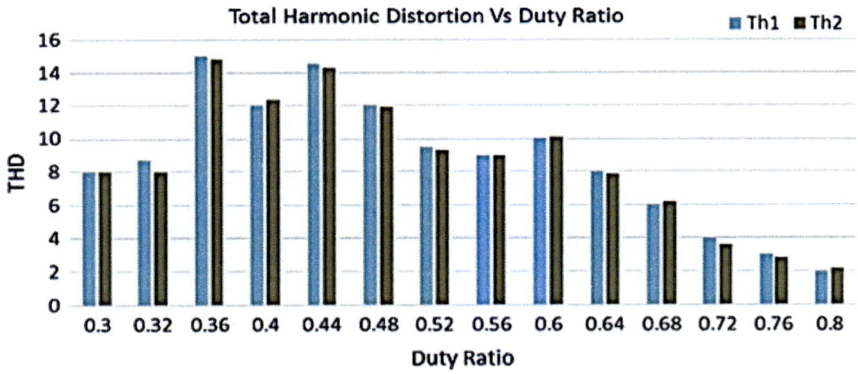

Fig. 7 Graph between THD and duty ratio

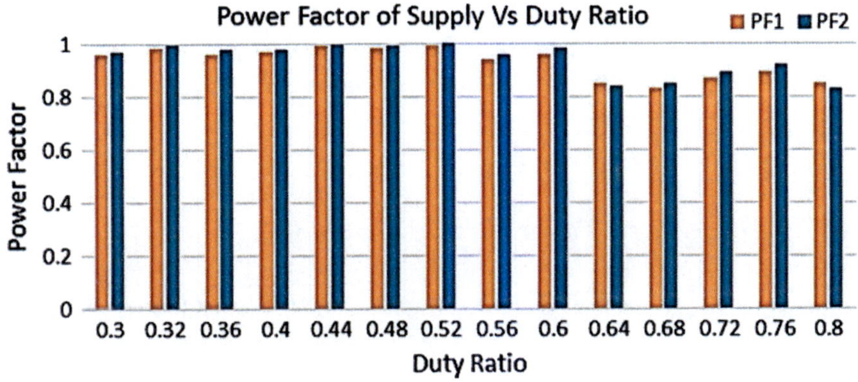

Fig. 8 Graph between power factor and duty ratio

4 Results and Discussion

4.1 Supply Current Versus Duty Ratio

On changing the duty ratio, the supply current will also vary. This variation of supply current with duty ratio is depicted in Fig. 3. Here, I_{s1} and I_{s2} are the source currents for MOSFET and IGBT, respectively. From the graph, it can be observed clearly that IGBT chopper topology draws less current as compared to MOSFET chopper topology for the same duty ratio and load torque. So, losses are less in case of IGBT chopper topology as compared to MOSFET chopper topology. Hence, IGBT chopper topology has more efficiency.

4.2 DC Link Current Versus Duty Ratio

The graph of DC versus duty ratio is depicted in Fig. 4. Here, I_{dc1} and I_{dc2} are the DC link currents for MOSFET and IGBT, respectively. From this graph, it is observed that IGBT chopper topology draws less DC link current as compared to MOSFET chopper topology. So in case of IGBT, less variation of electromagnetic torque is there that provides the smooth control.

4.3 Speed Versus Duty Ratio

The variation of WRIM speed is depicted in Fig. 5. Here, G_1 and G_2 represent the speed of rotor of WRIM for MOSFET and IGBT topologies, respectively. It can be illustrated from the graph that there is increment in speed with the increment in duty ratio. Speed control is linear as well as smooth in case of IGBT as increment or decrement in speed by changing duty ratio is less.

4.4 Source Reactive Power Versus Duty Ratio

The graph of supply reactive power versus duty ratio is depicted in Fig. 6. In this graph, Q_1 and Q_2 represent the reactive power drawn from source for MOSFET chopper topology and IGBT chopper topology, respectively. From the graph, it is illustrated that IGBT draws less reactive power from MOSFET for the same duty ratio. Less reactive power drawn means high power factor. The average value of reactive power for MOSFET chopper topology is 402.64 VAR, and average value of reactive power for IGBT chopper topology is 298.92 VAR.

4.5 Total Harmonic Distortion (THD) Versus Duty Ratio

The graph of THD versus duty is depicted in Fig. 7. In this graph, T_{h1} is total THD for MOSFET chopper topology and T_{h2} is THD for IGBT chopper topology. The average value of THD for MOSFET is 8.69%, and the value of THD for IGBT is 8.59%. From the graph, it is observed that THD is almost same for both the topologies.

4.6 Power Factor of Supply Versus Duty Ratio

The graph of power factor of supply versus duty ratio is depicted in Fig. 8. In this graph, PF_1 represents the power factor of supply for MOSFET chopper topology and PF_2 represents the power factor of supply for IGBT chopper topology. Power factor in case of IGBT is higher than MOSFET topology as observed from Fig. 8. The average power factor for MOSFET is 0.93, and average power factor for IGBT is 0.941.

5 Conclusions

This paper presented the analysis and thus the performance of speed control of WRIM by SPRS. In SPRS, Ćuk converter with MOSFET/IGBT has been used. These models have been created in Simulink/MATLAB. Effect on source current, power factor, DC link current, reactive power, THD and speed with the change in duty ratio is depicted by various graphs. Other than this, the parameters like source current, power factor, DC link current, etc., have been compared for MOSFET and IGBT chopper topologies. These results have shown power factor is 0.93 and 0.941 for MOSFET and IGBT, respectively. THD is almost same for both the topologies. Hence from results, it can be concluded that the IGBT with Ćuk converter is better option than IGBT for good performance.

References

1. Ajabi-Farshba R, Azizian MR (2014) Slip power recovery of induction machines using three-Level T-type converters. In: 5th IEEE conference on power electronics, drive systems and technologies (PEDSTC), pp 483–486
2. Yang X-H, Xi L, Yang XJ, Jiang JG (2008) Research on the application of PFC technology in cascade speed control system. In: 3rd IEEE conference on industrial electronics and applications, pp 1964–1969
3. Lavi A, Polge RJ (1966) Induction motor speed control with static inverter in the rotor. IEEE Trans Power Apparatus Syst PAS-85:76–84

4. Shepherd W, Stanway J (1969) Slip power recovery in an induction motor by the use of a thyristor inverter. IEEE Trans Ind Gen Appl IGA-5(1):74–82
5. Rahi OP, Kumar A (2016) Economic analysis for refurbishment and uprating of hydro power plants. Renew Energy 86:1197–1204
6. Shepherd W, Khalil AQ (1970) Capacitive compensation of thyristor controlled slip-energy recovery system. IEEE Proc 117(5):948–956
7. Taniguchi K, Takeda Y, Hirasa T (1987) High-performance slip-power recovery induction motor. IEEE Proc 134(4):193–198
8. Ram S, Rahi OP, Sharma V, Kumar P, Choudhary R, Vardhan G, Choudhary R (2016) Reactive power control of Induction motor drive using chopper operated slip power recovery scheme. In: 7th IEEE power India international conference (PIICON)
9. Ram S, Rahi OP, Sharma V, Kumar A (2015) Performance analysis of slip power recovery scheme employing two inverter topologies. In: Michael farady IET international summit(MFIIS), vol 2, pp 356–361
10. Ram S, Rahi OP, Sharma V, Kumar P, Vardhan G, Choudhary R (2016) Analysis of Induction motor drive using SPRS based on GTO/IGBT buck-boost chopper topologies. In: 7th IEEE India international conference on power electronics (IICPE)
11. Dubey GK (1995) Fundamentals of electrical drives, 2nd edn. Narosa Publication House, Delhi
12. Bhimbra PS (2005) Power Electronics, 3rd edn. Khanna Publishers, Delhi
13. Iqbal A et al (eds) (2020) Soft computing in condition monitoring and diagnostics of electrical and mechanical systems, vol 1096. In: Advances in intelligent systems and computing. Springer, Singapore. https://doi.org/10.1007/978-981-15-1532-3
14. Iqbal A et al (eds) (2020) Meta heuristic and evolutionary computation: algorithms and applications, vol 1096. In: Studies in computational intelligence. Springer, Singapore. https://www.springer.com/gp/book/9789811575709

Design and Analysis of Grounding Grid Performance Using ETAP 16.0.0 Software

Arun Kumar Bugaliya, Gourav Vardhan, and O. P. Rahi

Abstract This paper gives the different grounding mat design modifications and the impact on the design parameters of these modifications. Different modifications in different cases like case 1: grounding mat design without earth electrodes, case 2: grounding mat design with the electrodes at the corner of mat, case 3: grounding mat design with compression ratio without earth electrode, and case 4: grounding mat design with compression ratio and earth electrodes have been applied with the application of ETAP 16.0.0 version software. The primary framework of the grounding mat solidly depends upon, R_g, V_{touch}, V_{step}, and ground potential rise (GPR). These quantities are of great concern for safety of any grounding system. Here, our main concern is to minimize the design quantities by inserting electrodes and by changing the compression ratio to meet the standard safety regulations annunciated in various guidelines and standards. Here, in this paper, homogeneous soil model is considered for designing the grounding mat.

Keywords ETAP 16.0.0 version software · Grounding grid · Step voltage · Touch voltage · Compression ratio

1 Introduction

Grounding system of the substation should be designed in such a way that it should assure the secure and unswerving/unwavering operation of system and must also be committed to the safety measures of human beings in event of any kind of fault in the system [1]. Reduction in the ground resistance and ground potential increases the safety of power system equipment, but the safety of personnel can be ensured by emulating the potential spreading on the ground surface and by diminishing the step as well as touch voltage. With the regular increment in power system dimensions and voltage range, the frailty current has also rise with the same ratio as the dimensions

A. K. Bugaliya (✉) · G. Vardhan · O. P. Rahi
Department of Electrical Engineering, NIT Hamirpur, Hamirpur, HP, India
e-mail: arun02feb@gmail.com

A. Iqbal et al. (eds.), *Renewable Power for Sustainable Growth*, Lecture Notes in Electrical Engineering 723, https://doi.org/10.1007/978-981-33-4080-0_16

and range. Before designing the grounding grid, the complexity of different types of grid shapes and multi-layer soil model should be considered.

A method using non-uniform spacing between the grounding conductors of substation has been proposed by the Sverak [2]. The optimal arrangement of grounding grid's horizontal conductor is illustrated by Dawalibi [3–5]. By this optimal pattern, without exceeding the number of wires and potential, the leakage current is more uniformly distributed and that is the reason for obtaining low step voltage and low touch voltage [2]. The lack of side effects of a homogenous soil model has been discussed in IEEE standard [3].

In this paper, the authors have examined the effect of inserting grounding electrodes and also the effect of unequal span arrangement of horizontal conductors on touch potential and step potential for a particular substation built in Himachal Pradesh. In this paper, the four cases have been critically analyzed.

2 Methodology

Grid conductor size is one of the main concerns in grounding grid designing. Grid conductor helps to achieve the step voltage V_{step} and touch voltage V_{touch}. Grid potential rise (GPR) is considered as remote earth potential, while establishing the grounding grid for substation, some criteria are fixed that are defined in IEEE guidelines cum standard, i.e., IEEE-80-2000 [3]. To protect the electric equipment and personnel from shock, safety guidelines should be followed as given in IEEE-80-2000 [3]. Control equipment in substation and telecom equipment can be protected by effective grounding [7].

2.1 Grounding Grid Conductor Size

In this analysis, the grounding grid conductor used is stranded soft drawn copper. Grid conductor is required to be round so as to expose more cross-sectional area (up to 400 mm^2) with the surface. Coastal regions having low soil resistivity (similar to rainy season) are preferred to use tin copper conductor as grounding conductor. Sometimes electric joints may be formed under worst case scenario and to stop this, fusing copper-clad steel conductors are preferred. These worst situations are likely to occur in substation, e.g., increment in frailty current and increment in frailty period to which conductor are unprotected. For effective grounding, the grounding conductor is required to be made of soft drawn, annealed copper. Standard IEEE-80-2000 discusses the formula which is mandatory to calculate the cross-sectional area of grounding conductor. The formula, given for the calculation is as given below:

$$A = \frac{I_f}{\sqrt{\left(\frac{\text{TACP}*10^{-4}}{T_C \alpha_r \rho_r}\right)\ln\left(\frac{k_0+t_m}{k_0+t_a}\right)}} \quad (1)$$

where

T_c = Max. clearing time 1.0 s

α_r = Conductor resistivity's thermal coefficient at ref. temp, Tref = 0.0039

ρ_r = Ground conductor's resistivity at ref. temp., Tref in micro Ω-cm = 1.70

TACP = heat measurement p.u. vol. [7] joule/cm^3 °C = 3.42

t_m = Fusing temp = 1085 °C

t_a = Context temp = 55 °C.

Touch voltage V_{touch}, and it is the voltage divergence b/w a person and electrically conducting appliance which is touched by the person. It may be possible that the person is touching the equipment by one hand or by both hands and feet are in resting on ground. Same is given by the following equation:

$$V_{\text{Touch}} = (10^3 + 1.5 * C_r * \rho)\frac{0.116}{\sqrt{t_f}} \quad (2)$$

Here,

C_r = Reduction factor

C_S = 1, in absence of the protective surface layer (resistivity of both protective layer and soil is same)

C_S < 1, soil resistivity is lower than the protective layer.

t_f = Shock current duration, which is in 0.5–1.0 s range

ρ = Surface layer resistivity in Ω-m.

The fault hazard analysis is done for explaining the requirement of having touch voltage in limit [8].

2.2 *Permissible V_{step}*

Step voltage V_{step} is the voltage divergence b/w surface voltage that felt by a human being between their feet maintaining the distance of 1 m without being in touch with any kind of grounded tool.

$$V_{\text{Step}} = (10^3 + 6 * C_S * \rho)\frac{0.116}{\sqrt{t_f}} \quad (3)$$

Step and touch voltages also depend on the person's body impedance and type of shoes.

2.3 GPR

The ground potential rise is defined by [4]:

$$\text{GPR} = I_{\text{grid}} * R_{\text{grid}} \tag{4}$$

Here,

I_{grid} = Max. grounding grid current in amps, R_{grid} = Grounding resistance in Ω [8]

The following formula is used for the computation of substation R_{grid}

$$R_g = \rho * \left(\frac{1}{L_{\text{Total}}} + \frac{1}{\sqrt{20 * A}} * \left(1 + \frac{1}{1 + \sqrt{\frac{20h^2}{A}}} \right) \right) \tag{5}$$

Here,

R_g = Ground resistance in Ω, A = space covered by the gnd. grid in m^2, h = Grid's buried depth in m, ρ = Ground resistivity in Ω-m

L_{Total} = Total conductor buried in m

3 Grounding Grid Designing Process

Whenever, a specific voltage is applied to any material, due to its resistivity, a definite amount of current passes through it. There are many methods for the calculation of resistivity, one of them is four-electrode Wenner array method [9]. To calculate average soil resistivity, following equation is used:

$$\rho = 2\pi a R \tag{6}$$

Here,

ρ = Avg. soil resistivity, a = gap b/w electrodes, R = Testing inst. reading in Ω

Here, we have crushed rock surface having resistivity of 3300 Ω-m up to 0.1 m depth, than we are considering single layer soil model of resistivity 220 Ω-m. Conductor is at depth of 0.6 m. Grounding conductor is of MS FLAT and grounding electrode is of MS rod. Figure 1 is showing the soil model for the entire work carried out. Later the difference in different cases is of only number of grounding electrodes and optimal compression ratio [10].

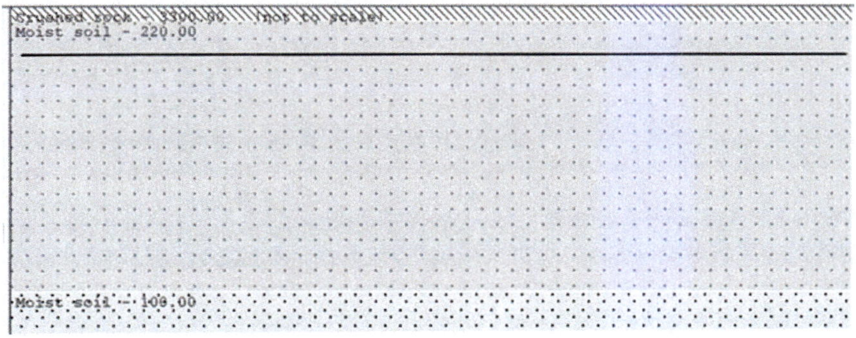

Fig. 1 Soil model

3.1 *Case1: Grounding Grid Design Without Earth Electrodes*

Here, first case is the basic conductor arrangement with zero grounding electrodes. The calculated step voltage is 365.3 V and calculated touch voltage is 634.7 V. As per IEEE rules both V_{Step} and V_{touch} should be less than the permissible limits of V_{Step} and, V_{touch}.

The ground potential rise is defined by (Figs. 2 and 3):

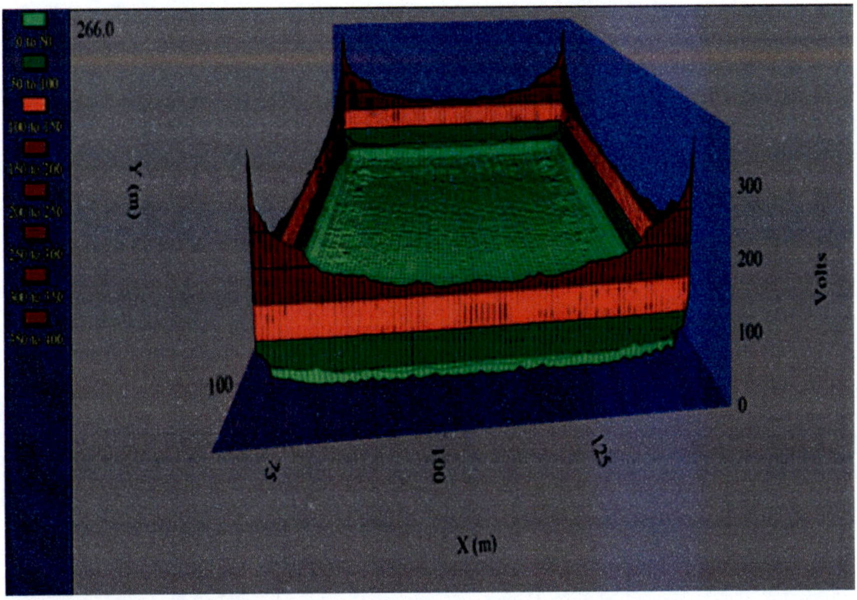

Fig. 2 Step voltage profile

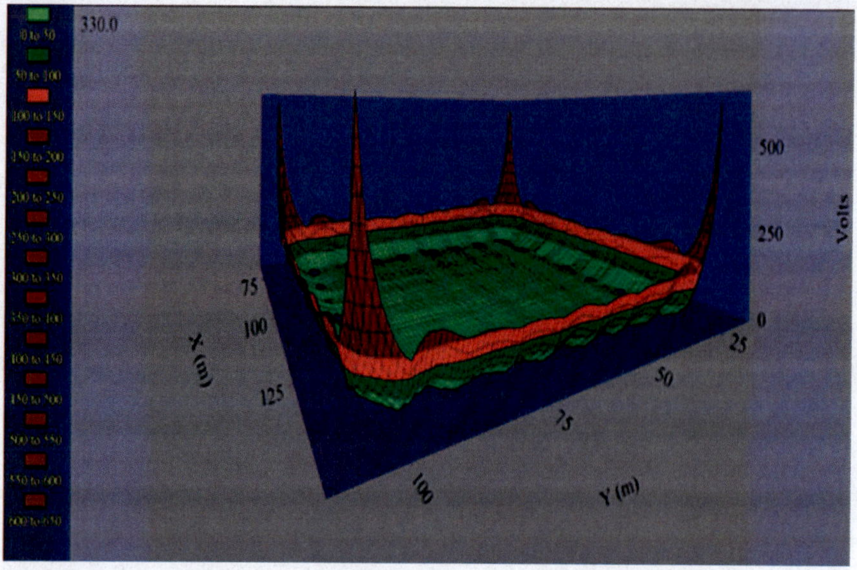

Fig. 3 Touch voltage profile

3.2 *Case2: Grounding Grid Design with Earth Electrodes*

Here, grounding electrodes are inserted at corners than again V_{Step} and V_{touch} calculated. The permissible V_{Step} is 2407.2 volts and calculated V_{Step} is 331.9 volts (Figs. 4 and 5).

The tolerable V_{touch} is 724.8 volts and calculated V_{touch} is 421.7 volts, which is less than that of case 1's V_{touch}.

4 Grounding Grid Designing Process

Optimal designing is the arrangement of conductors in substation's grounding grid in such a way that it equalizes potential of grounding surface and the leakage current distribution; this can be done by ensuring all grounding grid conductors fully exploit and thus reducing the V_{step} and V_{touch}. Diagram. 6 shows grounding grid alignment having irregular spacing between conductors. By this arrangement, potential gradient decreases and it is safe and economic way of designing. The problem arises in achieving the uneven spacing according to the complexity in figure where conductor span reduces from middle to edges of the grid. The *j*th conductor span from the middle conductor is:

$$D_j = D_{\max} C^j, \ (j = 0 \text{ to } M) \tag{7}$$

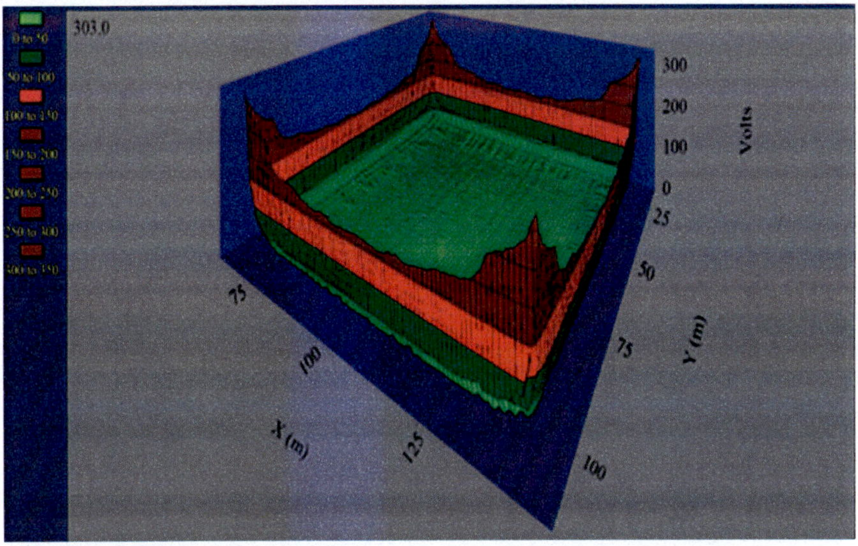

Fig. 4 Step voltage profile

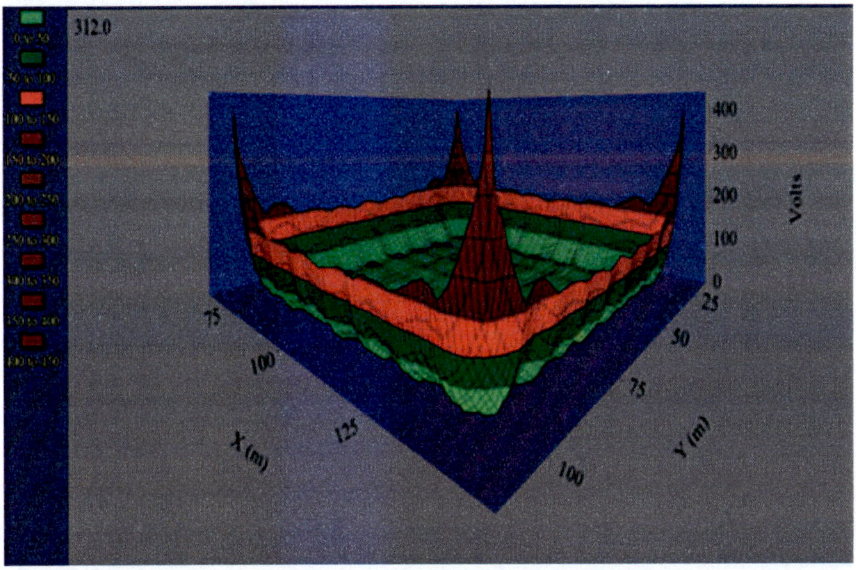

Fig. 5 Touch voltage profile

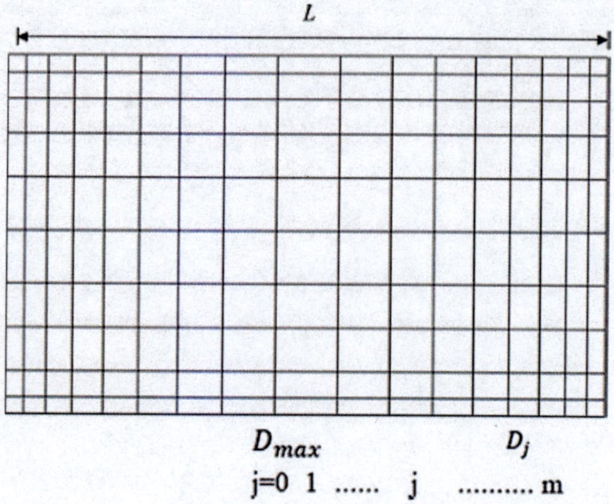

Fig. 6 Grounding grid with OCR

Here, C defining or is the notation for the compression ratio, that is unchanging factor, the value of C is between 'zero' and 'one', if C having value 'one', that means conductor span is uniform, conductors having equal spacing between them. If grid has odd conductors, at that time $M = \frac{N-3}{2}$ and if even, at that time $M = \frac{2N-4}{4}$ (Fig. 6).

If

L is grounding systems side length,

N conductors are placed on a direction of the grounding system, then the span for the middle conductor is given as:

$$D_{\max} = \frac{L(1-C)}{1+C-2C^{\left(\frac{N-2}{2}\right)}}(N \to \text{even}) \quad (8)$$

$$D_{\max} = \frac{L(1-C)}{2-2C^{(N-1)/2}}(N \to \text{odd}) \quad (9)$$

The OCR is taken as 0.75 as here the step voltage and touch voltage reaches minimum. At this OCR, next two cases have been analyzed.

4.1 *Case3: Grid with Optimal Compression Ratio and Without Earth Electrodes*

Here, in this third case, the basic grounding conductor arrangement having compression ratio of 0.75 with zero grounding electrodes has been considered. The calculated V_{step} is 301.4 V and calculated V_{touch} is 397.2 V. As per IEEE guide lines, both the V_{step} and V_{touch} should be less than the permissible limits of V_{step} and V_{touch} (Figs. 7 and 8).

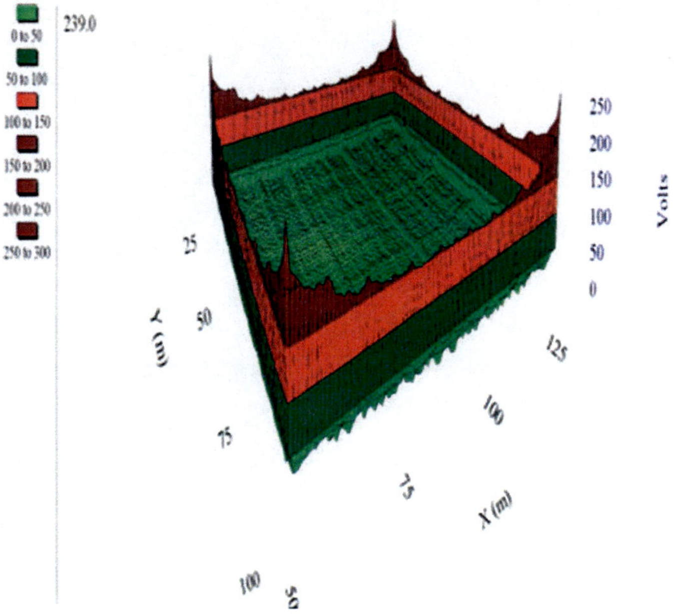

Fig. 7 Step voltage profile

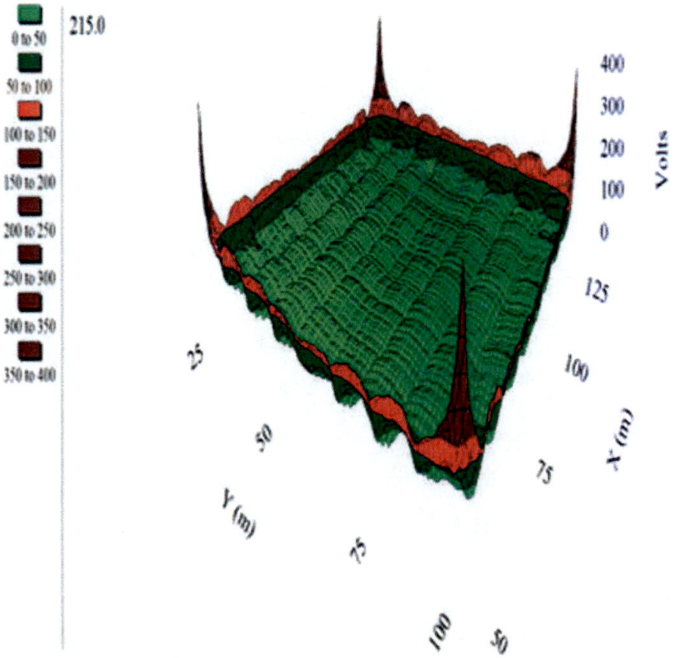

Fig. 8 Touch voltage profile

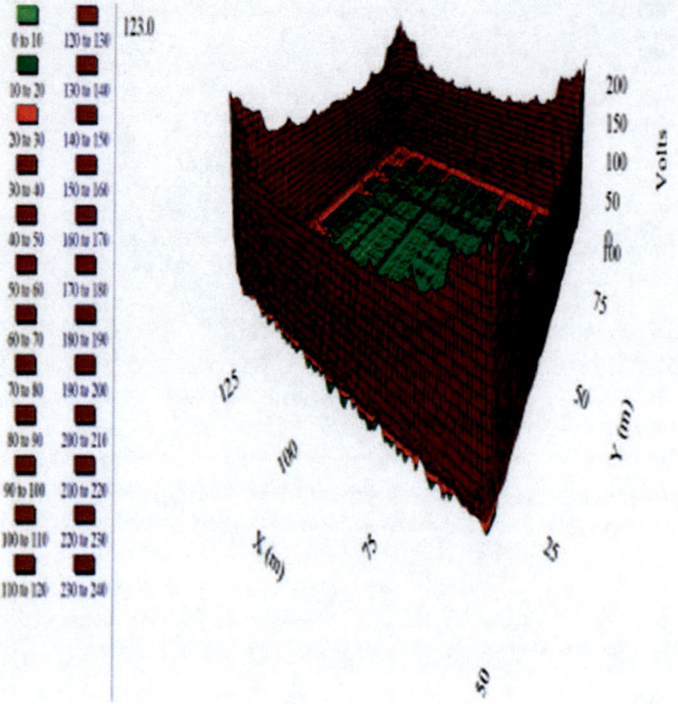

Fig. 9 Step voltage profile

4.2 *Case4: Grid with Optimal Compression Ratio and with Earth Electrodes*

Here, in this last but not the least fourth case, basic grounding conductor arrangement having compression ratio of 0.75 with grounding electrodes has been taken up for analysis. The calculated step voltage is 246.6 V and the touch voltage is 256 V. As per IEEE standards both the V_{step} and V_{touch} must be less than the tolerable limits of V_{step} and V_{touch} (Figs. 9 and 10; Table 1).

5 Conclusion

Here, authors have investigated an important grounding technique for 132 kV substation, for increased ground potential resistivity of surface which was quite high in this substation. The impacts of C_S on grid parameters have been examined with a single layer soil model for two cases including vertical conductors and excluding vertical conductors. The analysis was moved forward by taking optimal C_S from reference paper which ensure the minimum V_{Step}, V_{touch}, and R_g.

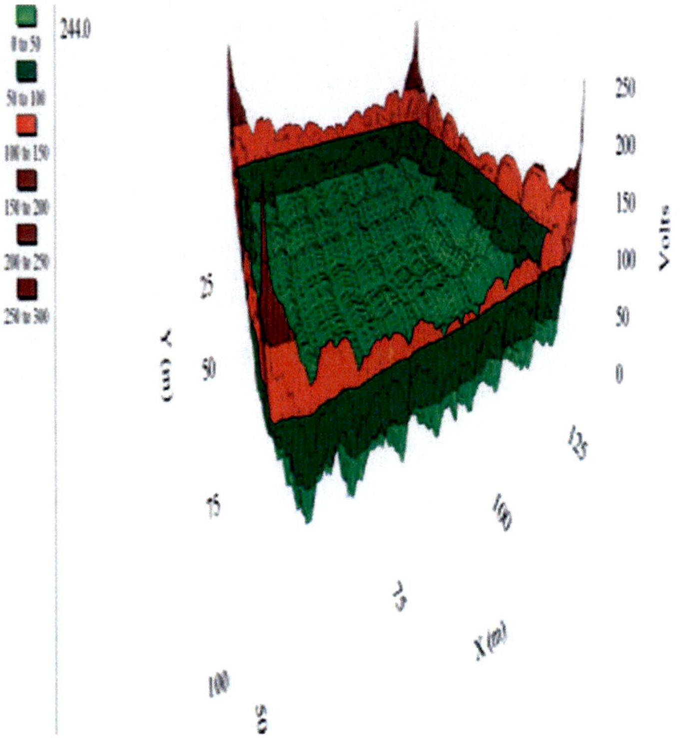

Fig. 10 Touch voltage profile

Table 1 Comparison table of the results of all the four cases

S. No.	Tolerable V_{step}	V_{step}	Tolerable V_{touch}	V_{touch}
1. Grid without earth electrodes	2407.2 v	365.3 v	724.8 v	634.7 v
2. Grid with earth electrodes	2407.2 v	331.9 v	724.8 v	421.7 v
3. Grid with optimal compression ratio and without earth electrodes	2407.2 v	301.4v	724.8 v	397.2 v
4. Grid with optimal compression ratio and with earth electrodes	2407.2 v	246.6 v	724.8 v	256 v

The results of this research work have shown the grounding grid mesh design for a 132/33 kV substation with maximum grid current of 31.7 kA. Finite element method was used to determine the tolerable limits of different parameters like step potential, touch potential, etc., on ETAP 16.0.0 software.

References

1. Patel V, Vaghamshi A, Sonagra K (2014) Substation grounding grid design case study using matlab. IJSRD 2(03)
2. Sverak J (1976) Optimized grounding grid design using variable spacing techniques. IEEE Trans Power App Syst 95(1):362–374
3. IEEE guide for safety in AC substation grounding. IEEE std.80-2000 (2000)
4. Huang L, Yan H (1995) Study of uneven spacing of grounding conductor. IEEE Trans Power Del 10(2):716–722
5. Dawalibi F, Mukhedkar D (1975) Optimal design of substation grounding in a 2 layer soil model, part I: analytical study. IEEE Trans Power App Syst PAS-94(2):252–261
6. Bhardwaj N, Sharma M, Rahi OP (2016) Seasonal influence on the substation grounding grid performance and its optimal design to neutralize the influence. In: IEEE conference, ICPEICES
7. Company SE, Saudi Electric Company Grounding. In: Riyadh (ed), Saudi Arabia
8. Earthing of power installations exceeding 1 kV A.C. In: EN 50522:2010, ed: British Standard (2012)
9. IEEE guide for measuring earth resistivity, earth Imp. and surface voltage of a ground system. IEEE 81-1983
10. Unde M, Kushare B, Effect of compression ratio on in layered soil. IJSRD, EISSN: 2319-1163, PISSN: 2321-7308

Grid-Interactive Microgrid Charging Infrastructural Network for Electric Vehicles with International Standards

Zeeshan Ahmad Arfeen, M. Saad Bin Arif, Md Pauzi Abdullah, Fatimah Khairiah, Esam Abubaker Ali Abubaker, and Heng Wei Wei

Abstract The electrification of road transport networks is an encouraging attempt to green highways and to lessen the impacts of global climatic scenarios. This research enlightens the current status, up-to-date employment, in compliance with numerous power charging topologies, and examines Electric vehicles (EV) influence and outlooks to society. Moreover, in this paper, the comparison of on and off board mode of recharging methods with one-direction and bi-directional energy flux are exclusively addressed. The work is well supported with Tables and Figures to outskirts the details. The international standards for EV charging and infrastructure for refuelling stations are also discussed and highlighted. In a sense, the paper gives fast insight and understanding of the current scenario of the charging pattern of EV to the readers.

Keywords Battery management system · DC bus · Power converter · Fast-charging station · Vehicle to grid

1 Introduction

Lowering power demand in road transport is the utmost challenging tasks to meet carbon dioxide (CO_2) reduction targets and Green House Gases emission (GHG). Because of the sector's reliance on conventional fuel generation sources and the

Z. A. Arfeen · M. S. B. Arif (✉) · M. P. Abdullah · F. Khairiah · H. W. Wei
School of Electrical Engineering, UTM, Skudai 81310, Johor, Malaysia
e-mail: saadbinarif@gmail.com

Z. A. Arfeen
e-mail: zaarfeen2@graduate.utm.my

M. S. B. Arif
Department of Electrical Engineering, Aligarh Muslim University, Aligarh, India

E. A. A. Abubaker
Department of Aeronautical, Automotive and Ocean Engineering, UTM, Skudai 81310, Johor, Malaysia

A. Iqbal et al. (eds.), *Renewable Power for Sustainable Growth*, Lecture Notes in Electrical Engineering 723, https://doi.org/10.1007/978-981-33-4080-0_17

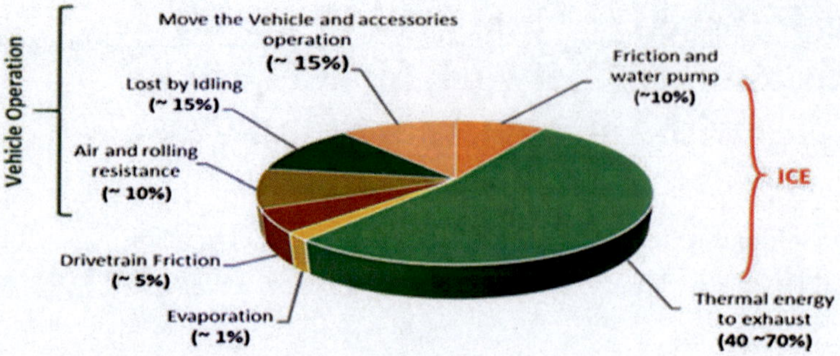

Fig. 1 Comparison of pattern ICEV versus EV [17]

colossal adverse effects for ecological change, air pollution and other social comportments, Electric vehicles (EVs) is a cost-effective alternative over conventional vehicles (CVs), which is eco-friendly, sustainable in the present phase [1]. EV exposure brings laurels to energy systems, like uplifting generation capacity during peak load intervals, load balancing, likely investment deferment; voltage-frequency regulation, spinning standby facility and reduce dependence on crude oil and green gas emissions, bringing economic and environmental benefits [2]. Apart from supporting the grid with the active power, EVs have the potential to improve the power quality by compensating the reactive power. The emissions over the complete running of EV, which include both the power generators and fuels used to energize vehicle-tank to car wheels and the direct tailpipe exhaustion generally called well to wheels is far less than ICEV [3]. The typical energy flow of a conventional internal combustion electric vehicle is shown in Fig. 1.

The electric vehicle refueling decks market is projected to exceed USD 27.7 billion by year 2027 from an projected USD 2.5 billion in twin year. DC Charging Station will bring in healthy gains adding significant momentum to global growth To date, there is about 400 electrified vehicle charging infrastructure fuel station in Malaysia. BMW, Schneider companies, aim to leverage on strategies by Green Technology Malaysia corporation. In this context about three thousands fast recharging stations for electric driven vehicles (EVs) set up in 2019 countrywide to cater to the growing concern in electric driven vehicles. The government of Malaysia also gives the privilege to free charge at the Charge EV charging stations nationwide currently. Solar-powered electric vehicle (EV) charging stations will be installed along the North-South Expressway (NSE) by early 2020. According to a leading newspaper report "The STAR" the cost of solar-EVCS in Malaysian Ringgit is approximately RM450000.

The hierarchy of the review is as under: Sect. 2 includes the recharging pattern, control strategy of zero-emission vehicles, International charging standards and infrastructure of refuelling stations. Section 3 is dedicated to power chargers along with its topologies off-board and onboard chargers. The ac and dc charging power level also placed in this section, and Sect. 4 is categorized with power streams of unidirectional and bi-directional energy flow. The concluding remarks are in the last Sect. 5.

2 Charging Pattern

Electric vehicles are generally based on the type of charging, either random or coordinated. In random charging the EVs start charging as they reside at house or car parking slots, heading to possible chaotic impacts on the national grid. Within this contemporary strategy, maximum energy losses, system congestion, overloading of the grid, voltage drops, exceed the thermal limits of the substation transformer, which result in accelerating the transformer's ageing [4]. On the other hand, coordinated charging builds on a modest union of procedures, such as time-delayed charging (pre-programmed charging) or intelligent charging, i.e. an optimized adaptive charging under the control of the local operator, based on load, cost or regulatory body [5], and other examples are given in [6, 7].

2.1 Barriers and Behaviors

Despite the environmental and economic benefits, EVs have a negative influence on the power grid if the way for intelligent charging is not developed. Random charging functionalities could place various types of technological problems. Suboptimal production dispatches, power demand load profile rises, system losses, transformer overloading, harmonics contents in current and voltage, large voltage variations, decreased network efficiency, and substantially raise the likelihood of power breakdown viz of system overloading arises. Nonetheless, due to the partial unpredictability of both consumers and power production from renewable sources, there are possible power imbalances between the generation output and the load. A robust and efficient power flow charging algorithm-based controller should be designed to exploit the practice of RES and cater to the impacts and burden of vehicle charging on the ac utility grid.

2.2 International Standards for Electric Vehicles

A few well-known standards are developed for EV recharging levels, for example SAE J1772, charge de moved CHAdeMO protocol and International Electrotechnical Commission IEC 61851-1 [8]. NIST—National Institute of Standards and Technology of the U.S Department of Commerce has a task to interlink and develop a framework that incorporates protocols and standardizations for data control management activity and to monitor the interoperability of intelligent grid appliances and networks [9]. Different standard .organization enlisted the U.S. NEC—National Electric Code: 690, SAE—Society of Automotive Engineers and IEEE-Engineers 1000~3~2, upkeep the standard to levelized the amount of harmonics and pure DC insertion. IWC- Infrastructure working council, a company, is working to build multiple codes and standards concerning the grid code interface. The standards defined by the SAE for EVs charging are. SAE J1772, J2293, J2836, J2847, J 2894 [10].

In the digital age, the dependence on fossil fuels and major concerns about the green gas environment, alternative fuel is the dire need for zero-emission vehicles. Considering the energy savings and long- mileage travel journey, the present status of the battery is immature in technology and becoming the bottleneck for its real implementation of EV. As the new battery coming in the market with features of having least cost, less weight, high power density, thus PHEV is becoming one of the promising answers for the novel power vehicle in the automation industry. As the power of a plug-in hybrid vehicles likely origin from the battery stacks, an energy supervisory strategy is desired to resolve the multi-power coupling issues [11].

3 EV Chargers

EV fast chargers play a vital role in the mass deployment and growth of the EV charging equipment. The lifetime, state of health (SoH) and the charging-discharging span have a closed affiliation with the features of battery quick chargers. An active battery power should possess momentous features like compact mass and volume, extremely reliable and effectual, elevated power density, and less price. They are generally classified as on and off-board power chargers (Fig. 2).

3.1 On/off-Board Energy Chargers

In a car-mounted battery charger, the customers have the freedom to avail of the provision of charging their vehicles whenever there is power source availability. The significant barriers regarded to an onboard power charger are the constraint of Level-1 power only. Moreover, in car-mounted onboard chargers the mass, area and cost restraints are the major causes to limit its power. In contrast, the off-board rapid-refuelling areas are gaining momentum in this regard. Despite the elevated price of infrastructure DC fueling decks, it offers tremendous promising characteristics such as lessen the mass of the zero emission vehicle; charging at raised energy levels; competence of rapid charging; dynamic recuperate networks for refuelling the EVs. However, active correspondence between utility companies and aggregators (station owners) make the fast charging station cost incentives. Table 1 gives the fast insight into the two modes of charging.

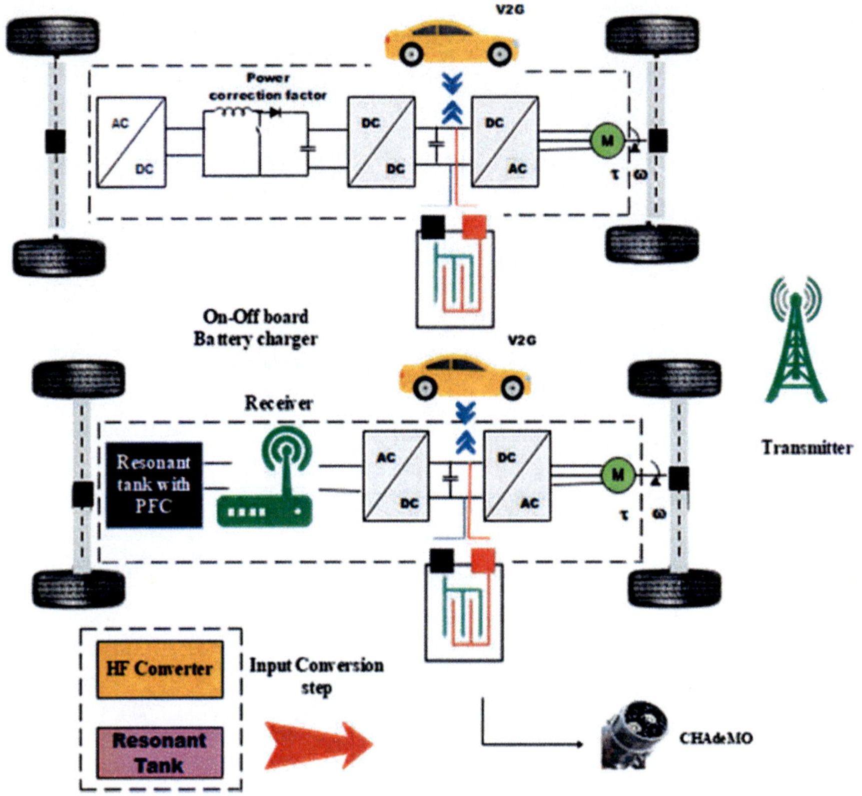

Fig. 2 local charging on board-off board with wireless charging of EVs [18]

3.2 Charging Power Levels

EV power chargers can be classified on the base of power rating. A variety of charging category as well as their particular range of energy are as follows—(i) charging Level-1: 120 V ac, 1-ϕ, 16 A and 2-KW power is utilized to charge car-mounted charger. Charging Level: 208–240 V ac, 1-ϕ, maximum of 12–80 A and (3–19.2) kW of power is confined to charge EV. Level-3 AC charging: In this specific type 400 V, 3-ϕ ac power is transformed into DC and after this power is directly supplied to charge EVs. Level-3 dc charging: power at 300–600 V DC is straight away supplied to accumulate EVs. On average, a maximum of 400A and 240 KW of power can be given by this type of fast charger [12].

Table 1 Comparison of onboard and off-board chargers

Types	Merits	Demerits
On-board chargers	Deprive of battery heating due to least power absorption	Slowly charging
	It doesn't involve more interconnection issues. Standard signal J1172 is applied for the action	Low power levels in KW
	Recharge everyplace within an electrical outlet, little infrastructure	Consideration of mass and size due to manufacturing of EV design; added components for charging
	Needed things by onboard, BMS are supplied voltage	Protocols concerning communication are not well-standards
	Max rated current and phase outline of the charging facility	Voltage, rated current capacity and phase configuration of the refueling area are necessary by onboard BMS
Off-board chargers	Superior rate of energy in (KW)	Battery heating arises, battery fading, battery recycle constraint
	Fewer concerns over size and weight	Independency is gone. Liable to refuel only at the station
	Better BMS systems	Risk of charging adoption
	Charge at elevated power levels	Troubleshooting—battery banks is not permissible by remote BMS
	Fringe benefits to both parties. (utility stakeholders and charging site owners)	Batteries to receive charge is limited
	BMS is matured and automated	Charging station personnel are the officials to check charging profiles
	Fast charging	Sophisticated infrastructure cost raised
		The inclusion of a battery management system causes more complexity and cost

3.3 Power Level-1, 2 and 3

As defined previously, the time-consuming method to achieve the charging dilemma of EV is by level-1 charging scenario. In this particular type of assembly might employ a standard connector J1722 for the ac socket of EV. In the second level charging scenario power is provided through dedicated facilities at public (station) and private places (shopping malls, hospitals, universities, etc.). The bright feasibility of level-2 has lesser power electronics interface converters, while the car mount facility is offered at this range of power level charging. Besides, adequate equipment is necessary to install at the public and household places for the level-2 charging. The level-3 charging process is applied to commercially and offers the quickest charging interval within half an hour. The working operation for level-III charging is equivalent to 480-V or needs a superior three-ϕ circuitry. A newly standard DC rapid charging popularly known as charge de move "CHADeMO" Consortium originated from Japanese state protocol is receiving recognition rapidly around the globe [13]. This defined standard is intended to raise the initiations of ground transportation and to lessen the anxiety concerning the distance among the users of EVs. CHADeMO can refill the battery up to 80% SoC around 30 min via optimal DC power cable, i.e., 50-kW (nearly 500 Volts, 100 A) (Table 2).

3.4 DC Recharging Systems

DC power networks demand dedicated electric wiring and equipment installations and can be fixed at public garages or refuelling areas. They have constant electric power than the AC networks and can charge EVs rapidly. All DC charging systems have permanently connected electric vehicle supply equipment (EVSE) that incorporates the dc charger. Connectors that are generally used in this category are SAE J 1772 Combo, CHAdeMO and IEC 62196 Mennekes Combo. Their taxonomy is according to the power levels provide in Table 3.

4 Power Streams

They are typically classified as unidirectional energy flow and bi-directional energy flow. These are discussed in detail in the following sections.

4.1 Unidirectional Energy Flow

In the case of electricity geared vehicles, usually, the rapid charging procedure of battery banks is done outside the car assembly via a fast power charger. Usually, two types of chargers are readily available in the market. Single directional energy flow has its merits which are pointed in the given Table 3. Some major contributions are; reduced hardware, simplified interconnection circuitry, and battery depletion of charges are less [14]. The main prospect to accept these single way chargers by grid utilities is the simple control architecture that managed the packed saturated feeders while the inclusion of mass penetration of vehicles. On another end, many issues such as performance, reliability, and safety are not fully covered. The cost of a one-way charger is less as compared to two-way chargers. Only an off-board outlet is required for fast charging the vehicles. Continuous charging and discharging cycling is the absence in single flow chargers so battery life prolonged in this case.

Table 2 Characteristics and infrastructure features of electric vehicles [19]

Electric vehicles	The capacity of Li batteries (Ah)	Mileage (km)	Topology of connectors	Charging power-I		Charging power-II		Charging power-III	
				120 V AC		240 V AC		DC Quick charging	
				Power demand (KW)	Expected time (h)	Power demand (KW)	Expected time for Charging, (h)	Power demand (KW)	Expected period (h)
Roadster tesla	54	340	J1772-SAE	2	>20	9.7–19.2	1.5–2	100	0.37
Toyota RAV-4	42	160	J1772-SAE	1.9	17	9.6	3.4	50	0.64
Mitsubishi i-MiEV	16	145	SAE- J1772 JARI/TEPCO	1.5	10	3.6	=4	50	0.26
Nissan leaf	24	160	J1772-SAE	1.8	11–15	6.6	2.5–3	50	0.3–0.4
Chevrolet volt	17	64	J1772-SAE	1.4	7–10	3.7	near 4	n/a	n/a
Toyota prius	5	25	SAE -J1772, TEPCO/JARI	1.5–1.8	3	3.2	near1.6	n/a	n/a

Table 3 DC Power levels to charge zero-emission vehicles

DC power level	Current rating (A)	Voltage rating (V)	Power rating (KW)
Stage-I	80	450	36
Stage-II	200	450	90
Stage-III	400	600	240

4.2 Bi-Directional Energy Flow

In this dual power flows several salient features are the stabilization of power, V → G technology, sufficient and controlled translation of power. Green vehicles with two-way chargers offer distinctive merit as a source of an engineering advancement as V → G [15]. When a fleet of electric vehicles are in office parking or rest car places, it can deliver power to intra-grid at its supreme level request of load balancing and thus improve the efficiency; this concept is first initiated the concept by Willett Kempton in his paper [16]. The pioneer believe that V → G concept its reality to conceivable stockpile and produce power from the fleets of car to poise stochastic intermittent renewable resources, enforce grid energy stability, clean and green environment and maximize profit, especially in peak hours. V → G, vehicle to vehicle V → V, and vehicle to home (V → H) are the promising solution of electricity-powered vehicles related to two-way flow. In the case of V2H, the extent of power accumulated in the battery pack will consider as a standby generation for home utilities with additional renewable resources. In the case of a vehicle to vehicle concept, local fleets of EVs are formed to mutual intake and outflow charge on the mutual strategy of the owner's will. Nonetheless, customer willingness is the prime blockade to indulge all these two-way energy concepts. The two-way energy flow implementing V2G features, accelerates the flexibility, sustainability, reliability and dynamics for the power system to control the power of EVs. Table 4 depicts a comprehensive assessment between one and two ways of charging energy flow hierarchy.

5 Conclusion

This manuscript examines the current values, the fleet status strategy by availing the benefits of the bidirectional functionality of fast battery chargers. The mass deployment and upgrading concerns in the real implementation of vehicle infrastructure are described. Energy level 1, 2, and 3 to charge the battery are investigated with IS-International Standards. Moreover, the paper gives a deep insight into the use of the car onboard and off-board charging systems with one and two-way energy direction. The progressive expansion, employment and adequate induction of green vehicles in recent days are proportionally hocked with the initiations of the latest

Table 4 Infrastructural comparison of unidirectional and bi-directional charging [20]

Salient features	Structure for single-directional mode-charging (G2V)	Structure for dual-directional mode-charging (V → G)
Present standing	Access	Access denied to all charging places
Power levels	Applicable to 1,2, 3 levels	Applicable to level 2 and 3
The direction of energy flow	Grid to vehicle only	Charge and discharge of the vehicle battery can be done
Switches topology	Diode bridge, one-way converters	Lower power MOSFET, normal power IGBT, excess power GTO
Battery influence	Battery life extends due to only charging phenomenon	Continuous charging/discharging cycling cause battery life shortens
Price matters	Low cost	High cost
Isolation and protection	Isolated as well as non-isolated components	• Increased protective events • Anti-islanding safety Iso
Control demands	The simple, firm grip of current charging; Time of use (TOU) power; dynamic billing applied to cope with ordinary control	• Complex control application • Additional drive controller • Massive precaution measures are employed
The necessity of hardware for the distribution system	A modest network for communication; Refreshing data usually not required; Utility equipment can be achieved with a massive deployment of zero-emission vehicles	Bidirectional battery charging communication system; The initial investment is required; The condition of essential data information to gain utility tasks
Issues and assignments	Power connection is mandatory; Reduced facilities	A duplex communication system is compulsory and standard power cable is needed. Advanced metering infrastructure is required. Modest interchange of data is necessary. Elevated cost. Greater stress and strain on equipment. Greater power losses. convoluted infrastructure, battery fading issue arise

(continued)

Table 4 (continued)

Salient features	Structure for single-directional mode-charging (G2V)	Structure for dual-directional mode-charging (V → G)
Merits	• Simple supervisory control and user-friendly • Doesn't imply interlinking matters • Dynamic modifications facility; • Q-power of charge's rate is readily available, even in case of reversal • The current and phase-angle method helps Q-the power to absorb or transmit without discharging of EV • Realize control of v and f • No issue of battery degradation	• Reduced utility losses • Reduced utility overloading • Stability of voltage • Improved load pattern • Reinforcement for real and virtual power • Ancillary facilities, i.e., compensation of v and f • Max power modelling • Filter harmonics current • Higher profit • more income generated • Spinning stocks • Max load shaving • access to valley filling criterion • Beneficial for utility and EV owner • Load issue enhancement • Harmonizing of generation-load demand • More absorption for RES • more customer favourite

International standards and recharging grid codes. Appropriate infrastructure, intelligent and adaptive power chargers, progression in battery chemistry and user-friendly software-based fueling stations are the working area by the researchers.

References

1. Boulanger AG, Chu AC, Maxx S, Waltz DL (2011) Vehicle electrification: Status and issues. Proc IEEE 99(6):1116–1138
2. Mwasilu F, Justo JJ, Kim E-K, Do TD, Jung J-W (2014) Electric vehicles and smart grid interaction: a review on vehicle to grid and renewable energy sources integration. Renew Sustain Energy Rev 34:501–516
3. Elgowainy A, Han J, Poch L, Wang M, Vyas A, Mahalik M, Rousseau A (2010) Well-to-wheels analysis of energy use and greenhouse gas emissions of plug-in hybrid electric vehicles. Argonne National Lab. (ANL), Argonne, IL (United States)
4. Rajakaruna S, Shahnia F, Ghosh A (2016) Plug in electric vehicles in smart grids. Springer
5. Mesarić P, Krajcar S (2015) Home demand side management integrated with electric vehicles and renewable energy sources. Energy Build 108:1–9
6. Iqbal A et al (eds) (2020) Soft computing in condition monitoring and diagnostics of electrical and mechanical systems, vol 1096. In: Advances in intelligent systems and computing. Springer, Singapore. https://doi.org/10.1007/978-981-15-1532-3

7. Iqbal A et al (eds) (2020) Meta heuristic and evolutionary computation: algorithms and applications, vol 1096. In: Studies in computational intelligence. Springer, Singapore. https://www.springer.com/gp/book/9789811575709
8. Yong JY, Ramachandaramurthy VK, Tan KM, Mithulananthan N (2015) A review on the state-of-the-art technologies of electric vehicle, its impacts and prospects. Renew Sustain Energy Rev 49:365–385
9. Hefner A (2010) Coordination and acceleration of standards for smart grid-distributed generation
10. Shaukat N, Khan B, Ali S, Mehmood C, Khan J, Farid U, Majid M, Anwar S, Jawad M, Ullah Z (2018) A survey on electric vehicle transportation within smart grid system. Renew Sustain Energy Rev 81:1329–1349
11. Peng J, Fan H, He H, Pan D (2015) A rule-based energy management strategy for a plug-in hybrid school bus based on a controller area network bus. Energies 8(6):5122–5142
12. Ahmad F, Alam MS, Asaad M (2017) Developments in xEVs charging infrastructure and energy management system for smart microgrids including xEVs. Sustain Cities Soc 35:552–564
13. Falvo MC, Sbordone D, Bayram IS, Devetsikiotis M (2014) EV charging stations and modes: international standards. In: 2014 International symposium on power electronics, electrical drives, automation and motion, pp 1134–1139. IEEE
14. Yilmaz M, Krein P (2012) Review of charging power levels and infrastructure for plug-in electric and hybrid vehicles and commentary on unidirectional charging. In: IEEE international electrical vehicle conference
15. Poullikkas A (2015) Sustainable options for electric vehicle technologies. Renew Sustain Energy Rev 41:1277–1287
16. Kempton W, Letendre SE (1997) Electric vehicles as a new power source for electric utilities. Transp Res Part D: Transp Environ 2(3):157–175
17. Tie SF, Tan CW (2013) A review of energy sources and energy management system in electric vehicles. Renew Sustain Energy Rev 20:82–102
18. Arfeen ZA, Khairuddin AB, Munir A, Azam MK, Faisal M, Arif MSB En route of electric vehicles with the vehicle to grid technique in distribution networks: Status and technological review. Energy Storage n/a (n/a):e115. https://doi.org/10.1002/est2.115
19. Vadi S, Bayindir R, Colak AM, Hossain E (2019) A Review on communication standards and charging topologies of V2G and V2H operation strategies. Energies 12(19):3748
20. Un-Noor F, Padmanaban S, Mihet-Popa L, Mollah MN, Hossain E (2017) A comprehensive study of key electric vehicle (EV) components, technologies, challenges, impacts, and future direction of development. Energies 10(8):1217

Design Kinematics and Control for a Differential Drive Mobile Robot

Shahida Khatoon, Md Istiyaque, Sajad Ahmad Wani, and Mohammad Shahid

Abstract The problem of differential drive mobile robot (DDMR) is to work in an organized way, in this research the main aims to design and control a DDMR which is on two different wheels. The construction and control for a DDMR is simple and easy. In this paper it has been shown that the classical PID-Controller of DC motor used for controlling the motion of the mobile robot platform. The work focuses on mechanics and map-reading or finding way of the mobile robot platform is a restriction free indoor environment. Mechanically the structure of the robot is fully prepared and drawing is presented in detail. Mobile robot motion is point to point movement and the mobile robot position error is reduced when it moves from a point to another points. The distance and orientation error of the robot is eliminated by translating and rotating the robot simultaneously. The control law is governed by kinematical model that provides the updated reference velocity to the PID-Controller connected to the DC motor. The tuning of the PID-Controller could do by the different gains to accomplish the desired motor speed for controlling the velocity. The SIMULINK model of the robot is prepared to confirm the effectiveness of the algorithms used and their implementation on a mobile.

Keywords Differential drive · Kinematics · Navigation · PID-controller and simulink model

1 Introduction

Robotics is the application of Science, Computer science and Engineering and is a rising research subject. Robotics was originally used in defense and military projects

S. Khatoon · M. Istiyaque (✉) · S. A. Wani
Jamia Millia Islamia University, New Delhi, India
e-mail: mohdistiyaque@gmail.com

M. Shahid
Department of Electrical Engineering, Galgotias College of Engineering and Technology, Greater Noida, India

A. Iqbal et al. (eds.), *Renewable Power for Sustainable Growth*, Lecture Notes in Electrical Engineering 723, https://doi.org/10.1007/978-981-33-4080-0_18

but slowly it finds its way in every face of human life like in cultural industry, education, for farming etc. [1–14]. Mobile robot is an electro-mechanical or virtual smart and quick. Among the differential kind of mobile robots DDMR is one of the simple and effective used structures [1]. Differential drive mobile robots have the ability to move freely in a prearranged workspace to achieve the desired trajectory [2]. This portability ability makes them superior for an extensive application in structured and nonstructural surroundings. The differential drive is made up of two wheels placed on the both end of the moving cart platform which is independently controlled for position and velocity [3]. In some cases, an extra wheel called the Castor wheel is used for balance in case of any instability [4]. Different cases arise in the rotation of DDMR. When both wheels rotate in the direction of the robot at same speed, the DDMR moves in a straight direction [5]. If one wheel rotates but the other is at rest, the DDMR moves in a circular path with the center the midpoint of the static wheel and vice versa [6].

The present research has been done on the mechanical structure of DDMR and its speed is controlled by using a PID-Controller and tuned accordingly. There are two types of errors suffered by a mobile robot platform, one is the systematic error and another is a not-systematic error. First error is caused by some inaccuracies present in the system and non-systematic error is caused by producing some external disturbances. That will be further decoded using PID controller. This research mainly focused on the systematic error for the straight motion in an obstacle-free indoor environment. The kinematic model is used to design and analyze the point to point motion of the differential drive mobile robot in an obstacle free environment [7]. It is a dynamic model for the position and motion estimation. Here the PID Controller is used based on kinematic model for the control of angular-velocity of both the motors present in the on either side of the mobile robot platform [8]. PID Controller is preferred because of its simple structure and theory [13, 14].

2 Mobile Robot Designing

In this paper, the model of the mobile robot used having two DC motors that have an optical encoder also which is connected to each motor. The robot positioning and velocity control is made possible by the kinematic model and the controller [9]. Mobile robot design generally consists of:

A. Kinematical Modelling
B. DC motor modelling
C. Implementation of PID-Controller with the model

A. **Mathematical Modelling of Mobile Robot**

The motion of a differential drive consists of two differently driven wheels that are placed on different side of DDMR Platform. The direction of motion of the Robot is change by changing the motion of both the wheels. Most of the mobile robots have a drive mechanism called Differential Drive having two wheels mounted on either side of the Robot. Each wheel is independently controlled by the DC motor with the help of PID controller connected [10].

The different parameters on which our present mobile robot is designed are as follows

U_i—left sided wheel Velocity
U_r—Right sided wheel Velocity
x, y—Relative robot position in Inertial frame
Theta—Relative Robot heading in Inertial frame
V—Mobile robot Targeted linear velocity
W—Angular velocity of Targeted mobile robot
D—Right and left wheel distance ($D = 15$ cm)
R—Radius of the wheel ($R = 1.5$ cm)

These parameters are important and useful as the input of basic kinematical design of the DDMR platform.

For the rolling motion of mobile robot to occur, the robot should move around a point known as Centre of Curvature (ICC). ICC is a point which lies midway along the common left and common right wheel axis. When we change the velocity of the DDMR the trajectory also changes. Since the rotation rate '*w*' of both wheels is same about the ICC, we can write as:

$$U_{\text{in}} = r\left(\frac{W_R + W_L}{2}\right)$$

$$W = r\left(\frac{W_R + W_L}{L}\right)$$

In this paper two coordinate systems explain the movement of the DDMR. One is Inertial frame of reference (X_i, Y_i) and other is Robot frame (X_r, Y_r). Inertial Reference frame is the frame fixed to the surroundings in which the robot moves and robot frame is a local frame attached with the robot [10]. The Robot position $P[x\ y\ q]$ is given by the Inertial frame of Cartesian coordinate system. Following equation expresses the relation between the inertial reference frame and robot frame of the present robot system:

The Robot under study is a Differential drive robot powered by two DC motors attached to each wheel of the robot. The motion of the wheels is controlled independently. The forward motion of the Robot is possible when both the wheels are driven in the direction of the robot. Turning right or left is possible when the velocity of the two wheels is not equal. When the motion of the two wheels is the same but in the different direction, the robot turns about a fixed point. A third wheel called a Castor wheel is needed for proper balance and stability of the mobile robot platform. Each wheel individually contributes to the motion of the robot as well as imposes constraints on its motion. Here it is assumed that the motion of the wheels of the robot is only in one direction at a time and does not slide or slip in other direction. This condition is best expressed by the Non-Holonomic constraint.

The angular velocity of the robot is calculated by the contribution of both the wheel independently. First, we consider each wheel's spinning speed at point P in the direction of $+X_r$. If any of the wheels of the robot spins while the other is at rest, then the second wheel contributes nothing to the linear speed ($U_2 = 0$). Since point P lies in between the two wheels, the robot moves instantaneously with half the speed, that is U_x. If this condition is applied to the other wheel the same thing happens, that is

$$U_x = \frac{1}{2}wl$$

Summation of these two speeds results from the final linear velocity of the mobile robot is given by

If only the right wheel rotates, the robot pivots around the left wheel and vice versa. The rotational velocity $w1$, when only right wheel spins can be computed as the wheel is moving instantaneously about the center of the left wheel of radius D. Therefore

$$W_l = \frac{R}{D}W_r$$

The same thing is true for the left wheel with the exception that it spins in the opposite direction at point P, that is $w_2 = -(R/D)\ wl$. The summation of these two speeds gives the final angular velocity of the mobile robot.

B. **Modelling of DC-Motor**

DC Motor is broadly used activating device or prime mover in the industry nowadays. Advanced technology has produced such DC motors with ironless rotors having low inertia and hence achieving a high ratio of torque-to-inertia. The DC motor is a form of Transducer converting Electrical energy into Mechanical energy. In this research work, Permanent Magnet Brushed DC (PMBDC) motor is utilized as the activating device for the motion control of the Differential Drive Robot [11]. PMBDC motor is the simplest form of DC motor having high torque-to-inertia ratio (Figs. 1 and 2).

Electrically PMBDC motor is modeled as a series combination of three electrical parameters, that is the Resistor (R), an Inductor (L) and an Electro-motive voltage source (V) as shown in Fig. 3. Mechanically the model is a combination of a moving mass (with inertia 'J' with units kg-square m) and a Viscous damping component B (units of B is N per meter per second) in this above simulink output the error is 98.7% that is removed using PID-controller in further section -C.

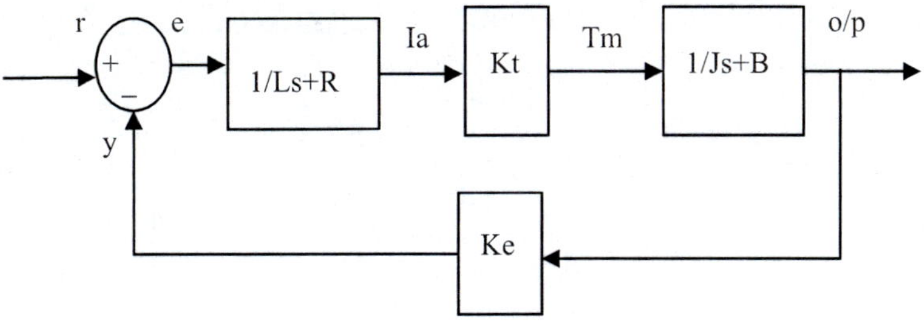

Fig. 1 DC Motor block diagram

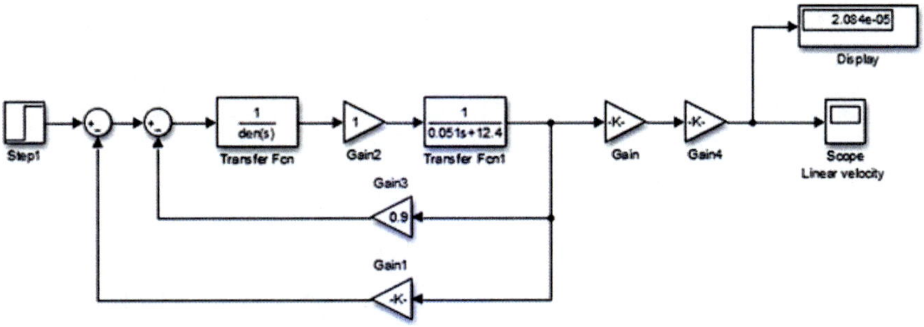

Fig. 2 Simulink model of DC motor with P-controller

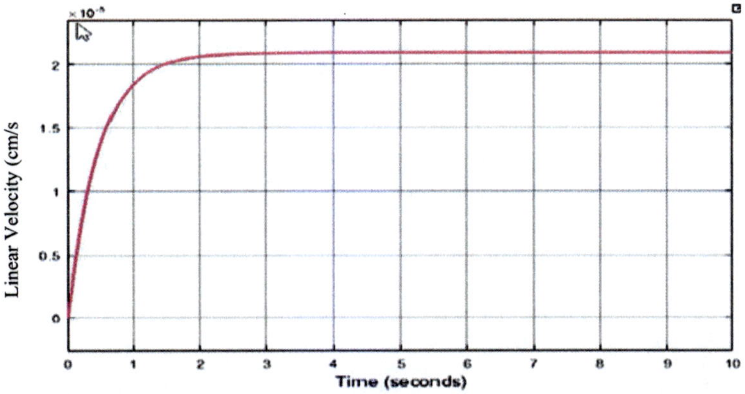

Fig. 3 Linear speed of close loop DC

C. Implementation of PID Controller with the model

The PID control known as Proportional-Integral- Derivative (PID) control. PID controller is considered to be simple, robust and the best controller in the control system designing. PID control works on a feedback mechanism used to control a

Process or Plant in a control system. It is a low class controller which is used to control the motion of a DC motor by assigning different values to the constants K_d, K_i and K_p. K_p measures the present error, K_d measures the prediction error and K_i is a measure of Past errors. The combination of all these Controllers gives control signal that is used to achieve the best control performance [12]. To acquire this optimal performance the three constants of the PID Controller must be tuned correctly and technically based on the Transfer Function of the DC motor with the help of its individual parameters.

The activating signal provided by the PID Controller, which is given to the DC motor, is given by the transfer function of the PID Controller (Figs. 4, 5, 6 and 7).

$$T(S) = K_p + \frac{K_I}{s} + K_D S$$

The graph above shows how the different parameters changes by increasing the values of K_p, K_d and K_i.

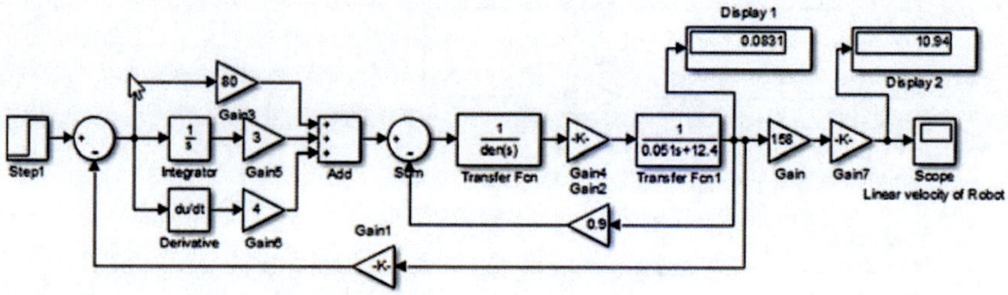

Fig. 4 Simulink model of closed loop DC motor with PID controlle

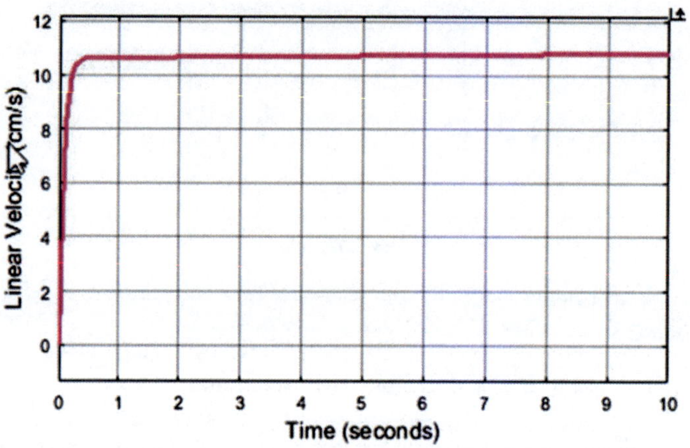

Fig. 5 Linear speed simulink model of closed loop using PID Controller controller

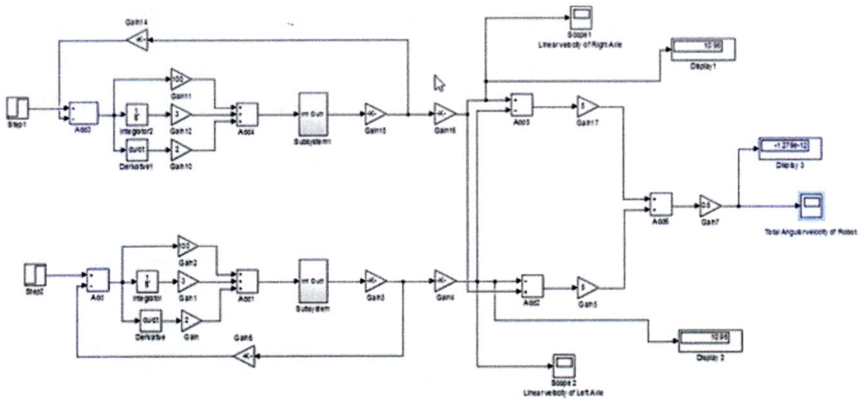

Fig. 6 Simulink modelo of DDMR of two wheel

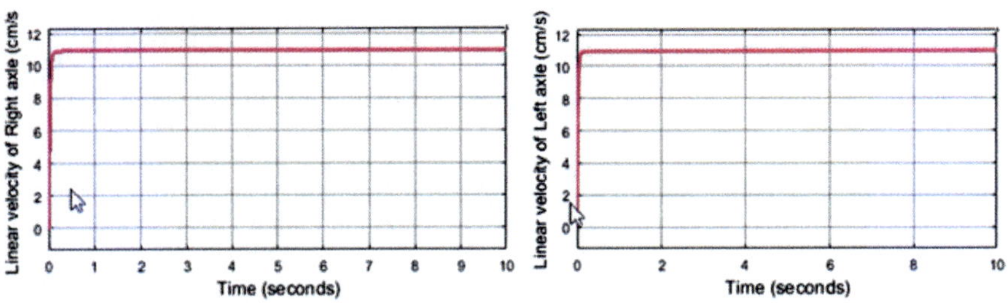

Fig. 7 Response curve of left and right using PID Controller

3 Conclusion

It is clear from the above work and the MATLAB output shown that the PID controller remove the error very well. The precise movement of the DDMR is depends on the DC motor used for the particular job, so that the robot can easily move in the different places where we wish to send without any disturbance. The controller designed in this paper for the robot is used for the minimizing application to move from one place to other, even could be well known and useful for the dangerous and risky mining condition it shows robust and easily controllable. That can of robot is also useful for the farming and transporting various plants for the same.

References

1. Tang Z, Xu X, Wang F, Jiang X, Jiang H (2018) Coordinated control for path following of two wheel independently actuated autonomous ground vehicle. IET J Inst Eng Technol 13(4):628–635 ISSN: 175-956X

2. Glotfelter P, Egerstedt M (2018) A parametric MPC approach to balancing the cost of abstraction for differential-drive mobile robots. In: 2018, IEEE international conference on robotics and automation (ICRA 2018), pp 732–737 10.1109/ICRA.2018.8461234
3. Rodriguez AA, Puttannaiah K, Lin ZY, Aldaco J, Li Z, Lu X, Mondal K, Sonawani SD, Ravishankar N, Das N, Pradhan PA (2017) Modeling, design and control of low-cost differential-drive robotic ground vehicles part II—multiple vehicle study. In 2017 IEEE conference on control technology and application (CCTA 2017), pp 161–166
4. Hsu L-Y, Chen T-L (2013) An optimal wheel torque distribution controller for automated vehicle trajectory following. IEEE Trans Veh Technol 62(6):2430–2440
5. Myint C, Win NN (2016) Position and velocity control for two-wheel differential drive mobile robot. Int J Sci Eng Technol Res (IJSETR) 5(9):2855–2949
6. Mahfouz AA, Aly AA, Salem FA, Mechatronics design of a mobile robot system. Int J Intell Syst Appl 23–36 https://doi.org/10.5815/ijisa.2013.03.03
7. Hipro BD, Zhongmin W (2017) Design and control for differential drivr mobile robot. Int J Eng Res Technol (IJERT) 6(10):327–334. ISSN: 2278-0181
8. Nurmaini S, Dewi K, Tutuko B, Differential drive mobile robot control design based on linear feedback control law. IAES international conference on electrical engineering, computer science and informatics, pp 1–7. https://doi.org/10.1088/1742-6596/755/1/011001
9. Vieira FC, Medeirds AAD, Alsiana PJ, Position and Orientation control of a Two-wheel differential drive nonholonomic mobile robot. Federal University of Rio Grande do Norte, Brazil
10. Crenganis M, Bologa O, Implementation PID controller for a mobile platform, Technical sciences of Romania, Lucian Blaga, University of Sibiu
11. Sailan K, Kuhnert KD, DC Motor angular position control using PID controller for the purpose of controlling the hydraulic pump. Ral Time System Institute, Siegen University, Germany
12. Amer SI, Eskander MN, Zales AM, Positioning and motion robot, Electronics Research Institute, Dokki, Caire, Egypt
13. Iqbal A et al (eds) (2020) Soft computing in condition monitoring and diagnostics of electrical and mechanical systems, vol 1096. In: Advances in intelligent systems and computing. Springer, Singapore. https://doi.org/10.1007/978-981-15-1532-3
14. Iqbal A et al (eds) (2020) Meta heuristic and evolutionary computation: algorithms and applications, vol 1096. In: Studies in computational intelligence. Springer, Singapore. https://www.springer.com/gp/book/9789811575709

Triple-Switch DC-to-DC Converter for High-Voltage Boost Application—Revista

Shima Sadaf, Nasser Al-Emadi, Mahajan Sagar Bhaskar, and Atif Iqbal

Abstract The presented manuscript gives a comparison of the recently proposed three-switch three-mode (TSTM-HS) and dual-duty three-mode (DDTM) DC-to-DC converters. DC microgrid is the easiest, fail-safe, economical, expandable, and high-yield solution which is gaining attention throughout the world and has become an important part of the distributed generation system due to its centralized renewable energy generation and decentralized storage. However, the non-conventional energy generation sources like fuel cells or solar PV cells are small-voltage power sources, **which necessitate the use of DC–DC converters having a high gain and high efficiency** in DC microgrids. For the voltage gain improvement, a TSTM-HS converter employs the voltage lift technique, and an extra switch is added to enhance the range of the duty cycle of the converter. Furthermore, there are three modes of operation with two kinds of duty cycles, and the use of extreme duty cycle across all the switches is avoided. On the other hand, three switches having two separate duty ratios, i.e., dual-duty are used to control the three working modes of the three-mode converters to obtain a broad duty ratio range.

Keywords DC–DC converter · DC microgrid · Dual-duty · Triple-switch · Three-mode · High-gain converter

S. Sadaf (✉) · N. Al-Emadi · A. Iqbal
Department of Electrical Engineering, Qatar University, Doha 2713, Qatar
e-mail: s.sadaf@qu.edu.qa

N. Al-Emadi
e-mail: alemadin@qu.edu.qa

A. Iqbal
e-mail: atif.iqbal@qu.edu.qa

M. S. Bhaskar
Renewable Energy Lab, Department of Communication and Networks, College of Engineering, Prince Sultan University, Riyadh 11586, Saudi Arabia
e-mail: sagar25.mahajan@gmail.com

A. Iqbal et al. (eds.), *Renewable Power for Sustainable Growth*, Lecture Notes in Electrical Engineering 723, https://doi.org/10.1007/978-981-33-4080-0_19

1 Introduction

In the recent past years, DC microgrids have gained large attention due to the use of highly efficient and reliable power electronics converters associated with it and their contribution to achieving the system safety, reliability and installation reliability [1–3]. Nonconventional energy sources like fuel cells, solar PV, etc., have also improved the DC microgrid systems incorporating DC–DC power electronic converters [4]. The DC voltage obtained by these renewable resources is quite low, i.e., 12–48 V, which necessitates the use of DC-DC power electronic converters with large voltage gain for the improvement of the output voltages in the DC microgrid connected system [5]. The traditional boost converter can theoretically provide a large gain in voltage at a greater value of the duty cycle. However, in practice, because of the capacitor and inductor effective series resistance, electromagnetic interference effect, the diode reverse recovery problem, and conduction losses across the switches, the classical boost converter is unable to attain the required large gain in voltage at extreme duty cycle [6]. High gain in voltage can be attained by employing the converters like half/full-bridge, push-pull, forward converters, and flyback of the isolated DC–DC type by adjusting the coupled inductors or transformers' turns ratio [7, 8]. However, these converter configurations possess sharp voltage rise across switches, loss of power, a saturation of the transformer core, etc., caused by the transformer leakage inductance [9]. These drawbacks can be overcome with the addition of a high-frequency transformer, snubber circuits, which in turn increases the converter price and size [10]. In order to attain a large gain in voltage together with decreased cost and size, the non-isolated or transformer-less converter can be a suitable option, if galvanic isolation is not needed [11]. Many different converter configurations such as quadratic boost and cascaded, the hybrid switched-capacitor/switched inductor along with a boost converter, voltage lift method, switched capacitor, voltage multiplier, etc., have been discussed in the literature so far considering numerous steps of boosting without incorporating any coupled inductors or transformers to avoid the leakage inductance, excessive ripples in the input current, and sharp rise in the voltage across the switches [12, 13]. However, circuit complexity, size, and cost increase due to an increase in the number of manifold stages involving capacitors and inductors. Furthermore, high transient current and high conduction losses are introduced across the switches in case of switched-capacitor topologies because of the multiple capacitor stages, which in turn decreases the efficiency, and also, high current capability inductor is required to overcome this issue [14, 15]. In order to overcome the above issues, for applications in DC microgrid, two recent converter topologies, three-switch three-mode high step-up (TSTM-HS) converter [16] and dual-duty three-mode (DDTM) DC/DC converter [17], respectively, have recently been proposed. In this paper, similarities and comparison studies of TSTM-HS and DDTM converters are presented.

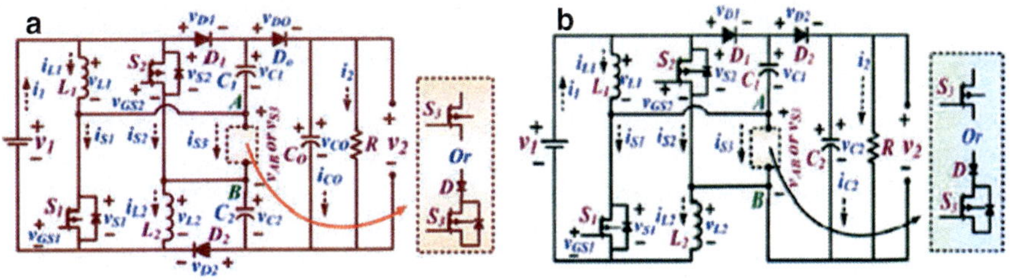

Fig. 1 Power circuitry of the **a** three-switch three-mode high step-up and **b** dual-duty three mode DC/DC converters

2 Three-Switch Three-Mode-HS Converter

The TSTM-HS converter is capable to attain a large gain in voltage without using any transformer or coupled inductor-based structure, operating with a broad duty cycle range where two different kinds of duty cycles are varied to obtain the output voltage. Furthermore, the load is supplied with the accumulated energy, and a high gain in voltage is attained by utilizing a lesser number of semiconductor devices. TSTM-HS converter's power circuit is shown in Fig. 1a. Both the switches S_1 and S_2 are supplied with the same gate pulse and the duty ratio k_1, whereas S_3 switch is supplied using a different gate pulse with duty cycle k_2 and delay k_1T seconds. Both the inductor L_1 and L_2 currents linearly increase during Mode 1 and Mode 2, while in Mode 3, currents through both the inductors L_1 and L_2 linearly decrease. The changes in the voltage gain are analyzed by keeping one of the two duty cycles constant at a time and varying the other. A large gain in voltage is observed to be attained by the three-switch three-mode-HS converter with the suitable duty ratio variations.

3 Dual-Duty Three-Mode (DDTM) Converter

The topology of the DDTM converter is developed without the use of any transformer, multiple switched capacitor/inductor, or voltage multiplier, and it has the capability of achieving a large gain in voltage with a broad duty cycle range along with a decreased switch voltage stress. The ease of duty ratio selection being the main advantage of DDTM converter, the increased range of operation because of the option of dual-duty ratios, and the converter can be regulated using various methods with the variation in voltage at the input. Furthermore, the occurrence of energy transfer from source to load without incorporating many energy transfer loops leads to an improved converter efficiency and performance. This property of achieving a high voltage gain with a broad range of duty cycles and higher efficiency makes the DDTM converter a better choice for DC microgrid applications. Figure 1b shows the DDTM converter's power circuit. The inductor L_1 and L_2 currents are increased linearly having constant slope values in Mode 1 and Mode 2, whereas the currents through the inductor L_1 and inductor L_2 in Mode 3 are decreased linearly with a constant slope.

4 Similarity and Comparative Study

Table 1 presents a detailed comparison of both the converters. A large gain in voltage with a broad duty cycle range, and a reduced component voltage stress is provided by the three-switch three-mode-HS converter. The main advantage of the three-switch three-mode-HS converter topology is its increased range of operation since it can work with a dual-duty ratio in all the three modes, which facilitates the working of the converter at a larger duty ratio (i.e., due to two duty ratios). Furthermore, efficiency and performance are improved due to the energy transfer without incorporating several energy transfer loops. The TSTM-HS converter possesses a broad range of duty cycle and a flexible voltage gain control by simply controlling the two different duty ratios and hence providing reliable operation. Therefore, devices with small voltage rating and internal resistance values can be used to develop the TSTM-HS converter. The DDTM converter topology provides high voltage gain for a given value of duty ratio and possesses a broader duty range, needs low voltage switches, and uses a lesser number of components in the circuit. Furthermore, both converters are observed to be providing voltage gain adjustment facility by suitably choosing the duty ratios, which makes it better than any converter operated on a single duty ratio.

In paper [16], the experimental investigation of the three-switch three-mode high step-up converter was carried out in the laboratory. The switches S_1 and S_2 were controlled by generating two gate pulses with half of the duty cycle (k_1), and the

Table 1 Comparison of TSTM-HS converter with DDTM converter

Parameters	TSTM-HS converter	DDTM converter
Voltage gain	$(3 - k_1 - 2k_2)/[1 - (k_1 + k_2)]$	$(2 - k_2)/[1 - (k_1 + k_2)]$
Max. switch voltage	$0.5(V_2 - V_1)$, $V_2 - 2V_1$	$0.5, (1 - V_1/V_2)$
PIV of load side diodes	$-(V_2 - V_1)/2$	$-(V_2 - V_1)/2V_2$
PIV of intermediate diodes	$-(V_2 - V_1)$	$-(V_2 - V_1)/V_2$
Average efficiency	92.06%	93.43%
No. of components	Diodes = 3, inductors = 2	Diodes = 2, inductors = 2
	Switches = 3, capacitors = 3	Switches = 3, capacitors = 2

switch S_3 was controlled by generating a gate pulse with 35% of the duty cycle and with a converter switching frequency of 50 kHz. Based on various tests performed, the TSTM-HS converter's average efficiency was observed to be 92.06%. Therefore, the results of the experimental investigations validated the theoretical analysis in good agreement, and hence, the TSTM-HS converter was found to be suitable for DC microgrid applications. In paper [17], to experimentally investigate the performance of DDTM converter, also a laboratory prototype was developed by considering duty ratio typical values as k_1= 50% and k_2= 35%. The highest value of efficiency was found to be 95.47%, and it was obtained at duty ratio values k_1= 50% and k_2= 35%, respectively, and after performing various tests, the DDTM converter's average efficiency was observed to be 93.43%. Therefore, the DDTM converter was found to be suitable and a better option for DC microgrid applications. The three-switch three-mode-HS converter's experimental results are shown in Fig. 2a, which indicate that the input voltage value of 36.3 V produced an output voltage of 401.1 V. A rise of 9.38 A in the input current is observed at the beginning of mode I because of capacitors C_1 and C_2 charging, whereas a rapid drop of the input current is observed as soon as the working of the converter was changed from mode I to mode II. On the other hand, Fig. 2b shows the experimental results for DDTM converter. The input/output average current and voltage values were observed to be 1.24A, 38.2 V, 13.61A, and 400.32 V, respectively. For the three-switch three-mode high step-up converter, the effect of duty cycles k_1 and k_2 on the gain in voltage was studied in [16]. Figure 3a shows the variation in voltage gain when k_1 is changed while k_2 is kept constant and changes in the voltage gain by changing k_2 while k_1 is kept constant. It is important to note that a large gain in voltage is attained by using the TSTM-HS converter by changing k_1 and k_2 appropriately. Figure 3b shows the voltage gain versus k_1 and k_2 plot for the DDTM converter. The gain in voltage attained by the DDTM converter is observed to be the same as that of TSTM-HS converter when the sum of duty ratios k_1 and k_2 is selected appropriately.

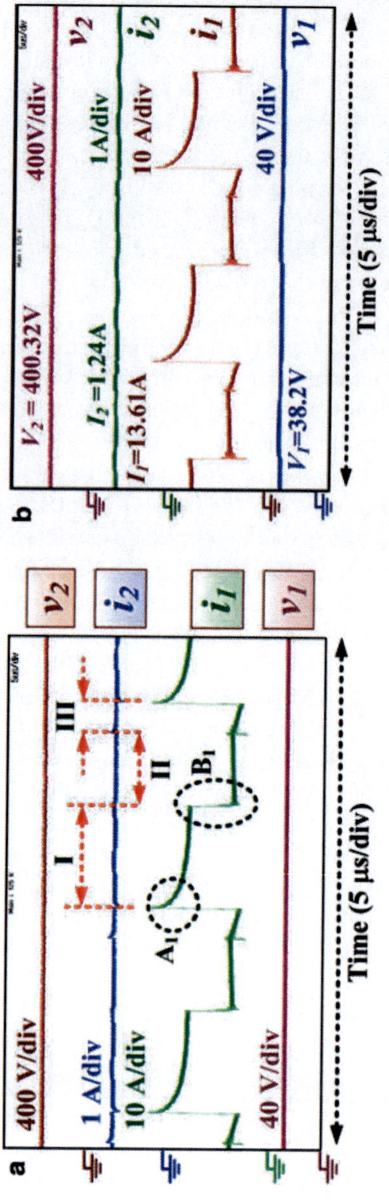

Fig. 2 Results of the experiments for a TSTM-HS converter and **b** DDTM converter

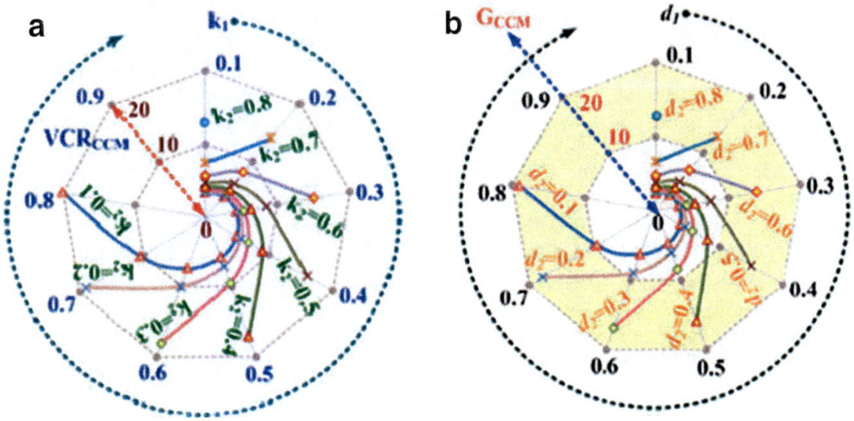

Fig. 3 Voltage gain plots with the two duty ratios for **a** TSTM-HS and **b** DDTM converters

5 Conclusion

The recently proposed TSTM-HS converter possessing a broad duty cycle range works in three modes, where two separate duty cycles are used to regulate the voltage gain. Furthermore, it avoids using a high duty cycle for individual switches to attain a large gain in voltage with low-voltage stress on semiconductor devices. Hence, for the DC microgrid applications in case of low to high voltage conversion, the TSTM-HS DC–DC converter is a viable solution. On the other hand, for applications in DC microgrid, the new DDTM converter can attain a large gain in voltage with a lesser components count as compared to the TSTM-HS converter. Therefore, the **DDTM converter is observed to be a better choice for improved efficiency with lower** voltage stress on semiconductor devices and to attain a large gain in voltage and a lesser number of components and hence reducing the circuit complexity, size and cost.

References

1. Luis EZ, Peter WL (2015) A high-efficiency unidirectional DC/DC converter for integrating distributed resources into DC microgrids. In: IEEE first international conference on DC microgrids 2015, ICDCM, pp 280–284. IEEE, Atlanta, GA, USA
2. Iqbal A et al (eds) (2020) Soft computing in condition monitoring and diagnostics of electrical and mechanical systems, vol 1096. In: Advances in intelligent systems and computing. Springer, Singapore. https://doi.org/10.1007/978-981-15-1532-3
3. Iqbal A et al (eds) (2020) Meta heuristic and evolutionary computation: algorithms and applications, vol 1096. In: Studies in computational intelligence. Springer, Singapore. https://www.springer.com/gp/book/9789811575709
4. Habumugisha D, Chowdhury S, Chowdhury SP (2013) A DC-DC interleaved forward converter to step—up DC voltage for DC microgrid applications. In: IEEE power and energy society

general meeting 2013, pp 1–5. IEEE, Vancouver, BC, Canada
5. Iqbal A, Bhaskar MS, Meraj M, Padmanaban SK, Rahman S (2020) Closed-loop control and boundary for CCM and DCM of nonisolated inverting n × multilevel boost converter for high-voltage step-up applications. IEEE Trans Ind Electron 67(4):2863–2874
6. Bhaskar MS, Reddy NS, Kumar RKP, Gupta YBSS (2014) A novel high step-up multilevel boost converter using double voltage-lift switched-inductor cell. In: International conference on circuits, power and computing technologies 2014, ICCPCT, pp 996–1001. IEEE, Nagercoil, India
7. Deng Y, Rong Q, Li W, Zhao Y, Shi J, He X (2012) Single-switch high step-up converters with built-in transformer voltage multiplier cell. IEEE Trans Power Electron 27(8):3557–3567
8. Padmanaban S, Bhaskar MS, Maroti PK, Blaabjerg F, Fedák V (2018) An original transformer and switched-capacitor (T & SC)-based extension for DC-DC boost converter for high-voltage/low-current renewable energy applications: hardware implementation of a new T & SC boost converter. Energies 11(4):783–805
9. Hsieh Y, Chen J, Liang T, Yang L (2011) A novel high step-up DC-DC converter for a microgrid system. IEEE Trans Power Electron 26(4):1127–1136
10. Tibola G, Lemmen E, Duarte JL, Barbi I (2017) Passive regenerative and dissipative snubber cells for isolated SEPIC converters: analysis, design, and comparison. IEEE Trans Power Electron 32(12):9210–9222
11. Tofoli FL, Pereira D, Paula WJ, Júnior D (2015) Survey on non-isolated high-voltage step-up dc-dc topologies based on the boost converter. IET Power Electron 8(10):2044–2057
12. Forouzesh M, Siwakoti YP, Gorji SA, Blaabjerg F, Lehman B (2017) Step-Up DC-DC converters: a comprehensive review of voltage-boosting techniques, topologies, and applications. IEEE Trans Power Electron 32(12):9143–9178
13. Sadaf S, Mahajan SB, Meraj M, Iqbal A, Alemadi N (2020) A novel modified switched inductor boost converter with reduced switch voltage stress. IEEE Trans Ind Electron 1–1
14. Iqbal A, Bhaskar MS, Meraj M, Padmanaban S (2018) DC-Transformer modelling, analysis and comparison of the experimental investigation of a non-inverting and non-isolated Nx multilevel boost converter (Nx MBC) for low to high DC voltage applications. IEEE Access 6:70935–70951
15. Kwon J, Kwon B (2009) High step-up active-clamp converter with input-current doubler and output-voltage doubler for fuel cell power systems. IEEE Trans Power Electron 24(1):108–115
16. Bhaskar MS, Alammari R, Meraj M, Padmanaban S, Iqbal A (2019) A new triple-switch-triple-mode high step-up converter with wide range of duty cycle for DC microgrid applications. IEEE Trans Ind Appl 55(6):7425–7441
17. Bhaskar MS, Meraj M, Iqbal A, Padmanaban S, Maroti PK, Alammari R (2019) High Gain transformer-less double-duty-triple-mode DC/DC converter for DC microgrid. IEEE Access 7:36353–36370

Initial Sizing and Sensitivity Analysis of a Personal Air Vehicle

Mohammad Irfan Alam

Abstract Megacities of the world are going through the unprecedented challenge of increasing congestion and frequent traffic jams. There is an immediate demand for a fast, safe, and efficient mode of mass mobility, especially in the densely populated cities of the world. Personal air vehicles could play a vital role in solving the above problems by blending the speed and efficiency of an airplane with the cost and convenience of a car. This paper discusses the initial sizing approach of a personal air vehicle. The objective is to size an electric vertical takeoff and landing (eVTOL) aircraft for the range of 200 km and payload of 200 kg. A dataset of various ongoing eVTOLs under development is used for the initial estimate of weight sizing. A sensitivity analysis is also carried out to study the effect of some design parameters which influence the selection of the final configuration. This study is based on the preliminary investigation and meant to serve as initial steps in the process of problem formulation for multidisciplinary design optimization of the system of transport.

Keywords Electric aircraft · On-demand urban-air-mobility (OD-UAM) · Air taxi

1 Introduction

Currently, megacities of the world are facing an unprecedented challenge to deal with alarming traffic situations due to the steep rise in the urban population. It has boosted demand for an alternate solution for urban travel. The projected population of some of the world's most populous cities indicates an average two-fold increase in the urban population over thirty years [1]. With the boom of the urban population, a significant amount of productive time is being wasted in traveling, due to congestion and long traffic jams in those megacities. Apart from the productivity loss, it also affects the work-life balance, social relationship, and health of individuals, especially living in those cities. Moreover, lives are under constant threat due to the increase in carbon footprints by increasing personal vehicles. Consequently, a dream of flying

M. I. Alam (✉)
Seoul National University, Seoul 08826, South Korea
e-mail: irfansnu@snu.ac.kr

A. Iqbal et al. (eds.), *Renewable Power for Sustainable Growth*, Lecture Notes in Electrical Engineering 723, https://doi.org/10.1007/978-981-33-4080-0_20

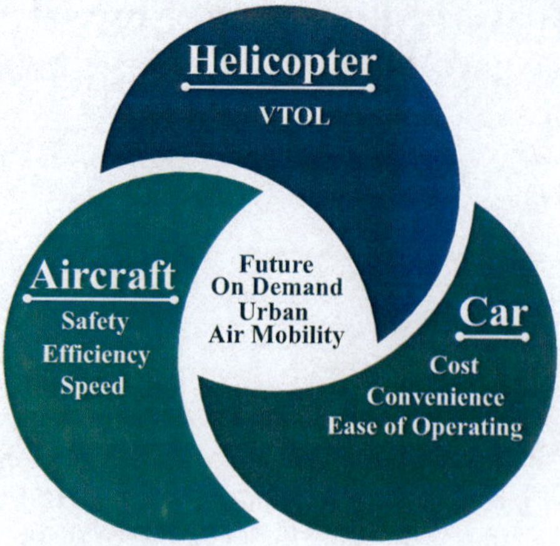

Fig. 1 On-demand urban air mobility

cars of the past has become a need in the present world. Therefore, to push the limit beyond the trivial way of mass mobility viz. Road transport, a new way of urban mobility through personal air vehicles, is the focus of research and development. A mode of travel that should be efficient, fast, safe, and reliable. In a nutshell, a personal air vehicle in the context of the future of on-demand air mobility should necessarily blend the features of a helicopter, an aircraft, and a private car, as illustrated in Fig. 1.

Now, the necessary technologies have reached the level, at which the third dimension of mass mobility for personal use is being explored [2]. The maturities of technical advancement and urgent demand for alternative urban transportation, have opened the enormous market potential of a private air vehicle [3, 4]. On realizing the fact, aviation tech industries and several startups are intensifying efforts to grab the opportunities. Ehang 184, Lilium, Kitty Hawk Cora, Joby S4, Vahana, Volocpter 2X are few to name among tens of efforts currently going on to achieve the common goal.

Polaczyk et al. [5] have discussed the worldwide efforts by the various industries involved in the design and development of personal air vehicles and their concepts. As a result of the ongoing efforts in the industries' front, PAV is now a focused area of research over the globe. Liu et al. [6] have discussed the PAV research activities focused on the United state and Europe.

In recent times, some interesting studies related to the initial sizing and conceptual design of all-electric small aircraft were carried out. Riboldi and Gualdoni [7, 8] have presented an approach for preliminary weight sizing of all-electric aircraft. Similarly, a conceptual design and performance analysis are also carried out [9, 10]. However, these studies are limited to conventional aircraft, i.e., without VTOL capability. Though, their model development and subsystem analyses could be useful at the later phase of electric VTOL design. Comparative analyses of different configurations and optimization studies for the possible application of on-demand urban air-mobility are also carried out [11, 12]. These studies play a vital role in decision making based on the given requirements at the early stage of design. Brelje and Martins [13] have presented a detailed overview of various design approaches and ongoing development efforts related to the electric and hybrid aircraft.

The standard initial sizing procedure of a conventional (CTOL) aircraft involves three main aspects. First, the estimation of an empty weight fraction driven by the fuel fraction (which is a dynamic mass), second, a sizing matrix plot (SMP) derived through the performance constraints and third, reliable historical datasets of various aircraft. However, the traditional approach for the initial sizing of an airplane is not directly applicable in the case of a personal air vehicle. In this study, a VTOL aircraft powered by a battery is being designed. The battery is a static mass, i.e., although it will discharge during operation, the weight will remain the same. Therefore, the SMP is not directly relevant due to the requirement of VTOL. There is a lack of reliable historical datasets of PAVs, thereby reducing the confidence of a designer on the initial sizing results.

It is worth mentioning that the initial sizing and conceptual design approaches vary in the literature primarily due to the variations in design requirements and possibly due to the lack of sufficient reliable datasets. Therefore, standardization of the methodology supported by reliable datasets to enhance the confidence of the developed concept at the initial stage of design is essential. Moreover, this step plays a critical role in accurate modeling for high-fidelity tools and multidisciplinary design analyses and optimization.

This paper is a result of a preliminary investigation for the conceptual design of PAVs, thereby serves as an initial process to achieve the final goal. It provides a basic sizing methodology backed by the relevant datasets of existing electric VTOL under active design and development. The present study explains the method step-wise, discusses the initial results and the PAV configuration.

2 Initial Sizing of PAVs

For any aircraft, its design requirements and mission specifications drive the design. Existing PAV configurations that are under development and testing vary from multicopter (e.g., Ehang 184) systems to the winged body (e.g., Lilium) depending upon mission requirements.

2.1 *Design Requirements and Mission Specifications*

Table 1 lists the design requirements for the present study. The vehicle needs to carry two passengers for a range of 200 km (a return journey on a single charge).

Figure 2 illustrates the complete mission specifications for the present study.

Based on the design requirements and mission specifications mentioned above, the initial concept of the vehicle could be achieved. Figure 3 represents the conceptualization based on the requirements.

Table 1 Design requirements

Parameters	Value
Range (R), (km)	200
Payload (m_p), (kg)	200
Maximum speed (v_{max}), (km/hr)	270

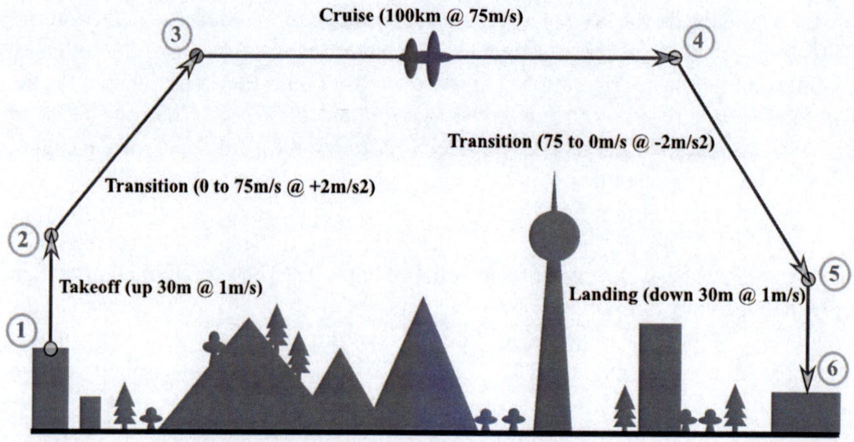

Fig. 2 Mission specifications

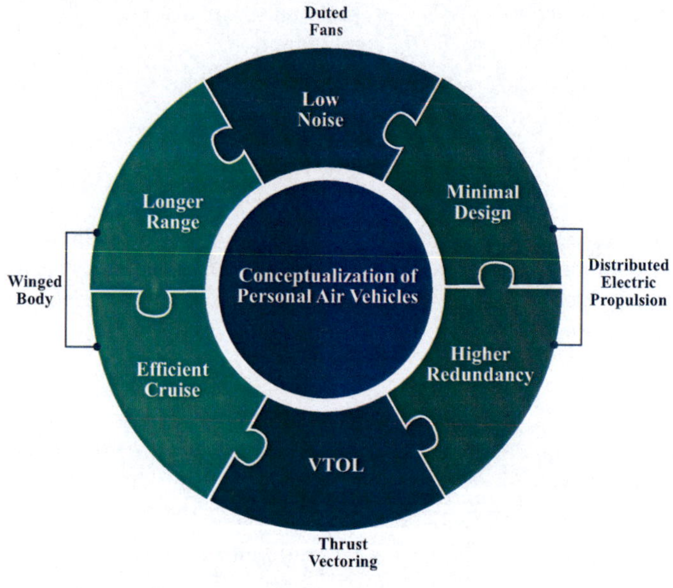

Fig. 3 Conceptualization of personal aerial vehicles

Multi-copter configurations exhibit excellent hover performance. However, those are inefficient and have speed limitations during the cruise. Winged body configurations have an efficient-cruise and can achieve higher speed. However, being a VTOL could be energy-intensive [11]. In the present study, a winged body configuration is selected to meet the design requirements. The optimal shape could deviate substantially from the chosen concept when a high fidelity multidisciplinary analysis is carried out. The next section describes the sizing approach.

2.2 Initial Sizing Methodology

Analogous to a standard sizing approach of a conventional aircraft, the maximum gross takeoff mass (m_t) can be expressed as a summation of masses of payload (m_p), battery (m_b), motor (m_m) and structure (m_s) as:

$$m_t = m_p + m_b + m_m + m_s \tag{1}$$

The payload mass (m_p) is obtained from the design requirements. The battery mass (m_b) can be expressed as:

$$m_b = \max\left\{m_{be},\ m_{bp}\right\} \tag{2}$$

where m_{be} and m_{be} are the battery masses based on energy and power respectively. m_{be} can be estimated as:

$$m_{be} = \frac{E_t}{\eta_b \cdot e_b} \tag{3}$$

where η_b and e_b are conversion efficiency and energy density of the battery, respectively. E_t is the total energy, which is the sum of the energy demand for hover (E_h) and energy demand for the cruise (E_c) as:

$$E_t = E_h + E_c \tag{4}$$

E_t can be further expressed in terms of hover power (P_h) and cruise power (P_c) as:

$$E_t = P_h \cdot t_h + P_c \cdot t_c \tag{5}$$

where t_h and t_c are hover duration and cruise duration of the PAV, respectively. Similar to m_{be}, m_{bp} be expressed as:

$$m_{bp} = \frac{1}{\eta_b} \max\left\{\frac{P_h}{e_p}, \frac{P_c}{e_p}\right\} \tag{6}$$

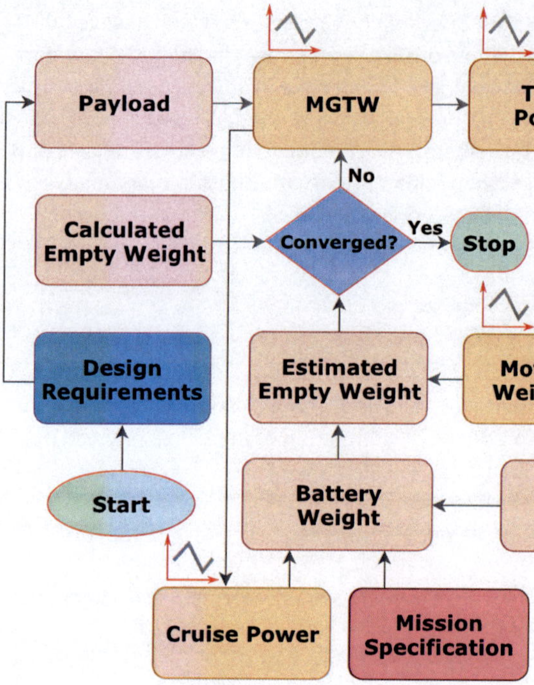

Fig. 4 An initial sizing methodology of a personal air vehicle

where e_p is power density of the battery. P_h is calculated based on standard actuator disk theory [11, 14] as:

$$P_h = \sqrt{\frac{\left(\frac{T}{K}\right)^3}{2\rho A_d}} \quad \text{where,} \quad K = \begin{cases} 1.26, & \text{if ducted fans} \\ 1, & \text{otherwise} \end{cases} \tag{7}$$

where ρ is air density, and A_d is total disk area required by the vehicle to hover. Figure 4 depicts the methodology used for the initial sizing of a vehicle. Figure 5 is used. The figure is plotted based on the data available for various PAVs under development and testing. Figure 6 gives the estimation of total hover power required based on the initial maximum takeoff mass of the vehicle. Figure 6 may not provide a reliable estimate due to the limited availability of data in the open literature. However, it can be used in the initialization of the sizing procedure because the feedback loop ensures the convergence of the vehicle mass-breakdown. Once hover power is estimated, the total disk area required can be calculated. For the given number of fans, its diameter can be obtained for the total disk area or vice-versa.

The payload mass is obtained from the design requirements. To get an estimate of maximum takeoff mass based on the payload of the vehicle.

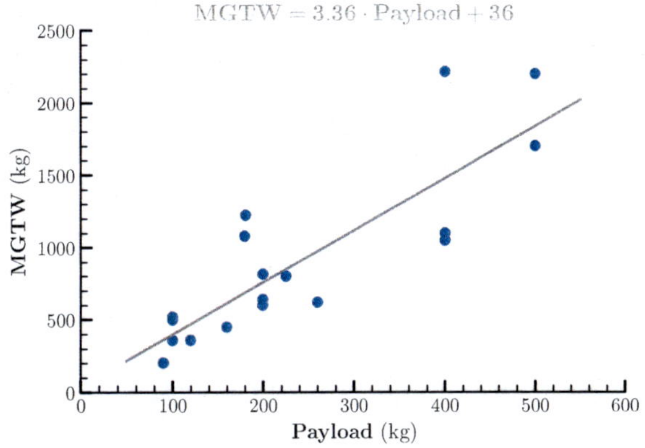

Fig. 5 MGTW estimation based on payload

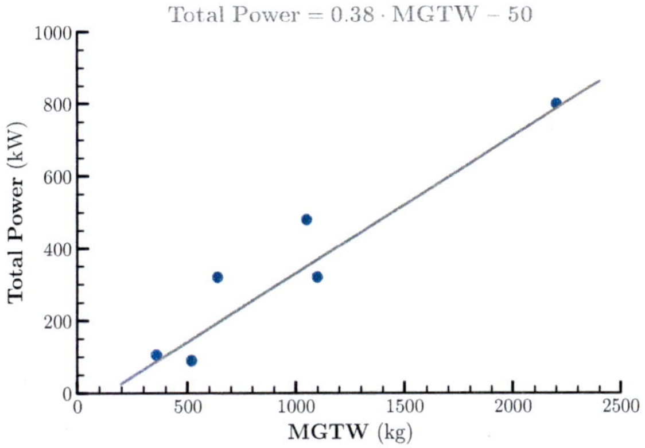

Fig. 6 Total vehicle power estimation

Figure 7 is plotted for the number of comparable sized existing CTOL electric airplanes. It was assumed that the vehicle would be as efficient as of them during the cruise condition. Therefore, the cruise power requirement can be estimated from the figure. At the last, motor mass is estimated based on the installed power of the vehicle. Though the data used is limited, it gives the fair estimate of currently available motors used in some of the small electric airplanes as 5 kW/kg (Fig. 8).

Table 2 lists the design parameters used in initial sizing of the PAV.

3 Results and Discussion

This section discusses the initial results of a personal aerial vehicle obtained through the initial sizing methodology. Figure 9 illustrates the system mass breakdown. Although, these results much depend on the design requirements and the mission

Fig. 7 Power required to cruise

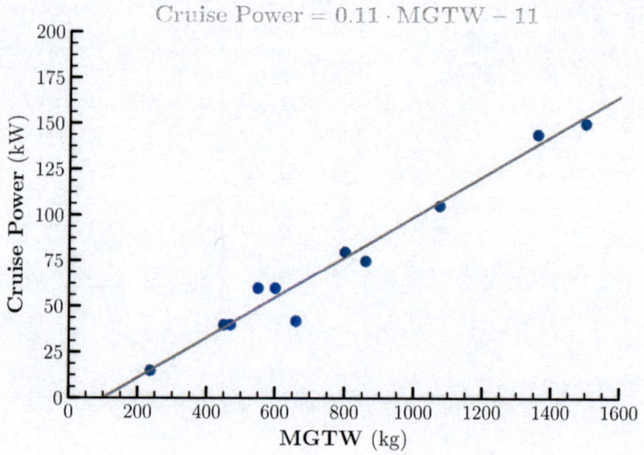

Fig. 8 Estimation of motor weight

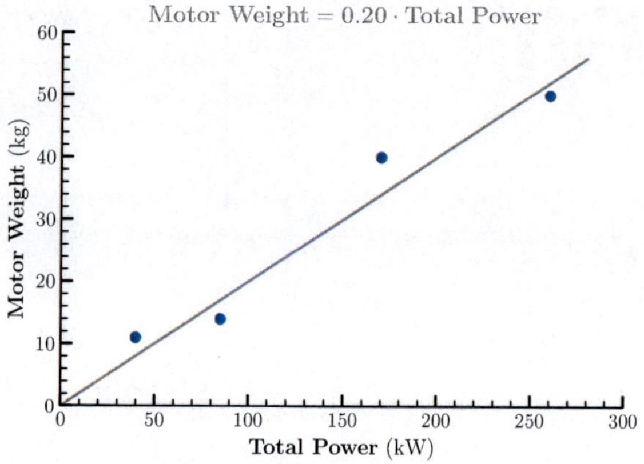

Table 2 Design parameters

Parameters	Value
Battery energy density (e_b), (Wh/kg)	400
Battery power density (p_b), (W/kg)	1000
Battery conversion efficiency (η_b)	0.85

specifications. It is worth mentioning that battery weight contributes around thirty percent of the MGTW in the present case, which is significantly higher in comparison to the conventional aircraft, due to the low energy density of battery compared to aviation fuels. Table 3 lists the results obtained.

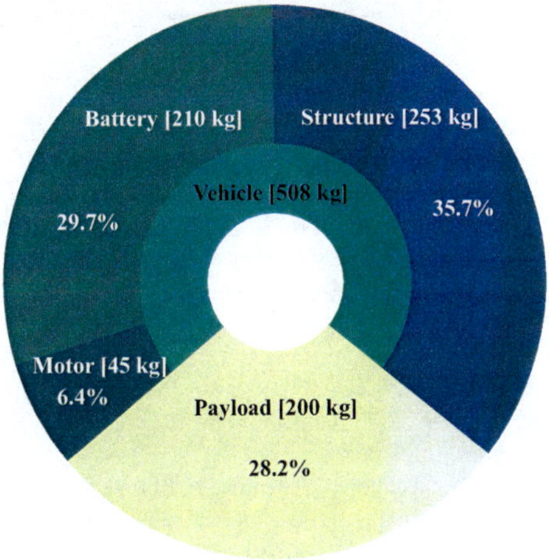

Fig. 9 System mass break-down

Table 3 Output parameters

Parameters	Value
Total energy required (E_b), (kWh)	71
Total power required (P_b), (kW)	220
Total number of fans	4
Total number of ducted fans	20
Diameter of fans (d_f), (m)	0.50
Diameter of ducted fans (d_d), (m)	0.15

4 Conclusions

Personal Air Vehicles have the potential to become a game-changer for Urban-Mobility. Once fully implemented, it could play a vital role in reducing travel time and carbon footprints. Although the system is far from fully developed for mass adaptation, and several barriers are ahead on the road, several companies have already demonstrated their models. The paper presented a methodology that can be used for the initial sizing of a vehicle based on given requirements.

Acknowledgements This study was supported by the Brain Korea 21+ Transformative Training Program for Creative Mechanical and Aerospace Engineers, School of Mechanical and Aerospace Engineering, Seoul National University, South Korea.

References

1. United Nations (2018) Department of Economic and Social Affairs, Population Division, The World's Cities in 2018–Data Booklet, (ST/ESA/ SER.A/417). https://www.un.org/en/events/citiesday/assets/pdf/the_worlds_cities_in_2018_data_booklet.pdf
2. Moore MD (2006) NASA personal air transportation technologies. SAE Technical Paper, p 0534. https://doi.org/10.4271/2006-01-2413
3. Holden J, Goel N (2016) Fast-forwarding to a future of on-demand urban air transportation. Uber Elevate, 27 Oct 2016, pp 1–98. https://www.uber.com/elevate.pdf
4. Romli FI, Rashid H, Harmin MY (2018) Market potential for personal air vehicle (PAV) concept: a Malaysian case study. Int J Pure Appl Math 119(15):3755–3760
5. Polaczyk N, Trombino E, Wei P, Mitici M (2019) A review of current technology and research in urban on-demand air mobility applications. http://www.aere.iastate.edu/~pwei/proceedings/vfs19_nick.pdf
6. Liu Y, Kreimeier M, Stumpf E, Zhou Y, Liu H (2017) Overview of recent endeavors on personal aerial vehicles: a focus on the US and Europe led research activities. Prog Aerosp Sci 91:53–66. https://doi.org/10.1016/j.paerosci.2017.03.001
7. Riboldi CE, Gualdoni F (2016) An integrated approach to the preliminary weight sizing of small electric aircraft. Aerosp Sci Technol 58:134–149. https://doi.org/10.1016/j.ast.2016.07.014
8. Riboldi CE, Gualdoni F, Trainelli L (2018) Preliminary weight sizing of light pure-electric and hybrid-electric aircraft. Transp Res Procedia 29:376–389. https://doi.org/10.1016/j.trpro.2018.02.034
9. Patterson MD, German BJ, Moore MD (2012) Performance analysis and design of on-demand electric aircraft concepts. In: 12th AIAA aviation technology, integration and operations (ATIO) conference. https://doi.org/10.2514/6.2012-5474
10. Patterson MD, German B (2014) Conceptual design of electric aircraft with distributed propellers: multidisciplinary analysis needs and aerodynamic modeling development. In: 52nd Aerospace sciences meeting, p 0534
11. Bacchini A, Cestino E (2019) Electric VTOL configurations comparison. Aerospace 6(3):26. https://doi.org/10.3390/aerospace6030026
12. Brown A, Harris W (2018) A vehicle design and optimization model for on-demand aviation. In: AIAA/ASCE/AHS/ASC structures, structural dynamics, and materials conference, p 0105
13. Brelje BJ, Martins JR (2018) Electric, hybrid, and turboelectric fixed-wing aircraft: a review of concepts, models, and design approaches. Progr Aerosp Sci 104:1–19. https://doi.org/10.1016/j.paerosci.2018.06.004
14. Seddon J, Newman S (2001) Basic helicopter aerodynamics, american institute of aeronautics and astronautics

Trajectory Tracking of Quadrotor Using LQR Controller

Mohammad Shahid, Md. Istiyaque, Shahida Khatoon, and Ibrahim

Abstract The quadrotors are more often used as an unmanned aerial vehicle in commercial as well as personal applications. But the quadrotor is inherently an unstable system due to its complex dynamics and need of dual controller for inner and outer loop control. The modelling and control of quadrotor UAV has been presented in this paper. The modelling of the quadrotor is done using Newton-Euler equations. And for controlling the quadrotor intelligent Hybrid controller and Linear Quadratic Regulator (LQR) control schemes are used. The results obtained after applying both controllers are compared and an optimal controller is proposed.

Keywords Quadrotor · Hybrid controller · ANFIS · LQR · UAV

1 First Section

The UAV have been defined in many ways from time to time in literature. In common man's perception, UAVs is a class of small aircraft that flies without any human pilot. They may be fully autonomous or remotely operated by a human pilot. An autonomous vehicle can be defined as an airplane that is operated and controlled by onboard computer. These types of flying vehicles are generally designed for a set of tasks or a specific mission. Generally, UAVs are powered or unpowered, tethered or untethered aerial vehicles. UAVs are generally used in military applications but in recent years they have been gaining interest in civil applications [1].

In this paper, a quadrotor which is classified as a rotorcraft is chosen for the investigation. The quadrotor has a very high non-linear and its variables are coupled in nature and closely interdependent. The quadrotor is a multivariable, complex and highly non-linear dynamical system and generally has four control inputs with six

M. Shahid (✉)
Department of Electrical Engineering, Galgotias College of Engineering and Technology, Greater Noida, India
e-mail: msjmi09@gmail.com

Md. Istiyaque · S. Khatoon · Ibrahim
Department of Electrical Engineering, Faculty of Engineering and Technology, Jamia Millia Islamia, New Delhi, India

A. Iqbal et al. (eds.), *Renewable Power for Sustainable Growth*, Lecture Notes in Electrical Engineering 723, https://doi.org/10.1007/978-981-33-4080-0_21

degree of freedom so as to act as an under-actuated system [2]. The under-actuated systems have degree of freedom more than the number of controlling inputs. These systems are very difficult to control due to nonlinear coupling between control inputs and degree of freedom [3]. The model uncertainties and parameter variations are some significant difficulties that has been encountered during quadrotor controller design. However, a very short period of time is permitted to stabilize the quadrotor with the desired precision. The actuator speed of all propellers' must be matched to attain attitude and altitude tracking and control in all three axes. Therefore, a more precise controller scheme is required to fulfil its high maneuver capabilities. Due to this reason, the modelled quadrotor needs an external controller.

The complete quadrotor control scheme is formed by combining these two control loops:

(a) Inner Control Loop
(b) Outer Control Loop.

The attitude angles are stabilized by the inner control loop whereas, the altitude is controlled using outer control loop. Due to the fact that only four degree of freedom from six can be stabilized using input voltages, the quadrotor is considered as under-actuated system. In this paper the roll, pitch and yaw as an internal loop and altitude from outer loop are considered as control states.

In the work, carried out in the paper, the quadrotor dynamics is modelled using Newton-Euler approach. The two modern and intelligent control strategies are applied to design the control systems for the quadrotor model and their performances are compared.

2 Quadrotor Modelling

The quadrotor control is quite difficult because of its nonlinear dynamics. Therefore, it is proposed to develop a simplified UAV model with least number of output states retaining main features of real aircraft.

2.1 *Concept and Generalities*

Quadrotor is a four-rotor fixed pitch type aerial vehicle. In Fig. 1, the four rotors indicate generated thrust (T_i). The navigation and control of aircraft is done by varying the speed of rotors. The direction of rotation of front and rear rotor is anti-clock wise, while the direction of rotation of left and right rotor is clock wise. This is necessary to remove the need of tail rotor in these types of helicopters. Figure 1 shows the quadrotor in hovering condition. The quadrotor structure is shown with blue colour. The vertical thrust and the force due to gravity are shown by red arrows. The earth fixed frame is shown with black colour [4].

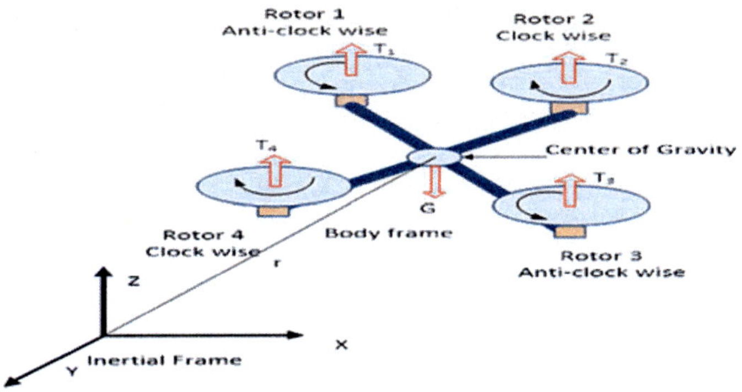

Fig. 1 Simplified diagram of quadrotor model

In the above diagram, the T_1, T_2, T_3 and T_4 are the thrusts produced by the propeller action and x, y, z are the coordinate axes in inertial frame. The collective sum of these thrust results in hovering of the quadrotor aerial vehicle. The red arrow pointing downward represents the effect of force of gravity on the modelled quadrotor. The earth fixed frame is represented by black colour. For roll movement, the speed of right side (or left) rotor is increased (or decreased). Similarly, for pitch motion, the front (or back side) rotor speed is increased (or decreased). And for yaw movement, the speed of front rotor is decreased (or increased) while increasing (or decreasing) the speed of rear motor concurrently. It is important to note that the total amount of force will remain constant during entire process.

The inputs of the quadrotor are designed in such a way so as to give desired performance as that of real aircraft. The derived inputs to the quadrotor motors are labelled as U_1, U_2, U_3 and U_4. Each input has different effects on quadrotor dynamics. The variation in U1 changes the altitude of quadrotor while U_2, U_3 and U_4 are responsible for the change in Euler's angle (Pitch, Roll and Yaw movement) of quadrotor [5].

The quadrotor dynamic model is a set of equations which are highly complex [6]. The equations of motion are derived using aerodynamics and forces acting on the aircraft. The complex quadrotor equations are linearized and converted into state space form for better understanding.

2.2 *State Space Model*

Defining the state space vector X of quadrotor:

$$X = [x_1 x_2 x_3 x_4 x_5 x_6 x_7 x_8 x_9 x_{10} x_{11} x_{12}]^T \tag{1}$$

This has to be mapped with quadrotor's degree of freedom in the following manner:

$$X = [\varnothing\ \dot{\varnothing}\ \theta\ \dot{\theta}\ \psi\ \dot{\psi}\ x\ \dot{x}\ y\ \dot{y}\ z\ \dot{z}]^{\mathrm{T}} \tag{2}$$

where,

$$\left.\begin{array}{ll} x_1 = \dot{\varnothing} & x_7 = x \\ x_2 = \dot{x}_1 = \dot{\varnothing} & x_8 = \dot{x}_7 = \dot{x} \\ x_3 = \theta & x_9 = y \\ x_4 = \dot{x}_3 = \dot{\theta} & x_{10} = \dot{x}_9 = \dot{y} \\ x_5 = \psi & x_{11} = z \\ x_6 = x_5 = \psi & x_{12} = \dot{x}_{11} = \dot{z} \end{array}\right\} \tag{3}$$

The state vector defines the linear and angular velocity and position of the quadrotor in space.

The control input vector U is given by:

$$U = [U_1 U_2 U_3 U_4] \tag{4}$$

where the inputs are mapped as:

$$U_1 = b\left(\omega_1^2 + \omega_2^2 + \omega_3^2 + \omega_4^2\right) \tag{5}$$

$$U_2 = bl\left(-\omega_2^2 + \omega_4^2\right) \tag{6}$$

$$U_3 = bl\left(-\omega_1^2 + \omega_3^2\right) \tag{7}$$

$$U_4 = d\left(-\omega_1^2 + \omega_2^2 - \omega_3^2 + \omega_4^2\right) \tag{8}$$

$$\omega_r = (-\omega_1 + \omega_2 - \omega_3 + \omega_4) \tag{9}$$

The object frame is converted to earth fixed frame using transformation matrix. The transformation matrix used to convert object frame is shown below:

$$R_{xyz} = \begin{bmatrix} C_\varphi C_\theta & C_\varphi S_\theta S_\psi - S_\varphi C_\psi & C_\varphi S_\theta C_\psi + S_\varphi S_\psi \\ C_\varphi S_\theta & S_\varphi S_\theta S_\psi + C_\varphi C_\psi & S_\varphi S_\theta C_\psi - C_\varphi S_\psi \\ -S_\theta & C_\theta S_\psi & C_\theta C_\psi \end{bmatrix} \tag{10}$$

where $S_\theta =$ sin (θ), $C_\psi =$ cos (ψ), etc., and R_{xyz} is the transformation matrix.

The state space model of quadrotor UAV obtained after simplification is described below:

$$f(X,U)=\left(\begin{array}{c}\dot{\varnothing}\\ \dot{\theta}\,\dot{\varnothing}a_1+\dot{\theta}a_2\omega_r+b_1U_2\\ \dot{\theta}\\ \dot{\varnothing}\dot{\psi}a_3+\dot{\varnothing}a_4\omega_r+b_2U_3\\ \dot{\psi}\\ \dot{\varnothing}\dot{\theta}a_5+b_3U_4\\ \dot{z}\\ g-(\cos\theta\cos\varphi)\frac{U_1}{m}\\ \dot{x}\\ u_x\frac{U_1}{m}\\ \dot{y}\\ u_y\frac{U_1}{m}\end{array}\right)\Bigg\} \tag{11}$$

where,

$$a_1=\frac{I_{yy}-I_{zz}}{I_{xz}} \tag{12}$$

$$a_2=\frac{J_{xx}}{I_{xx}} \tag{13}$$

$$a_3=\frac{I_{zz}-I_{xx}}{I_{yy}} \tag{14}$$

$$a_4=\frac{J_{yy}}{I_{yy}} \tag{15}$$

$$a_5=\frac{I_{xx}-I_{yy}}{I_{zz}} \tag{16}$$

$$b_1=\frac{l}{I_{xx}} \tag{17}$$

$$b_2=\frac{l}{I_{yy}} \tag{18}$$

$$b_3=\frac{1}{I_{xx}} \tag{19}$$

$$u_x=C_\theta C_\psi+S_\varnothing S_\psi \tag{20}$$

$$u_y=C_\varnothing C_\theta S_\psi+S_\varnothing S_\psi \tag{21}$$

where x, y and z are three positions and θ, ψ, and φ are the pitch, roll, and yaw respectively.

Where b denotes the thrust coefficient and d denotes the drag coefficient, U_i are the input variable to the input vector, ω_i are the angular speed of the of the propeller, and I_{xx}, I_{yy} and I_{zz} are the moments of inertia of the quadrotor. Many assumptions are made and disturbances are ignored while developing the model [7].

Transfer function obtained for pitch, yaw, roll and altitude are as follows:

Pitch channel

$$G_1 = \frac{56.65s + 4391}{s^3 + 105s^2 + 870s + 4430} \tag{22}$$

Yaw Channel

$$G_2 = \frac{105}{s^2 + 413s} \tag{23}$$

Roll channel

$$G_3 = \frac{65s + 4560}{s^3 + 109s^2 + 1023s + 2935} \tag{24}$$

Z-Axis

$$G_4 = \frac{1.63}{s^2 + 5s} \tag{25}$$

2.3 Modelling of Brushless Direct Current Motor

The model of the BLDC motor can be described as linear second order system. The armature current and motor torque are related by a constant K as:

$$M(t) = K.i(t) \tag{26}$$

The back emf V_{EMF} is proportionate to the angular velocity of BLDC motor:

$$V_{\mathrm{EMF}}(t) = e(t) = K.\omega(t) = K\frac{d\theta}{dt} \tag{27}$$

We can write the equation based on Newton's law combined with Kirchhoff's law as follows:

$$J\frac{d^2\theta}{dt^2} + \frac{d\theta}{dt} = Ki \tag{28}$$

$$L\frac{di}{dt} + Ri = V - K\frac{d\theta}{dt} \tag{29}$$

where R is a resistance and L is an inductance of the armature and J is a moment of inertia of the rotor. The transfer function derived for motor is shown below [8]:

$$G_4 = \frac{K}{\left[(R + Ls)(Js + b) + K^2\right]} \tag{30}$$

where, the implemented values are $V(s)$= 12 V; $J = 1 \times 10^{-2}$; $b = 3 \times 10^{-4}$; $K = 23 \times 10^{-1}$; $R = 1$; *and* $L = 0.5$.

3 Controller Design

To stabilize the quadrotor different control schemes are applied and investigated. Among many two effective control schemes are discussed and compared in present paper. The control algorithms discussed in this paper are: The Linear Quadratic Regulator, and a newly developed intelligent PD-ANFIS based hybrid controller. These applied techniques and controlled system responses are discussed in the following section:

3.1 LQR Control of Roll Moment

It is a convincing substitute to the techniques used to solve MIMO control problems. It is usually called as LQ or LQR. The LQR technique is used to generate best gain matrix K, exempting the pole placement methodology. In LQR, the system errors and control efforts are balanced optimally. The cost functions are defined by the designer that specifies the importance of control efforts and errors.

For a state space model:

$$x = A\dot{x} + Bu \tag{31}$$

With cost function:

$$J = \int_0^\infty (x^T Qx + u^T Ru + x^T Su)dt \tag{32}$$

The linearized quadrotor model (22–25) is utilized to calculate the system matrices to apply LQR control algorithm. The values of weight matrix, gain matrix and the new system matrix A_lqr used for roll, pitch, yaw and altitude control are calculated using equation (33–35) respectively.

$$Q = W * C' * C \tag{33}$$

$$K = lqr(A, B, C, D) \tag{34}$$

$$A\_lqr = A - B * K \tag{35}$$

The values obtained for input matrix B and Control matrix C are as it is used in new system matrices B_lqr and C_lqr respectively. As already discussed earlier that the aircraft is considered in hover condition, in which the disturbances associated are considered negligible. That's why the value of matrix D is considered zero during attitude control using LQR.

3.2 Hybrid Controller

The results obtained by applying ANFIS controller are not acceptable in case of a highly unstable nonlinear unmanned aerial vehicle. A fast control is required to control the quadrotor so that it can stabilize the system within very short time and can prevent the flying vehicle from falling to the ground. In investigation it is found that if two controllers are applied simultaneously to the quadrotor dynamics, it can achieve highly appreciable results [9]. So, an intelligent Hybrid Controller is developed as an initial experiment. In this controller, two control strategies are combined namely: (a) P-D and (b) PD trained ANFIS control scheme. The strategy is to utilize P-D controller for altitude and yaw control and the remaining roll and pitch moments are controlled by using intelligent PD trained ANFIS controller [10]. On combining the controllers provided appreciable results. Figure 3, describes the architecture of developed intelligent hybrid PD trained ANFIS controller (Fig. 2).

To stabilize the UAV, the LQR and intelligent PD trained ANFIS controller are applied and the results are analyzed. The comparison of responses concludes that the LQR control scheme provides better control for quadrotor UAV. Figure 3 Shows the comparative result after applying LQR and intelligent hybrid controller for altitude and attitude control.

4 Results and Discussion

See Fig. 3.

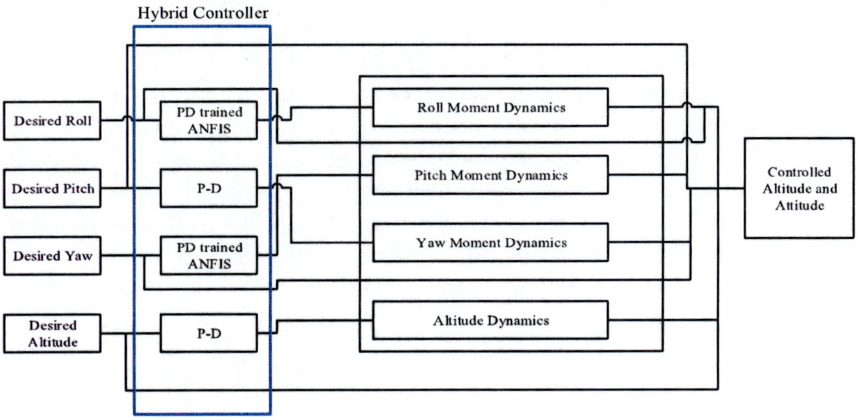

Fig. 2 Block diagram representation of hybrid control scheme applied to the quadrotor model

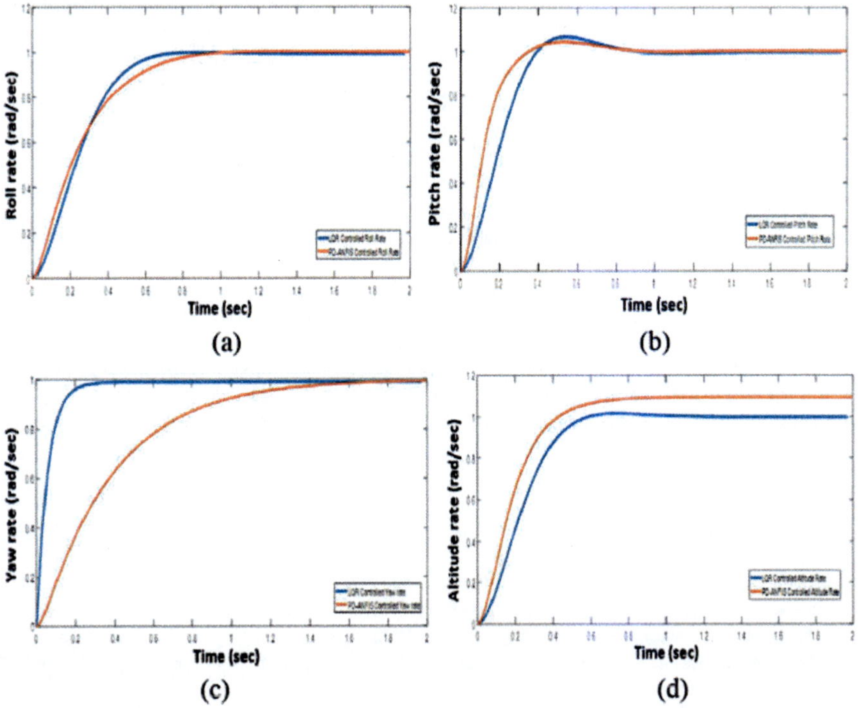

Fig. 3 Comparative results after applying LQR and intelligent hybrid controller to the (**a**) roll moment (**b**) pitch moment (**c**) yaw moment (**d**) altitude of the quadrotor

5 Conclusion

The present paper describes the modelling of quadrotor UAV using Newton-Euler formulation. The unstable behavior of the quadrotor convinced the researcher to search for an external controller to stabilize the vehicle. In investigation, it is observed that two controllers (i) LQR and (ii) intelligent hybrid controller have provided appreciable results. The results are compared for altitude and attitude stabilization of quadrotor. The comparative study proposed that the applied LQR control scheme is best suited for the quadrotor.

References

1. Azzam A, Wang X (2010) Quad rotor arial robot dynamic modeling and configuration stabilization. In: 2010 2nd International Asia conference on informatics in control, automation and robotics (CAR 2010), vol 1, pp 438–444. https://doi.org/10.1109/car.2010.5456804
2. Hou H, Zhuang J, Xia H, Wang G, Yu D (2010) A simple controller of minisize quad-rotor vehicle. In: 2010 IEEE international conference on mechatronics and automation, pp 1701–1706. https://doi.org/10.1109/icma.2010.5588802
3. Kim J, Kang M-S, Park S (2009) Accurate modeling and robust hovering control for a quad-rotor VTOL aircraft. J Intell Robot Syst 57(1):9. https://doi.org/10.1007/s10846-009-9369-z
4. Mian AA, Daobo W (2008) Modeling and backstepping-based nonlinear control strategy for a 6 DOF quadrotor helicopter. Chin J Aeronaut 21(2008):261–268
5. Rodi A, Mester G (2011) The modeling and simulation of an autonomous quad-rotor microcopter in a virtual outdoor scenario. 8(4):107–122
6. Lai L, Yang C, Wu C (2006) Time-optimal control of a hovering quad-rotor helicopter. J Intell Robot Syst 45(2006):115–135. https://doi.org/10.1007/s10846-005-9015-3
7. Bouabdallah S (2007) Design and control of quadrotors with application to autonomous flying
8. Aung W (2007) Analysis on modeling and simulink of DC motor and its driving system used for wheeled mobile robot. World Acad Sci Eng 26(December):299–306
9. Santoso F, Garratt MA, Anavatti SG (2019) Hybrid PD-Fuzzy and PD controllers for trajectory tracking of a quadrotor unmanned aerial vehicle: autopilot designs and real-time flight tests. IEEE Trans Syst Man Cybern Syst PP:1–13. https://doi.org/10.1109/tsmc.2019.2906320
10. Khatoon S, Nasiruddin I, Shahid M (2017) Design and simulation of a hybrid PD-ANFIS controller for attitude tracking control of a quadrotor UAV. Arab J Sci Eng 42(12):5211–5229. https://doi.org/10.1007/s13369-017-2586-z
11. Iqbal A et al (eds) (2020) Soft computing in condition monitoring and diagnostics of electrical and mechanical systems, vol 1096. In: Advances in intelligent systems and computing. Springer, Singapore. https://doi.org/10.1007/978-981-15-1532-3
12. Iqbal A et al (eds) (2020) Meta heuristic and evolutionary computation: algorithms and applications, vol 1096. In: Studies in computational intelligence. Springer, Singapore. https://www.springer.com/gp/book/9789811575709

Performance Analysis of PV Array Connection Schemes Under Mismatch Scenarios

Mohd Faisal Jalil, Mohammad Shariz Ansari, Sourav Diwania, and Mohammed Aslam Husain

Abstract The power produced through the photovoltaic (PV) plants is exposed to mismatch/partial shading scenarios, the actual position of the panel and connection scheme of the PV array. Consequently, peak power point tracking becomes complex owing toward the existence of compound maxima in power–voltage characteristics. There are different connection schemes of PV array, available in literature particularly series–parallel (S-P), honeycomb (H-C), bridge link(B-L). Then total cross tied (T-C-T) regularly used during partial shading scenarios. For each of the abovementioned connections, the scheme has relative benefits and drawbacks. In this paper, all conventional connection schemes are reconfigured by the odd–even arrangement. A 6 × 6 PV array is considered for investigation for two severe partial shading scenarios. The odd–even reconfiguration arrangement scatters the influence of intense shading over complete PV array and hence reduces the mismatch losses. The global maxima move close to the typical functional point as all strings contribute to overall instantaneous power generation. The results obtained present the improvement with reference to fill factor, peak power yield and the amount of multiple maxima.

Keywords Partial shading condition · PV array configuration · Reconfiguration

M. F. Jalil (✉) · M. S. Ansari · S. Diwania
KIET Group of Institutions, Delhi-NCR, Ghaziabad, India
e-mail: mfaisaljalil@gmail.com

M. S. Ansari
e-mail: shariz.ansari@kiet.edu

M. A. Husain
Rajkiya Engineering College, Ambedkar Nagar, Banda, UP, India

A. Iqbal et al. (eds.), *Renewable Power for Sustainable Growth*, Lecture Notes in Electrical Engineering 723, https://doi.org/10.1007/978-981-33-4080-0_22

1 Introduction

The solar energy's potential is tremendous; it is capable of supplying the world's demand for energy many times over [1–12]. The International Energy Agency forecasts that photovoltaic (PV) will contribute about 16% of worldwide electricity supply by 2050. Solar power is produced by transforming energy from sunlight into electricity, either straight using solar cells (PV) or secondarily using concentrating solar power (CSP) systems. In the PV power plant, PV modules transform sunlight into the electric current as per the photovoltaic effect. Whereas, the CSP system employs lenses or mirrors and tracking schemes to focus sunlight onto a fluid. The fluid is heated to 250–1000 °C and used to generate steam. In 2018, the global electricity production capacity of PV is 505 GW and CSP is 5.5 GW. In this work, we will be focusing on PV's issues and challenges. At the end of 2018, PV accounted for 2.4% of the world's electricity production. Although the global electricity supply from renewables is 26.2% with 15.8% hydropower, 5.5% wind power, 2.4% solar power and remaining from other sources.

The power produced by the PV power plant is reliant on several constraints such as irradiance, temperature, soiling and aging effect. The most significant parameter among these is irradiance. When modules in a PV array are receiving even irradiance, there will be a unique global peak in power–voltage characteristics for maximum power output. On the occurrence uneven irradiance due to neighboring construction, deposition of bird wastes and dust, shades of trees, etc., there will be several peaks in power–voltage characteristics, therefore operation at optimum power point becomes difficult and conventional MPP algorithms become ineffective under these conditions [1, 2]. There are several researchers used mostly commonly embraced single diode models for modeling PV module as it is sufficiently accurate but under low illumination or in uneven insolation, the implementation of this single diode model requires improvements [3]. Some researchers preferred two diode model for modeling PV panel under these uneven irradiances or partial shading conditions with improved results in comparison with the previous model at cost complexity as it requires estimation of seven parameters. The operation of the PV module in second and fourth quadrant is optimally presented with the help of the Bishop model [4]. Several researchers used this model but there is no generalized technique available for parameter estimation of Bishop model. The operation of PV array in these uneven conditions is presented for simple series and parallel connections, and outcomes show that series connections in the PV array must be avoided to obtain higher-power output in these conditions [5, 6]. There are several analyses on PV array connection schemes available in literature aiming to obtain optimal power output under partial shading conditions [7, 8]. Analyzed S-P, B-L, H-C and TCT for shades on equal PV array area but in different patterns. The power is different for different patterns irrespective of the same shade area. The TCT connection scheme is the most efficient in terms of maximum power output, and this is since the TCT connection scheme reduces mismatch that is occurring due to the non-uniformity in irradiance of PV array. Some authors further reconfigured array connection schemes to reduce these mismatch losses [9]. There are two types of reconfigurations, namely electrical reconfiguration and physical reconfiguration. The physical reconfiguration is more popular among researchers as it is simple and economical in comparison with electrical reconfiguration. The authors in [10] proposed the odd–even technique as fixed reconfiguration to structure the PV array electrically connected in TCT configuration. They only implemented the odd–even technique for TCT configuration. There are certain PV array configurations especially asymmetrical in which HC configurations give higher maximum power yield in contrast among distinct connection arrangement. This paper presents effectiveness investigation of PV plant configurations series–parallel (SP)–odd–even series–parallel (OESP), bridge link (BL)–odd–even bridge link (OEBL), honeycomb (HC)–odd–even honeycomb (OEHC), then total cross tied (TCT)–odd–even total cross tied (OETCT) with reference to peak power output.

2 PV Array Connection Schemes

The connection schemes S-P, B-L, H-C then T-C-T stand are presented in Fig. 1a–d [11, 12]. All these connection schemes have their own relative benefits and shortcomings. The peak power output of the PV plant depends on these configurations and shading pattern. Among these configuration, maximum power output is possible for T-C-T for most of the conditions, but in some conditions other configurations present better performance. This section presents a comparative analysis among abovementioned configurations and when these configurations are reconfigured as odd–even scheme of PV modules. Every module in a PV array has unique position. For example, module-23 presents module available in third column and second row of the PV array. For odd–even arrangement, electrical connections remain unaltered as in respective configuration, whereas physical arrangement is decided as odd–even technique [10] (Figs. 2, 3, 4, 5, 6, 7, 8, 9, 10, 11, 12 and 13; Table 1).

3 Conclusion

In this paper, comparative analysis of PV array connection schemes and reconfigurations, namely series–parallel (SP)–odd–even series–parallel (OESP), bridge link (BL)–odd–even bridge link (OEBL), honeycomb (HC)–odd–even honeycomb (OEHC) and then total cross tied (TCT)–odd–even total cross tied (OETCT) have been done for two severe partial shading scenario. An investigation is carried on 6 × 6 PV plant with MATLAB/Simulink. The instantaneous energy generation of the PV plant is same for all connection schemes and reconfigurations under uniform illumination conditions. For considered shading scenario, TCT connection scheme provides leading maximum power output of the array followed by HC, BL and SP connection schemes. Results show if TCT, HC, BL and SP connection schemes are

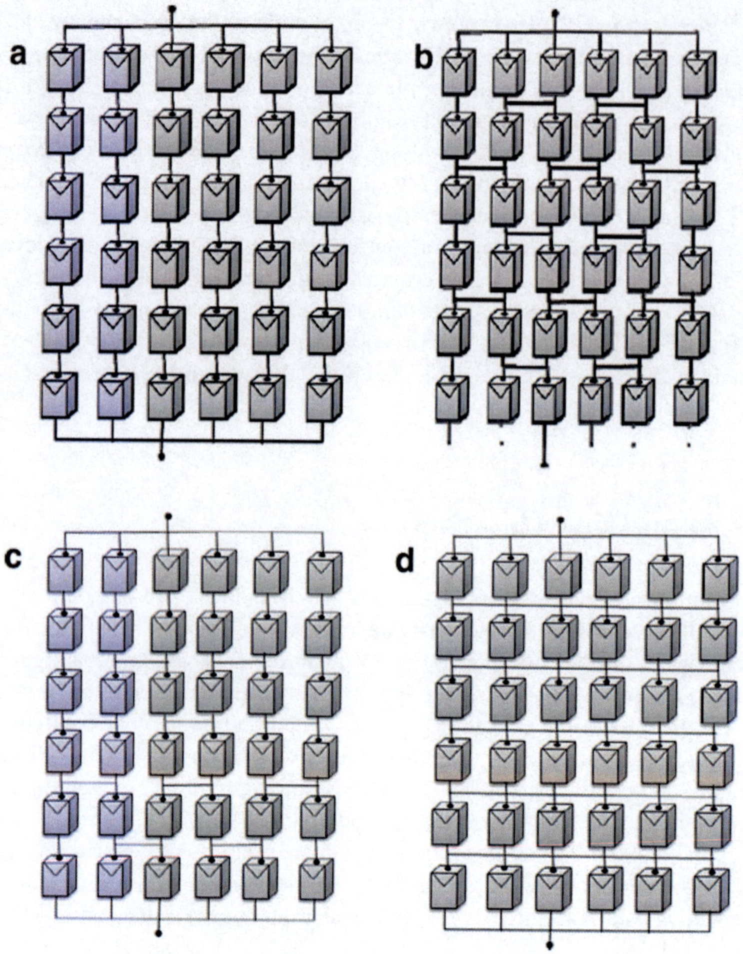

Fig. 1 **a** 6 × 6 PV array as per series–parallel connections ,**b** 6 × 6 PV array as per bridge link connections, **c** 6 × 6 PV array as per honeycomb connections, **d** 6 × 6 PV array as per total cross tied connections

Fig. 2 Shading scenario for PV array for case 1

11	12	13	14	15	16
21	22	23	24	25	26
31	32	33	34	35	36
41	42	43	44	45	46
51	52	53	54	55	56
61	62	63	64	65	66

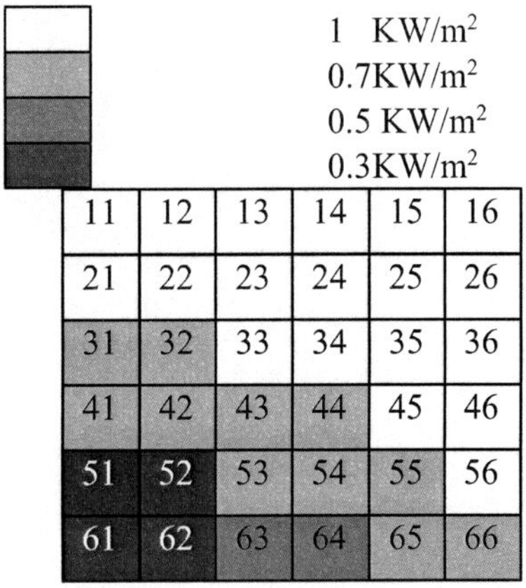

Fig. 3 Shading scenario for PV array for case 2

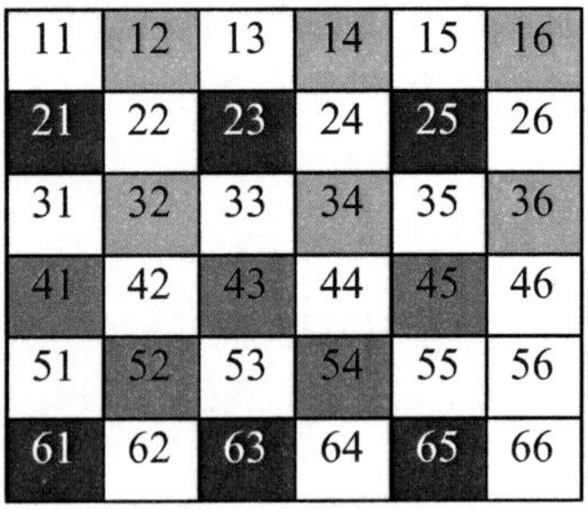

Fig. 4 Shade dispersion for shading scenario considered in case 1

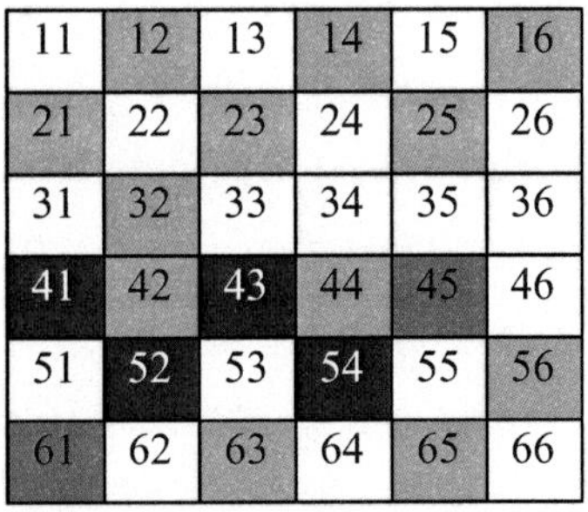

Fig. 5 Shade dispersion for shading scenario considered in case 2

reconfigured as per odd–even arrangement, and then there is increment in peak power yield through the PV plant. As per the results, if it reconfigures the PV array without changing the electrical connection scheme, then performance of the PV array is improved.

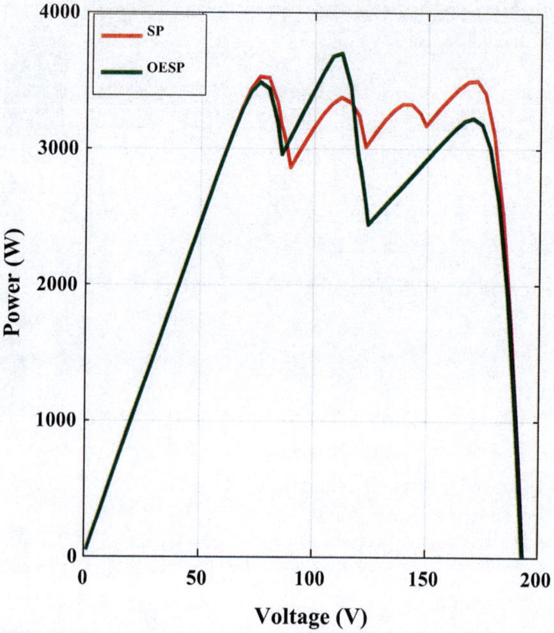

Fig. 6 P-V curves concerning SP and OESP for mismatch scenario in situation 1

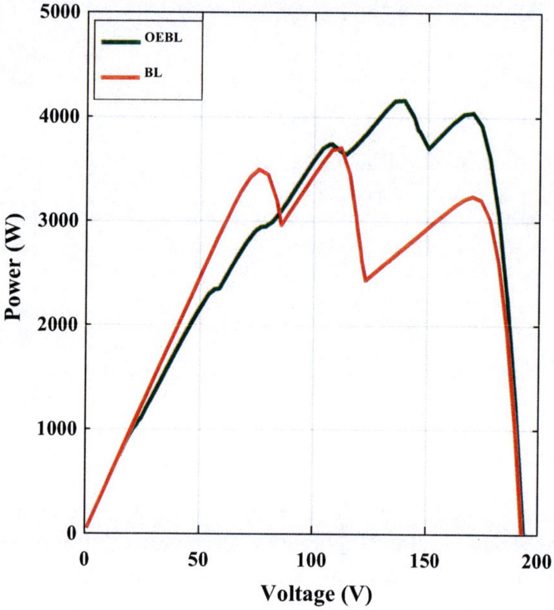

Fig. 7 P-V curves concerning BL and OEBL for mismatch scenario in situation 1

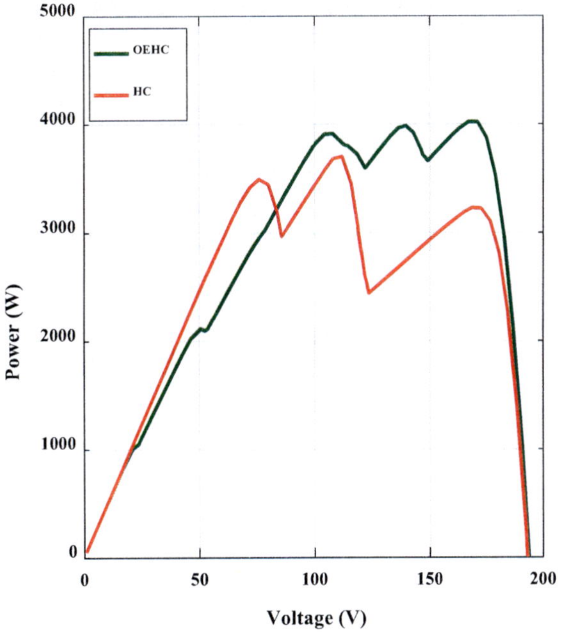

Fig. 8 P-V curves concerning HC and OEHC for mismatch scenario in situation 1

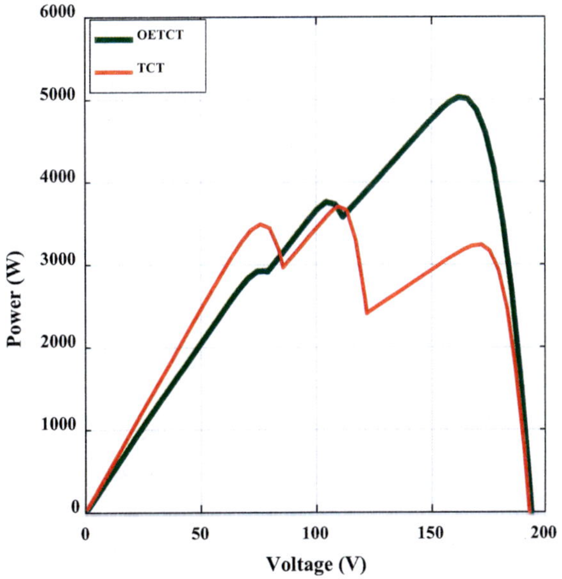

Fig. 9 P-V curves concerning TCT and OETCT for mismatch scenario in situation 1

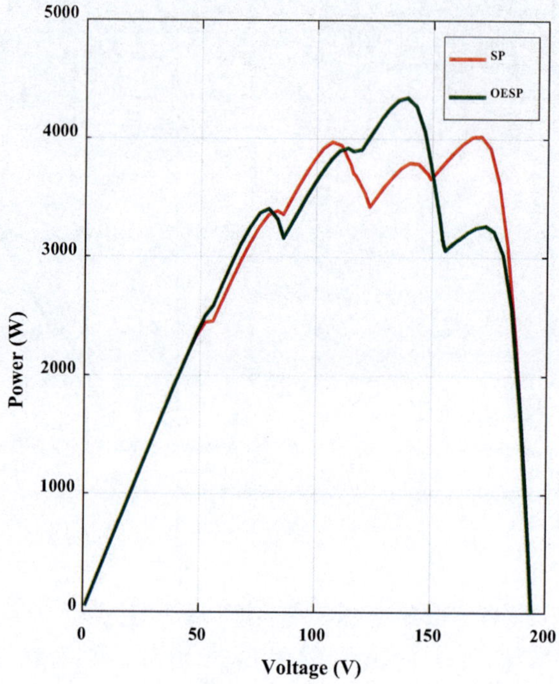

Fig. 10 P-V curves concerning SP and OESP for mismatch scenario in situation 2

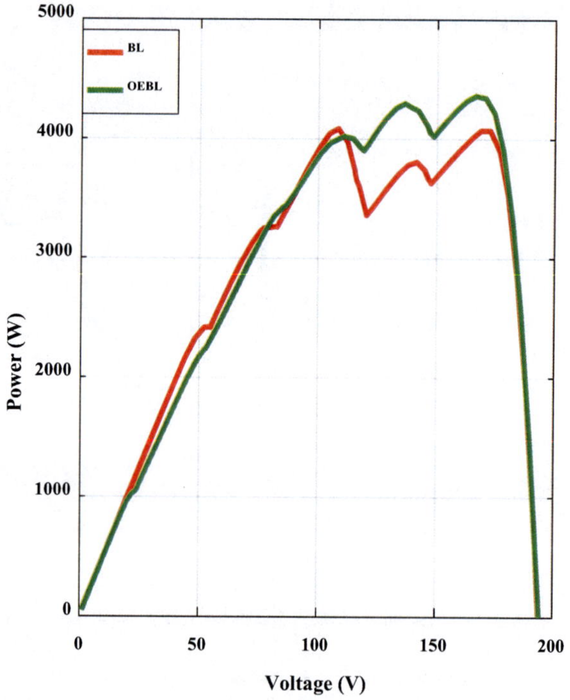

Fig. 11 P-V curves concerning BL and OEBL for mismatch scenario in situation 2

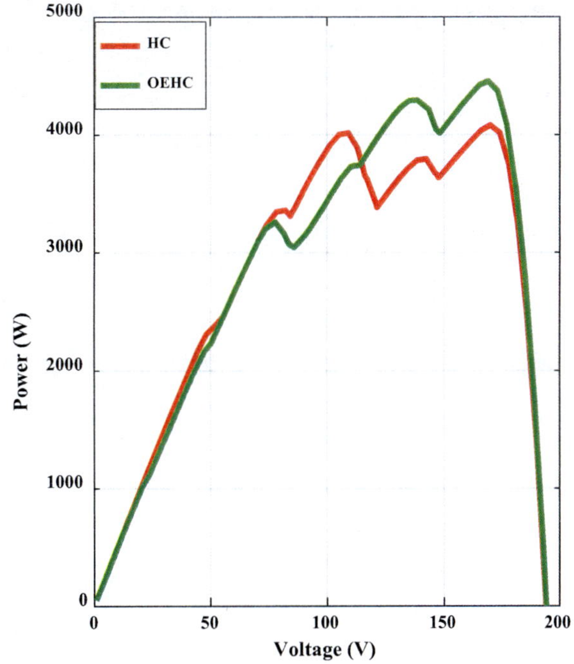

Fig. 12 P-V curves concerning HC and OEHC for mismatch scenario in situation 2

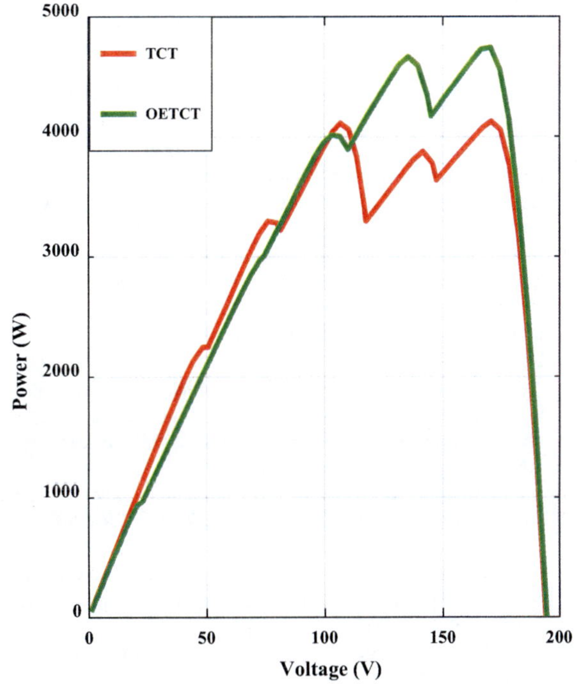

Fig. 13 P-V curves concerning TCT and OETCT for mismatch scenario in situation 2

Table 1 Results in terms of instantaneous maximum power of PV array

	Configuration	I_{max} (A)	V_{max} (V)	P_{max} (W)
Case 1	SP	44.02	79.96	3520
	OESP	33.02	112.1	3702
	BL	32.97	112.2	3700
	OEBL	29.78	139.8	4164
	HC	32.99	112.2	3702
	OEHC	23.51	170.9	4019
	TCT	33.83	109.5	3705
	OETCT	31.05	161.9	5028
Case 2	SP	23.48	171.1	4018
	OESP	31.24	139.2	4349
	BL	37.61	108.6	4084
	OEBL	26.22	166.3	4361
	HC	23.94	170.3	4078
	OEHC	26.30	169.4	4456
	TCT	24.09	171.1	4123
	OETCT	44.04	107.7	4744

References

1. Belhachat F, Larbes C (2018) A review of global maximum power point tracking techniques of photovoltaic system under partial shading conditions. Renew Sustain Energy Rev 92:513–553. https://doi.org/10.1016/j.rser.2018.04.094
2. Ansari MS, Manaullah Jalil MF (2016) Investigation of renewable energy potential in union territory of lakshadweep islands. In: 2nd IEEE international conference on innovative applications of computational intelligence on power, energy and controls with their impact on humanity, CIPECH 2016, pp 209–215. https://doi.org/10.1109/CIPECH.2016.7918768
3. Villalva MG, Gazoli JR, Filho ER (2009) Comprehensive approach to modeling and simulation of photovoltaic arrays. IEEE Trans Power Electron 24:1198–1208
4. Bishop J (1988) Computer simulation of effects of electrical missmatches in PV-Cell interconnection circuits. Sol Cells 25:73–89
5. Khatoon S, Ibraheem Jalil MF (2014) Analysis of solar photovoltaic array under partial shading conditions for different array configrations. In: Proceedings of the international conference on innovative applications of computational intelligence on power, energy and controls with their impact on humanity, CIPECH 2014, pp 452–456. IEEE. https://doi.org/10.1109/CIPECH.2014.7019127
6. Ramabadran R, Mathur B (2009) Effect of shading on series and parallel connected solar PV modules. Mod Appl Sci 3:32–41. https://doi.org/10.5539/mas.v3n10p32
7. Pendem SR, Mikkili S (2018) Modeling, simulation, and performance analysis of PV array configurations (Series, Series-Parallel, Bridge-Linked, and Honey-Comb) to harvest maximum power under various partial shading conditions. Int J Green Energy 15:795–812. https://doi.org/10.1080/15435075.2018.1529577
8. Khatoon S, Ibraheem Jalil MF (2015) Feasibility analysis of solar photovoltaic array configurations under partial shading conditions. In: 2015 Annual IEEE India conference (INDICON), pp 1–6. IEEE. https://doi.org/10.1109/INDICON.2015.7443701

9. Iysaouy L El, Lahbabi M, Oumnad A (2019) A novel magic square view topology of a PV system under partial shading condition. Energy Procedia 157:1182–1190. https://doi.org/10.1016/j.egypro.2018.11.284
10. Nasiruddin I, Khatoon S, Jalil MF, Bansal RC (2019) Shade diffusion of partial shaded PV array by using odd-even structure. Sol Energy 181:519–529. https://doi.org/10.1016/j.solener.2019.01.076
11. Iqbal A et al. (eds) (2020) Soft computing in condition monitoring and diagnostics of electrical and mechanical systems, vol 1096. In: Advances in intelligent systems and computing. Springer, Singapore. https://doi.org/10.1007/978-981-15-1532-3
12. Iqbal A et al. (eds) (2020) Meta heuristic and evolutionary computation: algorithms and applications, vol 1096. In: Studies in computational intelligence. Springer, Singapore. https://www.springer.com/gp/book/9789811575709

Modelling and Analysis of Bridgeless PFC Boost Convertor

Md. Shamsuddin and Piyush Agrawal

Abstract All type of electronics devices works on the DC power supply and use AC-to-DC conversion device. Passive elements are used in the AC-to-DC conversion device which creates the harmonics in the system. Harmonic current always introduces poor power factor in operating device. The low power factor created in any type of the electrical or electronics devices deteriorates its performance. PFC boost converter helps in improving the power factor. Power factor correction boost converter significantly improves both power factor and harmonic distortion Two types of circuit model (a) conventional and (b) bridgeless type are discussed in this paper. The difference between the conventional power factor correction converter and bridgeless power factor correction is that diode bridge is present in conventional power factor correction circuit, but no diode bridge is present in the type (b) PFC.

Keywords PFC · Bridgeless PFC · THD

1 Introduction

In recent years, as the demand has increased, the need for better power factor and less total harmonics distortion in the current drawn from the power supply has become an area of concern [1–9]. The need for electricity supply has always been there, and with increased demand and generation, power quality has always been an issue. Though there is a economic slowdown in most of the world, still the demand of electricity continues to grow. It grows at 2.1% per year according to IEA. The development of power electronic equipment's has led to more advanced converters and given many other advantages. The drawback is being increased harmonic contents and poor power factor. Many ways have been proposed to reduce harmonic like use of different types of filters and FACTS devices. In the development of PFC, mostly, researches are related to the continuous conduction mode (CCM) because CCM topology is simple

Md. Shamsuddin (✉) · P. Agrawal
Department of EEE, IILM Academy of Higher Learning, Greater Noida, India
e-mail: shamsuddinalam99@gmail.com

A. Iqbal et al. (eds.), *Renewable Power for Sustainable Growth*, Lecture Notes in Electrical Engineering 723, https://doi.org/10.1007/978-981-33-4080-0_23

and has high power gain [8, 9]. It may be used for universal voltage range. In recent year, electronic companies are working to reduce the size and weight of the electronics converter which are also used in the residential, commercial and industries. All street is supplied from different types of three-phase supply which creates unbalancing in a housing scheme. Unbalanced current flow throws the neutral of star connection configuration [1–3]. The waveform of the supply voltage gets distorted because of reactive power. The bridgeless PFC has no bridge loss because bridge rectifier is not present in this topology, and the bridgeless topology is mostly used in the dual boost PFC rectifier. Dual boost topology also reduced the conduction losses because of reduction of the semiconductors in the line current path.

2 Boost Converter (PFC)

Angle $\cos(\phi)$ between the voltage and current is known as the power factor. Unity power factor is also considered as the best power factor. Good power factor provides the better operating quality for the device. The main work of the PFC circuit makes the input power supply to work as a pure resistor. When the ratio between current and voltage is constant, then the input will be resistive and PF is unity [4, 5]. When the ratio is not constant between the current and voltage, then due to the existence of nonlinear load, input will contain phase displacement and harmonics distortions, and then, the power factor also becomes poor.

2.1 Low Power Factor

Power factor depends on the type of load. In case of purely resistive load, power factor is good. In case of inductive load, power factor is not good because in case of inductive load, current lags with the 90° from voltage. In condition of current lag or lead, the harmonics are generated in the device, and it makes cause of the low power factor.

Low PF does not provide better operation for the device.

The harmonics current is caused by a nonlinear load or by a solid-state rectifier [6–8].

2.2 Need of Power Factor Correction (PFC)

Nowadays, in the industry, residential and commercial places used a huge variety of electronic system which require supply. The electronic system uses DC-to-DC converter as they require DC power supply.

In Fig. 1, waveforms with and without PFC are shown.

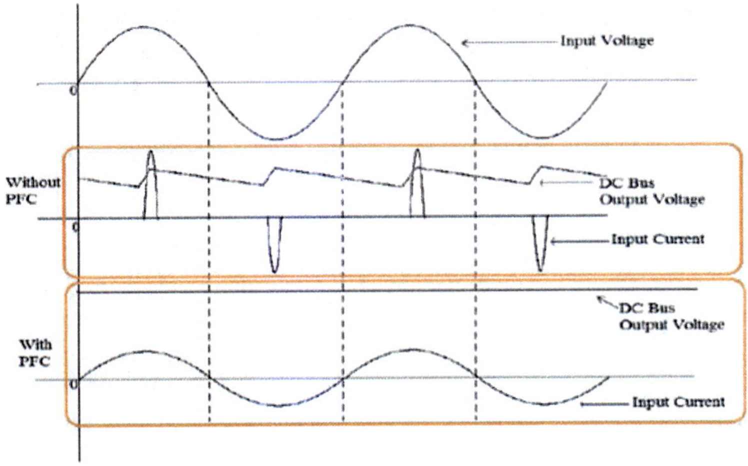

Fig. 1 Waveform using PFC and without PFC

2.3 *Types of PFC*

- Passive PFC;
- Active PFC.

Passive PFC as shown in Fig. 2 uses filter at the input. In the circuit of the passive power factor correction, only passive elements are used, for example, inductor (L) and capacitor (C). In the passive PFC circuit, the use of extra electronics component such as the bridge rectifier is a disadvantage. Passive elements also help to improve the nature of line current. [9] The use of PFC circuit improves the power factor to about 0.7–0.8. Low-pass filter is used in passive PFC circuit.

Benefits of Passive PFC

- Passive PFC has a normal structure.

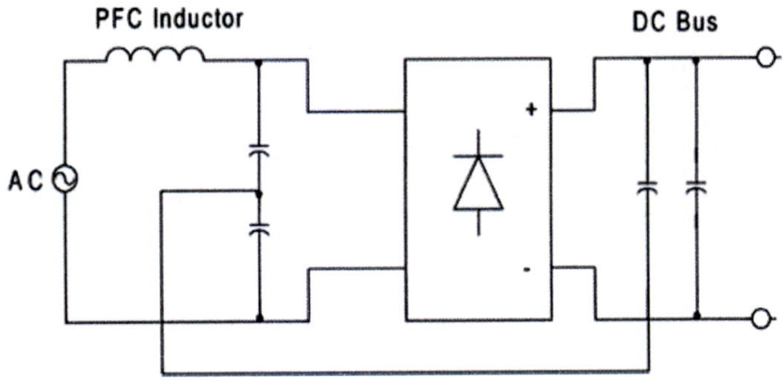

Fig. 2 Passive PFC

- It is good in quality and well built.
- In the passive PFC, only filter is required.
- Has no high switching loss.
- In the passive PFC, the used element does not create high-frequency EMI.

2.4 *Active PFC*

Figure 3 shows active PFC which gives much better result than the passive PFC. THD obtained from it is much lower than passive THD. The size and weight of active PFC are also less compared to passive PFC. The cost of active PFC circuit is less as it does not use the passive elements. Active PFC circuit works on the high switching frequency.

Benefits of Active PFC

- It is smaller than the passive PFC.
- 0.95 PF is generally achieved through this method.
- Active PFC also reduces harmonics current which is present in the device.

3 Conventional PFC Boost Converter

The use of electrolyte capacitor is for rectification and filter of harmonics in case of conventional PFC as shown in Fig. 4. It improves the distorted input and harmonics content. The important thing is that it decreases the harmonics and provides the better power factor in the input current for the power supply. The conventional method is the most widely used method in the improvement of power factor application. In the conventional PFC boost converter, full wave rectifier bridge is used. It works as the AC-to-DC converter [6]. This method is better for small to average power supply. As the increment in the power level, then heat dissipation in bridge rectifier becomes important for this application. It is also protected from heat problem in specific surface area (Figs. 5, 6 and 7).

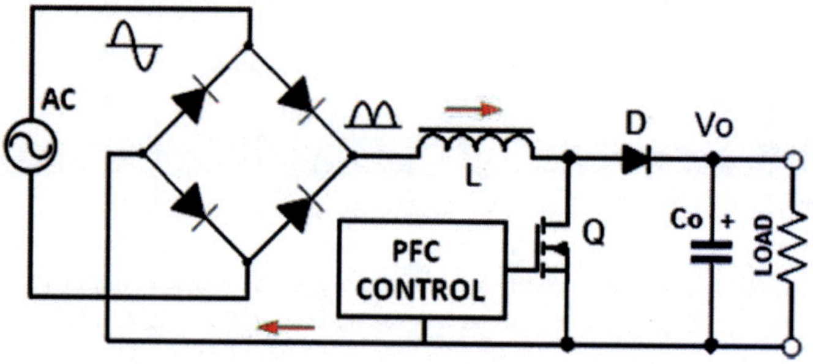

Fig. 3 Active power correction

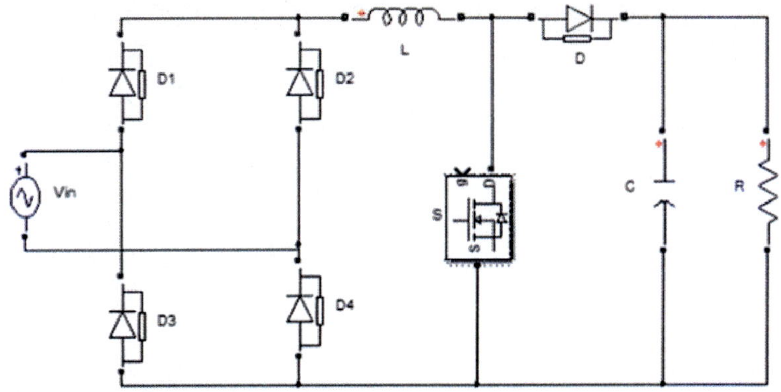

Fig. 4 PFC boost converter (conventional)

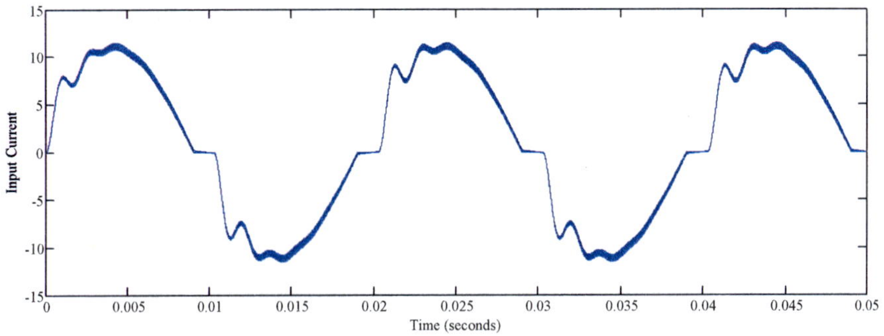

Fig. 5 Input current

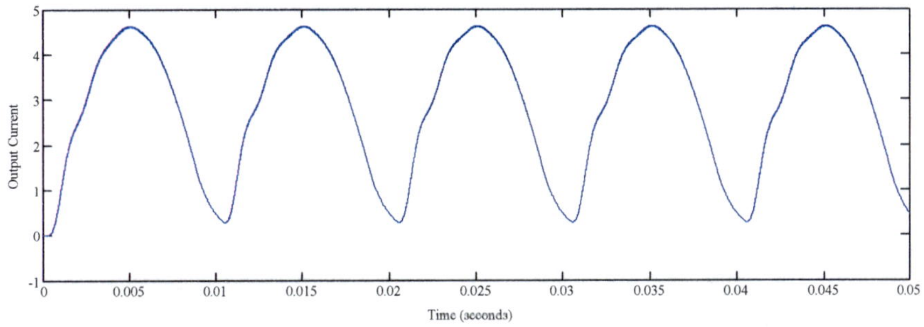

Fig. 6 Output voltage

Table 1 shows variation in power factor and THD with change in load at constant voltage 24 V. Load change at constant voltage O/P power is increased and PF is improved.

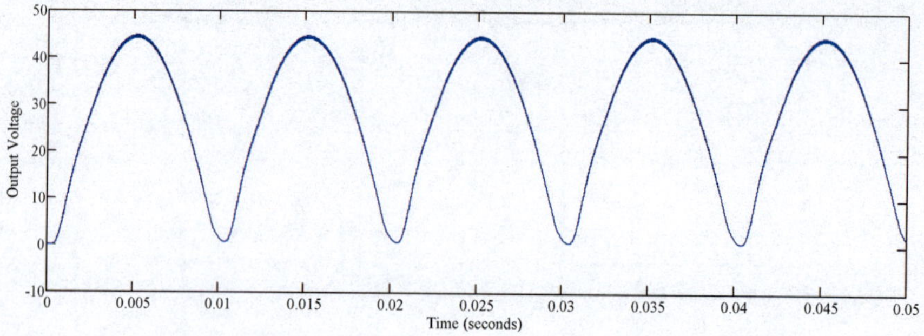

Fig. 7 Output current

Table 1 Change in PF and THD at 24 V

Input voltage 20 Vac		
Power factor	THD (%)	Output power
0.9618	15.80	101.6
0.9693	14.12	112.2
0.9762	12.50	125.3
0.9826	10.96	141.8
0.988	9.53	163.4

4 PFC Boost Converter (Bridgeless)

In the converter circuit, no bridge rectifier is used for the AC-to-DC converter. In this convertor, heat generates in the small amount. The bridgeless PFC converter is shown in Fig. 8. The inductor is present on the ac source side. Only two diodes are present, D1 and D2. Both diodes are connected in the parallel, and inductor is also connected in series. [6] Capacitor (C) which is used in the parallel also improves the PF of the system. There are two MOSFETs which are denoted as S1 and S2 used in the converter circuit.

5 Operations

Operation mode of this converter

- Boost converter;
- Return path ac input signal.

Mode-1

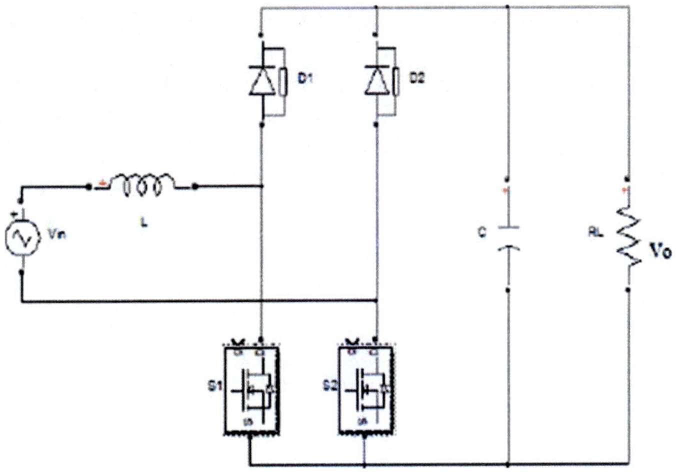

Fig. 8 Basic bridgeless PFC boost converter

Power MOSFET used in the bridgeless PFC is denoted by the switch S1, and it is an important part of the circuit. Power MOSFET is used for amplifying signal and switching purpose. MOSFET is high when AC voltage has positive cycle on gate of S1, and then, S1 is turned on and inductor stores the energy from the input source. After the closing of MOSFETS1, the stored energy passes through the diode D1, and its return path is diode 2 and MOSFET 2 as shown in Fig. 9.

Mode-2

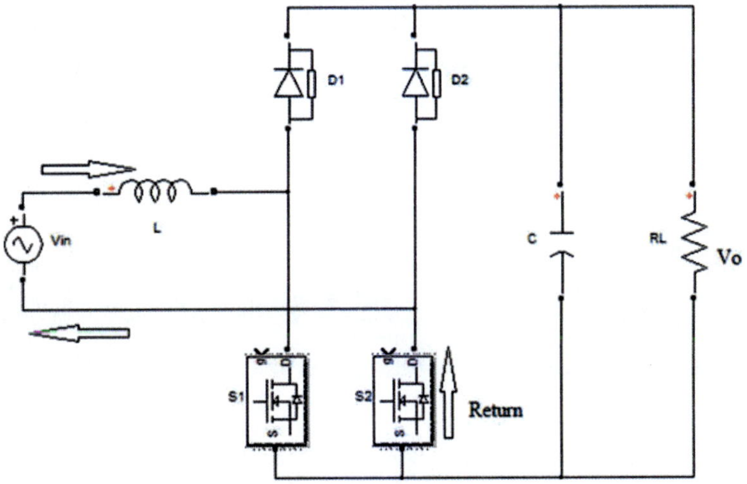

Fig. 9 Positive ½ line cycle

Turning on the power MOSFET S2, inductor stores the energy. When S2 is turned off, then inductor passes the stored energy through the (D1) and off the circuit through the load. The power MOSFET (S1) may be driven ON/OFF only on positive ½ line cycle and power MOSFET (S2) driven ON/OFF only in case of negative ½ line cycle as shown in Fig. 10.

The output current and output voltage are shown in Figs. 11 and 12, respectively. Table 2 shows variation in power factor and THD with change in load at constant voltage 24 V. Change in load at constant voltage PF is increased and harmonics decrease.

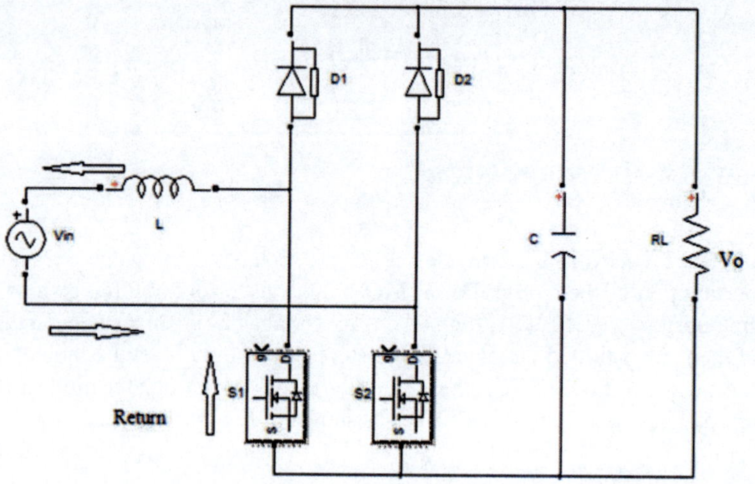

Fig. 10 Negative ½ cycle

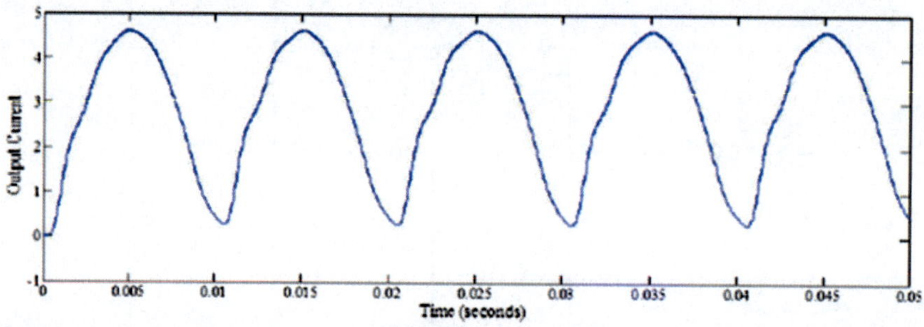

Fig. 11 Output current

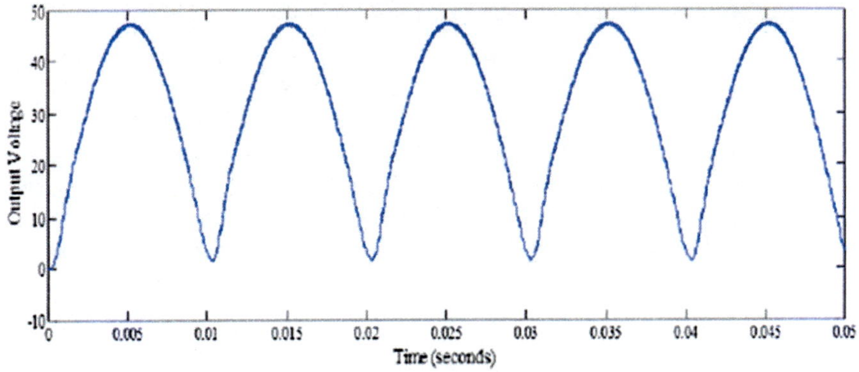

Fig. 12 Output voltage

Table 2 Change in PF and THD at 24 V

Input voltage 20 Vac		
Power factor	THD (%)	Output power
0.9656	13.07	120.3
0.9727	11.41	132.8
0.9792	9.78	148.2
0.9852	8.20	167.6
0.9903	6.67	192.9
0.9946	5.23	227.2
0.9979	3.87	276.3
0.9996	2.62	351.7
0.9987	1.54	482.6

5.1 Conventional Versus Bridgeless PFC Boost Converter

In bridgeless power factor correction converter, diodes are replaced with the power MOSFET, and inductor current passes through the two semiconductors, whereas in conventional PFC, circuit has a full bridge and inductor current passes through the all diode in the bridge. Efficiency is also improved in the bridgeless PFC converter because it has only two slow diodes but the conventional PFC has four slow diodes. The bridgeless PFC converter reduces the semiconductor devices, and it provides better power factor as compared to conventional PFC converter. Total harmonic distortion is reduced in the bridgeless PFC device.

6 Conclusion

Total harmonics distortion is reduced after use of PFC and PF is increased. Power factor correction method is also divided into the two parts (1) active and (2) passive. Bridgeless and modified bridgeless PFC are part of the active filtering method, and it also improves the power factor and current quality. After using the bridgeless power factor correction, power factor increases near about to unity power factor, and it is also utilized the grid power. The bridgeless PFC gives better result than conventional PFC.

References

1. Asim M, Tariq M, Mallick MA, Ashraf I, Kumari S, Bhoi AK (2018) Critical evaluation of offline MPPT techniques of solar PV for stand-alone applications. Lect Notes Electr Eng 435:13–21
2. Raghav R, Raza A, Asim M (2015) Distributed power flow controller-an improvement of unified power flow controller. Development 2(5)
3. Asim M, Riyaz A, Tiwari S, Verma A (2018) Performance evaluation of fuzzy controller for boost converter with active PFC. J Intell Fuzzy Syst 35(5):5169–5175
4. Asim, M, Parveen H, Mallick MA, Siddiqui A (2015) Performance evaluation of PFC boost converters. IJIREEICE 3(11):107–110
5. Ahamad I, Asim M, Sarkar PR, Khan FA (2016) Comparison of conventional PFC boost converter and bridgeless PFC boost converter. 4(5):210–212
6. Akhtar I, Asim M, Yadav RK, Agarwal P, Kirmani S (2020) Design of effective grid-connected solar system. Lect Notes Netw Syst 98:606–614
7. Shahabuddin M, Riyaz A, Asim M, Shadab MM, Sarwar A, Anees A (2019) Performance based analysis of solar PV emulators: a review. In: 2018 International conference on computational and characterization techniques in engineering and sciences CCTES, pp 94–99
8. Parveen H, Asim M, Ahamad I (2016) Efficiency analysis of bridgeless PFC boost converter. Int J Sci Res Dev 4(01)
9. Sarwar A, Shahid A, Hudaif A, Gupta U, Wahab M (2017) Generalized state-space model for an n-phase interleaved buck-boost converter. In 2017 4th IEEE Uttar Pradesh section international conference on electrical, computer and electronics (UPCON) (pp 62–67). IEEE

Performance Analysis of Different Sliding Mode Controller on Single Link Inverted Pendulum

Ajit Kumar Sharma and Bharat Bhushan

Abstract This article presents comparative study of three types of control involving sliding mode approaches, which are quasi-sliding law, robust reaching law and equivalent law. In order to stabilize the pendulum at upright position, these strategies have been deployed on inverted pendulum simultaneously fetching the cart to a chosen position. Sliding mode surfaces and control functions of each strategy are formulated. Control performance of these sliding mode controllers has been compared on the basis of performance measures in terms of time of convergence, disturbance rejection, chattering and stability.

Keywords Sliding mode control · Inverted pendulum · Exponential reaching law · Quasi-sliding mode

1 Introduction

The main requirement of any system depends on its control action taken by controller [1–12]. The control action is specifically designed to achieve the object. And it becomes more challenging for researcher to achieve optimal control performance if system complexity increases [1–12]. And due to gruesome complexity, conventional controllers are not much more comfortable to deal with plant uncertainties. For the same reason, sliding mode control (SMC) approach is applied as a controller which has ability to work satisfactorily under consideration of nonlinear factors of system as well. Sliding mode control evolved from pioneering works in the 1960s in the former Soviet Union [1, 2]. SMC is developed in 1960 by former Soviet Union. Since then for all linear and nonlinear system, SMC provides robust control technique, which also includes time variant systems, MIMO systems and large-scale

A. K. Sharma (✉) · B. Bhushan
Electrical Engineering Department, Delhi Technological University, Delhi, India
e-mail: sharmaajit01@gmail.com

B. Bhushan
e-mail: bharat@dce.ac.in

A. Iqbal et al. (eds.), *Renewable Power for Sustainable Growth*, Lecture Notes in Electrical Engineering 723, https://doi.org/10.1007/978-981-33-4080-0_24

systems. And due to its interesting features, SMC is extensively used in dynamic control of uncertain systems. Design and implementations of SMC are simple and cost effective. Further, SMC enforces order reduction and robustness against system uncertainties and disturbances [3, 4].

Among the other control methods, SMC method is considered as variable structured one [5]. System state space reaches at described surface and always remains there, whenever any discontinuous control structure reverses the system. The surface mentioned in the state space is known as sliding surface. Multiple control structures are designed to obtain the movement of state trajectory only move in the direction of switching. Hence, not a single control structure can define or cover absolute trajectory. Instead the absolute trajectory travels in the direction of the margins of control structures. And these movements of system which slides around boundaries are known as sliding mode [6]. And the ideal sliding motion is when the system dynamics restricted to move towards the sliding surface. Generally, we follow two rules to design SMC. First rule comprises system response which is achieved by designing a sliding surface. Hence plant dynamics are coercible its state variable to satisfy the other set of equation which define its switching surface. The second rule is to design a switched feedback gain which forces the state trajectory of plant towards the sliding surface. These two rules are formulated on the generalized Lyapunov stability theory [7].

Although, SMC is an efficient tool of control system, still it has few drawbacks. In ideal situations, switching of control must occur at infinitely high frequency. The trajectory of the system dynamics is being pushed by the switching to glide along the subspace of restricted sliding mode. But practically, the control computation takes so much time delay, and also, due to physical limitations of switching devices, it makes impossible to switching of control at high frequency. And the results this high frequency provides oscillations known as chattering. Chattering may result in energy loss, system instability, and sometimes it may lead to plant damage. Another issue reported in literature is sliding mode control scheme can completely reject the matched uncertainties, but fails in case of unmatched uncertainties [8–12].

In this study, three methods of control regarding sliding mode, i.e.: Quasi-sliding mode, exponential reaching law and equivalent control have been compared. Inverted pendulum has been chosen as control problem to test the efficacy of these approaches. The objective of control action is to bring the pendulum at a stabilized point in upright position point while maintaining the desired position of cart. In each of these control approaches, sliding mode surface is designed, and thereafter, control functions are formulated to meet control objectives. The performance measures in terms of chattering, time taken to converge, disturbance rejection and stability are recorded. Thereafter, a comparison analysis is performed on the basis of these performance measures.

This paper is organized as follows. In section, the objectives of the work have been introduced. In Section II, different sliding mode control is explained. System modelling is illustrated in Section III. Finally, the results obtained from the simulation of the present work have been presented in Section IV. Section V represents the conclusion which is followed by references.

2 Controller Design

For any plant uncertainty and external disturbance, SMC provides best solution due to its robust control techniques. And to deal with this plant uncertainty and external disturbances, a high gain control input is applied. However, this high gain control produces input chattering, which generates a high frequency in plant dynamics may produce unexpected instabilities [18, 19].

2.1 Reaching Law Control SMC

In reaching law approach, controller describes about reaching phase and sliding phase of system [16]. Reaching phase maintains system stable, and the sliding phase provides equilibrium to system. The objective of sliding mode may be represented in Fig. 1.

Let the plant be

$$\ddot{\theta}(t) = -f(\theta, t) + bu(t) + d(t)$$

where $f(\theta, t)$ and $bu(t)$ are plant parameters and $d(t)$ is external disturbance

The sliding function may be described as

$$s(t) = ce(t) + \dot{e}(t) \tag{1}$$

where 'c' is a constant which must satisfy Hurwitz condition, $c > 0$.

The tracking error and its derivative are defined as

$$e(t) = x(t) - \theta(t)$$

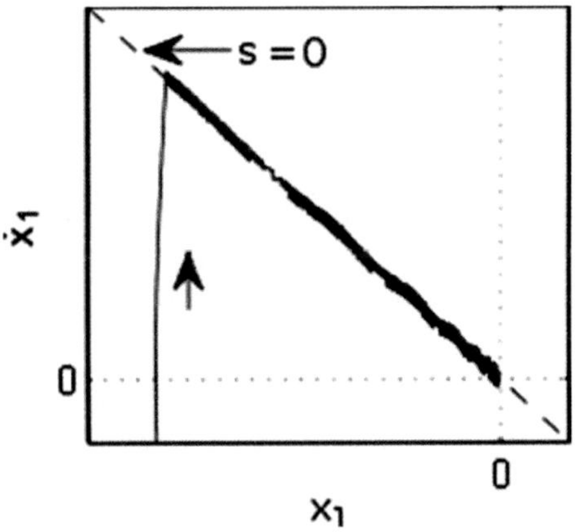

Fig. 1 Idea of sliding mode

$$\dot{e}(t) = \dot{x}(t) - \dot{\theta}(t)$$

where $x(t)$ is ideal positional signal.

Therefore, we have

$$\begin{aligned}\dot{s}(t) &= c\dot{e}(t) + \ddot{e}(t) \\ &= c\big(\dot{x}(t) - \dot{\theta}(t)\big) + \big(\ddot{x}(t) - \ddot{\theta}(t)\big) \\ &= c\big(\dot{x}(t) - \dot{\theta}(t)\big) + (\ddot{x}(t) + f(\theta, t) - bu(t) - d(t))\end{aligned} \quad (2)$$

Again on the basis of exponential reaching law, we know

$$\dot{s} = -\epsilon\ \mathrm{sgn}(s) - ks \quad (3)$$

where $\epsilon > 0, k > 0$

From Eq. (2) and (3),

$$\big(\dot{x}(t) - \dot{\theta}(t)\big) + (\ddot{x}(t) + f(\theta, t) - bu(t) - d(t)) = -\epsilon\ \mathrm{sgn}\, s - ks \quad (4)$$

The design of the controller functioning as sliding mode has been developed as

$$u(t) = \frac{1}{b}\big(\epsilon\ \mathrm{sgn}(s) + ks + c\big(\dot{x}(t) - \dot{\theta}(t)\big) + \ddot{x}(t) + f(\theta, t) - d(t)\big) \quad (5)$$

From the above equations as the disturbance ‘d’ is an unknown quantity among all other plant quantities, so to solve above problem, d is replaced by a conservative known quantity d_c.

Then, the governing equation for the controller based on sliding mode can be written as

$$u(t) = \frac{1}{b}\big(\epsilon\ \mathrm{sgn}(s) + ks + c\big(\dot{x}(t) - \dot{\theta}(t)\big) + \ddot{x}(t) + f(\theta, t) - d_c\big) \quad (6)$$

where d_c is chosen for guarantee in reaching condition.

Substituting Eq. (6) into Eq. (2) and by simplifying it,

$$\dot{s}(t) = -\epsilon \mathrm{sgn}(s) - ks + d_c - d \quad (7)$$

Reaching condition is ensured by term d_c. Hence, it gives guarantee that d is bounded, and therefore, d_c.

$$d_L \leq d(t) \geq d_U \quad (8)$$

where the bounds d_L and d_U are known.

Hence,

(i) When $s(t) > 0, \dot{s}(t) = -\epsilon \operatorname{sgn}(s) - ks + d_c - d$ and we want $\dot{s}(t) < 0$ so $d_c = d_L$
(ii) When $s(t) < 0, \dot{s}(t) = -\epsilon \operatorname{sgn}(s) - ks + d_c - d$ and we want $\dot{s}(t) > 0$ so $d_c = d_U$

Therefore, if we define $d_1 = \frac{d_U - d_L}{2}$,

$$d_2 = \frac{d_U + d_L}{2}$$

$$d_c = d_2 - d_1 \operatorname{sgn}(s) \tag{9}$$

2.2 Quasi-sliding Mode SMC

Although SMC is a robust control technique but having drawback of chattering. The system components are getting damaged due to an undesirable phenomenon of oscillations, known as chattering. In order to reduce the chattering, around a switching surface, a boundary layer has been proposed by using continuous control. In this proposed method to reduce chattering, system forces its state to stay in range of its neighbourhood.

Let the plant be

$$\ddot{\theta}(t) = -f(\theta, t) + bu(t) + d(t)$$

where $f(\theta, t)$ and $bu(t)$ are plant parameters and $d(t)$ is external disturbance

The sliding function may be described as

$$s(t) = ce(t) + \dot{e}(t) \tag{10}$$

where 'c' is a constant which must satisfy Hurwitz condition, c > 0.

The tracking error and its derivative are defined as

$$e(t) = x(t) - \theta(t)$$

$$\dot{e}(t) = \dot{x}(t) - \dot{\theta}(t)$$

where $x(t)$ is ideal positional signal.

Therefore, we have

$$\begin{aligned}\dot{s}(t) &= c\dot{e}(t) + \ddot{e}(t) \\ &= c\dot{e} + \left(\ddot{x}(t) - \ddot{\theta}(t)\right)\end{aligned}$$

$$= c\dot{e} + (\ddot{x}(t) + f(\theta, t) - bu(t) - d(t)) \quad (11)$$

Hence, the proposed controller will be

$$u(t) = \frac{1}{b}(c\dot{e} + \ddot{x}(t) + f(\theta, t) + \eta \text{sgn}(s)) \quad (12)$$

Lyapunov function is defined as

$$L = \frac{1}{2}s^2$$

Therefore,

$$\begin{aligned} \dot{L} = s\dot{s} &= s(\ddot{x}\ddot{x}(t) - f(\theta, t) - bu(t) + c\dot{e}) \\ &= s(\ddot{x}(t) - f(\theta, t) - (-f(\theta, t) + c\dot{e} + \ddot{x}(t) + \eta\ \text{sgn}(s))) \\ &= s(-d(t) - \eta\ \text{sgn}(s)) \\ &- sd(t) - \eta|s| \end{aligned} \quad (13)$$

To control the chattering, a saturated function sat(s) function is used, instead of sgn(s).

$$\text{sat}(s) = \begin{cases} 1 & s > \Delta \\ -1ks & |s| \leq \Delta, k = \frac{1}{\Delta} \\ -1 & s < -\Delta \end{cases} \quad (14)$$

where Δ is defined as boundary layer, switch control is used for outside boundary layer and linear feedback is used for inside boundary layer.

Since sliding variable and state variables does not converge to zero due to its smooth control law, but in its place converge to domains in a neighbourhood of the origin due to the effect of the disturbance. The said smooth control law is known as quasi-sliding mode control.

2.3 *Equivalent Control SMC*

SMC controller usually comprises the switching control u_{sw} and the equivalent control u_{eq}. System state is managed by switching control and equivalent control and maintains the system on sliding surface.

If we neglect the uncertainty of plant and external disturbance, the plant may be defined as

$$x^{(n)} = f(x, t) + bu(t) \quad (15)$$

The tracking error vector will be defined as

$$e = x_d - x = [e\ e^{\cdot}\ e^{(n-1)}]^T \tag{16}$$

And switch function

$$s(x, t) = ce = c_1 e + c_2 e^{\cdot} + \cdots e^{(n-1)} \tag{17}$$

where $c = \left[c_1 c_2 \cdots c_{n-1} 1\right]$

Let $s^{\cdot} = 0$

Then,

$$s^{\cdot}(x, t) = c_1 e^{\cdot} + c_2 e^{\cdot\cdot} + \cdots e^n$$

$$= c_1 e^{\cdot} + c_2 e^{\cdot\cdot} + \cdots c_{n-1} e^{n-1} + x_d^{(n)} - x^{(n)}$$

$$\sum_{i=1}^{n-1} c_i e^{(i)} + x_d^n - f(x, t) - bu(t) = 0 \tag{18}$$

Hence, the control input is designed as

$$u_{eq} = \frac{1}{b}\left[\sum_{i=1}^{n-1} c_i e^{(i)} + x_d^n - f(x, t)\right] \tag{19}$$

To satisfy the reaching condition of sliding mode,

$$s(x,).\dot{s}(x, t) \leq -\eta|s|, \eta > 0$$

The switching control law

$$u_{\text{sw}} = \frac{1}{b} K \operatorname{sgn}(s) \tag{20}$$

where $K = D + \eta$

Hence, sliding mode controller

$$u = u_{\text{eq}} + u_{\text{sw}} \tag{21}$$

And the stability equation will be

$$s\dot{s} = s(-K\ \operatorname{sgn}(s)) - s.d(t) = -\eta|s| \leq 0 \tag{22}$$

3 System Modelling

Figure 1 portrays a single link inverted pendulum which consists of a rod hinged on a moving cart. Table 1. gives the different parameters of inverted pendulum (Fig. 2).

The system's motion equation is determined by Euler–Lagrange method and Newton–Euler formalism while keeping system motion is frictionless.

$$(m + M)\ddot{x} + ml\cos\theta\theta - ml\sin\theta\dot{\theta}^2 = F \tag{23}$$

$$\left(ml^2 + I\right)\ddot{\theta} + ml\ddot{x}\cos\theta - mgl\sin\theta = 0 \tag{24}$$

By solving for $\ddot{\theta}$, and $\ddot{x}$ from Eq. (23) and (24), the state space equation of inverted pendulum is defined as

$$\ddot{\theta}, t(t) = -f(\theta, t) + bu(t) \tag{25}$$

Where

Table 1 Parameters and their units

Parameter	Meaning	Unit
θ	Pendulum angle	rad
x	Cart displacement	m
m	Pendulum mass	kg
M	Cart mass	kg
l	Pendulum's length	m
g	Gravitational acceleration	m/s^2

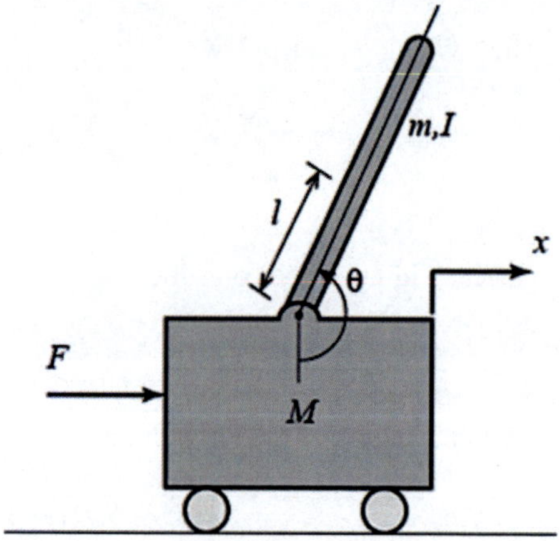

Fig. 2 Single link inverted pendulum

$$f(\theta, t) = \frac{m(m+M)gl}{(M+m)I + Mml^2}$$

And,

$$b = \frac{ml}{(M+m)I + Mml^2}$$

4 Implementation of Algorithm in MATLAB

Simulation is done on MATLAB/Simulink environment to evaluate the results for different controller cases. Simulation results give a brief about the algorithm need to be implemented in the system.

The following parameters were chosen for simulation: $M = 1$; m = 0.010; $L =$ 0.50, $g = 9.8$ m/s^2.

4.1 Results of Reaching Law Control SMC

Let the ideal position of pendulum be $r(t) = \sin t$, the initial state is [−0.15 0.15], and other constants are given as $c = 15$, $\epsilon = 0.5$, $k = 10$.

In Fig. 3, the simulation model of controller and pendulum is shown. Figure 4 gives the control input for plant with a constant reaching law, and the cart position and speed at equilibrium point are shown in Fig. 5.

4.2 Results of Quasi-sliding Mode Control SMC

Figure 6 shows the simulation model of controller and pendulum. On the basis of sliding mode control, the upper bound has been represented as, M=1 using sgn function The control input is shown in Fig. 7. Correspondingly, cart position and speed are shown in Fig. 8.

4.3 Results of Equivalent Control SMC

Let $d(t) = 50\ \sin(t)$, $\eta = 0.10$, r (i.e. the ideal position of signal) $= \sin(2\pi t)$, if c = 25, then D can be obtained = 50. The simulation results of control input and cart position and speed are shown in Figs. 10 and 11 (Fig. 9; Table 2).

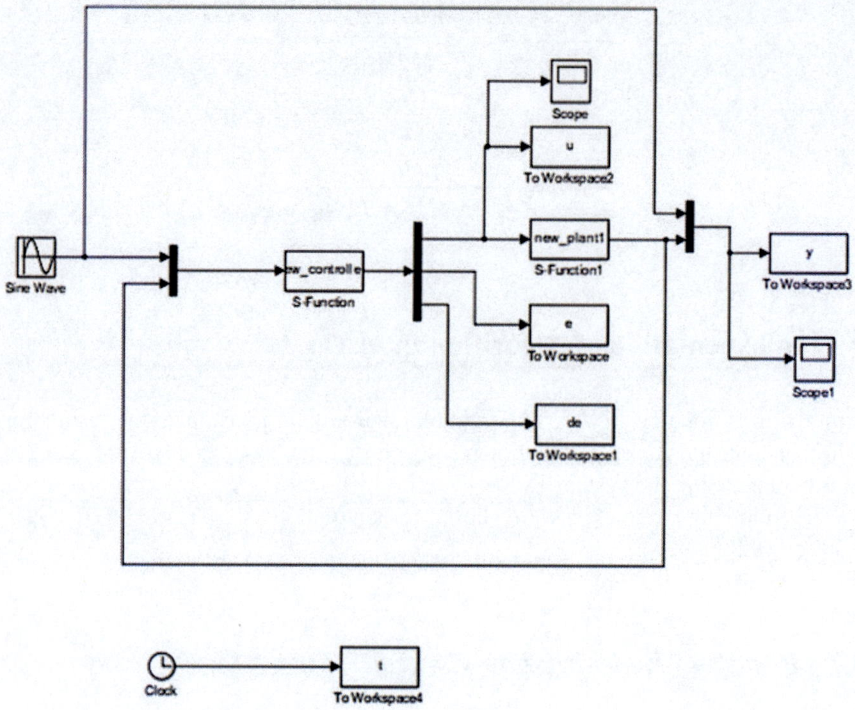

Fig. 3 Simulink model of system

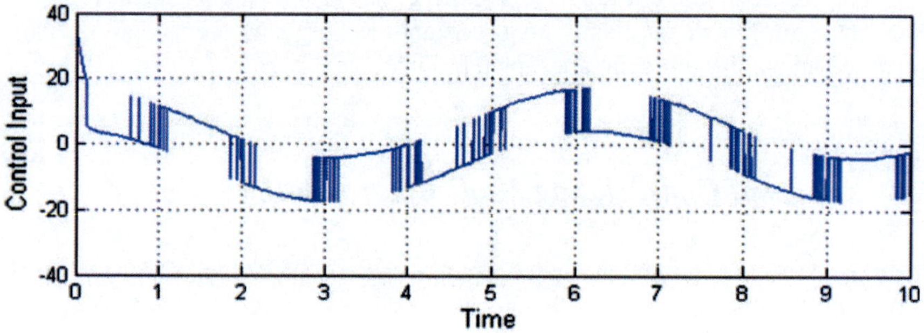

Fig. 4 Control input

From above results, it can be inferred that as far as stability is concerned all three control modes display satisfactory performance. Sliding mode controller with quasi law stands out on account of chattering, while equivalent mode has moderate effect of chattering. Reaching law is poorest among three and worst affected by chattering. As already stated that chattering is undesirable feature due to un-modelled dynamics and discrete time implementation of plant, this problem reduces the control accuracy

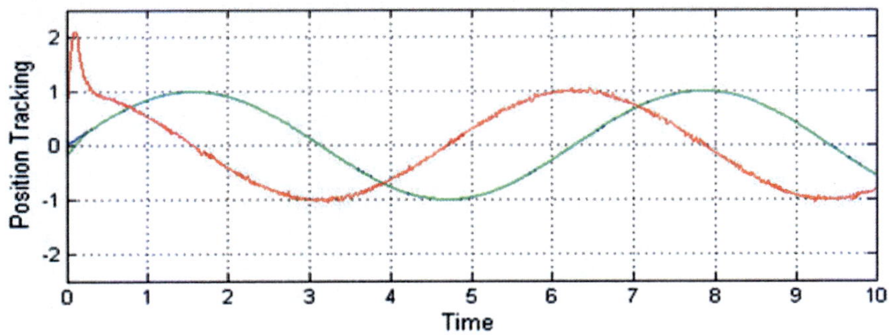

Fig. 5 Cart position and speed tracking

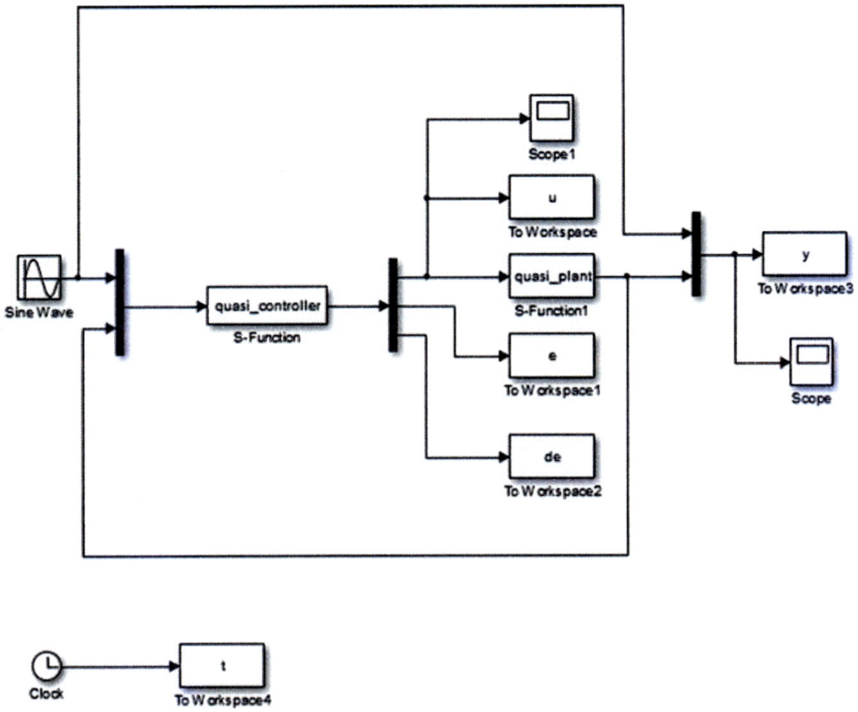

Fig. 6 Simulink model of system

and strength of mechanical components. Both reaching law and equivalent mode demonstrate superior convergence, while quasi mode has some delay in convergence. Disturbance rejection is least in equivalent law, while reaching law and quasi mode have moderate disturbance rejection.

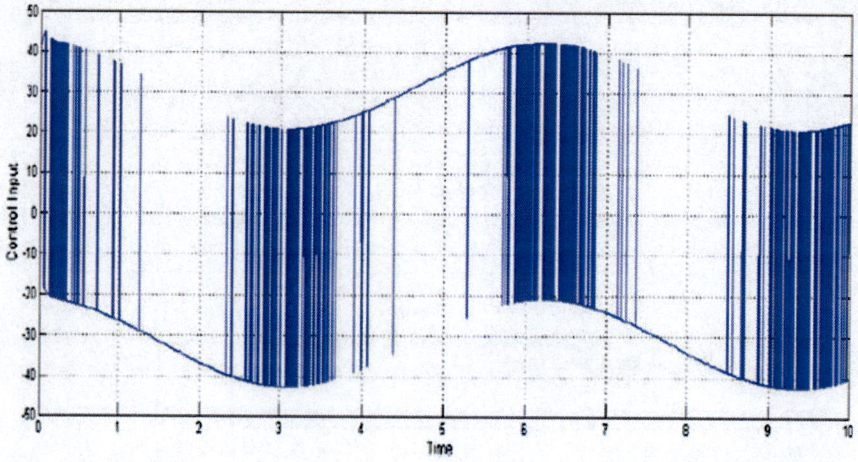

Fig. 7 Control input

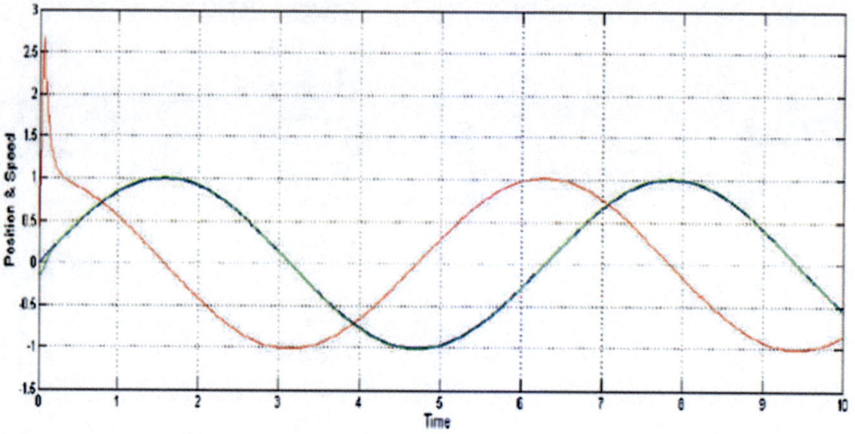

Fig. 8 Cart position and speed tracking

5 Conclusion

This paper highlights the difference among three different control technique of sliding mode control, their control strategies and performance measures. Inverted pendulum was chosen as control problem to analyse the performance of sliding mode control employing reaching law, equivalent law and quasi law. The objective of control action was to stabilize the pendulum at upright position point while bringing the cart to a desired position. Sliding surface and control functions were designed for each of these control strategies. Performance results of sliding model controllers were obtained on MATLAB platform. Finally, performance measures

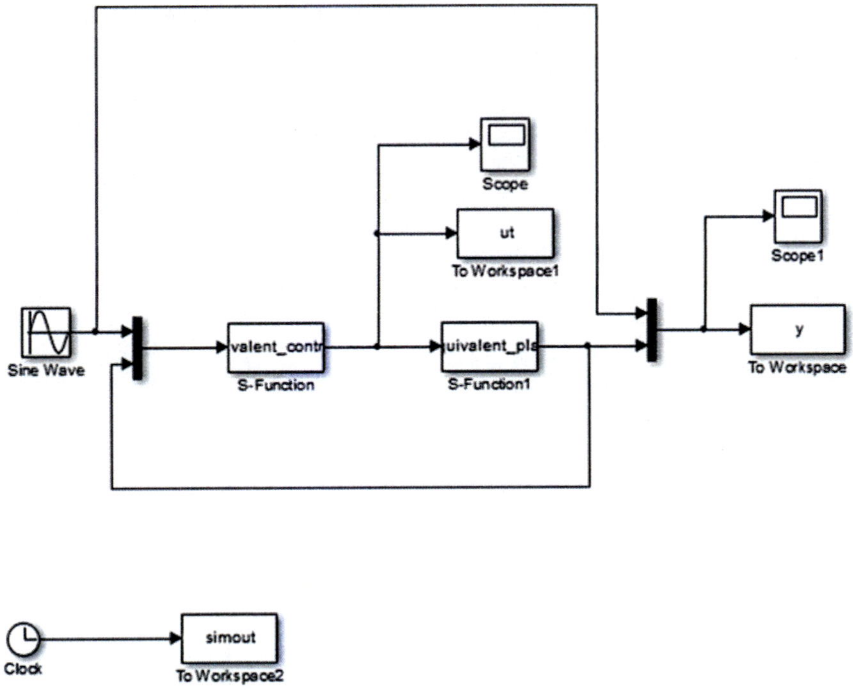

Fig. 9 Simulink model of system

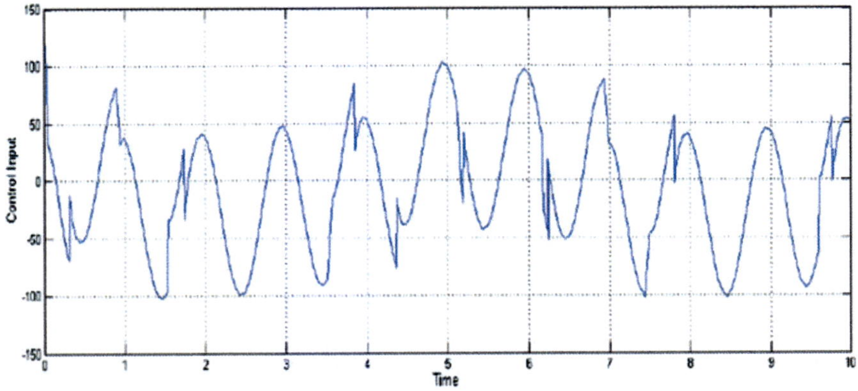

Fig. 10 Control input

were analysed on the basis of convergence rate, disturbance rejection, chattering and stability performance. From observations, it was noted that on stability sliding mode controllers gave similar stability performance. However, chattering phenomenon, disturbance rejection and convergence rate were different. Sliding mode controllers with quasi law gave standout results in chattering and moderate disturbance rejection

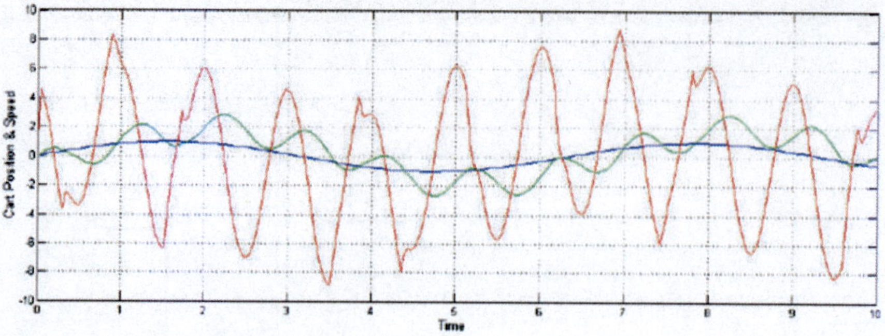

Fig. 11 Cart position and speed tracking

Table 2 Differences between all three controllers

Analysis	Reaching	Quasi	Equivalent
Stability	High	High	High
Chattering	More	Less	Moderate
Disturbance rejection	Moderate	Moderate	Less
Convergence	Fast	Medium	Fast

and convergence. Sliding mode controller with equivalent law had fast convergence, moderate chattering, but its performance was poor in disturbance rejection. Sliding mode controller with reaching law demonstrated fast convergence, moderate disturbance rejection. However, it performed worst on account of chattering among three strategies.

References

1. Itkis U (1976) Control systems of variable structure. Wiley, New York
2. Utkin VI (1977) Variable structure systems with sliding modes. IEEE Trans Autom Control 22:212–222
3. Utkin VI (1992) Sliding modes in control optimisation. Springer, Berlin
4. Edwards C, Spurgeon SK (1998) Sliding mode control: theory and applications
5. Hung JY, Gao W, Hung JC (1993) Variable structure control: a survey. IEEE Trans Ind Electron 40(1):222
6. Edwards C, Spurgeon S (1998) Sliding mode control: theory and applications. Taylor and Francis, London
7. Brockett RW (1983) Asymptotic stability and feedback stabilization. Differential geometric control theory. Birkhäuser, Boston
8. Bartolini G, Ferrara A, Usai E (1998) Chattering avoidance by second-order sliding mode control. IEEE Trans Autom Control 43(2):241–246
9. Latosiński P, Bartoszewicz A (2018) Discrete time sliding mode controllers with relative degree one and two switching variables. J Franklin Inst 355
10. Lee SH, Park MJ, Kwon OM (2019) Synchronization criteria for delayed Lur'e systems and randomly occurring sampled-data controller gain. Commun Nonlinear Sci Numer Simul 68

11. Iqbal A et al (eds) (2020) Soft computing in condition monitoring and diagnostics of electrical and mechanical system. In: Advances in intelligent systems and computings, vol 1096. Springer, Singapore. https://doi.org/10.1007/978-981-15-1532-3
12. Iqbal A et al (eds) (2020) Meta heuristic and evolutionary computation: algorithms and applications. In: Studies in computational intelligence, vol 1096. Springer, Singapore. https://www.springer.com/gp/book/9789811575709

A Resilient Hybrid Output Converter with Inherent Cross-Regulation Avoidance Feature

Raj Kumar Yadav, Adil Sarwar, Dipti Saxena, Mohammed Asim, and Chandra Prakash

Abstract Power electronic converters are required for integrating renewable energy sources with the grid. Concept of hybrid micro-grid allows coordinated operation and integration of AC and DC micro-grid. But it requires separate converters for feeding power to the grid which increases cost and control complexity. This paper proposes a hybrid multi-output converter (simultaneously AC and DC output) for standalone application and can be extended for grid application. DC and AC loads can be supplied simultaneously from a single DC source. The DC–DC operation utilizes quadratic boost circuitry to obtain higher gain at lower duty ratio, thereby reducing conduction loss. While a multilevel AC output is obtained which has a lower total harmonic distortion. The converter is tested under different dynamically changing load conditions. The parallel operation of DC–DC and DC–AC part of the proposed converter ensures cross-regulation-free operation.

Keywords Hybrid multi-output converter · Multilevel inverter (MLI) · Quadratic boost converter · Resilience

R. K. Yadav · C. Prakash
Department of Electronic instrumentation and Control Engineering, College of Engineering, Ajmer, India

A. Sarwar (✉)
Electrical Engineering Department, AMU, Aligarh, India
e-mail: adil.sarwar@zhcet.ac.in

D. Saxena
Electrical Engineering Department, MNIT, Jaipur, India

M. Asim
Electrical Engineering Department, Integral University, Lucknow, India

A. Iqbal et al. (eds.), *Renewable Power for Sustainable Growth*, Lecture Notes in Electrical Engineering 723, https://doi.org/10.1007/978-981-33-4080-0_25

1 Introduction

The conventional energy resources which played a major role in meeting the electrical requirement of the globe are being seen as a threat to humanity because of the pollution which produces in due process of electricity generation [1–7]. At present, renewable energy is looking one of the most prominent alternative sources of energy. This pollution is contaminating the air, soil and water bodies which are ultimately affecting human beings. Hence, the importance of renewable energy has seen a tremendous growth in the previous decade and seems to grow even more in the coming decade. The government initiatives of many countries also propagate it. These policies give stress on reducing generation from fossil fuels and look for alternate sources like renewable energy sources. Especially, solar energy is abundantly available throughout the world and can be utilized either by tapping heat content of sunlight or converting the solar energy directly into DC electricity by way of photovoltaic effect. PV has gained importance because of many factors such as low maintenance, ease of availability of sun, and technology availability. Although there are many emerging technologies to mitigate the problems of conventional sources, but the photovoltaic (PV) technology is one of the cleanest and popular with ease of generating power [1, 2]. Unlike wind energy solar PV technology can be installed anywhere, at any place around the world, even very close to the load center. This could eliminate the cost of transmission of electricity by way of costly transmission system. Moreover, it has many other advantages such as low operational and maintenance costs as well as long life. In most of the cases, maximum power point tracker is added to improve the efficiency of overall system. The efficiency of the system improves considerably.

Nowadays, there is a rapid increase in the use of power electronic converters in many devices such as aircraft, electric cars, and ships [6, 7]. With the use of these converters, the size, weight, maintenance, and cost of the system have come down, while the efficiency reliability and safety have increased. Both AC and DC loads characterize these power systems. In most of these, including the conversion of primary power, auxiliary systems are needed to boost up the voltage for the effective operation of the overall system [3].

These systems consist of different types of load—the interface of DC and AC with various kinds of energy sources including both non-conventional and conventional using converters of power electronics [2]. The figure shown in Fig. 1 displays the internal architecture of the system where an individual DC source ($V_{\mathrm{dc,is}}$), e.g., solar panel, fuel cell, etc., supplies both the AC and DC loads($V_{\mathrm{ac,out}}$ and $V_{\mathrm{dc,out}}$, respectively). The architecture of Fig. 1a uses different power converters for both the conversion type, i.e., DC–AC and DC–DC, while Fig. 1b shows both the conversions utilizing a single power converter stage. The converter in Fig. 1b is known as the hybrid converter in this paper. Due to the inherent shoot-through protection capability, this hybrid converter has the very important features of better reliability and higher power processing density [4]. These characteristics enable the hybrid converter perfect to use in smaller and compact systems with both the loads, i.e., AC

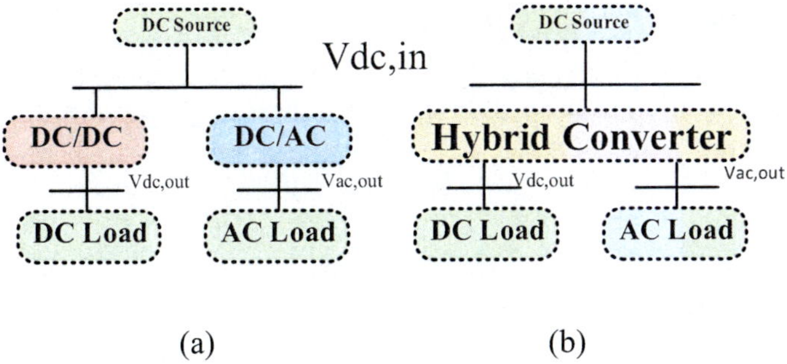

Fig. 1 **a** Architecture for power converter to supply AC and DC loads. **b** Hybrid converter-based architecture for supplying DC and AC loads

and DC loads. As an example, consider that both an AC fan and a LED lamp can be powered simultaneously by a hybrid converter at an initial stage of individual DC input.

Non-conventional energy sources supply purified and clean energy, so it is mostly connected to systems in smart residencies. These intended energy sources are highly limited or bounded and have very low power ratings which are approximate of the order of a hundred watts and terminal voltage due to space limits. To solve these issues, various topologies of hybrid converter were developed to simultaneously power DC and AC loads [4]. The disadvantages of conventional hybrid converters are discussed below.

In applications where high power handing is required, higher current rating operation for longer duration results in an increase in conduction losses and a reduction in the efficiency.

The main motto of the researchers is to design and develop new topologies of a power converter having lesser number of power semiconductor devices and reduced complexity. There is an amalgamation of multiple renewable energy sources to AC and DC system buses in hybrid grid system. This integration needs two individual converters, i.e., a DC–DC boost converter and a multilevel inverter.

Literature [4] suggests a topology of the converter that provides simultaneously DC and AC output, but there is a disadvantage of the operating range in this network topology due to which the range should be maintained in $\mathrm{M} + \mathrm{D} \leq 1$ which has been discussed above. Moreover, the two-level AC output produced by the hybrid converter in [4] has a higher THD and requires larger passive filters which makes the converter bulky and costly.

The shortcomings of [4] can be overcome by the proposed hybrid converter in this paper. The main objective of the paper has been summarized as follows:

(1) To outline a hybrid converter that will be capable of imparting AC and DC outputs simultaneously without the limit of $m + D \geq 1$.
(2) To use more gain DC–DC converter topology in parallel with the inverter for more efficient operation.

2 Proposed Converter Circuit

The proposed topology has a high gain DC–DC converter in parallel with a modified H-bridge inverter as shown in Fig. 2.

The proposed converter operation is explained in the following subsection.

2.1 *Quadratic Boost Converter*

Conventional boost circuit is used to step up the voltage at a high duty ratio. At high duty ratio, generally inductor gets saturated. Moreover, the gain obtained in case a conventional boost converter is limited to 4–5 times the input voltage. So, the DC–DC converter part of the circuit consists of a quadratic boost converter (QBC) configuration. The efficiency of quadratic boost converter is less when the duty ratio is low, but when the duty ratio is high, there is a significant increase in efficiency. As shown in Fig. 3 when the switch is on, then diodes D_1 and D_2 are off. The input current flows through inductor L_1 and diode D_2.

The equations when the switch is on are given by

$$\Delta i L_1 = \frac{V_{\text{in}}}{L_1} \tag{1}$$

$$\Delta i L_2 = \frac{V_{c1}}{L_2} \tag{2}$$

$$\Delta V_{c1} = -\frac{i L_2}{C_1} \tag{3}$$

$$\Delta V_{c2} = -\frac{V_{\text{out}}}{R_{c2}} \tag{4}$$

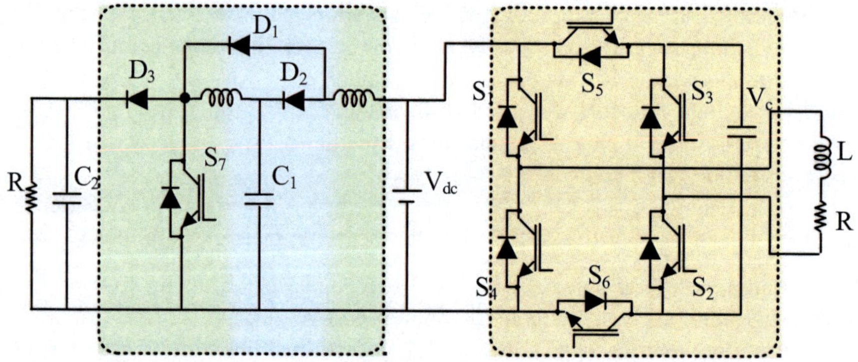

Fig. 2 Proposed topology

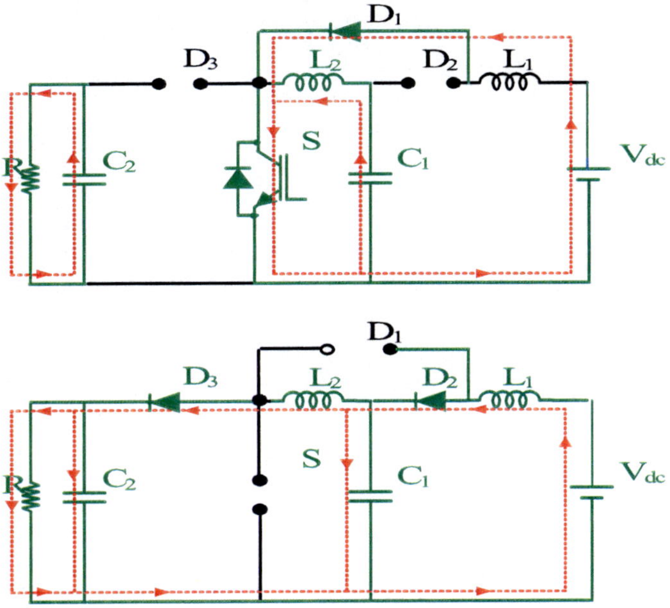

Fig. 3 Working mode of operation

When the switch is turned off, diodes D_1 and D_3 are turned ON, and diode D_2 is in OFF state. The capacitors C_1 and C_2 are all charged by inductor L_1 and L_2, respectively. The load energy demand is supplied by the inductor, as shown in Fig. 3.

The differential equations related to the state variables when the switch is turned OFF are

$$\Delta i L_1 = \frac{V_{\text{in}}}{L_1} - \frac{V_{c_1}}{L_1} \tag{5}$$

$$\Delta i L_2 = \frac{V_{c_1}}{L_2} - \frac{V_{\text{out}}}{L_2} \tag{6}$$

$$\Delta V_{c_1} = \frac{I L_1}{C_1} - \frac{I L_2}{C_1} \tag{7}$$

$$\Delta V_{\text{out}} = \frac{I L_2}{C_2} - \frac{V_{\text{out}}}{R_{c2}} \tag{8}$$

The average model equation of QBC

$$\Delta i L_1 = \frac{V_{\text{in}}}{L_1} - \frac{V_{c_1}}{L_1}(1 - D) \tag{9}$$

$$\Delta i L_2 = \frac{V_{c_1}}{L_2} - \frac{V_{\text{out}}}{L_2}(1 - D) \tag{10}$$

$$\Delta V_{c_1} = \frac{I L_1}{C_1}(1 - D) - \frac{I L_2}{C_1} \tag{11}$$

$$\Delta V_{\text{out}} = \frac{I L_2}{C_2} - \frac{V_{\text{out}}}{R_{c_2}} \tag{12}$$

The net inductor voltage must be zero in a switching period for steady state operation of an inductor in a DC–DC converter.

$$\Delta i L_{1(\text{closed})} + \Delta i L_{1(\text{open})} = 0 \tag{13}$$

$$\frac{V_{\text{in}}}{L_1} DT + \frac{V_{\text{in}}}{L_1}(1 - D)T - \frac{V_{c_1}}{L_1}(1 - D) = 0$$

$$\frac{V_{\text{in}}}{L_1} T = \frac{V_{c_1}}{L_1}(1 - D)T$$

$$\text{V}_{c_1} = \frac{V_{\text{in}}}{(1 - D)} \tag{14}$$

$$\Delta i L_{2(\text{on})} + \Delta i L_{2(\text{off})} = 0 \tag{15}$$

$$\frac{V_{c_1}}{L_2} DT + \frac{V_{c_1}}{L_2}(1 - D)T - \frac{V_{\text{out}}}{L_2}(1 - D)T = 0$$

$$V_{c_1} D + V_{c_1}(1 - D) - V_{c_1} D = V_{\text{out}}(1 - D)$$

$$V_{c_1} = V_{\text{out}}(1 - D) \text{ and } \frac{V_{\text{in}}}{(1 - D)} = V_{\text{out}}(1 - D)$$

$$V_{\text{out}} = \frac{V_{\text{in}}}{(1 - D)} \tag{16}$$

3 Modified H-Bridge Inverter Operation

Due to the advantage of higher-power quality waveforms, low electrical switching losses, and higher voltage generation capability, the use of multilevel inverter has been increasing day by day. There are several advantages of multilevel inverter. The minimized harmonics distortion in output voltage waveform is the most important being. In order to obtain the desired output voltage, several DC power supplies are interconnected via power electronic switches [5]. The output voltage waveform of the converter approaches nearly a sinusoidal waveform as level increases. In comparison with other topologies, the cascaded MLI is far better as it does not need any clamping diode and flying capacitor. The primary advantage of using H-bridge inverter is its modularity, while the disadvantage is that each cell of the H-bridge requires an isolated DC source. As the number of units connected in cascade increases, the number of levels of output also increases. The shortcoming of an H-bridge configuration (requirement of multiple DC power supply) can be partially mitigated by using modified H-bridge configuration and a capacitor in place of DC power supply (Fig. 4).

The operation of the cascaded three levels MLI is shown in Table 1. The most important task or a key issue in designing MLI is to reduce its harmonics distortion. To obtain a stepped output voltage resembling a sinusoidal waveform, multiple DC power supply are interconnected using power electronic switches in a multilevel inverter.

Fig. 4 Modified H-bridge 3-level MLI

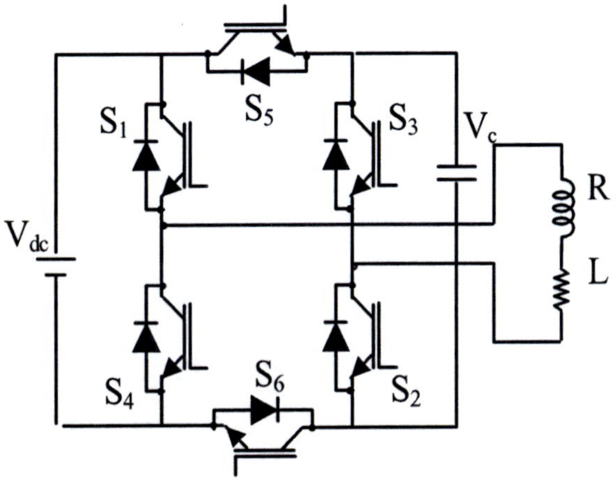

Table 1 Switching states for modified H-bridge inverter

S_1	S_2	S_3	S_4	S_5	S_6	V_{out}
1	1	0	0	1	0	V_c
1	1	0	0	0	1	V_{dc}
1	1	0	0	1	1	$(V_{dc}-V_c)$
0	0	1	1	1	0	$-V_c$
0	0	1	1	0	1	$-V_{dc}$
0	0	1	1	1	1	$-(V_{dc}-V_c)$

4 Simulation and Results

The simulation has been performed in MATLAB®/Simulink environment interactive development platform. The various parameters taken in the simulation have been shown in Table 2.

The simulation has been performed under various dynamically load changing conditions. Figure 5 shows the output AC and DC voltage when there is a step change in the DC load. The load is halved after 0.02 s (one cycle) of the operation. There is no effect of change in load on the output voltages implying that the cross-regulation effect has been avoided (Fig. 6).

The effect of change in a purely resistive load has also been seen. (1) by reducing the load resistance as shown in Fig. 7 and (2) By increasing the load resistance as shown in Fig. 8. In both cases, the output voltage (AC as well as DC) has a resilient feature.

Table 2 Simulation parameters

S. No.	Parameters	Value in simulation
1	V_{dc}	50 V
2	C_1 and C_2	2200 μF
3	Load (R and L)	10 Ω, 10 mH
4	L_1, L_2	10 mH, 20 mH

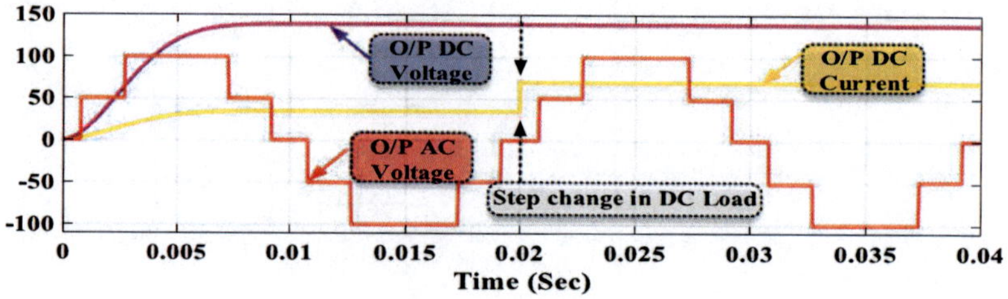

Fig. 5 Effect of changing DC load

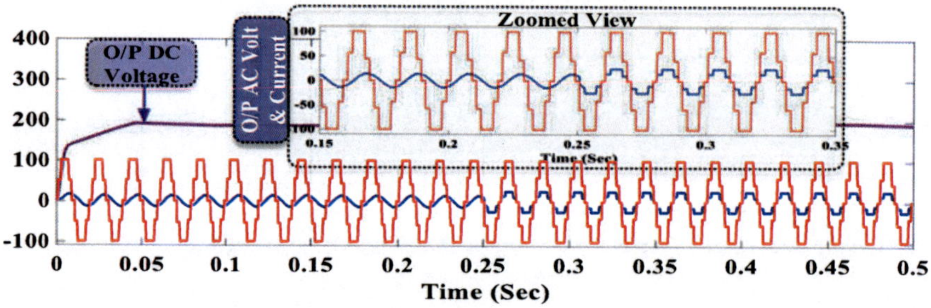

Fig. 6 Effect of changing AC load (from RL to R)

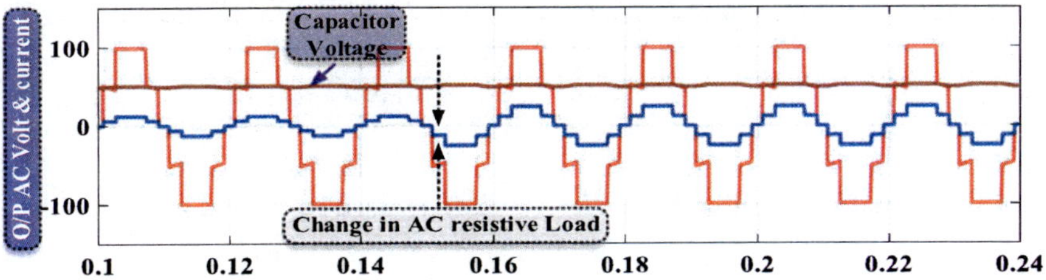

Fig. 7 Effect of decreasing the resistive load

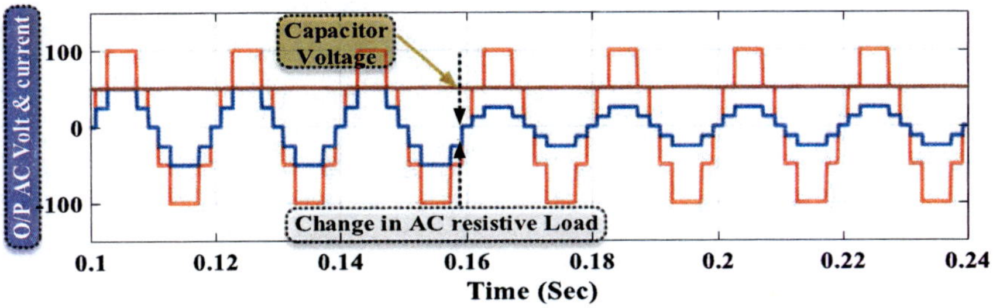

Fig. 8 Effect of increasing the resistive load

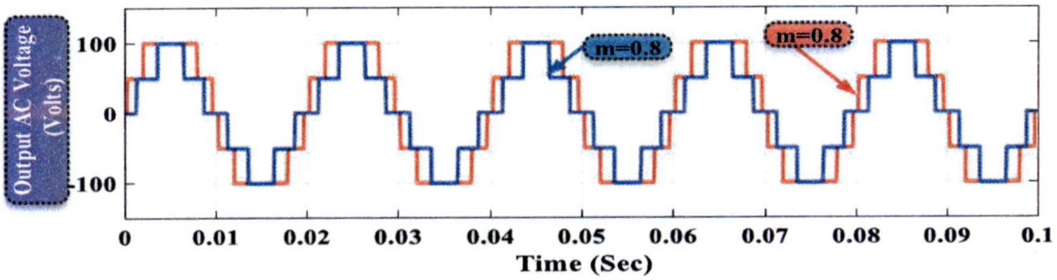

Fig. 9 Effect of change in modulation index

Nearest level control has been used to control the output AC voltage. The figure shown in Fig. 9 shows the effect of changing the modulation index from 0.6 to 0.8. The stresses across the various switches have been shown in Figs. 10, 11, and 12. The stress is very low in all the rest of the switches, while the stress is very higher in switches S_1 and S_3.

5 Conclusion

A hybrid output converter has been successfully modeled, and simulation results have been presented under different dynamically changing load conditions which prove the resilient nature of the converter topology. Modified H-bridge configuration

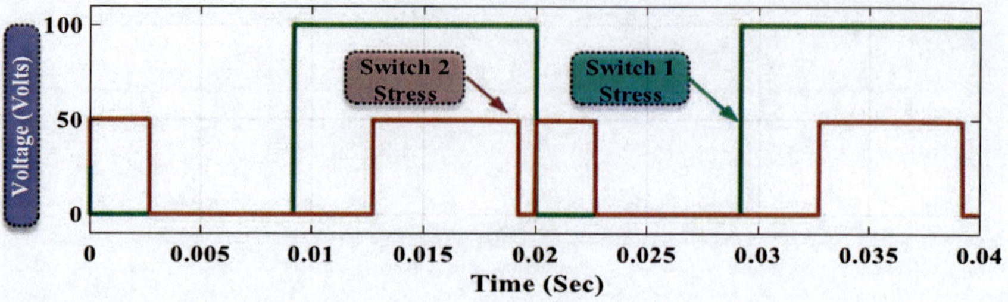

Fig. 10 Voltage stress across the switches S_1 and S_2

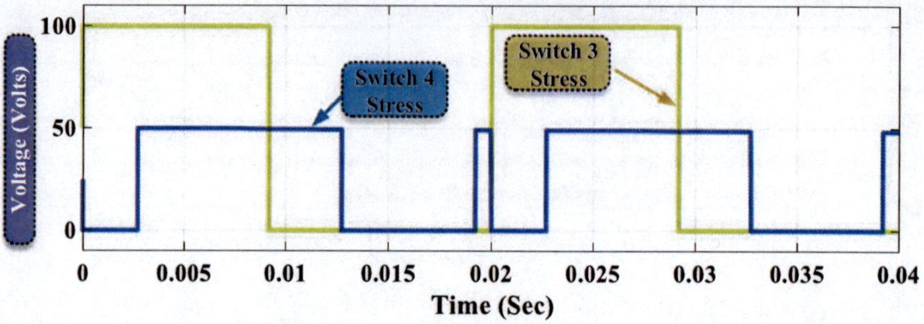

Fig. 11 Voltage stress across the switches S_3 and S_4

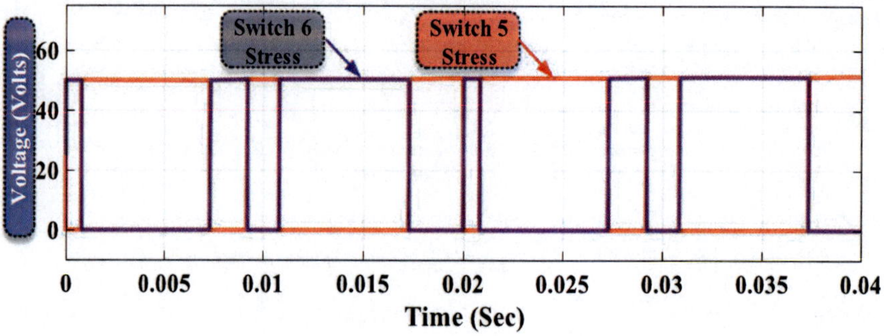

Fig. 12 Voltage stress across the switches S_5 and S_6

has been utilized for AC output voltage generation, while simultaneously quadratic boost converter topology has been utilized for DC output voltage generation. AC and DC output voltage boosting is the attractive feature of the proposed topology. Moreover, stepped wave AC output has a superior waveform quality with lower total harmonic distortion.

References

1. Asim M, Tariq M, Mallick MA, Ashraf I, Kumari S, Bhoi AK (2018) Critical evaluation of offline MPPT techniques of solar PV for stand-alone applications. In: Advances in smart grid and renewable energy 2018. Springer, Singapore, pp 13–21
2. Shahabuddin M, Riyaz A, Asim M, Shadab MM, Sarwar A, Anees A (2018) Performance based analysis of solar PV emulators: a review. In: 2018 international conference on computational and characterization techniques in engineering & sciences (CCTES), Lucknow, India, pp 94–99
3. Tomy A, Thomas AJ (2017) Sepic derived hybrid converter with simultaneous AC and DC outputs. In: 2016 IEEE annual India conference on INDICON 2016
4. Adda R, Ray O, Mishra SK, Joshi A (2013) Synchronous-reference-frame-based control of switched boost inverter for standalone DC nanogrid applications. IEEE Trans Power Electron 28(3):1219–1233
5. Siddique MD, Mekhilef S, Shah NM, Sarwar A, Iqbal A, Memon MA (2019) A new multilevel inverter topology with reduce switch count. IEEE Access 7:58584–58594
6. Ahamad I, Asim M, Sarkar PR, Khan FA (2016) Comparison of conventional PFC boost converter and bridgeless PFC boost converter. Int J Innovative Res Electr Electron Instrum Control Eng 4(5)
7. Asim M, Parveen H, Mallick MA, Siddiqui A (2015) Performance evaluation of PFC boost converters. Int J Innovative Res Electr Electron Instrum Control Eng 3(11):107–110

Review of Various Load Frequency Controllers

Alok Kumar, Mohammed Asim, Mirza Mohd. Shadab, and Iram Akhtar

Abstract In a power system, various fluctuations can occur in voltage and frequency due to the disturbances and sudden changes in the load (due to load change or fault). Whenever there is a change in the load, voltage and frequency get affected/disturbed. Voltage regulator is designed to compensate the voltage fluctuations and to compensate or maintain the frequency of the system. A load frequency controller (LFC) is designed. A large number of LFC are proposed and designed in order get the most effective, accurate, efficient and adaptive control. In this paper, various controller designs are studied that are proposed for load frequency controller.

Keywords Frequency · LFC · Settling time

1 Introduction

In a power system, high numbers of power elements and devices, it is nonlinear in nature and is based on multi-input multi-output (MIMO) system. The system performance can be improved by various control techniques and are used to compensate the disturbances along the path of the system [1–7].

Different control techniques have evolved in the past and in order to maintain the power system in its secure state and protect from various disturbing phenomenon. Various techniques such as excitation system control, generator prime mover control, clearing the faults faster, power compensation, current injector and voltage compensators. [1].

Nowadays, various modern techniques such as FACTS controllers, HVDC controllers and static VAR compensators (SVC) [3] are used as they provide more

A. Kumar (✉) · M. Asim · M. Mohd. Shadab
Department of Electrical Engineering, Integral University, Lucknow, India
e-mail: alokk1159@gmail.com

M. Asim
e-mail: asimamu@gmail.com

I. Akhtar
Department of Electrical Engineering, Jamia Milia Islamia, Delhi, New Delhi, India

A. Iqbal et al. (eds.), *Renewable Power for Sustainable Growth*, Lecture Notes in Electrical Engineering 723, https://doi.org/10.1007/978-981-33-4080-0_26

accurate, efficient and fast response. Automatic generation control (AGC) consists of control of voltage and frequency both. This paper discusses various proposed LFC controllers that are used. To understand the basics of load frequency control, a basic diagram representation of an AGC is shown in Fig. 1.

Various objectives [2] of AGC are as follows:

- The system frequency is maintained constant near to nominal value.
- Uphold the value of interchange of power between the areas.
- Each generator that is connected in the system is to be maintained at most economical values.

A number of load frequency controllers are proposed based on various techniques such as direct synthesis approach, various adaptive techniques such as by using interaction estimations, fuzzy logic control approach for load frequency control and neuro fuzzy simulation-based hybrid intelligent PI control.

The basic components that are to be considered in a load frequency controller are turbine, governor, generator and load, controller and droop. All these components are used to maintain the system frequency [3, 4]. The various parameters of these

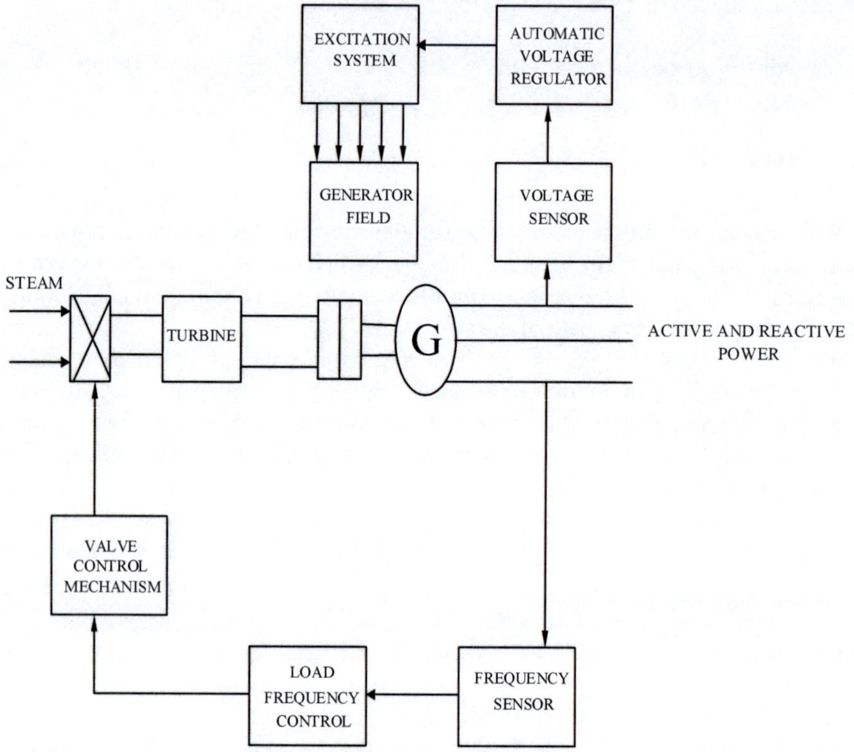

Fig. 1 Automatic voltage regulator

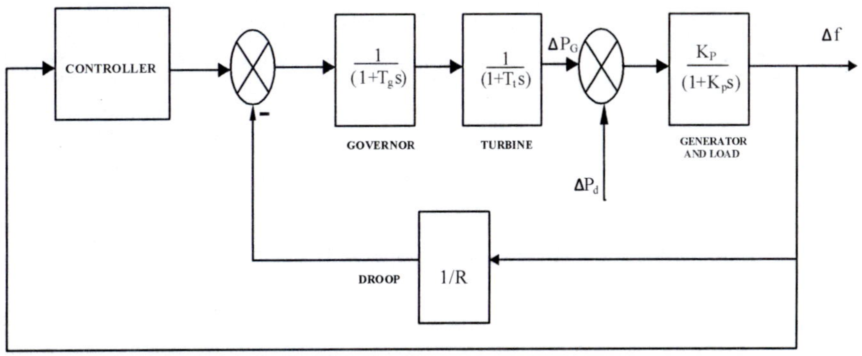

Fig. 2 Single-area system with load frequency controller

components are considered in Laplace form. The block diagram of a general load frequency controller is shown in Fig. 2.

In this paper, various load frequency control models are compared on various previous-based proposed techniques and the various control parameters of the proposed techniques are compared.

2 Types of Load Frequency Control Methods

2.1 Internal Model Control (IMC)

In this controller design technique, the control is based on Q parameterization technique. This technique has been developed for various discrete and continuous time systems. It can be used for single-input single-output, multi-input multi-output systems and discrete systems.

This technique provides a good option instead of using a robust controller or an optimal controller. It can be used in process control industry to reduce error and also reduce noises.

2.2 Soft Computing

In a complex power system or plant, the traditional control systems cannot provide accurate results and are therefore it is difficult to control the plant using conventional methods. Soft computing technique uses approximation in order to solve very difficult problems.

Soft computing is a technique that is a part of computational intelligence (CI) family. It consists of a large number of innovative search techniques. These techniques

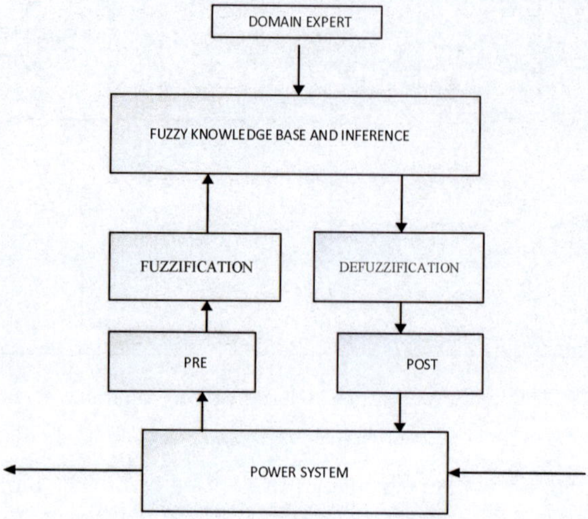

Fig. 3 Basic diagram of a fuzzy logic controller

are obtained from various branches like physics, chemistry, mathematics, etc. Some of the soft computing techniques that are widely used are genetic algorithm, particle swarm optimization, optimization, fuzzy logic, etc. [6, 7].

2.3 Fuzzy Logic Controller

Dynamic characteristics of a power system are complex and variable, and various conventional methods of control cannot provide the result that is desired. Therefore, intelligent controllers can be replaced with conventional controllers to get fast and good dynamic response in load frequency control problems.

The fuzzy logic controllers can be used to solve a wide range of control problems if the system robustness and reliability are more important as conventional controllers are slow and have low efficiency in nonlinear system applications. It minimizes the fluctuation on system output.

A fuzzy logic controller mainly consists of three parts—fuzzification, rule base table and defuzzification [7] (Fig. 3).

2.4 Genetic Algorithm (GA)

There are various evolutionary algorithms, and genetic algorithms are the part of evolutionary algorithms. Natural selection and genetics is the basis of operation of this algorithm. These are generally used to obtain a high-quality solution for search and optimization problem. This newly intelligent soft computing control technique can be used in load frequency controller [5].

Genetic algorithm consists of five stages.

2.4.1 Initial Population

Initially, the set of individuals are called a population. Every problem can be solved with the help of each individual. The parameters or properties of an individual are called genes. A chromosome is formed with the help of string, which is formed when various genes are joined together.

2.4.2 Fitness Function

It is a function that determines the fitness of an individual. Fitness is an individual ability to compete with others. Each individual is assigned with a fitness score.

2.4.3 Selection

In this phase of the genetic algorithm, the individual that is the fittest in the population is selected and its genes are transmitted to the future generation. On the basis of the fitness score, parents are selected which is pair of individual. In the population, individuals which have high fitness or fitness score are likely to be picked up for reproduction.

2.4.4 Crossover

It is the most important phase in a GA. A crossover point is chosen at random from within the genes for each pair of parents to be mated.

2.4.5 Mutation

In various offsprings that are formed in the process, various genes that have low random probability can be subjected to mutation. It means that the bits in the string can be flipped.

2.5 Seeker Optimization Algorithm (SOA)

These algorithms are designed based upon the type of the plant and its various parameters and components. A multi-area system is completely different from a single-area system because the control operation required is complex and the design of a load frequency controller is also complex

Seeker optimization algorithm is also based on population just like other optimization techniques such as PSO and GA. In 2007, seeker optimization technique was introduced by Dai et al.

Each member of the seeker population is known as seeker. The flowchart below shows the algorithm and process of a seeker optimization algorithm. In this method, every seeker is the member of the population. Each seeker tends to move to its best-suited position according to his opinion. The best position of the seeker is updated by optimum solution [6].

In order to find the optimum solution very accurately and fast, these algorithms are used. However, premature convergence can be found in many studies which is a drawback in these algorithms.

These problems can be improved by using a hybrid or a very highly accurate technique and can be implemented to systems. The controller can be designed by various algorithms, optimization techniques and classical methods with new systems implemented. The PID controller parameters are to be selected for optimal values by optimal solution and load frequency control (Fig. 4).

2.6 Direct Synthesis Approach

In this method, a new PID controller is designed which is based on direct synthesis approach in frequency domain. Various parameters of the controller are obtained using frequency response matching with the DS controller. This method provides the linear equations, and the solution of these equations gives the values of the controller parameters. This design method is developed for both single-area system as well as multi-area power systems (Fig. 5).

Let us consider some examples and compare them with various proposed methods.

Example 1 A non-reheat turbine system for single area is considered, the design methods of some predefined controllers are compared, and the results are shown in Table 1.

Example 2 A reheat turbine system for single area is considered, the some of the design methods are compared, and the results are shown in Table 1 (Fig. 6; Table 2).

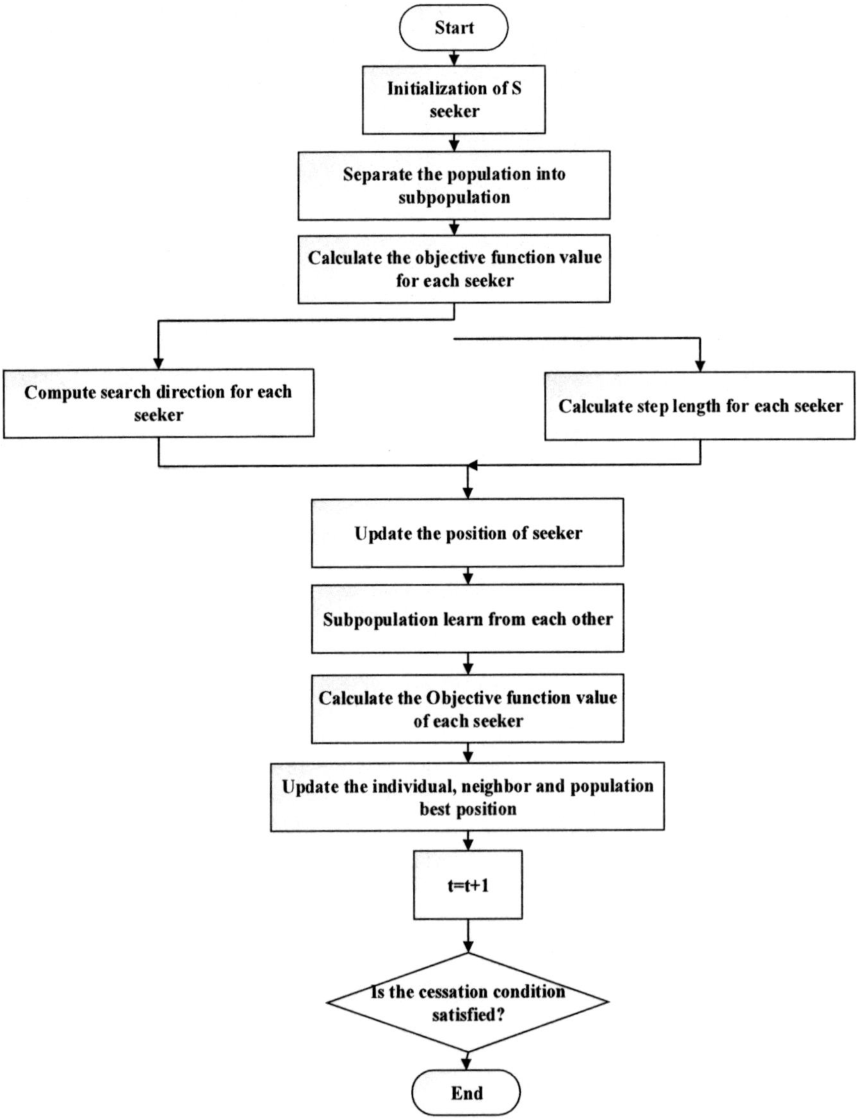

Fig. 4 Seeker optimization algorithm

3 Conclusion

The various load frequency control technique from the literature are studied and compared. It is seen that various techniques such as direct synthesis approach, seeker optimization control, robust decentralized control, fuzzy logic controller, adaptive load frequency controller, etc., are being used in industries nowadays. These methods

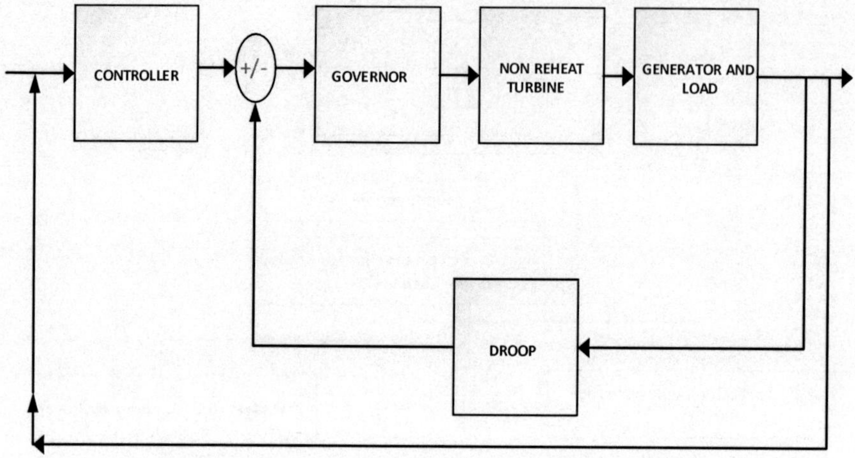

Fig. 5 Unity negative output feedback configuration

Table 1 Comparison of various parameters by Nishat and Pan, Padhan and Majhi and Tan

S. No	Method	Settling time	Peak value	ISE
1	Nishat and Pan	Low	Low	Low
2	Padhan and Majhi	Moderate	Moderate	Moderate
3	Tan	High	High	High

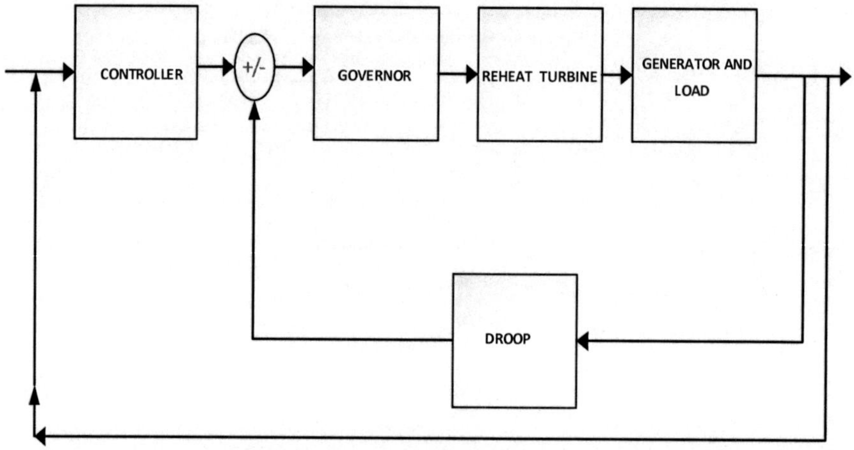

Fig. 6 Unity negative output feedback configuration

Table 2 Comparison of various parameters by Nishat and Pan, Padhan and Majhi and Tan

S. No.	Method	Settling time	Peak value	ISE
1	Nishat and Pan	Moderate	Low	Low
2	Padhan and Majhi	Low	Moderate	Moderate
3	Tan	High	High	High

focus on the best possible control of the load frequency that varies in a power system significantly. Depending upon the type of the system, the design of the controller varies. It is clear that by changing some of the parameters and using optimal solution for these values, the response of the whole system can be improved significantly. The system parameters such as settling time, magnitude and the integral square error have improved, and a better response is obtained.

References

1. Tan W (2009) Tuning of PID load frequency controller for power systems. Energy Convers Manage 50:1465–1472
2. Anwar MN, Pan S (2015) A new PID load frequency controller design method in frequency domain through direct synthesis approach. Electr Power Energy Syst 67:560–569
3. Gundes AN, Chow L (2013) Controller synthesis for single-area and multi-area power systems with communication delays. In: American control conference (ACC), Washington, DC, USA 17–19 June 2013
4. Kumar A, MdNishat A A PID and PIDA controller design for an AVR system using frequency response matching. Int J Innovative Technol Exploring Eng (IJITEE), 8(10)
5. Asim M, Riyaz A, Tiwari S, Verma A (2018) Performance evaluation of fuzzy controller for boost converter with active PFC. J Intell Fuzzy Syst 35(5):5169–5175
6. Dai C, Zhu Y, Chen W (2007) Seeker optimization algorithm. In: Wang Y, Cheung Y, Liu H (eds) Computational intelligence and security. CIS 2006. Lecture notes in computer science, vol. 4456. Springer, Berlin, Heidelberg. https://doi.org/10.1007/978-3-540-74377-4_18
7. Shahabuddin M, Asim M, Sarwar A (2020) Parameter extraction of a solar PV cell using projectile search algorithm. In: 2020 international conference on advances in computing, communication & materials (ICACCM), Dehradun, India, pp 357–361. https://doi.org/10.1109/ICACCM50413.2020.9213005.

Techno-Economic Analysis of Diesel/Wind/PV/Battery Hybrid Energy System for Androth Island

Mohammad Shariz Ansari, Md. Faisal Jalil, Manaullah, and Sourav Diwania

Abstract The net present cost (NPC) and the cost of energy (COE) are very high in the remote islands because energy is mainly generated from diesel generators. Pollution is also increasing due to the usage of diesel generators. So, the objective in this work is to minimize the NPC, COE, and pollution by using renewable energy sources, for example, solar and wind in addition to diesel. Androth Island of union territory of Lakshadweep (UTL) in India has been considered for the study. The average wind speed and average daily radiation at Androth Island are 5.34 m/s and 5.68 kWh/m^2/day, respectively. The modeling, simulation, and optimization have been performed by hybrid optimization model for electric renewable (HOMER) software. The model that gives an optimal solution includes diesel–PV–battery. This model gives the least NPC and COE. For sensitivity analysis, the wind speed (m/s) and the fuel rate ($) have been considered as sensitivity variables. The optimal system is obtained for the sensitivity variable of 5.38 m/s of wind velocity and 1$/L of diesel price. This optimal system also reduces pollution by 90%.

Keywords Optimal system · NPC · COE · Renewable energy · HOMER

1 Introduction

Energy is a basic requirement for every single social action, generation everything being equal and the arrangement of all services. Then again, energy is additionally and still to be the greatest emergency to individuals, since right now, most of the energy

M. S. Ansari (✉) · Md. F. Jalil · S. Diwania
KIET Group of Institutions, Delhi-NCR, Ghaziabad, U.P, India
e-mail: shariz.ansari@kiet.edu

Md. F. Jalil
e-mail: faisal.jalil@kiet.edu

Manaullah
Jamia Millia Islamia, Delhi, New Delhi, India

A. Iqbal et al. (eds.), *Renewable Power for Sustainable Growth*, Lecture Notes in Electrical Engineering 723, https://doi.org/10.1007/978-981-33-4080-0_27

utilized on earth originates from customary petroleum products, and some of them will be depleted in a very long while as per the ongoing investigating and devouring rate [1–3]. Also, there are still about 1.5 billion occupants overall as yet have no accesses to power [4]. For these remote islands and towns, energy is being regularly supplied by diesel generators. In any case, they felt progressively increasingly tense since they as often as possible face inflated fuel costs as a result of the gigantic climb in the cost of diesel and shipment cost [5]. Additionally, the negative natural effects from crafted by diesel hurt the nearby biological system and cause water, noise, air, and soil contamination. In the issue, they experience the ill effects of energy shortage or power outages often. Luckily, remote regions are typically rich in locally accessible sustainable power resources [6]. Because of the increasing expense of diesel fuel and the quickly decreasing expense of sustainable power resources, the supply of energy by renewables is presently getting to be more focused with conventional energy, accordingly encouraging broadly use of sustainable power sources for stand-alone system, for example, PV–battery, wind–battery or hybrid system. Until now, look into on RESs is normally done in the area of modeling of the system, sizing of the element, simulation, economic evaluation, and especially optimization of the system. For doing such type of research, the computer tools and simulation models are normally required. 37 PC tools for understanding the RESs have been looked into in [7], and an audit of the strategy of optimizing the hybrid renewable energy sources was done in [8, 9]. Out of these simulation devices, HOMER software is a standout among the most broadly utilized for stand-alone RESs [10]. Utilizing HOMER for RES techno-economic analysis, simulation and modeling have been the topic of considerable prior investigations, for instance, the likelihood of accomplishing energy sovereignty in an island utilizing wind turbine, PV, biogas generator, and battery was assessed in [5]. A techno-economic study of RES for Rural Electrification in South Algeria has been done in [11].

2 Profile of Androth Island

The Androth Island is one of the inhabited islands in the union territory of Lakshadweep. It is situated at east of Kavaratti Island. The distance from Androth Island and Kochi is 64 nautical miles (119 km). The area of Androth Island is 34.90 km^2. It is the major island in union territory of Lakshadweep. It is shown in Fig. 1. The maximum span of the island is 4.66 km, and width is 1.43 km. It lies between 10°48′ and 10°50′ N latitude and 73°38′ and 73°42′ E longitude. It has east–west and north–south trend with a long tail on the east. Out of all ten inhabited islands, Androth Island has least lagoon area [12].

Fig. 1 Location of Androth Islands of UTL

3 Optimal Hybrid Model for Androth Island

3.1 Inputs to the HOMER

For simulation and optimization in HOMER, six kinds of information are required. These data include meteorological data, load profile data, economic data, technical data, equipment characteristics data, and search space data.

3.2 Meteorological Data for Androth Island

Monthly average solar global irradiance (SGI) data for Androth Island is downloaded from NASA database. Monthly average global horizontal radiation has been recorded by NASA. The monthly average solar global irradiance for Androth Island is shown in Table 1. Table 1 gives the clearness index and daily radiation in kWh/m^2/day. This table shows the maximum clearness index is 0.679 and minimum lucidity index is 0.453. This maximum and minimum lucidity index occurs in the month of March and June, respectively. The maximum and minimum daily radiation shown by Table 1 are 6.92 kWh/m^2/day and 4.68 kWh/m^2/day, respectively. The maximum solar radiation at selected location falls in March and minimum falls in June. The yearly average radiation is 5.68 kWh/m^2/day [13].

Monthly average wind speed data at Androth Island in (m/s) is downloaded from NASA database. Monthly average wind speed has been recorded by NASA for duration of 10 years. Average monthly wind velocity is also shown in Table 1. Maximum speed of wind is 8.93 m/s, and it is accessible in June. Minimum speed of wind is 3.48 m/s, and it is accessible in February. The normal wind velocity is 5.38 m/s.

Table 1 Monthly average solar global irradiance (SGI) and wind speed data for Androth Island

Month	Clearness index	Daily radiation ($kWh/m^2/day$)	Average wind speed (m/s)
Jan.	0.659	5.79	4.22
Feb.	0.678	6.43	3.48
Mar.	0.679	6.92	3.82
Apr.	0.647	6.81	4.17
May	0.57	5.97	5.19
Jun.	0.453	4.68	8.93
Jul.	0.456	4.72	7.95
Aug.	0.5	5.22	7.66
Sep.	0.557	5.71	6.1
Oct.	0.548	5.29	4.28
Nov.	0.585	5.21	3.95
Dec.	0.629	5.36	4.86
Average	**0.580**	**5.68**	**5.34**

3.3 *Load Profile of Androth Island*

The load shape of the selected island is an important input data for the optimization of proposed model in HOMER, because for an optimal sizing, different energy sources depend on the load shape. Figure 2 displays the load shape of Androth Island. Figure 2a shows the daily residential load shape, and Fig. 2b displays the daily commercial load profile. In Androth Island, energy is mainly utilized to fulfill the residential and industrial loads only. Hence, only these two types of loads will be painstaking in HOMER. The energy consumptions by the residential and commercial load at Kavaratti Island are 6.52 MU per year and 1.05 MU per year, respectively, in [12]. The conversion of these data into daily energy consumption gives 15,397.26 kWh/day for residential load and 2876.71 kWh/day for the commercial load. The peak loads of residential and commercial loads are 3270.43 kW and 481.00 kW, respectively. HOMER uses these data for simulation and optimization.

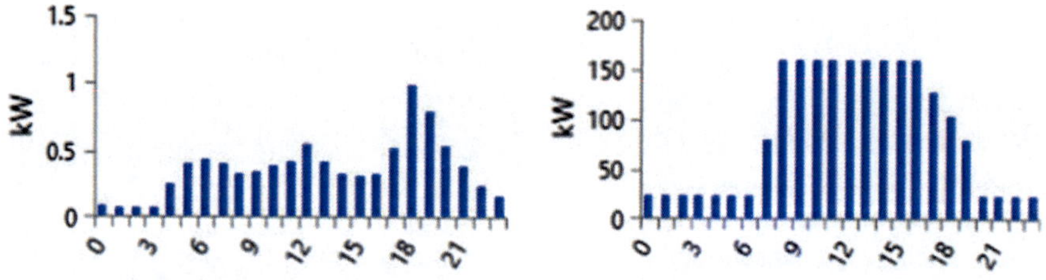

(a) Daily residential load profile (b) Daily commercial load profile

Fig. 2 Load profile of Androth Island

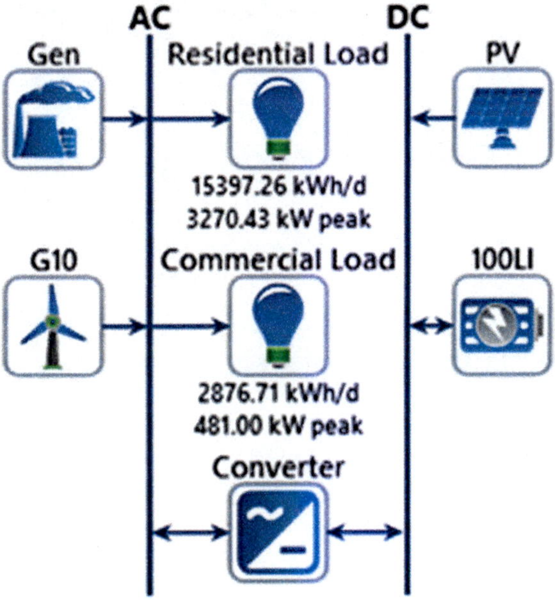

Fig. 3 Proposed system model for Androth

4 System Architecture of Proposed Model for Androth Island

In Androth Island power is generated mainly from diesel generators, but there are huge potential of solar PV and wind is also available. Since diesel is costly and it is depleted very fast, so the aim is to decrease the use of diesel and maximize the usage of renewable energy. The author has proposed a model shown in Fig. 3 that consists of wind turbines, solar PV, diesel generators for emergency use, batteries, converters, and loads. The rating and cost of these components have been described in previous chapter. HOMER simulates the proposed model to give the optimal result.

5 Results of Proposed Model for Androth Island

5.1 Sensitivity and Optimization Results

The sensitivity and optimization results from HOMER are presented in Fig. 4. The upper part of the figure gives the result of different sensitivity variables, and lower part gives the optimization result for a particular selected sensitivity variable.

The sensitivity variables for diesel price have been taken as 0.9$/L, 1$/L and 1.1$/L. The sensitivity variable for wind speed has been taken as 3.0 m/s, 8 m/s, 10 m/s, and 5.38 m/s. The average speed of wind for the considered island has

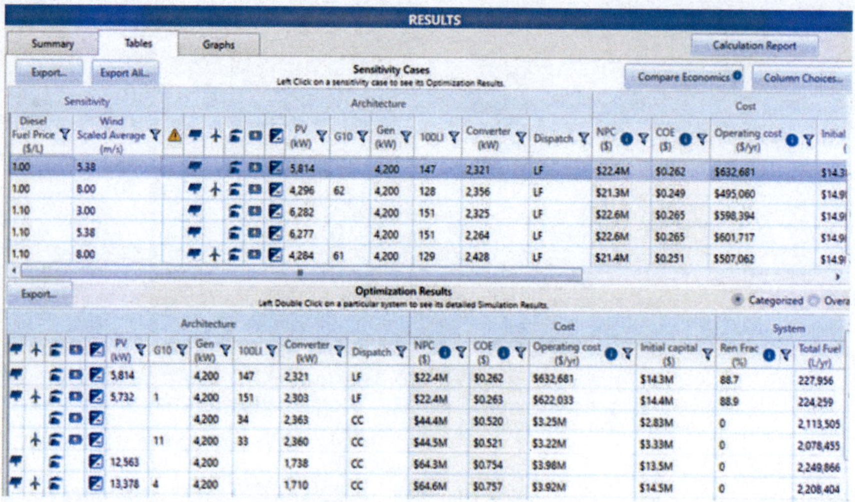

RESULTS

Summary | Tables | Graphs | Calculation Report

Export... | Export All...

Sensitivity Cases

Left Click on a sensitivity case to see its Optimization Results.

Compare Economics | Column Choices...

Diesel Fuel Price ($/L)	Wind Scaled Average (m/s)	PV (kW)	G10	Gen (kW)	100LI	Converter (kW)	Dispatch	NPC ($)	COE ($)	Operating cost ($/yr)	Initial
1.00	5.38	5,814		4,200	147	2,321	LF	$22.4M	$0.262	$632,681	$14.3
1.00	8.00	4,296	62	4,200	128	2,356	LF	$21.3M	$0.249	$495,060	$14.9
1.10	3.00	6,282		4,200	151	2,325	LF	$22.6M	$0.265	$598,394	$14.9
1.10	5.38	6,277		4,200	151	2,264	LF	$22.6M	$0.265	$601,717	$14.9
1.10	8.00	4,284	61	4,200	129	2,428	LF	$21.4M	$0.251	$507,062	$14.9

Export...

Optimization Results

Left Double Click on a particular system to see its detailed Simulation Results.

Categorized | Overa

PV (kW)	G10	Gen (kW)	100LI	Converter (kW)	Dispatch	NPC ($)	COE ($)	Operating cost ($/yr)	Initial capital ($)	Ren Frac (%)	Total Fuel (L/yr)
5,814		4,200	147	2,321	LF	$22.4M	$0.262	$632,681	$14.3M	88.7	227,956
5,732	1	4,200	151	2,303	LF	$22.4M	$0.263	$622,033	$14.4M	88.9	224,259
		4,200	34	2,363	CC	$44.4M	$0.520	$3.25M	$2.83M	0	2,113,505
	11	4,200	33	2,360	CC	$44.5M	$0.521	$3.22M	$3.33M	0	2,078,455
12,563		4,200		1,738	CC	$64.3M	$0.754	$3.98M	$13.5M	0	2,249,866
13,378	4	4,200		1,710	CC	$64.6M	$0.757	$3.92M	$14.5M	0	2,208,404

Fig. 4 HOMER sensitivity and optimization result for Androth Island

been taken from NASA by HOMER, and it is 5.38 m/s. HOMER has used all the combination of these sensitivity variables for simulation and optimization. Total 32,106 solutions were simulated, and all of them are feasible. Out of them, total 4033 were omitted in which 2634 omitted for lacking a converter and 1255 omitted for having an unnecessary converter. The actual value of sensitivity variables at the considered island is 5.38 m/s. Using these sensitivity variables for the optimum solution, HOMER uses only solar PV, diesel generators, and storage systems. The optimum solution for this combination comes out to be total net present cost as $22,368,262.22 and levelized cost of energy as 0.262 $/kWh.

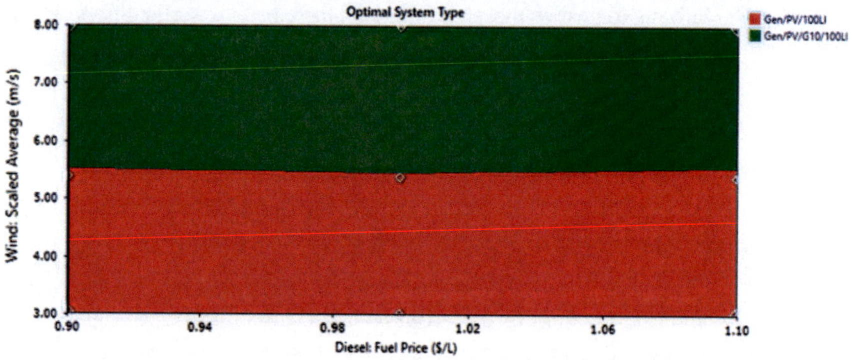

Fig. 5 Optimal system type plot for Androth Island

Optimal system type plot for the proposed model of the Androth Island is shown in Fig. 5. It shows that which combination of hybrid system is optimal at a particular value of wind speed and diesel price. Green color displays the hybrid system consisting of all sources, and red color shows all sources except wind. From the optimal system type plot, it is clear that for slow wind speed at Androth Island, diesel/PV/battery system is giving optimal solution, and if wind velocity is high, then system consisting all the sources is giving an optimal solution. Table 1 shows that the typical wind velocity is 5.38 m/s. For this wind speed, optimal system consists of diesel/PV/battery systems only.

5.2 *Net Present Cost (NPC)*

Figure 6 and Table 2 give the category wise net present cost in bar chart form and in tabular form, respectively. Total net present cost comes out to be $22,368,262.22 that consists of operating cost, capital cost, salvage value, replacement cost, and resource cost for the considered model.

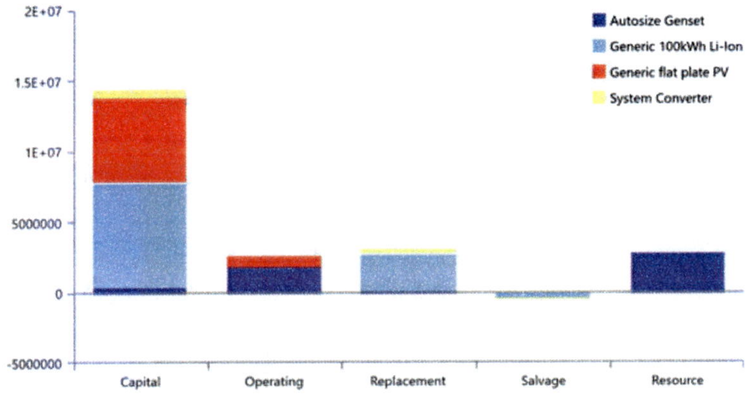

Fig. 6 Cost summary of proposed model at Androth Island

Table 2 Net present cost of proposed model at Androth Island

Name	Capital	Operating	Replacement	Salvage	Resource	Total
Autosize genset	$420,000	$1.82 M	$115,934	−$84,975	$2.84 M	$5.12 M
Generic 100 kWh li-Ion	$7.40 M	$18,919	$2.59 M	−$331,751	$0.00	$9.67 M
Generic flat plate PV	$5.90 M	$754,343	$0.00	$0.00	$0.00	$6.66 M
System converter	$689,906	$0.00	$287,874	−$53,582	$0.00	$924,198
System	$14.4 M	$2.60 M	$2.99 M	−$470,308	$2.8 M	$22.4 M

Table 3 Emission by diesel-only system and hybrid system

Pollutant	Emission by only diesel generator (kg/year)	Emission by PV–diesel–battery system for diesel price $1/L (kg/year)	Emission reduction (kg/year)
Carbon dioxide	7,787,491	593,161	7,194,330
Carbon monoxide	49,088	3739	45,349
Unburned hydrocarbons	2142	163	1979
Particulate matter	298	22.7	275.3
Sulfur dioxide	19,070	1453	17,617
Nitrogen oxides	46,113	3512	42,601

Total NPC includes $14,400,000.00 as capital cost, $2,600,000.00 as operating cost, $2,990,000.00 as replacement cost, −$470,308.00 as salvage value, and $22,368,262.22 as resource cost. Among these, the maximum cost is capital cost which is $14,400,000.00 that accounts for Gen sets capital cost ($ 420,000), Li-ion batteries capital cost ($7,400,000.00), PV capital cost ($5,900,000.00), and converters capital cost ($689,906.00). The total operating cost is $2,600,000.00, total replacement cost is $2,990,000.00, total salvage value is −$470,308.00, and total resource cost is $2,840,000.00

6 Emission of Different Pollutants at Androth Island

Table 3 shows the emission of different types of pollutant materials by only diesel system and by a hybrid system. These pollutant materials are CO_2, CO, unburned hydrocarbon, particulate matters, SO_2, and nitrogen oxide. From Table 3, it is clearly seen that a huge amount of pollution is reduced if we use a hybrid system. In a diesel-only system, the CO_2 emission is 7,787,491 kg/year, and in the hybrid system, the CO_2 emission is 593,161 kg/year. It means a reduction in CO_2 emission is 7,194,330 kg/year which is a great reduction. The reduction of other pollutant materials such as carbon monoxide, sulfur dioxide, nitrogen dioxide is mentioned in Table 3.

7 Comparative Analysis of Both the System

Table 4 gives comparative analysis of both the systems for different value of sensitivity variables, i.e., for different value of diesel price and wind speed. This table shows that NPC of $21.1 M and COE of 0.262 $/kWh are minimum for a hybrid

Table 4 Comparative analysis of diesel-only system and hybrid system (Androth Island)

Sensitivity variables		Hybrid system					Diesel only		
Diesel fuel price ($/L)	Wind speed (m/s)	System	NPC (M$)	COE ($/kWh)	Renewable fraction (%)	CO_2 emission (kg/year)	NPC (M$)	COE ($/kWh)	CO_2 emission (kg/year)
0.9	3	PV–diesel–battery	22.1	0.262	88.8	593,161	61.1	0.717	7,787,491
0.9	5.21	PV–diesel–battery	22.1	0.262	88.8	593,161	61.1	0.717	7,787,491
0.9	8	Wind–PV–diesel–battery	21.1	0.257	91.9	430,782	61.1	0.717	7,787,491
1	3	PV–diesel–battery	22.4	0.262	88.7	596,700	64.9	0.762	7,787,491
1	5.21	PV–diesel–battery	22.4	0.262	88.7	596,700	64.9	0.762	7,787,491
1	8	Wind–PV–diesel–battery	21.3	0.249	93.2	358,554	64.9	0.762	7,787,491
1.1	3	PV–diesel–battery	22.6	0.265	90.3	510,843	68.7	0.806	7,787,491
1.1	5.21	PV–diesel–battery	22.6	0.265	90.2	519,327	68.7	0.806	7,787,491
1.1	8	Wind–PV–diesel–battery	21.4	0.251	93.3	355,800	68.7	0.806	7,7,491

system consisting of all the sources for diesel price 0.9 $/L and wind speed 8 m/s. But the typical wind velocity is limited to 5.38 m/s, and the actual price of diesel as of today is $1/L at the considered island. That is why we are considering wind speed of 5.38 m/s and diesel price of $1/L for the actual value of NPC and COE that comes out to be $22.4 M and 0.262 $/kWh, respectively, for a PV–diesel–battery hybrid system.

Table 4 also shows that the NPC and COE for the same sensitivity variable of diesel-only system and that comes out to be $64.9 M and 0.762 $/kWh, respectively. Comparing these two values for hybrid and diesel-only system at the same wind speed and diesel price, i.e., 5.38 m/s and $1/L, respectively, it is clearly seen that the NPC is reducing from $64.9 M to $22.4 M, and COE is reducing from 0.762 $/kWh to 0.262 $/kWh. It means If we use a hybrid system in place of diesel-only system, then there will be a great reduction in NPC and COE. If we use a mix system in place of diesel-only system, then CO_2 emission will be reduced from 7,787,491 kg/year to 519,327 kg/year which is again a great reduction.

8 Conclusions

The sensitivity and optimization results obtained from HOMER are shown in Fig. 4. In order to obtain the optimal result, the proposed model is being simulated for thousands of combinations by HOMER. The combination which gives the optimal result includes diesel–PV–battery and converters. This combination gives the least NPC and COE which is $22,368,262.22 and $0.262, respectively. For sensitivity analysis, the wind speed (m/s) and fuel rate ($) have considered as sensitivity variables. The above optimal solution, i.e., least NPC as $22,368,262.22 and least COE as $0.262 has been determined for 5.38 m/s of wind speed and $1/L of diesel price. Figure 5 shows the optimal system type plot. This plot shows that for different rate of wind velocity and for different rate of diesel price which type of system is economical, i.e., PV–diesel–battery system is economical, or PV–wind–diesel–battery system is economical. From the optimal system type plot, it is clearly visible that for a higher value of wind velocity, wind–PV–diesel–battery system is economical, and for a lower value of wind speed, diesel–PV–battery system is economical. Table 2 and Fig. 6 give the NPC and cost\summary that contains operating cost, capital cost, salvage value, replacement cost, and resource cost. Among these, the maximum cost is capital cost which is $14,400,000.00 that accounts for Gen sets capital cost ($ 420,000), Li-ion batteries capital cost ($7,400,000.00), PV capital cost ($5,900,000.00), and converters capital cost ($689,906.00). The total operating cost is $2,600,000.00, total replacement cost is $2,990,000.00, total salvage value is −$470,308.00, and total resource cost is $22,368,262.22.

References

1. Legros G, Havet I, Bruce N, Bonjour S (2009) The energy access situation in developing countries. USA. https://doi.org/10.1503/cmaj.160656
2. Iqbal A et al (eds) (2020) Soft computing in condition monitoring and diagnostics of electrical and mechanical systems. In: Advances in intelligent systems and computing, vol 1096. Springer, Singapore. doi:https://doi.org/10.1007/978-981-15-1532-3
3. Iqbal A et al (eds) (2020) Meta heuristic and evolutionary computation: algorithms and applications. In: Studies in computational intelligence, vol 1096. Springer, Singapore. https://www.springer.com/gp/book/9789811575709
4. Elsen C, Darses F, Leclercq P (2011) An anthropo-based standpoint on mediating objects: evolution and extension of industrial design practices. Paris, France
5. Manaullah, Ansari MS (2019) Feasibility analysis of renewable energy options for the union territory of Lakshadweep Islands. Int J Glob Energy Issues 42:63–80. https://doi.org/10.1504/IJGEI.2019.100690
6. Nasiruddin I, Khatoon S, Jalil MF, Bansal RC (2019) Shade diffusion of partial shaded PV array by using odd-even structure. Sol Energy 181:519–529. https://doi.org/10.1016/j.solener.2019.01.076
7. Manaullah, Ansari MS (2019) Techno-economic feasibility of PV-wind-diesel battery hybrid energy system for Lakshadweep Island in India. Indian J. Power River Val Dev 69:141–148 (2019)
8. Ansari MS, Jalil MF (2017) Investigation of renewable energy potential in union territory of Lakshadweep islands. In: CIPECH 2016. IEEE, Ghaziabad, India, pp 209–213. https://doi.org/10.1109/CIPECH.2016.7918768
9. Anoune K, Sizing a PV-wind based hybrid system using deterministic approach Sizing a PV-Wind based hybrid system using deterministic approach. Energy Convers Manag 169:137–148. https://doi.org/10.1016/j.enconman.2018.05.034
10. Abid M, Ansari MS (2014) Solar photo voltaic power generation in union territory of Lakshadweep Island: projected level dissemination using technology diffusion models. In: CIPECH 2014. IEEE, Ghaziabad, India, pp 448–452. https://doi.org/10.1109/CIPECH.2014.7019036, Electronic ISBN: 978-1-4799-5871-9
11. Chermat F, Khemliche M, Badoud A, Latreche S (2018) Techno-economic feasibility study of **investigation of renewable energy system for rural electrification in South Algeria. Eng Technol** Appl Sci Res 8:3421–3426
12. A joint initiative of Government of India and Administration of UT of Lakshadweep: 24 × 7 POWER FOR ALL (Lakshadweep Islands) (2016)
13. Stackhouse P, Whitlock C, Zhang T (2004) Solar renewable energy data sets from NASA satellites and research. In: World Renewable Energy Congress VIII (WREC 2004), pp 279–283

Transformer-Based DC-to-AC Grid-Connected Multi-Level Inverter Topology

Shahbaz Ahmad and Farhad Ilahi Bakhsh

Abstract The traditional line inverters contain line current of square shape in which the harmonics of higher order are present. Due to this, higher order harmonic content EMI is produced due to which large amount of heat in the core of the transformer is triggered. In this paper, a transformer-based grid-connected multi-level inverter topology is presented where transformer with multiple tappings is used to get the output voltage nearly sinusoidal. The output of the inverter is connected to the grid. During each periodic cycle, switching of power electronic devices is done in a very systematic order using the controller circuit. The voltage is given as a command signal to the controller circuit since MOSFET is used as a switch. The complete performance of the DC-to-AC converter is controlled by varying the quantity of the output voltage by utilizing the tap-changing transformer. The analysis of the circuit is done in the MATLAB/Simulink which shows the extreme curtailment in total harmonic distortion.

Keywords DC-to-AC inverter · Tap-changing transformer · MATLAB · Multi-level

1 Introduction

The power electronics-based converters are extensively used in various applications such as commercial, telecommunication, transportation, generation, distribution and transmission of electrical energy. AC-to-DC converters are broadly intended for various uses such as battery charging in different appliances [1–9]. Generally, in grid-connected inverter, AC-to-DC converter is operated at switching angle greater than 90° [2–5]. Thus, the power is flowing from DC source to the grid. However, the

S. Ahmad (✉)
Department of Electrical Engineering, Aligarh Muslim University, Aligarh, India
e-mail: shahbazdept@gmail.com

F. I. Bakhsh
Department of Electrical Engineering, National Institute of Technology Srinagar, Srinagar, India

A. Iqbal et al. (eds.), *Renewable Power for Sustainable Growth*, Lecture Notes in Electrical Engineering 723, https://doi.org/10.1007/978-981-33-4080-0_28

conventional converters or rectifiers have ample amount of higher order harmonics content in the current (line). Therefore, the line current which is square shaped and has high total harmonic distortion will be their on the AC side. This circuit topology has a simple construction, but it is not frequently used for DC to AC power conversion because it contains harmonics in the line current. To overcome this, an inductor is required of large value on the load side which increases the cost of the frequently used inverter [6, 7].

Here, a transformer-based grid-connected inverter is used. The transformer is used have multiple tappings. On primary side or multiple tap side of the transformer battery, source and the switches are placed, whereas on the other side of the transformer, grid is connected. The DC source and the switches are placed in such a manner that when proper switching of the switches is done with the help of controller circuit in both the positive and negative cycles of AC grid, the output voltage is nearly sinusoidal. Due to this, THD in the output voltage will be very low as compared to the conventional system. MOSFET of low current rating is used as a switch as the number of switch used is large in number.

2 Description

Figures 1 and 2 present the block and circuit diagrams of the projected inverter, respectively. Here, source (DC) is put in the middle on the multiple tapping side of the transformer. The switches T_1, T_2 up to T_n and T'_1, T'_2 up to T'_n are power semiconductor switches whose proper switching is done so that it can operate for the desired duration of time. The tapping of every switch is done in such a manner so that the output voltage becomes nearly sinusoidal as shown in Fig. 3, which is connected across the grid.

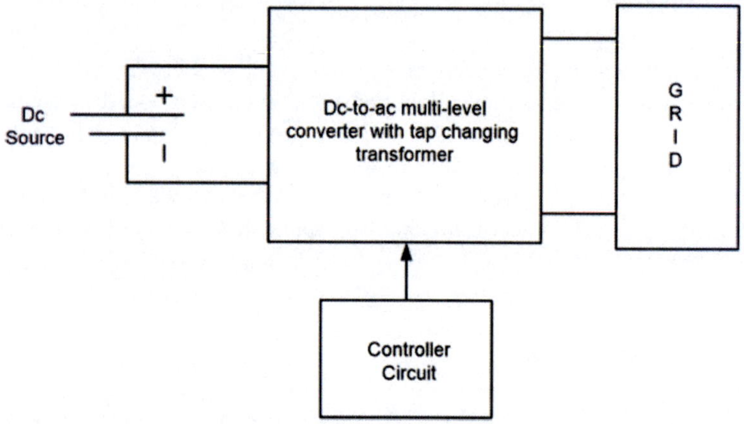

Fig. 1 Block diagram of the proposed grid-connected multi-level inverter

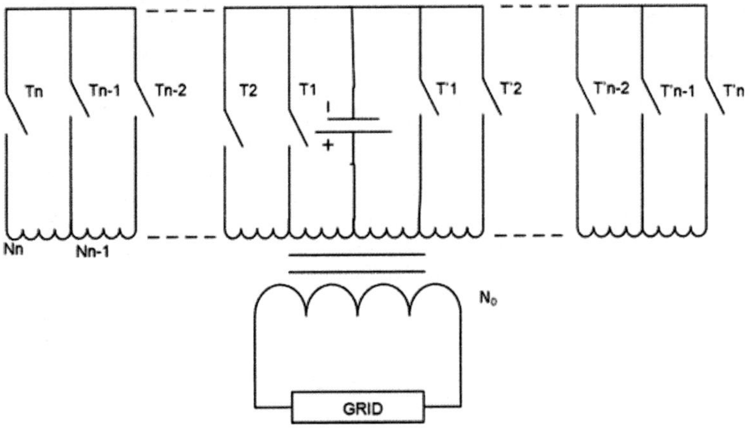

Fig. 2 Internal diagram of the proposed grid-connected multi-level inverter

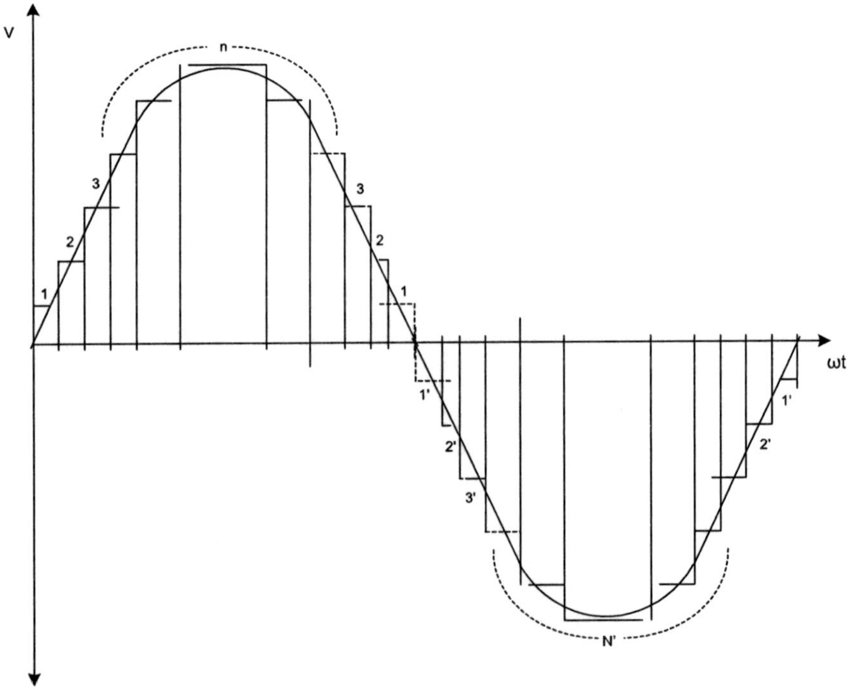

Fig. 3 Output voltage waveform

3 Working

3.1 First Half Cycle

For the conduction during ωt = 0 to ωt = 180°, the switches T_1, T_2 up to T_n conduct. The duration of conduction of each switch for one periodic cycle and the input voltage to the grid is shown in Fig. 4.

The time period of conduction of switch T_1, T_2 up to T_n is α_1, α_2 up to α_n, respectively

$$V_1 = \frac{N_0}{N_1} V_{dc}$$

where V_1 is output voltage generated.

When switch T_1 conducts, V_1 will be the output voltage whose duration depends upon the time period of switch T_1 which is α_1. When switch T_2 conducts, V_2 will be

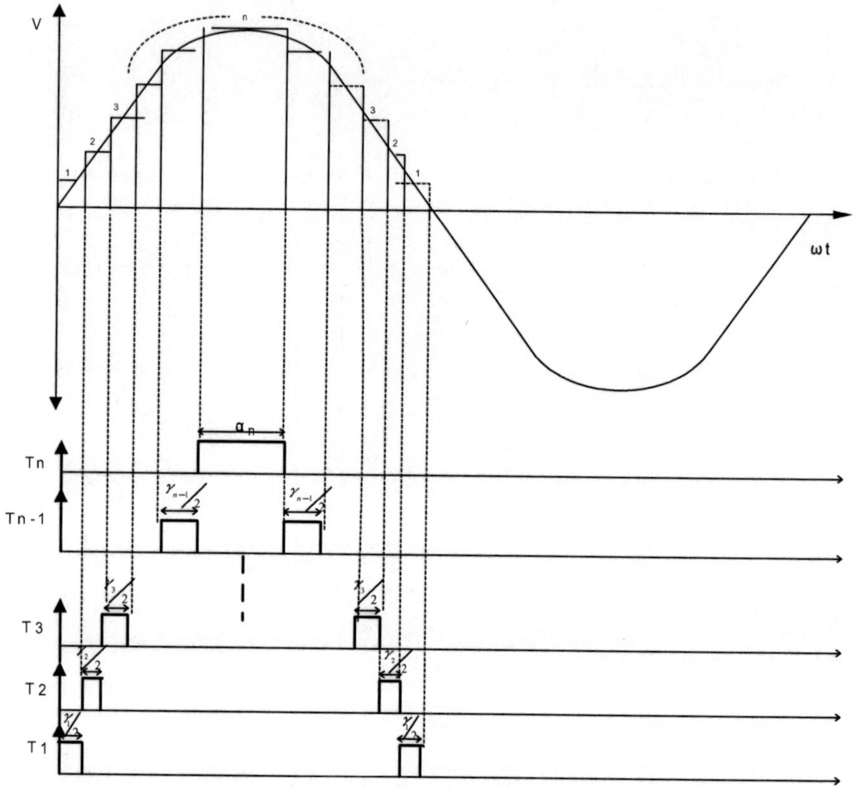

Fig. 4 Output voltage and gate pulses from for ωt = 0 to ωt = 180°

the output voltage whose duration depends upon the time period of switch T_2 which is α_2, and in the similar manner, the other switches will operate. The output voltage is given by

$$V_0 = V_2 = \frac{N_0 V_{dc}}{N_2}$$

where $N_{n-1} < N_n$.

3.2 During the Second Half Cycle

For the conduction during ωt = 180° to ωt = 360°, the switches T'_1, T'_2 up to T'_n conduct. The duration of conduction of each switch for one periodic cycle and the input voltage to the grid is shown in Fig. 5.

The time period of conduction of switch T'_1, T'_2 up to T'_n is α'_1, α'_2 up to α'_n, respectively. When switch T'_1 conducts, $-V_1$ will be the output voltage whose

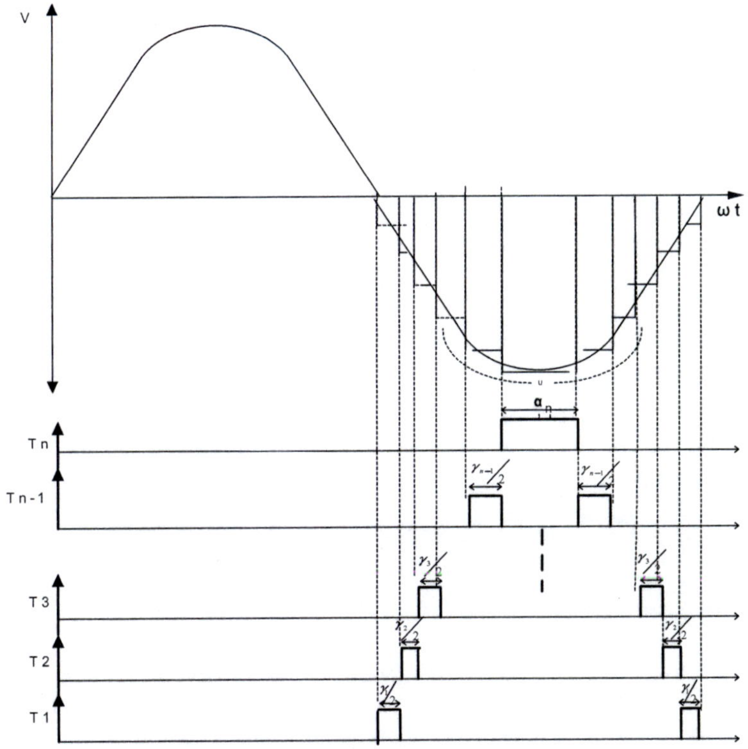

Fig. 5 Output voltage and gate pulses from for ωt = 180° to ωt = 360°

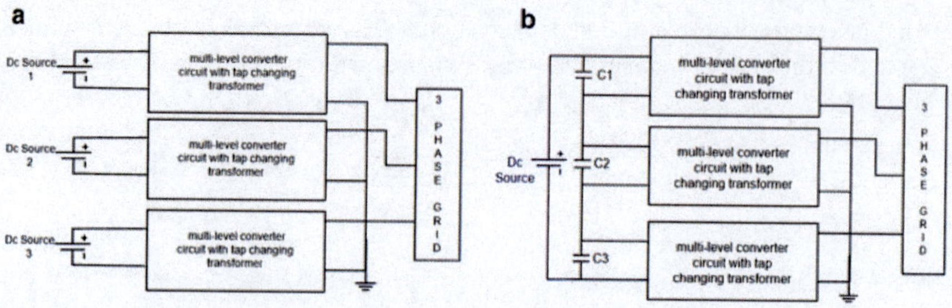

Fig. 6 Block diagram of the proposed three-phase inverter with (**a**) separate sources (dc) (**b**) three capacitors and a source (DC)

duration depends upon the time period of T'_1 which is α'_1. When T'_2 conducts, $-V_2$ will be the output voltage whose duration depends upon the time period of T'_2 which is α'_2, and in the similar manner, the other switches will operate.

In this manner, each half cycle requires 'n' switches. Therefore, total number of switches used in each periodic cycle will be '$2n$' for generating n level of output voltage. With the increase in the levels, the output voltage becomes more close to sinusoidal which reduces the total harmonic distortion. The system described above can be implemented for three-phase system presented in Fig. 6.

4 MATLAB Simulation

4.1 Simulation Model

The MATLAB simulation model of the transformer-based grid-connected multi-level inverter is shown in Fig. 7, where multiwinding transformer is taken for the Simulink library, in which required number of tappings is done, which is used for variation of amplitude of the voltage level in the respective duration of switch used. The power MOSFET is taken as a switch which is represented as M1, M2 up to M9. The duration of conduction of each switch depends upon the pulse or the combination of pulses given to the switches. The pulse transformer is represented by P1, P2, up to P14.

4.2 Simulation Results

The output voltage and grid current waveforms obtained from the developed simulation model are shown in Figs. 8 and 9, respectively. It can be observed from Figs. 8 and 9 that the waveforms of output voltage and grid current are almost sinusoidal in shape whose Fourier analysis is shown in Figs. 10 and 11, respectively.

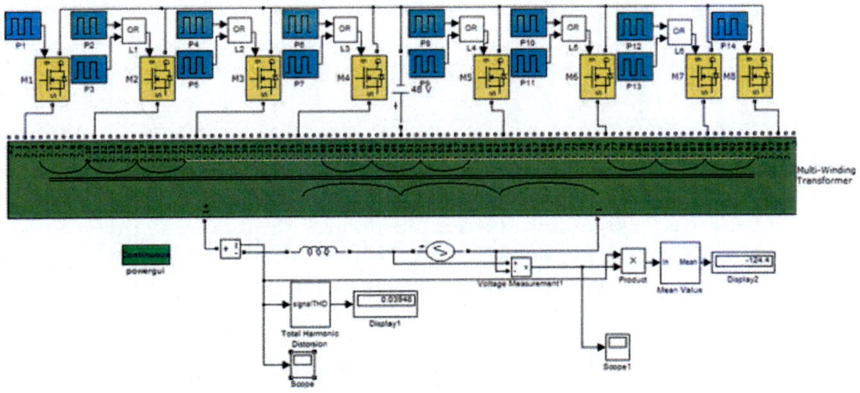

Fig. 7 MATLAB/Simulink model of the proposed inverter

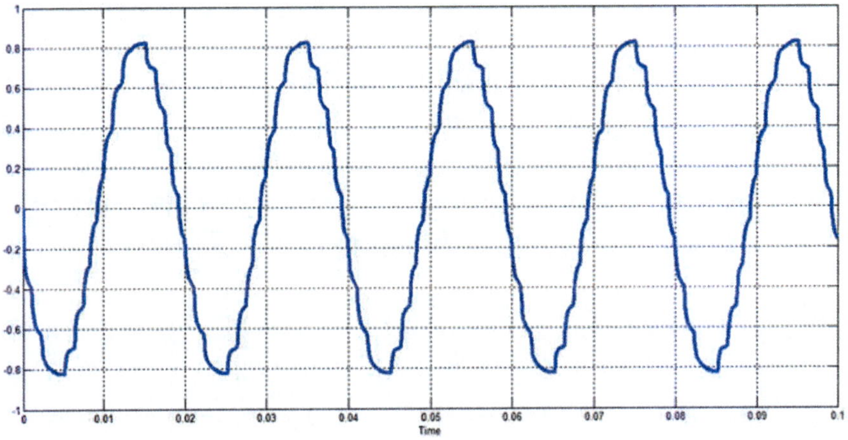

Fig. 8 Waveform of the output voltage

The output voltage value observed is 226.7 V with THD 8.98%, the grid current is 0.272 A, and its THD is 3.91%. The power fed to grid is 57.44 W.

5 Conclusion

A transformer-based multi-level inverter model and its method of controlling are shown in this paper. The output is attained by switching eight power MOSFET in a strategical way with the help of tappings in the transformer. The duration of each level of voltage is controlled by operating each switch for relevant time period which results in output voltage waveform to become nearly sinusoidal. Hence, the total

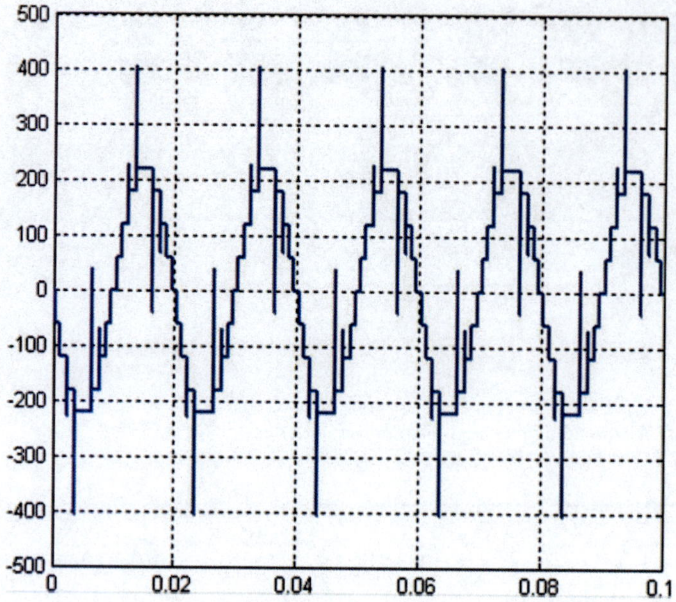

Fig. 9 Waveform of the system grid current

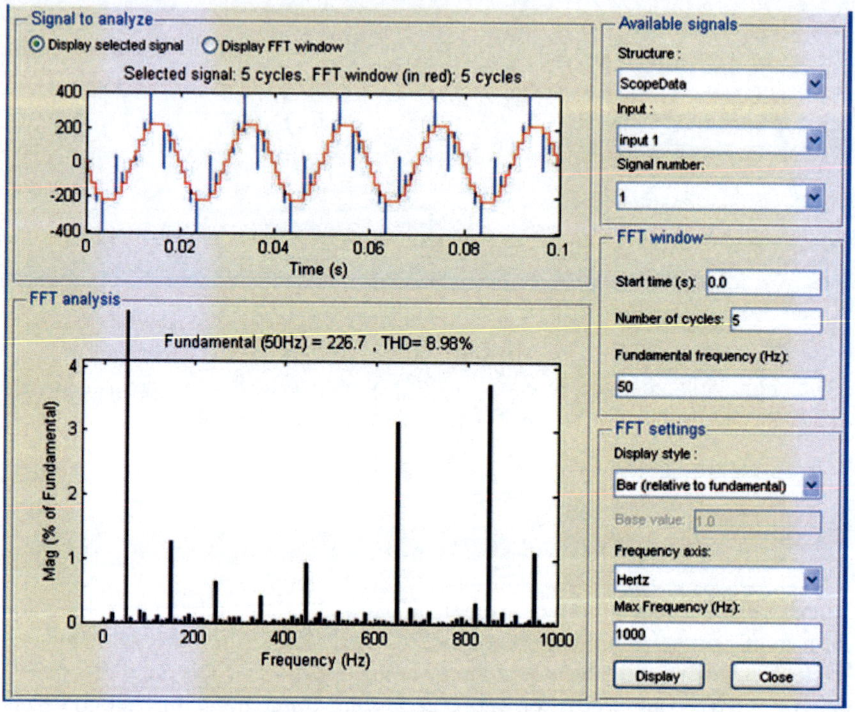

Fig. 10 Output voltage Fourier analysis

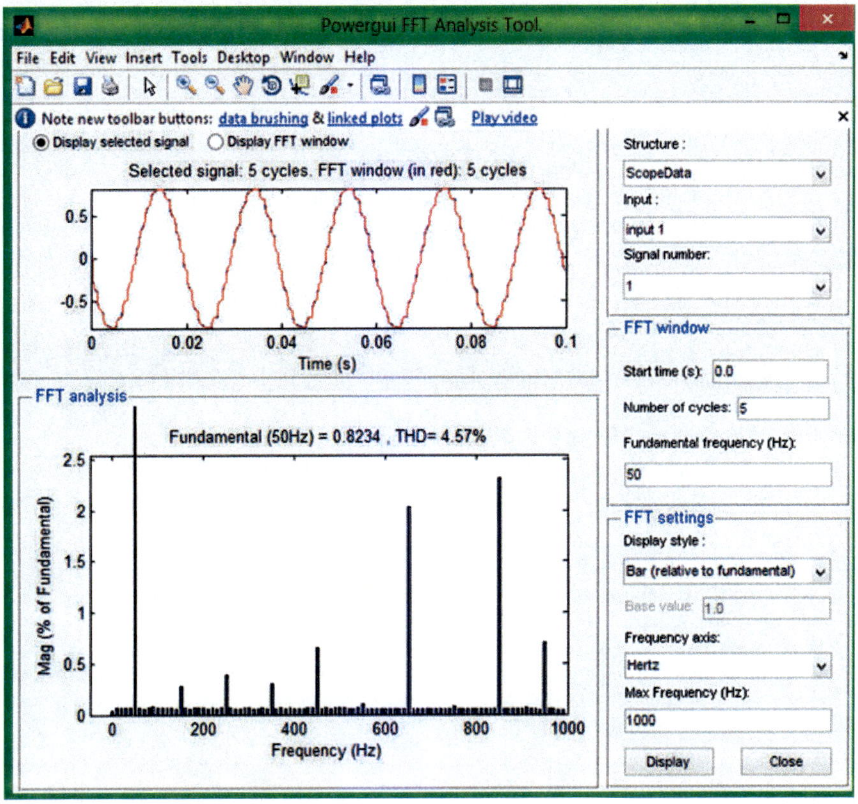

Fig. 11 Grid current Fourier analysis

harmonic distortion is reduced substantially, i.e., around 9% compared to 48.3% in case of conventional inverters. Moreover, the grid current waveform is also almost sinusoidal, and its THD is about 4.5%. The MATLAB simulation model developed is showing the acceptable performance.

Appendix

The conduction period of power MOSFET:

$T_1 = T_1' = 2 * 7.77 * 10^{-4}$ s, $T_2 = T_2' = 2 * 1.11 * 10^{-3}$ s, $T_3 = T_3' = 2 * 1.22 * 10^{-3}$ s, $T_4 = T_4' = 2 * 2.88 * 10^{-3}$ s.

References

1. Asghar MSJ (2004) Power electronics. PHI Ltd, New Delhi
2. Rukonuzzaman M (2019) Islanding detection and control of grid connected inverter. In: International conference on energy and power engineering (ICEPE), Dhaka, Bangladesh, 14–16 Mar 2019
3. Kan S, Ruan X, Dang H, Li Z, Huang X (2019) Second harmonic current reduction in front-end DC–DC converter for two stage single-phase photovoltaic grid connected inverter. IEEE Trans Power Electron 34(7)
4. Zhang W, Armstrong M, Elgendy MA (2019) Mitigation of DC current injection in transformer-less grid-connected inverters using a voltage filtering DC extraction approach. IEEE Trans Energy Convers 34(1):426–434
5. Chi P-C, Purnama I, Hsieh Y-C, Lin J-Y, Chiu H-J (2017) Simple four-quadrant grid-tie control scheme with unity power factor rectifier mode for single-phase DC/AC converters. Renew Power Gener IET 11(11):1483–1493
6. Yazdani S, Ferdowsi M, Shamsi P (2020) Internal model based smooth transition of a three-phase inverter between islanded and grid-connected modes. IEEE Trans Energy Convers 35(1)
7. Ahmad S, Bakhsh FI, Jamil Asghar MS (2013) A novel DC-to-AC controlled multi-level inverter. In: International conference on multimedia, signal processing and communication technologies (IMPACT), AMU, India, pp 315–319, 23–25 Nov 2013
8. Iqbal A et al (eds) (2020) Soft computing in condition monitoring and diagnostics of electrical and mechanical systems. In: Advances in intelligent systems and computing, vol 1096. Springer, Singapore. https://doi.org/10.1007/978-981-15-1532-3
9. Iqbal A et al (eds) (2020) Meta heuristic and evolutionary computation: algorithms and applications. In: Studies in computational intelligence, vol 1096. Springer, Singapore. https://www.springer.com/gp/book/9789811575709

A Stratified Optional Mathematical Model

Tanveer Ahmad Tarray, Zahoor Ahmad Gani, and Baziga Youssuf

Abstract This paper advocates a stratified optional mathematical model constructed via Singh and Mathur's model with comparative and Neyman provision and grander increase in productivity. The delinquent verbalized stays cracked through Lagrange multipliers method, and the optimum allocation is grown. Mathematical artworks stand prearranged in sustenance of the existing learning and the consequences verbalized over LINGO.

Keywords RRM · SRM · Proportional allocation · Neyman allocation · Mean square error

1 Introduction

For the literature, see Tarray et al. [3].

2 Mangat [1] Randomized Response Model

$$\alpha = \pi + (1 - \pi)(1 - P)$$

and

$$\hat{\pi} = (\hat{\alpha} - 1 + P)/P$$

where

$$\hat{\alpha} = n'/n$$

T. A. Tarray (✉) · Z. A. Gani · B. Youssuf
Islamic University of Science and Technology, Awantipora, India
e-mail: tanveerstat@gmail.com

A. Iqbal et al. (eds.), *Renewable Power for Sustainable Growth*, Lecture Notes in Electrical Engineering 723, https://doi.org/10.1007/978-981-33-4080-0_29

.

$$V(\hat{\pi}) = \frac{\alpha(1-\alpha)}{nP^2}$$

$$= \frac{\pi(1-\pi)}{n} + \frac{(1-\pi)(1-P)}{nP}.$$

3 Singh and Mathur [2] Optional Randomized Response Model

and

$$\alpha_1 = \pi T + (1-T)\alpha$$

$\hat{\pi}_1 = n'/n$

$$\text{MSE}(\hat{\pi}_1) = \frac{\alpha_1(1-\alpha_1)}{n} + (1-T)^2(1-P)^2(1-\pi)^2.$$

4 Proposed Model

In this proposed model, we extent Singh and Mathur [2] study to stratified sampling with

$$Y_i = T_i \pi_{Si} + (1-T_i)\,\alpha_i, \quad i = 1, \ldots, k$$

with

$$\hat{\pi}_{mi} = \frac{n'_i}{n_i},$$

and

$$\hat{\pi}_{\text{ST}} = \sum_{i=1}^{k} w_i \hat{\pi}_s = \sum_{i=1}^{k} w_i \left(n'_i/n_i\right)$$

Theorem 1

$$B\left(\hat{\pi}_{\mathrm{ST}}\right) = \sum_{i=1}^{k} w_i(1 - T_i)(1 - P_i)(1 - \pi_{Si})$$

Proof we have

$$B(\hat{\pi}_{ST}) = E(\hat{\pi}_{ST}) - \pi_S$$

$$= \sum_{i=1}^{k} w_i(1 - T_i)(1 - P_i)(1 - \pi_{Si})a$$

Theorem 2

$$V\left(\hat{\pi}_{\mathrm{ST}}\right) = \sum_{i=1}^{k} w_i^2 \frac{Y_i(1 - Y_i)}{n_i},$$

where Y_i is given already.

Theorem 3

$$\mathrm{MSE}\left(\hat{\pi}_{\mathrm{ST}}\right) = \sum_{i=1}^{k} w_i^2 \frac{Y_i(1 - Y_i)}{n_i} + \left[\sum_{i=1}^{k} w_i(1 - T_i)(1 - P_i)(1 - \pi_{Si})\right]^2.$$

4.1 Proportional Allocation

Theorem 4 For $n_i = n\ (N_i/N)$),

$$\mathrm{MSE}\left(\hat{\pi}_{\mathrm{ST}}\right)_P = \frac{1}{n}\left[\sum_{i=1}^{k} w_i Y_i(1 - Y_i)\right] + \left[\sum_{i=1}^{k} w_i(1 - T_i)(1 - P_i)(1 - \pi_{Si})\right]^2$$

4.2 Neyman Allocation

Theorem 5 Subject to $n = \sum_{i=i}^{k} n_i$

$$\frac{n_i}{n} = \frac{w_i\sqrt{Y_i(1 - Y_i)}}{\sum_{i=1}^{k} w_i\sqrt{Y_i(1 - Y_i)}}$$

Theorem 6 Mean square error $\hat{\pi}_{\text{ST}}$ is as

$$\text{MSE}(\hat{\pi}_{\text{ST}})_N = \frac{1}{n}\left[\sum_{i=1}^{k} w_i\sqrt{Y_i(1-Y_i)}\right]^2 + \left[\sum_{i=1}^{k} w_i(1-T_i)(1-P_i)(1-\pi_{Si})\right]^2$$

5 Relative Efficiency

We write MSE of $\hat{\pi}_1$ as

$$\text{MSE}(\hat{\pi}_1) = \frac{[Y(1-Y)]}{n} + [(1-T)(1-P)(1-\pi_S)]^2 \quad (1)$$

with $Y = T\pi_S + (1-T)\{\pi_S + (1-P)(1-\pi_S)\}$ and $\pi_S = w_1\pi_{S1} + w_2\pi_{S2}$.

5.1 *Proportional Allocation*

$(\hat{\pi}_{\text{ST}})_P$ reduces to

$$\text{MSE}(\hat{\pi}_{\text{ST}})_P = \frac{1}{n}\left[\sum_{i=1}^{2} w_i Y_i^*(1-Y_i^*)\right] + [(1-T)(1-P)(1-\pi_S)]^2 \quad (2)$$

where $\pi_S = (w_1\pi_{S1} + w_2\pi_{S2})$ and $Y_i^* = T\pi_{Si} + (1-T)\{\pi_{Si} + (1-P)(1-\pi_{Si})\}$, for $i = 1, 2, \ldots, k$ and $\pi_{S1} \neq \pi_{S1}$.

From (1) and (2), $\text{PRE}\left((\hat{\pi}_{\text{ST}})_P, \hat{\pi}_1\right) = \frac{\text{MSE}(\hat{\pi}_1)}{\text{MSE}(\hat{\pi}_{\text{ST}})_P} \times 100$

Discoveries are publicized in Table 1.

5.2 *Neyman Allocation*

Beneath hypothesis for two strata

$$\text{MSE}(\hat{\pi}_{\text{ST}})_N = \frac{1}{n}\left[\sum_{i=1}^{2} w_i\sqrt{Y_i^*(1-Y_i^*)}\right]^2 + [(1-T)(1-P)(1-\pi_S)]^2 \quad (3)$$

$$\text{PRE}\left((\hat{\pi}_{\text{ST}})_N, \hat{\pi}_1\right) = \frac{\text{MSE}(\hat{\pi}_1)}{\text{MSE}(\hat{\pi}_{\text{ST}})_N} \times 100$$

Table 1 PRE of $\hat{\pi}_{\mathrm{ST}}$ via $\hat{\pi}_1$

n	π_{S1}	π_{S2}	T	w_1	w_2	π_S	P				
							0.6	0.63	0.66	0.69	0.72
10	0.08	0.13	0.50	0.90	0.10	0.09	100.03	100.03	100.04	100.04	100.05
				0.80	0.20	0.09	100.05	100.06	100.07	100.08	100.09
				0.70	0.30	0.10	100.06	100.07	100.09	100.10	100.12
				0.60	0.40	0.10	100.07	100.08	100.10	100.11	100.13
				0.50	0.50	0.11	100.08	100.09	100.10	100.12	100.14
				0.40	0.60	0.11	100.07	100.08	100.10	100.11	100.13
				0.30	0.70	0.12	100.06	100.07	100.09	100.10	100.12
				0.20	0.80	0.12	100.05	100.06	100.07	100.08	100.09
				0.10	0.90	0.13	100.03	100.03	100.04	100.04	100.05
20	0.18	0.23	0.70	0.90	0.10	0.19	100.04	100.05	100.05	100.06	100.07
				0.80	0.20	0.19	100.08	100.09	100.10	100.11	100.12
				0.70	0.30	0.20	100.10	100.11	100.13	100.14	100.15
				0.60	0.40	0.20	100.12	100.13	100.14	100.16	100.18
				0.50	0.50	0.21	100.12	100.14	100.15	100.17	100.18
				0.40	0.60	0.21	100.12	100.13	100.14	100.16	100.18
				0.30	0.70	0.22	100.10	100.11	100.13	100.14	100.15
				0.20	0.80	0.22	100.08	100.09	100.10	100.11	100.12
				0.10	0.90	0.23	100.04	100.05	100.05	100.06	100.06
30	0.28	0.33	0.90	0.90	0.10	0.29	100.09	100.09	100.09	100.09	100.09
				0.80	0.20	0.29	100.15	100.16	100.16	100.16	100.17
				0.70	0.30	0.30	100.20	100.20	100.21	100.21	100.22
				0.60	0.40	0.30	100.23	100.23	100.24	100.24	100.25
				0.50	0.50	0.31	100.24	100.24	100.25	100.25	100.26
				0.40	0.60	0.31	100.22	100.23	100.24	100.24	100.24
				0.30	0.70	0.32	100.20	100.20	100.20	100.21	100.21
				0.20	0.80	0.32	100.15	100.15	100.15	100.16	100.16
				0.10	0.90	0.33	100.08	100.08	100.09	100.09	100.09

Results are compiled in Table 2.

$$\mathrm{PRE}\left(\left(\hat{\pi}_{ST}\right)_N, \left(\hat{\pi}_{ST}\right)_P\right) = \frac{\mathrm{MSE}(\hat{\pi}_{\mathrm{ST}})_P}{\mathrm{MSE}(\hat{\pi}_{\mathrm{ST}})_N} \times 100$$

for unalike ethics of n, P, w_1, w_2, π_{S1}, π_{S2}, and T. Discoveries are publicized in Table 3.

Table 2 PRE of $\hat{\pi}_{ST}$ via $\hat{\pi}_1$

n	π_{S1}	π_{S2}	T	w_1	w_2	π_S	P				
							0.6	0.63	0.66	0.69	0.72
10	0.08	0.13	0.50	0.90	0.10	0.09	406.05	377.03	349.39	323.11	298.19
				0.80	0.20	0.09	586.06	544.48	504.90	467.32	431.74
				0.70	0.30	0.10	796.13	739.67	685.98	635.06	586.91
				0.60	0.40	0.10	974.12	904.57	838.49	775.87	716.73
				0.50	0.50	0.11	1032.78	958.18	887.37	820.32	757.07
				0.40	0.60	0.11	936.08	867.71	802.87	741.54	683.73
				0.30	0.70	0.12	740.16	685.78	634.26	585.56	539.72
				0.20	0.80	0.12	532.21	493.15	456.17	421.25	388.41
				0.10	0.90	0.13	363.24	336.76	311.72	288.09	265.89
20	0.18	0.23	0.70	0.90	0.10	0.19	291.46	273.15	255.91	239.79	224.81
				0.80	0.20	0.19	421.67	395.49	370.87	347.86	326.51
				0.70	0.30	0.20	574.17	538.82	505.60	474.57	445.80
				0.60	0.40	0.20	704.25	661.05	620.49	582.63	547.54
				0.50	0.50	0.21	748.50	702.61	659.55	619.40	582.21
				0.40	0.60	0.21	680.06	638.39	599.33	562.92	529.23
				0.30	0.70	0.22	538.99	506.09	475.26	446.55	420.00
				0.20	0.80	0.22	388.42	364.90	342.88	322.38	303.44
				0.10	0.90	0.23	265.67	249.77	234.89	221.05	208.27
30	0.28	0.33	0.90	0.90	0.10	0.29	166.93	164.54	162.33	160.29	158.44
				0.80	0.20	0.29	243.35	239.95	236.81	233.92	231.29
				0.70	0.30	0.30	334.03	329.47	325.24	321.35	317.81
				0.60	0.40	0.30	413.22	407.64	402.48	397.74	393.43
				0.50	0.50	0.31	443.06	437.14	431.66	426.62	422.03
				0.40	0.60	0.31	406.04	400.67	395.69	391.11	386.93
				0.30	0.70	0.32	324.40	320.16	316.24	312.62	309.32
				0.20	0.80	0.32	235.49	232.47	229.67	227.09	224.74
				0.10	0.90	0.33	162.13	160.10	158.21	156.48	154.90

6 Conclusion

This manuscript speeches the delinquent of assessing the quantity π_S inhabitants fitting subtle group using optional randomized response technique. An optional RRM using the estimator $\hat{\pi}_1$ model has been proposed. The anticipated RRM model is supplementary competent than Singh and Mathur [2] optional RRM.

Table 3 PRE of $\hat{\pi}_{ST}$ via $\hat{\pi}_{ST}$

n	π_{S1}	π_{S2}	T	w_1	w_2	π_S	P				
							0.6	0.63	0.66	0.69	0.72
10	0.08	0.13	0.50	0.90	0.10	0.09	405.94	376.91	349.26	322.97	298.04
				0.80	0.20	0.09	585.78	544.17	504.58	466.97	431.36
				0.70	0.30	0.10	795.62	739.12	685.40	634.43	586.22
				0.60	0.40	0.10	973.41	903.80	837.67	774.99	715.77
				0.50	0.50	0.11	1031.99	957.34	886.46	819.35	756.02
				0.40	0.60	0.11	935.39	866.97	802.08	740.70	682.82
				0.30	0.70	0.12	739.68	685.27	633.71	584.98	539.09
				0.20	0.80	0.12	531.95	492.87	455.87	420.93	388.07
				0.10	0.90	0.13	363.14	336.66	311.60	287.97	265.76
20	0.18	0.23	0.70	0.90	0.10	0.19	291.33	273.01	255.77	239.64	224.66
				0.80	0.20	0.19	421.33	395.14	370.52	347.49	326.12
				0.70	0.30	0.20	573.58	538.21	504.96	473.91	445.11
				0.60	0.40	0.20	703.42	660.19	619.59	581.70	546.58
				0.50	0.50	0.21	747.57	701.65	658.56	618.37	581.15
				0.40	0.60	0.21	679.26	637.56	598.47	562.03	528.30
				0.30	0.70	0.22	538.43	505.51	474.66	445.93	419.36
				0.20	0.80	0.22	388.12	364.59	342.55	322.04	303.09
				0.10	0.90	0.23	265.55	249.65	234.77	220.92	208.14
30	0.28	0.33	0.90	0.90	0.10	0.29	166.79	164.40	162.18	160.15	158.29
				0.80	0.20	0.29	242.98	239.58	236.43	233.54	230.90
				0.70	0.30	0.30	333.37	328.79	324.56	320.67	317.12
				0.60	0.40	0.30	412.28	406.70	401.53	396.78	392.45
				0.50	0.50	0.31	442.02	436.09	430.60	425.55	420.95
				0.40	0.60	0.31	405.13	399.75	394.76	390.17	385.99
				0.30	0.70	0.32	323.77	319.53	315.59	311.97	308.67
				0.20	0.80	0.32	235.14	232.11	229.31	226.73	224.38
				0.10	0.90	0.33	161.99	159.96	158.08	156.35	154.77

References

1. Mangat NS (1994) An improved randomized response strategy. J R Statist Soc 56(1):93–95
2. Singh HP, Mathur N (2003) An optionally randomized response technique. Alig J Statist 23:1–5
3. Tarray TA, Sigh HP, Masood S (2019) An endowed randomized response model for estimating a rare sensitive attribute using poisson distribution. Trends Appl Sci Res 12:1–6

Internet of Things Applications in Electric Vehicles—A Review

Aminul Hoque Emdadul Hoque, Mohammad Rafi Ahmed, Anas Barber, Annaufal Rizqullah, Atif Iqbal, and Amith Khandakar

Abstract The Internet of things (IoT) is a revolutionary field that contributes to the Fourth Industrial Revolution along with artificial intelligence (AI), machine learning cloud computing, and many other fields that contribute to constructing smart ecosystems. Such rapid developments in these fields affect the lifestyle and daily usage of things around. Since the IoT mainly refers to develop smart behavior of the devices used in the environment and managing the energy consumptions in an efficient and optimized method. However, electric vehicles (EVs) would play a great role in power consumption of daily life specially to meet the environmental concerns and energy insufficiency. This would result in a huge investment in the infrastructure along with new related legislations. Such machines require very huge electrical energy storage which surely needs smart power management and monitoring systems. Batteries of EVs are very huge electrical energy storage which needs an accurate charging estimation to protect the powered equipment in the EVs. This paper discusses the field applications of IoT in EVs that would enhance the experience of using EVs in daily life in terms of environmental issues and improving the experience of having EVs.

Keywords Internet of things · Electric vehicle · BMS · O-FD · O-MI

A. H. E. Hoque · M. R. Ahmed · A. Barber · A. Rizqullah · A. Iqbal (✉) · A. Khandakar
Department of Electrical Engineering, Qatar University, Doha, Qatar
e-mail: atif.iqbal@qu.edu.qa

A. H. E. Hoque
e-mail: ae1304311@qu.edu.qa

M. R. Ahmed
e-mail: ma1603851@qu.edu.qa

A. Barber
e-mail: ab1605615@qu.edu.qa

A. Rizqullah
e-mail: ar1605250@student.qu.edu.qa

A. Khandakar
e-mail: amitk@qu.edu.qa

A. Iqbal et al. (eds.), *Renewable Power for Sustainable Growth*, Lecture Notes in Electrical Engineering 723, https://doi.org/10.1007/978-981-33-4080-0_30

1 Introduction

The Internet of things (IoT) has become increasingly popular in the world over the last two decades. There is a tremendous need for IoT in many applications since things are becoming smarter, automated, and efficient. The use of IoT is extended in both civil and military purposes where it covers different aspects as complimentary and necessity. Data produced by IoT devices such as sensors, cameras, vehicles, and other devices would further enhance the research and developments in this field. These applications will benefit all sectors including companies, citizens, and companies in daily life by using and building the IoT ecosystem [1–26].

The purpose of this review paper is to connect the dots and clarify the big picture of the use of IoT in a revolutionary area which is electric vehicles (EVs). Research papers are prepared and published in many fields to tackle many problems in a different area, where this review paper is concerned with the main purpose of the new lifestyle of transportation which is reducing pollution as a general issue and carbon emission as specialized. Furthermore, charging is the main concern for EV owners or people who are planning to be one of the owners of EVs. Energy storing unit as a whole is an issue for researchers to tackle the charging efficiency and speed, quality or time, and cost. Both subjects are discussed to shed light on these two main issues.

2 Environmental Contribution

2.1 IoT Tools to Monitor Green and Brown Energy in EVs

Electric vehicles are (EVs) are dramatically increasing and becoming popular and contributing to reducing carbon emissions [1, 2]. The main concern of the EVs is to resolve the emission of CO_2 which is the environmental concern. Such an issue is not completely resolved just by using EVs. The charging process is a part that should be considered whether it is charged from green and renewable energy or not. EV batteries are charged with renewable green energy sources such as solar and battery-stored energy or the grid produced by gas turbines [3, 4]. Such mixed energy sources for the EVs needed to be monitored because of the environmental concerns reduce the emission carbon to the air. Such a monitoring system is becoming easy to be applied by IoT-based systems. Setups are applied to monitor the feedings of the energy to EVs battery in addition to developing an algorithm that is used to process this, especially where the data could be collected in the buildings that provide data that cannot be calculated directly. The IoT infrastructure provides the potential in this domain as a complementary function for EVs drivers [6].

The two months test was conducted as a pilot study to prove the methodology to differentiate the mixed energy in the UK. It is coming to conclude that different households could be connected to take collaborative decisions [5] (Fig. 1).

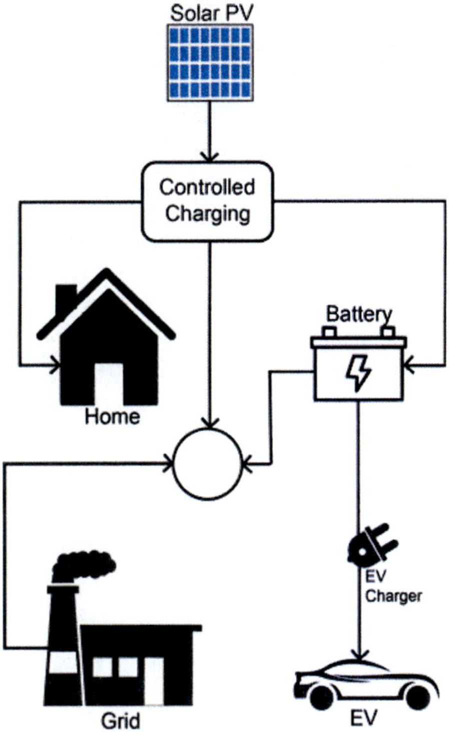

Fig. 1 IoT based smart energy management system for EV charging

The tests have successfully distinguished the flow of mixed energy supplied from many sourced and mechanism of the separating process [6].

2.2 Gathering Environmental Data to IoT Cloud

An increase of 0.4% carbon dioxide, nitrogen dioxide annually, results in vapors chlorofluorocarbons 5% annual gas which contributes to global warming and damaging the ozone layer. It is becoming a chance to use non-polluting and soundless cars whereas reopening EVs were a matter of discussion in 1960, nowadays, research has to tackle the phenomenon of the energy crisis. Countries are searching for alternatives to the crude oil after 1973 when the problem of oil reserve appeared [7, 8].

EVs are great potentials to work as data providers. A group of sensors could be added to the vehicle and form develops a communication protocol that allows a vehicle to vehicle (V2V) communication. Every vehicle shall be consisting of many sensor types to get data on geographic location. All data could be stored and on a cloud, and then a smart and interactive map could be developed to show the real-time info of the cities on the map [9] (Fig. 2).

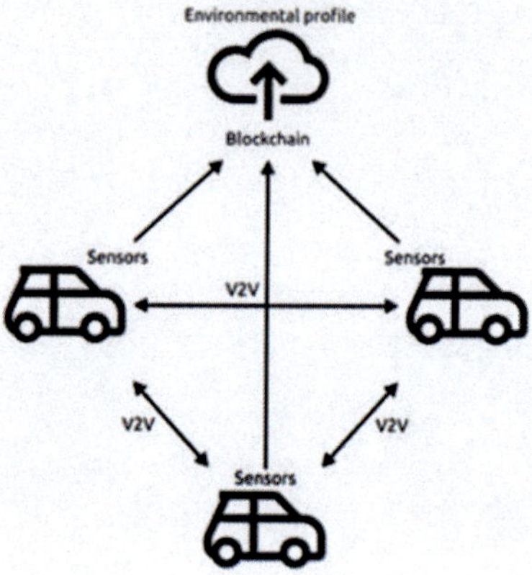

Fig. 2 Conceptual model of the system

3 IoT Systems as Smart Solutions for EVs

3.1 Smart Parking and Charging System

Nowadays, adopting IoT ecosystems is becoming very popular, especially in perusing to build things smart such as homes, buildings, mobility, and environments. However, non-interoperability in the IoT is one of the obstacles that have seamless communication between IoT devices [15]. There are many different domains and standards used as interoperability which causes the problem of communication with each other [10, 11]. IoT application , therefore, tends to vertical communication type of model that does not allow data sharing horizontally with IoT ecosystems. To solve this problem, an independent IoT massaging domain was published by Open Group in 2014 which are Open Messaging Interface (O-MI) and Open Data Format (O-DF) along with IoT project bIoTope in the EU [15]. These domains were extended practically for reaching interoperability in the smart mobility domain, electric vehicle charging in the context of smart cities. Many levels of challenges resolved as a part of a proof of concept of the smart EV charging ecosystem with a mobile app interface. Since the expectation of the EVs is very high with the rise in the sales of EVs. The feedback of using EVs is to have a more sustainable tool along with reducing the greenhouse emissions; however, the impact is dependent on the primary energy source which will include the grid, distribution network, charging infrastructure, etc. [12, 13].

Interoperability is consisting of many different domain bases, and generally, data exchange interoperability could be defined as the ability to exchange and process information whenever whatever by whoever needed [14].

The proposed paper is to bring the O-MI and O-DF in practical use and an IoT market place relying on those standards to complete the ecosystem of IoT. Furthermore, it evaluated the real implementation of the bIoTope ecosystem within the smart city domain depending on the EV charging case [15].

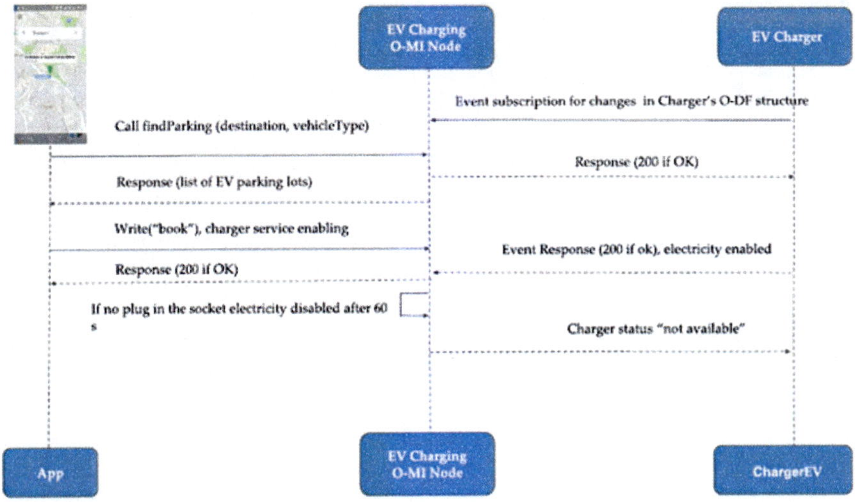

Fig. 3 EV charging service sequence diagram

Generally, smart cities would have many challenges to face, where non-interoperability between the technologies used in the area since many different entities is involved: users, telecommunication networks, buildings, etc.

With the help of the tools available, it is possible to develop an IoT infrastructure that would help to find the needed parking lots and charging slot details which include the distance of the parking, charging quality, adaptors, the possibility to reserve and so on (Fig. 3).

3.2 *IoT Based Battery Management System*

IoT offers the technology of interconnectivity between devices to devices with the help of the Internet. Along with the advancement of technology, smart devices can communicate and share beneficial data with each other. The application of IoT allows people, public administration, and companies to benefit day-to-day life [15, 16, 25, 26].

The world is advancing toward making the concept of smart city a reality. This implies that public administration services would be digitized to increase the quality of people's life. Smart grid is a part of the revolution of smart city, and IoT can immense the improvement of the smart grid. IoT, i.e., network-based interconnection system links with the wireless network using the interface of sensor and electronic identifiers to establish communication between a man and machine or from one machine to another.

With the widespread adoption of EV, development of batteries and the bidirectional power flow system (V2G), power peak, and problem can be solved. IoT would play a very important role in the monetarization of this whole process. Firstly, an EV user can use a mobile application to figure out the parking and charging station in a convenient location. With the mobile app, the user can view the availability of the charger and reserve a time for the EV after providing information car and the battery specification. The user can either choose to supply power from the stored energy in an EV battery or can charge the EV. The IoT-based battery monitoring system can be integrated into this technology so that the user can monitor the battery while charging or discharging, the estimated time it would take to charge the battery and the state of charge of the battery. The user arriving at the designated time at the station is required to provide a passcode to get access to the charger, and the rest of the update can easily be seen from the mobile app. Electric vehicles (EV) charging system is already being integrated into the smart city with the help of IoT-based technology [17].

3.3 *IoT as a Solution to Expand the Shared Charging Stations*

One of the main obstacles that are the topic of challenge discussion is to overcome the charging issues in the EVs where the time and quality matter since it directly deals with the main and most expensive part of the EV [25, 26]. Shared electricity is one of the challenges in compounds and condominiums because of unprepared infrastructure. IoT system developments are a key solution along with sensing devices and communication systems [19]. Few homeowners will only use the general reluctance of the EV charging stations, and there would be also issues of health, safety, and environment (HSE) [20]. Additionally, authorization issues may be faced as a part of granting permission from local governors [23]. Lopez-Behar study showed four main issues in sharing EV charging station which are parking availability, rules, and regulations, unavailable charging infrastructure, and building limitations [22].

To tackle these problems, Fig. 4 illustrates an overview of the compounds and the solution as an EV charging platform. This solution has many advantages that are avoiding different identification cards by the charging infrastructure because every charging infrastructure has different cards since there will be no need for preregistration with the local service providers [19]. Additionally, digital currencies could have introduced using peer-to-peer framework [3], and the cost would be reduced since there is no need for third-party management. The main feature of the proposed solution is that it is user authentication by Bluetooth Low Energy (BLE) communication using cellphones that process the charging energy for the EV [23]. Moreover, the IoT system will monitor the energy consumption which consists of sensors and a microcontroller board to transmit the data to a server that would play the role of the management unit [23].

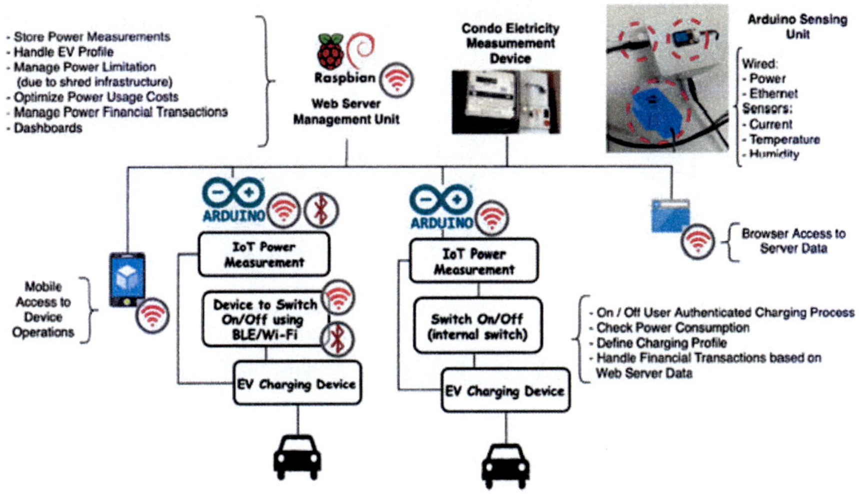

Fig. 4 Overview of the proposed electric vehicle (EV) charging platform in shared spaces

4 Conclusion

IoT is a key solution for many challenges in the field of EVs and adapting smart solutions as mainly the paper summarized the charging management concept in terms of green and brown energy that would evolve the experience of preserving the environment by reducing the usage of the grid in charging the EVs on peak time. The environmental issues are the main concern in the adaptation of these tools which depend on the primary source of the energy. The user should be provided with key solutions that concern the daily need of any consumer. As moving to the BMS, smart parking and charging are the main contribution of these concerns. Furthermore, future research also should convey some more key solutions and specialized in regional characteristics and the behavioral of different individuals in different societies that could be achieved with artificial intelligence and machine learning.

Acknowledgements It was possible for this publication to be made by UREP grant# [24-091-2-018] from the Qatar National Research Fund (a member of Qatar Foundation). The statements made herein are solely the author's responsibility.

References

1. Treese W (2017) The open market Internet index for 11 November 1995. Available online http://treese.org. Accessed on 2 Sept 2017
2. Burnett GE (2009) On-the-move and in your car. Int J Mob Hum Comput Interact 1(1):60–78
3. Bellavista P, Cardone G, Corradi A, Foscini L (2013) Convergence of MANET and WSN in IoT urban scenarios. IEEE Sens J 13:3558–3567

4. Bunce L, Harris M, Burgess M (2014) Charge up then charge out? Drivers' perceptions and experiences of electric vehicles in the UK. Transp Res Part A Policy Pract 59:278–287
5. Bourgeois J et al (2015) Harvesting green miles from my roof: an investigation into self-sufficient mobility with electric vehicles. UbiComp
6. Gubbi J et al (2013) Internet of Things (IoT): a vision, architectural elements, and future directions. Future Gener Comput Syst 29.7:1645–1660
7. Chowdhury N, Price BA, Smith A, Kortuem G, van der Linden J, Moore J (2016) EV charging: separation of green and brown energy using IoT. In: Proceedings of the 2016 ACM international joint conference on pervasive and ubiquitous computing: adjunct, Heidelberg, Germany
8. Casals LC et al (2016) Sustainability analysis of the electric vehicle use in Europe for CO2 emissions reduction. J Clean Prod 127:425–437
9. Jochem P, Babrowski S, Fichtner W (2015) Assessing CO2 emissions of electric vehicles in Germany in 2030. Transp Res Part A Policy Pract 78:68–83
10. Rus C, Marcus R, Pellegrini L, Leba M, Rebrisoreanu M, Constandoiu A (2019) Electric cars as environmental monitoring IoT network. In: IOP conference series: materials science and engineering, vol 572, p 012091
11. Hernandez-Munoz J, Vercher J, Munoz L, Galache J, (2011) Smart cities at the forefront of the future Internet. In: The future Internet. Springer, Berlin/Heidelberg, Germany, pp 447–462
12. Mulligan CE, Olsson M (2013) Architectural implications of smart city business models: an evolutionary Perspective. IEEE Commun. Mag 51:80–85
13. Zorzi M, Gluhak A, Lange S, Bassi A (2010) From today's intranet of things to a future internet of things: a wireless and mobility-related view. IEEE Wirel Commun 17:44–51
14. Kubler S, Yoo MJ, Cassagnes C, Främling K, Kiritsis D, Skilton M (2015) Opportunity to leverage information-as-an-asset in the IoT—the road ahead. In: Proceedings of the 3rd international conference on future Internet of Things and Cloud, Rome, Italy, 24–26 Aug 2015, pp 64–71
15. Karpenko A, Kinnunen T, Madhikermi M, Robert J, Främling K, Dave B, Nurminen A (2018) Data exchange interoperability in IoT ecosystem for smart parking and EV charging. Sensors 1–19 (13 Dec 2018)
16. Guo Dengfeng K, Shan X (2010) The Internet of Things hold up smart grid networking technology. North China Electr 2:59–63
17. ITU (2005) The Internet of Things. ITU Internet Rep. 2005, p 212
18. Fang X, Misra S, Xue G, Yang D (2012) Smart grid—the new and improved power grid: a survey. IEEE Commun Surv Tutorials 14(4):944–980
19. Roberts BP, Sandberg C (2011) The role of energy storage in development of smart grids. Proc IEEE 99(6):1139–1144
20. Al-Fuqaha A, Guizani M, Mohammadi M, Aledhari M, Ayyash M (2015) Internet of things: a survey on enabling technologies, protocols, and applications. IEEE Commun Surv Tutor 17:2347–2376
21. Stat of the Week: Percent of Households That Rent By Country. Available online https://evadoption.com/stat-of-the-week-percent-of-households-that-rent-by-country/. Accessed on 21 May 2019
22. Axsen J, Goldberg S, Bailey J, Kamiya G, Langman B, Cairns J, Wolinetz M, Miele A (2015) Electrifying vehicles: insights from the canadian plug-in electric vehicle study. Simon Fraser University, Vancouver, BC, Canada
23. Lopez-Behar D, Tran M, Mayaud JR, Froese T, Herera O, Merida W (2019) Putting electric vehicles on the map: a policy agenda for residential charging infrastructure in Canada. Energy Res Soc Sci 50:29–37
24. IoT and Blockchain Paradigms for EV Charging System
25. Iqbal A et al (eds) (2020) Soft computing in condition monitoring and diagnostics of electrical and mechanical systems. In: Advances in intelligent systems and computing, vol 1096. Springer, Singapore. https://doi.org/10.1007/978-981-15-1532-3
26. Iqbal A et al (eds) (2020) Meta heuristic and evolutionary computation: algorithms and applications. In: Studies in computational intelligence, vol 1096. Springer, Singapore. https://www.springer.com/gp/book/9789811575709

Space Heating for Kashmir Valley: Issues, Challenges and Remedies

Ahmed Sharique Anees, Salman Ahmad, and Zahoor Ahmad Ganie

Abstract In winters especially in Kashmir, the power cut is frequent and the quality of power is very poor, i.e., low voltage, so for to survive the harsh winter, the people here adopt traditional heating system like bukharis, angithi and hamams and gas bukharis. No doubt they protect them from harsh winter but they are associated with lots of disadvantages. These traditional heating arrangements produce toxic gases which produce hazardous effect on our health. In this regard, we have developed a prototype system which eliminates these toxic gases from rooms, when their concentration rises drastically. This paper also discusses the feasibility of using passive ground source heating system (GSHS), which is an alternate environment-friendly and safe solution.

Keywords Passive space heating · Ground source heat system · Ground heat exchanger

1 Introduction

In many regions, winter is associated with snow and freezing temperatures. In the past, the heat generated by burning fossil fuels and firewood is used to be only option in order to survive the harsh cold during winters. But with the advancement in the time, humans developed other more efficient, more safe and most importantly more environment-friendly heating systems. These include electric blowers, electric blankets, electric heaters, electric furnaces, etc. These heating systems not only provide better heating facilities but also ensure high level of safety during its operation, but the capital and running costs of electrical heating equipments are high, and those expenses can seldom be tolerated by a person living under poverty line [1–24]. As per the data published by CIA World Fact book on 10-01-2017 that nearly half of the population of the world falls below poverty line [1]. A study conducted by EU SILC survey has found a correlation between the population living under poverty line

A. S. Anees (✉) · S. Ahmad · Z. A. Ganie
Department of Electrical Engineering, Islamic University of Science and Technology, Awantipora, Pulwama, Jammu and Kashmir, India
e-mail: shariq.anees@islamicuniversity.edu.in

A. Iqbal et al. (eds.), *Renewable Power for Sustainable Growth*, Lecture Notes in Electrical Engineering 723, https://doi.org/10.1007/978-981-33-4080-0_31

Table 1 Different cases along with no. of person died

Case No.	Dated	No. of Death	District	Reasons of Death
CASE 1	06-01-2019	05	Srinagar	Gas heater
CASE 2	14-01-2018	02	Pulwama	Angithi
CASE 3	20-01-2019	01	Bandipora	Gas heater
CASE 4	29-12-2018	04	Pattan	Gas heater
CASE 5	01-03-2010	03	Rajouri	An earthen stove

and the populations which cannot afford to keep their homes adequately warm [2]. Following the same trend, the majority of peoples in Kashmir living below poverty line makes use of conventional heaters like Bukharis and gas heaters for heating their homes. These heating arrangements despite saving cost are still quite disadvantageous and dangerous as far as safety of the people is concerned. Some of the examples in the form of case studies of the past untowards incidents are mentioned in Table 1, which caused the loss of many lives [3–7].

2 Literature Review

Ahmed Abdullah Ibrahim (2018) employed a MQ-7 carbon monoxide gas sensor and an MQ-135 air quality sensor. When the value of the readings touched the safety limits, the buzzer beeped indicating danger. He made the whole system so compact that it could be carried from one place to the other [8]. H. Saha et al. (2007) employed microelectromechanical systems (MEMS) structure and nano-crystalline zinc oxide, and these structures power consumption range of 100–150 Mw and temperature range of 150–2000 °C have been achieved with moderately high sensitivity and response time of less than 15 s [9]. YU-KAI HE et al. (2006) used four-detector system which is employed and majority of the interference factors like power source anti-jamming, mismatch of the detectors, gas cell material's absorption and dusts influence, etc., can be compensated [10]. Emil CORDOS et al. (2006) designed a single sensing unit developed which could detect both carbon monoxide and methane. They even presented the detection method and circuitry [11]. S.O. Podgornyi et al. (2016) conducted their research on cadmium telluride (CdTe) nano-films. They successfully developed CO micro-impurities sensors in the laboratory. They investigated the surface physiochemical properties of CdTe nano-films and evaluated their applicability in CO diagnostics. [12]. M. Mello et al. (2006) integrated a photodetector based on gallium nitride (GaN) and an UV light source of detection of carbon monoxide. Thus, the fraction of light in the UV spectral region is modified which is detected by the GaN photodetector as a function of the concentration of the gas [13]. M. N. Mohammad et al. (2019) aimed to design a system for vehicles which could sense and detect the harmful gases and alarmed the passengers and particularly the driver once the carbon monoxide content exceeded the safe level so that the driver could operate the ventilation system immediately. The setup also had the provision that a text message (SMS) was delivered to the authorized user through GSM [14]. Pratik Makh et al. (2018) aimed to develop a low-cost sensor system to detect bronchitis by sensing the carbon monoxide concentration in the exhaled air. The humidity and temperature of the exhaled breath were also taken into consideration with the help of sensors. They processed and analyzed the collected data using an Arduino microcontroller. Thus, they determined whether the subject was suffering from bronchitis or not [15]. M.L. Homer et al. (2010) developed two types of polymer-based sensors for the detection of carbon monoxide [16].

Another promising long-term solution for space heating via renewable energy is by utilizing ground heats. The soil heat is transferred to the fluid flowing inside the metallic pipe. Thus, the air enters in the room at some temperature higher than the actual temperature of the space. An active ground heat system with pump is proposed in [17]. The ground heat profile for designing of an effective system to harness most of its heat energy is drawn in [18]. The steady periodic temperature profile is utilized for designing of such systems. The experimental results and perdition results have been compared in [19]. Another method called no dig method for underground heat exchanger is proposed in [20], for an input of 0.89 kW, the output was found to be 2.9 kW, and the energy saving as compared with the vertical ground heating system was found to be 29% [21]. However, most of the methods proposed in the literature are of active types which include an electric pump and therefore continuous power supply is required [22].

3 Traditional Heating Systems

In this section, different traditional methods with their advantages and disadvantages are discussed.

3.1 Kangri

A small earthen pot enclosed in wooden basket is used for individual heating as shown in Fig. 1. The fuel for Kangri is wooden coal covered with wood powder and ash. Ash is used to slow down the rate of burning coal. A small steel spoon is also attached to make opening for to increase heating. This is a movable heating arrangement but this makes fire in case of mishandle, due to burning of coal it produces CO gas and also consumes oxygen.

Fig. 1 Traditional small movable Bukhari

3.2 Fixed Bukhari

This is a metal cylinder or box type. In this heating arrangement, small piece of wood or coal is used as fuel. There is an opening in the top portion for inserting fuel and in the bottom section another opening for removing the by-product. This container is connected through the pipe, which removes the flue gases from the container to the outside room which needs to be heated as shown in Fig. 2, but this is a fixed-type heating arrangements. When the container gets heated up, this in turns heats the surrounding as well.

Fig. 2 Traditional fixed-type Bukhari

Fig. 3 Outside and inside view of the traditional Hamam

3.3 Hamams

These hamams are used for centralized heating of the houses. Hamams also run on the firewood as shown in Fig. 3. These have an advantage that they have an opening on the outside through which fuel can be inserted. Thus, they remain cutoff from the house, but they have a disadvantage that they require a proper installation which lands at a very huge cost. Below the floor, some piped arrangements are laid down which runs through the wall and floor when flue gases pass through these pipe, and it gets get heated up and radiates its heat to the surrounding.

3.4 Gas Heater

This is more sophisticated than above methods, but it is costly as well. In this, LPG gas is burned to warm up the space directly. The air gets heated up instantly and the toxic gases mixed up to the living space air which becomes very dangerous if proper ventilation would not be provided. There are several models of gas heaters are available in the market.

The main disadvantage of the above-discussed traditional home heating system is the productions of toxic gases like carbon monoxide, methane, butane, etc. These gases are responsible for causing suffocation in the surrounding atmosphere and may even cause a serious casualty if proper measures are not taken. Inhaling CO can become fatal; it reduces the oxygen-carrying capacity of blood cell. CO is a poison gas which is also called as silent killer having these characteristics colorless, odorless and tasteless.

4 Prototype System and Its Working

The schematic diagram of the proposed system is shown in Fig. 4. The system is designed based on the low cost so that it should be affordable to the poors and will

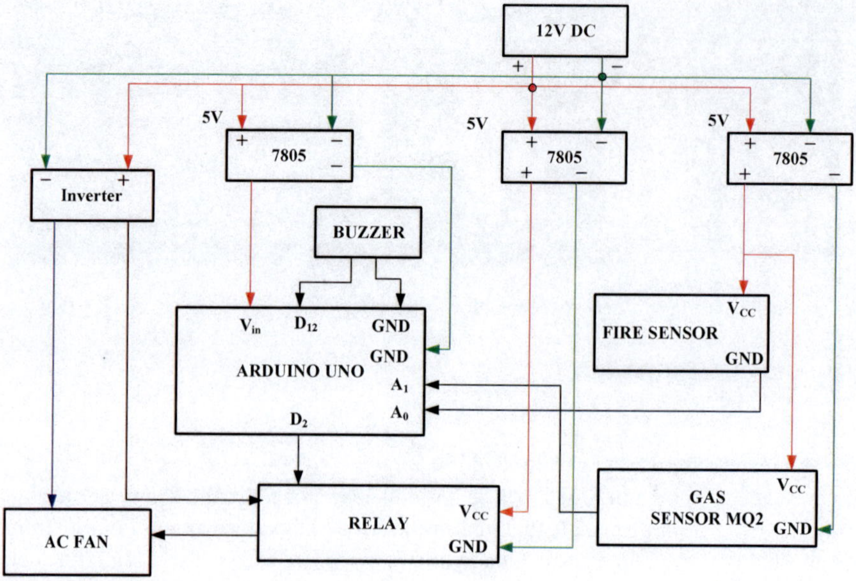

Fig. 4 Block diagram of CO detection system

save many lives. Operating these cheap heating systems requires extremely careful handling, and many a times there occurs the ventilation failure which poses a threat to life. So by employing this system, one could afford being a bit careless while operating cheap heating systems like Bukhari and gas heaters. The actual prototype developed in the laboratory is shown in Fig. 5.

The main components of the developed prototype are shown with the arrows. These components include arduino UNO, buzzer, voltage regulator, fire sensor, etc., when the concentration of carbon monoxide increases in air due to lack of proper ventilation. The MQ2 gas sensor senses concentration of the toxic gases. Now if the value comes out to be more than the preset value, a signal is generated which is received by the relay through microcontroller. In this way, the load circuit which was open till now is completed by the action of relay. And the exhaust fan runs and evacuates all the toxic gases from the atmosphere along with alarm. In addition to this, we have a flame detector sensor on this setup as well.

If the fire is detected, the digital output sends a high signal to the microcontroller and buzzer beeps making us aware about the fire.

5 Renewable Energy-Based Passive GSHS System

It is also known as earth air heat exchanger or underground air tunnel. In this technique, air is admitted into the pipes which are seated deep down the earth (3.5 m)

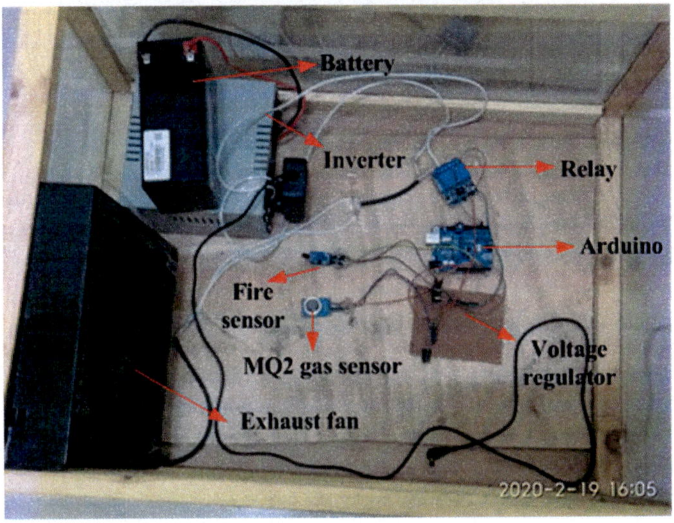

Fig. 5 Prototype model for CO detection system

in vertical or horizontal fashion. The air circulates through these pipes for charge or discharge and then is sent to the building for heating/cooling as shown in Fig. 6. Figure 6 demonstrates the open loop air-based ground heat exchanger be taken by natural wind or fan. To put this system in practical use, we have to first dig the earth about 3.5 m deep, as we know the soil temperature below a certain depth remains constant throughout the year. Long horizontal PVC pipes of diameter 0.10 m are buried in flat land with dry soil. With the help of fan, primary air is blown into the pipes.

In order to eradicate/remove the primary air supply at a continuous level, we can back fed the already conditioned air from the room with the help of fan through pipes at a depth of (3.5 m), then this air is readmitted into the room, and the cycle continues.

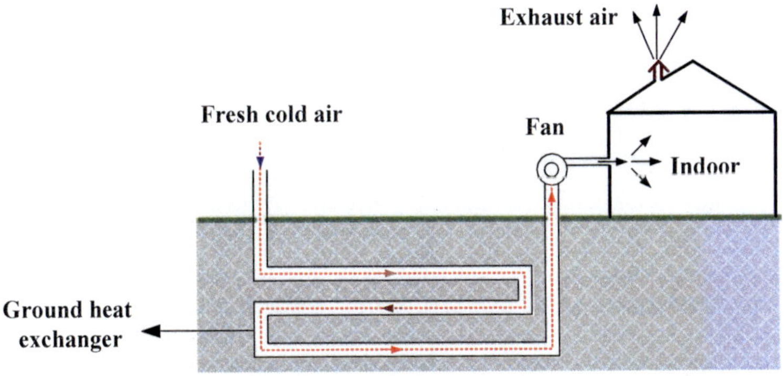

Fig. 6 Schematic diagram of the air-based GHE preheating/precooling

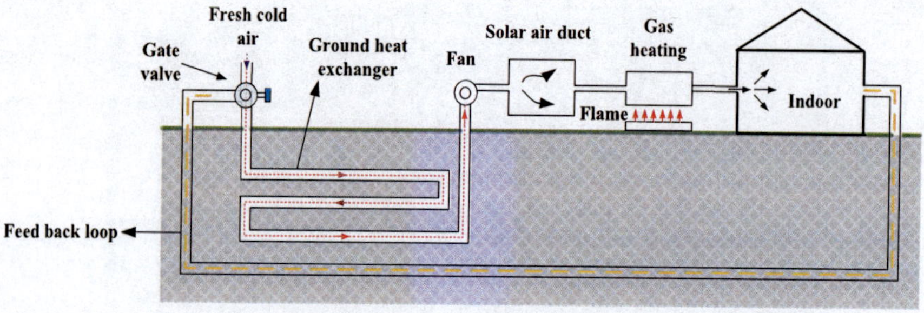

Fig. 7 Hybrid closed loop model of GHE

This constitutes the closed loop system as shown in Fig. 7. As the potential of ground heat energy is relatively low as compared to other means, there is an inevitable need to increase its output efficiency rates. For achieving this goal, we have proposed coupling the passive ground heating system unit with a solar air heating duct and LPG gas system for maximum possible efficiency. The passive GSHS or semi-passive GSHS consumes very small amount of electrical energy which is only 20–50% of any other electrical heating method, which is the main advantage. Humidity of the space can also be controlled by the GSHS. The ground temperatures more or less remain constant throughout the year because of that the thermal output and proper functioning of GSHS will not be affected by the outside weather condition.

In order to determine the feasibility of GSHS temperature of room, surrounding and underground need to be monitored. We notice that there is sufficient temperature differences between room and underground which can be utilized for cooling and heating purposes. The temperature at different location and levels is shown below in Table 2.

The thermal properties of the ground heat exchanger depend on the length of the heat exchanger, thermal conductivity of soil, diffusivity of the soil and type of soil and movement of underground water. The main steps in installing GSHS are to estimate the length of ground heat exchanger (GHX) to meet the heating and cooling demand requirement. The length of GHX can be calculated as given in (1).

Table 2 Temperature at different location and levels

S. No.	Temperature sensor location	Temperature in °C
1.	Room temperature	26–28
2.	Surrounding temperature	27–31
3.	At 75 cm underground temperature	19–21
4.	At 150 cm underground temperature	17
5.	At 220 cm underground temperature	15

$$l = \frac{l_h R_{\text{etr}} + l_y R_{\text{gtr}1y} + l_m R_{\text{gtr}1m} + l_h R_{\text{gtr}6h}}{T_{mf} - (T_s + T_c)} \tag{1}$$

where l is the length of heat exchanger, T_{mf} is mean fluid temperature, T_s temperature of underground soil, and T_c is temperature compensation associated with GHX. Also l_h, l_y and l_m are peak,/ground load for six hours, 1 month and 1 year, respectively. Rgtr6h, Rgtr1m and Rgtr1y are ground thermal resistance for six hours, 1 month and 1 year, respectively. R_{etr} is the effective thermal resistance of GHX. The heat exchange rate is found from using (2) or (3).

$$H_{\text{er}} = M_a^{\bullet} C_{\text{sh}} (T_{\text{oft}} - T_{\text{ift}}) \tag{2}$$

or

$$H_{\text{er}} = \rho V^{\bullet} C_{\text{sh}} (T_{\text{oft}} - T_{\text{ift}}) \tag{3}$$

where $M_a^{\bullet}$ is the mass flow rate, and C_{sh} is the specific heat. The T_{oft} and T_{ift} are output and input temperatures. Using the above expressions, the actual heat transfer rates can be calculated. The effectiveness which is defined as actual heat transfer rate and maximum heat transferred can be found using (4).

$$\varepsilon = \frac{H_{\text{er\_GHX}}}{H_{\text{er\_max}}} \tag{4}$$

6 Conclusion

In this paper, a carbon monoxide detection system is developed by employing an MQ2 gas sensor. If the concentration of harmful gases is found to be more than the threshold value, a signal is generated and sent to run the exhaust fan, and the toxic gases are evacuated. Further a provision for fire detection is also installed. The buzzer beeps in order to make an aware sound about the flame. The whole setup works on a 12 V DC battery which eliminates the requirement of a continuous main supply for working of the proposed system. In other words, this system is self-sustaining and beneficial as it is cost effective. In cold regions like Kashmir where there have occurred a number of asphyxiation-related incidents in the recent past, this setup would be extremely helpful to prevent such mishaps. For long-term solutions from renewable energy, the passive and low-energy means must be adopted as neat alternatives for heating/cooling. Also in this paper, a feasibility analysis of the passive GSHS is carried out to figure to long-term solution for space heating by utilizing ground heat.

Acknowledgements The work is financially supported by seed grant No. IUST\TEQIP\19\36-53 dated 15\04\2019 from TEQIP-III Islamic University of Science and Technology, Awantipora, Jammu and Kashmir.

References

1. http://world.bymap.org/Poverty.html
2. https://www.reddit.com/r/europe/comments/alonn7/people_who_cannot_afford_to_keep_their_home/
3. https://www.dailyexcelsior.com/5-of-a-family-die-due-to-asphyxiation/
4. https://kashmirobserver.net/2018/local-news/2-labourers-die-asphyxiation-pulwama 27774
5. https://www.greaterkashmir.com/news/kashmir/construction-company-employee-diesdue-to-suffocation-in-bandipora/
6. https://kashmirreader.com/2018/12/29/pattan-youth-dies-of-asphyxiation-3-others-critical/
7. https://zeenews.india.com/news/jammu-and-kashmir/three-of-jawans-family-die-of-asphyxiation-in-jk_592394.html
8. Ibrahim AA (2018) Carbon dioxide and carbon monoxide level detector. In: IEEE 2018 21st international conference of computer and information technology (ICCIT), 21–23 Dec 2018. https://doi.org/10.1109/iccitechn.2018.8631933
9. Saha H, Basu S, Bhattacharyya P, Basu PK (2007) MEMS based nano crystalline zinc oxide methane gas sensors. In: IEEE internationational workshop on physics of semiconductors, 16–20 Dec 2007. https://doi.org/10.1109/iwpsd.2007.4472608
10. He Y-K, Wang R-L, Yang Y-Q (2006) A novel carbon monoxide detection system based on infrared absorption used in mine. In: IEEE international conference on machine learning and cybernetics, 13–16 Aug 2006. https://doi.org/10.1109/icmlc.2006.258392
11. Cordos E, Ferenczi L, Cadar S, Costiug S, Pitl S, Aciu A, Ghita A (2006) Methane and carbon monoxide gas detection system based on semiconductor sensor. In: IEEE international conference on automation, quality and testing robotics, 11 Dec 2006. 10.1109AQTR.2006.254633
12. Podgornyi SO, Demeshko IP, Podgornaya OT, Lukoyanova OV, Skutin ED, Fedotova KI (2016) Cadmium telluride nano films in CO detection. In: IEEE dynamics of systems, mechanisms and machines (dynamics), 15–17 Nov 2016. https://doi.org/10.1109/dynamics.2016.7819060
13. Mello M, de Risi A, Passaseo A, Lomascolo M, De Vittorio M (2006) Carbon-monoxide (CO) real time monitoring in combustion engines by an optical detection system. In: IEEE Ph.D. research in microelectronics and electronics, 12–15 June 2006. https://doi.org/10.1109/rme.2006.1689987
14. Mohammed MN, Ghanesen Y, Al-Zubaidi S, Ali MAM, Ismael Al-Sanjary O and NS Zamani, Investigation on carbon monoxide monitoring and alert system for vehicles, 2019 IEEE 15th International Colloquium on Signal Processing & Its Applications (CSPA), Penang, Malaysia, 2019, pp. 239–242, https://doi.org/10.1109/CSPA.2019.8696001
15. Makh P, Mudhalwadkar R, Babtiwale A (2018) Sensor system development for bronchitis detection from exhaled breath. In: Second international conference on inventive communication and computational technologies (ICICCT), 20–21 Apr 2018. https://doi.org/10.1109/icicct.2018.8473092]
16. Homer ML, Shevade AV, Zhou H, Kisor AK, Lara LM, Yen SPS, Ryan MA (2010) Polymer based carbon monoxide sensors. Sensors 2010 IEEE, 1–4 Nov 2010. https://doi.org/10.1109/icsens.2010.5690345
17. Jakhara S, Misrab R, Bansalc V, Soni MS (2015) Thermal performance investigation of earth air tunnel heat exchanger coupled with a solar air heating duct for northwestern India. Energy Build 87:360–369

18. Gao J, Li A, Xu X, Gang W, Yan T (2018) Ground heat exchanger: applications, technology integration and potential for zero energy buildings. Renew Energy 128:337–349
19. Sivasakthivel T, Murugesan K, Sahoo PK (2015) Study of technical, economical and environmental viability of ground source heat pump system for Himalayan cities of India. Renew Sustain Energy Rev 48:452–462
20. Zhu N, Hu P, Xu L, Jiang Z, Lei F (2014) Recent research and applications of ground source heat pump integrated with thermal energy storage systems: a review. Appl Therm Eng 71:142–151
21. Verma V, Murugesan K (2014) Optimization of solar assisted ground source heat pump system for space heating application by Taguchi method and utility concept. Energy Build 82:296–309
22. Sivasakthivel T, Murugesan K, Sahoo PK (2014) A study on energy and CO_2 saving potential of ground source heat pump system in India. Renew Sustain Energy Rev 32:278–293
23. Iqbal A et al (eds) (2020) Soft computing in condition monitoring and diagnostics of electrical and mechanical systems. In: Advances in intelligent systems and computing, vol 1096. Springer, Singapore. https://doi.org/10.1007/978-981-15-1532-3
24. Iqbal A et al 2020 (eds), Meta heuristic and evolutionary computation: algorithms and applications. In: Studies in computational intelligence, vol 1096. Springer, Singapore. https://www.springer.com/gp/book/9789811575709

Deregulation and Its Effect on Indian Power Industry

Vimal Singh Bisht, Navneet Joshi, Jagdish Singh Mehta, and James Kunjwal

Abstract Nowadays, electricity is considered as a commodity for consumers, and quality of power is a very important issue which cannot be neglected. Through this paper, we will understand how deregulation has played an important role in improving the state of power sector in India. This paper provides an idea of how Indian power sector have reformed itself with growing demand of power; it presents a detailed understanding of Indian power industry before and after of deregulation. It helps us to understand the current structure of power industry in different states of India. Salient features of the Indian Electricity act, 2003, are also discussed in the paper.

Keywords Deregulation · Indian power scenario · Electricity act 2003 · Market models

1 Introduction

The term deregulation focuses on unbundling of the major components present in the power system, i.e., generation, transmission, and distribution. Deregulation also focuses on availability of these components for sale, thus giving market a competitive look. For achieving all these features, new set of rules for operation and sales need to be defined [1–17].

Laai [1] Concept of deregulation comes into existence in the year 1982 when Chile restructured its power sector for the first time followed by United Kingdom in the year 1990. Deregulation is basically changing the monopoly rule and regulation of regulated industry that affects how companies do business and how consumer may buy electricity, thus allowing other independent generators and transmitters to access

V. S. Bisht (✉)
Department of Electronics Engineering, GEHU, Bhimtal, India
e-mail: bishtvimal@yahoo.com

N. Joshi
Mathematics Department, GEHU, Bhimtal, India

J. S. Mehta · J. Kunjwal
Mechanical Engineering Department, GEHU, Bhimtal, India

A. Iqbal et al. (eds.), *Renewable Power for Sustainable Growth*, Lecture Notes in Electrical Engineering 723, https://doi.org/10.1007/978-981-33-4080-0_32

the transmission and distribution grid to serve the customers. Deregulation leads to a competition eventually resulting in a low price of electricity to the consumers.

When we talk about India power sector, it is an industry of almost a century year old, and in these years, it has changed itself at a very slow pace. With the change in power industry around the world, Indian power sector had also changed itself and it all started in 1998, as a regulatory commission was formed under the Electricity Regulatory Commission Act 1998. The purpose of this act was to introduce competition in the power industry of the country.

Followed to this in the year 2003, Electricity Act 2003 comes into force from June 10, 2003, and bulk trading of electricity was introduced in this act. It also provided open access to transmission and distribution system, thus providing opportunities for new players to enter in Power industry.

Now over the years, the power supply industries in India have experienced a gradual and shatter growth. The ratification of the Electricity Act 1948 led to the solidification of the power industry, both at the state level and regional level. This step led to the public sector ownership of almost the entire industry, and the private sector has effectively enrolled themselves in the power generation and in meeting todays demand supply, hence participating in the betterment and smooth functioning of the countries power demand.

2 Power Scenario in India

As India is the second largest country in the world with respect to its population, there is a huge need of developing a strong power sector to meet the demand of 1.4 billion people. India has done significantly well in reforming its power sector leading to meet its growing demands. India is working a lot on renewable energy sources particularly on solar energy to meet its energy demand as a result there is significant drop in energy imports from Bhutan. Indian power sector will offer motivating example for many countries in the world [16, 17].

The initial power scenario in India was owned and governed by numerous government agencies. Over the years, there has been a radical change in the power sector in India, and we entered into the league of privatization since post 1991. Earlier in 1990s, there was no involvement of private sectors in India. But due to continuous increase in demand of power and failure of government in meeting it compelled the government to go for the privatization of power sector. This policy leads to the new reforms in power generation. Generation growth pattern of the country for conventional sources of energy is given below in (Fig. 1). It can be clearly seen that there has been significant growth since 2009–10 with the current year target of 1330 BU. The power scenario in India comes under the concurrent list in the constitution of India that is both state and the central government have the full authority to codify the sector. Largely, the whole power industry can be divided in to three parts: generation, transmission, and distribution (Table 1).

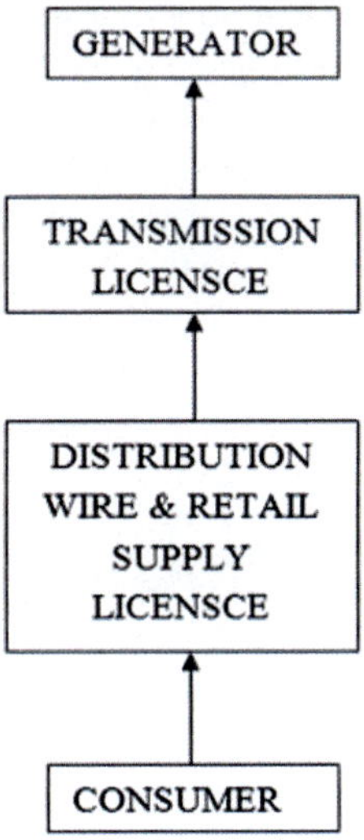

Fig. 1 Upstream and downstream monopoly (1947–1998)

Table 1 Generation pattern in past years for conventional generating sources [15]

Years	Generation in billion units	Growth in percentage (%)
2009–10	771.551	6.6
2010–11	811.143	5.56
2011–12	876.887	8.11
2012–13	912.056	4.01
2013–14	967.150	6.04
2014–15	1048.673	8.43
2015–16	1107.822	5.64
2016–17	1160.141	4.72
2017–18	1206.306	3.98
2018–19	1249.337	3.57
2019–20	1252.611	0.26
2020–21	91.913	−22.85

2.1 Generation

These are also termed as "GENCOS" in the deregulated markets. The function of this body is to generate power using various renewable and non-renewable sources of energies available around us. Thus, it is this body which is responsible for meeting of the energy demand of the nation. This is a body which is responsible for generation of power. In India, there are more than 30 government & private generating companies.

The generation levels of the power in the country have been considerably increased from 1110.458 BU (in the year 2014–15) to 1390.467 BU in the year 2019–20 [9]. If we talk about year 2019–20, we have reduced thermal generation limit by focusing more on renewable sources of energy. India has great potential to improve renewable generation limits by focusing more on solar generation. In past few years, government has launched various schemes to promote generation via solar energy. As per the recent data, the total installed capacity of the country is 3,70,348 MW out of which maximum is of thermal generation which is 230,600 MW, i.e., 62.8%, of the total [9]. To meet its installed capacity, India also imports electricity from Bhutan, and for this year (2020–21), a target of 7.230 BU is set for Bhutan [9].

2.2 Transmission

As the resources of power generation in India is located at different places and are limited to certain regions, therefore, there is a need to evacuate this power produced by different generating stations. Power Grid and Central Transmission Utilities are responsible for the interstate power transmission, whereas State Transmission Utilities are responsible for intra-state transmission. Power Grid is a company formed in the year 1989 and is fully owned by government of India. Power Grid is responsible for operation and establishing of national and regional grids for easy transmission of power by providing better quality, security & reliability. Improvements of this sector are very important since demand and generation are both expected to grow at a very rapid rate in the near future. The main challenge for the sector is to minimize the transmission losses and planning of future expansion. Currently as on December 2016, the total installed capacity of lines was 362,121 circuit kilometers, out of which 104,640 circuit kilometers is achieved during 12th plan, i.e., up to December 2016.

2.3 Distribution

This entity acts as a very important Link between consumers and utilities in power industry. In deregulated environment, it is also termed as "DISCOS." As power lies in concurrent list, its duty of state is to distribute the power to the urban and rural areas of the state with the assistance of the central government. There are approximately 60 distribution utilities (both government and private players) in India. Its main function is to provide quality power to consumer ends and controlling its theft. Distribution companies continuously update their metering system to control theft of electricity. In India, there are approximately 200 million customers fetching load from 400 GW connected load. It is the only source of revenue for the entire power industry.

3 Indian Electricity Act. 2003

The act was notified in the gazette of India on 2/06/2003, and it came in to enforcement from 10/06/2003. 2003 act is a very comprehensive act which replaced Indian Electricity Act 1910, Electrical Supply Act 1948, and Electricity Regulatory Commission Act 1998. 2003 Electricity Act mainly focuses on consolidation of laws related to generation, transmission, and distribution, i.e., introducing completion in the power market. This act replaced some very old aged existing policies existing in the country. The act has been amended twice once as amendment 2003 and one as amendment 2007. Though the act has been enacted nicely and is very beneficial, there are still many section those to be enacted. The key features of the act are listed below:

- Power generation was de-licensed, and captive generation was allowed freely, provided that they should comply with the technical standards relating to connectivity with grid.
- One cannot transmit, distribute, or trade electricity without a valid license.
- Generating companies would be free to take up distribution licensees.
- There will be an open access to the transmission system.
- The state governments were asked to restructure their electricity board, and State Electricity Regulatory Commission (SERC) was made mandatory.
- Metering of the power supplied was made necessary, and strict provisions were made to control theft of electricity.
- Special emphasis was given to rural electrifications and distributed generation, and distribution was permitted in rural and remote areas.
- It was said that Central Electricity Authority (CEA) will be responsible to prepare National Electricity Plan, i.e., preparing all technical standards for commissioning of new electrical plant and connecting it to the grid.

4 Restructuring Across States

It started off with breaking vertically unbundling of the utilities. Vertically unbundling means separation of generation, transmission, and distribution. Odisha was the first state in India to witnessed restructuring of its power sector, i.e., in the year 1996, when Odisha Electricity Regulatory Commission (OER) was established. Odisha has become the first state of India where government provides no subsidy to the power sector 1996. Before, this government was giving an amount of 250 crore (approx) as subsidy. Also Odisha is the only state where no budget is provided to the distribution companies which is approx. minimum of 100 crore in any state of India and more than 5000 crore to the distribution companies since deregulation Odisha has witnessed major drop in losses and high growth of revenues.

Haryana was the second state of the country to follow Odisha. In the year 1998, Haryana Electricity Regulatory Commission was formed (HERC) consisting of Haryana Power Generation Corporation Limited (HPGCL),

Haryana Vidyut Prasaran Nigam Limited Uttar Haryana Bijli Vitran Nigam Limited (UHBVN), and Dakshin Haryana Bijli Vitran Nigam Limited (DHBVN). Andra Pradesh and Karnataka also joined the league in year 1999 followed by Uttara Pradesh in 2000 which formed three bodies UPRVUNL and UPJVNL which are set up for generation, and for distribution and transmission, UPPCL was formed. An additional company was formed for rural electrification which work under these companies.

Uttarakhand and Rajasthan deregulated there power sectors in the year 2001. Delhi and Madhya Pradesh were next to follow in the year 2002. In Delhi, distribution was first to get privatized whose 51% of the stake was sold to two private companies, whereas transmission system is still with the government. Assam and Gujrat joined the reform in year 2004, followed by Maharashtra in year 2005, and West Bengal in the year 2007.

5 Indian Power Market Structure

Indian market has been reformed significantly since 1947 and is still changing to provide better services to the consumers. If we look in to 1947–1948 era, we use to follow upstream and download model shown (Fig. 1). Here, we can clearly see that there is a huge monopoly in the industry. Consumers have no options other than going to distribution sector. This causes customer to pay a higher tariff, and in turn, the quality of supply was also poor [16, 17].

With the starting of 1998, a partial competition was introduced in the market with competition in the upstream and monopoly in downstream, i.e., captive generators can also do generation. Though there was still no choice for the consumer to buy electricity, different generating companies were there which can be seen below in (Fig. 2). This somehow helped in meeting the generation and demand gap. It was for the first time that private generators were in the competition. Involvement of private GENCOS reduces the monopoly up to certain amount; however, over all, there was still monopoly in the market because consumer still has to buy it from a single distributor.

This monopoly was suppressed up to a larger extent by the starting of 2003, when the Electricity Act was introduced. Form 2003–2016, it was distribution companies who had the choice, i.e., they can buy power directly from the generating companies and sell it to the consumers. In this model, distribution company moved to the generator who were providing quality power at reasonable price. This causes generating companies to lower their tariffs and to provide quality power, so that they can sustain in the business. Working of model can be understood from (Fig. 3). This environment can still be seen as partial competitive environment as consumers still do not have any choice other than going to distribution companies, which was the case since establishment of the industry.

Now with the starting of 2016, we entered in to a new era of possibilities, wherein consumers along with distributors had choices available. This model broke the

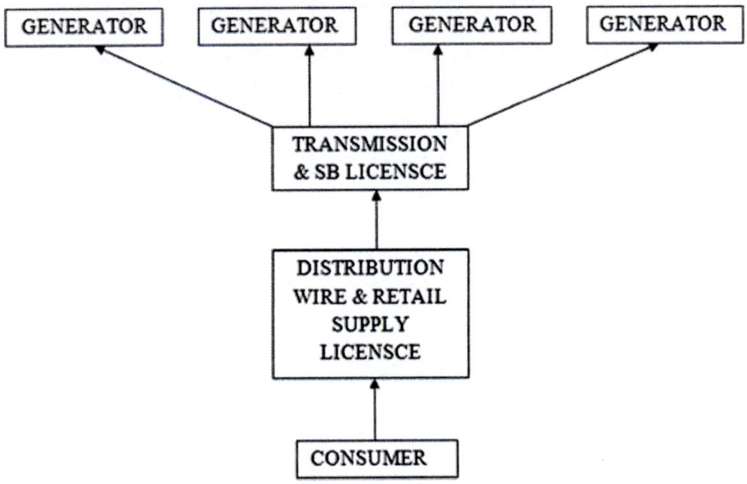

Fig. 2 Upstream partial competition, downstream monopoly (1998–2003)

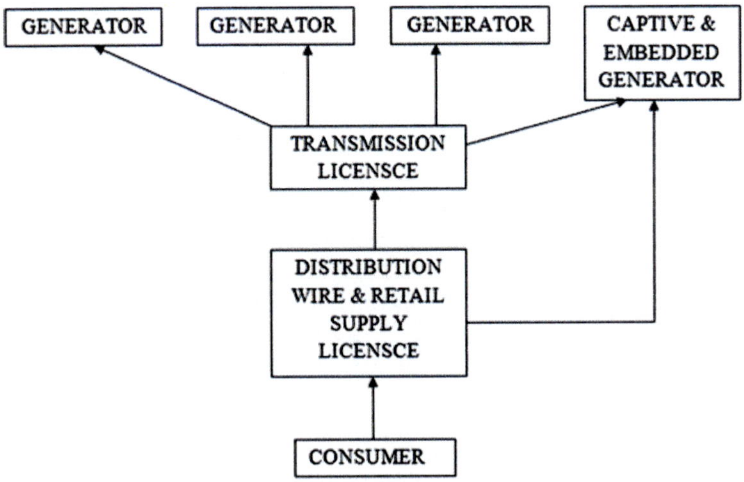

Fig. 3 Upstream competition, downstream partial competition (2003–2016)

monopoly of eras at once. Now, consumers have direct choice of going to generators directly bypassing the distributors. It was for the first time that power exchangers were introduced in the market. Independent power producers were also allowed to generate and distribute directly or through power exchanger to the consumers. This structure of market has introduced a very healthy competition among the market players. Working of model can be understood from (Fig. 4). But we still have lot to do; we need to upgrade this model to a model, wherein we can further subdivide the distribution system in to smaller division to increase the profit.

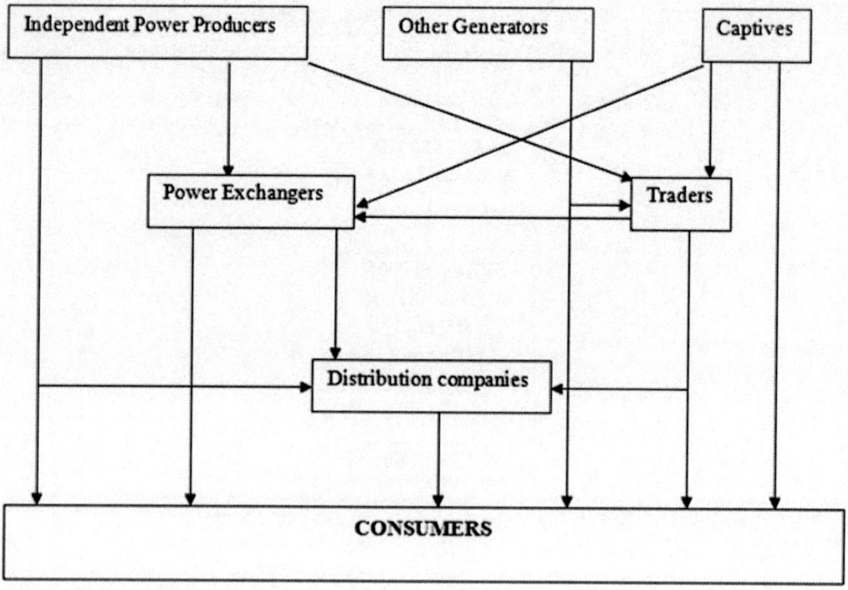

Fig. 4 Current market scenario (2016 onward)

6 Conclusion

Despite of various constraints, the Indian power sector has achieved significant success. Deregulated environment have facilitated consumers by providing more options of buying quality power at efficient rates. It is due to the deregulated environment that distributed generations are gaining popularity, and thus, overall generation capacity of the entire system is getting improved. Generating companies can directly transmit electricity to the customers, i.e., they can bypass the distribution companies. Transmission scenario has lot to change as there is very minor changes since 1991 to present condition; thus, emphasis should be given to improve power quality and minimization of losses. New metering systems are used to control theft. Deregulation of power sector alone does not mean that it is end to every problem; there are still many unturned stones. As demand of power in India is increasing at a very rapid rate, we need to upgrade our power system entities at same rate, so that we can match our demand and generation capacity.

References

1. Laai LL (2002) Power system restructuring and deregulation. Wiley, England
2. Sood YR, Padhy NP, Gupta HO (2002) Wheeling of power under deregulated environment of power industry—a bibliographical survey. IEEE Trans Power Syst 17(3):870–878

3. Government of India (2003, June) The Electricity Act. 2003. Gadzet of India
4. Kothari DP, Nagrath IJ, Power system engineering. TMH Publication, New Delhi
5. Sood YR (2010, October) Feasibility assessment of simultaneous bilateral and multilateral transaction. Int J Electr Power Energy Syst 32(8):879–885
6. Wood AJ, Wollenberg BF (1996) Power generation, operation and control, 2nd edn. Wiley, New York
7. Padhy NP, Sood YR (2004) Advancement in power system engineering education and research with power industry moving towards deregulation. In: Proceedings of IEEE, power engineering society general meeting, vol 1, pp 71–76, 6–10 June 2004
8. Galiana FD, Conejo AJ, Gil HA (2003) Transmission network cost allocation based on equivalent bilateral exchanges. IEEE Trans Power Syst 18(4):1425–1431
9. Sood YR (2007) Evolutionary programming based optimal power flow and its validation for deregulated power system analysis. Elsevier Electrical Power and Energy Systems 29:65–75
10. Sun C, Lu S (2010) ST combined economic emission HS using improved quantum-behaved particle swarm optimization. Expert Syst Appl 37:4232–4241
11. Singh A (2004) Power sector reform in India: current issues and prospects. Energy Policy 34:2480–2490
12. Singh A (2010) Towards a competitive market for electricity and consumer choice in the Indian power sector. Energy Policy 38:4196–4208
13. Lee WJ, Nolan GJ Power system deregulation and SMD: status and future
14. Projections (2005) IEEE industry applications society, vol 11, pp 52–60
15. Government of India. Ministry of Power, Gov.nic.in
16. Iqbal A et al (eds) (2020) Soft computing in condition monitoring and diagnostics of electrical and mechanical systems. In: Advances in intelligent systems and computing, vol 1096. Springer, Singapore. https://doi.org/10.1007/978-981-15-1532-3
17. Iqbal A et al (eds) (2020) Meta heuristic and evolutionary computation: algorithms and applications. In: Studies in computational intelligence, vol 1096. Springer, Singapore. https://www.springer.com/gp/book/9789811575709

The Impact of Energy, Chemical Reaction, Power Radiation, and MHD-Free Convection in the Presence of Thermophoresis

Navneet Joshi, Vimal Singh Bisht, Abhijit Singh Bhakuni, and Manoj Kumar Singh

Abstract The present study signifies the impact of study of power radiation, chemical reaction, thermophoretic, and MHD flow for a dense viscous fluid having viscosity and thermal conductivity dependent on temperature. The study is performed in the presence of radiation for lower stagnation point considering an isothermal cylinder. Using similarity transformation for governing flow are nonlinear partial differential equation is obtained which is then transformed into ordinary differential equation considering various physical parameters these equations were solved and the results are demonstrated. The effect of different parameters is studied and is represented in the form of graphs. The effects of S.F.C and Nusselt number on the isothermal cylinder are also represented in tabular form.

Keywords Chemical reaction · Power radiation · MHD · Lower stagnation point · Natural or free convection

1 Introduction

The temperature difference-driven free convention flow finds great application in industries. Buoyancy is also of great importance from an environmental point of view. If the difference between the land temperature and air temperature rises, it may lead to complex flow patterns [1–13]. Such flows are importance in various branches of engineering. Nuclear power plants, missiles, thermal power plant, gas turbines, and few are examples of engineering domains. Hossain et al. [3] examine the effect of temperature on transmission of fluid on curly surface when it is heated

N. Joshi (✉)
Mathematics Department, GEHU, Bhimtal, India
e-mail: navneet.nimt@gmail.com

V. S. Bisht · A. S. Bhakuni
Electronics Engineering Department, GEHU, Bhimtal, India

M. K. Singh
School of Computing Department, GEHU, Bhimtal, India

A. Iqbal et al. (eds.), *Renewable Power for Sustainable Growth*, Lecture Notes in Electrical Engineering 723, https://doi.org/10.1007/978-981-33-4080-0_33

vertically, whereas the radiation conduction impact on mixed transmission of horizontal cylinder was investigated by Hossain et al. [3]. Merkin [4] and [5] studies the elliptic cross section of the cylinder for free conviction at its boundary layer. An elliptic cross-sectional cylinder's free transmission boundary layer was examined by Merkin et al. [6]. The evanescent-free transmission close to the lower immobile end of a cylindrical surface prone to an abrupt change in the surface temperature was examined by Harris et al. [2]. Molla et al. [7] examined the changing viscosity's effect on cylindrical surface while fluid passes through it. Ahmad et al. (2008) investigated the constant laminar-free transmission boundary on a horizontal cylinder having elliptic cross section where both horizontal (blunt elliptic cylinder) and vertical (slender elliptic cylinder) axes are subjected to a uniform face examining the fluid to be in compressible and viscous. Sravanthi et al. (2019) examined the radiation absorption and impact of chemical reaction on Rivlin–Ericksen flow past a vertical plate which is porous in nature. Bhatt et al. [11] performed the recent study on nanofluid considering the stretching cylinder model in a porous medium. It has been seen that study of mathematical models plays an important role and the variety of applications is very high. Hence, studying CFD problem gives a chance to researchers in related field and problems easily match with those existing in industry.

The objective of work is to analyze the various parameters of chemical reaction effects which are temperature dependency of viscosity and the thermal conductivity dependency of thermophoretic with temperature on the free transmission flow MHD which is very near to the torpidity point of a circular cylinder which is isothermal in nature.

2 Problem Formulation

Let us consider a porous isothermal cylinder having radius r and the temperature T_w Assume temperature of fluid to be T_∞. It is assumed that "O" is the centre of curvature and "a" is the radius of curvature of the body surface. Further a uniform magnetic field of strength "B_0" is applied normally to the surface of isothermal cylinder. Choosing O as the orthogonal coordinate system at a point near to the stagnation level which is x and y.

The governing equations of the flow will be as follows:

Continuity Equation:

$$u_x + v_y = 0 \tag{1}$$

Momentum Equation (2):

$$uu_x + vu_y = g\beta(\mathrm{T} - \mathrm{T}_\infty)ax + \frac{1}{\rho}\left[\mu(\mathrm{T})u_y\right]_y$$

$$-\frac{\sigma B_0^2}{\rho}u - \frac{\mu}{\rho k'}u + g\beta_c(C - C_\infty) \quad (2)$$

Energy Equation:

$$\rho C_P(uT_x + vT_y) = (k(T)T_y)_y - (q_r)_y \quad (3)$$

Diffusion Equation:

$$uC_x + vC_y = DC_{yy} - K_1(C - C_\infty) - [V_T(C - C_\infty)]_y \quad (4)$$

From the equations mentioned above, u represents fluid velocity in the x-axis direction and v represents velocity in the direction of y-axis. V_t is thermophoretic velocity, T is fluid temperature, gravity acceleration is represented by g, thermal expansion is represented by β, concentration expansion is β_c, electrical conductivity defined as σ, magnetic field applied is having strength of B_0, assuming pressure to be constant specific heat is represented as C_p, viscosity of fluid depending upon temperature is $\mu(T)$, and thermal conductivity depending upon temperature is $k(T)$.

The radiation heat flux given by Brewster [2] is:

$$q_r = -\frac{4\sigma_1}{3k_1}\frac{\partial T^4}{\partial y} \quad (5)$$

Assuming the boundary conditions initially as:

$$u = 0\,(\text{initial}),\ v = 0\,(\text{initial}),\ T = T_w(\text{wall temperature})\ \text{at}\ y = 0$$

$$u \text{ tending to zero, } T \text{ tending to infinity as } y \text{ tending to infinity} \quad (6a)$$

$\mu(T)$—Viscosity, $k(T)$—Thermal conductivity varies with temperature as follows:

$$\mu(T) = \frac{\mu_\infty}{1 + \gamma(T - T_\infty)} \quad (7)$$

$$k(T) = k_\infty(1 + (b(T - T_\infty))) \quad (8)$$

Introducing the similarity variables and non-dimensional parameters:

$$\psi = Gr^{\frac{1}{4}}axvf(\eta)\ \ \eta = Gr^{\frac{1}{4}}ay\ \ \theta = \frac{T - T_\infty}{T_w - T_\infty} \quad (9)$$

$$Gr = \frac{g\beta(T_w - T_\infty)}{a^3\nu^2}\ \ G_m = \frac{g\beta_c(C_w - C_\infty)}{a^3\nu^2}\ \ \varphi = \frac{C - C_\infty}{C_w - C_\infty}$$

where θ and f are non-dimensional reduced temperature and stream function, respectively, and also the stream function is defined by ψ:

$$f''' = A\left[f'^2 - ff'' + \frac{\varepsilon\theta' f''}{(1+\varepsilon\theta)^2} + \frac{1}{\sqrt{\mathrm{Gr}}} f'\left(\mathrm{Ha} + \frac{1}{(1+\varepsilon\theta)K}\right) - \theta - \frac{Gm}{Gr}\phi\right] \tag{10}$$

$$\theta'' = -\frac{3N}{4+3N(1+\omega\theta)}\left[\mathrm{Pr}\, f\theta' + \omega\theta'^2\right] \tag{11}$$

$$\varphi'' = (\tau\theta' - f)Sc\varphi' + \tau Sc\varphi\theta'' + L\varphi Sc \tag{12}$$

Here, variable thermal conductivity and variable viscosity parameter are defined, respectively, as $(\omega) = b(T_w - T_\infty)$ and $(\varepsilon) = \gamma(T_w - T_\infty)$, Hartman number $(\mathrm{Ha}) = \frac{\sigma B_0^2}{\mu_\infty a^2}$, radiation parameter $(N) = \frac{k_1 k_\infty}{4\sigma_1 T_\infty^3}$, Prandtl number $(\mathrm{Pr}) = \frac{\mu_\infty C_p}{k_\infty}$, $A = (1+\varepsilon\theta)$, thermophoretic parameter $(\tau) = -{}^{m}\!/_{T_0}(T_w - T_\infty)$.

Taking initial boundary conditions as:

$$f(\eta) = 0,\ (\text{initial})\ f'(\eta) = 0\ (\text{initial})\ \theta(\eta) = 1\ (\text{initial})\ \text{at}\ \eta = 0 \tag{13}$$

$f'(\eta)$ tending to zero, $\theta(\eta)$ tending to zero, $\phi(\eta)$ tending to zero, as η tending to infinity.

S.F.C are the most important characteristics from engineering point of view and can be written as follows:

$$\mathrm{Nu} = \frac{Gr^{-\frac{1}{4}}}{ak_\infty(T_w - T_\infty)} q_w,\ C_f = \frac{Gr^{-\frac{3}{4}}}{a^3 x v_\infty}\tau_w$$

Using the variable Eqs. (7)–(9) and initial and boundary conditions (13), we get local Nusselt number and skin friction:

$$\mathrm{Nu} = -(1+\omega)\theta'(0)\ C_f = \frac{1}{(1+\varepsilon)} f''(0)$$

3 Result and Discussion

Using ordinary differential equation, from 9 to 11, the numerical values have been calculated for velocity temperature and concentration. By adopting Runge–Kutta method for fourth order and fifth order, the values of dependent variables are calculated at terminal point. The admissible tolerance is taken, viz. O (10^{-6}).To achieve missing boundary condition, we have used shooting technique $\Delta\ \eta = 0.05$. The values of dependent variables are computed at the final instant by making use of fourth- and fifth-order Runge–Kutta technique. For different dimensionless band of parameters, velocity $f'(\eta)$, temperature $\theta(\eta)$, and concentration (φ) have been examined and displayed in Figs. 1, 2, 3, 4, 5, 6, 7, 8, 9, 10, 11, 12 and 13. Figures 1, 2, 3, and 4 represent the distinct Hartman number impact (Ha = 0, 0.5, 1.5), chemical reaction parameter (L = 0.2, 0.6, 0.8), thermophoretic parameter (τ = 0, 0.5, 1), and Schmidt number (Sc = 0.5, 1, 1.5). From the figure drawn below, it can be concluded that there is a decrease in velocity with the increases in magnetic, chemical, thermophoretic, and Schmidt number parameters, respectively. Figure 8 shows the impact of Prandtl number on dimensionless temperature. Therefore, we can see that with the increase in Prandtl number its momentum decreases sharply. Figure 7 exhibits that with an increase in temperature magnetic parameter increases. It is seen from Fig. 9 that increase in viscosity parameter will result in increase of velocity, but on moving away from the surface of cylinder it decreases. From Figs. 11, 12 and 13, we can see with an increase in chemical reaction parameter and Schmidt number, there is a decrease in concentration level. Figure 10 exhibits that with the increase in magnetic parameter, there is an increment in concentration. The numeric values for S.F.C and Nusselt number are represented in Table 1 for the distinct parameter of dimensionless parameters. It is examined that the dimensionless wall velocity gradient $f''(0)$ increases with increasing variable viscosity parameter, while

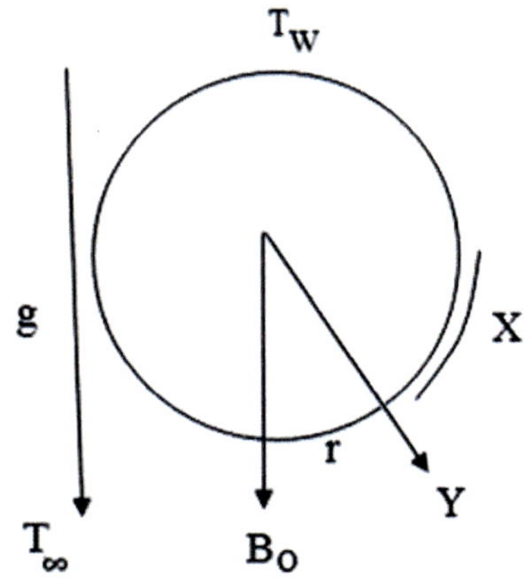

Fig. 1 Geometrical representation of the problem

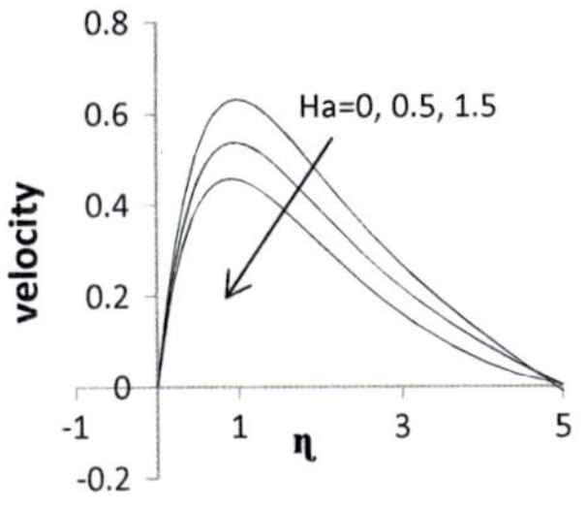

Fig. 2 Velocity profile for various estimate of Ha at ε = 0.5, K = 1, Pr = 0.71, ω = 0.5, Sc = 0.5, Gr = Gc = 1

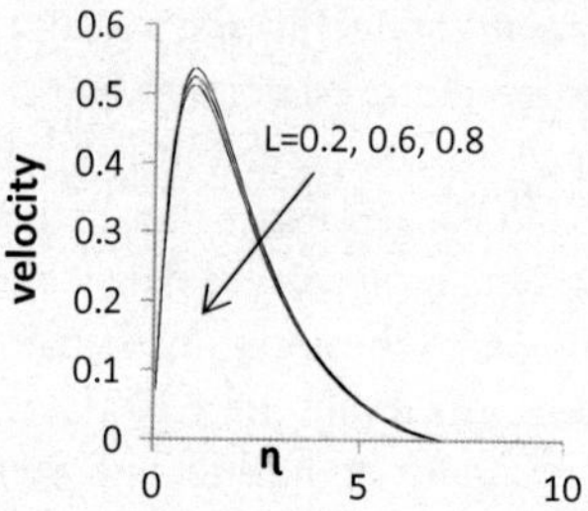

Fig. 3 Velocity distribution for various estimate of L at $M = 0.5$, $Sc = 0.5$, $\varepsilon = 0.5$, $Gr = Gc = 1$, $K = 1$, $Pr = 0.71$

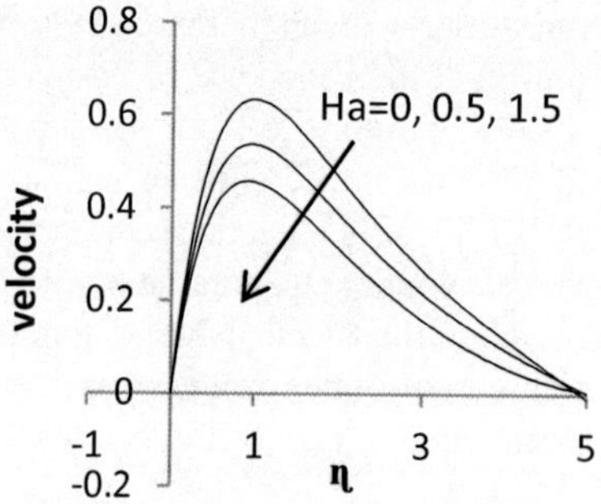

Fig. 4 Velocity distribution for various estimate of τ at $K = 1$, $Pr = 0.71$, $\varepsilon = 0.5$, $Gr = Gc = 1$, $\omega = 0.5$, $L = 0.2$, $Sc = 0.5$, $\tau = 0.5$

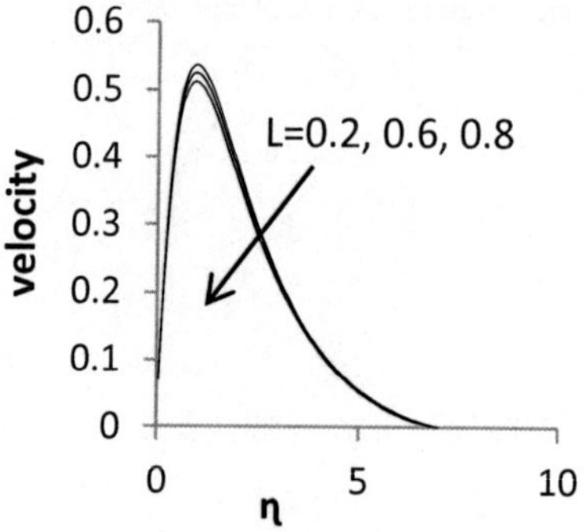

Fig. 5 Velocity distribution for various estimate of Sc at $K = 1$, $Pr = 071$, $Gr = Gc = 1$, $\varepsilon = 0.5$, $\omega = 0.5$, $\tau = 0.5$

it depreciates with increase in other parameters discussed in the work. Moreover, the value of $-\theta'(0)$ depreciates with the increment in magnetic parameter, chemical reaction parameter, and Schmidt number, while there is an increment with increase in viscosity parameter, Prandtl number, and thermophoretic parameter.

Fig. 6 Velocity distribution for various estimate of L at $\varepsilon = 0.5$, $K = 1$, $\Pr = 0.71$, $\omega = 0.5$, $Sc = 0.5$, $\tau = 0.5$, $Gr = Gc = 1$

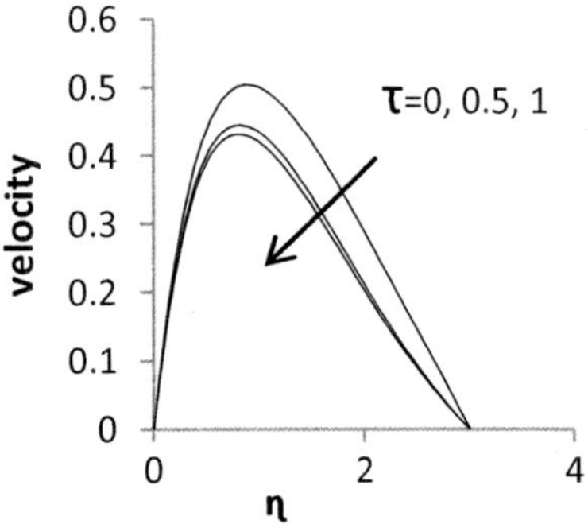

Fig. 7 Temperature profile for various estimate of Ha at $\varepsilon = 0.5$, $Gr = Gc = 1$, $\omega = 0.5$, $L = 0.2$, $Sc = 0.5$, $\tau = 0.5$, $K = 1$, $\Pr = 0.71$

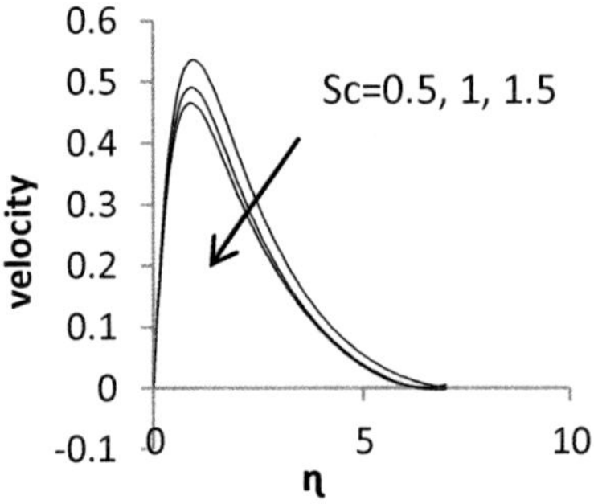

Fig. 8 Temperature profile for various estimate of Pr at $K = 1$, $Gr = Gc = 1$, $\omega = 0.5$, $L = 0.5$, $\tau = 0.5$, $Ha = 0.5$, $\varepsilon = 0.5$

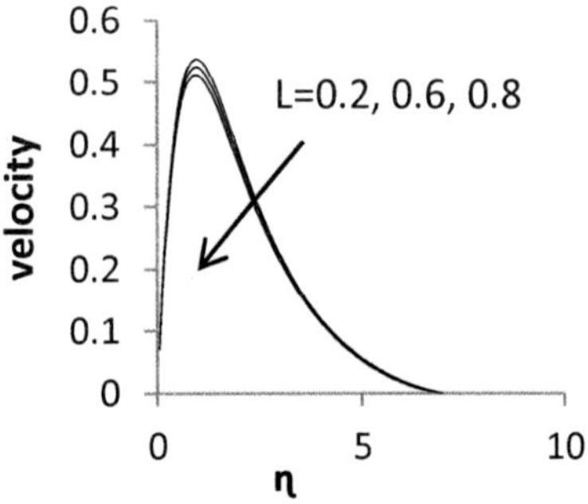

Fig. 9 Velocity distribution for various estimate of ε at $K = 1$, $Gr = Gc = 1$, $\omega = 0.5$, $L = 0.5$, $\tau = 0.5$, $Ha = 0.5$

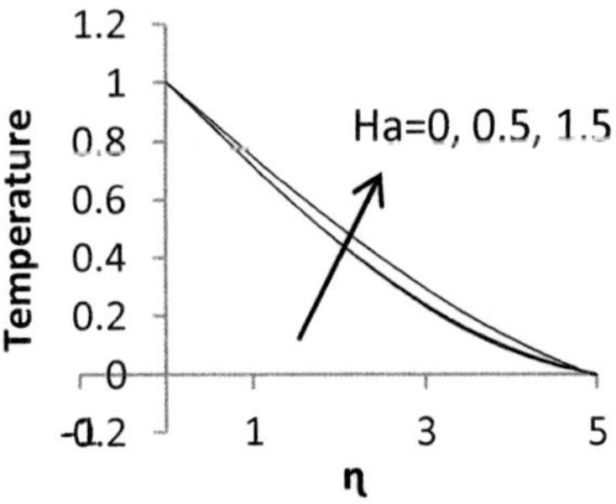

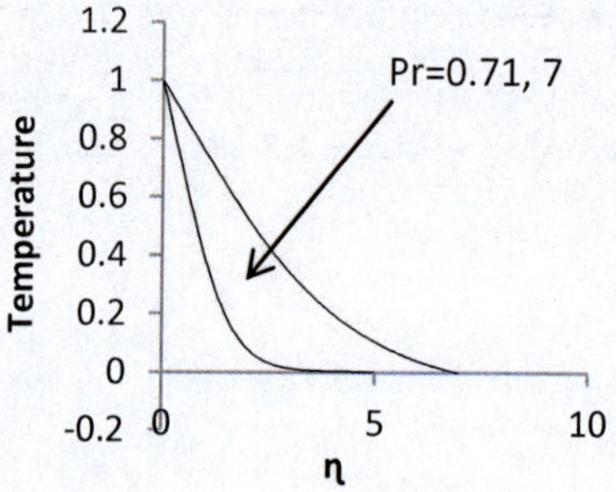

Fig. 10 Concentration distribution for various value of Ha at $K = 1$, $\mathrm{Gr} = \mathrm{Gc} = 1$, $\omega = 0.5$, $L = 0.5$, $\tau = 0.5$, $\varepsilon = 0.5$

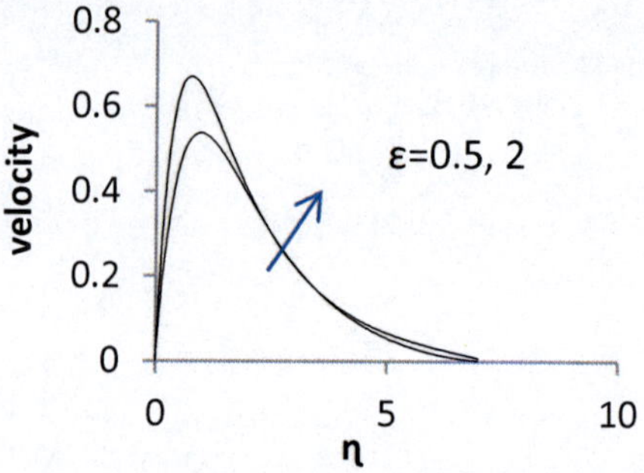

Fig. 11 Concentration distribution for various value of L at $K = 1$, $\mathrm{Gr} = \mathrm{Gc} = 1$, $\omega = 0.5$, $\tau = 0.5$, $\varepsilon = 0.5$, $\mathrm{Ha} = 0.5$, $\mathrm{Pr} = 0.71$

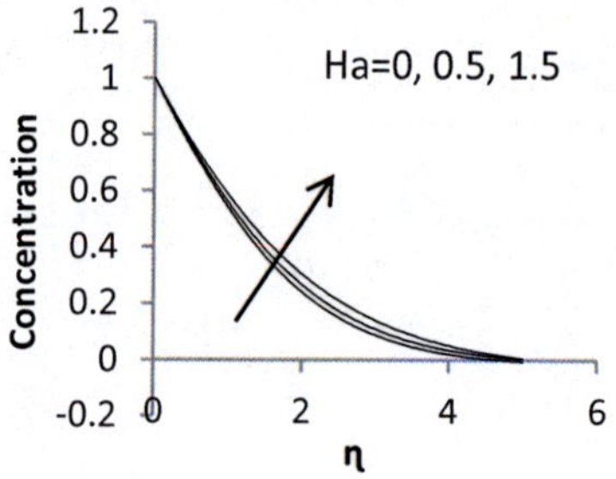

Fig. 12 Concentration distribution for various estimate of τ at $K = 1$, $\mathrm{Pr} = 0.71$, $\varepsilon = 0.5$, $\mathrm{Gr} = \mathrm{Gc} = 1$, $\omega = 0.5$, $L = 0.2$, $\mathrm{Sc} = 0.5$, $\mathrm{Ha} = 0.5$

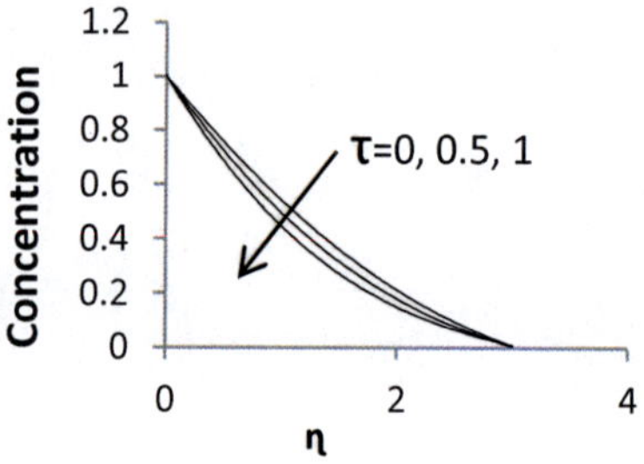

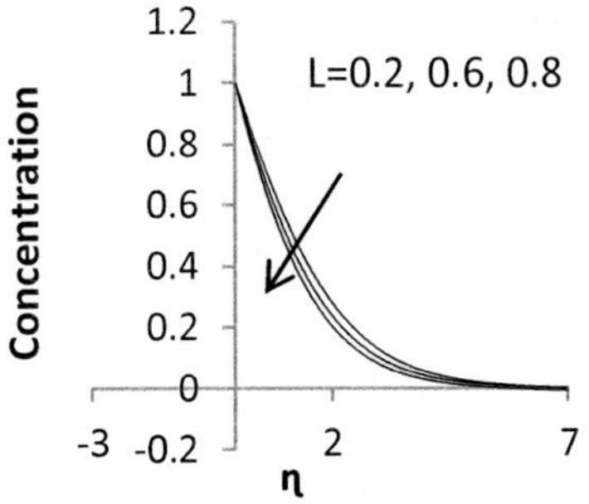

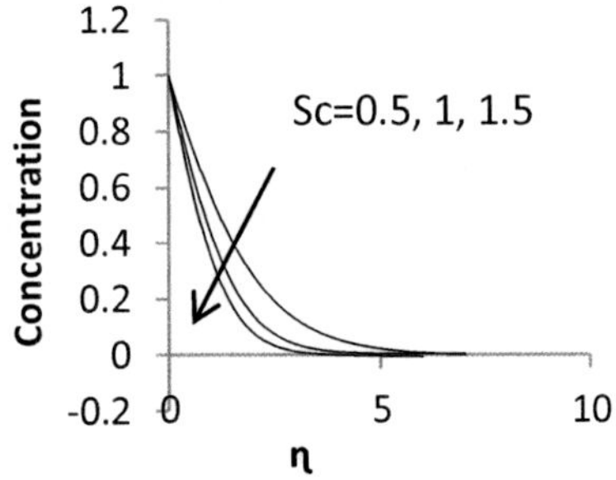

Fig. 13 Concentration profile for various estimate of Sc at $K = 1$, $\mathrm{Pr} = 0.71$, $\varepsilon = 0.5$, $\mathrm{Gr} = \mathrm{Gc} = 1$, $\omega = 0.5$, $L = 0.2$, $\mathrm{Ha} = 0.5$, $\tau = 0.5$

Table 1 S.F.C $(f''(\eta))$ and Nusselt number $(-\theta'(\eta))$ at $\eta = 0$ for various range of parameters at $\mathrm{Gr} = \mathrm{Gc} = 1.0$

ε	M	L	τ	Pr	Sc	$f''(\eta)$	$-\theta'(\eta)$
0.5	0.5	0.2	0.5	0.71	0.5	1.5268794	0.239101
2	0.5	0.2	0.5	0.71	0.5	2.450375	0.25850
0.5	0.5	0.2	0.5	0.71	0.5	1.5268794	0.239101
0.5	1	0.2	0.5	0.71	0.5	1.38861	0.23850
0.5	0.5	0.6	0.5	0.71	0.5	1.505791	0.234971
0.5	0.5	0.8	0.5	0.71	0.5	1.4666431	0.232180
0.5	0.5	0.2	0	0.71	0.5	1.41977	0.34140
0.5	0.5	0.2	1	0.71	0.5	1.39454	0.34149
0.5	0.5	0.2	0.5	7	0.5	1.488114	0.57245
0.5	0.5	0.2	0.5	0.71	1	1.46104250	0.233490
0.5	0.5	0.2	0.5	0.71	1.5	1.4183281	0.228750

References

1. Brewster MQ (1992) Thermal radiative transfer and properties. Wiley, New York
2. Harris SD, Ingham DB (2000) Transient free convection near the lower stagnation point of cylinder surface subjected to a sudden change in surface temperature. Int Commun Heat Transfer 27:1091

3. Hossain MA, Kutubuddin M, Pop I (1999) Radiation-conduction interaction on mixed convection past a horizontal circular cylinder. Int J Heat Mass Transfer 35:307
4. Merkin JH (1976) Free convection boundary layer on an isothermal horizontal circular cylinder. In: ASME/AIChE heat transfer conference, St. Louis, MO, 9–11 Aug 1976
5. Merkin JH (1977) Mixed convection a horizontal circular cylinder. Int J Heat Mass Transfer 20:73
6. Merkin JH (1977) Free convection boundary layer on cylinders of elliptic cross section. ASME J Heat Transfer 99:453
7. Molla MM, Hossain MA, Gorla RSR (2005) Natural convection flow from an isothermal horizontal circular cylinder with temperature dependent viscosity. Heat Mass Transfer 41:594
8. Uddin Z, Kumar M (2011) Radiation effects on MHD heat and mass transfer free convection flow near the lower stagnation point of an isothermal cylinder embedded in porous domain with the presence of radiation. Jordan J Mech Ind Eng 5(2):133–138
9. Mabood F, Sateyi (2019) Multiple slip effects on MHD unsteady flow heat and mass transfer impinging on permeable stretching sheet with radiation. 3052790. https://doi.org/10.1155/2019/3052790
10. Sravanthi CS, Gorla RSR (2019) Absorption and chemical reaction effects on Rivlin-Ericksen flow past a vertical moving porous plate. Int J Appl Mech 24(3):675–689
11. Bhatt Bipin C, Pathak Govind, Singh Khilap (2019) Influence of non-uniform heat absorption/generation on Magneto hydrodynamic flow via a permeable stretching cylinder in porous medium due to nanofluid. Int J Res Anal Rev (IJRAR) 6(2):389–396
12. Iqbal A et al (eds) (2020) Soft computing in condition monitoring and diagnostics of electrical and mechanical systems. In: Advances in intelligent systems and computing, vol 1096. Springer, Singapore. https://doi.org/10.1007/978-981-15-1532-3
13. Iqbal A et al (eds) (2020) Meta heuristic and evolutionary computation: algorithms and applications. In: Studies in computational intelligence, vol 1096. Springer, Singapore. https://www.springer.com/gp/book/9789811575709

Performance Assessment of Unipolar Control Technique-Based Cascaded H-Bridge Multilevel Inverter

Md Abdullah Ansari, Naved Khan, Wasif Dilshad, and Khaliqur Rahman

Abstract In this paper, a single phase five level cascaded H bridge multilevel inverter followed by unipolar control schemes have been analyzed for full range of linear modulation indices. Pulse width modulation (PWM)—based modulation scheme is used for the switching of power electronic switches. Inverter performance has been analyzed and discussed by taking performance parameters as output voltage transferred to load, harmonics in output voltage, switching stress across the switches, power loss, and efficiency of inverter. The proposed topology is simulated in MATLAB/Simulink.

Keywords Amplitude modulation · Power loss · Unipolar · Total harmonics distortion

1 Introduction

With the rapid exhaustion of conventional energy sources, forthcoming power generation sector is shifting towards non-conventional resources like solar, wind, etc. As these resources are abundant in nature, but their optimal utilization is not an easy task due to their non-uniform presence [1–15]. Also, DC output obtained from these resources need to be transformed to AC which is done by inverters [1]. Main problem with the trivial two level inverters is their high total harmonic distortion (THD), switching losses, low efficiency, etc. [2]. This problem has been solved by cascading multiple units to make multilevel inverters (MLI) [3].

Md A. Ansari (✉) · N. Khan · W. Dilshad · K. Rahman
Electrical Engineering Department, Aligarh Muslim University, Aligarh, UP 202002, India
e-mail: maansari11@myamu.ac.in

N. Khan
e-mail: nk130685@gmail.com

W. Dilshad
e-mail: wasif.dilshad138@gmail.com

K. Rahman
e-mail: er.khaliqurrahman@gmail.com

A. Iqbal et al. (eds.), *Renewable Power for Sustainable Growth*, Lecture Notes in Electrical Engineering 723, https://doi.org/10.1007/978-981-33-4080-0_34

MLIs are being extensively utilized in the field of medium as well as high voltages and energy applications. The research areas include flexible AC transmission (FACTS), high voltage DC transmission (HVDC), renewable energy like solar, wind, and many more [4, 5]. They achieve desired output voltage from combination of DC voltage sources [6]. Since, prime objective is to obtain nearly sinusoidal voltage as output which is directly related to the number of voltage levels [7]. Higher the output voltage levels, lower is total harmonic distortion (THD) which results in better performance of inverter [8].

Classical topologies like flying capacitor (FC), neutral point clamped (NPC), and cascaded H bridge (CHB) gained popularity due to their simple circuit arrangement. Besides their popularity in the industries, they suffer from many flaws. In NPC, number of clamping diodes increase quadratically with increase in voltage levels [9]. And also furthermore, capacitor voltage balancing problem also intensifies with increasing voltage levels. Similarly, FC topology requires large size capacitors whose number also increases significantly with higher voltage levels along with voltage balancing problem also. CHB topology requires isolated DC sources for producing multi leveled output [10–12]. In symmetrical mode (SM), all DC sources are identical whereas in asymmetrical mode (AM), some or all of the DC sources are different [13–15]. In asymmetrical mode of operation, THD achieved is lower as compared with SM. And also, higher voltage levels can be achieved for different DC source ratios.

2 Description of Proposed Topology

2.1 Generalized Design of Proposed Topology

Figure 1 shows the basic circuit diagram of a single phase n-cell CHB-based MLI. Each cell is responsible for producing three levels, i.e., $+V_{DC}$, 0 and $-V_{DC}$. $+V_{DC}$ is produced when S1 and S2 is closed, whereas $-V_{DC}$ is obtained when S3 and S4 are closed. And 0 is obtained when either of switches S_1 and S_3 or S_4 and S_2 are closed. Hence, n-cell produces $2n + 1$ output voltage levels when all DC sources are equal. Phase voltage V_a obtained at the output is sum of all voltages obtained at each cell, i.e.

$$Va = V_1 + V_2 + \ldots + V_n \tag{1}$$

where V_1, V_2, ..., V_n represents the output voltage of each cell.

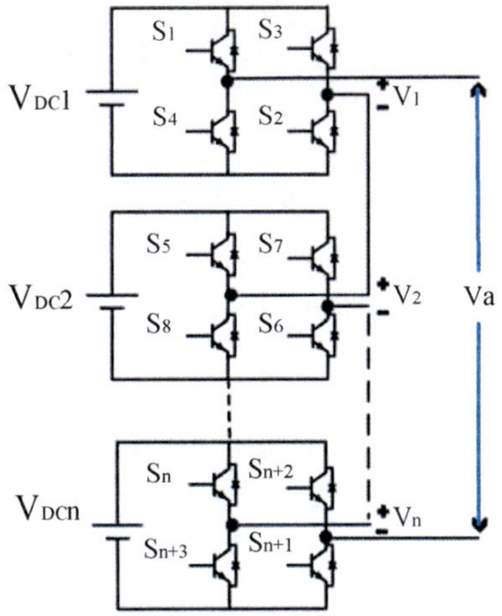

Fig. 1 Single phase n-cell CHB

2.2 *Proposed Topology*

Figure 2 depicts the circuit diagram of proposed topology. The whole assembly comprises of two DC sources along with eight unidirectional switches. Table 1 shows switching table corresponding to nine levels. State "1" represents switch in closed position whereas "0" state represents the switch is in open position. As described above, V_{DC1} is obtained at Va when S_1 and S_2 are closed, along with switch pair (S_5, S_7) being closed which is depicted by symbol "1" in the table.

Figure 3 shows the corresponding voltage level states obtained from analyzing Table 1. Diagonal switches are responsible for positive voltage at the output, whereas off-diagonal switches are responsible for negative voltage and same row switches produce zero voltage at the output.

Generalized values for n-cell single phase CHB is given below:

- Number of maximum voltage levels $(Va) = 3^n$
- Total number of IGBT switches $= 4n$

Fig. 2 Proposed topology

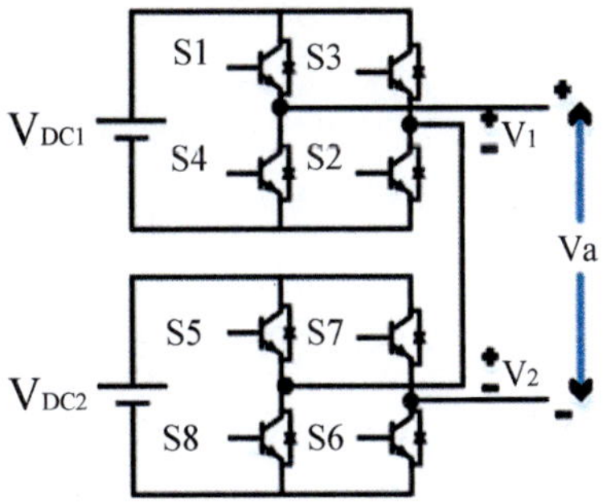

Table 1 Switch triggering sequence of proposed topology

State	V_o	S_1	S_2	S_3	S_4	S_5	S_6	S_7	S_8
1	$V_{DC1} + V_{DC2}$	1	1	0	0	1	1	0	0
2	V_{DC1}	1	1	0	0	1	0	1	0
3	V_{DC2}	1	0	1	0	1	1	0	0
4	$V_{DC1} - V_{DC2}$	1	1	0	0	0	0	1	1
5	$V_{DC2} - V_{DC1}$	0	0	1	1	1	1	0	0
6	0	0	0	0	0	0	0	0	0
7	$-V_{DC1}$	0	0	1	1	1	0	1	0
8	$-V_{DC2}$	1	0	1	0	0	0	1	1
9	$-V_{DC1} - V_{DC2}$	0	0	1	1	0	0	1	1

- Total number of DC sources = n.

2.3 Control Scheme

In order to activate the switches, there must be proper switching scheme to obtain the desired output voltage level at the desired point of time.

Sinusoidal PWM type is dealt in this paper with unipolar switching scheme. In unipolar scheme, the individual switches are triggered by comparing V_{ref} and V_c of reference and carrier signals, respectively.

Phase-shifted PWM technique is used in which phase shifted sinusoidal reference signals are compared with carrier signals in order to generate gate signals which is shown in Fig. 4.

Circuit parameters of the proposed topology is described below.

3 Performance Analysis

3.1 Output Voltage

The net RMS value of voltage received at the output is defined as:

$$V_{o,\text{rms}} = \sqrt{\frac{\int_0^T V_o^2(t)\mathrm{d}t}{T}} \tag{2}$$

where $V_o(t)$ is the output voltage varying periodically with period T.

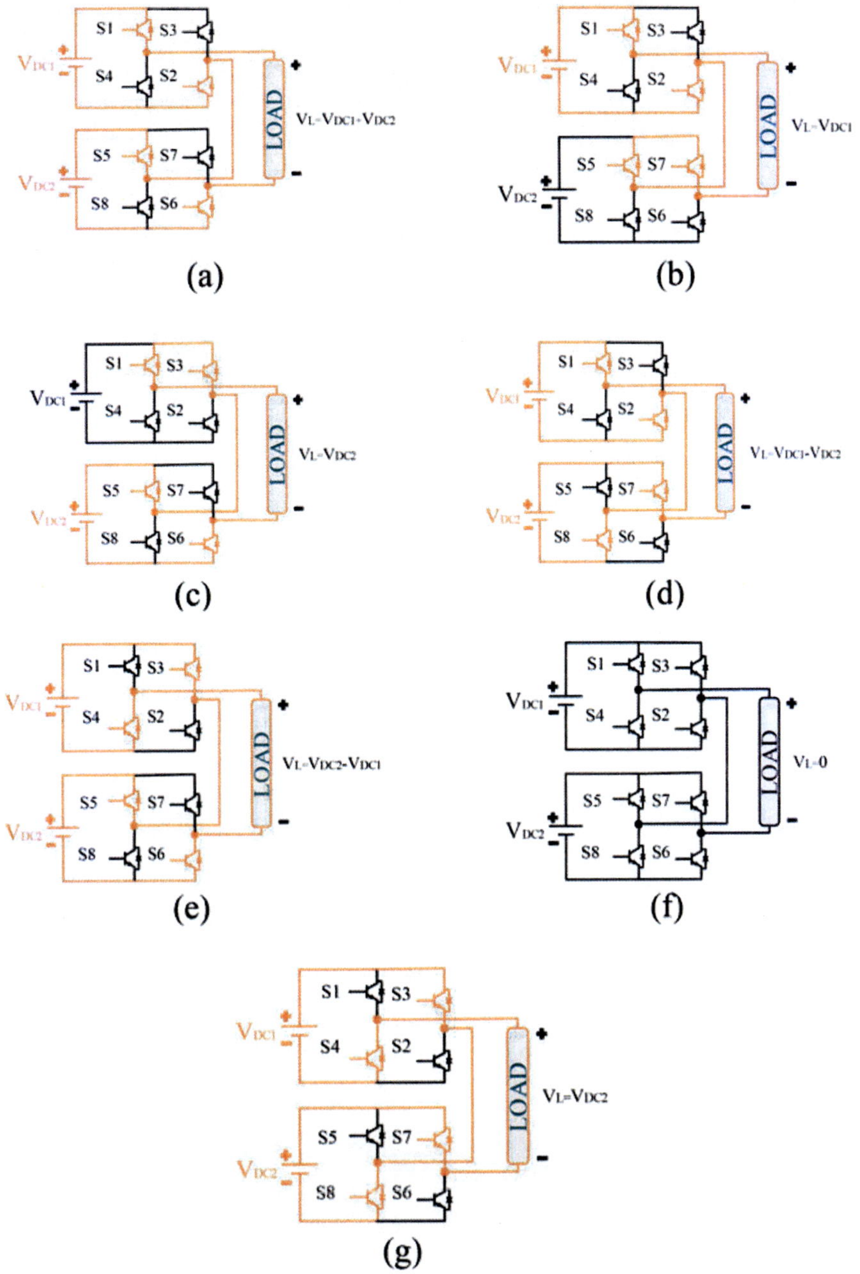

Fig. 3 **a–g** showing different voltage levels of proposed topology w.r.t. Table 1

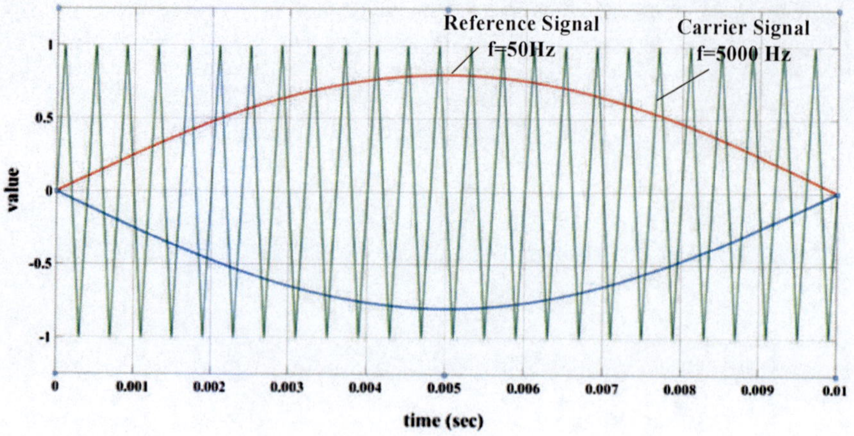

Fig. 4 Phase shifted PWM technique

Component	Specification	Number
Carrier signal	Triangular $F_c = 5000$ Hz	2
Reference signal	Pure Sinusoidal $F = 50$ Hz	2
Voltage sources	$V_1 = 100$ V, $V_2 = 100$ V	2
Switches with anti-parallel diodes	IGBT, Ron $= 1\text{e} - 3\ \Omega$	8
Load	$R = 10\ \Omega$, $L = 250$ mH	1

3.2 *Total Harmonic Distortion (THD)*

It represents the quantity of harmonics present in the signal other than the fundamental frequency component.

$$\text{THD} = \sqrt{\frac{V_{\text{rms}}^2 - V_1^2}{V_1}} \tag{3}$$

where V_{rms} and V_1 are the RMS and fundamental component of output voltage, respectively.

3.3 Switching Stress

Switching stress is a figure of merit which describes the total active switch stress developed in the power electronics switches (S):

$$S = \sum_{j=1}^{N} V_j I_j \quad (4)$$

where V_j is the peak voltage applied across switch j,

I_j is the peak current applied across switch j

For best performance, switching stress is kept minimum.

3.4 Power Loss

Power loss is the combined effect of conduction Loss (P_c) as well as switching Loss (P_{SW}). Conduction Loss is the composite loss of switch and anti-parallel diode due to ON state drop in voltage and the equivalent switch resistance.

$$P_{\text{total loss}} = P_c + P_{SW} \quad (5)$$

Conduction loss is calculated by taking ideal IGBT switch with anti-parallel diode.

3.5 Efficiency of Inverter

Efficiency is the net useful power delivered to the output. It is given by the formula (6).

$$\eta = \frac{P_o}{P_o + P_{\text{loss}}} \times 100\% \quad (6)$$

4 Results and Discussion

The output voltage obtained corresponding to amplitude modulation index of 0.95 is illustrated in Fig. 5.

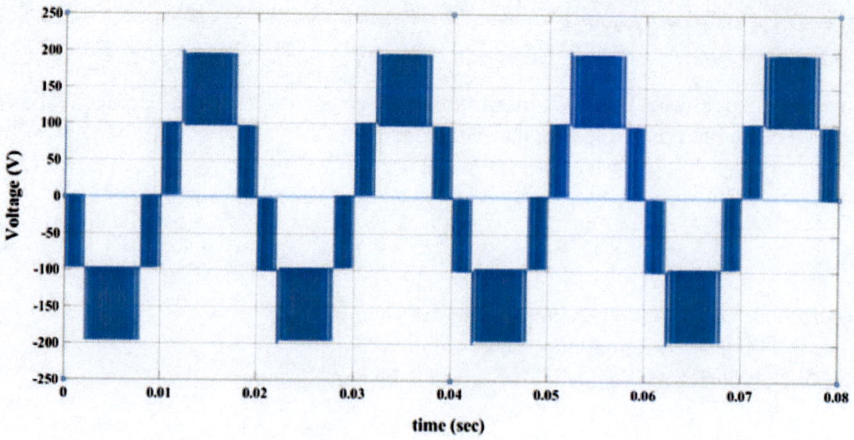

Fig. 5 Output five level waveform for ma = 0.95

Fig. 6 Output fundamental (RMS) voltage versus amplitude modulation index (Ma)

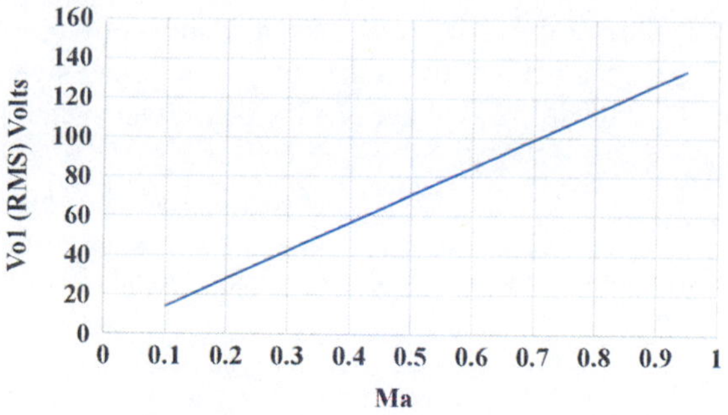

4.1 Output Voltage Transferred to Load

Since, output voltage being stair-cased form contains fundamental along with other harmonic components, the useful output is only the fundamental part. And hence, RMS value of fundamental is calculated which is shown in Fig. 6.

4.2 THD of Output Voltage

Coefficients of frequency terms other than fundamental frequency constitutes to noise or harmonics in the system. Quantitative term used for determining their contribution is known as THD. Since,

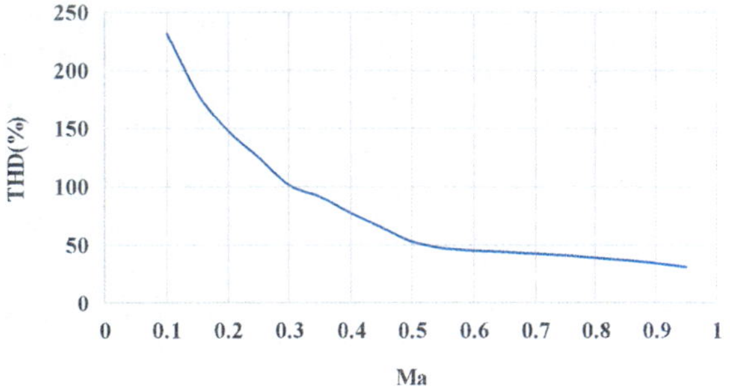

Fig. 7 THD (%) versus amplitude modulation index (Ma)

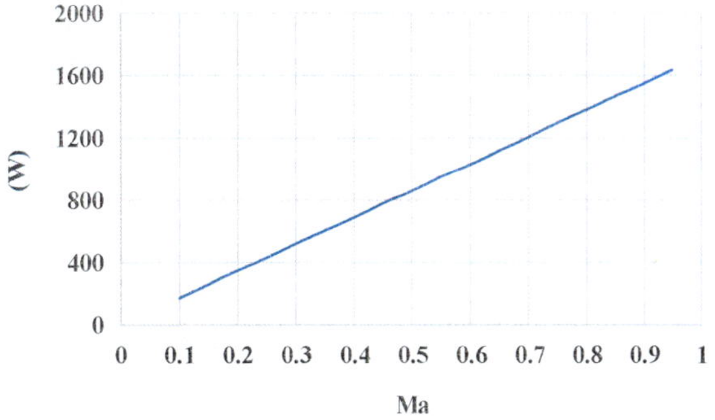

Fig. 8 Switching stress versus amplitude modulation index (Ma)

$$V_0 = \sum_{n=1,3,5}^{\infty} \frac{4V_s}{n\pi} \sin(n\omega t) \tag{11}$$

From Fig. 7, it is clear that THD decreases with increase in the value of Ma.

4.3 Switching Stress

As described previously, it is the overall stress developed in the switches during current flow in the circuit. It can be inferred from Fig. 8 that the switching stress depends on on-state current flowing through the switches.

4.4 Power Loss

It is the actual loss that occurs due to switching as well as conduction of switches. Average conduction loss is calculated by taking product of instantaneous voltage and

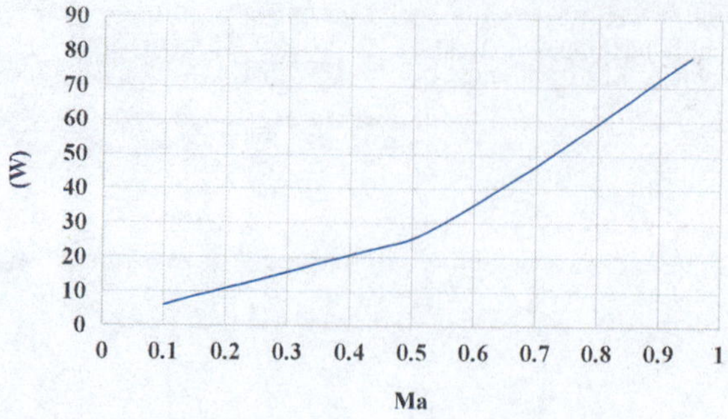

Fig. 9 Average switching loss versus amplitude modulation index (Ma)

current averaged over a complete cycle. Individual switching duration determines the switching loss which is depicted in Fig. 9.

4.5 *Efficiency*

Finally, the efficiency of the inverter is calculated by calculating input power and output power.

$$P_{in} = \sum_{k=1}^{2} I_{s,k} V_{s,k} \tag{12}$$

where V_s and I_s are the average input voltage and input current of both source V_1 and V_2, respectively.

$$P_o = V_o I_o \cos(\varphi) \tag{13}$$

V_o and I_o are the RMS values of fundamental output voltage component and current, respectively. It is so because only the fundamental power is useful and rest harmonic power is lost as heat and is not useful. Figure 10 explains the relationship of efficiency with Ma.

5 Conclusion

In this paper, detailed analysis of single phase five level cascaded H bridge inverter followed by unipolar control technique was performed. It was found that as the value of amplitude modulation index was increased, output RMS voltage, average switching losses, switching stress and efficiency increased, whereas value of THD decreased.

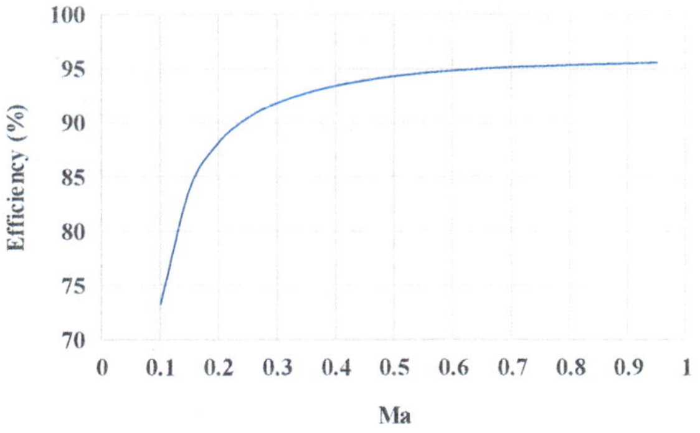

Fig. 10 Efficiency versus amplitude modulation index (Ma)

References

1. Ansari MA, Lone JA, Tariq M (2020) Performance analysis and comparative evaluation of two-winding multi-tapped transformer based nine-level inverter. In: International conference on power electronics, control and automation (ICPECA), New Delhi, India, pp 1–5
2. Ansari A, Mohammad A, Tariq M, Riyaz A (2020) Residential energy conservation using efficient home appliances. Int J Innov Technol Explor Eng 9(3):3457–3465
3. Azeem H, Yellasiri S, Jammala V, Naik BS, Panda AK (2019) A fuzzy logic based switching methodology for a cascaded h-bridge multi-level inverter. IEEE Trans Power Electron 34(10):9360–9364
4. Siddique MD, Mekhilef S, Shah NM, Memon MA (2019) Optimal design of a new cascaded multilevel inverter topology with reduced switch count. IEEE Access 7:24498–24510
5. Rahman K, Meraj M, Mahajan SB, Iqbal A (2019) Single phase ZAC-source AC-AC converter with high buck and boost voltage conversion capability. IEEE Trans Ind Electron 1–1
6. Behbahanifard H, Abazari S, Sadoughi A (2019) New scheme of SHE-PWM technique for cascaded multilevel inverters with regulation of DC voltage sources. ISA Trans
7. Siddique MD, Mekhilef S, Shah NM, Sarwar A, Memon MA (2019) A new single-phase cascaded multilevel inverter topology with reduced number of switches and voltage stress. Int Trans Electr Energ Syst 1–21
8. Gupta VK, Mahanty R (2015) Optimized switching scheme of cascaded H-bridge multilevel inverter using PSO. Int J Electr Power Energ Syst 64:699–707
9. Salam Z, Majed A, Amjad AM (2015) Design and implementation of 15-level cascaded multi-level voltage source inverter with harmonics elimination pulse-width modulation using differential evolution method. IET Power Electron 8(9):1740–1748
10. Rahman K, Tariq A, Bakhsh FI (2012) Modeling and analysis of multilevel inverters using unipolar and bipolar switching schemes. In: IEEE-IEEE international conference on advances in engineering, science and management (ICAESM 2012), pp 466–471
11. Rahman K, Aware MV, Iqbal A, Al-Ammari R, Abu-Rub H (2014) Common-mode voltage control through vector selection in three-to-five phase matrix converter. In: IEEE international symposium on industrial electronics, pp 2087–2092
12. Rahman K, Iqbal A, Al-Ammari R (2013) Space vector model of a three-phase to five-phase AC/AC converter. In: IEEE AFRICON conference
13. Khoshkbar Sadigh A, Abarzadeh M, Corzine KA, Dargahi V (2015) A new breed of optimized symmetrical and asymmetrical cascaded multilevel power converters. IEEE J Emerg Sel Top Power Electron 3(4):1160–1170

14. Iqbal A et al (eds) (2020) Soft computing in condition monitoring and diagnostics of electrical and mechanical systems. In: Advances in intelligent systems and computing, vol 1096. Springer, Singapore. https://doi.org/10.1007/978-981-15-1532-3
15. Iqbal A et al (eds) (2020) Meta heuristic and evolutionary computation: algorithms and applications. In: Studies in computational intelligence, vol 1096. Springer, Singapore. https://www.springer.com/gp/book/9789811575709

A Scientific Study on Effect of Polarization in Calculation of Rain Attenuation Using ITU-R Model

Arun Kumar, Natwar Singh Rathore, and Alok Kumar Pandey

Abstract This paper addresses the current need to work on higher frequency levels for radio wave communication because of need for higher speed requirement in communication system and also because the current frequency spectrum is congested. While establishing radio communication links of higher frequencies it is important to study various problems associated with them. Rain induced attenuation at higher frequency is a major problem. Therefore, a study has been done on rain attenuation and effect of polarization is calculated for six different regions of the world. The simulation results are tested on ITU-R model and various findings throughout the simulation work have been concluded.

Keywords ITU-R · Polarization · Rain rate · Radio wave · Specific attenuation

1 Introduction

Radio waves play the major role in communication [1–16], or it can be said that radio waves make the wireless communication possible. Many scientists with likes of Heinrich Rudolf Hertz, James Clerk Maxwell, David Edward Hughes etc. conducted various experiments and gave their findings to the world which sets the base for modern research which are still ongoing for further improvements in this field. Another aspect in current communication is that the radio spectrum which is currently in use is deeply concentrated which calls for higher frequencies band, as in [1]. For the use of higher frequencies, it is essential to study the effect of polarization due to rain attenuation because it fades signal and affect the satellite communication. In order to predict the rain attenuation various models are being developed keeping

A. Kumar (✉) · N. S. Rathore · A. K. Pandey
KIET Group of Institutions, Delhi-NCR, Ghaziabad, India
e-mail: arun.kumar.en@kiet.edu

N. S. Rathore
e-mail: natwar.singh@kiet.edu

A. K. Pandey
e-mail: alok.pandey@kiet.edu

A. Iqbal et al. (eds.), *Renewable Power for Sustainable Growth*, Lecture Notes in Electrical Engineering 723, https://doi.org/10.1007/978-981-33-4080-0_35

in mind various parameters including geographical location, system constraints etc. ITU-R model is the most popular and worldwide accepted model being in use to predict rain induced attenuation [2]. It predicts the rain attenuation of a region based on its geographical location, type of polarization used, frequency level and rain rate of that region. It is important to know rain attenuation in radio wave communication, so a theoretical study has been done on various region of the world with ITU-R model. Rain attenuation is a very major problem at higher frequencies, as in [3]. Because of absorption and phenomenon of scattering by droplets of rain, attenuation of radio wave happens, as in [4]. Researchers have been working upon this problem for long (since twentieth century). Various models have been developed to predict rain attenuation. Some of these models are Crane Global Model, Simple Attenuation Model ITU-R, Two-Component Model etc., as in [5, 12–14]. Due to wider acceptability of ITU-R model, it has been used by us to study rain attenuation phenomenon.

2 System Description

ITU-R model for rain attenuation prediction has been developed by International Telecommunication Union (ITU) to predict rain attenuation. The ITU has divided the world in 15 different regions based on their geographical location, as in [6]. Parameters like frequency and polarization type are being used to predict the rain attenuation of that region [8–10]. The specific attenuation denoted by γ_R (dB/km) is evaluated from rain-rate R (mm/h) using the power-law relationship [11]. It is represented as:

$$\gamma_R = k R^{\alpha} \tag{1}$$

where, the values of the coefficients k and α are found as functions of frequency f (GHz). The value of frequency ranges from 1 to 1000 GHz. The details of the evaluated equations, which were calculated from curve-fitting to power-law coefficients, are described in [7, 11].

3 Proposed Algorithm

A simple algorithm based on ITU-R model has been developed during our study that could be used to predict rain attenuation. This algorithm can be used to develop program to reduce computational time taken to calculate rain attenuation. The proposed algorithm steps are shown as below:

Step 1: Initialize the algorithm with the command “Start”.
Step 2: Declare the input variables and percentage availability for rain-rate region.

Step 3: Declare the 15 array of size 7 of all 15 rain regions with their respective rain rate at different percentage availability.
Step 4: Declare the 4 arrays of 2 each for horizontal polarization coefficients and vertical polarization coefficients for the all required frequency level (data of 10 frequencies are considered in this paper) in sorted type.
Step 5: Now initialize the rain region and percentage availability.
Step 6: Check rain region from 1 to 15 from predefined array of rain regions.
Step 7: For the given percentage availability, check the rain rate in rain region obtained from step 6.
Step 8: From 1 to 10 frequency levels, find Specific Attenuation for each frequency Specific Attenuation = Coeffecient-1 * Rain-Rate Coeffecient-2.
Step 9: Display the result
Step 10: End

The flow-chart for the proposed algorithm is depicted in Fig. 1.

Above Algorithm can be used to develop the program to find the rain attenuation at a faster rate for many regions at different frequency range.

4 Experimental Data

Using the proposed algorithm, a C program has been developed and specific attenuation for 6 regions have been calculated based on the ITU-R model.

The specific attenuation in (dB/km) due to horizontal polarization (S_H) and vertical polarization (S_V) for regions A, D, G, K, N and Q have been calculated for different frequencies, with detailed analysis (Tables 1, 2, 3, 4, 5 and 6).

5 Results

The result of above experimental data has been plotted between specific attenuation Vs frequency for all six regions in MATLAB platform. The plot obtained for the given regions are shown below.

In Fig. 2 for A-region, it is noticed that up to frequency 15 GHz, the attenuation due to horizontal and vertical polarization is nearly equal and above 15 GHz for horizontal polarization it rises faster than vertical polarization. In the Fig. 3 for D-region, it is noticed that up to frequency 7 GHz the attenuation due to horizontal and vertical polarization is nearly equal and above 7 GHz for horizontal polarization it rises faster than vertical polarization.

In the Fig. 4 for G-region, it is noticed that the up to frequency 5 GHz, the attenuation due to horizontal and vertical polarization is nearly equal and above 5 GHz for horizontal polarization it rises faster than vertical polarization. In the Fig. 5 for K-region, it is noticed that up to frequency 5 GHz, the attenuation due to

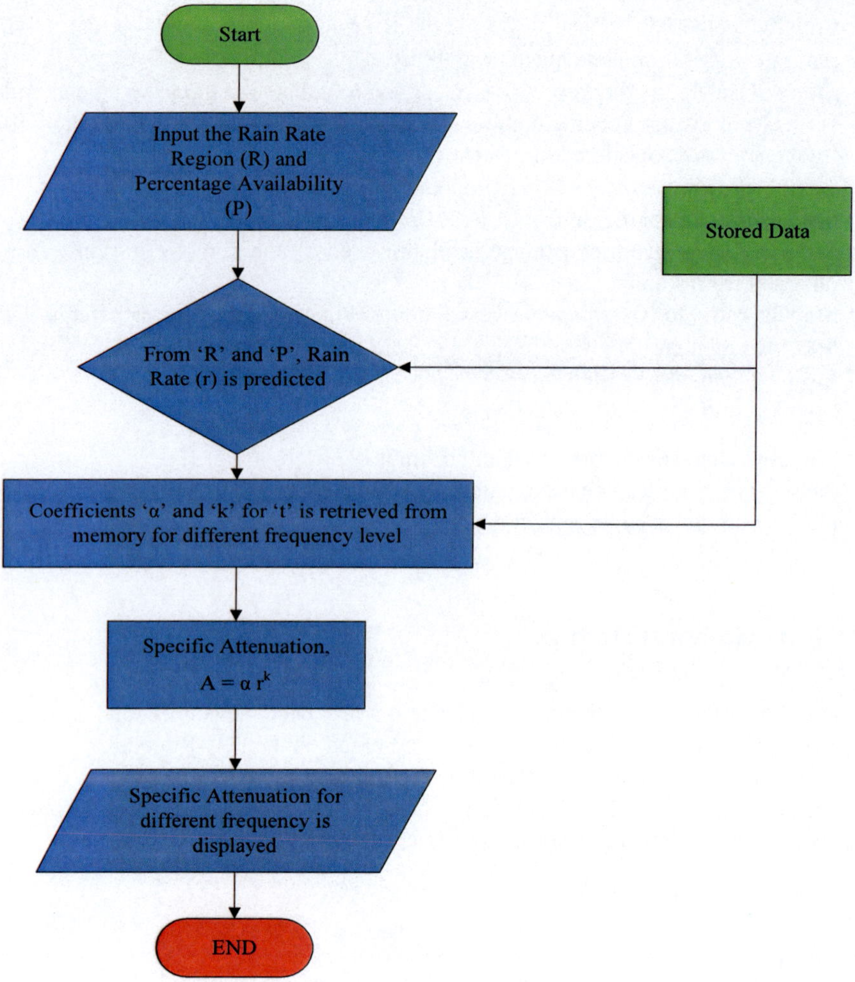

Fig. 1 Flowchart for the proposed algorithm

horizontal and vertical Polarization is nearly equal and above 5 GHz for horizontal polarization it rises faster than vertical polarization.

In the Fig. 6 for N-region, it is noticed that up to frequency 3 GHz, the attenuation due to horizontal and vertical polarization is nearly equal and above 3 GHz for horizontal polarization it rises faster than vertical polarization. In the Fig. 7 for Q-region, it is noticed that from 1 GHz, the attenuation due to horizontal polarization rises faster than vertical polarization.

Table 1 Specific attenuation (dB/km) for A-region

f (GHz)	A-region	
	Specific attenuation (dB/km)	
	Due to horizontal polarization (S_H)	Due to vertical polarization (S_V)
1	0.000194	0.000184
5	0.007367	0.005670
10	0.166175	0.141413
15	0.463250	0.439026
20	0.825032	0.744803
25	1.254450	1.103227
30	1.727168	1.529171

Table 2 Specific attenuation (dB/km) for region D

f (GHz)	Region D	
	Specific attenuation (dB/km)	
	Due to horizontal polarization (S_H)	Due to vertical polarization (S_V)
1	0.000449	0.000387
5	0.031972	0.021330
10	0.492960	0.404713
15	1.22404	1.083136
20	2.05812	1.745650
25	2.977001	2.507305
30	3.923300	3.368211

Table 3 Specific attenuation (dB/km) for region G

f (GHz)	Region G	
	Specific attenuation (dB/km)	
	Due to horizontal polarization (S_H)	Due to vertical polarization (S_V)
1	0.000699	0.000572
5	0.069404	0.042937
10	0.87534	0.705152
15	2.044679	1.744934
20	3.33508	2.737095
25	4.698596	3.867924
30	6.050667	5.110802

Table 4 Specific attenuation (dB/km) for region K

f (GHz)	Region K	
	Specific attenuation (dB/km)	
	Due to horizontal polarization (S_H)	Due to vertical polarization (S_V)
1	0.000969	0.000764
5	0.122842	0.071888
10	1.336218	1.061490
15	2.983808	2.479343
20	4.759211	3.812257
25	6.576043	5.323143
30	8.325411	6.948472

Table 5 Specific attenuation (dB/km) for region N

f (GHz)	Region N	
	Specific attenuation (dB/km)	
	Due to horizontal polarization (S_H)	Due to vertical polarization (S_V)
1	0.002138	0.001541
5	0.490744	0.250962
10	3.728094	2.862957
15	7.463659	5.813100
20	11.275698	8.515949
25	14.863460	11.550470
30	18.056126	14.638242

Table 6 Specific attenuation (dB/km) for region Q

f (GHz)	region Q	
	Specific attenuation (dB/km)	
	Due to horizontal polarization (S_H)	Due to vertical polarization (S_V)
1	0.002572	0.001816
5	0.678663	0.336278
10	4.740169	3.611422
15	9.25031	7.096316
20	13.798459	10.278690
25	17.989515	13.846835
30	21.643408	17.427540

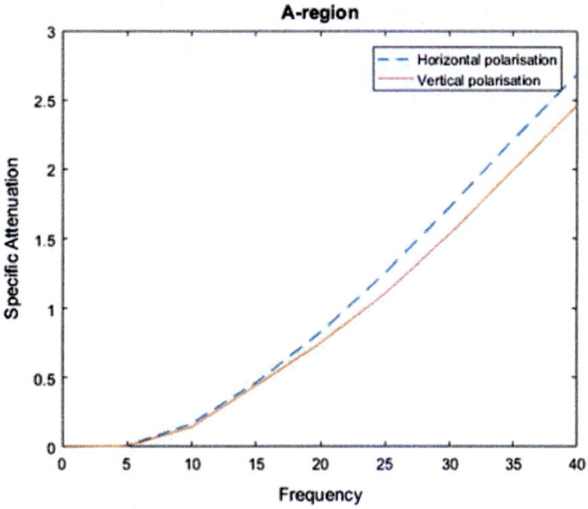

Fig. 2 Specific attenuation versus frequency for A-region

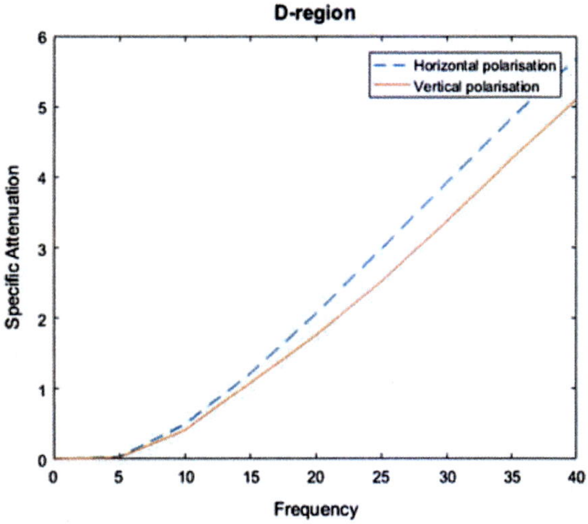

Fig. 3 Specific attenuation versus frequency for D-region

6 Conclusion

The type of polarization for radio wave propagation is chosen keeping in mind various parameters. This paper is based on ITU-R model, first a simple algorithm is developed to find rain attenuation in the model. The effect of polarization has been shown through various simulations in six regions of the world i.e. region A, D, G, K, N and Q. It has been observed that rain attenuation increases sharply at 5 GHz. Therefore, based on simulation results from ITU-R model, it is observed that the wave with vertical polarization has lesser rain attenuation than the horizontal

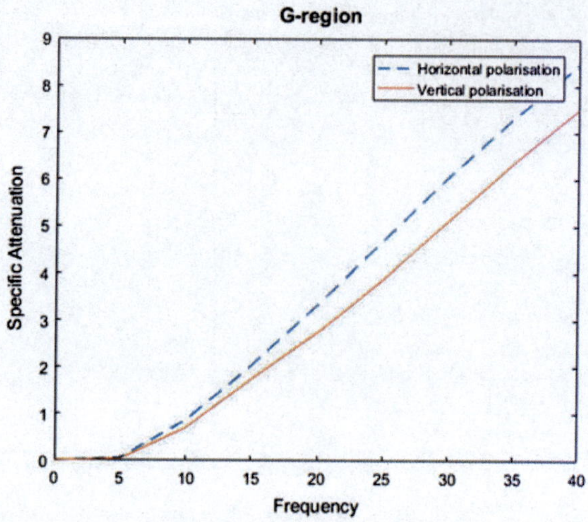

Fig. 4 Specific attenuation versus frequency for G-region

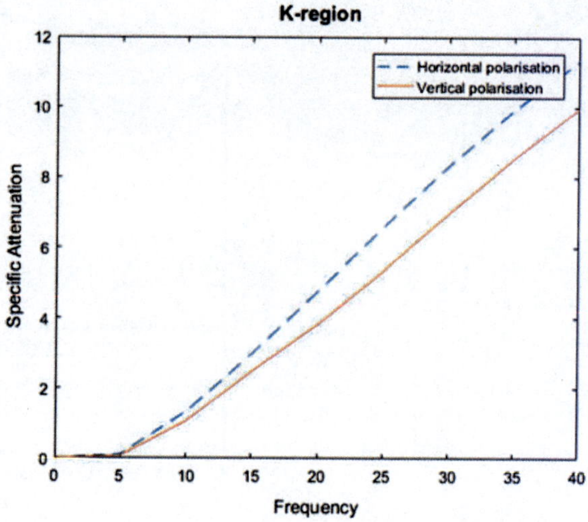

Fig. 5 Specific attenuation versus frequency for K-region

polarization. The proposed method is very simple, accurate, and fast, which makes it highly suitable in real time practice.

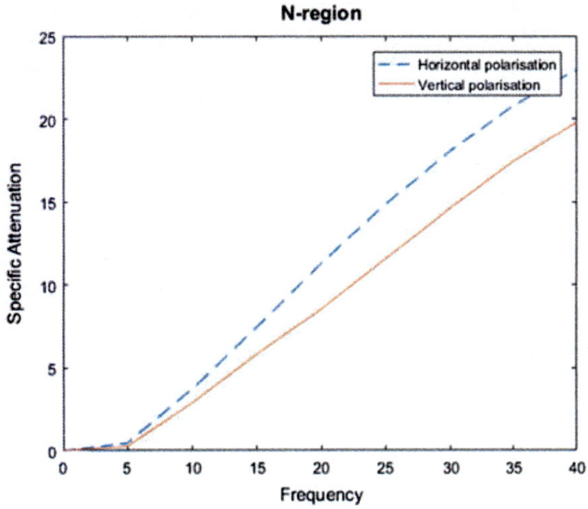

Fig. 6 Specific attenuation versus frequency for N-region

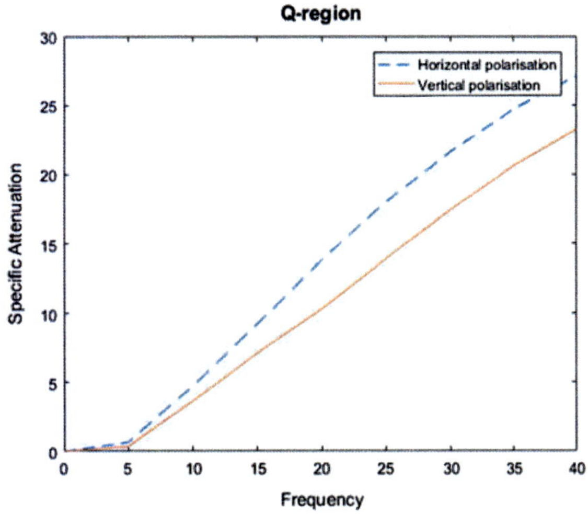

Fig. 7 Specific attenuation versus frequency for Q-region

References

1. Kestwal MC (2014) Prediction of rain attenuation and impact of rain in wave propagation at microwave frequency for tropical region, Uttarakhand, India
2. Panchal P, Joshi R (2016) Performance analysis and simulation of rain attenuation models at 12–40 GHz band for an earth space path over Indian cities. *In:* 7th international conference on communication, computing and virtualization 2016
3. T. C. Ramadorai, ""Rain attenuation and prediction in the Satellite-Earth Path," in *Proceedings of the Workshop on HF VHF and Microwave Communications*, New Delhi, India, February 1987
4. Meeks R, Beafore FJ (2011) "Propagation", the science and study of radio wave reflection, refraction, diffraction, absorption, polarization, and scattering

5. Sridhar M, Padma Raju K, Srinivan Rao Ch (2012) Estimation of rain attenuation based on ITU-R model in Guntur (A.P), India. ACEEE Int J Commun 3(3)
6. Characteristics of precipitation for propagation modelling. Rec. ITU-R PN.937-1
7. Specific attenuation model for rain for use in prediction methods. Rec. ITU-R P.838-3
8. Shrestha S, Choi DY (2017) Rain attenuation statistics over millimeter wave bands in South Korea. J Atmos Solar Terr Phys 152:1–10
9. Shrestha S, Choi D-Y (2018) Diurnal and monthly variations of rain rate and rain attenuation on Ka-band satellite communication in South Korea. Progress Electromagn Res 80:151–171
10. Sujan S, Nadeem I, Sun-Woong K, Seung-Jo H, Choi DY (2017) Rain specific attenuation and frequency scaling approach in slant-path for Ku and Ka-band experiments in South Korea. In: ICEIC 2017 international conference on electronics, information, and communication, Phuket, Thailand, January 2017, pp 625–628
11. Rakshit G, Adhikari A, Maitra A (2017) Modelling of rain decay parameter for attenuation estimation at a tropical location. Adv Space Res 59(12):2901–2908
12. Shrestha S, Choi D-Y (2017) Rain attenuation over terrestrial microwave links in South Korea. IET Microwaves Antennas Propag 11(7). Article 1031
13. Schuld M, Petruccione F (2018) Machine Learning. In: Supervised learning with quantum computers. Springer, Cham, Switzerland, p 21
14. Biscarini M, Marzano F, Montopoli M, De Sanctis K, Iess L, Montagna M, Mercolino M, Lanucara M (2016) Optimizing data volume return for Ka-band deep space links exploiting short-Term radiometeorologica model forecast. IEEE T Antenn Propag 64:235–250
15. Iqbal A et al (eds) Soft computing in condition monitoring and diagnostics of electrical and mechanical systems. In: Advances in intelligent systems and computing, vol 1096. Springer, Singapore,. https://doi.org/10.1007/978-981-15-1532-3
16. Iqbal A et al (eds) Meta heuristic and evolutionary computation: algorithms and applications. In: Studies in computational intelligence, vol 1096. Springer, Singapore. https://www.springer.com/gp/book/9789811575709

Performance Analysis of Five-Level-Cascaded H-Bridge Inverter

Mohd Faizan, Mohd Suhaib Ashraf, Syed Sultan Ahmed, Hamza Mashhood, and Imtiaz Ashraf

Abstract Multilevel inverter structures have the several points of interest of running at high DC bus voltages for higher strength applications that is executed the utilization of an arrangement association of exchanging devices. Alternatively, reduction in the harmonics of the output voltage is accomplished via switching between distinctive voltage stages. The two important typical topologies of multilevel inverters are clamped diode inverter and the cascaded H-bridge inverter. In this article, we will talk about the multilevel-cascaded inverter. The multilevel-cascaded inverter uses single-phase complete bridge inverter sequence strings to create multilevel phase legs. A particular benefit of this topology is that every bridge has modular modulation, power, and security requirements. Also, topologies of multilevel inverters can produce a better quality of performance when operating at a lower frequency of switching. In this paper, the traditional PWM inverter is modeled for MATLAB Simulink. With the help unipolar and bipolar switching schemes, the performance analysis of conventional PWM inverters and the cascaded H-bridge multilevel PWM inverters is executed and the consequences are.

Keywords Multilevel inverter · PWM inverter · Total harmonic distortion (THD) · MATLAB

M. Faizan (✉) · M. S. Ashraf · S. S. Ahmed · H. Mashhood · I. Ashraf
Department of Electrical Engineering, Zakir Hussain College of Engineering and Technology, Aligarh Muslim University, Aligarh, UP 202002, India
e-mail: mohdfaizan1130@gmail.com

M. S. Ashraf
e-mail: mohdsuhaibashrafamu93@gmail.com

S. S. Ahmed
e-mail: Sultanahmed41830@gmail.com

H. Mashhood
e-mail: hmhamza798@gmail.com

I. Ashraf
e-mail: iashraf@rediffmail.com

A. Iqbal et al. (eds.), *Renewable Power for Sustainable Growth*, Lecture Notes in Electrical Engineering 723, https://doi.org/10.1007/978-981-33-4080-0_36

1 Introduction

One of the frequent region converter topologies utilized in higher strength medium voltage drives is the CHMI [1]. It consists of many devices of H-connect single-stage strength cells. The H-bridge cell on their AC aspect are sometimes connected in cascaded to realize medium voltages working and lower total harmonic distortion. For action, the quantity of strength cells in a cascaded H-bridge inverter is primarily determined by its working voltage and price of production. For example, a nine-level inverter may be used in drives with the range of line-to-line voltage of 3300 V, where the cascaded multilevel inverters incorporate a total of twelve power cell exploitation elements of the 600 V category [2–4]. Using similar power cells results in a compact system that is a way of decreasing cost. Multilevel inverters create an AC voltage utilizing voltage steps acquired utilizing DC substance, condenser banks [5].

The structure of CHB MLI comprises of grouping of multiple single-stage inverters [6, 7]. Cascaded H-bridge inverter waveform of bipolar adjustment, the sine curve regulating waveform, triangular transporter waveform, and the gate signals, respectively, for higher switches S_1 and S_3. In the similar inverter side, the upper and base switches work in a reciprocal manner, with switched on and the different became off [8–10]. So we want to think about two unbiased gate signals V_{g1}, V_{g3}, created by using evaluating sinusoidal wave to triangular carrier wave. Normally, unipolar adjustment call for two sign curve waves, that are of the similar magnitude and recurrence out of segment at 180° [11–13]. Compared with an unusual triangular transporter wave, the two modulating waves generate gating signals for the higher turns, separately S_1 and S_3. It tends to determine that the two higher devices do not transfer simultaneously, which is particular from the bipolar pulse width modulation where each of the four devices are switches at equal interval. The inverter yield voltage switches either within the positive half pattern among zero and $-V_d$ or within the negative half pattern of the essential frequency among zero and V_d. So this scheme is referred to as unipolar modulation [14].

Multilevel inverters (MLI) use different modulation techniques to acquire a improved yield voltage with lower THD. Sine curve pulse broad regulating, specific consonant disposal, space vector broad modulation methods for cascaded H-bridge MI [2–4, 20, 21]. In the following way, multilevel inverters have various benefits as compared to conventional bipolar inverter in the following ways.

- Capability for high voltage operation with a decrease dv/dt per switch.
- Higher efficiency.
- Low electromagnetic interference.

2 Control Topologies

Figure 1 shows the five-level multilevel inverter utilizing two full H-connect. The yield voltage can be achieved by utilizing just two full bridge inverter [15]. The

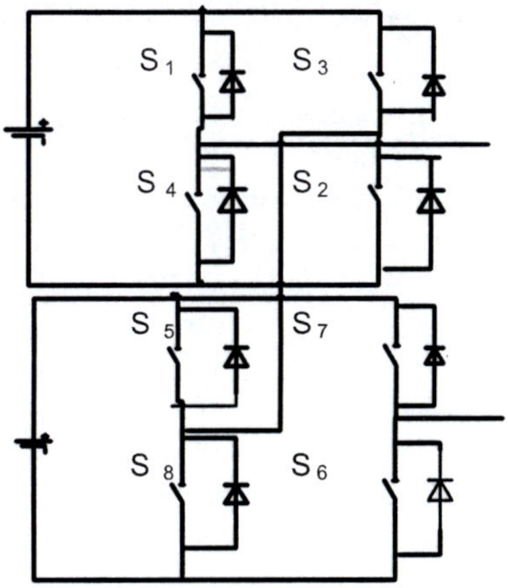

Fig. 1 Five-level multilevel inverter

troublesome yield voltage level can be acquired by shutting the suitable switches as appeared in exchanging in the table. The yield voltage of both the scaffolds is either included or subtracted by shutting separate changes to get 5 diverse voltage levels, i.e., $V_1, V_2, 0, -V_1, -V_2$ [16, 17]. Table 1 shows the switching scheme for five-stage inverter. Here level one indicates the switches is ON, level zero indicates the switches is OFF. It comprises of two arrangement associated H-bridge MLI with a two DC sources. The voltage heights of V_{dc}, $V_{dc}/2$, 0, $V_{dc}/2$ and nV_{dc} pulses are produced using a multicarrier level-shifted SPWM technique. The inverter cell required for n level is $(n - 1)/2$.

Table 1 Switching scheme and output voltage CHB MLI

V_s	S_1	S_2	S_3	S_4	S_5	S_6	S_7	S_8
0	0	0	0	0	0	0	0	0
$+V_{dc}$	1	1	0	0	0	0	0	0
$+2V_{dc}$	1	1	0	0	1	1	0	0
$+V_{dc}$	1	1	0	0	0	0	0	0
0	0	0	0	0	0	0	0	0
$-V_{dc}$	0	0	1	1	0	0	0	0
$-2V_{dc}$	0	0	1	1	0	0	1	1
$-V_{dc}$	0	0	1	1	0	0	0	0
0	0	0	0	0	0	0	0	0

2.1 Switching Scheme with Operation

Two H-bridges are connected in cascaded with two separate DC sources. All the switches are OFF for 0 V in Fig. 2a. For V_{1dc}, S_1 and S_2 are switched ON for next voltage level in Fig. 2b. For V_{2dc}, (S_1, S_2) and (S_5, S_6) are switched ON for getting maximum voltage level in Fig. 2c. For $-V_{1dc}$, S_3 and S_4 are switched ON for negative voltage level in Fig. 2d. For $-V_{2dc}$, (S_3, S_4) and (S_7, S_8) are switched ON for getting maximum negative voltage level in Fig. 2e.

This topology requires a lesser number of equipments since there are no more diodes and capacitors for clamping. But it has a separate DC wellspring of H-bridge [18]. And it hires two different DC applications and produces five degrees of the yield voltage.

In this process, a sinusoidal wave is compared to four triangular service waves of identical amplitude, organized one over the other. A CHB MLI output voltage is the complete individual DC supplies given to every one of the symmetric H-bridges as set out in condition [19].

$$V_O = V_{\text{DC1}} + V_{\text{DC2}} \tag{1}$$

Using Eq. (2), the authentic range of steps is given inoutput voltage.

$$N_{\text{step}} = 2n + 1 \tag{2}$$

Therefore, 'n' is number of H module.

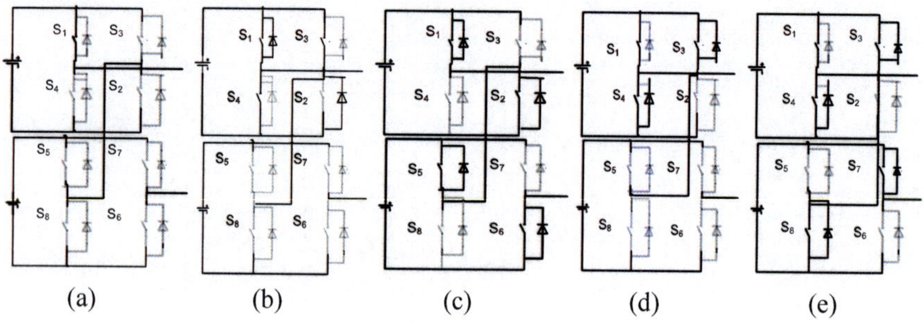

Fig. 2 Five-level-cascaded H-bridge MLI (Switching topology)

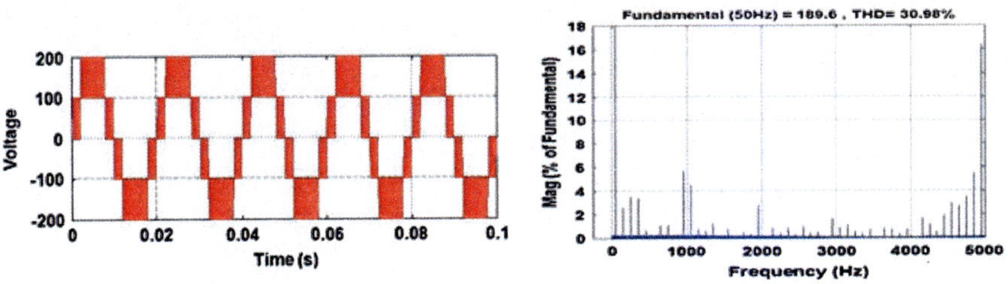

Fig. 3 Unipolar output voltage harmonic spectrum = 0.95

3 Simulation Model

3.1 Model Description of Unipolar Five-Level Inverter

In unipolar switching scheme, the switch sets S_1, S_2 and S_3, S_4 will not operate in pair. Instead, the switches of the main leg, i.e., S_1 and S_4 are worked by using comparing the triangular provider wave (V_c) with the sinusoidal reference signal (V_{ref}). The switches of the opposite leg, i.e., S_2 and S_3 are worked by means of comparing V_c with $-V_{\mathrm{ref}}$. Following logic is utilized to work these switches:

(a) If Vref is greater than V_c, S_1 is on and If Vref is greater than V_c, S_4 is on.
(b) If −Vref is greater than V_c, S_3 is on and If $-V_{\mathrm{ref}}$ is greater than V_c, S_2 is on.

The reference wave, i.e., sinusoidal wave is given a suitable shift (π/2 electrical in our case), when compared with a carrier wave in the second 1 single-stage bridge inverter.

4 Simulation Results

See Figs. 3 and 4.

5 THD Calculation

This paper deals with the sinusoidal PWM which further sub-categorized as unipolar and bipolar. In unipolar scheme, the individual switches are triggered by using contrasting reference and carrier signals, Vref and Vc, respectively. Whereas diagonal switches area is activated simultaneously in bipolar scheme. Firstly, we describe all the parameters of success and then quantify the same. They are (1) Output voltages transferred to load, (2) Total harmonic distortion in output voltages, (3) Switching stress.

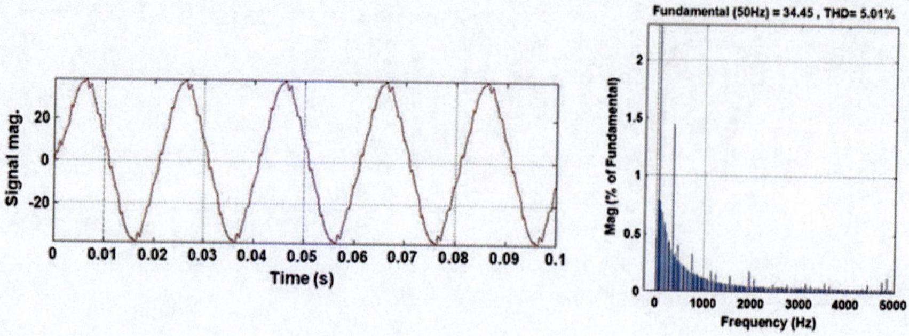

Fig. 4 Unipolar output current harmonic spectrum = 0.95

Fig. 5 Output (RMS) voltage versus amplitude modulation index

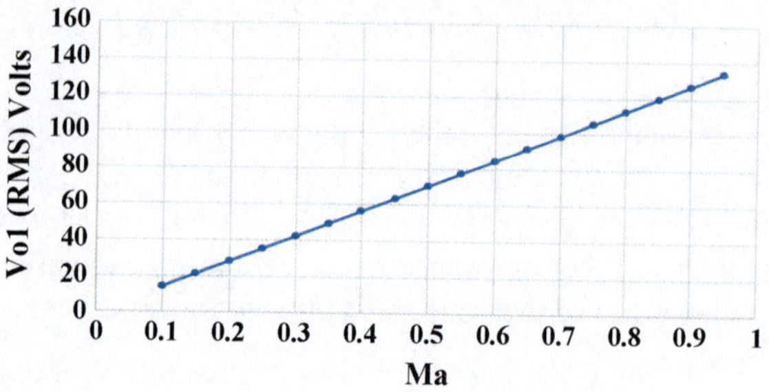

5.1 *Output Voltage Transferred to Load*

The net voltage rms obtained at the output is set to:

$$V_{o,\mathrm{rms}} = \sqrt{\frac{\int_0^T V_o^2(t)\mathrm{d}t}{T}}$$

where $V_o(t)$ is the voltage of output which varies periodically with time T.

Because, together with other harmonic elements, the output voltage is stair-cased from include fundamental, the useful output is only the basic part. And thus, the fundamental RMS value is determined as shown in Fig. 5.

5.2 *THD of Output Voltage*

Other than fundamental component harmonics present in the signal.

$$\mathrm{THD} = \sqrt{\frac{V_{\mathrm{rms}}^2 - V_1^2}{V_1}}$$

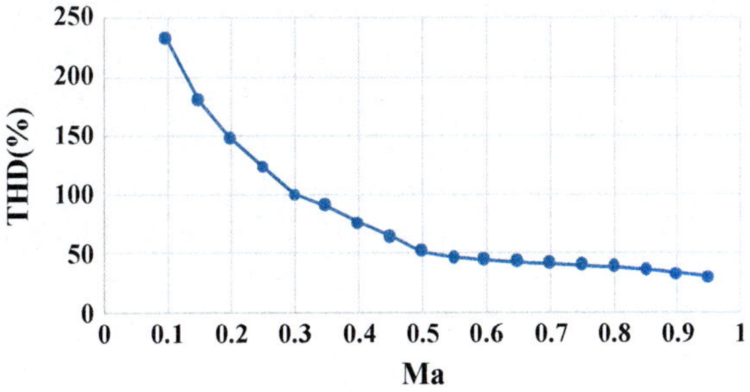

Fig. 6 THD in percentage of fundamental versus Ma

where V_{rms} and V_1 are the respective RMS and fundamental part of output voltage. The term quantitative used to determine their contribution is known a THD. Because,

$$V_0 = \sum_{n=1,3,5}^{\infty} \frac{4V_s}{n\pi} \sin(n\omega t)$$

From Fig. 6. it is clear THD decrease linearly with increases in the value of Ma.

6 Discussion

Two complete bridges have a five-level-cascaded inverter. Each inverter stage will produce three extraordinary voltage outputs, $V_{2\text{dc}}$, $V_{1\text{dc}}$ or zero, utilizing various connections of the four switches S_1 to S_4. To achieve the five-level output of every bridge is utilized in sequence. The proposed circuit makes use of a multilevel sine PWM method to decrease the harmonics in the voltage output wave as in contrast to the switching perspective variation approach. Similarly, the output waveform is obtained in the form of a staircase for five-stage H-connect multilevel inverter as shown in the simulation results of the indicated Fig. 4. The product of voltage values of 200, 100, 0, −100 and −200 V is studied from waveforms. The stress voltage throughout the quantity switches of the cascaded H-bridge MLI for a modulation index of 0.95. It is found that the stress voltage is 100 V for all the changes from S_1 to S_8.

7 Conclusion

The performance analysis of unipolar control scheme of five-level H-bridge inverters is done. The performance parameters are considered as output power and THD in output voltage. Five-level H-bridge inverter gives more adequate results than the three-level inverter, and also find its application in the field of the grid system and integrating PV arrays. On comparing with the flying capacitor and diode-clamped inverters, multilevel uses less number of components. It has low cost and light in weight than those of the inverters (i.e., flying capacitor and diode clamped). We also can get smooth switching with the assist of new switching methods. The results are the output voltages and its total THD are near 30.98% and the current output wave is much closer to the sign curve wave and has a total harmonic distortion current nearly about 5.01%.

References

1. Asghar SM (2004) Power electronics. Prentice-Hall, New Delhi
2. Rahman K, Tariq A, Bakhsh FI (2012) Modelling and analysis of multilvel inverters using unipolar and bipolar switching schemes. In: IEEE, ICAESM, Nagapattinam, Tamil Nadu, India, 30–31 March 2012, pp 466-471
3. Afarulrazi AB, Zarafi M, Utomo WM, Zar A (2010) FPGA Implementation of unipolar SPWM for single phase inverter. In: 2010 international conference on computer applications and industrial Electronics (lCCAiE 2010), Kuala Lumpur, Malaysia, December 5-7 2010, pp 671–676
4. Iqbal A, Tariq M, Rahman K, Al-qahtani A (2014) Control of packed U-cell multilevel five-phase voltage source inverter, vol 2014(1). Hamad bin Khalifa University Press (HBKU Press), pp. ITPP0493
5. Jiang M-C, Wang W-S, Fu H-K, Kuei W-C (2011) A novel single phase soft-switching unipolar PWM inverter. In: 8th international conference on power electronics ECCE Asia, The ShIlla JeJu, Korea, May30-June 3 2011, pp 2874–2879
6. (2019) Low switching frequency based asymmetrical multilevel inverter topology with reduced switch count. IEEE Access 7:86374–86383
7. Shanthi B, Natarajan SP (2009) Comparative study on unipolar multicarrier PWM strategies for five level flying capacitor inverter. In: international conference on control, automation, communication and energy conservation, Tamilnadu, India,4-6 June 2009, pp 1–7
8. (2019) A new single-phase cascaded multilevel inverter topology with reduced number of switches and voltage stress. Int Trans Electr Energ Syst 30(2)
9. Sharma R, Ball JAR (2009) Unipolar switched inverter low frequency harmonicscaused by switching delay. IET Power Electron 2(5):508–516
10. Rahman K, Meraj M, Mahajan SB, Iqbal A Single phase ZAC-source AC-AC converter with high buck and boost voltage conversion capability. IEEE Trans Ind Electron. https://doi.org/10.1109/tie.2019.2956386
11. Rahman K, Al-Emadi N, Iqbal A, Rahman S (2018) Common mode voltage reduction technique in a three-to-three phase indirect matrix converter. IET Electr Power Appl 12(2):254–263
12. Gupta AK, Khambadkone AM (2006) A space vector PWM scheme for multilevel inverters based on two-level space vector PWM. IEEE Trans Ind Electron 53(5):1631–1639
13. Rahman K, Iqbal A, Al-Emadi N, Ben-Brahim L (2017) Common mode voltage reduction in a three-to-five phase matrix converter fed induction motor drive. IET Power Electron 10(7):817–825
14. Franquelo LG et al (2006) Three-dimensional space vector modulation algorithm for four-leg multilevel converters using ABC coordinates. IEEE Trans Ind Electron 53(2):458–466
15. Li L, Czarkowski D et al (1998) Multilevel selective harmonic elimination PWM technique in series-connected voltage inverters. In: Conference record of IAS annual meeting, pp 1454–1461
16. A new single phase single switched-capacitor based nine-level boost inverter topology with reduced switch count and voltage stress. IEEE Access 7:174178–174188

17. (2019) Optimal design of a new cascaded multilevel inverter topology with reduced switch count. IEEE Access 24498–24510
18. Du Z et al (2006) Active harmonic elimination for multilevel converters. IEEE Trans Power Electron 21(2):459–469
19. Zambra DAB, Pinheiro JR (2013) Comparison of phase-shift and step wave modulation technique applied to symmetrical cascaded multilevel inverter. IEEE Latin America Trans 11(5):1156–1162
20. Iqbal A et al (eds) (2020) Soft computing in condition monitoring and diagnostics of electrical and mechanical systems. In: Advances in intelligent systems and computing, vol 1096. Springer, Singapore. https://doi.org/10.1007/978-981-15-1532-3
21. Iqbal A et al (eds) (2020) Meta heuristic and evolutionary computation: algorithms and applications. In: Studies in computational intelligence, vol 1096. Springer, Singapore. https://www.springer.com/gp/book/9789811575709

Realization of a Flyback DC–DC Converter for Experimentation-Assisted Teaching in Power Electronics

Mohammad Fahad, Shahrukh Khan, Mohammad Muktafi Ali Khan, and Adil Sarwar

Abstract The applications and importance of the power electronics sector are witnessing rapid growth in the power engineering sector. However, the multidisciplinary nature of power electronics leads to various challenges. The education paradigm needs to shift from teaching mode toward learning and experimenting mode with a wide use of software and hardware tools. In view of these factors, this paper proposes an experimental hardware test bench for flyback converter realization.

Keywords Power electronics education · Flyback converter realization · Hardware test bench

1 Introduction

The power electronics sector has transformed the traditional power generation, transmission, and distribution and has spawned numerous new applications like electric vehicles, smart home systems, more electric aircraft, and widespread, low-cost and reliable renewable power, as well as smart grid concept [1–8]. Power electronics teaching requires the incorporation of the latest simulation software and hardware tools and discussion of advanced and interesting applications to attract present-day students [2, 3]. Efficacious results on student enrollment for the courses of electric machines and electric drives were observed due to a motivating learning environment. Power electronics, however, being a multidisciplinary field encompassing circuit theory, signal processing, analog and digital electronics, electromagnetics, electric machines, etc., makes it a daunting task. Lack of adequate practical experimentation is not a solution, and it will only lead to under-skilled future engineers. Moreover,

M. Fahad (✉) · S. Khan · M. M. A. Khan · A. Sarwar
Aligarh Muslim University, Aligarh, India
e-mail: mfahad3@myamu.ac.in

S. Khan
e-mail: skhan9@myamu.ac.in

A. Sarwar
e-mail: adil.sarwar@zhcet.ac.in

A. Iqbal et al. (eds.), *Renewable Power for Sustainable Growth*, Lecture Notes in Electrical Engineering 723, https://doi.org/10.1007/978-981-33-4080-0_37

Table 1 Simulation parameters

Parameters	Magnitude
Operating frequency	10 kHz
Input DC source voltage magnitude	12 V
Magnetizing inductance	0.001 H
Transformer turns ratio	12:5
Capacitance	500 μF
Load	1 Ω

many new issues are gaining traction like electromagnetic interference (EMI) and electromagnetic compatibility (EMC). Theoretical teaching only will develop superficial skills in students to meet the indispensable and practical issues to be dealt with in power electronic devices like thermal characteristics, EMI and EMC, reliability and fault tolerance issues, etc. A major development, the Bologna declaration proposed by 29 European countries, aims to promote higher education with the considerations of globalization. The new proposed education system includes shifting the education paradigm from teaching-based to learning-based. This paper proposes an experimental educational hardware setup for flyback converter realization. A flyback converter has the benefit of isolated operation of input and output unlike conventional buck and boost DC–DC converters. DC-DC converters have been the subject of abundant research with authors developing numerous topologies with high efficiency and gain parameters [9, 10]. The output voltage ripples have the potential to be diminished by employing interleaved operation of the flyback converter as demonstrated for a buck–boost in [4]. PV-fed systems can also be developed by adding a suitable inverter after the buck–boost operation [5, 7, 8].

2 Description of the Flyback Converter

The flyback converter was developed by NASA in the 1960s for space vehicle applications. The operation procedure is analogous to the buck–boost DC–DC converter with the added assets of electrical isolation between the input and output. As an inductor is used to store energy in a buck–boost converter, similarly the magnetizing inductance of a transformer is used to store energy in the flyback converter, the basic function remaining the same [6]. Figure 1 compares the buck–boost converter and the flyback converter. The switch Sw above is a power electronic switch which can be a MOSFET, IGBT, etc. Flyback converters are generally used for power applications below 100 W power range including power supplies for televisions and computer displays. Also, with the use of multiple secondary windings and corresponding diode and capacitors, multiple outputs can be procured from a single input DC source.

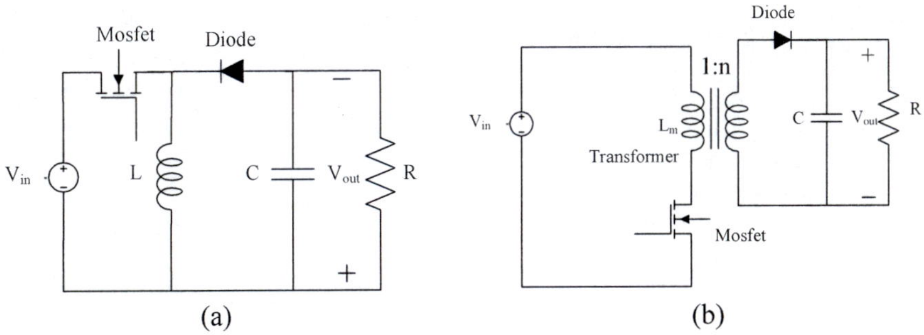

Fig. 1 **a** Buck–boost converter schematic, **b** flyback converter schematic

3 Operation and Analysis

The assumptions made for the analysis of flyback converter are the following:

1. A very large capacitance value is chosen such that output voltage V_o remains effectively constant.
2. During the steady-state conditions, all the voltages and current waveforms attain a periodic nature with equal rise and fall values over the period.
3. The gating pulse for the switch has a high frequency of the order of kHz with DT the closed switch period and $(1 - D)T$ the open switch period, T as the time period.
4. The switch and diode exhibit ideal behavior.

During the closed switch period, the magnetizing inductance of the transformer (L_m) accumulates energy supplied by the DC source. Immediately upon the turning off of the switch, the diode D_1 conducts and the energy accumulated is delivered across the load with the corresponding inductor voltage and current scaling up to a value multiplied by the transformer turns ratio.

Analysis for switch closed period

During subinterval DT, while the switch Sw conducts, the equivalent circuit is highlighted in Fig. 2a. The respective values for the inductor voltage v_l, capacitor current i_c and DC input source current i_{in} are as follows.

$$v_l = V_{\text{in}} \tag{1}$$

$$i_c = -\frac{v}{R} \tag{2}$$

$$i_{\text{in}} = i \tag{3}$$

With sufficient values of inductance and capacitance leading to a negligible inductor current ripple and capacitor voltage ripple to boot, the magnetizing current

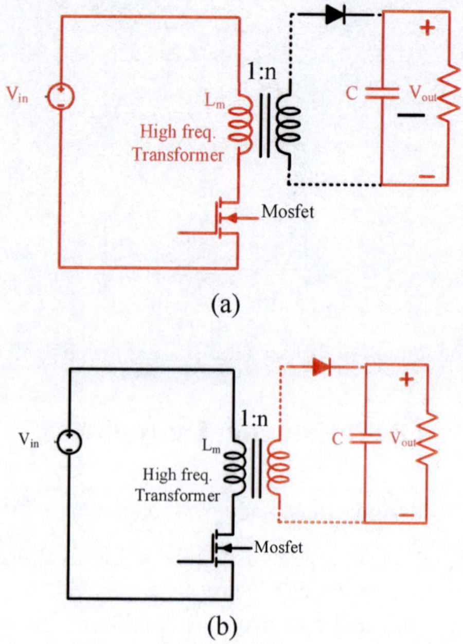

Fig. 2 Two operating modes of a flyback converter **a** switch in ON state and the capacitor supplying the load output voltage, **b** switch in OFF state and the secondary winding supplying energy to the load

i and output capacitor voltage v can be represented by their average DC values I and V. The above equations are transformed into

$$v_l = V_{\text{in}} \tag{4}$$

$$i_c = -\frac{V}{R} \tag{5}$$

$$i_{\text{in}} = I \tag{6}$$

Analysis for the switch open

When the switch is turned OFF throughout the second subinterval, the energy is conveyed to the secondary side of the transformer and the diode starts conducting, as shown in Fig. 2b. The voltage across the magnetizing inductance v_l, the capacitor current i_c and the current supplied by the DC input source are, respectively, given by

$$v_l = -\frac{v}{n} \tag{7}$$

$$i_c = \frac{i}{n} - \frac{v}{R} \tag{8}$$

$$i_{\text{in}} = 0 \tag{9}$$

We obtain the following relations.

During steady-state conditions, the volt-second area for the inductor requires to be zero, therefore we have

$$\langle v_l \rangle = D(V_{\text{in}}) + (1 - D)\left(-\frac{V}{n}\right) = 0 \tag{10}$$

Solving for the input–output voltage conversion ratio

$$\frac{V}{V_{\text{in}}} = n\frac{D}{1 - D} \tag{11}$$

4 Simulation

The simulation was implemented in PLECS® v4.1 environment. The simulation parameters are enumerated in Table 1.

5 Experimental Test Bench Setup

The Schematic of the setup is depicted in Fig. 3. IC 7805 and IC 74122 were used to generate pulses. Power MOSFET IRF260 was employed as the switch. The Diode used on the secondary winding was MUR5060.

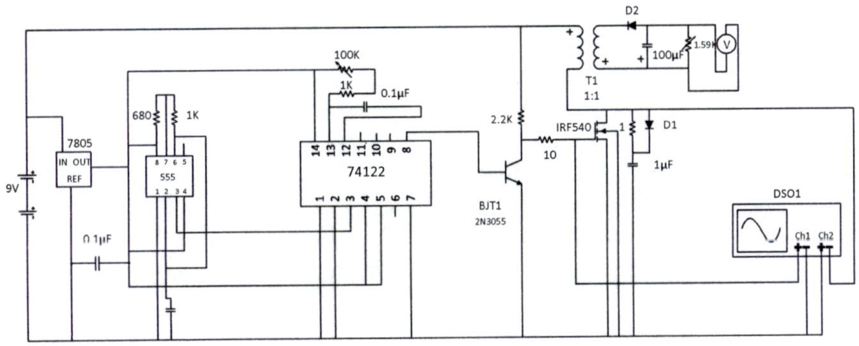

Fig. 3 Hardware setup schematic

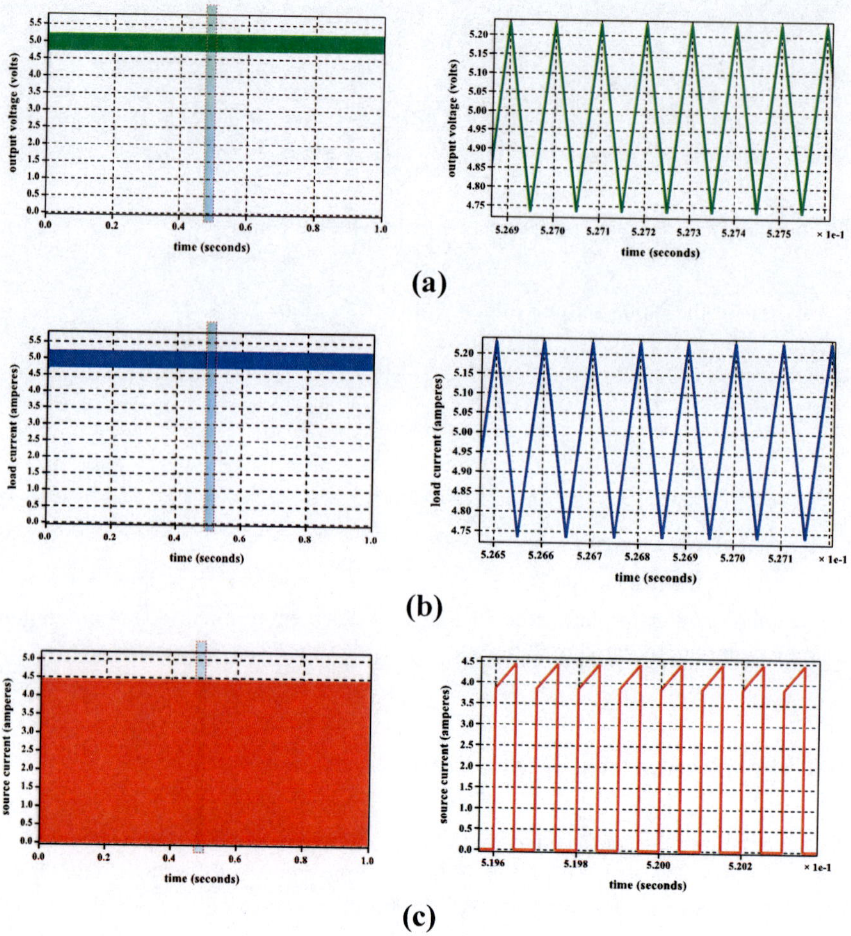

Fig. 4 **a** Load voltage, **b** load current waveform, **c** DC source current

6 Results and Discussion

6.1 Flyback Converter Simulation Results

The output waveforms of the simulation are depicted in Figs. 4 and 5.

6.2 Experimental Verification

The hardware setup for experimental verification is depicted in Fig. 6.

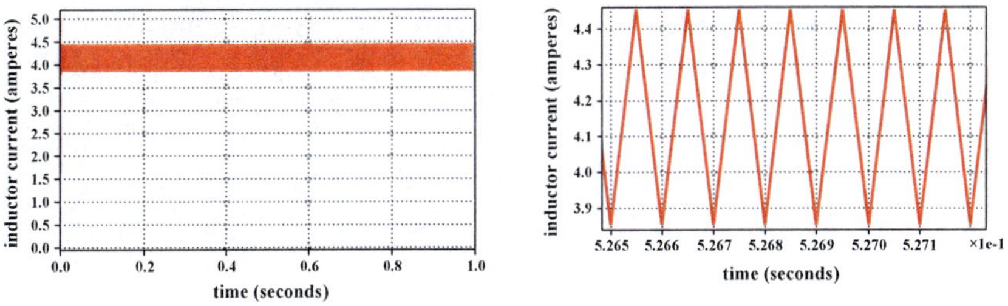

(a)

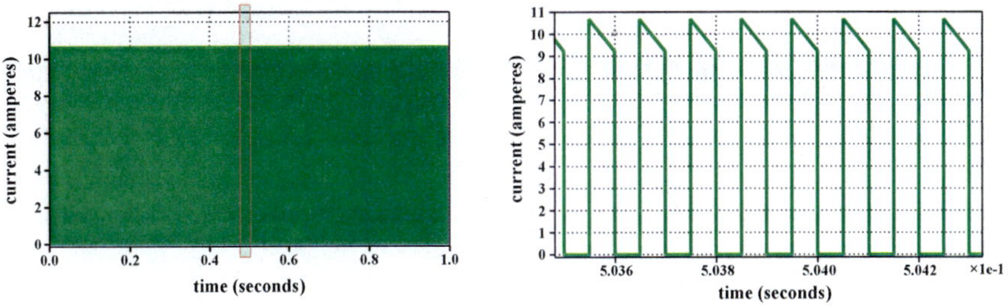

(b)

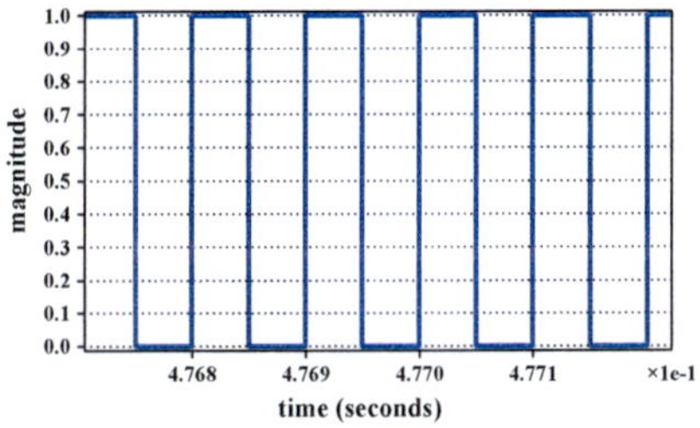

(c)

Fig. 5 Simulation waveforms **a** inductor current waveforms, **b** diode current, **c** switch gating pulses—duty cycle

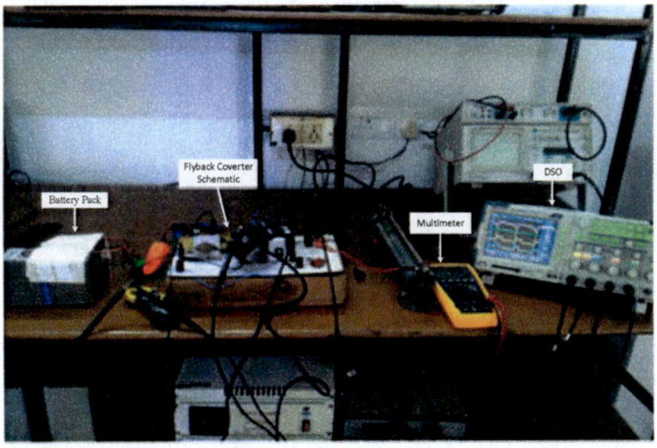

Fig. 6 Hardware setup

In Fig. 6, the numerous output waveforms including load voltage, input DC source current, duty cycle pulses, inductor current, and the output of the ICs 555 and 74,122 are presented. Figure 6a depicts the input and output voltages and the pulses generated by 74,122 and 555 ICs. Spikes present in the waveforms can be spotted due to open-loop operation. Implementing Closed-loop operation can eliminate the spikes. Also, Figs. 7b–e illustrate the enlargement of current ripple for a higher frequency.

7 Conclusion

This paper proposes an experimental hardware setup for flyback DC–DC converter realization on a hardware test bench. The need of experimental-based education in the power electronics engineering sector is demonstrated, and the description and working rules of flyback converter are presented. The circuit is simulated to describe the behavior of the converter. The hardware results are adequate, and the flyback converter operation is conducted on the test bench [7].

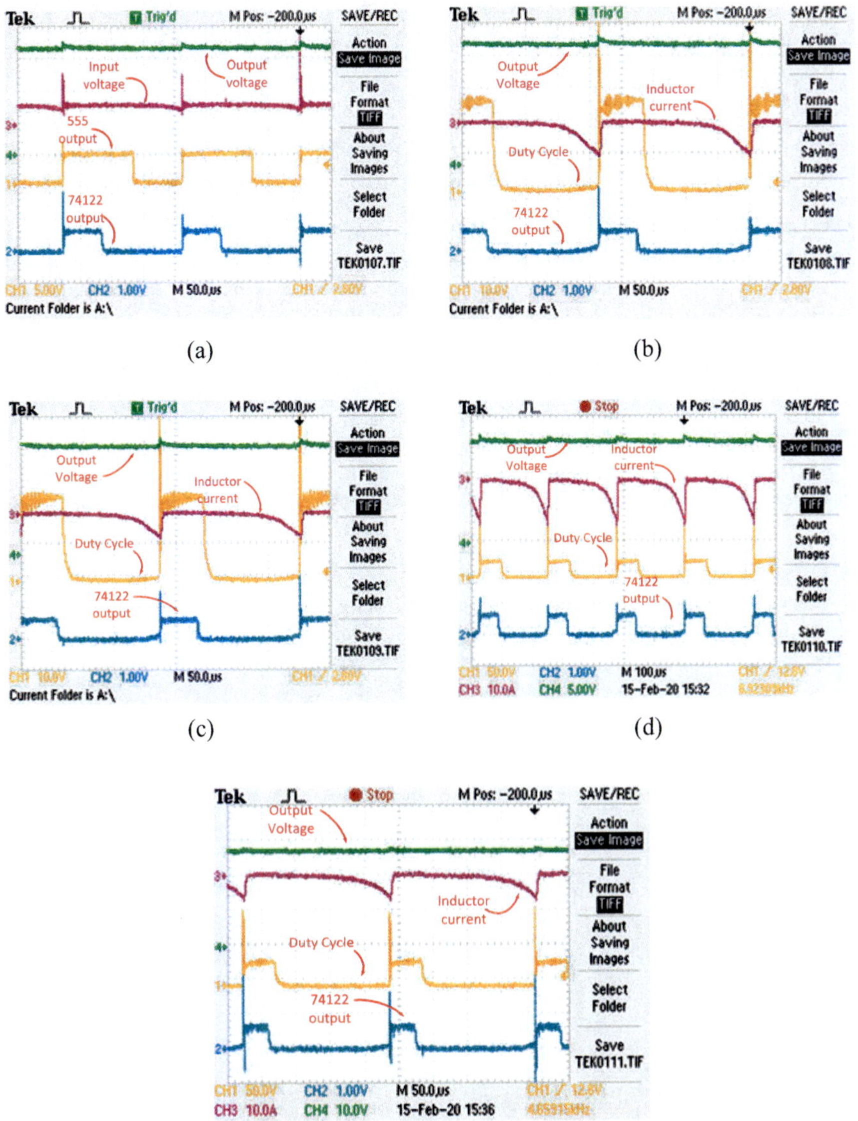

(a) (b)

(c) (d)

(e) load change from 27.5 Ω to 88 Ω

Fig. 7 Output waveforms obtained from experimental setup

References

1. Committee CI (2015) Draft recommended practice for the design and application of power electronics in electrical power systems. In: IEEE Std 1662-2016 (Revision IEEE Std 1662-2008), pp 1–68

2. Leite V, Barbosa J, Teixeira H, Araújo RE (2005) A generic converter for experimentation based power electronics learning. In: 2005 European conference on power electronics and applications, vol 2005, pp 1–10
3. Ayasun S, Nwankpa CO (2005) Induction motor tests using MATLAB/Simulink and their integration into undergraduate electric machinery courses. IEEE Trans Edu 48(1):37–46
4. Sarwar A, Shahid A, Hudaif A, Gupta U, Wahab M (2017) Generalized state-space model for an n-phase interleaved buck-boost converter. In: 2017 4th IEEE Uttar Pradesh section international conference on electrical, computer and electronics, UPCON 2017, vol 2018, pp 62–67
5. Iqbal MT, Tariq M, Ahmad MK, Bin Arif MS (2017) Modeling, analysis and control of buck converter and Z-source converter for photo voltaic emulator. In: 1st IEEE international conference on power electronics, intelligent control and energy systems, ICPEICES 2016
6. Kircioğlu O, Ünlü M, Çamur S (2016) Modeling and analysis of DC-DC SEPIC converter with coupled inductors
7. Iqbal A et al (eds) (2020) Soft computing in condition monitoring and diagnostics of electrical and mechanical systems. In: Advances in intelligent systems and computing, vol 1096. Springer, Singapore. https://doi.org/10.1007/978-981-15-1532-3
8. Iqbal A et al (eds) (2020) Meta heuristic and evolutionary computation: algorithms and applications. In: Studies in computational intelligence, vol 1096. Springer, Singapore. https://www.springer.com/gp/book/9789811575709
9. Ahmad J, Zaid M, Sarwar A, Lin C-H, Ahmad S, Sharaf M, Zaindin M, Firdausi M (2020) A voltage multiplier circuit based quadratic boost converter for energy storage application. Appl Sci 10:8254. https://doi.org/10.3390/app10228254
10. Ahmad J, Lin C-H, Zaid M, Sarwar A, Ahmad S, Sharaf M, Zaindin M, Firdausi M (2020) A new high voltage gain DC to DC converter with low voltage stress for energy storage system application. Electron 9:2067. https://doi.org/10.3390/electronics9122067

Nine-Step Multilevel Inverter Output Analysis Using the EP Approach

Isarar Ahamad and A. J. Ansari

Abstract Multilevel voltage source inverter has a number of advantages over conventional inverter. By combining the different sets of DC voltage sources, stepped AC voltage waveform is generated by Cascaded H-bridge inverter at a higher level with fewer harmonic. When the level of inverter increases, the THD becomes low and inverter produces a good voltage wave form and reduced the cost. Multilevel technology is very attractive for photovoltaic application due to the necessity of various sources on DC side. This paper provides a nine-level topology using equal phase (EP) method. EP method is a method for finding switching angle, phase delay, and pulse width for good waveform and lowering the THD. The nine-level inverter model is implemented on MATLAB/Simulink.

Keywords Multilevel · Economic · THD · Simulink model

1 Introduction

The sinusoidal output waveform adds additional step when DC voltage sources at disparate ranges increase [1–12]. For higher-power applications such as compound system, solar system, and flexible AC transmission systems, multilevel inverters are used. In [1, 2], by rising the voltage level, it reduces the need for filters, and efficiency is high due to fewer harmonics. For low-power applications, switching frequencies are not as limited as in high-power applications. In [3], the different control methods can be implementing for reducing harmonics in staircase waveforms. The multilevel inverter is basically classified into diode clamped, flying capacitor, and cascaded multilevel inverter. The control of cascaded multilevel inverter is very easy as compared to diode clamp and flying capacitor. The control methods are such as modulation of multicarrier pulse widths or multiple hysteresis bands control. In [4,

I. Ahamad (✉) · A. J. Ansari
EED, Integral University, Lucknow, India
e-mail: israr@iul.ac.in

A. J. Ansari
e-mail: ajansari@iul.ac.in

A. Iqbal et al. (eds.), *Renewable Power for Sustainable Growth*, Lecture Notes in Electrical Engineering 723, https://doi.org/10.1007/978-981-33-4080-0_38

Fig. 1 H-bridge multilevel inverter

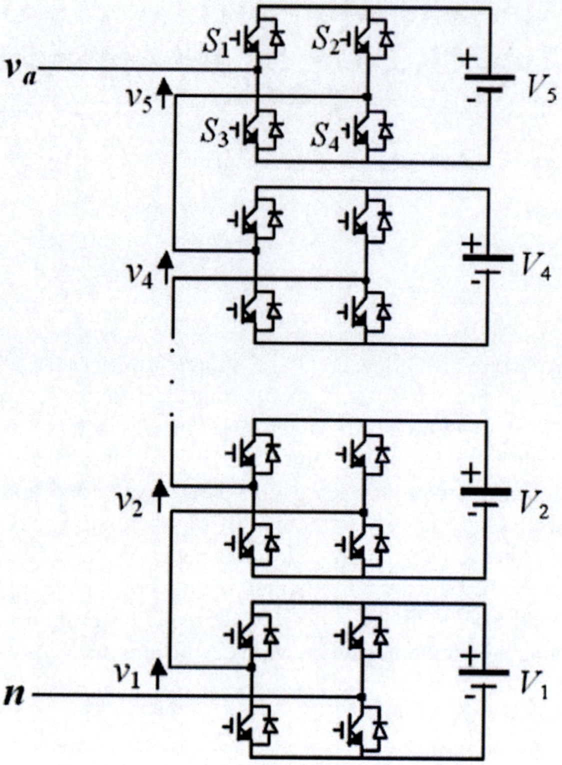

5], due to their cell structure, for solar PV systems, the multilevel inverter is ideal. That solar array offers varying levels of DC voltage. A multilevel inverter (MLI) achieves lower costs, high output power, and less THD. The fuel cells in high-power applications are used with the solar PV module, and we can conveniently implement the H-bridge inverter in cascaded manner as displayed in Fig. 1.

2 Modeling of Multilevel Inverter Using EP Method

The firing angles are taken in 0–Π range in equal step process displayed in Fig. 2. The key switching angles are obtained by formula below.

$$\text{Number of H bridges}(N) = (m - 1)/2$$
$$\theta_i = i \times 1800/m$$

and $i = 1, 2, \ldots, (m - 1)/2$
m = level of output voltage,
θ = switching angle.

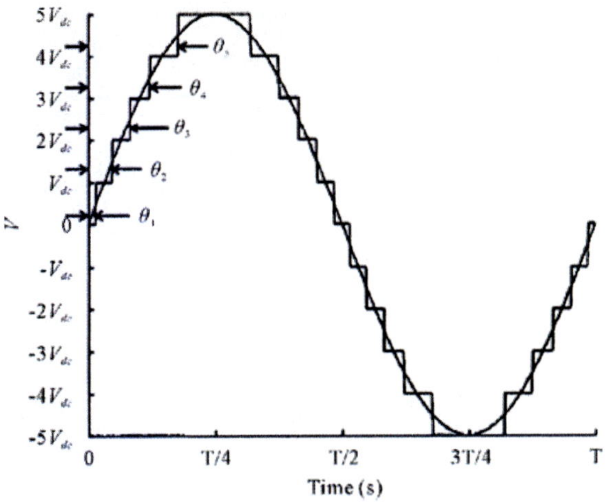

Fig. 2 Multilevel inverter output voltage and its switching angle

3 Working of Nine-Stage Multilevel Inverter

The inverter model is shown in Fig. 3 with each section of the loop, and the inverter uses basically four legs and each leg contains H-bridge with a DC source.

To know the system in [6] operation, look at basic simulation of single-phase topology shown in Fig. 5. The H-bridges in series connected by four different inverter DC generating units. Eventually, in [1], the method of calculating the switching angles, switching times, and pulse width are used in equal phase technique. Figure 6 displays nine-level output waveform. From Fig. 7, we can see the nine-level inverter waveform and spectrum of total harmonic.

4 Switching Technique of Nine Level Inverter

Modulation techniques like conventional sine pulse width modulation (SPWM) and PWM space vector technique for particular harmonic removal and active harmonic removal can be used to control inverter modeling [7, 8] by the equal phase method. In this proposed system, this equivalent phase (EP) method is applied. This switching system provides a switching time, switching angle, and pulse width of semiconductor switches with different firing pulse to ON and OFF the switches for the required switching frequency. Pulse width can be calculated and implemented [9] at the following switches given in Tables 1 and 2. The flowchart is given in Fig. 4 to obtain the procedure to obtain the switching angle for cascaded H-bridge multilevel inverter. This flowchart is an iteration process to find the best output value for the inverter where harmonics become less.

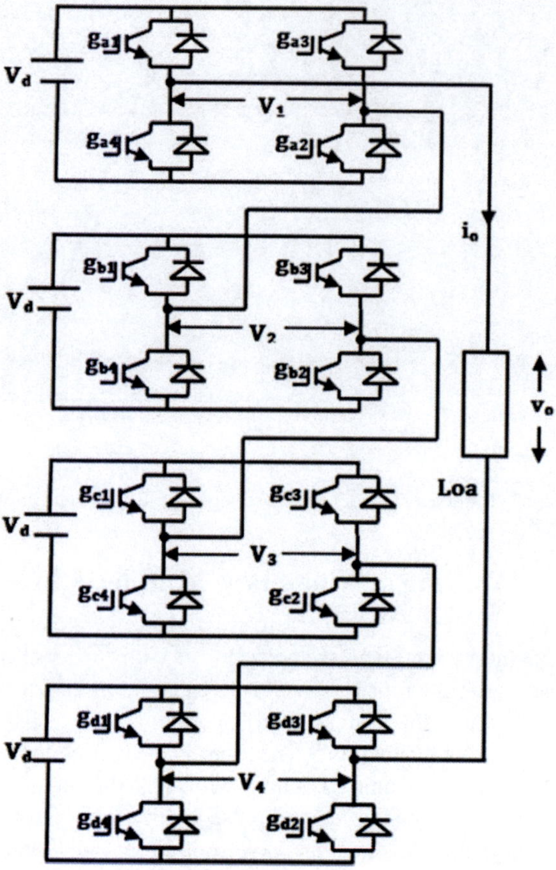

Fig. 3 Nine-level cascaded H-bridge multilevel inverter

5 Simulation Result

There are typically multiple modulation control systems [10] on the basis of switching frequency the multilevel inverter is used. The suggested model is based on MATLAB/Simulink. The switching angles vary in the order of 0–Π in the equal phase process.

Each connection contains one single H-bridge inverter. Like this there is need of four H-bridges with sixteen switches. From each switch, we give the following time delay, switching angles, and pulse width as shown in Tables 1 and 2 to obtain the nine-level inverter.

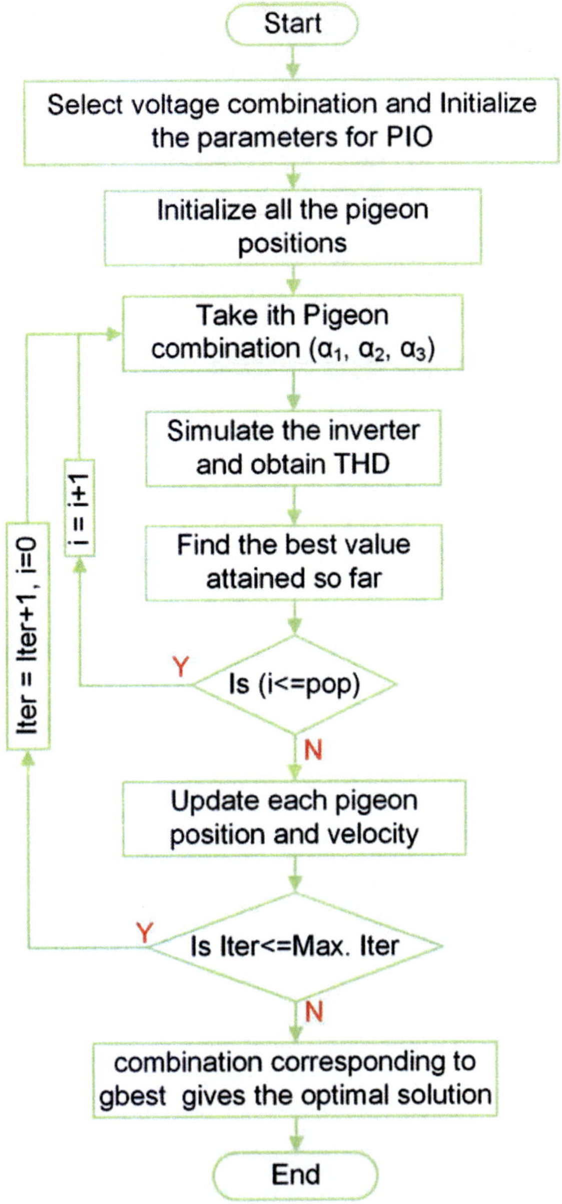

Fig. 4 Obtaining switching angle for multilevel inverter

6 Conclusion

The paper is designed and implemented for a resistive load of nine-level H-bridge multilevel inverter in cascaded manner by equal phase (EP) method. Multilevel inverters give an practical solution in effective manner for increasing power and

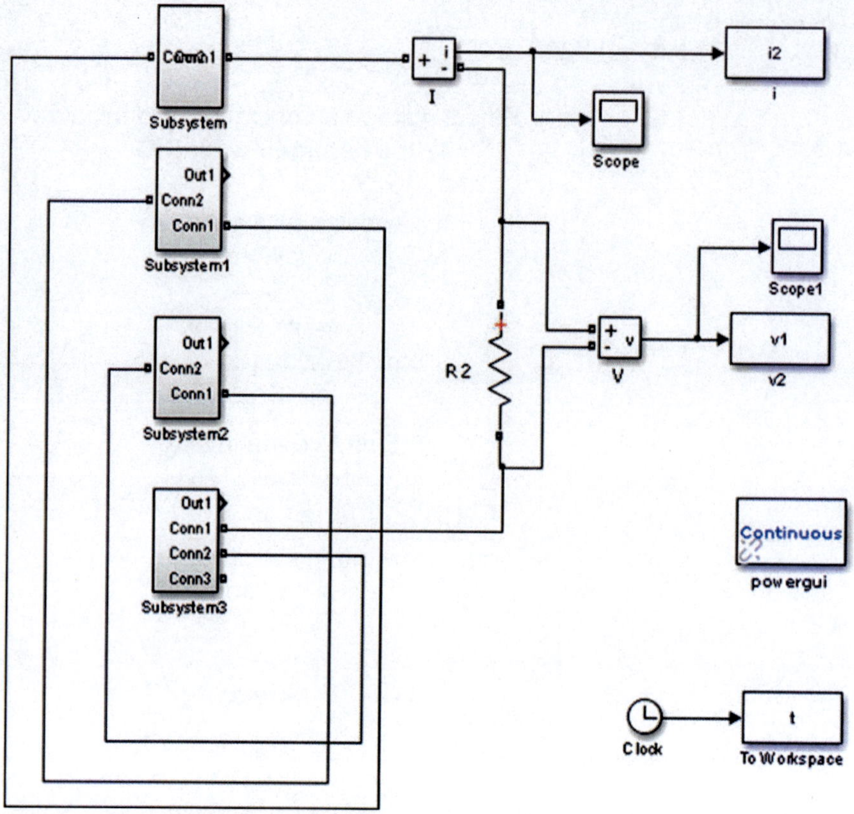

Fig. 5 Simulink model of nine-level inverter

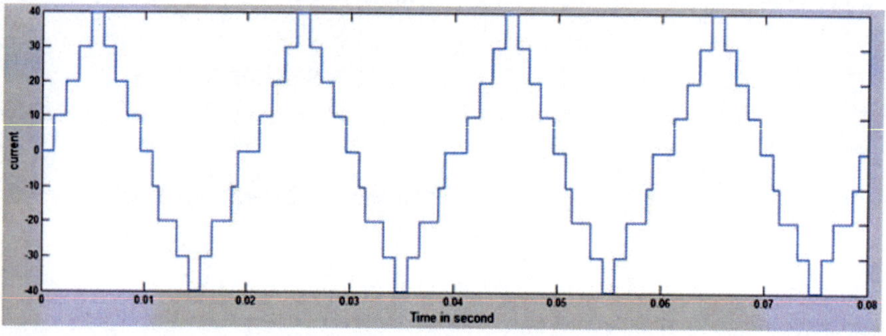

Fig. 6 Output wave form using EP method

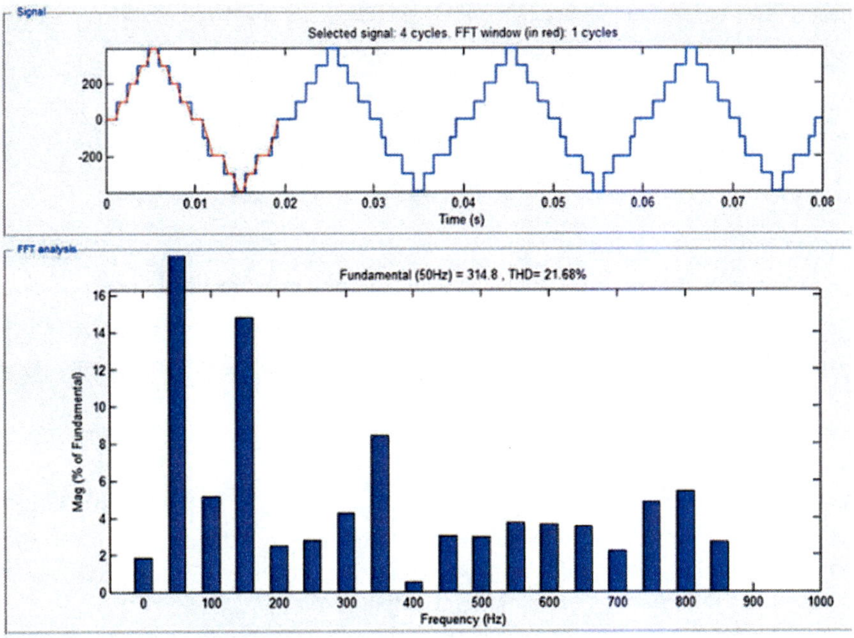

Fig. 7 THD waveform using EP method

Table 1 Switching state to obtain positive pulse

Pulse No.	Angle	Time	$S_1, S_2, S_6, S_{10}, S_{14}$	S_5	S_9	S_{13}
0	21.8	0.0012				
V	43.5	0.0024	PW = 41.76			
2 V	65.3	0.0036		PW = 29.41		
3 V	87	0.0048			PW = 17.64	
4 V	109	0.0060				PW = 5.88
3 V	130.5	0.0072				
2 V	152.3	0.0084				
V	173.6	0.0096				

reduce total harmonics distortion (THD). The nine-level inverter simulation model is simulated along with a harmonic analysis. The results of the simulation model indicate that when the level of output increases, output response of developed multilevel H-bridge inverter is improved and the harmonic distortion is decreased.

Table 2 Switching state to obtain negative pulse

Pulse No.	Angle	Time	$S_4, S_8, S_{11}, S_{12}, S_{15}$	S_9	S_7	S_3
0	195.8	0.0108				
−V	217.6	0.0120	PW = 41.76			
−2 V	239.4	0.0132		PW = 29.41		
−3 V	261.2	0.0144			PW = 17.64	
−4 V	282.9	0.0156				PW = 5.88
−3 V	304.6	0.0168				
−2 V	326.4	0.0180				
−V	348.2	0.0192				
0	360	0.02				

7 Future Scope

The paper can be implemented to multilevel inverter level by reducing DC sources and switches and other components to make reliable, improved efficiency, and inverter cost effective for obtaining the same level output result.

Acknowledgements The author would like to acknowledge Integral University for providing the MCN number "IU/R&D/2020–MCN000838".

References

1. Rodriguez J, Lai JS, Peng FZ (2002) Multilevel inverters: a survey of topologies, controls, and applications. IEEE Trans Ind Electron 49(4):724–738
2. Franquelo VJ, Rodriguez J, Leon I, Kouro S, Portillo R, Prats MAM (2008) The age of multilevel converters arrives. IEEE Ind Electron Mag 2(2):28–39
3. Duarte JL, Lokos J, van Horck FBM (2013) Phase-shift-controlled three-level converter with reduced voltage stress featuring ZVS over the full operation range. IEEE Trans Power Electron 28(5):2140–2150
4. Minai AF, Usmani T, Mallick MA (2015) Optimum sizing and estimation of a 30 kWp hybrid solar photovoltaic system with multilevel inverter. Int J Res Sci Innov (IJRSI) 31–36. ISSN 2321-2705
5. Asim M, Tariq M, Mallick MA, Ashraf I, Kumari S, Bhoi AK (2018) Critical evaluation of offline MPPT techniques of solar PV for stand-alone applications. In: Advances in smart grid and renewable energy. Springer, Singapore, pp 13-21
6. Usha Nandhini MS, Vinothini N, Prakash R (2012) Modified H-bridge nine level inverter with low switching frequency. IJACT 3(2):1–5
7. Minai AF, Tariq A (2011) Analysis of Cascaded multilevel inverter. In: International conference on power electronics (IICPE). IEEE, pp 1–6
8. Ahamad I, Asim M, Sarkar PR, Khan FA (2016) Comparison of conventional PFC boost converter and bridgeless PFC boost converter. Int J Innov Res Instr Control Eng 4(5):5
9. Naderi R, Rahmati A (2008) Phase-shifted carrier PWM technique for general cascaded inverters. IEEE Trans Power Electron 23:1257–1269

10. Malinowski M, Gopal Kumar K, Rodriguez J, Pérez MA (2010) A survey on cascaded multilevel inverters. IEEE Trans Ind Electron 57(7):2197–2206
11. Iqbal A et al (eds) Soft computing in condition monitoring and diagnostics of electrical and mechanical systems. In: Advances in intelligent systems and computing, vol 1096. Springer, Singapore. https://doi.org/10.1007/978-981-15-1532-3
12. Iqbal A et al (eds) Meta heuristic and evolutionary computation: algorithms and applications. In: Studies in computational intelligence, vol 1096. Springer, Singapore. https://www.springer.com/gp/book/9789811575709

MATLAB-Based Modeling and Simulations for the Low- and High-Temperature Module Power Generation of PV Panels in Kuala Lumpur and Genting Highlands, Malaysia

Qamar Ul Islam and Fatemeh Khozaei

Abstract As a source of green energy, solar energy provides a clean way for the production of heat and electricity. PV solar panels, in particular, have extensively applied for the production of electricity with reasonable operating costs and fewer environmental in comparison to fossil fuels. Previous studies suggest that the PV panel's temperature decreases its power generation. The current study compares the productivity of photovoltaic panels in the capital city of Kuala Lumpur with the coolest city of Malaysia namely Cameron Highlands. Typical KC200GT Photovoltaic Panel was used as a common PV model, and the analysis was performed in MATLAB Editor. When giving the information on solar irradiation and panel temperature, the model output illustrated the I-V, P-V, and P-I characteristics. The result of the study confirms that the photovoltaic panel power output is significantly higher in the coolest city of Malaysia (Cameron Highlands) comparing the capital city (Kuala Lumpur).

Keywords Efficiency · MATLAB simulation · Temperature · Solar energy · Algorithm

1 Introduction

In any discussion about the future of the earth, global warming usually tops the list. The negative effects of climate change are clear to everyone, and for decades, the developed countries have been attempted to find a solution accordingly. For the last centuries, we have been dependent heavily on the sources of energy such as coal and oil while burning such fuels creates a great number of green gases [1–39]. When it

Q. U. Islam (✉)
Department of Electrical Engineering, College of Engineering and Technology, School of Engineering and Technology, Baba Ghulam Shah Badshah University, Rajouri, J&K, India
e-mail: qamarulislam@bgsbu.ac.in

F. Khozaei
Department of Architecture, Kerman Branch, Islamic Azad University, Kerman, Iran
e-mail: fkhozaei@iauk.ac.ir

A. Iqbal et al. (eds.), *Renewable Power for Sustainable Growth*, Lecture Notes in Electrical Engineering 723, https://doi.org/10.1007/978-981-33-4080-0_39

comes to fighting global warming and replacing fossil fuels with renewable energies, most countries despite their other available sources count on solar energy [13] as one of the cleanest and accessible sources of energy. For collecting solar energy and converting solar radiational to electricity, various photovoltaic panels have invented and used globally. Each panel consists of several cells that create an electric current by exposing it to the sunlight. The photovoltaic effect was initially discovered in the 1830s and paved the way for the creation of photovoltaic panels. PV cells such as semiconductor materials collect the sun radiations and convert the light photons to electricity. Solar energy can be applied for not only the generation of electricity but heating and desalination of water as well as cooking.

While the application of solar panels is increasing day after day, the current PV panels demand high investment cost, and returning of profit for the investors takes over 5 years. In terms of power generation and efficiency as well as energy storage systems, there are still a great number of improvements are required. These factors besides the lack of balance between supplied and demanded amounts of electricity [13] are some of the barriers to the vast usage of solar energy in the current time, though it must be noted that the PV panels that are available in the market currently are noticeably more affordable than their earlier generations that bring the hope for near future.

The scholars in the last few years have attempted to optimize the efficiency of PV panels. The current PV panels convert the sun lights more to heat than electricity and their power efficiency is commonly lower than expectations. Beside the PV panel, performance as well as efficiency decreases by heating which causes increasing the temperature of panels [6]. Studies even suggest that a little increase in the temperature of PV panels can notably decrease their performance and efficiency (Kumar et. al. 2007). Accordingly minimizing the heat gain of PV panels should be taken into account for assuring improved efficiency [5]. The importance of PV panels cooling becomes even more clear when canonizing the fact that the available PV panels in the market currently only have an efficiency of 6–20% and hence about 80% of received solar energy is wasted converting to heat. It has repeatedly confirmed that roughly increasing the temperature of PV panels for each 1 °C can predict decreasing of .5 efficiency [2, 16, 18] and the temperature over 40 °C can significantly decrease the Panel's efficiency [27]. As with the increase in the solar irradiance daily solar panel temperature also increases gradually [4, 24] therefore photovoltaic panels require an efficient cooling system to keep an appropriate operating temperature.

To get closer to daily use, it is necessary to minimize heat loss. Various strategies have been used for cooling photovoltaic panels such as air cooling, heat sink, and water cooling [5, 25]. However, the another strategy has been finding the best locations for harvesting solar radiation. Cities with a great number of shining days and at the same time cool weather can be considered the most ideal place for such a purpose.

Considering the highest and lowest temperatures in the various cities of a country provides a picture of the efficiency and outputs of the modules. The current study attempts to compare the photovoltaic panel efficiency in the general for the highest and lowest annual temperature for the capital of Malaysia, Kuala Lumpur, and the coolest region of the country so-called Cameron Highlands. Using MATLAB software version R2018a, the study hence simulates the output of the photovoltaic panel and presents voltage, current, and power curves.

2 Literature Review

The photovoltaic panels temperature can be considered as an essential parameter that affects their conversion performance and efficiency, the higher the operating temperature, the lower would be the output [1–39]. Various strategies for cooling PV panels have tested, and the result for efficiency increase has stated. Evaluating the performance of PV panels applying aluminum fins El Mays et. [20] found an improved efficiency of up to 17.7%. Chandrasekar and Senthilkumar [10] studied the temperature of a finned photovoltaic panel associated with cotton wicks, which resulted in increasing efficiency up to 14% by lowering the surface temperature of PV panels. Studies suggest that the cooling of PV panel boards can expand their effectiveness by 14.3% and hardware costs recoup in 8–7 years with this framework [31].

Some studies have relied on the numerical models for examining cooling strategies on PV panels efficiency [25]. Applying air-cooled heat sinks on the rear of PV boards comprise of ribbed dividers with explicit measurements can increase the efficiency of panels up to 16% [25]. Studies suggest that reducing the temperature of PV panels for about less 30 °C can enhance the efficiency of the panels up to 14% [23]. Proell et al. [26] have given another idea to improve the warm productivity of PV/T by concentrating daylight with explanatory concentrator reflectors to accomplish higher brilliant force have been done at ZAE Bayern in Garching Germany.

Gotmore and Borkar [14] researched the presentation improvement of photovoltaic boards utilizing latent convection cooling under common convection. They attempted a temporary exploratory investigation on photovoltaic boards with and without balance cooling to consider the impact of the working temperature on the voltage–current and yield intensity of the created board. They evaluated this method successfully as it was evidenced by the effective dropping of PV panels temperature and consequently 5.5% the efficiency of the panel was increased.

Cuce et al. [9] tentatively analyzed the impacts of uninvolved cooling on sun-oriented cell execution parameters and their outcomes indicated that the force cell vitality and vitality efficiencies of the PV cell increment significantly with the strategy proposed cooling. Expansion of about 20% in the yield intensity of the PV cell is acquired with a radiation state of 800 W/m^2. The most extreme cooling level is watched for a power level of 600 W/m^2. Mojumder et al. (2016) dissected the presentation examination of an incorporated air-cooled photovoltaic warm gatherer framework with cooling blades. Their conclusive outcomes uncovered that the greatest warm effectiveness and most extreme productivity were separately 56–19 and 13–75% utilizing four blades at a mass progression of 0.14 kg/s and sun-oriented radiation of 700 W/m^2. To the moment yet the expense of the hardware is generally less affordable than cooling of PV panels by air.

Some studies have attempted to examine the effect of air cooling of PV panels using a large number of fans. For example, Kaiser et al. [19] used this technique and air conduit size and stream speed were taken as an independent variable while the efficiency of PV boards was taken as a dependent variable. Tonui and Tripanagnostopoulos [28] attempted to examine cooling a photovoltaic board using a 0.1 m wide ventilated channel behind it. Subsequently photovoltaic/warm frameworks PV/T is made which speaks to the ideal arrangement utilizing photovoltaic boards with the beneficial job of delivering heat vitality. This component is extremely essential because two kinds of vitality are the most utilized power and warmth. Additionally, each ideal capacity carries a resource for the other.

While suggesting this method can positively perfume in terms of PV board increased efficiency, adding water cooling strategies and splashing water on the boards can enhance the system performance even more effectively [1, 6, 11, 12]. The circulation of water at the rear side of a typical PV panel is also found as a good means for reducing its temperature [17, 30]. Among all these strategies, water shower cooling is considered to be the most effective one [23].

While numerous studies have been conducted on the PV modules performance, it must be noted that their output in the laboratory and under the daily circumstances of the outdoor environment cannot be completely equal. Radiation loss, reflection, shadow, dirt, and dust can reduce the efficiency of PV panels. Besides temperature incompatibility, cable AC/DC, and inverter role in the performance of PV panels should not be overlooked [5]. Accordingly, some studies have attempted to measure the efficiency of PV panels in the outdoor environment and precisely examine the effects of the sun and surrounding environmental temperatures [8].

Simulation of PV panels modules using MATLAB software has been considered a precise way for examining their performance and the effect of various factors such as shadow, heat, and cooling. This method has been used by numerous scholars [32–37], and the current study applies the same approach for comparing the PV panels efficiency and performance in two different cities of Malaysia. The further parts of this paper will discuss the research methodology, data analysis, and findings of the research. It will also explain in detail the reason behind case studies as well.

3 Research Methodology

The warmest and coolest areas of Malaysia, as well as Kuala Lumpur, were examined. The current output of the PV panels was examined in the highest and lowest temperature recorded. The highest temperature of Kuala Lumpur was recorded during May

Table 1 Power output of PV module at different temperature in Kuala Lumpur & Cameron highlands, Malaysia

Temperature description	Kuala Lumpur	Cameron highlands
Highest Temperature (T_h)	84 °F (302.039 K)	57.2 °F (287.15 K)
Power in highest temperature (P_h)	429 W, 13.41 A, 32 V, 100 W/m^2	461 W, 13.56 A, 34 V, 100 W/m^2
Lowest temperature (T_l)	82 °F (300.928 K)	53.6 °F (285.15 K).
Power in lowest temperature (P_l)	431.6 W, 13.49 A, 32 V, 100 W/m^2	465.6 W, 13.69 A, 34 V, 100 W/m^2

as 84 °F (302.039 K) and the coldest month was recorded to be December with the average 82 °F (300.928 K) **avg. the coldest**. On average, the warmest month is May 57.2 °F (287.15 K). On average, the coolest month is July in July is 53.6 °F (285.15 K).

4 Findings

As shown in Table 1, the module output power increases as the PV module temperature decreases. This simulation study results support the previous studies indicating the curve Pmax changes from the 431.6 to 465.6 W as temperature changes from 82 °F (300.928 K) to 53.6 °F (285.15 K). Accordingly, the result of this study supports the previous research that the geographical location of photovoltaic panels has a significant effect on the PV panel's power generation (Figs. 1 and 2).

5 Conclusion

Solar energy is a clean source for the production of heat and electricity. In recent years, PV solar panels have been used widely for harvesting solar energy and creation of electricity with almost minimum operating costs and fewer environmental impacts than conventional power plants that use fossil fuels. Previous studies suggest that the PV panel's temperature decreases its power generation. The reliance on transformation productivity and the temperature of photovoltaic cells is a pivotal report space for scientists. The studies affirm a reduction in productivity while the temperature of PV cells rises [7]. Most examinations attest that the delivered electricity changes directly with the working temperature. The normal productivity decrease is around 0–45% for each degree over 25 °C. The electrical force reference delivered by photovoltaic boards is resolved for the normal cell temperature of 25 °C and the extent of sun-powered radiation 1000 W/m^2. Right now, cells produce the greatest force estimated in Watt top Wp. This worth is reasonable for makers however in genuine

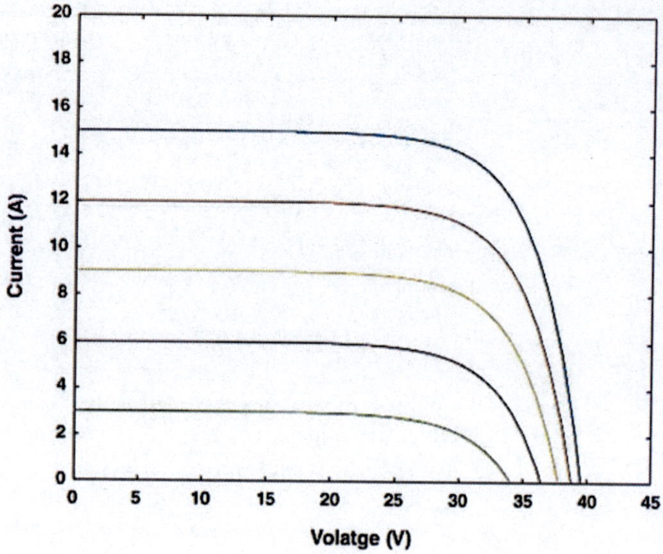

Fig. 1 I-V characteristics of PV module for highest temperature condition at $T = 84$ °F (302.039 K) and $T_r = 298$ K with irradiance ($G = 20$, 40, 60, 80 and 100 W/m^2) in Kuala Lumpur, Malaysia

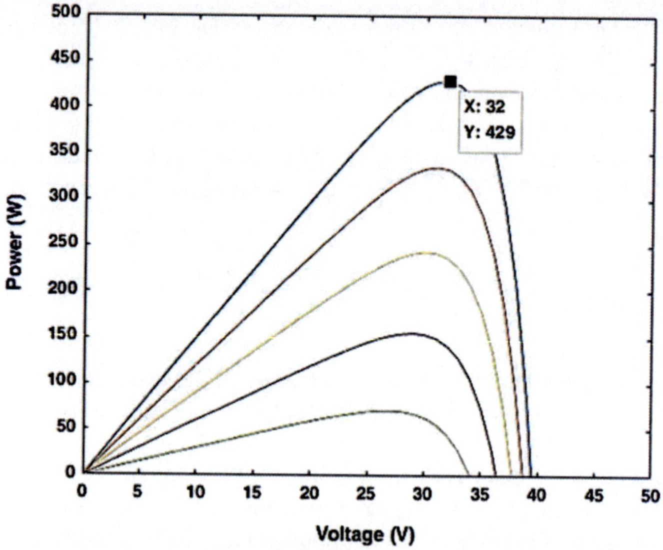

Fig. 2 P-V characteristics of PV module for highest temperature condition at $T = 84$ °F (302.039 K) and $T_r = 298$ K with irradiance ($G = 20$, 40, 60, 80 and 100 W/m^2) in Kuala Lumpur, Malaysia

working conditions the quality of PV cells relies upon changes in temperature and level of sun oriented radiation.

The current study aimed to examine the power efficiency differences and the photovoltaic module between the capital and coolest city in Malaysia. Typical KC200GT Photovoltaic Panel was used, and the analysis was performed in MATLAB Editor. Such a generalized PV model is easy to be used for the implementation of MATLAB modeling and simulation platform. Given solar irradiation and panel temperature, the model returns an I-V, P-V, and P-I characteristics. The result of the study confirms that the photovoltaic panel power output is significantly higher in the coolest city of Malaysia (Cameron Highlands) comparing the capital city (Kuala Lumpur). The result of the study accordingly is in line with previous studies stating increasing module temperature decreases the panel power output. Further studies might concentrate on the best cooling systems for PV panels in Malaysia.

References

1. Abdolzadeh M, Ameri M (2009) Improving the effectiveness of a photovoltaic water pumping system by spraying water over the front of photovoltaic cells. Renew. Energ. 34:91–96
2. Alboteanu et al (2013) Cooling system for the photovoltaic module. Recent Res Environ Geol Sci 133-138
3. Andreev VACM, Grilikhes VA, Rumiantsev VD (1997) Photovoltaic conversion of concentrated sunlight. Wiley
4. Bahaidarah et al (2013) Performance evaluation of a PV (photovoltaic) module by back surface water cooling for hot climatic conditions. Energy 59:445-453
5. Bayrak F, Oztop HF, Selimefendigil F (2019) Effects of different in parameters on temperature and efficiency for cooling of photovoltaic panels under natural convection. Sol Energ 188:484–494
6. Bigorajski, Chwieduk D (2018) Analysis of a micro photovoltaic/thermal e PV/T system operation in a moderate climate. Renew Energ. https://doi.org/10.1016/j.renene.2018.01.116 in press
7. Borkar DS, Prayagi SV, Gotmare J (2014) Performance evaluation of photovoltaic solar panel using thermoelectric cooling. Int J Eng Res 3(9):536–539
8. Chen et al (2015) Simulation and experimental analysis of the optimal buried depth of the vertical U-tube ground heat exchanger for a ground-coupled heat pump system. Renew Energ 73:46-54
9. Cuce E, Bali T, Sekucoglu SA (2011) Effects of passive cooling on the performance of silicon photovoltaic cells. Int J Low-Carbon Technol 1–10
10. Chandrasekar M, Senthilkumar T (2016) Passive thermal regulation of flat PV modules by coupling the mechanisms of evaporative and fin cooling. Heat Mass Transf 52:1381–1391. https://doi.org/10.1007/s00231-015-1661-9 UndStoffuebertragung
11. Chandrasekar et al (2015) A review on the thermal regulation techniques for nonintegrated flat PV modules mounted on building top. Energ Build 86:692-697
12. Elnozahy et al (2015) Performance of a PV module integrated with standalone building in hot arid areas as enhanced by surface cooling and cleaning. Energ Build 88:100-109
13. Fakouriyan S, Saboohi Y, Fathi A (2019) Experimental analysis of a cooling system effect on photovoltaic panels' efficiency and its preheating water production. Renew Energ 134:1362–1368
14. Gotmare JA, Borkar DS, Hatwar PR (2015) Experimental investigation of PV panel with fin cooling under natural convection. Int J Adv Technol Eng Sci 447–454

15. Ibrahim, Fudholi A, Sopian K, Othman MY (2014) Efficiencies and improvement potential of building integrated photovoltaic thermal (BIPVT) Syst. Energ Convers Manag 77:527–534
16. Jakhar S et al (2016) Historical and recent development of concentrating photovoltaic cooling technologies. Renew Sustain Energ Rev 60:41-59
17. Jakhar S et al (2017) An integrated photovoltaic thermal solar (IPVTS) system with earth water heat exchanger cooling: energy and exergy analysis. Sol Energ 157:81-93
18. Kane et al (2017) Optimization of thermoelectric cooling technology for active cooling of a photovoltaic panel. Renew Sustain Energ Rev 75:1295–1305
19. Kaiser et al (2014) Experimental study of cooling BIPV modules by forced convection in the air channel. Appl Energ 135:88-97
20. El Mays A, Ammar R, Hawa M, Akroush MA, Hachem F, Khaled M et al (2017) Improving photovoltaic panel using a finned plate of aluminum. Energ Procedia 119:812–817. https://doi.org/10.1016/j.egypro.2017.07.103
21. Michael JJ, Iniyan S, Goic R (2015) Flat plate solar photovoltaic thermal (PV/T) systems: a reference guide. Renew Sustain Energ Rev 62–38
22. Moharram KA et al (2013) Enhancing the performance of photovoltaic panels by water cooling. Ain Shams Eng J 4:869-877
23. Nižetić S et al (2016) Water spray cooling technique applied on a photovoltaic panel: the performance response. Energ Convers Manag 108:287-296
24. Peng et al (2017) Cooled solar PV panels for output energy efficiency optimization. Energ Convers Manag 150:949-955
25. Popovici, Hudis SV¸ Mateescu TD, Chereche NC (2016) Efficiency improvement of photovoltaic panels by using air cooled heat sinks. Energ Proc 85:425–432
26. Proell, Osgyan P, Karrer H, Brabec CJ (2017) Experimental efficiency of a low concentrating CPC PVT flat plate collector. Sol Energ 147:463–469
27. Rajput, Yang J (2018) Comparison of the heat sink and water type PV/T collector for polycrystalline photovoltaic panel cooling. Renew Energ 116:479-491
28. Tonui JK, Tripanagnostopoulos Y (2007) Improved PV/T solar collectors with heat extraction by forced or natural air circulation. Renew Energ 32:623–637
29. Tselepis S, Tripanagnostopoulos Y Economics analysis of hybrid photovoltaic/thermal solar systems and comparison with standard PV modules. In: Proceedings of the international conference PV in Europe, pp 7–11
30. Wang Y et al (2017) Experimental study on direct-contact liquid film cooling simulated dense-array solar cells in a high concentrating photovoltaic system. Energ Convers Manag 135:55-62
31. Yang L-H, Liang J-D, Hsu C-Y, Yang T-H, Chen S-L (2019) Enhanced efficiency of photovoltaic panels by integrating a spray cooling system with a shallow geothermal energy heat exchanger. Renew Energ 134:970–981
32. Bellia H, Youcef R et al A detailed modeling of photovoltaic module using MATLAB. NRIAG J Astron Geophys 3(1):53–61
33. Revati D, Natarajan E I-V and P-V characteristics analysis of a photovoltaic module by different methods using matlab software. Mater Today Proc
34. Reddy GS, Reddy TB et al A MATLAB based PV module models analysis under conditions of nonuniform irradiance. Energ Procedia 117:974–983
35. Nguyen XH, Nguyen MP Mathematical modeling of photovoltaic cell/module/arrays with tags in Matlab/Simulink. Environ Syst Res 4(1):24
36. Da Silva RM, Fernandes JLM Hybrid photovoltaic/thermal (PV/T) solar systems simulation with Simulink/Matlab. Solar Energ 84(12):1985–1996
37. Fekkak B, Menaa M et al Control of transformerless grid-connected PV system using average models of power electronics converters with MATLAB/Simulink. Solar Energ 173:804–813
38. Iqbal A et al (eds) (2020) Soft computing in condition monitoring and diagnostics of electrical and mechanical systems. In: Advances in intelligent systems and computing, vol 1096. Springer, Singapore. https://doi.org/10.1007/978-981-15-1532-3
39. Iqbal A et al (eds) (2020) Meta heuristic and evolutionary computation: algorithms and applications. In: Studies in computational intelligence, vol 1096. Springer, Singapore. https://www.springer.com/gp/book/9789811575709

Tuning of Controllers for a Boost Converter Used to Interface Battery Source to BTS Load of a Telecommunication Site

Mohd Khursheed, M. A. Mallick, and A. Iqbal

Abstract To enhance the reliability of supply to the remote load like base transceiver station (BTS) load, battery storage sources are used. However, selection of suitable number of battery units and regulation of power using a DC-DC converter is still a challenging task. A very few papers are reported in this direction. To fill this gap, method to select the number of battery units required for a given BTS load and the designing of current and voltage controller for interfacing the battery source via boost converter to the load is discussed. Bode plot analysis is carried out for the tuning of inner current and outer voltage controllers to regulate the power supply. The performance has been analyzed using MATLAB/Simulink.

Keywords Battery storage · Converter · BTS load · Current controller · Voltage controller

1 Introduction

Balancing and stabilizing power [1–17] is an exigent task in microgrids. There are many methods available which deal with this challenge, which include shedding of load, sources unloading, and combining battery storage with distributed sources for managing power when energy is in abundance or in shortage [4]. The units for storage are situated near every distributed source and linked to every power electric converters or they can be connected to microgrid through a central battery storage unit. For energy storage, there is a need of converting it into another form, like electrochemical energy, chemical energy, mechanical or thermal energy [11]. The stored energy can be readily used during the system being operated in islanding mode, when the major power intent of stored energy is controlling the voltage thereby maintaining system frequency when it is being operated in islanding mode [9]. Additionally, when the

M. Khursheed (✉) · M. A. Mallick
Department of Electrical Engineering, Integral University, Lucknow, India 224024
e-mail: Khursheed20@gmail.com

A. Iqbal
Department of Electrical Engineering, Qatar University, Doha 122001, Qatar

A. Iqbal et al. (eds.), *Renewable Power for Sustainable Growth*, Lecture Notes in Electrical Engineering 723, https://doi.org/10.1007/978-981-33-4080-0_40

Table 1 Parameters of DC-DC boost converter

S. No.	Parameters	Value
1.	Rated output voltage	48
2.	Source rated power	1.6 kW
3.	Inductor	2000
4.	Capacitor	3000
5.	Current controller	0.16 + 120/s
6.	Voltage controller	0.3 + 80/s

grid is in connected mode, the storage system must have the capability to give a quick response to the changes in transient power [15]. Many applications such as managing customer energy and ancillary services, to integrate the non-conventional sources (RES) with the microgrid, also require the battery storage system [5].

1.1 Functions of Battery Storage System

The main functions of battery storage in renewable sources-based supply systems are [16, 17]:

- A battery storage and power converter helps in maintaining power to microgrid loads by supplying the active and reactive component of power to the microgrid loads.
- Battery storage improves voltage stability, i.e., maintaining bus voltage of microgrid during fluctuations in supply and consumes power [10].
- Moreover, loads having low power factor and unbalance loads can be compensated using battery storage.
- Battery storage provides energy during periods of peak demand.
- The inertia of power electronic converters is very small and consequently responses quickly to any disturbance in AC microgrids. By incorporating battery storage units, the output response could be enhanced [3].
- Battery storage is very important in microgrids as the renewable energy sources have a stochastic and intermittent nature.

2 Specifications of Battery Storage System

The battery storage model which is easily accessible in the Sim Power Systems Block library has been considered. This model has a little bit different charging and discharging cycle characteristic, whereas the other parameter of the model keep unchanged [1]. A Simulink model of 12 V, 80 Ah lead-acid batteries for an array has been used for the modeling the centralized BESS [2, 9]. In the final simulation, a stiff battery source has been considered for only supplying power to BTS load without taking into account the charging and discharging action [7].

3 Design of Battery Energy Storage System (BESS) for BTS Load

Output voltage of single battery = 12 V
Ampere hour of each battery = 80 AH

Backup of 8 h is to be provided solely by battery energy storage system

Output voltage of BTS = 48 V
Current requirement of BTS = 30 A
The battery bank would be able to supply current of 30 A, 48 V for 4 h.
Total no. of batteries to be connected in series, $N_s = 48/12 = 4$
Current supplied by each battery for 8 h $I_B = 80/8 = 10$ A
Total no. of batteries to be connected in parallel, $N_p = 30/10 = 3$

Thus, the battery bank comprises four batteries in series and three in parallel, respectively.

Total no. of batteries required for BTS operation = $4 \times 3 = 12$ lead-acid batteries.

4 Bidirectional DC-DC Converter

A bidirectional DC-to-DC converter has been used for interfacing BESS to microgrid [8]. The main purpose of BESS is to provide additional safety and to improve the stability of overall microgrid system [12]. As shown in Fig. 1, the bidirectional converter comprises of two converters: one is buck converter and another is boost converter, respectively. For both the converters, IGBT is used as a switching

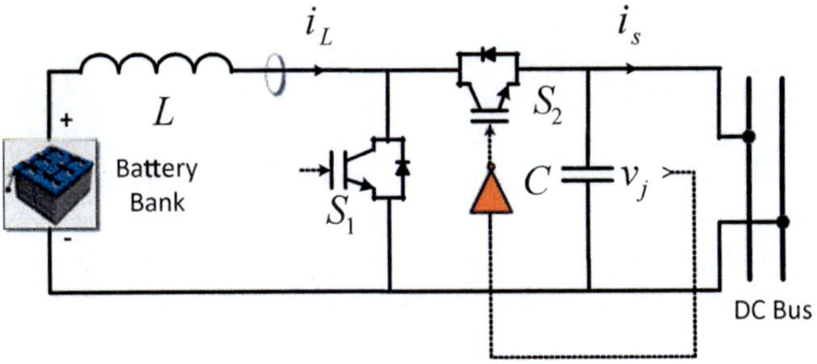

Fig. 1 Boost convertor circuit diagram

device. These switches are triggered using pulse width modulated (PWM) signals. With the help of proper controlling of converters, bidirectional power flow is maintained. When surplus power is available, the converter steps down the output thereby charging the battery. While during shortage of power, the converter steps up the output thereby feeding the DC bus [11].

5 Control of Boost Converter

For getting desired output voltage even when there is changes in input source voltage, load current and converter parameters are required. Any change that occurs is due to harmonics in supply, sudden load fluctuations, and switching of solid-state devices. The controllers are required for minimizing sensitivity to changes in load and hence attenuation in input–output transmission required for large system bandwidth [5, 12]. The controllers are designed for ensuring stability of the closed-loop system [6]. Moreover, controller should exhibit robust performance for all plant variations encountered during operation.

5.1 Double-Loop Control of DC-DC Boost Converter

The multi-loop control technique with inner current control loop for DC converters gives better result as compared to voltage mode control, because of high current protection ability; higher stability margin assured by the fast control loop (i.e., current loop) thus better Vin rejection. Two control units are employed in addition to active balancing circuits. The outer control loop provides the reference current to inner control loop [14]. The outer control loop is meant for controlling the voltage across the capacitor whereas the inner control loop controls the current through inductor.

The following two control loops are considered for the design of closed-loop DC-DC boost converter.

- Inner current control loop (fast response)
- Outer voltage control loop (slow response).

Bandwidth (BW) of the current control loop normally taken higher with respect to voltage loop in order to make the controller design simpler [13]. This design technique is normally used for fuel cell use. Fuel cells are highly sensitive toward the ripple current. For steady and stable working of the converter and fuel cells, average value of the input current must be regulated to be DC which requires small band widths of voltage loop and large band width of current loop. Two-loop average current control boost converter schematic diagram are shown in Fig. 2.

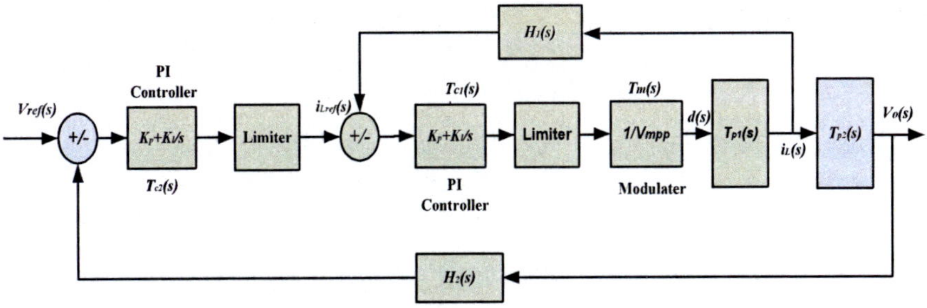

Fig. 2 Two-loop average current control system

6 Designing of Inner Current Control Loop (ICCL)

Figure 3 shows the control strategy for controlling current using inner loop. For designing must be such that it has faster dynamics (higher BW) as compared to the outer voltage loop (low BW). As inductor current changes rapidly with respect to the output voltage, it results in simple design (Table 1).

Output current to control transfer function for the boost converter is given by,

$$T_{p_1}(s) = \frac{i_L(s)}{d(s)} = \frac{CV_o + 2(1-D)I_L}{LCs^2 + \frac{L}{R}s + (1-D)^2} \tag{4}$$

Bode plot of transfer function $T_{p_1}(s) = \frac{i_L(s)}{d(s)}$ has been shown in Fig. 4. Bode plot of $T_{p_1}(s)$ gives the value of PM = 90° at f_c= 1.83 kHz (ω_c= 11,500 rad/s) for improving the frequency gain and to minimize the error between the reference current and actual inductor current a PI controller is used though keeping a positive phase margin at particular crossover frequency.

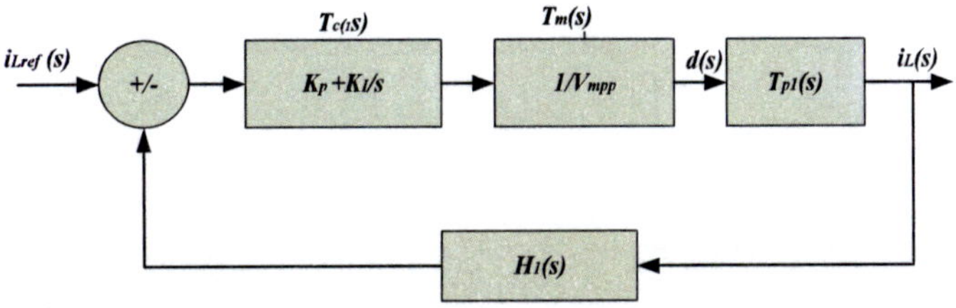

Fig. 3 Inner current control loop

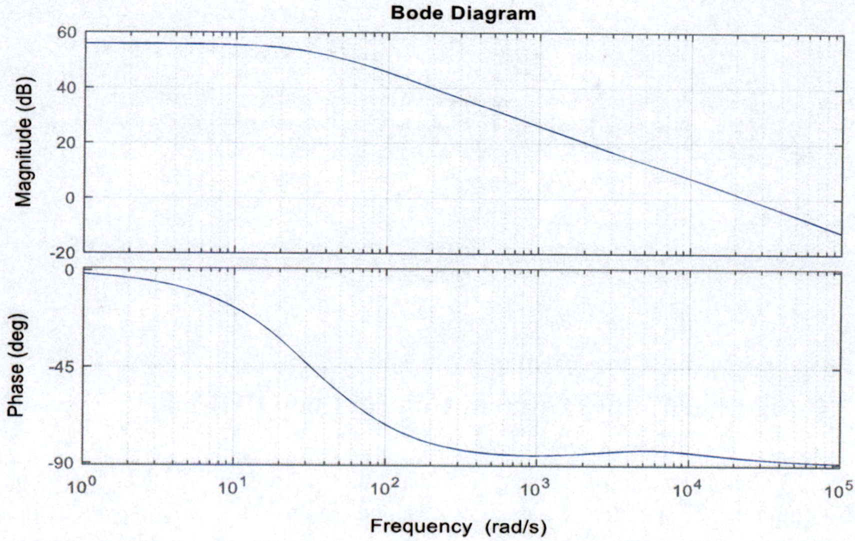

Fig. 4 Bode plot for transfer function $\frac{i_L(s)}{d(s)}$

6.1 Feedback Gain Design

Assuming 100% efficiency, the average inductor current $I_L = 62.5$ A, taking the reference value of the inductor current = 62.5 A. Thus, the gain $H_1(s)$ of the feedback of current loop can be calculated as

$$H_1(s) = \frac{62.5}{62.5} \tag{5}$$

Thus, feedback loop with unity gain is to be designed.

6.2 Modulator Design

A unipolar or bipolar, either triangular or sawtooth wave has been employed for carrier signal. The switching frequency of the converter must be similar to carrier signal frequency. For modulation signal, unipolar triangular wave has been considered. The maximum value of inductor current is given as $(V_o\text{-}V_{in})/L$. During first half cycle $(T_s/2)$ of the wave, the ripple current attains a maximum $V_{P\text{-}P}$ of the triangular modulating signal.

Considering noise, etc., the maximum $V_{P\text{-}P}$ voltage of the modulating signal has been taken as 1.25 V. Thus, the modulator transfer function is given as

$$T(s) = \frac{1}{V_{m,p-p}} = \frac{1}{1.25} = 0.8 \quad (6)$$

Bode plot of $\frac{i_L(s)}{i_{L_{\text{ref}}}(s)}$ for the current controller loop of boost converter is depicted in Fig. 5. From the Bode plot of $\frac{i_L(s)}{i_{L_{\text{ref}}}(s)}$, the value of bandwidth obtain is equal to 191 Hz (1200 rad/s).

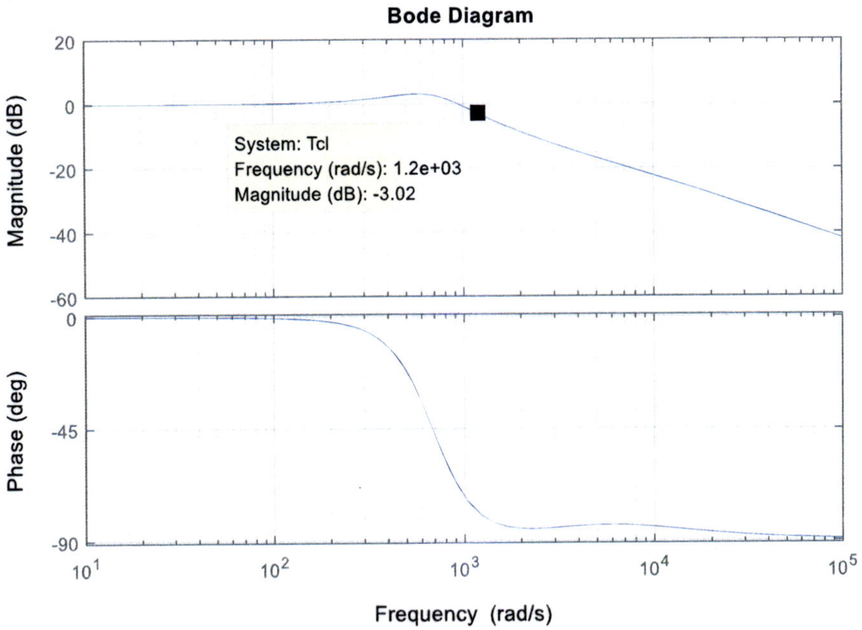

Fig. 5 Bode plot of transfer function of $\frac{i_L(s)}{i_{L_{\text{ref}}}(s)}$

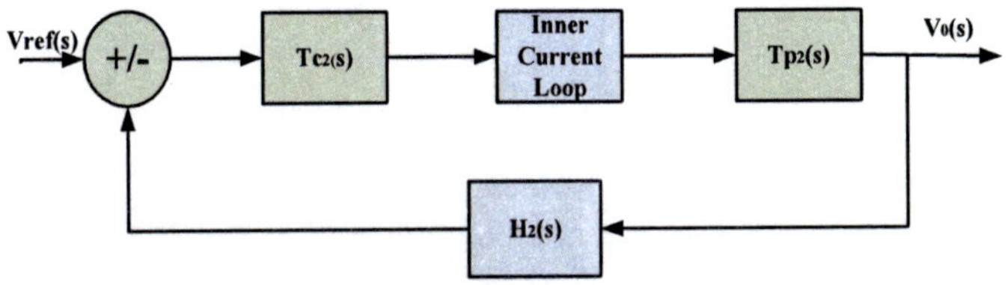

Fig. 6 Block diagram of outer VCL

7 Designing of Voltage Control Loop (VCL)

Figure 6 gives the design and working of VCL. The inner CCL reference current ($i_{L,\mathrm{ref}}$) has been generated by this loop. As the inner current loop is very fast, it corrects the current error quickly. Thus, the effect of current loop can be neglected while deigning the VCL. The transfer function of current loop has been ignored in the calculation and the duty cycle has been substituted with a constant value D (steady-state value).

Thus putting $d\hat{}(s) = 0$ in Eq. (6), and this gives,

$$T_p(s) = \frac{v_o(s)}{i_L(s)} = \frac{(1-D)}{Cs + \frac{1}{R}} \tag{7}$$

Bode plot for$T_{p2}(s)$ is given in Fig. 9. The PM $= 141^{\circ}$ is obtained on $f_c =$ 802 Hz (5040 rad/s). Bode plot of transfer function $\frac{v_o(s)}{v_{o_{\mathrm{ref}}}(s)}$ of voltage controller loop is depicted in Fig. 8. From the Bode plot, the value of bandwidth of $\frac{v_o(s)}{v_{o_{\mathrm{ref}}}(s)}$ is 15.4 Hz (96.8 rad/s). From Figs. 7 and 10, it is clear that the bandwidth of current controller transfer function $\frac{i_L(s)}{i_{L_{\mathrm{ref}}}(s)}$ is higher than that of voltage controller $\frac{v_o(s)}{v_{o_{\mathrm{ref}}}(s)}$.

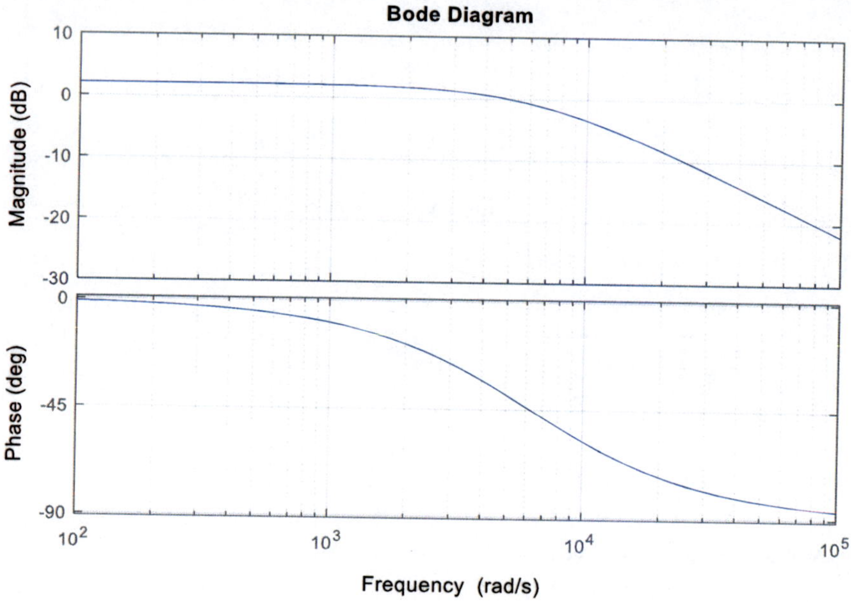

Fig. 7 Bode plot of transfer function $\frac{v_o(s)}{i_L(s)}$

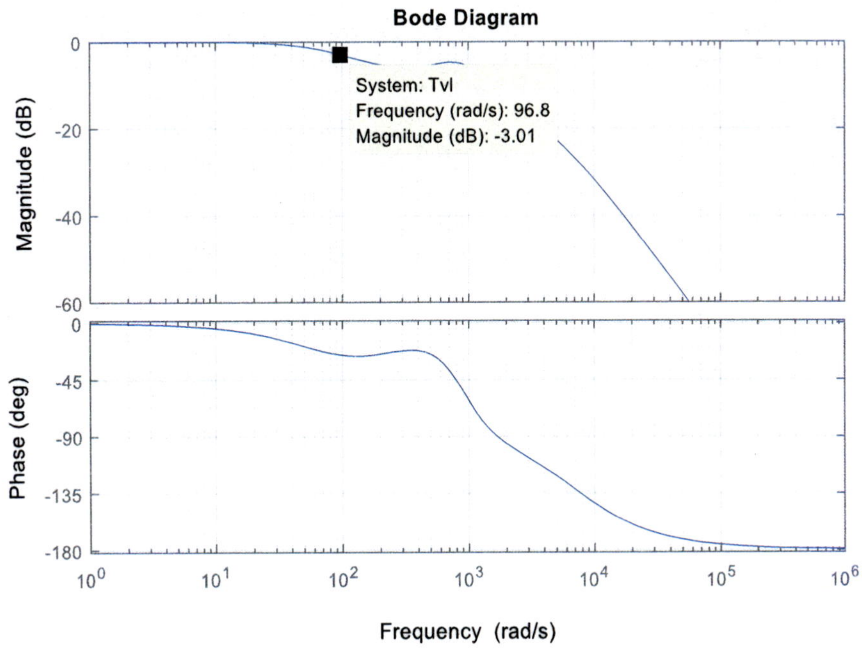

Fig. 8 Bode plot for transfer function $\frac{v_o(s)}{v_{o_{\text{ref}}}(s)}$

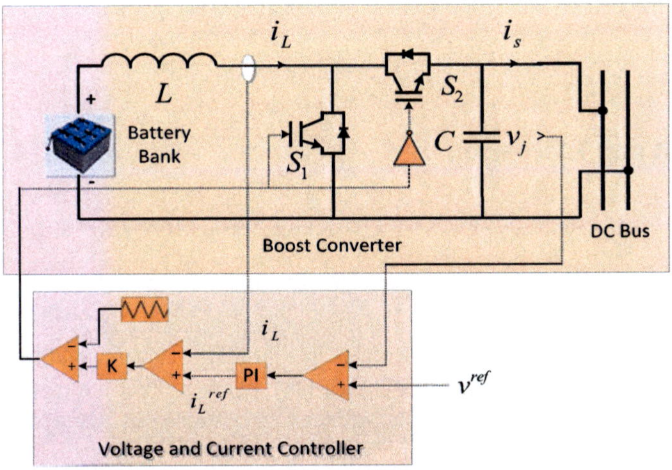

Fig. 9 Connection diagram of closed-loop control scheme for boost converter with voltage and current control loop

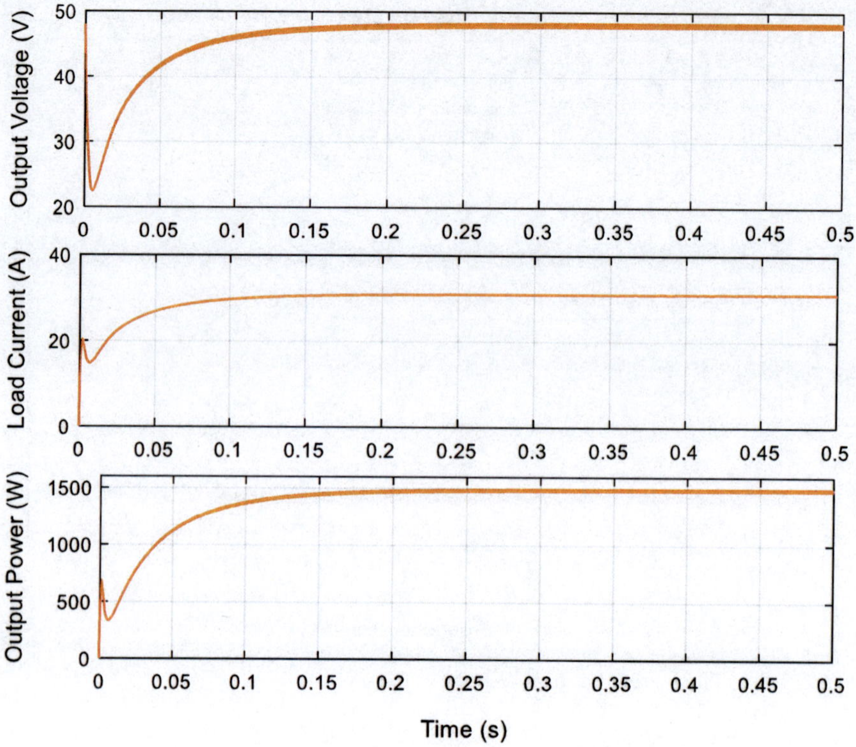

Fig. 10 Output voltage, load current, and output power supplied by boost converter to the BTS load

8 Result and Discussion

Current control of boost converter for checking the performance of voltage and current mode control has been shown in Fig. 9. The input source is a battery supplying power to BTS load through the closed-loop controlled DC-DC boost converter. The value of BTS load resistance is 1.6 O and the voltage reference is $V_{\text{ref}} = 48$ V. The closed-loop system is examined. It operates in continuous current conduction mode and at a constant output voltage of 48 V. It has been seen that the output voltage is independent of any change in input voltage and output load. At rated BTS load condition, the boost converter is supplying a power of 1440 W as shown in Fig. 10.

9 Conclusion

This paper describes tuning of the double-loop controller that is outer voltage controller and inner current controller implemented through the boost converter of the battery source feeding supply to the BTS load.

The voltage control loop maintains the capacitor voltage of respective stages, whereas the current controller maintains the inductor current. The voltage controller provides the reference current for CCL, whereas the current control loop (CCL) provides the gate signal to the corresponding IGBT switch.

Moreover, the designing procedure of the current and voltage controller is also given. The required closed-loop transfer function for each controller is derived and their frequency response characteristic has been examined using Bode plots. It has been concluded from the result that the bandwidth of voltage loop is low (slow) with respect to the high BW (fast) of current controller.

Simulation is done in MATLAB–Simulink environment. From the simulation result, it has been concluded that current is tracking the reference value and the output voltage is independent of any variation in the input voltage and that of any change in load.

Acknowledgements The author would like to acknowledge Integral University for providing MCN No: IU/R&D/2020- MCN 000834.

References

1. Matthias D, Andrew C, Sinclair G, McDonald JR (2006) Dynamic model of a lead acid battery for use in a domestic fuel cell system. J Power Sour 161(2):1400–1411
2. Inthamoussou FA, Pegueroles-Queralt J, Bianchi FD (2013) Control of a super capacitor energy storage system for microgrid applications. IEEE Trans Energ Convers 28(3):690–697
3. Delille G, Francois B, Malarange G, Fraisse J-L (2009) Energy storage systems in distribution grids: new assets to upgrade distribution network abilities. Proceedings of 20th international conference and exhibition on electricity distribution-part I, pp 1–4
4. Chen H, Cong TN, Yang W et al (2009) Progress in electrical energy storage system: a critical review. Prog Nat Sci 19(3):291–312
5. Chen J, Chu C (1995) Combination voltage-controlled and current-controlled PWM inverters for UPS parallel operation. IEEE Trans Power Electron 10(5):547–558
6. Datta M, Senjyu T, Yona A, Funabashi T, Kim C-H (2009) A coordinated control method for leveling PV output power fluctuations of PV–diesel hybrid systems connected to isolated power utility. IEEE Trans Energ Convers 24(1):153–162
7. Wang M-H, Tan S-C, Lee C-K, Hui SY (2018) A configuration of storage system for DC microgrids. IEEE Trans Power Electron 33(5)
8. Kwon M, Choi S (2018) Control scheme for autonomous and smooth mode switching of bidirectional DC–DC converters in a DC microgrid. IEEE Trans Power Electron 33(8)
9. Farrokhabadi M, König S, Cañizares CA (2018) Battery energy storage system models for microgrid stability analysis and dynamic simulation. IEEE Trans Power Syst 33(2)
10. Tremblay O, Dessaint LA, Dekkiche A (2007) A generic battery model for the dynamic simulation of hybrid electric vehicles. Proceedings of IEEE vehicle power and propulsion conference, pp 284–289

11. Mauracher P, Karden E (1997) Dynamic modelling of lead/acid batteries using impedance spectroscopy for parameter identification. J Power Sour 67(1–2):69-84
12. Ahmadi R, Ferdowsi M (2014) Improving the performance of a line regulating converter in a converter-dominated DC microgrid system. IEEE Trans Smart Grid 5(5):2553–2563
13. Sudhoff SD et al (2003) Stability analysis methodologies for DC power distribution systems. In: Proceedings of 13th international ship control systems symposium (SCSS), Orlando, FL, USA, p 235
14. Sul S (2011) Design of regulators for electric machines and power converters. In: Control of electric machine drive system. Wiley, Piscataway, NJ, USA, pp 170–173
15. Wiegman HLN (1999) Battery state estimation and control for power buffering applications. Ph.D., The University of Wisconsin, Madison, p 216. AAT 9956303
16. Iqbal A et al (2019) Performance analysis of closed loop control of diesel generator power supply for base transceiver (BTS) load. Int J Innovative Technol Explor Eng 8(9):2483–2495
17. Mohd K et al (2011) Performance evaluation of an indirect vector controlled drive using synchronous current controller. Int J Eng Res Appl 1(4):2062–2071

Systematic Study of Maximum Power Point Tracking Methods Used in Photovoltaic-Based Systems

Sajad Ahmad Tali and Faroze Ahmad

Abstract Photovoltaic (PV) systems use maximum power point tracking (MPPT) devices for extracting maximum power from solar panel irrespective of dynamic change in temperature and irradiance. In this paper, concise and organized assessment of numerous maximum power point (MPP) methods is discussed that encompasses the perturb and observe (P&O), open-circuit voltage (OCV), incremental conductance (IC) approach, neural network approach (NN) and the fuzzy logic method (FLC). The MPPT strategies are assessed on the bases of simplicity, speed, implementation cost and performance under partial shading condition (PSC); therefore, the right literature review is crucial while designing photovoltaic systems. The comparison of various MPP tracking algorithms is likewise given. Further, mathematical modeling of current equations of solar cell is also done. Moreover, the workability of perturb and observe MPPT is tested in MATLAB/Simulink and results carried out are also given.

Keywords PV systems · MPPT methods

1 Introduction

The development of the country is largely dependent on successful production and supply of electricity. In "1950," India started the electrification program in rural areas to improve the standard of life and improve economic development [1–6]. Harvesting solar energy for clean electricity helps in improving various sectors such

S. A. Tali (✉) · F. Ahmad
Department of Electronics and Communication Engineering, Islamic University of Science and Technology, Awantipora 192122, J&K, India
e-mail: talisajad89@gmail.com

F. Ahmad
e-mail: drferoz07@gmail.com

A. Iqbal et al. (eds.), *Renewable Power for Sustainable Growth*, Lecture Notes in Electrical Engineering 723, https://doi.org/10.1007/978-981-33-4080-0_41

as domestic, healthcare and education sectors. Photovoltaic (PV) systems have extensive capability in electrifying the modern world. It is essential to maximize the efficiency without getting affected due to variations in solar irradiance. Many techniques and methods are implemented on PV system to make it stable and efficient. When PV panel is operated consistently at MPP, maximum power is drawn from it irrespectively of dynamic changes in the environment [5, 6]. Therefore, diverse power transfer techniques have been developed to maintain the MPP correctly. The second section discusses various MPPT algorithms in detail. In the third section, MPPT methods discussed are compared keeping in view speed, oscillations around MPP and complexity. In the fourth section, mathematical modeling of the current equations of solar cell is demonstrated in MATLAB/Simulink. Finally, in the fifth section P&O method is tested in MATLAB/Simulink and results carried out have been given in graphs.

2 MPPT Methods

The different MPPT algorithms discussed in the literature are P&O, IC, OCV, NN, FLC; amongst them, IC, P&O and FLC are widely suitable ones [1–6].

2.1 Perturb and Observe Method

The purpose of this approach is to attain MPP. It is the most commonly used algorithm, and MPP is obtained by the repetitive method. This algorithm calculates the power by measuring the output voltage ($V_{pv}(k)$) and the current ($I_{pv}(k)$) and at the kth instant and the multiplication of $I_{pv}(k)$ and $V_{pv}(k)$ results in instantaneous power. The instantaneous power is equated with power stored in memory. Then, change in power (ΔP) and voltage (ΔV) is calculated. P&O works on the certainty; if ΔP and ΔV are positive, the algorithm moves in same direction; otherwise, course is reversed and voltage is raised or reduced accordingly until MPP is reached. P&O is simple to enforce and inexpensive, but it keeps on oscillating continuously around MPP ensuing in wastage of energy and its response becomes slow during speedily changing atmospheric conditions [1] (Fig. 1).

2.2 Incremental Conductance Method

This method overcomes limitations of P&O method. It is based on Eqs. (1) though (3) that determine that the value of derivative can be equal to zero, higher than zero or lesser than zero. When the condition of having incremental conductance equals to instantaneous conductance, it stops unsettling the operating point. The merit of

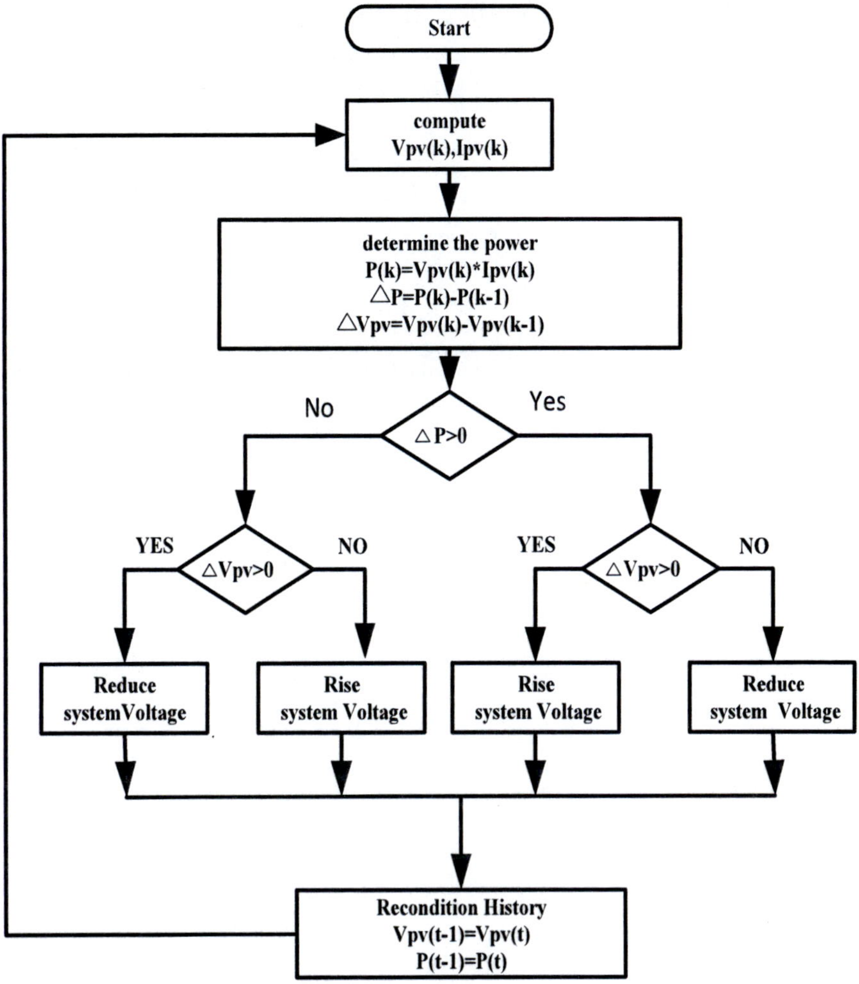

Fig. 1 Flowchart of P&O algorithm

IC method over P&O method is its better performance with decreased oscillations around MPP even when atmospheric conditions are changing at the faster rate. The drawbacks of this technique are it is complex and costly and cannot track under partial shading conditions [2].

$$P_{\mathrm{PV}} = V_{\mathrm{PV}} \times I_{\mathrm{PV}} \tag{1}$$

$$\frac{dP_{\mathrm{PV}}}{dV_{\mathrm{PV}}} = I_{\mathrm{PV}} + V_{\mathrm{PV}} \times \frac{dP_{\mathrm{PV}}}{dV_{\mathrm{PV}}} \tag{2}$$

$$\frac{dP_{PV}}{dV_{PV}} = \begin{cases} 0, \text{duty cycle is kept fixed} \\ +\text{ve, increase duty cycle} \\ -\text{ve, decrease duty cycle} \end{cases} \quad (3)$$

2.3 *Open-Circuit Voltage Method*

The OCV method works on the truth that MPP voltage is the fixed percentage of the open-circuit voltage (V_{OC}); hence, it is changeless. The position of the MPP gets changed around two percent (2%) when the temperature and the solar insulation vary. Generally, V_{MPP} is K_1 times (V_{OC}). The value of V_{MPP} is obtained from Eq. (4), but this equation is not valid for partial shading condition and from Eq. (5) K_1 is calculated. Choosing the value of "K_1" is tedious; for polycrystalline, its value ranges from 0.73 to 80.

$$V_{MPP} = K_1 * V_{OC} \quad (4)$$

$$K_1 = \frac{V_{MPP}}{V_{OC}} \cong \text{Constant} < 1 \quad (5)$$

The OCV method is simple to design and is having lesser "cost"; however, load gets no power while measuring open-circuit voltage because current through the photovoltaic system is zero at that point of time. Therefore, energy produced gets wasted momentarily. It has also the slower response as V_{MPP} is proportional to the V_{OC} but may not locate correct MPP at rapidly changing conditions and this method is not valid for partial shading condition [3].

2.4 *Neural Network Method*

Neural network (NN) approach is trained to produce accurate MPPT operation under changing characteristics of the PV system without requiring the complete knowledge. For MPPT operation, the network constitutes three well-defined layers forming input layer, middle layer taken as the hidden layer and output layer. The inputs to the network are taken as the short-circuit current (I_{sc}) and the open-circuit voltage (V_{oc}) corresponding to PV system that provides the output of the network in the form of duty cycle (D) together with the achievement of MPP due to hidden layer. This duty cycle output further drives the conversion process in power converter. It performs better, "provided" it is initially taught at large. The NN is the strong and comparably simple design and adjusts itself under changing temperature and illumination in order to provide efficient output without redesigning the network [4].

Table 1 Differentiation of MPPT methods [1–4]

Method	Convergence speed	Complexity	Efficiency	Oscillations around MPP
P&O	Low	Low	Low	Yes
IC	Fast	Medium	High	Yes
OCV	Medium	Low	Low (86%)	No
NN	Fast	High	High (98%)	No
FLC	Fast	High	High	No

2.5 Fuzzy Logic Method

The FLC can be utilized to obtain the maximum power that can be produced by the solar panel under dynamically changing climatic condition. The inputs of the FLC are alteration in voltage (ΔV) and alteration in power (ΔP) of the PV panel, and corresponding output can be ΔV, ΔP, duty cycle (D) or change in the duty cycle (ΔD). FLC is simple to implement and has fast speed with decreased oscillation, and it does not depend on the mathematical modeling. However, it is unable to find GMPP under PSC [2, 3].

3 Analysis of MPPT Methods

The comparison of various MPPT methods discussed is given in Table 1.

4 Mathematical Modeling in MATLAB/Simulink

See Figs. 2, 3, 4, 5, 6 and 7.

5 Simulation and Results

The complete PV system along with MPPT is indicated in Fig. 8, with the aim to test P&O method. Although P&O has simple design and can efficiently track the MPP, P&O MPPT technique has regular oscillations near the MPP as shown in Fig. 9 as a result of which a fraction of energy is lost. Therefore, this oscillatory nature needs to be addressed without increasing the complexity of the P&O algorithm, as well as its economics.

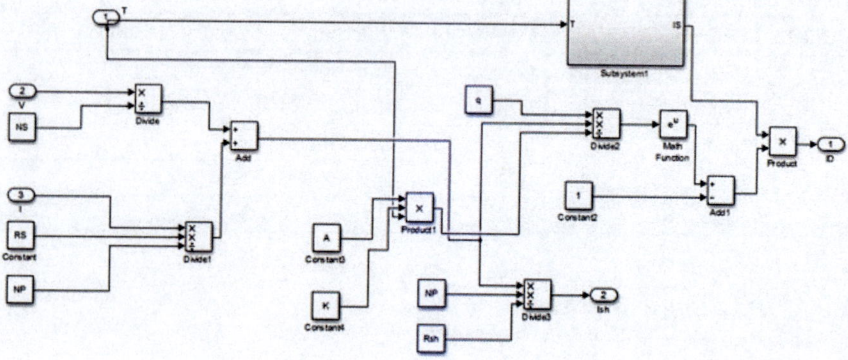

Fig. 2 Diode current and shunt leakage current block

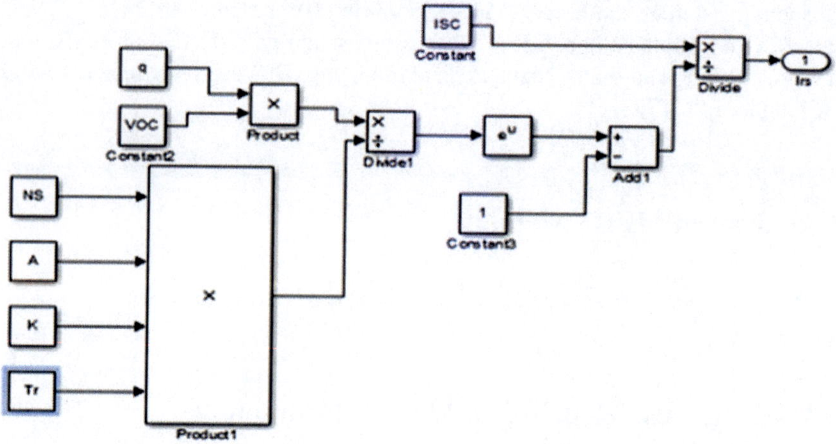

Fig. 3 Reverse saturation current block

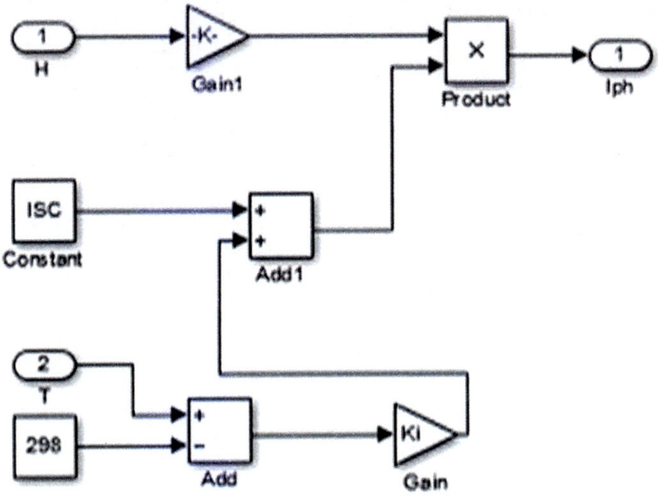

Fig. 4 Photocurrent block

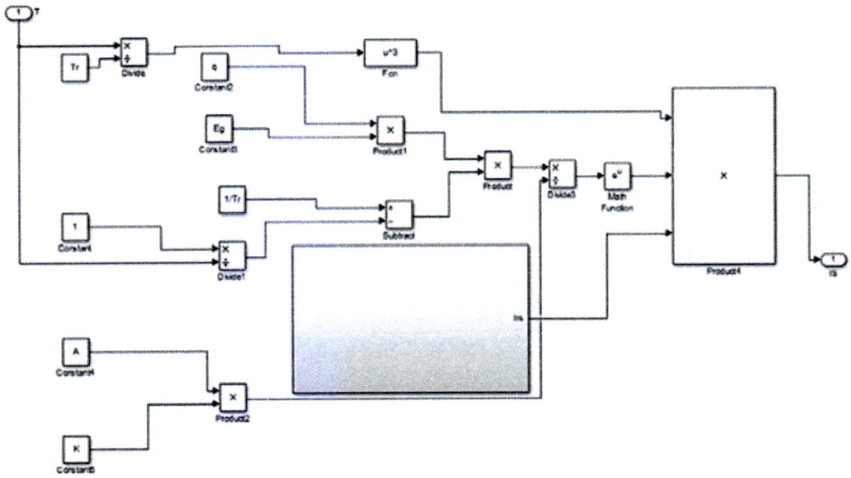

Fig. 5 Saturation dark current block

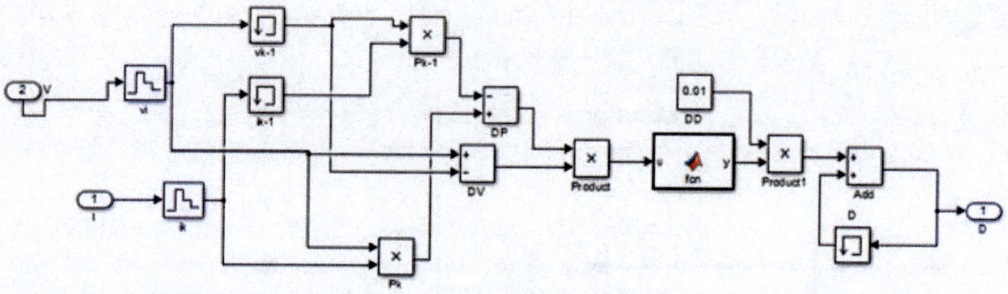

Fig. 6 MPPT block

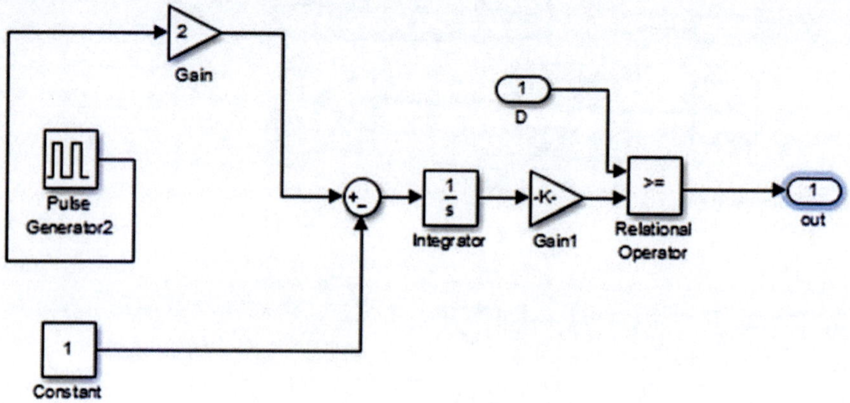

Fig. 7 PWM block

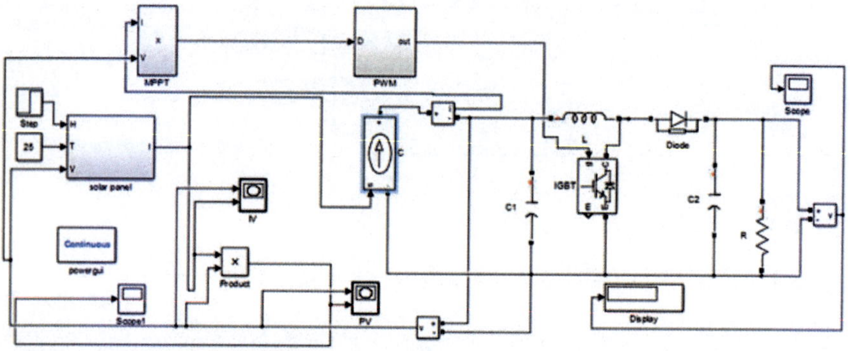

Fig. 8 Photovoltaic system block in MATLAB/Simulink

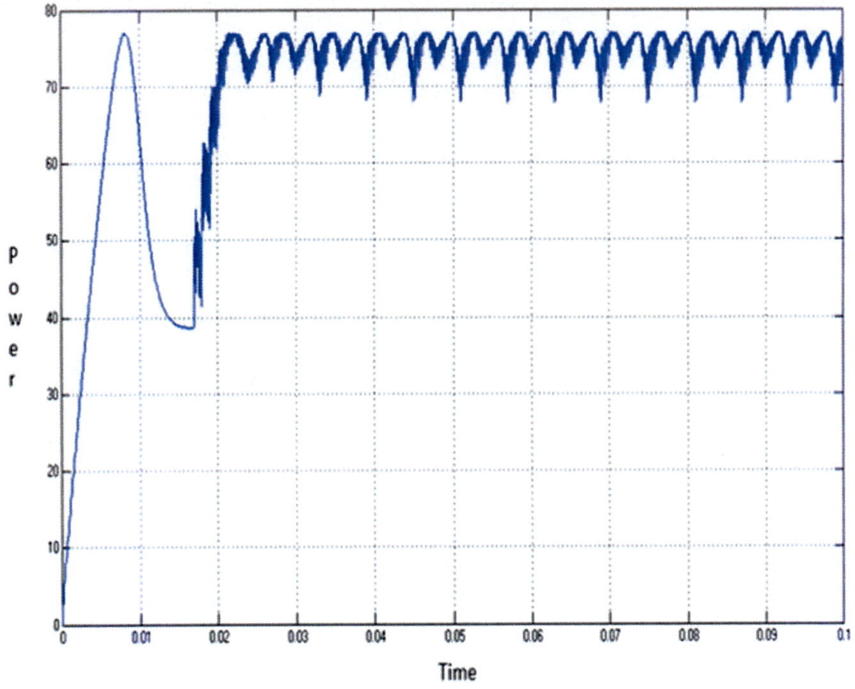

Fig. 9 "Solar panel," output power with P&O algorithm at irradiance range of 1000-1500 W/m^2

References

1. Abdul-Kalaam R, Muyeen SM, Al-Durra A (2016) Review of maximum power point tracking techniques for photovoltaic system. Glob J Control Eng Technol 2:8–18
2. Faranda R, Leva S (2008) Energy comparison of maximum power point tracking techniques for photovoltaic systems. WSEAS Trans Power Syst 3:447–455
3. Esram T, Chapman PL (2007) Comparison of photovoltaic array maximum power point tracking techniques. IEEE Trans Energ Convers 22(2):439–449
4. Kamarzaman NA, Tan CW (2014) A comprehensive review of maximum power point tracker algorithms for photovoltaic systems. Renew Sustain Energ Rev 37:585–598
5. Iqbal A et al (eds) (2020) Soft computing in condition monitoring and diagnostics of electrical and mechanical systems. In: Advances in intelligent systems and computing, vol 1096. Springer, Singapore. https://doi.org/10.1007/978-981-15-1532-3
6. Iqbal A et al (eds) (2020) Meta heuristic and evolutionary computation: algorithms and applications. In: Studies in computational intelligence, vol 1096. Springer, Singapore. https://www.springer.com/gp/book/9789811575709

Hybrid Video Watermarking Based on LWT, SVD and SWT Using Fused Images for Data Security

Haweez Showkat, Rohun Nisa, and Asifa Baba

Abstract We propose a new scheme to protect IPR of owners and stop illegal modification and distribution of videos. We propose embedding of a fused watermark into videos to solve ownership disputes. The embedding watermark is an image that consists of two watermarks fused using SWT, so as to make the system more secure. Embedding and extraction are done using LWT and SVD. Experimental outcomes show that the extracted watermarks are stout against various noise attacks and are evaluated using correlation coefficient. The PSNR is also calculated to evaluate the perceptual video quality.

Keywords Lifting wavelet transform · Stationary wavelet transform · Singular-valued decomposition · Digital video watermarking

1 Introduction

The huge advancements made in the field of multimedia communication have completely changed the scenario of transactions and perceptions, which has in turn led to untoward apprehensions regarding the information shared and received [1–19]. It is very convenient to replicate any content, modify it with perfection and then redistribute it without any objection. Due to these apprehensions, digital watermark came into existence. Watermark helps in the authentication of information received, whether it is a text, audio, image or video. In the process of watermarking, the exclusive information is embedded in the information to be sent. This watermark can then be extracted on the receiver side only by detector [4] to prove its ownership.

The watermarking should have certain properties [1–19]:

H. Showkat (✉) · R. Nisa · A. Baba
Islamic University of Science and Technology, Awantipora, J&K, India
e-mail: haweezsk@gmail.com

R. Nisa
e-mail: rohunnisa@islamicuniversity.edu.in

A. Baba
e-mail: asifababa@gmail.com

A. Iqbal et al. (eds.), *Renewable Power for Sustainable Growth*, Lecture Notes in Electrical Engineering 723, https://doi.org/10.1007/978-981-33-4080-0_42

1. The watermark should not decrease the perceptual quality of an image, audio or video. The PSNR should be above 40 db for commercial implementation.
2. The watermark should be robust enough to withstand any attack by the intruder.
3. The watermark should be exclusive for copyright protection.
4. The process should be such that it can be used in real-time systems.
5. Capacity, i.e., no. of bits to be embedded should be taken into consideration.
6. The watermark should not be detectable by human eye, to keep the embedding information secret.
7. The watermark must be alive even in the analogue form of an image, video or audio [6].

There has been much advancement in the field of watermarking since it came into existence. First watermarking was used to be done in the spatial domain. The operations are done directly on pixels in this domain. The most common method used in spatial domain was least significant bit (LSB) method. It is easy to implement and understand, but if an attack is launched on each pixel, the watermark can be easily removed. Then the trend of transforming the spatial domain into the frequency domain came into being. The main techniques used in frequency domain are as follows:

Discrete Fourier transform (DFT): It produces complex value coefficients in the frequency domain for real value inputs. The problem in this method was that it had symmetry constrains.

Discrete cosines transform (DCT): The cosine and sine coefficients are produced from real value inputs. It removes the symmetry constrains of DFT.

Discrete wavelets transform (DWT): The image is divided into four frequency sub-bands: low-frequency band, low–high-frequency band, higher low-frequency band and high-frequency band. It represents frequency as well as location characteristics. Many other techniques are used in frequency domain that includes LWT and inverse wavelet transform (IWT). There is another trend of using neural networks for watermarking, which has a lot of scope in future [8].

2 Preliminaries

2.1 Lifting Wavelet Transform (LWT)

These are second age group wavelets. LWT is better than DWT in many of its characters like in compression, denoising, watermarking, memory requirements, etc.

They reduce aliasing effects of the image and increase the smoothness. They also have better frequency localization features. They help for removing boundary artefacts for good reconstruction features [5]. They are less complex as they reduce complexity of processing to about the half as compared to DWT. LWT is a good option when speed is a critical consideration, like in case of live videos.

LWT has three simple steps that include split, predicting and updating.

Split Step:

This step splits the input signal $X(a)$ into even and odd fragments, respectively.

$$X_e(a) = X(2a)$$
$$X_o(a) = X(2a + 1)$$

Prediction Step:

These steps predict the odd sample models from an even sample model. The variation between the predicted value P_σ $[X_e(a)]$ and the actual value $X_o(a)$ is termed as an elaborated signal of $D(a)$.

Update Step:

In this step, update operator U_s is used. It helps the signal $D(a)$ so as to revise the even samples. In this step, lifting of bands takes place.

2.2 *Singular Value Decomposition (SVD)*

SVD is used to convert a single matrix into three matrixes, left matrix (U), singular matrix (S) and right matrix (V). S matrix specifies the relation between matrix U and matrix V, which are the two orthogonal matrices. An image is also considered as a matrix, which is transformed into vectors U, S and V.

$$I = \mathrm{USV}^T$$

Suppose we have an image matrix I of order $n * n$, maximum diagonal elements in S will be n [15]. It particularly specifies the luminance of the image. Generally, SVD has good embedding advantages, i.e., after embedding the quality of the image is not reduced. It represents that part of an image which is not visual to the human eye, so embedding does not decrease the perceptual quality. It is known to keep the symmetric properties of the images intact. Using SVD helps in making images robust to geometric noises like cropping, scaling, rotation, etc. It also gives non-invertible nature to the watermark.

2.3 *Stationary Wavelet Transform (SWT)*

SWT has same the way of operation like DWT except that it does not involve down sampling. The process generates four images that have about the half of original information. The images generated have the same size that of the original. Out of four generated images, one is an approximated image and other three are detailed images. The algorithm is decimated, which is applied to rows and columns, respectively, at each point. It has the properties of translation and invariance so as to remove up sampling and down sampling steps in DWT. Up sampling is done using a factor of 2 $(r - 1)$ where r refers to level. The algorithm of SWT is known as "algorithm à trous" in French (meaning of trous is holes in English) as it also inserts zeros in filters.

Figure 1 depicts the steps of the algorithm. The Gi and Hi represent low- and high-pass filters, respectively.

3 Previous Work

Divjot et al. [1], in his paper, propose a scheme in which the watermarking of video has been done by applying DWT and extracting high-frequency bands of both watermark and the video. Then the singular values are modified using singular values of both video frames and the watermark, considering the scaling factor. Then IDWT is applied and video is recovered back for sending it to the sender. This proposal is evaluated based on PSNR and CC which give satisfactory results. Heman et al. [2] This paper proposes a watermarking scheme which uses DWT and SVD for embedding and taking out of fused watermark. The fusion of images is done using DWT. This method outperforms many other schemes without decreasing the perceptual quality. Praful et al. [9] propose watermarking format using DWT-SVD. It uses high-frequency band for embedding and extraction. Zhu Shoo-long et al. [7] The paper proposes

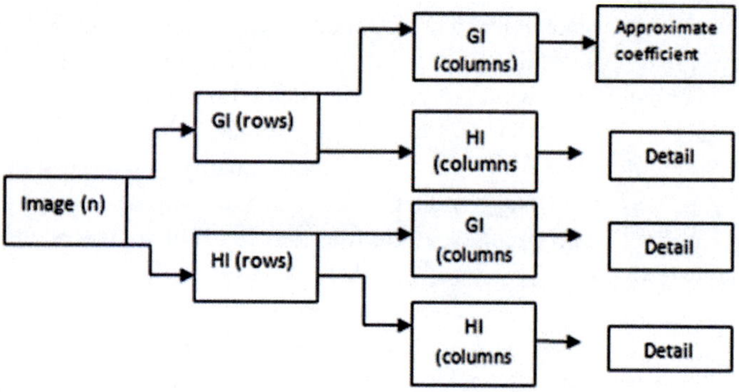

Fig. 1 Steps of SWT algorithm

the scheme for fusing the images using wavelet fusion, so as to improve geometrical properties of the fused images. It decomposes the images according to their frequency content and then fused under certain criteria. Then images are reconstructed having information of all the images, resulting into a fused image. Mirajkar et al. [12] This paper proposes the union of images using stationary wavelet transform (SWT). This method is compared with other schemes, and it outperforms many of them.

4 Proposed Methodology

Videos and watermarks are first converted into the greyscale regime so as to work upon them.

4.1 Embedding of Watermark

1. The frames are extracted from input video, so that each frame is worked upon individually.
2. Apply LWT on cover frame (A), so as to decompose it into its corresponding sub-bands, LL, LH, HL and HH.
3. Singular value decomposition is applied on the high-frequency band.

   $$[\mathrm{U, S, V}] = \mathrm{svd(HH)}$$

4. Take two images as inputs which have to be taken as watermarks. Fuse the two watermarks to obtain E, using stationary wavelet transform whose algorithm is given in fusion section.
5. Apply LWT on fused watermark to obtain L_L, L_H, H_L, H_H.
6. Apply SVD on the high-frequency band H_H of fused watermark image.

   $$[\mathrm{Uh, Sh, Vh}] = \mathrm{svd\ (H\_H)}$$

7. Transform the singular values of the frames with singular values of the fused watermark using scaling factor β, so as to obtain singular values of watermarked image.

   $$\mathrm{Sigma\,N} = \mathrm{S} + (0.\beta * \mathrm{Sh})$$

8. Apply SVD to SigmaN to obtain the singular values.

   $$[\mathrm{Uw, Sw, Vw}] = \mathrm{svd\ (Sigma\_N)}$$

9. Obtain the higher band of watermarked frame.

   $$\mathrm{Aw} = \mathrm{U} * \mathrm{Sw} * \mathrm{V}^{\mathrm{T}}$$

10. Apply ILWT to obtain watermarked cover image A^*, using sub-bands LL, LH, HL and Aw.

11. Finally, we can recover the watermarked video.

4.2 Extraction of Watermark

1. Extract frames out of the received video at the receiving end.
2. Apply LWT on each frame A*, to obtain the sub-bands A, B, C and D.
3. Apply SVD on higher band D.

$$[U1,\ S1,\ V1] = \mathrm{svd}(D)$$

4. Take the original image (A), apply LWT on it so as it gets the sub-bands LL, LH, HL and HH, just like was done in embedding section.
5. Apply SVD on high-frequency sub-band.

$$[U,\ S,\ V] = \mathrm{svd}\ (HH)$$

6. Decompose watermark (E) and apply SVD on its high-frequency band.
7. Obtain the matrix X.

$$X = Uw * S1 * Vw^{T}$$

8. Take out the singular values of the high-frequency band of watermark image.

$$\mathrm{recovered\_hh} = (X - S)/\beta.$$

9. Apply ILWT using L_L, L_H, H_L and recovered_hh, to obtain fused watermark.
10. Apply wavelet-based anti-fusion (explained in Sect. 4.4) on the obtained watermark so as to get the two watermarks back for further evaluation.

The algorithm used for fusion and anti-fusion of the watermarks is as follows.

4.3 Fusion of Two Watermarks

1. Take the two image watermarks that have to be fused, im1 and im2.
2. Apply SWT on both the images and extract their sub-bands.

$$[A1L1, H1L1, V1L1, D1L1] = \mathrm{swt}\ (im1)$$
$$[A2L1, H2L1, V2L1, D2L1] = \mathrm{swt}\ (im2)$$

3. Start union by adding two lower bands and multiplying it with the scaling factor.

$$AfL1 = \beta * (A1L1 + A2L1)$$

4. Take absolute values of sub-band details and subtract one from the other.

$$D = (\mathrm{abs}\ (H1L1)\text{-}\mathrm{abs}\ (H2L1)) == 1$$

HfL1 = D * H1L1 + (~D) * H2L1
D = (abs (V1L1)-abs (V2L1)) ==1
VfL1 = D * V1L1 + (~D) * V2L1
D = (abs (D1L1)-abs (D2L1)) ==1
DfL1 = D * D1L1 + (~D) * D2L1

5. Obtain the merged image.

   Fused image = iswt (AfL1, HfL1, VfL1, DfL1)

4.4 Anti-Fusion

1. Take the fused watermarked image (E) from the extraction end. Decompose it into sub-bands using SWT.

   [R1L1, S1L1, T1L1, U1L1] = swt (E)

2. Extract the lower sub-band.

   REL1= (R1L1 − (β * A2L1))/β

3. Extract the other sub-bands by taking the absolute values of sub-bands obtained and subtracting one from other.

   F = (abs (S1L1)-abs (S1L1))>=0
   SEL1 = F * S1L1 + (~F) * S1L1
   F = (abs (T1L1)-abs (T1L1))>=0
   TEL1 = F * T1L1 + (~F) * T1L1
   F = (abs (U1L1)-abs (U1L1))>=0
   UEL1 = F * U1L1 + (~F) * U1L1

4. Apply ISWT to the obtained sub-bands so as to get 1st watermark image.

   imw1 = iswt (REL1, SEL1, TEL1, UEL1)

5. Take the extracted fused watermark image (E) again. Apply SWT to obtain its sub-bands.

   [R1L1, S1L1, T1L1, U1L1] = swt (E)

6. Obtain the first lower sub-band where β is the scaling factor.

   REL1 = (R1L1 − (β * A1L1))/β

7. Extract the other sub-bands by taking the absolute values of sub-bands obtained and subtracting one from other.

   F = (abs (S1L1)-abs (S1L1))>=1
   SEL1 = F * S1L1 + (~F) * S1L1
   F = (abs (T1L1)-abs (T1L1))>=1

TEL1 = F * T1L1 + (~F) * T1L1
F = (abs (U1L1)-abs (U1L1))>=1
UEL1 = F * U1L1 + (~F) * U1L1

8. Apply ISWT to the obtained sub-bands so as to extract 2nd watermark image.

imw2 = iswt (REL1, SEL1, TEL1, UEL1)

5 Parameters for Simulation

The algorithm is simulated on MATLAB, Core2Duo CPU, 4 GB RAM machine.

Wavelet filters for image transformations in LWT are 'haar'. The two watermarks used for embedding are of 'jpeg' format. The source for video taken is YouTube. Its parameters are shown in Table 1.

Table 1 Parameters of cover video

	Parameters	Test video
1	Width * height	640 * 360
2	Duration of video	8 s
3	No. of frames extracted	120
4	Rate of frame extraction	15

Table 2 Proposed algorithm results after evaluation of peak signal to noise ratio (PSNR) and correlation coefficient (CC)

	Attack	PSNR	CC of fused watermark	CC of WM1	CC of WM2
1	NO ATTACK	74.3547	0.9931	0.9795	0.9914
2	GUASSIAN (0.005)	33.8738	0.9923	0.9787	0.9893
3	GUASSIAN (0.01)	31.6654	0.9921	0.9785	0.9888
4	GUASSIAN (0.02)	30.1126	0.9916	0.9780	0.9875
5	SALTAND PEPPER (0.005)	42.5350	0.9924	0.9787	0.9893
6	SALTAND PEPPER (0.01)	39.6595	0.9924	0.9787	0.9893
7	SALTAND PEPPER (0.02)	36.8698	0.9923	0.9787	0.9891
8	ROTATION (15)	26.3210	0.9923	0.9787	0.9891
9	ROTATION (30)	25.7430	0.9923	0.9786	0.9889
10	ROTATION (45)	25.5949	0.9923	0.9786	0.9890
11	CROPPING (1/2)	29.2719	0.9925	0.9786	0.9890
12	CROPPING (2/3)	29.4307	0.9924	0.9786	0.9889

6 Evaluation Parameters

6.1 Peak Signal to Noise Ratio (Psnr)

In this age of HD and 3D quality videos, the perceptual quality of videos is of great importance to the viewers specially while broadcasting. So it is important that while embedding the watermark in the cover video, the quality of the video should not decrease significantly. To evaluate the video quality perception, we use PSNR. To calculate PSNR, we first have to calculate mean square error (MSE)

$$\text{MSE} = \frac{1}{n}\sum_{i=1}^{n}(x_1 - x_2)^2$$

$$\text{PSNR} = 10 * \log_{10}\left(\max_i^2 /\text{MSE}\right)$$

Max_i tells us about the maximum limit of pixel data. $\times 1$ is pixel value in the cover frame, and $\times 2$ is corresponding value of pixel in the watermarked frame.

PSNR value of around 40 db is taken as good for implementation of the given scheme.

6.2 Correlation Coefficient (CC)

To evaluate how much difference is there in the watermark embedded and extracted, we use correlation coefficient. It gives the perspective of how robust a watermark is towards the noise attacks. The highest value of CC is 1 and 0 is the minimum value. There is a trade-off between the visibility and sturdiness of the watermark.

7 Experimental Results

Evaluation for our scheme is based on PSNR and CC. Scaling coefficient (β) is taken as 0.5 (Table 2 and Figs. 2, 3, 4).

8 Conclusion

The proposed scheme outperforms many other schemes in literature [1]. This scheme involves videos which is one of the major sought after research domain and most venerable to spread illegally. This scheme also takes into account salt and pepper

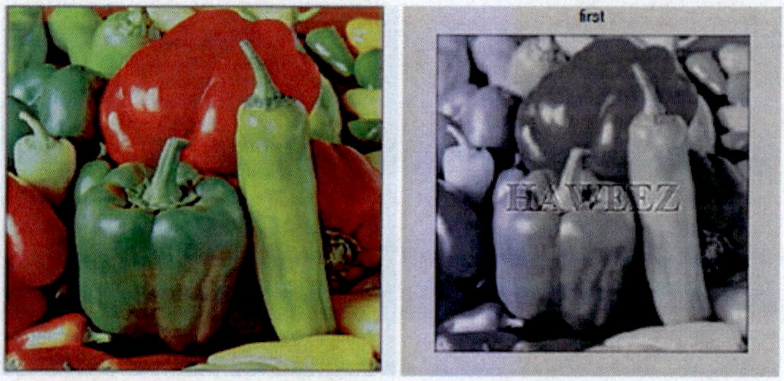

Fig. 2 Embedded and extracted watermark1 (WM1)

Fig. 3 Embedded and extracted watermark2 (WM2)

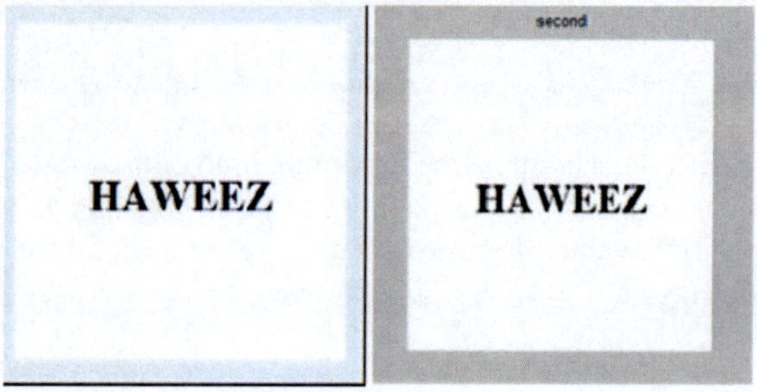

Fig. 4 Embedded and extracted fused watermark

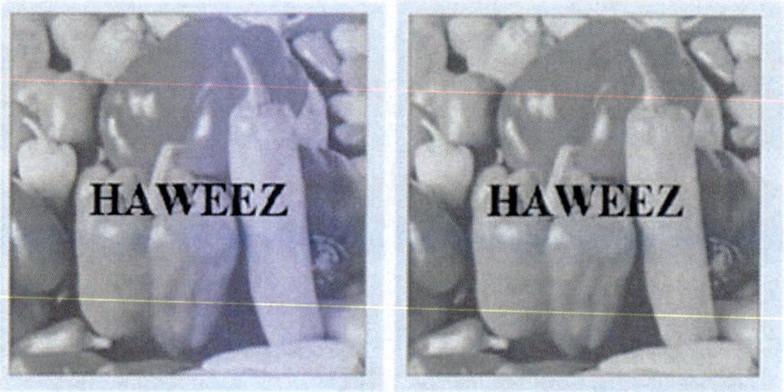

noise and geometrical noises, which are generally ignored in literature. The scheme ensures security as the results yield high correlation coefficient in both embedded watermarks as well as high PSNR of the video after various attacks. Applications of this scheme are enormous. One of them being that large amount of invisible information of the patient can be saved in a medical video or image.

References

1. Thind DK, Jindal S (2015) A semi blind DWT-SVD video watermarking. Procedia Comput Sci 46:1661–1667
2. Hemdan EE, El-Fishawy N, Attiya G (2013) Hybrid digital image watermarking technique for data hiding. In: IEEE 30th national radio science conference 2013, pp 220–227
3. Zheng W, Mo S, Jin X, Qu Y (2018) Robust and high capacity watermarking for image based on DWT-SVD and CNN. In: 13th IEEE conference on industrial electronics and applications 2018. IEEE
4. Wu HT, Cheung YM (2010) Reversible watermarking by modulation and security enhancement. IEEE Trans Instrum Meas 59:221–228
5. Ftima SB, Talbi M, Ezzedine T (2017) LWT-SVD secure image watermarking technique. In: International conference of electronics, communication and aerospace technology 2017, pp. 510-517
6. Csapodi M, Vandewalle J, Preneel B (1999) CNN algorithms for video authentication and copyright protection. J VLSI Sig Proc 23:449–463
7. Long ZS (2002) Image fusion using wavelet transform. In: Symposium on geospatial theory, processing and applications, Ottawa
8. Manjunath A, Ravikumar HM (2010) Comparision of DWT, LWT, SWT and S-transformin power quality analysis. Eur J Sci Res 39:569–576
9. Saxena P, Garg S, Srivastava A (2012) DWT-SVD semi-blind image watermarking using high frequency band. In: 2nd international conference on computer science and information technology 2012, Singapore, pp 28–29
10. Watanadelok PB, Rattanapitak W, Udomhunsakul S (2009) Multi-focus image fusion based on stationary wavelet transform and extended spatial frequency measurement. IEEE Trans Image Process 77–81
11. Liu R, Tan T (2002) An SVD-based watermarking scheme for protecting rightful ownership. IEEE Trans Multimed 4(1):121–128
12. Mirajkar PP, Sachin DR (2013) Image fusion based stationary wavelet transform. Int J Adv Eng Res Stud 2:99–101
13. Zhi-Ming Z, Rong-Yan L, Lei W (2003) Adaptive watermark scheme with RBF neural networks. In: IEEE international conference neural networks and signal processing, December 2003, pp 14–17
14. Khan MI, Rahman MM, Sarker MIH (2013) Digital watermarking for image authentication based on combined DCT, DWT and SVD transformation. Int J Comput Sci Issues 10(3)(1)
15. Showkat H (2014) SVD-DWT based digital video watermarking using fused images and low-middle frequency bands. Int J Adv Res Comput Sci Softw Eng 4(8)
16. Hartung F, Girod B (1998) Watermarking of uncompressed and compressed video. Sig Process 66(3):283–301
17. Vidyasagar MP, Han S, Chang E (2005) A survey of digital image watermarking techniques. In: 3rd IEEE international conference on industrial informatics
18. Xu D, Wang R, Wang J (2011) A novel watermarking scheme for H.264/AVC video authentication||. Sig Process 267–279
19. Aslantas V, Dogan LA, Ozturk S (2008) DWT-SVD based image water-marking using particle swarm optimizer. In: IEEE international conference on multimedia and expo 2008, IEEE Proceedings, Germany, pp 241–244

Modeling and Simulation of Solar PV-Based Grid-Tied Multilevel Inverter

Mohd Suhail Khan, Mirza Mohammad Shadab, Mohammed Asim, and Javed Ahmad

Abstract The conventional or normal line commutated converters/inverters have higher-order harmonics as it has square-shaped waveform of line current. Power transformers or other equipments became heated, and this heating occurs due to the existence of higher or excessive sequence of harmonics. Pulse width modulation-based converters/inverters having MOSFET/IGBT as switching devices possess high switching losses, as compared to thyristor and SCR. Thyristor-based forced commutated inverters/converters having issues of commutation, and it is not good for PWM applications. A new multilevel inverter topology is proposed in this paper, in which two half-wave, three-pulse thyristor-based converters are used to work as a three-level (multilevel) inverter, by controlling firing angles of switches. The multilevel inverter is powered by photovoltaic solar panels. This topology has an advantage of having no separate circuit for synchronization due to natural commutation. The proposed model is simulated and analyzed for two different conditions initially by connecting it with RLE load and after that the RLE load is replaced by the solar PV array in a three-phase three-pulse converter circuit. Both groups of thyristors are triggered at an angle more than 90° for inverting mode of operation. The proposed model is analyzed for different firing angle to reduce THD and to analyze the efficacy or potency of the proposed model.

Keywords MLI · THD · AC–DC converters

M. S. Khan (✉) · M. M. Shadab · M. Asim
EED Integral University, Lucknow, India
e-mail: mohdsuhail1991@yahoo.com

M. Asim
e-mail: asimamu@gmail.com

J. Ahmad
National Taiwan University of Science and Technology (NTUST), Taipei, Taiwan
e-mail: D10807822@mail.ntust.edu.tw

A. Iqbal et al. (eds.), *Renewable Power for Sustainable Growth*, Lecture Notes in Electrical Engineering 723, https://doi.org/10.1007/978-981-33-4080-0_43

1 Introduction

For a higher-power level above, 1 kW in rating normally a three-phase system is used [1–10]. As compared to single-phase converter, three-phase converters are provided for higher output voltage due to which the wave voltage reflected in the output voltage is high [1–10]. Multilevel inverters are widely used for medium-voltage high-power converters like fans, pumps and material transport depending on their specific features and restriction like power or voltage level, reliableness, price and other technological characteristics such as their unique characteristics and limitation [1, 2]. It results in a very simple tool to smooth the load voltage and load current. This is the reason which makes three-phase converters that are used in primarily excessive power variable speed drives. In high-power applications such as in the AC transmission line system, multilevel inverters are used. The increase in voltage level decreases the need for filters, on the other hand, higher efficiency is obtained as a result of a decrease in harmonics. A new idea has been developed to increase the level with a lower number of switches. This is made up of topology which is closely related to each application [3, 4]. Earlier, single-phase AC to DC converter is used as grid-tied inverters for the use of solar energy and feeding it to the grid. However, as a result of these cases, there is a high level of THD. In this paper, three-phase three-pulse circuits are used to connect loads and a DC voltage source, later on which acts as an inverter by controlling the switching angle α [5, 6]. MATLAB/Simulink model is examined. In this paper. the three-phase three-pulse converters are used and then total harmonics distortion is found at different condition for converter circuit. The two three-phase half-wave three-pulse MLI circuit has been used as a three-level inverter by varying the switching angle above 90°. Two three-phase, half-wave, three-pulse converter circuits have been shown in Fig. 1 [7, 8]. In the circuit, two half-wave

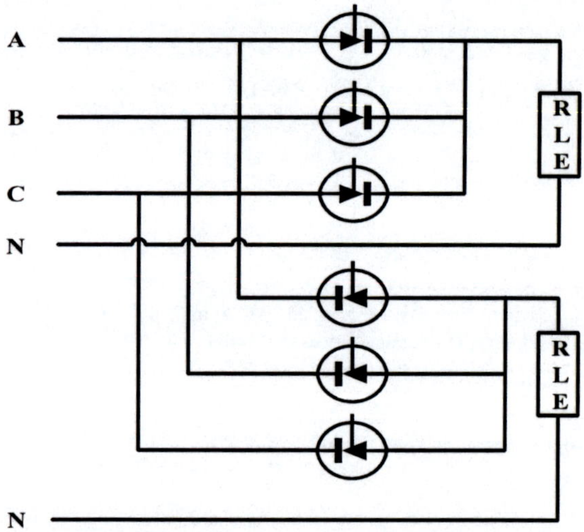

Fig. 1 Circuit diagram of two three-phase, three-pulse converters

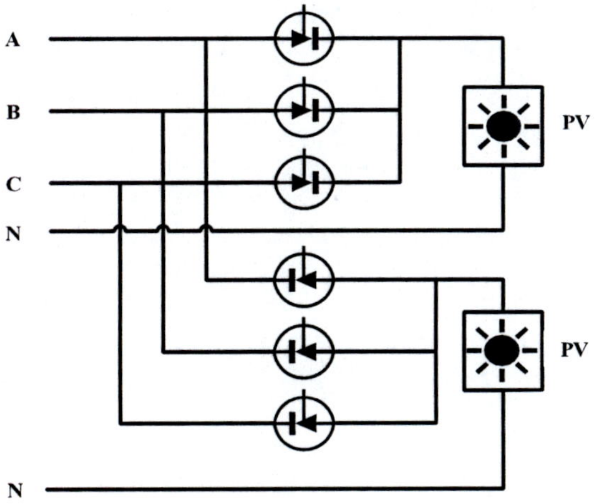

Fig. 2 Circuit diagram of the two three-phase, three-pulse converters with the solar array

converters have been connected in antiparallel mode via the same AC source and with independent loads. Both groups of thyristors are triggered at an angle greater than 90° for inverting mode of operation and after that the battery is replaced by the solar PV array in a three-phase three-pulse converter circuit which is shown in Fig. 2. Using various combinations of switching angles (α), THD and power transfer analyses are taken for both the condition, and the results are compared for both the circuits [7, 8].

2 Proposed Scheme

RLE load with a fully controlled converter, a multilevel converter projects two ways operation, namely rectification and inversion mode of operation. When the firing angle of the converter is hike above 90°, the converter act as an inverter and it acts like a rectifier for angle lower than 90°. By controlling the switching angle (α) and DC voltage source at the load end side, the line commutated control rectifier can operate in reverse mode operation. It is a phase-controlled converter with RLE load. Here, for DC-to-AC conversion, a multilevel circuit topology has been proposed where batteries are replaced by the solar PV array. Analyzation of the circuit has been for three levels of line current. For better performance, it can be increased to excessive levels. However, it adds the cost of the converter as the level of inverter increases. Therefore, a worthy deal should be created between the line total harmonic distortion and the cost of additional hardware. Two three-phase, three-pulse converters topology has been used for three levels which act as an inverter by linking a DC voltage source at the load end side and by managing the switching angle α.

3 Control Strategy

The load current one of two can be continuous or discontinuous. In instance of continuous current operation, the current of conducting thyristors does not drop to zero level, it depends on the condition of a DC voltage source, phase angle of load or inductor and the switching angle. For three-level line current, two three-phase, half-wave, three-pulse AC-to-DC converters are used. Two three-phase, half-wave, three-pulse AC-to-DC converters are shown in Fig. 2 in which two half-wave converters are connected in antiparallel through a same source with independent load. Both groups of thyristors are triggered or activated at angle above than 90° for inversion operation. T_1 is switched at firing angle 95° and subsequent switches were triggered with phase delay of 120°. One of the biggest advantages of three-pulse converter is that each converter has continuous current mode of operation, stepped sinusoidal wave resembles the waveform of line current and the harmonics content can be reduced drastically with the selection of switching angle.

4 Simulation of Three-Phase Three-Pulse Converter

The Simulink model of two three-phase half-wave three-pulse MLI has been shown in Fig. 3. It consists of three AC voltage sources one for each phase, six thyristors (T_1 to T_6). For each thyristor, the triggering pulses from a six-pulse generator with the help of a Simulink library. The DC source voltage is 24 V having negative polarity for inversion mode of operation. The R-L load is given as $R = 5\Omega$ and $L = 0.1$ H, respectively, for a two three-phase half-wave three-pulse MLI.

Initially, a MATLAB/Simulink model is obtained by simulating the circuit by connecting it with a RLE load. The solar PV array is connected in the circuit with the help of a controlled voltage source which converts the input signal from photovoltaic array into equivalent voltage source. T_1 is switched at firing angle 95° and subsequent switches were triggered with a phase delay of 120° deg. Total harmonic distortion and power transfer study is done for various combinations of switching angles.

5 Analysis and Simulation Results

Using MATLAB/Simulink, the simulation of the preferred model is done. The pulse generator block of the Simulink block set gives the triggering pulses. Using various combinations of switching angles (α), THD and power transfer study is done. It has been seen that the power transfer to the grid initially increases for increasing switching angle, thereafter it decreases after 115° (for battery array connected system) and for the solar PV array, connected system power transfer to grid constantly increases. The THD is almost constant up to 125°. Therefore, a suitable compromise is made for selecting the switching angle. Figure 4 shows the output waveform of the line current

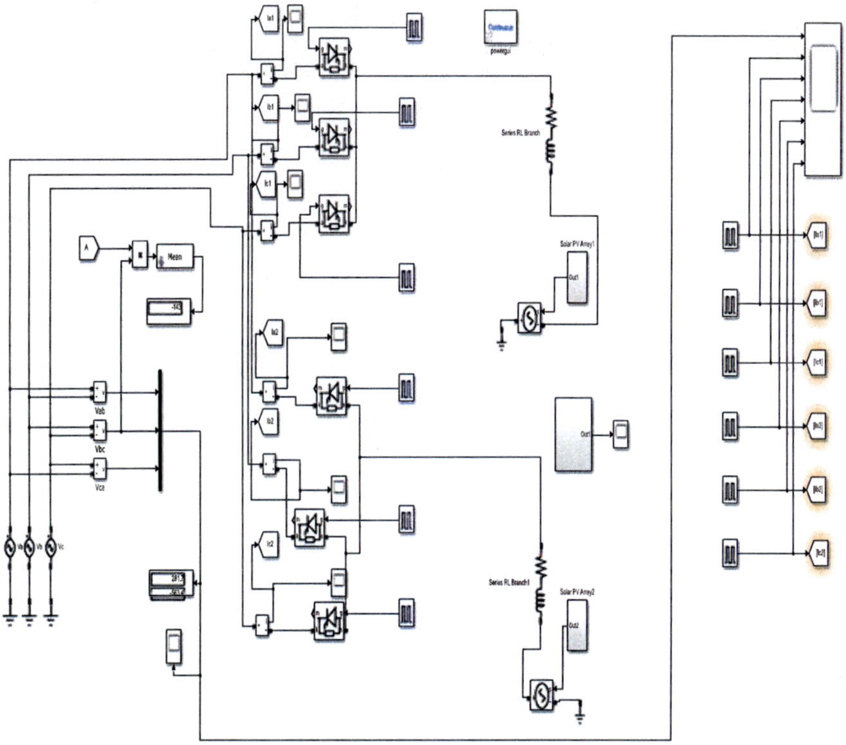

Fig. 3 Simulink model of Two three-phase, three-pulse converters

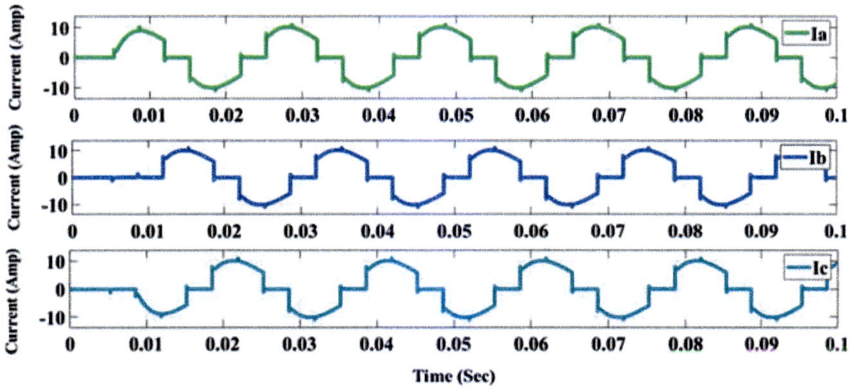

Fig. 4 Output waveform of the line current

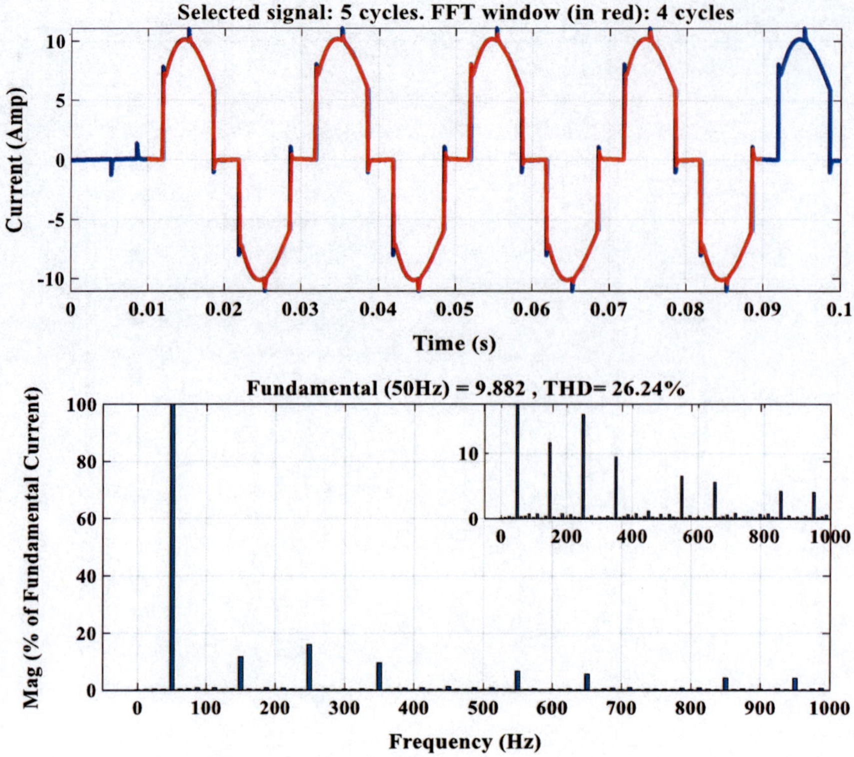

Fig. 5 Line current with THD and harmonics at switching angle α = 95° for the two three-phase, three-pulse multilevel inverters

of the proposed model. It has been shown in Fig. 5 that for the three-phase three-pulse converter, THD of line current comes out to be 26.70% at switching angle $\alpha =$ 95°. A comparison of the total harmonic distortion and power transfer analysis for various combinations of switching angles (α) for battery and solar array connected system has been shown in Table 1 and 2, respectively. Comparison of switching angle versus THD for battery and solar cell connected three-phase three-pulse converter system has been shown in Fig. 6, whereas comparison of switching angle versus power transfer for battery and solar cell connected three-phase three-pulse converter system has been shown in Fig. 7.

6 Conclusion

In this paper, modeling and simulation analysis of two half-wave, three-pulse converter are successfully done. Total harmonic distortion and power transfer examination are done for various combinations of switching angles. It has been shown

Table 1 Comparison of results (Switching angle vs. THD)

Switching angle (Deg)	THD (%)	
	With battery	With Solar Array
95	28.49	26.70
100	32.01	26.63
105	37.79	26.75
110	44.54	26.45
115	49.25	26.59
120	53.43	26.48
125	59.11	26.71

Table 2 Comparison of results (switching angle vs. power transfer)

Switching angle (Deg)	Power transfer (W)	
	With battery	With solar array
95	−64.55	−338.2
100	−146.1	−691.4
105	−187.2	−999.7
110	−199.8	−1260
115	−200.3	−1472
120	−194.5	−1634
125	−180.2	−1749

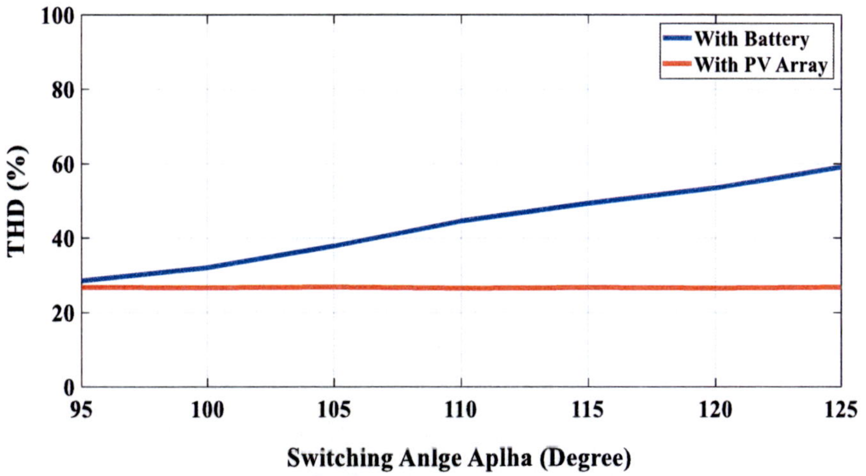

Fig. 6 Comparison of switching angle versus THD for battery and solar cell connected three-phase three-pulse converter system

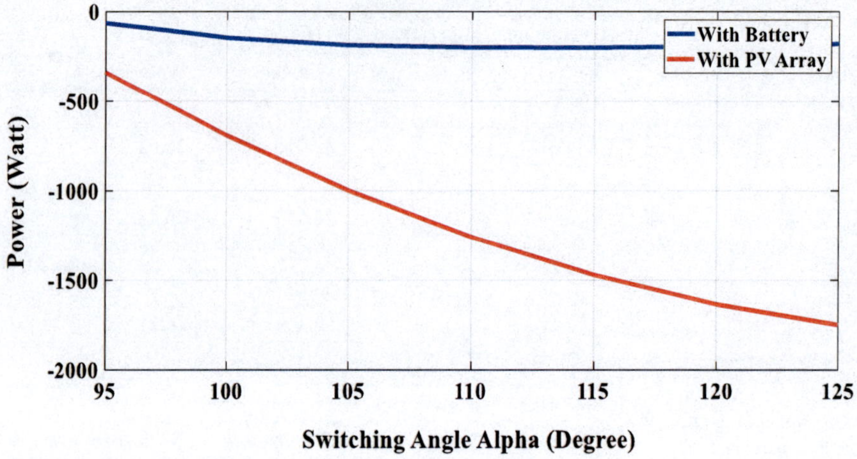

Fig. 7 Comparison of switching angle versus power transfer for battery and solar cell connected three-phase three-pulse converter system

that for three-pulse converter (solar PV array connected) system, the THD of line current comes out to be 26.70% at switching angle $\alpha = 95°$, and for battery-connected system, it is 28.49%. In this paper from the MATLAB/Simulink results shown, it has been clearly observed that in case of a battery-connected system as the firing angle rises the power delivered by the battery drop gradually and after a certain rise in the firing angle power which is delivered by a battery becomes constant. However, in the case of solar PV array, the power drops rapidly with the rise in the firing angle of the converter. From THD results, it is found that in case of the battery-based system, power delivered increases with the increase in firing angle; however, in the case of solar PV array, it almost remains constant with the change in the firing angle. For solar PV-based system, it is advisable to set the firing angle as high as possible because it increases the power delivering capacity of the solar PV array without increasing the THD of the output current.

References

1. Rashid MH Power electronics circuits, devices, and applications, 3rd edn. Pearson
2. Sabari Nathan L, Karthik S, Ravi Krishna S The 27-level multilevel inverter for solar PV applications. IEEE. 978-1-4673-0934-9/12/2012
3. Sarwar A, Asghar MSJ (2011) Simulation and analysis of a multilevel converter topology for solar PV based grid connected inverter. Smart Grid Renew Energ 2:56–62. https://doi.org/10.4236/sgre.2011.21007. Published Online February 2011
4. Sarwar A, Jamil Asghar MS Multilevel converter topology for solar PV based grid-tie inverters. In: 2010 IEEE international energy conference

5. Asim M, Tariq M, Mallick MA, Ashraf I, Kumari S, Bhoi AK (2018) Critical evaluation of offline MPPT techniques of solar PV for stand-alone applications. In: Advances in smart grid and renewable energy. Springer, Singapore, pp 13–21
6. Asghar MSJ (2004) Power electronics. PHI Learning, India, Chapter 6
7. Shadab MM, Tufail M, Alam Q, Mallick MA (2014) Simulation and analysis of a grid connected multi-level converter topologies and their comparison. J Electr Syst Inform Technol 1(2):166–174. Elsevier. https://doi.org/10.1016/j.jesit.2014.07.007
8. Shadab MM, Mallick MA, Tufail M, Jamil Asghar MS Development of a novel three phase grid-tied multilevel inverter topology
9. Raghav R, Raza A, Asim M (2015) Distributed power flow controller-an improvement of unified power flow controller. Development 2(5)
10. Asim M, Mallick MA, Malik A, Saaqib M (2015) Modelling and simulation of 5 parameter model of solar cell. Int J Electr Electr Comput Syst 4

Improvement of Voltage Stability Margin in a Radial Distribution System

Afroz Alam, Mohammad Zaid, Umair Shahajhani, and Adil Sarwar

Abstract Voltage instability is one among the main causes of blackout in a radial distribution system (RDS). There are many voltage-based indices for analyzing instability in voltages. In this work, voltage stability margin (VSM) is used to analyze voltage instability of a 69-bus system because VSM helps in identifying the bus on which voltage collapse may occur and accordingly, the loadability limit of a particular feeder section is decided. The VSM of all feeders is evaluated by determining the loading indices of individual feeders. Based on the values of VSM; the selection of the weakest feeder is made. On the bus of the weakest feeder having worst voltage profile, a distributed energy resource has been placed which has significantly improved voltage profile and VSM of all the feeders.

Keywords Voltage stability margin (VSM) · Distribution system · Feeder · Loadability limit

1 Introduction

Voltage stability is an important factor that allows the limit of loadability of a transmission or distribution system [1–9]. With the variation of load, voltage stability of a system also varies. The type of load also plays a vital role in voltage stability [1]. Variation in load is very common in the distribution system. Sudden variation in load leads to instability in voltage across a feeder which may result in voltage collapse on that feeder. Therefore, various voltage stability indices should be evaluated to monitor the VSM of a distribution system. Most of the indices which were utilized in earlier days have nonlinear characteristic. A novel index for its measurement has been proposed in [2]. The proposed VSM method has linear characteristic

A. Alam (✉) · M. Zaid · U. Shahajhani · A. Sarwar
Department of Electrical Engineering, Z.H.C.E.T., Aligarh Muslim University, Aligarh, UP, India
e-mail: afroz.iit@gmail.com

M. Zaid
e-mail: mohammad.zaid@zhcet.ac.in

A. Iqbal et al. (eds.), *Renewable Power for Sustainable Growth*, Lecture Notes in Electrical Engineering 723, https://doi.org/10.1007/978-981-33-4080-0_44

and provides better results as compared to other indices. It also gives a good approximation of the distance to voltage collapse. VSM is tested on a 69-bus system, and the result showed that by improving VSM of the system, voltage profiles also improve. The main role of VSM is to evaluate that particular bus on which voltage collapse may occur. The previous indices have been evaluated by considering losses in the system. Therefore, it is necessary to evaluate the system parameters near the voltage collapse point. In [3], authors have proposed a new index to calculate the VSM of the system and also to simultaneously reduce the losses in the system. In [3], authors have evaluated the VSM based on evaluation of local parameters instead of real-time measurements. The concept of VSM is mainly based upon the loadability limit of a bus or a feeder; lower the value of VSM, weaker is the bus and is more susceptible to voltage collapse, whereas higher the VSM, higher is loadability limit and hence the stability of the bus will be higher. Out of the several ways of calculating VSM, a generalized method is discussed in [3]. Allocation of distributed generation (DG) in a network leads to various advantages in a network which includes minimization in power losses and enhancement of voltage profile [4, 5, 8, 9].

In this paper, VSM is evaluated by considering loading indices of feeders as discussed in [2]. The concept of VSM is utilized to improve the voltage profile of the 69-bus system by the placement of type-1 DGs (generating active power only). Placement of DGs in a system leads to various advantages including minimization in power losses and augmentation of voltage profile.

This paper is divided into 6 sections. Section 2 describes the overall system configuration of the 69-bus system. Section 3 gives a brief detail about VSM and the loading index (L_i). Section 4 outlines the problem formulation to calculate VSM by considering the loading index of different feeders. In Sect. 5, results are discussed. Finally, the conclusion is discussed in Sect. 6.

2 System Configuration

A 69-bus test system consisting of 8 feeders and 68 branches is shown in Fig. 1. Average loading of the system is (4293.7 + j4041.9) kVA. Loading index for each feeder is evaluated. Based on loading index, VSM of each feeder is calculated. The feeder having least VSM is regarded as the worst feeder. To improve VSM and voltage profile, DGs of 4 and 2 kW are manually placed at 27th bus and 65th bus, respectively. The load flow analysis is done using MATLAB version 2015.

3 VSM

The VSM is characterized as the distinction in loadability breaking point and current working level. Lower value of VSM shows that the bus is experiencing voltage instability. It indicates the margin of critical point and operating point of voltages.

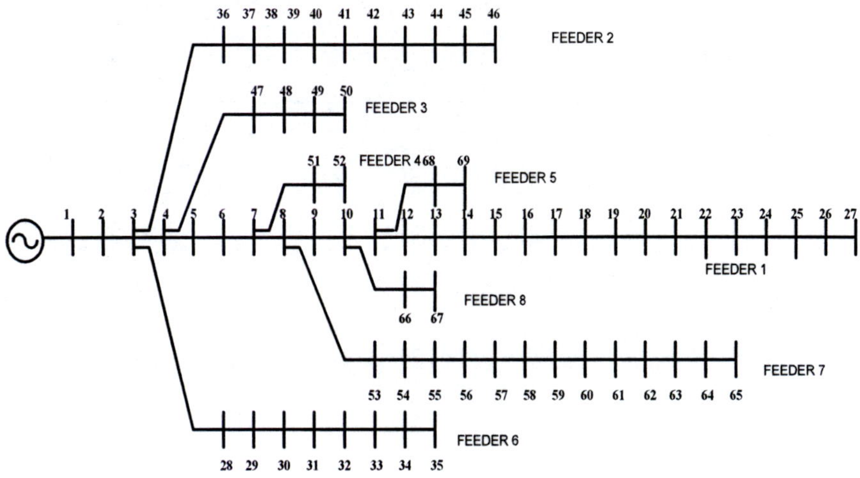

Fig. 1 IEEE 69-bus system

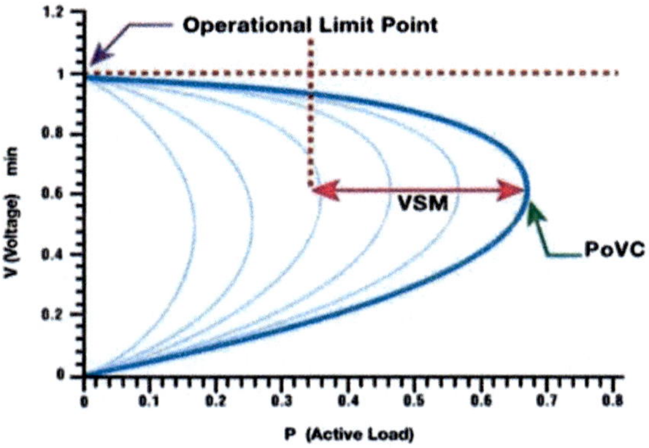

Fig. 2 Voltage stability margin [2]

Figure 2 shows a typical representation of VSM, where PoVC represents the voltage collapse point. Beyond PoVC, system enters into an unstable region. In any system, VSM values are used for the selection of weakest bus of the system. In the presented work, VSM is calculated by multiplying loading indices of individual feeders.

4 Problem Formulation

Consider a feeder as shown in the following figure (Fig. 3).

This feeder may be regarded as a small part of a big system. However, for the sake of simplicity, we may assume the first bus as having a generating source as shown in the above figure. It is a well-known fact that loads are not fixed as these vary from

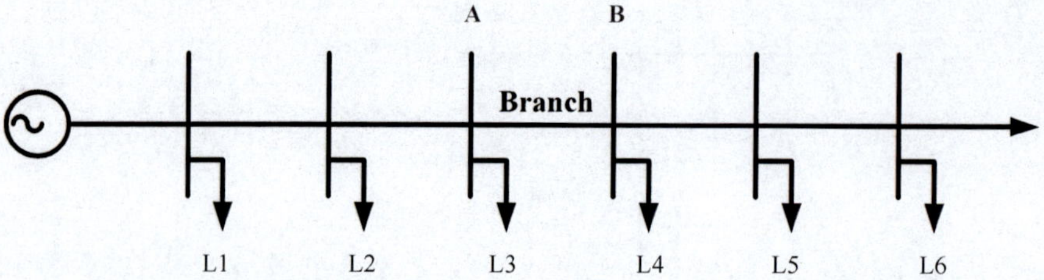

Fig. 3 A typical radial distribution system

a minimum value to a maximum value through the operating time interval. Now, consider the ith branch connected between the buses A and B. As explained in [2], the loading index (L_i) of each branch is evaluated by using Eq. (1).

$$L_i = \left(2\frac{V_B}{V_A}\cos\delta_{AB} - 1\right)^2 \tag{1}$$

where V_A and V_B are the voltages at buses A and B, respectively, and δ_{AB} can be obtained by subtracting the phase angle of voltage at Ath from phase angle of voltage at Bth bus and vice versa. After evaluating loading indices of all sections, the VSM of each feeder can be determined by multiplying these indices as given by Eq. (2).

$$\text{VSM} = \prod_{i \in \Omega} L_i \tag{2}$$

Here, Ω is the set of all sections of a feeder. By performing load flow analysis, voltage profile of all the buses of a feeder is evaluated. The bus which has the worst voltage profile is identified, and a DG is manually placed on that bus. In this paper, only type 1 (supplying only active power) DGs of sizes 4 and 2 kW are placed at the buses having poor voltages. The allocation of DG improves the VSM of the feeder.

Consider a branch (say b1) between bus 47 and bus 48 of feeder 3 of the system shown in Fig. 1. Similarly, consider branches b2 and b3 connected between buses 48 and 49 and 49 and 50, respectively. The loading index of b1 is evaluated by using Eq. (1) as follows:

$$L_{b1} = \left[2\frac{0.9978}{0.9982}\cos(-0.0013 + 0.0001447) - 1\right]^2$$

One may note that the voltages at bus 48 and bus 47 are 0.9978 and 0.9982, respectively. Also, δ_{48} and δ_{47} are −0.0013 and −0.0001447, respectively. Similarly, loading indices of branch b2 and b3 are determined as 0.9998 and 0.9998, respectively. Hence, by multiplying the loading indices of b1, b2, and b3, the VSM of feeder 3 can be determined as

$$\begin{aligned} \text{VSM} &= [L_{b1} * L_{b2} * L_{b3}] \\ &= [0.9995 * 0.9998 * 0.9998] \\ &= 0.9991 \end{aligned}$$

Power losses of the system have been calculated as follows

$$P_L = \sum_{p=1}^{N_{br}} I_P^2 R_P$$

where

P_L = power loss,
N_{br} = number of feeders,
I_P = current in branch p,
R_P = resistance of branch p.

5 Results and Discussion

VSM provides a more accurate distance to the bus on which voltage collapse may occur. The concept of VSM is applied to the 69-bus system; the VSM of each feeder is calculated. Results obtained for the 69-bus system are divided into the following 3 cases:

Case 1: When no DG is placed

When no DG is connected, power flow is in one direction only. In this case, the VSM of individual feeders is obtained as 0.7541, 0.8318, 0.9990, 0.9998, 0.9998, 0.8714, 0.7996, and 0.9998. Since feeder 1 has the lowest value of VSM, i.e., 0.7541, it is the weakest feeder and prone to voltage collapse. Figure 4 shows that voltages at buses 27 of feeder 1 and 65 of feeder 7 have the poor voltage profile.

Case 2: When one DG is placed at 27th bus

When DG of 4 kW is connected at the 27th bus of the 69-bus system, the VSM of all individual feeders is obtained as 0.8766, 0.8596, 0.9991, 0.9998, 0.9998, 0.8846, 0.8510, and 0.9998. Now, it can be observed from Fig. 5 that VSM of the weakest feeder has been significantly improved from its previous value. Also, there is marginal improvement in the voltage profile of the 27th bus.

Case 3: When two DGs are placed at 27th bus and 65th bus

When multiple DGs of size 4 kW on the 27th bus and 6 kW on the 65th bus are placed, the VSM of individual feeders is determined as 0.9007, 0.8835, 0.9991, 0.9998, 0.9998, 0.9075, 0.8747, and 0.9998. It can be observed that VSM has further

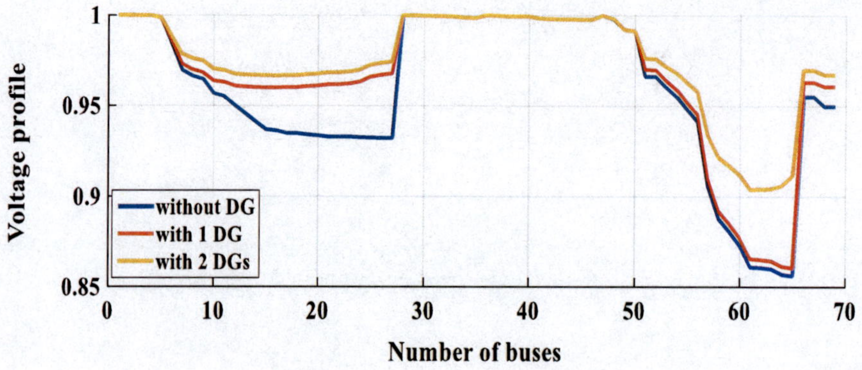

Fig. 4 Voltage profile of 69-bus system

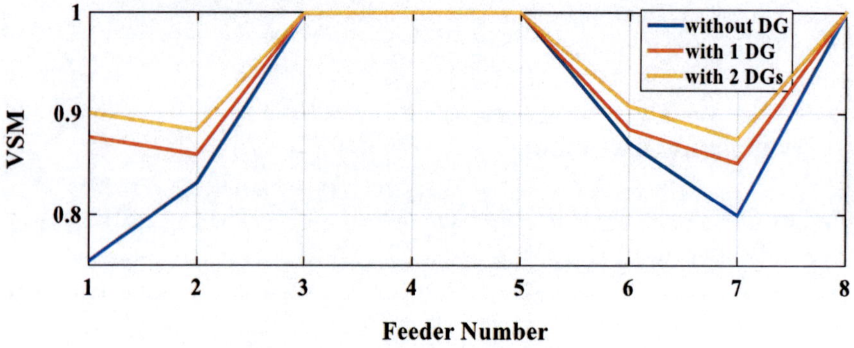

Fig. 5 VSM of feeders

been improved from its previous value. Also, there are marginal improvements in the voltage profiles of 27th and 65th buses.

The 69-bus system shown in Fig. 1, consists of 8 radial feeders and a baseload of (4293.7 + j4041.9) kVA. In the given system, branch 26 (which connects buses 26 and 27) of feeder 1 having VSM as 0.7541 and branch 64 (connected between 64 and 65) of feeder 7 having VSM 0.7996 are found to have the lowest value of VSM in their respective feeders. Therefore, feeder 1 and feeder 7 are considered as the vulnerable feeders of the system, whereas bus 27 is having voltage as 0.9322 pu and bus 65 having voltage as 0.8560 pu is found to have the lowest voltage bus in their respective feeders. Comparatively, bus 65 has the poorest voltage profile among all the buses of the system and hence considered as the worst voltage profile bus. Table 1 shows the system power losses for the cases discussed above. It can be inferred that apart from improvement in worst voltage value, the system power loss is also significantly reduced after placement of DGs. The system power loss without DG was 223.5 kW. When a single DG is placed at bus 27, the VSM of the system is

Table 1 Results for 69-bus system

No. of DG(s) placed	Average VSM of 69-bus system	Worst voltage (in pu)	The average real power loss of 69-bus system in kW
No DG is placed	0.9069	0.8560	223.5
1 DG of 4 kW placed at bus 27	0.9337	0.8603	91.1
2 DGs of sizes 4 and 6 kW placed at buses 27 and 65	**0.9456**	**0.9112**	**38.5**

improved to 0.9337 and the power loss is reduced to 91.1 kW; similarly when two DGs are placed at requisite buses, the power loss is further reduced to 38.5 kW.

6 Conclusion

In this work, VSM is utilized to improve the bus voltages of a system. For this, first the feeder with poorest VSM is identified and DG is manually placed at the bus having the worst voltage profile on that feeder. For the test system (consisting of 8 feeders), loading indices of individual branches of a feeder are evaluated and multiplied together to get the VSM of that feeder. It can be concluded that, by the placement of DG(s), VSM of all feeders has improved from their previous values and the system power loss is also reduced significantly. Improving the voltage profile with the help of VSM is more effective as compared to the other voltage stability indices because it varies linearly with the load. In the future work, metaheuristic techniques for optimizing the sizes and positions of various DGs for the purpose of improving the VSM may be utilized.

References

1. Taylor CW (1994) Power system voltage stability. McGraw-Hill
2. Haque MH (2006) A linear static voltage stability margin for radial distribution systems. In: 2006 IEEE power engineering society general meeting, Montreal, Que, pp 1–6
3. Zabaiou T, Dessaint L, Kamwa I (2014) Preventive control approach for voltage stability improvement using voltage stability constrained optimal power flow based on static line voltage stability indices. 8:924–934
4. Vu K, Begovic MM, Novosel D, Saha MM (1999) Use of local measurements to estimate voltage stability margin. IEEE Trans Power Syst 14(3):1029–1035
5. Fateh A, Nor M, Sulaiman M (2018) Determining voltage stability margin values by measuring the hypotenuse under PV and QV curves. 1(1). IEEE

6. Zaid M, Alam A, Sarwer Z, Shahajhani U (2019) Optimal allocation of distributed energy resources in a distribution system. In: 2019 innovations in power and advanced computing technologies (i-PACT), Vellore, India, pp 1–6
7. Alam A, Gupta A, Bindal P, Siddiqui A, Zaid M (2018) Power loss minimization in a radial distribution system with distributed generation. In: 2018 international conference on power, energy, control and transmission systems (ICPECTS), Chennai, pp 21–25
8. Iqbal A et al (eds) (2020) Soft computing in condition monitoring and diagnostics of electrical and mechanical systems, vol 1096. In: Advances in intelligent systems and computing. Springer, Singapore. https://doi.org/10.1007/978-981-15-1532-3
9. Iqbal A et al (eds) (2020) Meta heuristic and evolutionary computation: algorithms and applications, vol 1096. In: Studies in computational intelligence. Springer, Singapore. https://www.springer.com/gp/book/9789811575709

Optimization Algorithm-Based Maximum Power Point Tracking Techniques for Solar PV Systems

Piyush Agrawal, Mohammed Asim, and Mohd Tariq

Abstract Recently, researchers have focused more on the optimization-based technique for MPPT for PV systems. Modern optimization-based techniques are very effective in finding global maxima from the pool of many local maxima. This paper gives an overview of modern optimization techniques based MPPT for PV system. The algorithms discussed in the paper are PSO, cuckoo search algorithm, and genetic algorithm (GA). The study is based on the most recent techniques available in literature. This paper should serve as a reference for future research in optimization algorithm-based MPPT techniques applied to PV systems.

Keywords Photovoltaic systems · MPPT · Particle swarm optimization · Cuckoo search algorithm · Genetic algorithm

1 Introduction

Energy demands have seen a significant rise. Coping with this, a sustainable manner government all over the world is expanding their share in PV technologies. With new technologies coming into existence for solar PV, it has become one of the most potential systems to overcome the rising energy demands. Solar PV is economical in operation, and they produce clean energy. Solar PV over the years have role in sustaining degradation by serving as an alternative to fossil fuels [1].

Energy harvesting from solar is less efficient. MPP tracking is an essential part of PV systems. MPPT techniques vary in their application type [2], complexity, sensor requirement, ease of application and designing, and in many other respects. With the amount of research conducted in this field, it becomes very difficult to point out the perfect technique. The survey of papers from our bibliography shows the MPPT topic that is being monitored and has received a considerable amount of attention in

P. Agrawal · M. Asim
Department of Electrical Engineering, Integral University, Lucknow, India

M. Tariq (✉)
Department of Electrical Engineering, ZHCET, AMU, Aligarh, India
e-mail: tariq.ee@zhcet.ac.in

A. Iqbal et al. (eds.), *Renewable Power for Sustainable Growth*, Lecture Notes in Electrical Engineering 723, https://doi.org/10.1007/978-981-33-4080-0_45

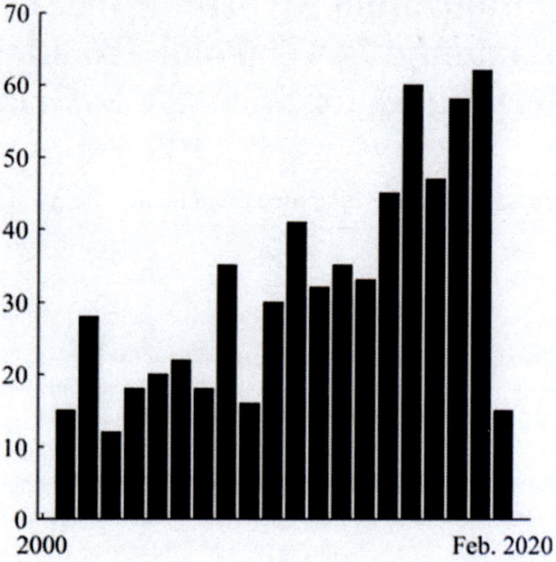

Fig. 1 Number of MPPT papers per year since 2000

the past two decades (Fig. 1). Recently, a shift over modern optimization algorithm-based MPPT techniques has been observed. Many papers are available in the literature for traditional MPPT techniques [1, 2]. But literature lacks a paper dedicated solemnly on modern optimization-based MPPT technique. With new optimization-based MPPT techniques being reported at a high pace, this manuscript would aid benefits to researchers.

This manuscript discusses the latest additions in the field of MPPTs. We have avoided the discussion on non-popular or the techniques showing lesser improvements from the previously obtained MPPTs. The algorithm selected for this paper is PSO, cuckoo search, and GA. Paper draft is as follow: in Sect. 2, the problem formulation is discussed, section three discusses the MPPT algorithms, discussion on result are presented in the next section, and the conclusion is given in last.

2 Problem Statement

The problem of MPPT can be understood by the means of Fig. 2. The plot of PV array is shown in Fig. 2, and the objective of any MPPT technique is to find and make the operation of PV array for which power extracted from the array is maximize. This MPP is obtained under specific condition, as generation of PV array is a function of temp. and irradiance. Optimization-based MPPTs are particularly preferable, and they have proven way better than conventional, as in partial shading conditions many local maxima may exist. In contrast, if optimization algorithm-based MPPT

Fig. 2 Power-characteristics of PV array

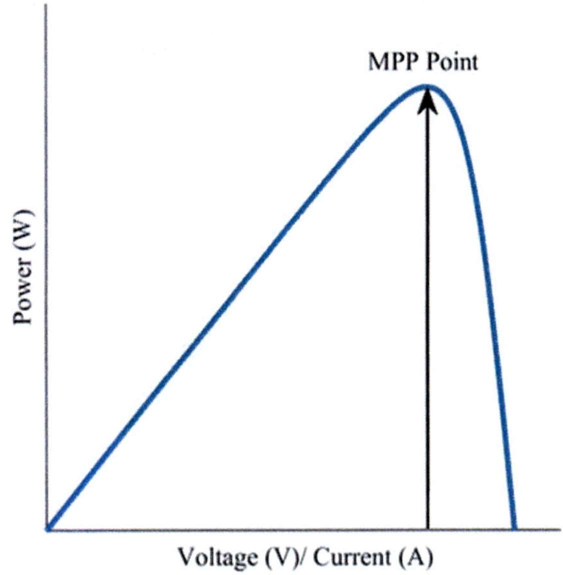

techniques are not tuned properly, they might require a larger time to reach to MPP [3].

A block diagram of solar PV system is shown in Fig. 3. The boost converter is selected for the application of MPPT techniques. PV array is connected in series with a boost converter, and duty ratio of converter is decided by the MPPT technique. It is feed to PWM generator for generating switching pulses of the converter. Modeling

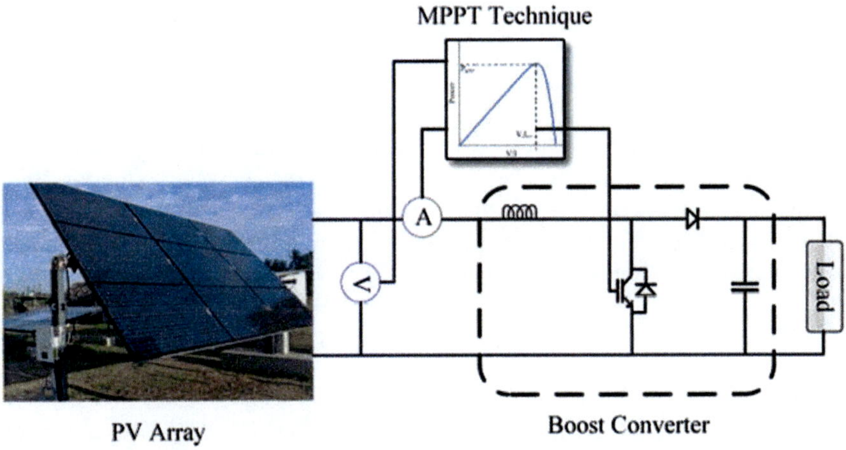

Fig. 3 Schematic diagram of a PV array

Table 1 Simulation parameters

Simulation parameters	Value
Switching frequency	50,000 Hz
L	1 mH
C	1 mF
Load	50 Ω

of the PV system and MPPT technique is done in MATLAB. Simulation parameters of modeling are given in Table 1.

3 MPPT Techniques

MPPT is important aspects of the PV, and this paper revolves around three modern optimization algorithms based on MPPT techniques.

3.1 PSO

PSO was given by Kennedy in 1995 [3]. This algorithm mimics the behavior of a group of animals. The term particle denotes a single individual in a group like a bird in a flock or individual fish in a school. Each particle in a group behaves in a distributed manner having individual intelligence and the intelligence of collective group, where if one individual finds food all the group instantly tries to follow it no matter how far from the food they individually located [4]. Optimization methods that work on swarm intelligence are known as behaviorally inspired methods, unlike GA and DE which are the evolutionary type of techniques. While solving the optimization problems from the PSO method, the particles are initially scattered in the domain. The particles have two characters namely position and velocity. All particles remember their best searched position, particles interact with each other and adjust their velocity and position keeping in mind their best searched position and best among the positions of individual particle known as global best position.

Particles update their speeds and positions according to Eqs. (1) and (2) until finding two extreme points as shown in.

$$V_i(n+1) = w * V_i(n) + rand * c_1(p_{best}(n) - x_i(n)) + rand * c_2(g_{best}(n) - x_i(n)) \tag{1}$$

$$x_i(n+1) = x_i(n) + v_i(n+1) \tag{2}$$

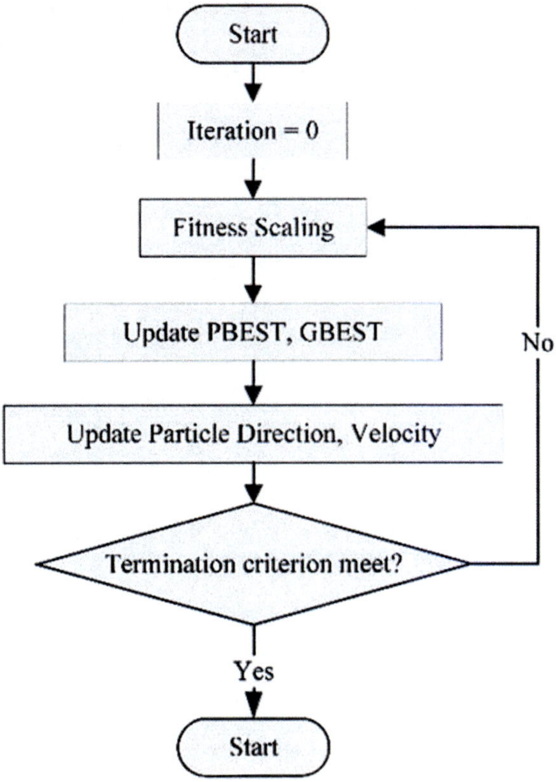

Fig. 4 Flowchart of PSO

where i is the particle index; w is the internal coefficient of velocity, c_1, c_2 are acceleration coefficients, and rand is a random constant generated between 0 and 1. The working of PSO is shown in Fig. 4, and simulation results obtained from PSO algorithms MPPT are depicted in Fig. 5.

4 GA

GA was introduced in by Henry Holland in the early 1960s [5]. GA is inspired by natural genetic and selection of the fittest theory of Darwin. Initially, a population is created by randomly generating solutions in the search domain. After that new population is created by crossover mutation and survival operators. Figure 6 represents the results obtained by the GA in MPP tracking. The algorithm GA is given below.

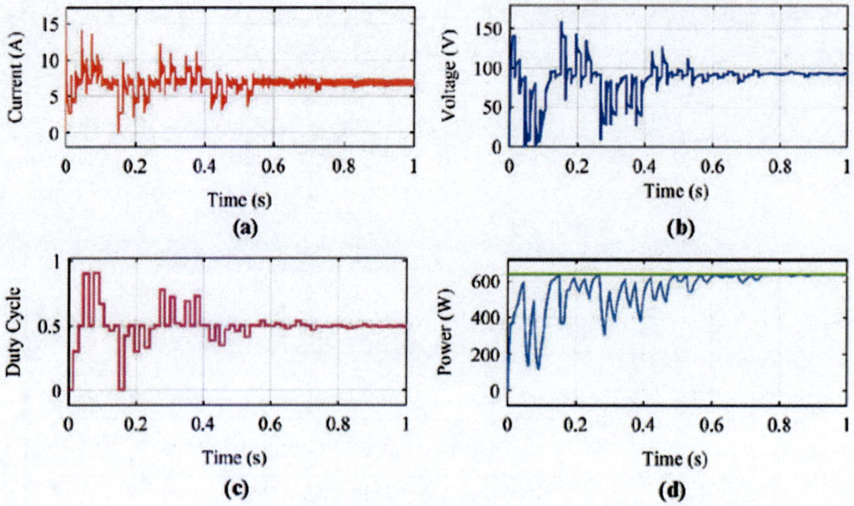

Fig. 5 Results obtained from PSO algorithm for MPPT, **a** PV output current, **b** output voltage, **c** duty ratio of converter, **d** tracked power

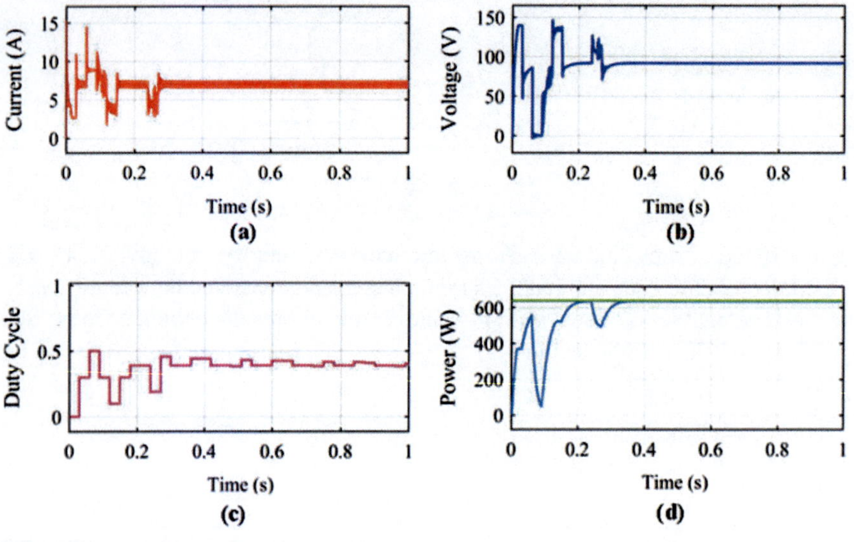

Fig. 6 Results obtained from GA for MPPT, **a** PV output current, **b** output voltage, **c** duty ratio of converter, **d** tracked power

```
while Generation • 5000 do
    Selection
    Crossover
    Mutation
    if Child has better fitness then
        Child Survives
    else
        Parent Survives
    end if
end while
Selection algorithm:
x = y
while x == y do
    n = rand
    m = rand
    Compute scaled fitness of chromosomes
    x = chromosome falling in the range of n
    y = chromosome falling in the range of m
end while
Crossover Algorithm:
n = 1
while n • 32 do
    if n • 16 then
        Copy bits from First Parent
    else
        Copy bits from Second Parent
    end if
    n = n + 1
end while
Mutation algorithm:
n=rand
if n ≤ 0.2 then
    bit = randi(32)
    child(bit) = rand
else
 child = child
end if
```

5 Cuckoo Search Algorithm

It is proposed in 2009 by Xin-She-Yang and Saush Deb [6]. This algorithm mimics the reproductive strategy of cuckoos. The cuckoo lays an egg in a host communal nest. If the host birds found out the peracetic eggs, parasitic birds either throw away the laid eggs or leave the nest. But some cuckoo's have the ability to mimic the eggs of host birds. Similar looking eggs create an illusion in host birds' minds and chances of cuckoo eggs to survive increases.

The cuckoo search algorithm was also applied for finding MPPT, and the algorithm was implemented using MATLAB. Results obtained by the cuckoo search algorithm are shown in Fig. 7.

6 Discussion

Many MPPT technique and grid integration were reported in literature [7–11]. With traditional techniques such as P&O and IC being popular among all, new optimization-based techniques have recently been reported to perform well in varied conditions. There is not a thumb rule to choose any MPPT for a particular system. It depends upon applicability and ease of implementation and system requirements. Ease of implementation makes an MPPT very popular, with traditional MPPT's such as P&O are easier to implement than the MPPT's involving modern optimization algorithms. But modern algorithm-based techniques show a greater performance for partial shading conditions. In this study, it is found out that the convergence rate of the cuckoo search algorithm was better than other search algorithms. All the techniques were implemented in MATLAB®/Simulink. Table 2 represents the major characteristics of the discussed MPPT techniques.

7 Conclusion

In this paper, optimization algorithm-based MPPT techniques are discussed. A comparison of these techniques is made. Modern optimization algorithms are very competitive in founding MPPT in very wide scenarios, where many local minima are present, but they suffer from the condition of sluggish behavior and complexity in construction. While traditional MPPT's are easy to implement, but they are not suitable in many minima conditions. In this report, it is found out that the cuckoo search algorithm is the most suitable optimization algorithm for the MPPT application. Concluding table and discussion serve as a guide for future research.

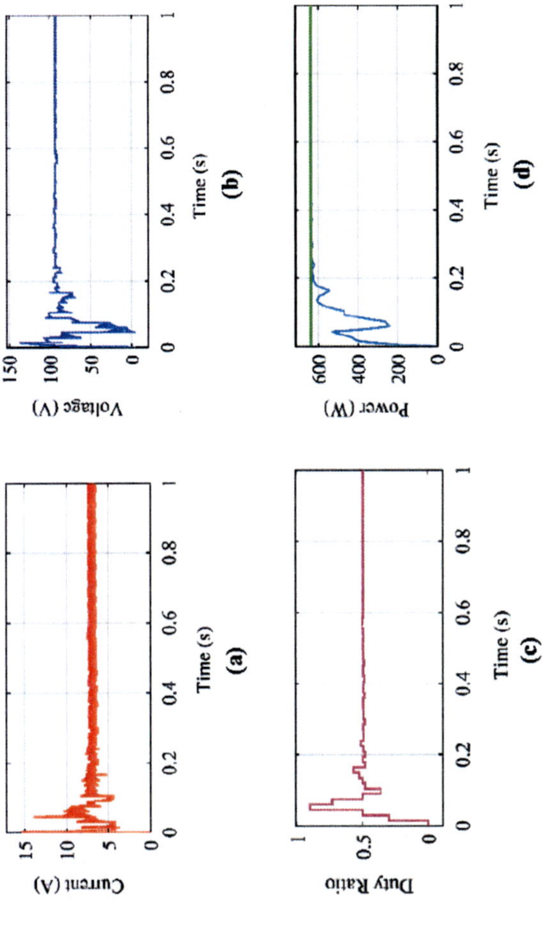

Fig. 7 Cuckoo search algorithm for MPPT, **a** PV output current, **b** output voltage, **c** duty ratio of converter, **d** tracked power

Table 2 Characteristics of different MPPT's

Technique	Tuning required	Speed	Complexity	Required signals
GA	Yes	Fast	High	Varies
PSO	Yes	Very fast	High	Varies
Cuckoo search	Yes	Very fast	Very high	Varies

Acknowledgements The authors would like to acknowledge Integral University for providing the MCN number IU/R&D/2020-MCN000821.

References

1. Iqbal MT, Tariq M, Ahmad MK, Arif MSB (2016) Modeling, analysis and control of buck converter and Z-source converter for photo voltaic emulator. In: 2016 IEEE 1st international conference on power electronics, intelligent control and energy systems (ICPEICES), Delhi, pp 1–6
2. Khan MJ, Mathew L (2017) Different kinds of maximum power point tracking control method for photovoltaic systems: a review. Arch Computat Methods Eng 24:855–867
3. Eberhart R, Kennedy J (1995) New optimizer using particle swarm theory. In: Proceedings of international symposium on micro machine and human science, pp 39–43
4. Shi H, Wen H, Hu Y, Jiang L (2018) Reactive power minimization in bidirectional DC-DC converters using a unified-phasor-based particle swarm optimization. IEEE Trans Power Electron 33(12):10990–11006
5. Tang KS, Man KF, Kwong S, He Q (1996) Genetic algorithms and their applications. IEEE Sig Process Mag 13(6):22–37
6. Gandomi AH, Yang X, Alavi AH (2013) Cuckoo search algorithm: a metaheuristic approach to solve structural optimization problems. Eng Comput 29:17–35
7. Asim M, Tariq M, Mallick MA, Ashraf I An improved constant voltage based MPPT technique for PMDC motor. Int J Power Electron Drive Syst 7(4)
8. Tariq M, Iqbal MT, Iqbal A, Meraj M, Roomi MM, Khan MSU (2016) Comparative analysis of carrier schemes for PWM in multilevel PUC inverter for PV applications. In: 2016 4th international conference on the development in the in renewable energy technology (ICDRET), Dhaka
9. Asim M, Tariq A, Tariq M (2011) Simulink based modeling, simulation and performance evaluation of an MPPT for maximum power generation on resistive load. In: 2nd international conference on environmental science and technology. IACSIT Press, Singapore
10. Tariq M, Iqbal MT, Meraj M, Iqbal A, Maswood AI, Bharatiraja C (2016) Design of a proportional resonant controller for packed U cell 5 level inverter for grid-connected applications. In: 2016 IEEE international conference on power electronics, drives and energy systems (PEDES), Trivandrum, pp 1–6
11. Bharatiraja C, Munda JL, Bayindir R, Tariq M (2016) A common-mode leakage current mitigation for PV-grid connected three-phase three-level transformerless T-type-NPC-MLI. In: 2016 IEEE international conference on renewable energy research and applications (ICRERA), Birmingham, pp 578–583

A Review of Power Factor Correction and Reduction in Total Harmonic Distortion for LED Drivers

Akanksha Verma, Rajesh Narayan Deo, Mahmood Alam Ansari, Megha Tomar, and Gautam Nath

Abstract The light-emitting diode (LED) driver is of great importance for efficient lighting. The main challenge for the LED driver is to reduce the THD and improve the PF of the circuit. Low power factor in LEDs makes it unsuitable for commercial use. Low power factor not only increases the losses but also is destructive for the electrical equipments. This paper deals with the technologies used for power factor correction and the requirement for using such technologies. It also covers the various problems related to power factor and the related works on power factor and THD improvement for LED driver.

Keywords LED driver · Total harmonic distortion (THD) · Power factor correction (PFC)

1 Introduction

LEDs need DC supply of voltage after the conversion of supplied current by the power line. Hence, the performance characteristics totally depend upon the design of LEDs [1–22]. LEDs are now being used in the lighting market, and the conventional lighting sources are replaced by LEDs for the purpose of decoration, display, and public lighting applications. Since the LED has long lifetime and is mercury-free, its popularity is increasing in lighting market [1–22]. The shape, size, color, and pattern of illumination can be varied accordingly in LEDs [1]. The V-I characteristic of LEDs resemble with that of a diode. It implies that a minute variation in the voltage can cause huge fluctuation in the current and its luminous outputs. So, LEDs should be powered by constant current source [2, 3].

AC sinusoidal voltage source is converted to rectified sinusoidal voltage either through half or full wave rectifier before taking it in use. Block diagram of light-emitting diode driver structure is shown in Fig. 1. This is the basic structure of the driver. Various converter topologies can be used in power factor correction. LEDs

A. Verma (✉) · R. N. Deo · M. A. Ansari · M. Tomar · G. Nath
Department of Electrical and Electronics Engineering, Galgotias College of Engineering and Technology, Greater Noida 201306, India
e-mail: akanksha24v@gmail.com

A. Iqbal et al. (eds.), *Renewable Power for Sustainable Growth*, Lecture Notes in Electrical Engineering 723, https://doi.org/10.1007/978-981-33-4080-0_46

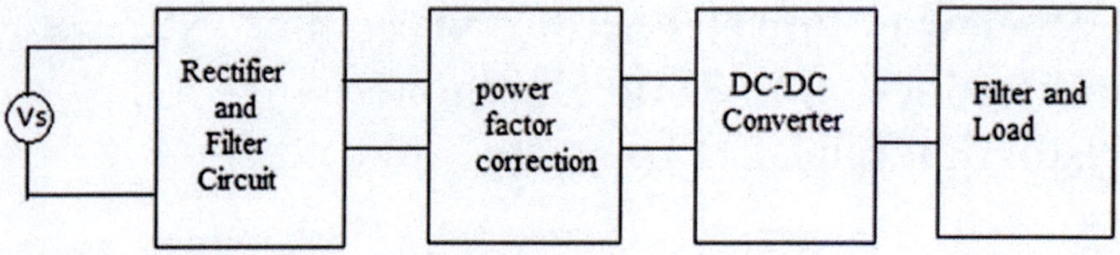

Fig. 1 Block diagram of LED driver

powered by AC power source are considered to be nonlinear loads. Due to this nonlinearity, the LEDs are considered to have low PF, and the total harmonic distortion (THD) is also large when compared to other lighting devices. Although LEDs save lot of energy, due to low power factor, the incandescent lamp is preferred over LEDs due to its high input power factor in the commercial use. Due to low power factor, the useful power gets wasted in the form of resistive losses in wiring. The other consequence of low power factor is THD. So design factors improve power factor and also trim down THD. Hence, it is beneficial to correct power factor for many types of load [21, 22].

Power factor is related to apparent power and real power. Real power is the capacity to operate in a given time limit. The product of voltage and current of the circuit is apparent power. Power factor may be used to find out the contribution of current to real power in the load. The load which has better power factor requires less current. A load with a power factor equal to unity demands least amount of current. Low pf and increased THD may result in high energy costs transmission losses and damage of equipments.

The power factor in case of LED is 30% less than that of incandescent lamps. The input current in case of LED contains high amount of harmonics due to which large portion of power is wasted in the form of reactive power.

2 Power Factor and Total Harmonic Distortion

Reason for low pf may be due to phase difference of voltage and current. The other reason may be distortion and harmonics in the current. To overcome this problem, it is requested to couple a power factor correction (PFC) stage to the bridge rectifier which will improve the current shaping to be more sinusoidal. Power factor correction means correcting the characteristics of the parameters that produce power factor <1. Power factor control is required to balance the effects of harmonics due to nonlinear loads like rectifier and smoothing capacitor. Design of LED driver that improves power factor and reduces THD is regarded as improving PF/THD performance. A power factor corrector includes amplifier and multiplier arranged in a manner to maintain the power factor.

Nonlinear loads are responsible for creating harmonic currents apart from the original AC current. Addition of linear component like capacitor and inductor would not help to reduce the harmonic currents. Active power factor correction is the best approach toward improvement of power factor and reduction of THD. The active PFC controls the input current to the load and maintains it as the voltage waveform (sinusoidal wave). Boost circuits, buck circuits and buck–boost circuits are common active PFC. Boost converters can be inserted between bridge rectifier and the input capacitors. Boost converter maintains the constant DC voltage at its output and draw current which is in phase with frequency and voltage. Active PFC can increase power factor up to 0.99 (99%).Switched mode converters are used for high-power LED application for better efficiency [4]. Circuit diagram of boost converter and buck–boost Converters is shown in Fig. 2. From Table 1, it is seen that boost and buck–boost converters are best for LED lighting applications.

Buck, boost and buck–boost are the basic topologies of DC–DC converter. From Table 1, it is clear that boost and buck–boost operate well in DCM. Input fed to these topologies is not pure DC. Hence, rectifier is needed here. Output of the rectifier is pulsating dc. This pulsating DC drives the main circuit. Pulsating DC has high value of ripples and harmonic contents. This results in low power factor and increased THD.

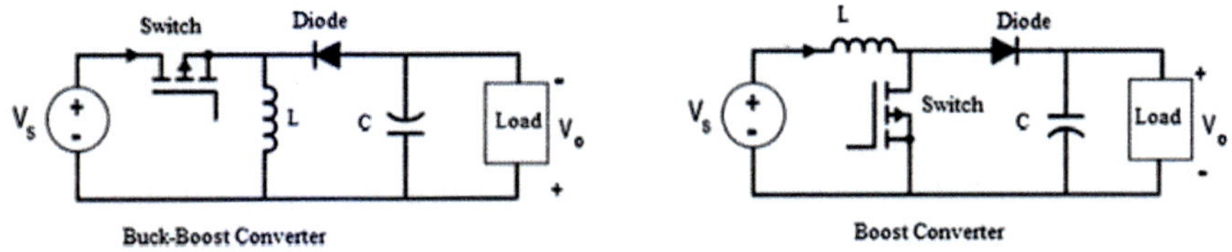

Fig. 2 Buck–boost and buck converter

Table 1 Comparison of converters

Converters	Waveform of line current	PFC in DCM
Buck	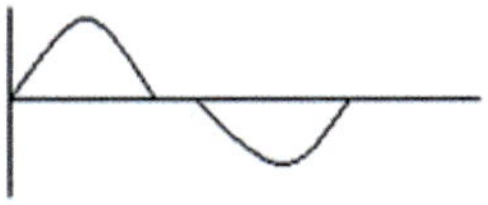	Poor
Boost	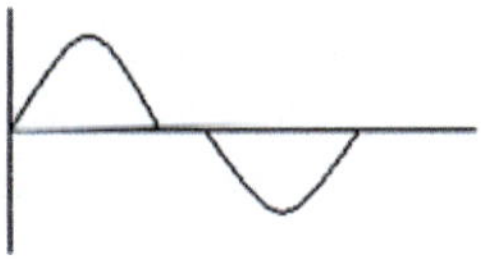	Better
Buck–boost	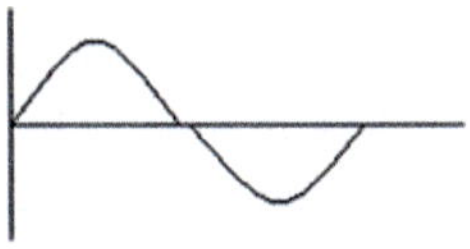	Excellent

3 Existing Techniques

Anwar et al. [5] worked on improvement of PF for LEDs having energy storage with the third port. It includes the omission of heavy electrolytic capacitors. The LED driver including flyback converter is taken in use. This LED driver operates in two manners for PFC. The first is uncontrolled rectifier, and the second is flyback converter. The flyback converter contains three switches, having transformer, capacitor with power decoupling and capacitor for output filter. The first switch has responsibility for controlling shape of input current. The second switch stores the power in the above-mentioned capacitor, and the third switch makes the path for current flow in LED.

Simulation results clearly show the improvement of power factor and the operation of capacitor used for decoupling in accordance to the controller. As the power factor is improved, the harmonics are reduced. The elimination of electrolytic capacitor will tend to maximize the life time of LEDs.

Aclkgoz et al. [6] worked on PFC control of LED lamp driver based on robust controller. The proposed LED driver contains transformed boost converter and LED strings. The transformer used is step-down transformer to step down AC voltage. Neuro-fuzzy controller is introduced which controls DC voltage. The total harmonic distortion in this proposed structure is approximately 1.10%, and power factor is 0.9999. Two neuro-fuzzy controllers are included; one is for voltage control and other is for the control of i_c current. In this, boost converter is used so that higher output voltage can be achieved. The proposed LED structure increases power quality, reduces THD and improves power factor.

El-Moniem et al. [7] worked on a PFC control for LED lamp driver with current sensorless. In the paper, PFC controller with boost mode is used. The mentioned structure is easy to design as current sensor is not used and the total cost is low. Instead of current sensor, two ADCs are used for sensing of input and output voltages, and zero-crossing detectors are used for accuracy. It has been stated from the experiment results that in case of the above control techniques of PFC circuit, the input current contains minimized THD (5%), and the power factor is increased up to 0.998.

Zaky et al. [8] illustrated control of PF for LEDs using sensorless predictive current controller. This paper eliminates sensorless predictive current controller as this will improve power factor of the LED driver and will bring reduction in cost. Under the proposed paper, the power circuit and the controller are isolated. The control part basically consists of two-loop control, so as to get maximum power factor having sinusoidal input current. Speed of operation is also fast. The effect of PFC on the input current is presented in the paper. Sensorless predictive current LED driver consists of a LED lamp string supplies by full bridge rectifier and step-down transformer. Sensorless control consists of two control circuits. A PI controller is also used for adjustment of LED lamp output voltage. After theoretical and experimental analysis, it is noted that THD of the current is only 3%, and PF has improved up to 0.9996.

Singh et al. [9] worked on buck–boost PFC converter with bridgeless concept for multiple string LED driver. The proposed design uses dimming technique for lighting adjustment. It consists of BL buck–boost PFC which supplies power to the flyback converter which is isolated. Low voltage from flyback DC to DC converter feeds the synchronous mode buck converter which is used to drive multistring LED. Thus, design provides improved power factor at low cost. The preferred converter for such LED driver has three stages. The stage 1 is BL buck–boost PFC converter which regulates the DC bus voltage and is used for improving PF and THD in source current. In this way, the life time of LED driver increases. The second stage, as seen earlier, consists of a constant DC voltage which feeds the isolated flyback, DC–DC converter, and in the same way, this becomes source to the buck LED driver working in synchronous mode and the cooling system. The cooling system is responsible for adjusting the junction temperature. The third stage consists of a constant current synchronous buck LED driver which has pulse-width modulation dimming characteristics. The above-mentioned buck converter controls the illumination. The topology of buck converter also increases efficiency of the LED driver.

Cheng et al. [10] proposed LED driver consisting of buck–boost and buck converter. In these converters, MOSFET acts as an active switch. The two switches can operate in ZVS on inductor current being freewheeled in converter to flow the current through the diodes of MOSFET. The buck–boost operates in DCM to increase PF at input line. Both the active switches reduce the switching losses. Use of auxiliary switch or use of snubber circuits is not suggested due to the complexity of the circuit and increment in cost. There are two stages in LED driver. The first one is buck–boost converter which performs PFC and the second stage is buck converter which steps down the voltage in order to drive the LEDs. Prototype of 60 W LED is constructed and tested. The experiments done clearly resulted in reduction in switching losses, and the pf was increased up to 0.99 and efficiency increased up to 93%.

Cheng et al. [11] proposed a LED driver in single stage with PWM dimming for streetlight application. This LED driver offers power factor improvement to a great extent. The single stage extent consists of AC to DC converter with inductors in coupled form and LLC resonant converter of half bridge type. AC to DC converter with coupled inductors operates in DCM. LLC resonant converters are used to reduce switching losses and hence increase efficiency of the circuit. A prototype for supplying 144 W rated LED is constructed and tested with input voltages from 100 to 200 V. The experiment results show a high efficiency up to 93.130% at 120 V_{AC} and maximum PF up to 0.99 at 100 V V_{AC}. The THD is also very low (up to 12.02% at 100 V_{AC}). It produced minimum output voltage ripple (less than 4%) and current ripple (less than 4%) at 110 input voltage and PWM dimming capability range from 20 to 100%.

Cheng et al. [12] worked on LED driver with single stage having PFC and soft switching. This design is useful for energy saving indoor lighting application. The proposed structure is integrated with dual buck–boost converter having coupled structures and series resonant converter in half bridge form along with a bridge rectifier circuit to form a LED driver of single stage. The suggested topology has greater efficiency (more than 91%) increased pf (>0.96) and low output current ripple factor

(<8%). Prototype circuit is tested for feeding 18 W LED with voltages of 100–120 V. The result shows a low-current THD (<10%) suitable for lighting in indoor applications.

Cheng et al. [13] proposed a novel LED with PFC in lighting for indoor applications. It consists of a buck converter in inverse mode and PFC. The proposed design offers low THD, improved pf and high efficiency. A prototype supply 18 W to the LED tube lamp driver is tested with input voltage of 110 V. The result demonstrated efficiency greater than 90%, improved power factor (>0.96) low THD (less than 27%), minimum voltage ripple factor (less than 1%) and minimum current ripple factor (less than 2%).

Zhang et al. [14] worked on a high PF isolated LED free from flicker. The first stage is boost converter for PFC which consists of boost inductor, input bridge diode and main switch. This converter is operated in boundary condition mode. The second stage is of flyback converter which is made of diode and transformer power switch. It works in state of quasi-resonant. Switching losses can be improved. After testing the prototype supplying 60 W power to the LED driver at 110 V, improvement in the stated parameters can be seen. Using the proposed technique, PF of 0.999 is maintained and THD of 4.1%. The overall circuit efficiency comes out to be 90.58%.

Wong et al. [15] proposed LED driver with single phase with least processing required and increased power factor. The design is such that the offline power supplying the LED should have the following characteristics: greater efficiency, large life span, improved PF and low THD. Minimal power processing allows to increase efficiency. In this technique, front-end converter is used, which operates in DCM and is modulated with PWM. Converter with front-end gives dual output. One output is attached to the load of LED by the switch and other is attached to the DC storage capacitor. The power flow in the circuit is controlled to obtain the minimal power processing, so as to reduce the DC storage capacitor. The proposed circuit has no requirement of electrolytic capacitor, and hence, the life span is increased. The prototype is constructed and tested, and the PF is found to be greater than 0.98 at the input voltages of 85, 110, 130 V, and harmonic content is also reduced.

Chowdhary et al. [16] worked on efficient LED driver having low power and galvanic isolation. The LED driver consists of active buck converter with PFC correction topology. The system consists of PFC unit fed by rectifier which is responsible for reducing the harmonic content of the current by controlling the sending end PF. It also consists of DC link inverter having buck converter with soft switching and LLC resonant inverter. PWM signals of 20 kHz control the switching converters. The software used for simulation of 12 W LED driver is PLECS. The results after simulation show the reduction in total harmonic distortion (<10.7904%), and the sending end PF is increased up to 0.9893.

Gritti et al. [17] introduced new technique for correction of power factor for a flyback converter. Primary regulation technique is applied to the converter to reduce the THD. In conventional PFC flyback converter, the THD was reportedly found to be higher. Some effort is being taken in the proposed technique to reduce the THD and improve power factor. A prototype of 35 W LED driver is tested to verify the proposed scheme. The results of the experiment clearly mark the reduction of THD.

Total harmonic distortion is found to be lower than 5% which seems better than the existing schemes. The power factor is increased up to 0.99.

Fang et al. [18] proposed flicker-free technique with high power factor by energy channeling ripple cancelation. In order to have two output voltages, the input power is divided into two segments. The output voltage consists of one main voltage and other auxiliary voltage. The main output is responsible for providing storage for energy and has twice of line frequency ripple voltage. Auxiliary voltage cancels the ripple voltage. In this way, flicker-free performance can be achieved. The proposed prototype of 8.5 W and 50 V/0.17 A is tested. The experiment results show 5.8% 120 Hz of twice line frequency ripple current and high PF (approximately 0.97) and full load efficiency of 86% under 110 V rms input.

Pandey [19] introduced an AC LED driver with bridgeless boost converter. The proposed scheme offers an improvement in PF and THD for LED driver circuit. This topology consists of boost and buck–boost converter. The proposed structure consists of one inductor and one capacitor, having low value. The conventional and proposed method is compared after simulation. It is found that in conventional circuit, THD is 68.52% and PF up to 0.98, whereas in the proposed circuit, THD is approximately equal to 0.93% and PF is about 0.991. In this circuit, passive components are not used in large number.

Baek et al. [20] illustrated a LED with single mode. This driver is based on buck converter in inverted mode and switches for controlling current path. This scheme includes circuitry for control, diode bridge and buck converter in inverted mode. Various switches attached to the LED strings are used in a parallel. Conventional LEDs use forward voltage with fixed value but the proposed LED driver uses forward voltage with variable values as per the level of input voltage. The proposed driver with 7 W power supply with four LED segments is constructed and tested. The experiment results show that at input voltage of 110 V, the PF is obtained to be 0.94, and efficiency is increased up to 94%.

Table 2 represents the techniques used for the correction of power factor and total harmonic reduction. It also compares the techniques on the basis of converters.

4 Conclusion

The discussed schemes provide power factor correction and reduction in THD. There is need of power factor control for LED driver to reduce the losses and cancel out the effect of harmonics which occur due to nonlinearities. Active power factor correction helps in the reduction in harmonics and improved power factor. Various prototypes have been built ranging from low to high power, supplying the LED driver with variable input line voltages from 100 to 200 V. The main focus of proposed structures is to reduce THD, improve power factor, reduce ripple factor and increase the efficiency of the circuit. A review of such schemes is done and found that the PF is increased up to 0.999 and THD is reduced up to 0.93%.

Table 2 Comparison of techniques with PFC and THD reduction

	Techniques applied	Converters used	PFC and THD
1	LED having energy storage with third port	Flyback	–
2	LED driver based on neuro-fuzzy controller	Boost	PF-0.9999 THD-1.10%
3	LED lamp driver with current sensorless	Boost	PF-0.998 THD-5%
4	LED with sensorless predictive current controller	–	PF-0.9996 THD-3%
5	Bridgeless concept for multistring LED driver	Buck–boost	–
6	LED driver with buck–boost and buck converter	Buck–boost and buck	PF-0.99
7	Single-stage LED driver with PWM dimming	LLC resonant	PF-0.99 THD-12.02%
8	Single-stage LED	Dual buck–boost	PF > 0.96 THD < 10%
9	Novel LED with PFC	Buck	PF > 0.96 THD < 27%
10	Isolated flicker-free LED	Boost and flyback	PF-0.999 THD-4.1%
11	Single-phase LED driver	Front-end converter	PF > 0.98
12	LED driver having low power and galvanic isolation	Buck converter and LLC resonant inverter	PF-0.9893 THD < 10.7904%
13	LED driver with flyback converter	Flyback	PF- 0.99 THD < 5%
14	Flicker-free LED	-	PF-0.97
15	AC LED driver with bridgeless boost converter	Boost and buck–boost	PF-0.991 THD-0.93%
16	Single-mode LED	Buck	PF-0.94

References

1. Chuang Y-C, Ke Y-L, Chuang H-S, Hu C-C (2010) Single stage power factor-correction circuit with flyback converter to drive leds for lighting applications. In: IEEE conference on industry application-society (IAS), pp 1–9, Oct 2010
2. Hasan J, Nguyen DH, Ang SS (2010) A RGB-driver for LED display panels. In: IEEE application power electronics conference on exposition 25th annual, pp 750–754, Feb 2010
3. Oh IH (2008) An analysis of current accuracies in peak and hysteretic current controlled power LED drivers. In: IEEE application power electronics conference and exposition 23th annual, pp 572–577, Feb 2008
4. Qu X, Wong SC, Tse CK (2007) Color control system for RGB LED light sources using junction temperature measurement. In: Proceedings of IEEE IECON, pp 1363–1368
5. Anwar S, Badawy MO, Sozer Y (2015) Power factor correction of LED drivers with third port energy storage. In: IEEE applied power electronics conference and exposition, pp 2206–2211

6. Açıkgöz H, Uçar F, Coteli R, Dandil B (2016) Power factor correction control of LED lamp driver based on robust controller. In: International engineering sciences and education conference. 01–03 Dec 2016
7. EL-Moniem MSA, Azazi HZ, Mahmoud SA (2014) A current sensorless power factor correction control for LED lamp driver. Alexandria Eng J 53:69–79
8. Zaky MS, Azazi HZ, Touti E (2018) PFC control for LED lamp driver using sensorless predictive current controller. Eng Technol Appl Sci Res 8(5):3373–3379
9. Jha A, Singh B (2017) Bridgeless buck-boost PFC converter for multistring LED driver, pp 1–8
10. Cheng H-L, Lin C-W (2014) Design and Implementation of a high power-factor LED driver with zero voltage switching-on characteristics. IEEE Trans Power Electron 29(9)
11. Cheng C-A, Chung T-Y (2016) A single-stage LED streetlight driver with PFC and digital PWM dimming capability. Int J Circuit Theor Appl
12. Cheng C-A, Chang E-C, Tseng C-H, Chung T-Y (2017) A single-stage LED tube driver with power-factor corrections and soft switching for energy saving indoor lighting applications. Appl Sci 7(2):1–14
13. Cheng C-A, Chang C-H, Cheng H-L, Chang E-C, Lai C-C (2018) A novel LED tube lamp driver with power factor correction for indoor lighting applications. In: IEEE 7th global conference on consumer electronics, pp 700–701
14. Zhang Z, Zhou H, Kang L, Zhang Z, Ren B, Jiang M (2018) A flicker-free high power isolated LED driver. In: The 30th Chinese control and decision conference, pp 4966–4970
15. Wu H, Wong S-C, Tse CK, Ron Hui SY, Chen Q (2018) Single-phase LED drivers with minimal power processing, constant output current, input power factor correction and without electrolytic capacitor. IEEE Trans Power Electron 33(7):6159–6170
16. Chowdhury D, Hussain MI, Gholam Zakaria M, Ziaur Rahman Khan M, Haider MZ (2016) An electrically isolated low power LED driver offering power factor correction with ameliorated mains current THD. In: 8th IEEE India international conference on power electronics, pp 1–6
17. Gritti G, Shao J (2017) A very low THD and high power factor primary side regulation LED driver. In: 19th European conference on power electronics and applications, pp 1–7
18. Fang P, Liu Y-F (2016) An energy channeling LED driver technology to achieve flicker-free operation with true single stage power factor correction. IEEE Trans Power Electron
19. Pandey RK (2016) An AC LED driver with improved total harmonic distortion and power factor. Int J Sci Eng Res 7(5)
20. Baek J, Chae S (2017) Single-stage buck-derived LED driver with improved efficiency and power factor using current path control switches. IEEE Trans Indust Electron 64(10):7852–7861
21. Iqbal A et al (eds) (2020) Soft computing in condition monitoring and diagnostics of electrical and mechanical systems, vol 1096. In: Advances in intelligent systems and computing. Springer, Singapore. https://doi.org/10.1007/978-981-15-1532-3
22. Iqbal A et al (eds) (2020) Meta heuristic and evolutionary computation: algorithms and applications, vol 1096. In: Studies in computational intelligence. Springer, Singapore. https://www.springer.com/gp/book/9789811575709

Case Study of Synchronization of Solar Power Converter

Prashant Kumar Singh, Rajesh Narayan Deo, Vishal Vajpayee, Vipin Verma, and Vivek Bajpai

Abstract In this paper, the modeling, hardware evaluation and simulation of PV solar power converter are discussed. The DC is converted into AC by grid-tied PV converter & then connected to a power grid. Voltage profile can be modified by connecting a known load of solar System. Hence, the system still experiences over wear to solar PV range connected is a way excess than local loads. The controlling of power factor also seems to enhance the voltage profile all over the feeder.

Keywords DC–AC inverter · Power converter · Solar PV

1 Introduction

Human population is growing day by day so does the increase in energy demand, and solar energy sources are one of the most efficient energy systems [1–3]. A Global Status Report in 2017 said that approximately, 62% of renewable resources are responsible for the total energy production [4]. As fossil fuels are the main reason for greenhouse effect, its goal is to decrease the use of fossil fuels. Solar energy sources are one of the most efficient energy systems because it's miles a carbon free era and also easy for installation, safe, dependable and maintenance free but solar PV output can vary unknowingly and indefinitely due to climatic conditions as the solar PV is a fluctuating source [5]. Overvoltage, device overloading and numerous operations of voltage regulations are one of the main problems when installing a solar PV. Generally, power flow in Solar PV is single directional and drops rapidly when it gets cloudy [6], whereas during a sunny day, power flow increases fiercely, and this may cause upward thrust of output energy and system starts to characteristic bidirectionally, hence can be reason for overload inside distribution feeder [4].

P. K. Singh (✉) · R. N. Deo · V. Vajpayee · V. Verma · V. Bajpai
Department of Electrical and Electronics Engineering, Galgotias College of Engineering and Technology, Greater Noida 201306, India
e-mail: rnps2007@gmail.com

A. Iqbal et al. (eds.), *Renewable Power for Sustainable Growth*, Lecture Notes in Electrical Engineering 723, https://doi.org/10.1007/978-981-33-4080-0_47

2 Solutions to Overvoltage

After performing a number of studies for range of studies for getting answers to accurate overvoltage of distribution-system, Tie and Gan said that voltage upward push is laid low with the size of cable. By the result of simulation, it has been proved that growing the dimensions decreases the rise in voltage of distribution network. One of the answers to correct overvoltage is to change the faucet of OLTC transformer [7] but because of fast need of sun PV, Chirapongsananurak and Hoonchareon stated that overvalue problems in distribution network could not be corrected any longer modifying faucet of OLTC transformer. Because of the abrupt trade of weather, work of OLTC transformer is not always sufficient to offset quick voltage medications. Moreover, the sturdiness of transformer gets reduced along with boom in range of transformer working. It is used to stock piece of power produced by sun PV which is saved in EESS, and it additionally absorbs any remaining active strength transferred to application framework via sun PV [8]. As the load may be at once carried out by requirement and therefore controlled signal may be produced by means of imperative attendant, reverse energy drift and high call is decreased [9]. So, the utilization of solar PV is not always constant at some point of the year [4]. Hence, combining the call for real power response has been proposed. Cosmic PV converter generates and absorbs receptive energy for voltage changes (Fig. 1).

3 Voltage Profile Challenges

To enhance the voltage profile for suspicious network, cosmic PV system is the best [10]. An overvoltage may be visible at the network if range of cosmic PV system installed is extra as compared to loads. Figures 2, 3, 4 and 5 suggest assessment of voltage profile for five scenarios. The end substation voltage is 11.017 kV for all the instances before putting in solar PV systems [11]. The voltage falls because of high strength loss attributable to the resistance [12]. Hence, adjustments in voltage profile are imitated by connecting cosmic PV structures of distinctive potential. Hence, suitable range needs to be ±5% of eleven kV.

3.1 Case 1

In Case 1, the voltage profile for three distinct situations, voltage pre-connecting cosmic PV system, voltage after 1 MW cosmic PV machine and voltage for a five MW cosmic PV system are connected (Fig. 2). These cosmic PV systems are connected at Sub 20. The setup of five MW cosmic PV increments voltage profile for community greater when correlated with at least one cosmic PV setup.

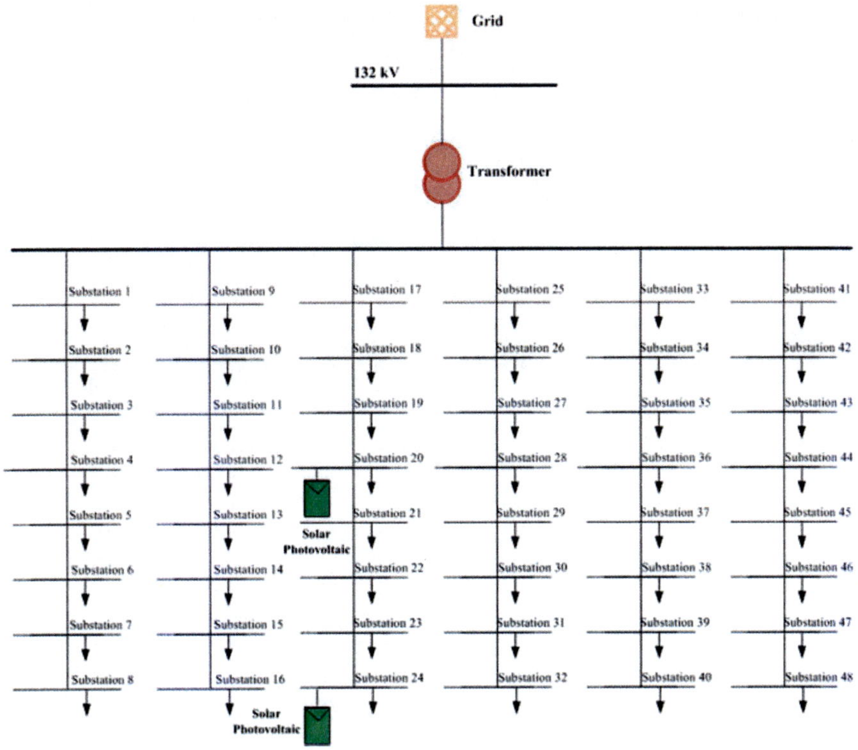

Fig. 1 A typical distribution network

Fig. 2 Voltage profile

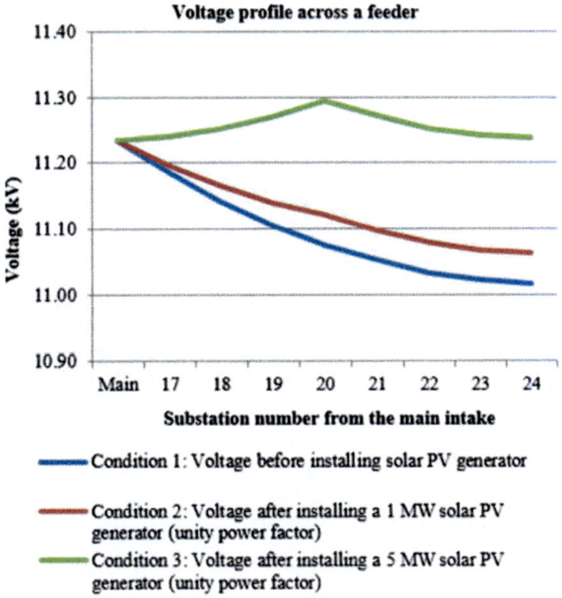

Fig. 3 Voltage profile

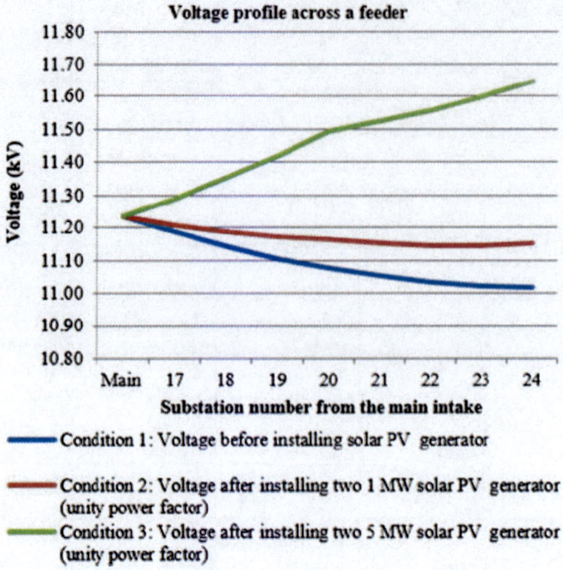

Fig. 4 Voltage profile

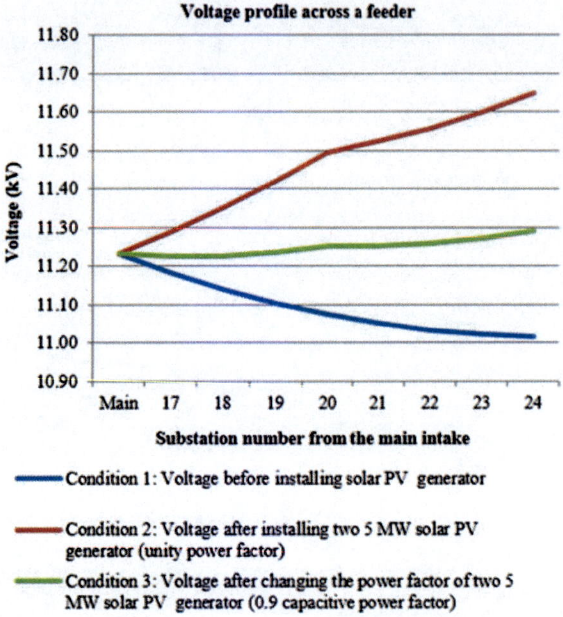

Fig. 5 Voltage profile

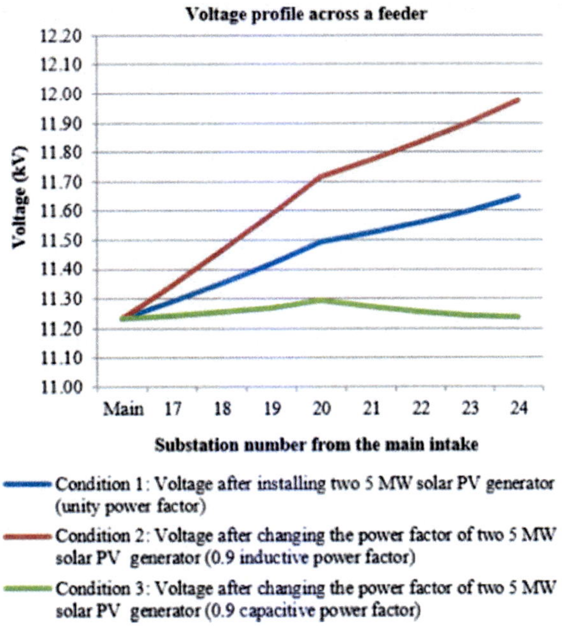

3.2 *Case 2*

For Case 2, assessment of voltage profile throughout the main feeders for three distinct conditions can be pre-connecting a cosmic PV device and connecting two of one MW and five MW cosmic PV systems (Fig. 3). Hence, because of huge cosmic PV potential of motor, community experience overloads.

3.3 *Case 3*

For Case 3, voltage profile in three distinct situations (Fig. 4). Cosmic PV is established at Sub 20 and Sub 24, respectively. The end substation voltage is 11.648 kV. Therefore, voltage at give up sub falls to 11.293 kV after modifying the electricity by 1–0.9.

3.4 *Case 4*

In Case 4, voltage profile in three distinct conditions (Fig. 5). Cosmic PV structures are mounted at the end of the substations, Sub 20 and Sub 24, respectively. The

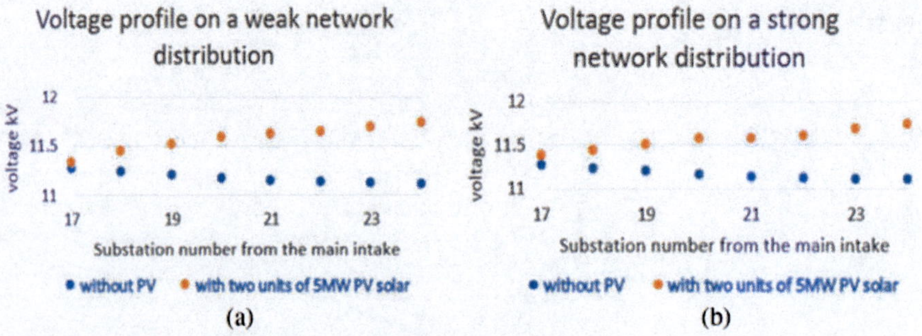

Fig. 6 **a** Network distribution (weak) and **b** network distribution (strong)

voltage rises extra than the overvoltage situations by changing the solidarity electricity element to inductive strength thing. According to Fig. 4, voltage in feeder = 11.974 kV afterward running on inductive energy *D*.

3.5 *Case 5*

In Case 5, effect of sun PV at the bus voltages for susceptible and intense electric network is discussed. The voltage profile for feeble grid before and after connecting two cosmic PV of 5 MW (Fig. 6a) and voltage profile at robust grid before and after putting in two devices of five MW cosmic PV systems at harmony energy issue (Fig. 6b) [4].

4 Solar Inverter Synchronization

In cosmic panel inverter, the energy supply is provided to domestic loads by converter, and unless until the domestic load needs no electricity delivered, then strength is given to network. The delivered voltage from the cosmic panel is granted to DC–DC inverter. The delivered voltage from DC–DC inverter is used throughout battery in domestic load [13] for strength network through PWM inverter. This kind of alternating current supply can be furnished to domestic load [11] (Figs. 7 and 8).

In Fig. 9, the first and second waveforms are of the grid and the PWM 1-phase converter, respectively; consequently by looking out waveforms, it can be stated that they are completely in segment [11].

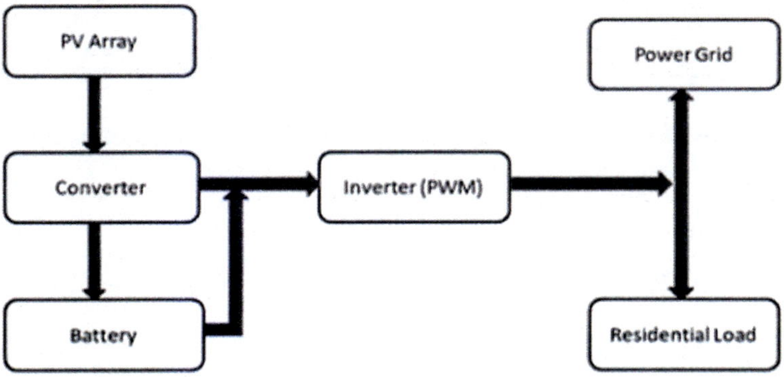

Fig. 7 Block diagram of typical solar cell converter sync

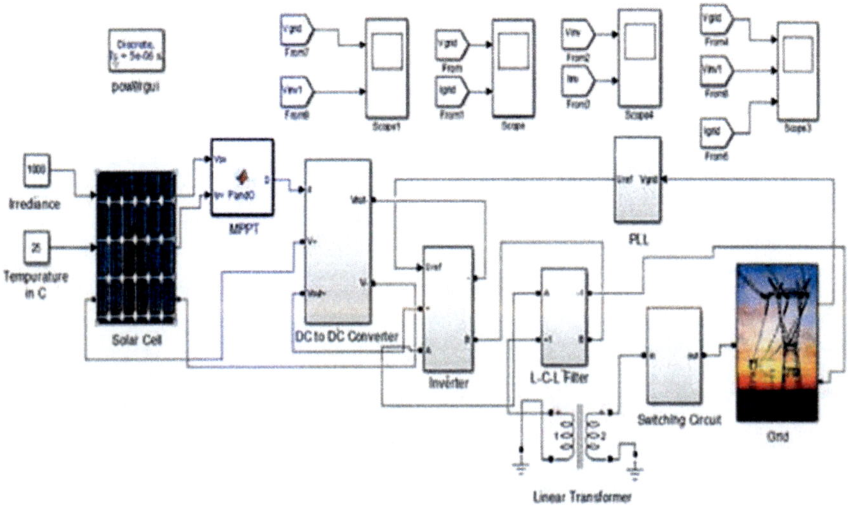

Fig. 8 Simulation of typical solar cell

5 Various Isolations and Their Effects on Solar Power Inverter

The single inductor-based twin enter/output converter inside the energy glide management gadget becomes produced by the usage of MOSFETS.

Of rating a hundred V, 23 A and the score of diode is 23 A [14]. Three 24 V 60 W cosmic photovoltaic systems turned into energy origin and individual DC load. The gate pulses are produced with the aid of the use of RTI board and XOR common

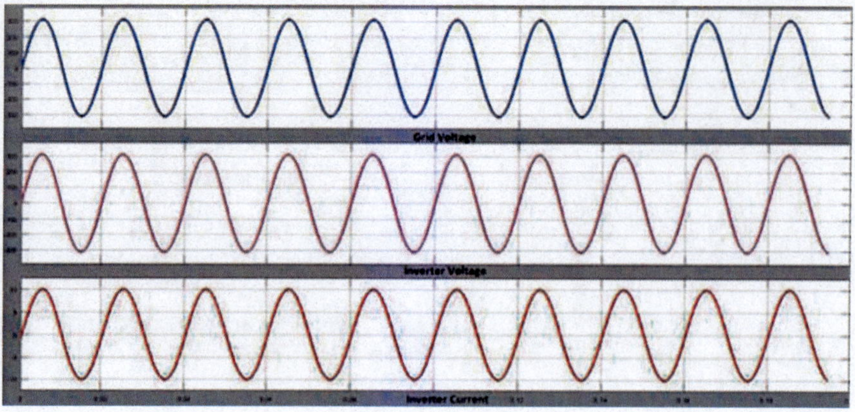

Fig. 9 Waveforms are of the grid

sense gate. These pulses are then provided to a quad-bundled comparator, taking action among the RTI board and optocouplers [15].

Case I

Effective comments of inverter because of increment in cosmic PV is from 18 to 36 W: Inverter's working on DIBM with cosmic photovoltaic and battery's gratifying an unmarried DC-load requirement.

Cosmic PV input turned into coupled to 36 W (as shown in Fig. 10) [16–19]. The maximum power point monitoring controller increases distributed maximum strength factor tracking to change cosmic solar photovoltaic to a brand new maximum power point. The voltage controller will increase volt as enter voltage permit to visible in equation is decreased because of boom in 'Din.' Because of increment in 'Din' and 'dS3' depletes, therefore energy consumption depletes [20].

Case II

Mode adjustments in DOBM because of cut back in unmarried direct current load from 12 to 37 W: The individual DC load energy requirement is 37 W that is given by cosmic photovoltaic [14–16, 20–23], whereas the cosmic DC load energy requirement is decreased to 12 W. Because of excess electricity, 'V' will increase. The maximum energy point tracking takeovers the control, resulting in lack of voltage manipulate (Fig. 11).

Case III

Method variation in DIBM because of decrement in cosmic photovoltaic from 18 to 36 W: Inverter is working at DOBM for single DC load energy required and cosmic PV. At 'tspv,' the solar photovoltaic input is reduced to half from 18 to 36 W. Since the solo DC load electricity need is 20 W, the maximum electricity point tracking control is reduced dispensed maximum power point tracking manipulate to shift to the brand new the maximum power point manipulate. The voltage controller will

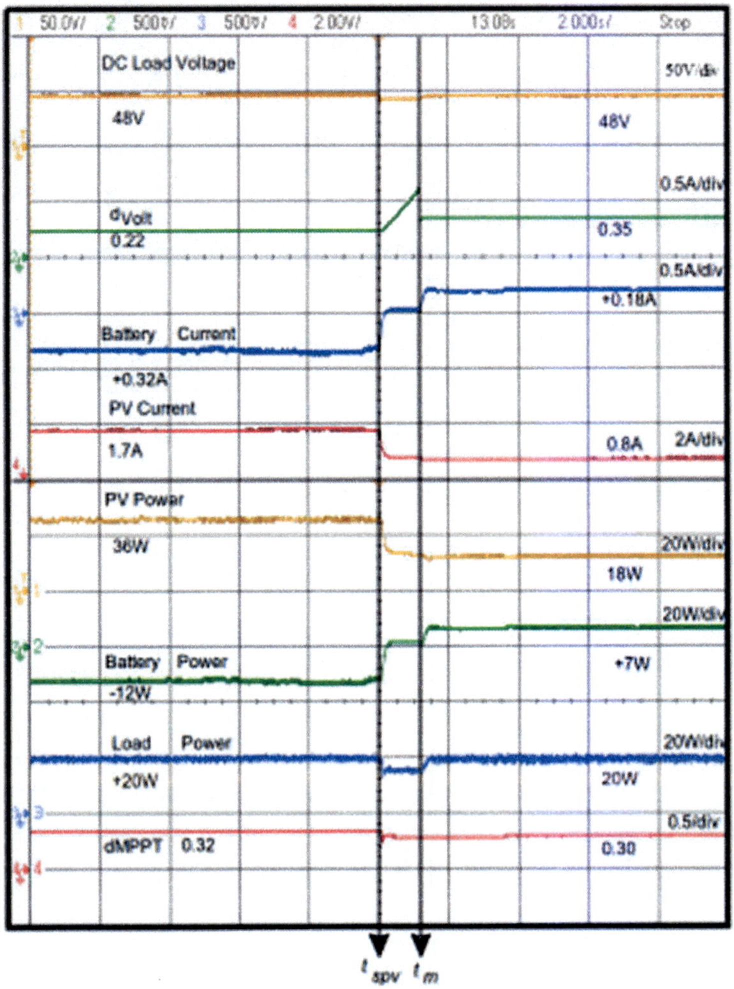

Fig. 10 Dynamic response of the converter to change in solar PV insolation [14]

increase the volt to decompensate for drop in V_0; however, this is not always enough [15, 16, 19, 23, 24]. The mode grow to be one-of-a-kind manage moves inverter from DOBM-DIBM (Fig. 12).

6 Improvement of Power Generation by PV

This phase portrays the set of the experimental constructive expansion for photo-voltaic electricity production network located on interacting version to a typical PC.

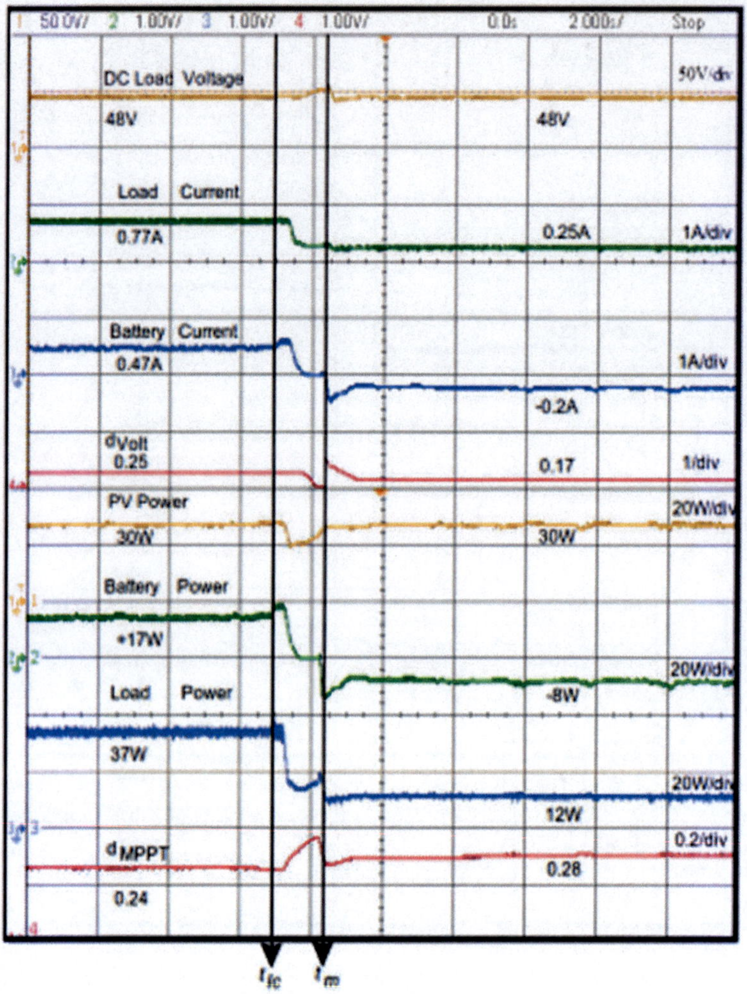

Fig. 11 Mode adjustments in DOBM

Therefore, the created photovoltaic converting device subsists of a photovoltaic panel, an electricity inverter, calculating load and circuits [25]. The set of the experimental take a look at is explained. The controller produces minimum pulse for the energy transistor [16, 25]. OS exercises are improved to degree one of a kind appliance and earn economical control process [25]. The peak value perspective of cosmic PV modules recognizes all through analysis (Figs. 13, 14 and 15).

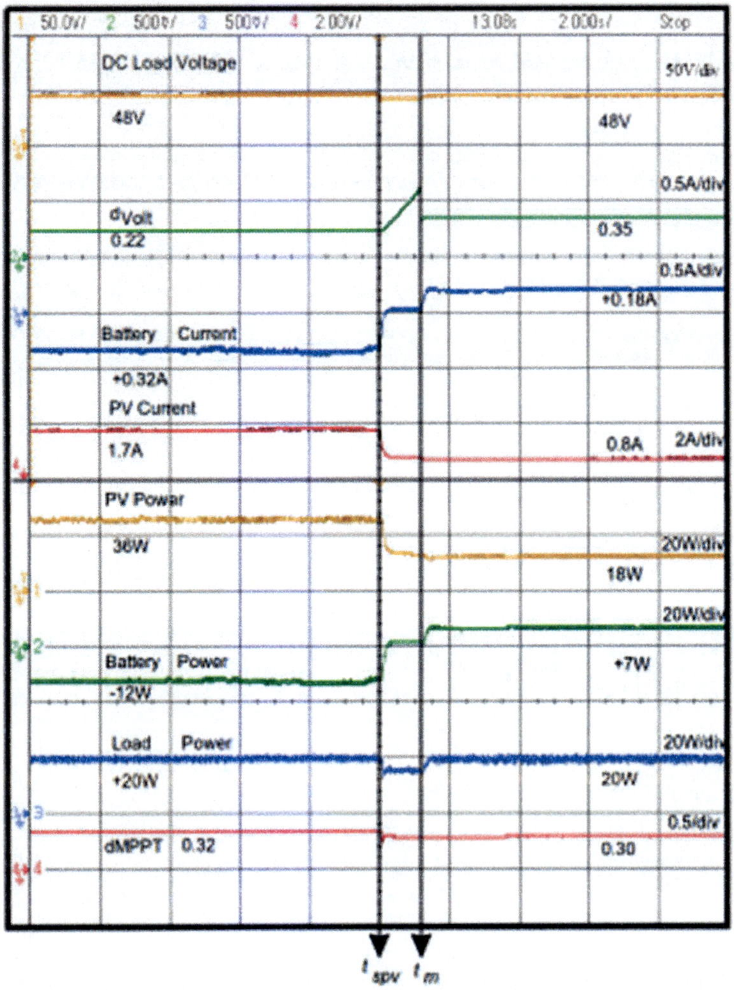

Fig. 12 Method variation in DIBM

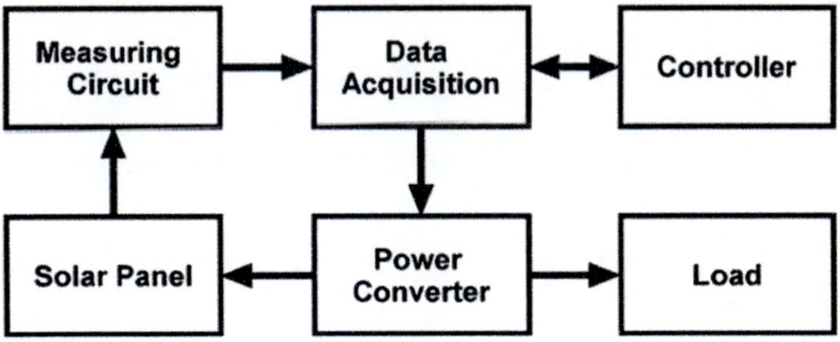

Fig. 13 Improvement of power generation by PV

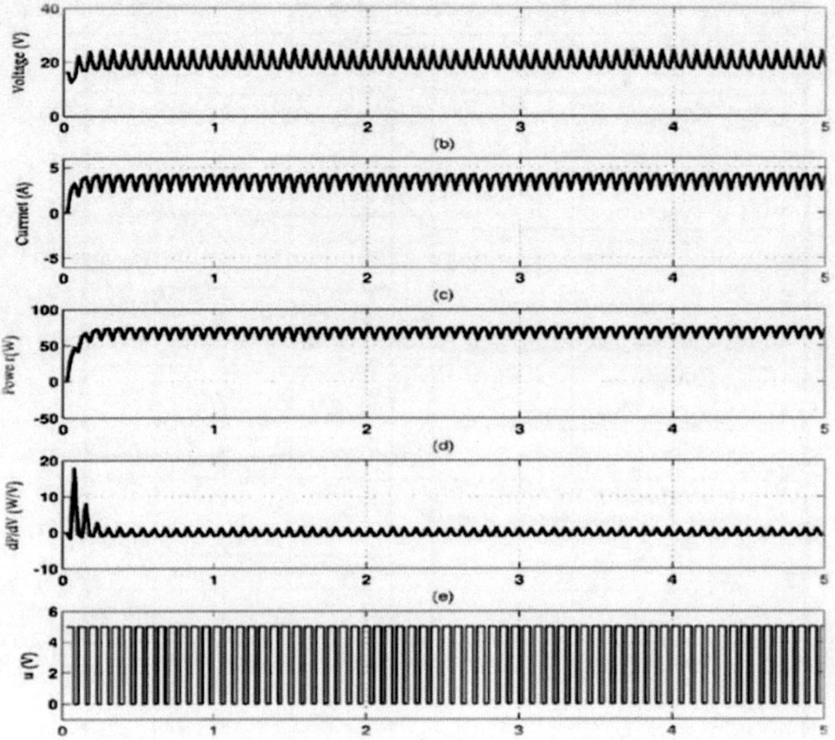

Fig. 14 Result of PV production system [18]

7 Conclusion

A number of nations are worrying to solve the difficulty of global warming. The utilization of sustainable power, especially sun PV, is also a solution to conquer this issue. However, the distribution community may experience problems like overvoltage, unsychronization, etc., which can be solved by the above-stated methods.

System Parameters				
PV	MPP PWR	2.5 kW		
	MPP Volt.	161.5 V		
Battery	Power	1.0 kW		
	Voltage	50 V		
Grid	Voltage	Single phase, 120 V_{RMS}		
A PV/Battery Universal Optimizer				
Operation Mode		PV-only	PV-battery	Battery-only
PV Output Voltage		161.5 V	161.5 V	-
Series Voltage		38.6 V	50.0 V	50.0 V
Max. Rated Power (PPR_{total})		0.5 kW (20%)	1.4 kW (55%)	0.8 kW (22%)
Absolute η_{DAB}		97.5%	97.5%	97.5%
$\eta_{DC\text{-}DC}$	PV Only	99.5%	99.4%	-
	Discharging	-	99.6%	98.1%
	Charging	-	99.0%	98.1%
A DC-AC Inverter				
Rated Power		2.5 kW	2.5 kW	2.5 kW
DC-Link Voltage		200 V	211.5 V	200 V
$\eta_{Inverter}$		98.5%	98.5%	98.5%
Total Efficiency				
η_{Total}	PV Only	98.0%	97.9%	-
	Discharging	-	98.1%	96.6%
	Charging	-	97.7%	96.6%

Fig. 15 Case research of a photovoltaic battery series inverter architecture

References

1. Yi JLM, Naayagi RT, Logenthiran T Modelling and implementation of single phase dual stage grid-tied solar power inverter. IEEE
2. Iqbal A et al (eds) (2020) Soft computing in condition monitoring and diagnostics of electrical and mechanical systems, vol 1096. In: Advances in intelligent systems and computing. Springer, Singapore. https://doi.org/10.1007/978-981-15-1532-3
3. Iqbal A et al (eds) (2020) Meta heuristic and evolutionary computation: algorithms and applications, vol 1096. In: Studies in computational intelligence. Springer, Singapore. https://www.springer.com/gp/book/9789811575709
4. Norhasmi NNF, Raveendran SK, Ramachandaramurthy VK Power factor control of solar photovoltaic inverter as solution to overvoltage
5. Allamehzadeh H Solar energy overview and maximizing power output of a solar array using sun trackers
6. Musunuri S, Nirantare P Multi-functional solar inverter for grid harmonic elimination
7. Camm EH, Williams E Solar power plant design and interconnection
8. Yang X, Zhai Y Design of inverter power supply for household solar power generation in pastoral area
9. Richard J Bravo Solar PV power plants harmonics impacts
10. Nirmal Mukundan CM, Jayaprakash P Cascaded H-bridge multilevel inverter based grid integration of solar power with PQ improvement
11. Sondawale SR, Jagtap PP, Tarnekar SG Synchronization of solar inverter with power grid
12. Pal D, Bajpai P Active and reactive power control in three phase solar PV inverter using modified IC method

13. Phannil N, Jettanasen C, Ngaopitakkul A Power quality analysis of grid connected solar power inverter
14. Anand I, Subramaniam S, Biswas D, Kaliamoorthy M (2018) Dynamic power management system employing single stage power converter for standalone solar PV applications
15. Shao H, Li X, Tsui CY, Ki WH (2014) A novel single-inductor dual-input dual-output dc-dc converter with pwm control for solar energy harvesting system. IEEE Trans Very Large Scale Integr (VLSI) Syst 22(8):1693–1704
16. Babaei E, Abbasi O (2016) Structure for multi-input multi-output dcdc boost converter. IET Power Electron 9(1):9–19
17. de Brito MAG, Galotto L, Sampaio LP, e Melo GDA, Canesin CA (2013) Evaluation of the main MPPT techniques for photovoltaic applications. IEEE Trans Indust Electron 60(3):1156–1167
18. Luque A, Hegedus S (2011) Handbook of photovoltaic science and engineering. Wiley
19. Jongerden MR, Haverkort BR (2009) Which battery model to use? IET Softw 3(6):445–457
20. Huang MH, Chen KH (2009) Single-inductor multi-output (simo) dc-dc converters with high light-load efficiency and minimized crossregulation for portable devices. IEEE J Solid-State Circuits 44(4):1099–1111
21. Nami A, Zare F, Ghosh A, Blaabjerg F (2010) Multi-output dc-dc converters based on diode-clamped converters configuration: topology and control strategy. IET Power Electron 3(2):197–208
22. Khaligh A, Cao J, Lee Y-J (2009) A multiple-input dc–dc converter topology. IEEE Trans Power Electron 24(3):862–868
23. Rehman Z, Al-Bahadly I, Mukhopadhyay S (2015) Multiinput dc–dc converters in renewable energy applications–an overview. Renew Sustain Energy Rev 41:521–539
24. Keyhani H, Toliyat HA (2013) A ZVS single-inductor multi-input multi-output dc-dc converter with the step up/down capability. In: 2013 IEEE energy conversion congress and exposition, pp 5546–5552, Sept 2013
25. Feshara HF, Ibrahim AM, El-Amary NH, Sharaf SM (2019) Performance evaluation of variable structure controller based on sliding mode technique for a grid-connected solar network
26. Haura R Jr, Casaro MM Single-phase dual-stage isolated multilevel inverter applied to solar energy processing

Generation of HVDC from Voltage Multiplier Using Opto-Isolator and Marx Generator

Asim Rahman Ansari, Mohd. Khursheed, Ahmed Riyaz, and Mintu Kumar

Abstract The main function of MARX generator is to produce a high potential pulse for testing the insulation of the electrical instruments such as transformers and electrical lines. However, power losses in conventional Marx generator due to capacitor and resistor are still a challenging task. A very few papers are reported in this direction. To fill this gap, method to implement the other static devices in place of conventional components is discussed. The presented topology comprises of a MOSFET, a capacitor, and two diodes. A 555 timer has been used to give pulses to the passive elements (capacitors) to be charged in parallel during ON interval. During OFF interval, charge storage devices are connected in series with the help of solid-state switches. Hence, MOSFET is used as a switching device; lossless charging of capacitors is done with the help of diodes. A comparison between different level generators is made and result is presented. The performance of the system is evaluated with the help of simulation results of PROTEUS software.

Keywords Marx generator · Current Marx generator · MOSFET · Opto-isolator · High voltage

1 Introduction

Power electronics devices and solid-state gadgets are turning out to be increasing more reasonable for high power application [1–6]. They could furnish the best force frameworks with smallness, unwavering quality, high redundancy rate, and long age. The ascending of high power generator utilizing solid-state gadgets kills constraints

A. R. Ansari (✉) · M. Kumar
Department of Electrical Engineering, FET, MJP Rohilkhand University, Bareilly, India
e-mail: ashi0786ansari@gmail.com

Mohd. Khursheed
Department of Electrical Engineering, Integral University, Lucknow 226026, India

A. Riyaz
Department of Electrical Engineering, BGSB University, Rajouri, India

A. Iqbal et al. (eds.), *Renewable Power for Sustainable Growth*, Lecture Notes in Electrical Engineering 723, https://doi.org/10.1007/978-981-33-4080-0_48

of regular segments and guarantees high power innovation to be generally utilized in industrial application [1].

2 Development and Advancement of Marx Generator

2.1 Traditional Instrument

A conventional Marx instrument comprises of a capacitor bank, potentiometers, and spark gaps arrangement (Fig. 1). The capacitors are charged in parallel manner via the series resistor 'R_s' by applying an input voltage V, and discharged in series through the spark gaps to get high pulse voltage, which is 'n' times the applied voltage V, where 'n' is number of charging stage of capacitors. The discharging time constant RC is very small in comparison to the charging time constant, that is of few milliseconds [1, 2].

The arrangement discussed above for obtaining high pulse voltage has been first suggested by Marx in nineteenth century. Generally, the charging opposition is picked to restrain the charging current around 50–100 mA, and the value of capacitance for generator is selected so that the time constant RC is in the range of 10–60 s.

The disadvantages of the conventional Marx circuit are as follows:

- Large size, high cost, and power losses due resistor and capacitor
- Low pulse repetition rate and short life time of the spark gap switches.
- Slow charging as the current flow through the charging resistor.

So as to tackle these issues, some new Marx circuits have been suggested. Improved topology uses semiconductor switches like MOSFET or IGBT.

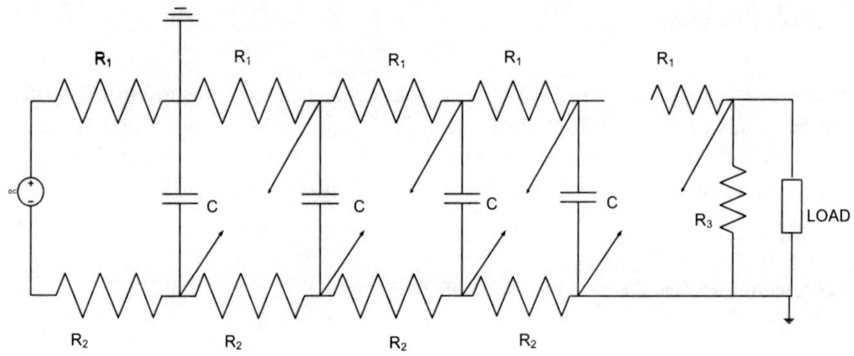

Fig. 1 Traditional Marx generator

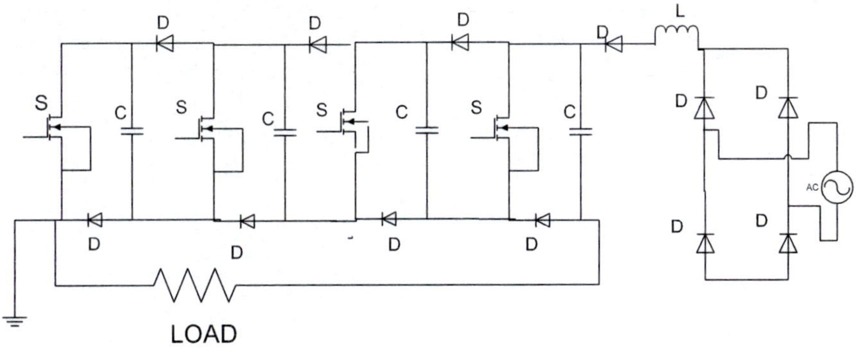

Fig. 2 Existing circuit of conventional high voltage generation

2.2 *Enhanced Marx Generator Topology*

With the advancement of power electronics devices, the use of solid-state device is increasing day by day in field of high-voltage application. They can offer the maximum power operation for the circuit with the conservative, high redundancy rate, and good span of life. This circuit utilizes solid-state switches for ex-MOSFET and passive elements, and also the opto-isolator is supported by switches (diodes). Subsequently, thus the limitation of conventional Marx generator [1, 2] is overcome by enhanced Marx generator as in Fig. 2.

Here, AC voltage is step down and applied to the rectifier to get constant DC output voltage to apply across the generator circuit. The capacitors charged in parallel manner when connected to the converter when all the capacitor are completely charged, the MOSFET switches are turn on at the same time to connect the capacitors and to discharge to get high dc voltage across the load. Few limitation of the circuit causes the voltage not to be exactly same as that of $n * V$ (where n is stage).

In this paper, different circuits with three, four, and five stages of Marx generator are discussed and results are obtained for load voltage and current. In this scheme, 555 timer works as an astable mode, i.e., pin 2 and 6 are short circuited and yield is joined with base of specific transistor. A 555 timer generates input pulses for capacitor. During on period of pulses capacitors use 12 V supply in parallel manner. During of time of pulses, MOSFET switches are triggered with the help of opto-coupler.

3 Design Details

Similar four levels are used to obtain the required high voltage in this work. For convenience, only the circuit parameters and component values of 4-level generator are mentioned in Tables 1 and 2 sequentially.

Table 1 Parameters of timer

S. no.	Parameters	Calculation	Value
1	T_{on}	$0.693\ (R1 + R2)\ C7$	0.7 ms
2	T_{off}	$0.693 * R2 * C7$	0.6 ms
3	Duty cycle	$T_{\text{on}}/(T_{\text{on}} + T_{\text{off}})$	53.8%

Table 2 Parameters of the Marx generator

S. no.	Parameters	Values
1	$C1$, $C2$	100 nf
2	$C3$, $C4$, $C5$, $C6$, $C7$	100 µf/35 V
3	$C8$	47 µf/160 V
4	$R1$	10 k
5	$R2 - R6$	3.9 k
6	$R7 - R18$	1 k, $V_{\max}$ (50–100 kV)
7	Diode	1N4007

The connection diagram for different level of Marx generator in PROTEUS software is shown in Fig. 3a–c. The presented topology is examined for 3-, 4-, and 5-level. For higher voltage, more levels can be included. Ideal switches and components are considered to realize the Marx circuit in PROTEUS software. The value of timer circuit parameter taken is given in Table 1 and that of Marx generator component is in Table 2. The simulation result of the proposed solid-state switch-based high voltage generator for different level is exhibited in Fig. 4.

4 Result and Discussion

In this paper, different level Marx generators have been discussed for voltage enhancement. It can be seen from simulation result, for a 3-stage Marx generator for an input of 15 V the load current is 0.0039 Amps exhibited in Fig. 4a and the output voltage is 39 V, depicted in Fig. 4b. On increasing one stage, i.e., 4-stage Marx generator, the load current of 0.0048 A exhibited in Fig. 4c and the output voltage is 48 V, depicted in Fig. 4d. When we have increased one more stage for high voltage, load current is 0.0057 A exhibited in Fig. 4e and the output voltage is not exactly five times and it is found to be 57 V as the value of load is 10 kΩ selected for all the simulations.

The number of MOSFETS switch used depends on the voltage output to be enhanced. Moreover, as the MOSFET drivers use strategy for self-provided power, the pulse repetition rate is higher, the power consumed from the source is decreased, and therefore overall economy can be achieved. Also, as high rating capacitors and resistors are replaced with that of lower value components so the voltage output

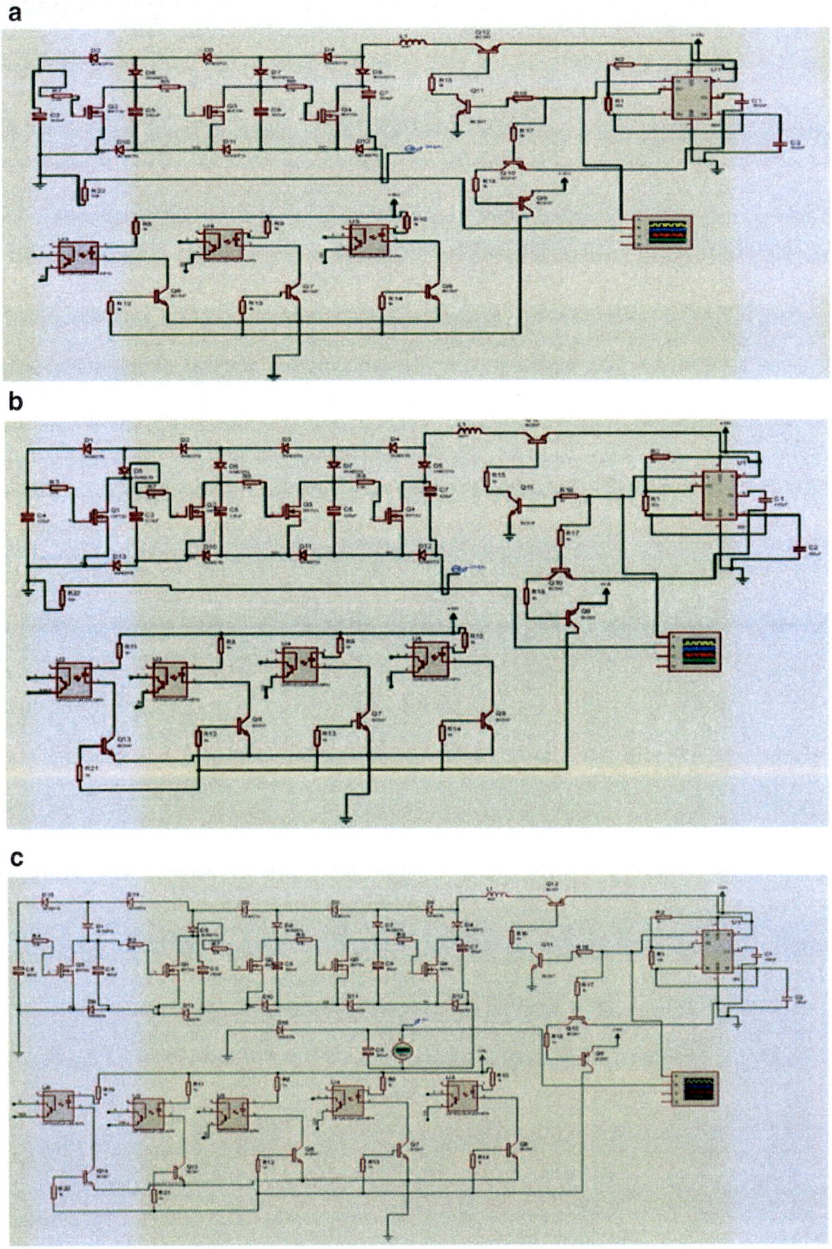

5

Fig. 3 **a** 3-level MOSFET switch-based Marx generator, **b** 4-level MOSFET switch-based Marx generator, **c** 5-level MOSFET switch-based Marx generator

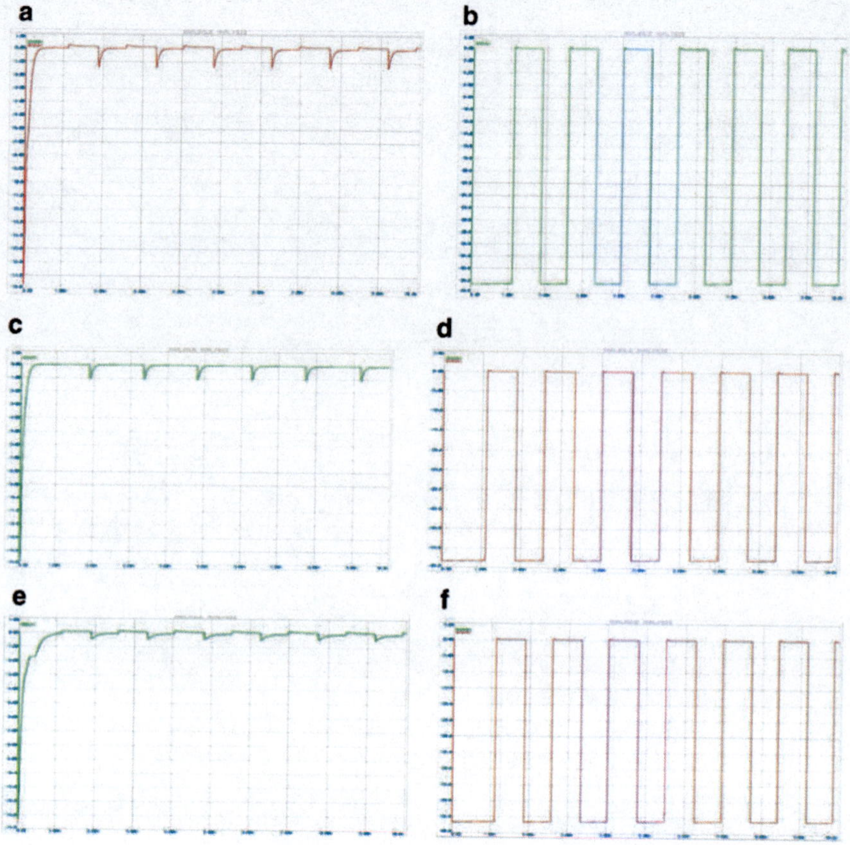

Fig. 4 **a** Output current of 3-level generator, **b** output voltage of 3-level generator, **c** output current of 4-level generator, **d** output voltage of 4-level generator, **e** output current of 5-level generator, **f** output voltage of 5-level generator

and pulse repetition rate are higher. Thus, there is no need of pulse transformer for obtaining the higher voltage (Table 3).

Table 3 Result data of different level Marx generator simulated in PROTEUS

S. no.	Parameters	3-stage	4-stage	5-stage
1	Input voltage V_i (V)	15	15	15
2	Ideal output voltage V_o (V)	45	60	75
3	Obtained output voltage V (V)	39	48	57
4	Voltage drop in passive elements (V)	06	12	18
5	Capacitance value (μF)	100	125	100
6	Efficiency = $(V/V_o) * 100$ (%)	86.66	80	76

5 Conclusion and Future Scope

- From above comparison, if the numbers of stages are more, load voltage increases which is equal to ideal output voltage.
- With higher pulse frequencies, the percentage voltage decrease in capacitor can be limited to 10%. Important point is this that current can be limited by the inductor of specific value. It causes step up of the voltage for the capacitors to charge up to its maximum input voltage. Value of inductor $L = \Delta V * \Delta T/\Delta I$, where $\Delta V =$ change in output voltage, $\Delta I =$ change in inductor current, and $\Delta T =$ maximum pulse width.
- From the simulation result, it has been concluded that the presented circuit enhances the input voltage effectively. This topology is very compact and easy to implement.

References

1. Wu Y, Liu K, Qiu J, Liu X, Xiao H (2007) Repetitive and high voltage Marx generator using solid-state devices. IEEE Trans Dielectr Electr Insul
2. Platts DA (1990) Gigawatt Marx bank pulsers, ultra-wideband radar: proceedings of the first los Alamos symposium. CRC Press
3. O'Loughlin J, Lehr J, Loree D (2001) High repetition rate charging a Marx type generator, presented at the 13th IEEE international pulsed power conference. Las Vegas, Nevada
4. Patel BM, Devmurari SH, Rathod MP (2014) Development of high voltage solid state marx generator for liquid applications. Int J Eng Res Technol (IJERT) 3(12). ISSN: 2278-0181
5. Iqbal A et al (eds) (2020) Soft computing in condition monitoring and diagnostics of electrical and mechanical systems, vol 1096. In: Advances in intelligent systems and computing. Springer, Singapore. https://doi.org/10.1007/978-981-15-1532-3
6. Iqbal A et al (eds) (2020) Meta heuristic and evolutionary computation: algorithms and applications, vol 1096. In: Studies in computational intelligence. Springer, Singapore.https://www.springer.com/gp/book/9789811575709

An Overview of Using Hydrogen in Transportation Sector as Fuel

Raghav Gupta and Mohd Faisal Jalil

Abstract After the industrial revolution, transportation became an important part of our lives, but for the kind of transportation we are talking about fuel is the part whose importance cannot be ignored. But with the starting of twenty-first century, availability of fossil fuels is degrading at a very faster pace. Along with this, automobile produces the highest amount of air pollution by burning fossil fuels. Hence, it is the need of the hour to obtain some substitute of these fossil fuels, among all the avaliable renewable resources Hydrogen is the best one avaliable to us. The best thing with hydrogen is the availability in bulk, and along with this for the generation of power we need a fuel cell that will work as an engine and generate electricity as input is provided to it. Another best part of hydrogen is that its product's water vapors as a by-product when it is combusted. Hence, we can say that usage of hydrogen as fuel will not harm nature as well. There are several methods of getting hydrogen from fossil fuels and water and a lot more for storage and transportation of hydrogen. This is a brief study comprising of all the aspects of hydrogen production, storage, and transport to develop an approach toward the usage of hydrogen as a fuel.

Keywords Hydrogen · Fossil fuels · Hydrogen vehicles · Production · Transportation · Storage · By-products

1 Introduction

As the world reaching to the point of eliminating fossil fuels, electric cars have an increase in demand from last few years [1–15]. The major ways to power up an electric vehicle is either a fuel cell or a battery which is powering the vehicle [14, 15]. Hydrogen can be produced using water which is available in abundance. At first glance, hydrogen seems like a superb way to drive a car. After going through a large number of information available on different sources regarding the production, storage, and transportation of hydrogen, this study is prepared. Several authors tried

R. Gupta (✉) · M. F. Jalil
KIET Group of Institutions, Delhi-NCR, Ghaziabad, India
e-mail: raghav.r1998@gmail.com

A. Iqbal et al. (eds.), *Renewable Power for Sustainable Growth*, Lecture Notes in Electrical Engineering 723, https://doi.org/10.1007/978-981-33-4080-0_49

to give solutions on these problems; along with this they also tried to give a solution to some other issues as well such as increasing the yield [1–3].

Along with the proposed solutions, there is also a comparative data. This will give a clear detailing about all the processes available and the economic benefits of different solutions available and proposed. Hydrogen fuel can be called as green fuel when the production of hydrogen is also done using some renewable sources. The work on hydrogen is already done by several scientists; some of them studied for this purpose are as follows [4].

2 Methods Available for Production

There are different methods of hydrogen production, but the major portion of the total is produced by these two methods [5–7, 14, 15]:

(1) Steam reforming
(2) Electrolysis
(3) Thermochemical method
(4) Fossil fuel method or biomass gasification
(5) Solar energy method
(6) Coal gasification
(7) Biomass pyrolysis (Fig. 1).

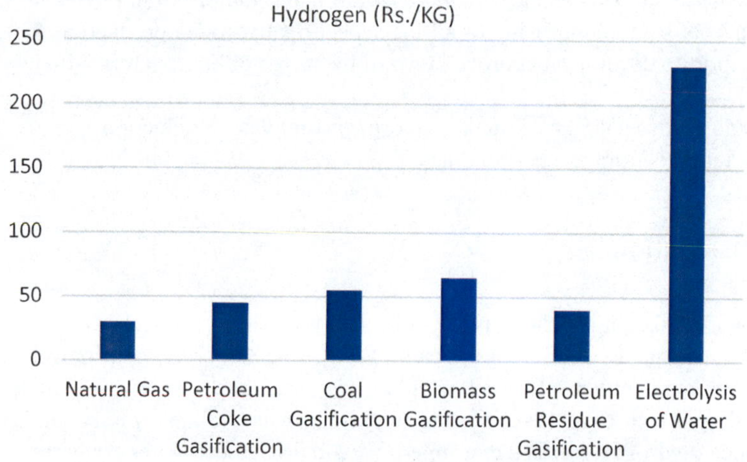

Fig. 1 Comparison of cost of hydrogen via different sources of production [8]

2.1 Steam Reforming Method

In the steam reforming method, the natural gas is a combined steam at a temperature of 710–1120 °C which is a difficult process. On the other hand, electrolysis is the process of passing a current through water which leads to the splitting of water. But the process becomes costly when production is done in bulk as we have to use any inert electrode which can sustain in both acidic as well as basic medium since during the production of hydrogen using electrolysis then for the production of hydrogen, we need acidic medium whereas for the production of oxygen basic medium is needed. But since the raw materials for steam reforming methods are also renewable, this method could not be used for the production of hydrogen in the future, and we need some alternate sources which could be utilized for the production of hydrogen [9].

2.2 Electrolysis

The crucial method used for obtaining hydrogen in bulk after the steam reforming method is electrolysis. During this process, two electrodes made of some passive materials are dipped in some electrolytes having some active salt for the increase in the process of electrolysis. For the process of electrolysis, DC source is used and hydrogen is evolved at cathode due to its positive nature and oxygen on the anode.

The important factor which to be remembered in electrolysis is the alkalinity of water and it should be maintained for the proper exchange of ions. Another important thing for the selection of material for electrodes should be used in such a way that the material should be able to sustain in both the nature that is acidic and basic nature since the production of hydrogen happens in acidic nature and evolution of oxygen happens in basic medium. The setup is generally utilized for the production of hydrogen [2, 5] (Fig. 2).

The reaction taking place on cathode is $4H^+ + 4e^- \rightarrow 2H_2$.

The reaction taking place on anode is $2H_2O \rightarrow O_2 + 4H^+ + 4e^-$

3 Problems and Solutions with Hydrogen

Being available in abundance and so useful, there are several problems which go hand in hand with hydrogen. The major problems with hydrogen will be discussed below along with the solutions [1].

Fig. 2 Electrolysis process

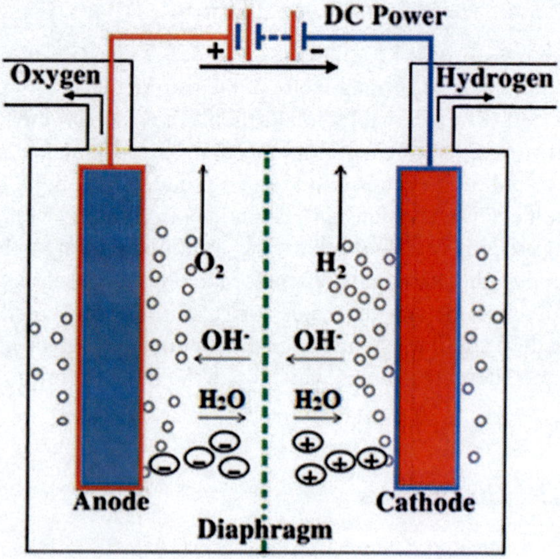

3.1 Cost of Production

The main constrain which we face with hydrogen as a fuel is cost. The main part of fuel is produced by electrolysis which utilizes platinum electrodes for electrolysis. As in the production from the water we need an acidic nature and the same for the oxygen we need basic nature [1]. Hence, for the production process, we need an electrode that can sustain in both of these environments. Platinum being an inert metal used for this purpose. But being so costly overall cost of hydrogen increases. The solution to this problem is proposed in such a way that we can prepare an electrode which will be a multilayered structure of nickel, graphene, and compound of iron, manganese, and phosphorous. This can be used as the substitute of the platinum electrode as nickel will provide more surface area and other constituents will enhance the conductivity. Along with this, the cost of the constituents of the newly form electrodes is so cheap and readily available, which will decrease the overall cost of the production of hydrogen. The other benefit of using this material is, it will be giving three times output on providing the same amount of energy to the electrolysis setup (Fig. 3).

3.2 Storage of Hydrogen

Storage of hydrogen becomes another problem as being a very light gas storage is not an easy task. Density is 1/4 of air and 1/9th of natural gas under the same condition.

For condensing the gas, it should be cooled at −253 °C with a specific gravity of 0, which is roughly 1/10th of gasoline.

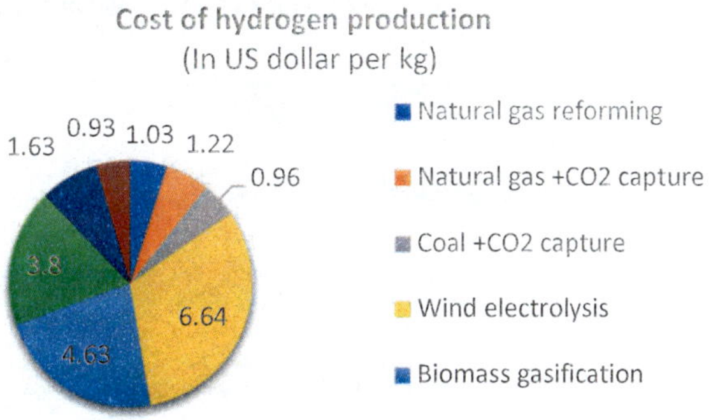

Fig. 3 Cost of hydrogen production using different sources [8]

The main issues which we are facing with electrical power are the storage which is difficult as well as costly and required high maintenance cost. So, we have an advantage of hydrogen over electricity which is that it can be stored.

The compressed gas storage method of storage of hydrogen is very expensive and bulky because it utilizes pressure compressed cylinders. Generally, these cylinders are made up of steel and hence transportation and management become very difficult.

Storing hydrogen in the form of liquid is one of the most practical approaches which is utilized for the storage purpose of hydrogen. Liquid hydrogen starts to boil at a temperature of −253 and hence for the storage temperature of the vessel utilized for the storage of hydrogen must be maintained below this temperature. Generally, we utilize vacuum-insulated vessels or super-insulated vessels for the storage of liquid hydrogen. The major disadvantage of using this method for the storage is the required amount of energy which has to be utilized for the conversion of the gas to a liquid state by compression and along with this, the other major cost is the refrigeration plant cost which is essential for the primary refrigeration such as liquid nitrogen plant. The overall net amount 25–39% of the boiling value of hydrogen is utilized during liquification of hydrogen (Fig. 4).

Another important and safe method of storage of hydrogen is storage as a metal hydride. In this method, hydrogen is reacted with any of the active metal which forms hydride after getting reacted with the metal. Some numerous metals and alloys form solid compounds after the reaction with hydrogen which forms metal hydride. And the other importance of this method is the property of hydride due to which extracting out of the hydrogen becomes very easy from metal hydrides.

3.3 *Transportation of Hydrogen*

For the transportation of hydrogen, two methods are generally used. They are liquid hydrogen transportation and metal hydride transportation. Both of these storage methods are preferred as per the need of time and availability of resources.

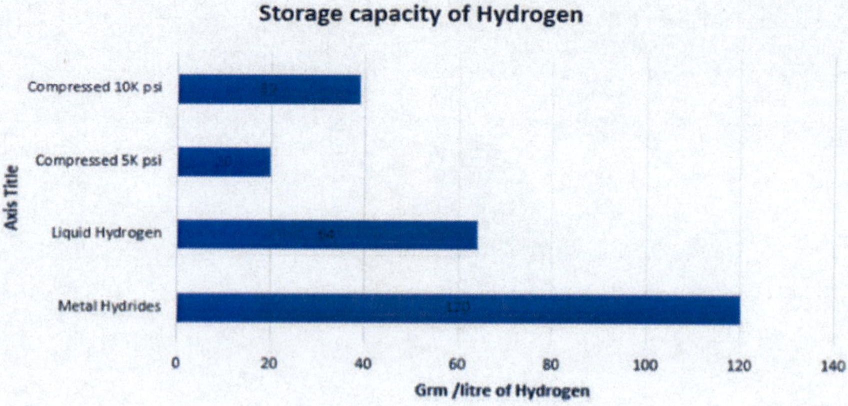

Fig. 4 Comparison between storage methods of hydrogen

The transportation of hydrogen through pipelines with a jacket of liquid nitrogen has been proposed. The cost of transportation may be greater than that of natural gas pipeline transportation but it is justified as it can be applied where the liquid fuel is required.

On the other hand, in metal hydride transportation, fuel will be delivered in the form of hydrides. The main constrain of metal hydride will be much greater than that of gaseous fuel [4] (Fig. 5).

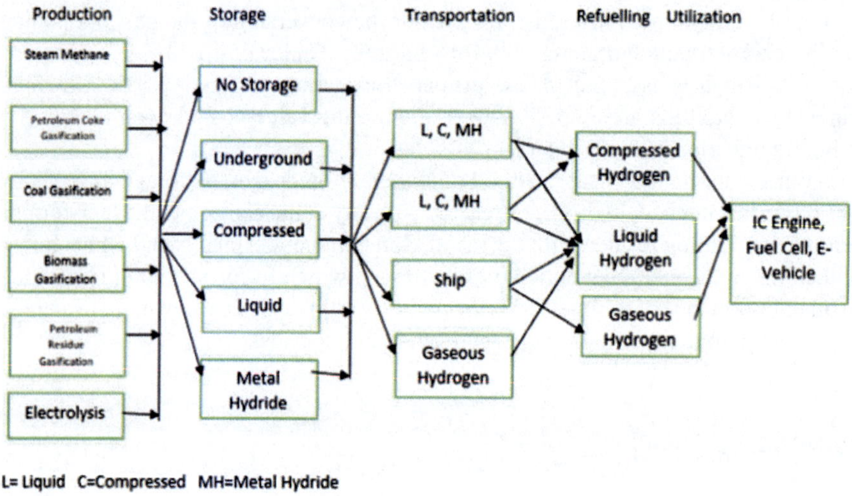

Fig. 5 Cost difference (Rs./kg of H_2)

4 Hydrogen Vehicle Over E-Vehicles

As we know that everything in this world is finite, the same goes for lithium as well. We have a limited quantity of lithium available in this world for our utilization. And as we have discussed above, fossils fuels will be finished in few more years, and what if all the vehicles need lithium-ion batteries for their operation [6]. Neither we have this much resource available nor there is any effective method for the safe disposal or recycling method of these lithium-ion batteries. Another problem with batteries is their life and charge holding capability. If a battery is left for 15 days without any usage also, it will destroy its charge up to 40% of its charging. Along with this, charging of batteries is another great issue which always remain side by side of the E-vehicles [10, 11].

From Table 1 [12], we can see that the charging time of a battery can be seen, and on the other hand, a fuel cell works as an engine until you keep on supplying the hydrogen and oxygen to it, it will keep on supplying the energy to the vehicle.

Through the above comparison [12, 13], we can easily see the benefits of fuel cell vehicles over electric vehicles. From all these points, we can see that hydrogen vehicles will be the vehicles of the future.

Another fact linked with lithium is that the mining of raw material such as lithium is also not a green activity which means the mining of lithium releases a large amount of carbon in the atmosphere which leads to the increase of greenhouse effect. The biggest difference between the two is that a battery stores energy, while a fuel cell generates energy by converting available fuel. A fuel cell can have a battery as a system component to store the electricity it is generating. A fuel cell directly "converts" an energy source into electricity through a chemical reaction—one step rather than multiple steps. This allows a fuel cell to remain efficient, quiet, and clean.

Table 1 Charging time of E-vehicles

Time for 100 km	Supply type	Power (kW)	Voltage	Max. current (A)
6.5–8.5 h	Single phase	3.2	230 V AC	17
3.5–4.5 h	Single phase	7.4	230 V AC	33
2.5–3.5 h	Three phase	11.5	400 V AC	17
1.5–2 h	Three phase	23	400 V AC	33
25–30 min	Three phase	44	400 V AC	64
25–30 min	Direct current	50	400–500 V DC	110–125
15 min	Direct current	120	400–500 V DC	310–350

5 Conclusion

If we can cut down the production cost of hydrogen then the economy could be boosted at a very higher rate. On taking about India, being a farming focused country, if we can bring down the fuel cost then the cost of fertilizers will be decreased to a great extent. Along with this, on talking about the mobility sector using hydrogen as a fuel, this will surely be a revolution as the energy output of hydrogen is about three times then our traditional fossil fuels which are already getting over at a very high rate. Surely there are several problems in using it as fuel but as per the outputs we are obtaining from the studies if the implementation of the newly opposed electrodes proposed by Rice University is done the production will be increased by three times. That is on giving the same amount of electricity as input our yield will get increased by three times. Hence, we must now focus on the utilization of hydrogen in several other sectors such as solar energy storage and liquid hydrogen-based engines. These all are the domains which will surely take this world to a new standard in various fields.

References

1. Dutta S (2014) A review on production, storage of hydrogen and its utilization as an energy resource. J Ind Eng Chem 20:1148–1156. https://doi.org/10.1016/j.jiec.2013.07.037
2. Al-Shara NK, Sher F, Iqbal SZ, Sajid Z, Chen GZ (2020) Electrochemical study of different membrane materials for the fabrication of stable, reproducible and reusable reference electrode. J Energy Chem 49:33–41. https://doi.org/10.1016/j.jechem.2020.01.008
3. Gherghele C, Gherghele V (2019) Hydrogen the fuel of the future. Marit Holl 68:34–37. https://doi.org/10.2307/3957240
4. Malhotra RK (2016) Report on transportation through hydrogen fuelled vehicles in India
5. Kwasi-Effah CC, Obanor AI, Aisien FA (2015) A review on electrolytic method of hydrogen production from water. Am J Renew Sustain Energy 1:51–57
6. Van De Kaa G, Scholten D, Rezaei J, Milchram C (2017) The battle between battery and fuel cell powered electric vehicles: a BWM approach. Energies 10. https://doi.org/10.3390/en10111707
7. Manoharan Y, Hosseini SE, Butler B, Alzhahrani H, Senior BTF, Ashuri T, Krohn J (2019) Hydrogen fuel cell vehicles; current status and future prospect. Appl Sci 9. https://doi.org/10.3390/app9112296
8. Abbasi T, Abbasi SA (2011) "Renewable" hydrogen: prospects and challenges. Renew Sustain Energy Rev 15:3034–3040. https://doi.org/10.1016/j.rser.2011.02.026
9. Davis C, Edelstein B, Evenson B, Brecher A, Cox D (2016) Hydrogen fuel cell vehicle. Int J Latest Trends Eng Technol 7:1–39. https://doi.org/10.21172/1.73.027
10. Offer GJ, Howey D, Contestabile M, Clague R, Brandon NP (2010) Comparative analysis of battery electric, hydrogen fuel cell and hybrid vehicles in a future sustainable road transport system. Energy Policy 38:24–29. https://doi.org/10.1016/j.enpol.2009.08.040
11. Notter DA, Gauch M, Widmer R, Wäger P, Stamp A, Zah R, Althaus HJ (2010) Contribution of Li-ion batteries to the environmental impact of electric vehicles. Environ Sci Technol 44:6550–6556. https://doi.org/10.1021/es903729a
12. Denton T (2016) Electric and hybrid vehicles
13. Wu D, Ren J, Davies H, Shang J, Haas O (2019) Intelligent hydrogen fuel cell range extender for battery electric vehicles. World Electr Veh J 10. https://doi.org/10.3390/wevj10020029

14. Iqbal A et al (eds) (2020) Soft computing in condition monitoring and diagnostics of electrical and mechanical systems, vol 1096. In: Advances in intelligent systems and computing. Springer, Singapore. https://doi.org/10.1007/978-981-15-1532-3
15. Iqbal A et al (eds) (2020) Meta heuristic and evolutionary computation: algorithms and applications, vol 1096. In: Studies in computational intelligence. Springer, Singapore. https://www.springer.com/gp/book/9789811575709

VP-ZA-LMS Control Algorithm for Double-Stage Grid Integration of Solar PV System

Md. Ibrahim, Ikhlaq Hussain, M. A. Mallick, and Mukul Chankaya

Abstract The present study presents the integration of double-stage solar PV system using variable parameter-based zero attracting least mean square control algorithm (VP-ZA-LMS) with the grid. The VP-ZA LMS gives fast transient and swift response as compared to other control algorithms like simple LMS and LMF. Extraction of maximum power from PV array is achieved using incremental conductance (INC) maximum power tracking. System is simulated in MATLAB software. Simulation results and performances under various operating conditions like steady state, varying insolation and load unbalancing are considered.

Keywords Solar photovoltaic (SPV) · Grid integration · VP-ZA-LMS · Power quality · Adaptive control · Power electronics

1 Introduction

Renewable energy is playing the key role in minimizing the gap between rapid acceleration in energy demand and depletion in fossil fuel-based generation. Among solar energy resources, solar energy is being available more or less at every location of the world [1–12]. Solar energy-based generation contributes in minimizing greenhouse gas emission (GHG). Solar photovoltaic (SPV) is an emerging renewable technology as it is a clean source of energy, i.e., there is no release of greenhouse gases into the atmosphere [2–4]. Next, the photovoltaic conversion process of sunlight into electricity does not require any fuel and hence no variable cost. In addition, it is a scalable power. To elaborate, it is deployable on an industrial scale, and it can be utilized for a single household. Due to ease of installation, less maintenance and ecofriendly nature, the SPV system is extensively being used to supply power to users. Reliability, voltage imbalance, weak grid system and poor power quality are some major

Md. Ibrahim (✉) · M. A. Mallick
Department of Electrical Engineering, Integral University Lucknow, Lucknow, India
e-mail: ibrahimalig3893@gmail.com

I. Hussain · M. Chankaya
Department Electrical Engineering, University of Kashmir, Srinagar, India

A. Iqbal et al. (eds.), *Renewable Power for Sustainable Growth*, Lecture Notes in Electrical Engineering 723, https://doi.org/10.1007/978-981-33-4080-0_50

expected challenges to be dealt while integration of solar photovoltaic (SPV). In the power system field, the SPV system is proven to be a leading technology owing to the fact that it is very effective in supplying power at remote positions (where transmission network cannot have range) because of its easy installation, low maintenance and has several other positives [5, 6]. One-stage and two-stage grid integration are possible for solar PV system. In double stage, DC–DC, i.e., boost converter is employed between PV array and VSC [7–9]. Double stage is more suitable for system with less input voltage [10]. In literature, many MPPT techniques are used in which INC technique is being extensively used due its simplicity and fast response [5, 6, 10].

Power qualities are the major concern for grid-tied SPV system which are mainly due to [11, 12]:

(1) Voltage and power fluctuation: Since the voltage is dependent on environment, i.e., temperature and solar irradiance, thus frequent environmental changes cause voltage fluctuation and hence power of the SPV system changes.
(2) Frequency fluctuations: It often takes place owing to power imbalance between load and supply.
(3) High harmonic contents in currents: Because of the use of converters and inverters, high contents of harmonics are produced in the grid current.

In order to maintain power quality, there are international standards and codes. According to IEEE-519 standards, the harmonic content in the grid current should not exceed 5%, as per IEEE-1547, voltage fluctuation which must be $\leq 0.5\%$.

2 Proposed System

The proposed system and control algorithm are shown in Figs. 1 and 2, respectively. The primary components are a PV array of 32 kW capacities, DC–DC boost converter, interfacing inductances and RC filter. INC MPPT technique is utilized to get MPP condition. The output of control algorithm is passed through hysteresis current

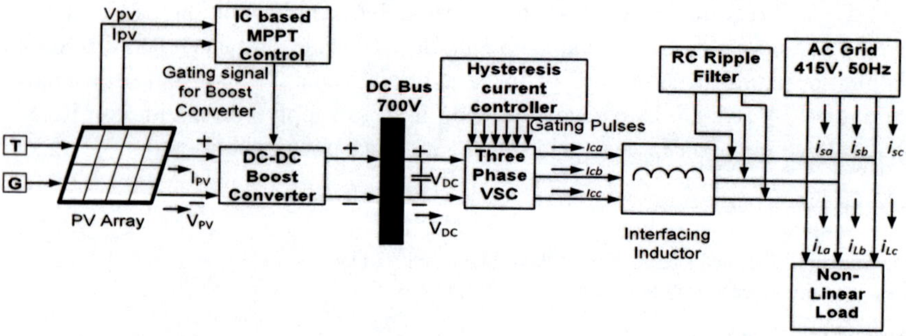

Fig. 1 Block diagram of proposed topology

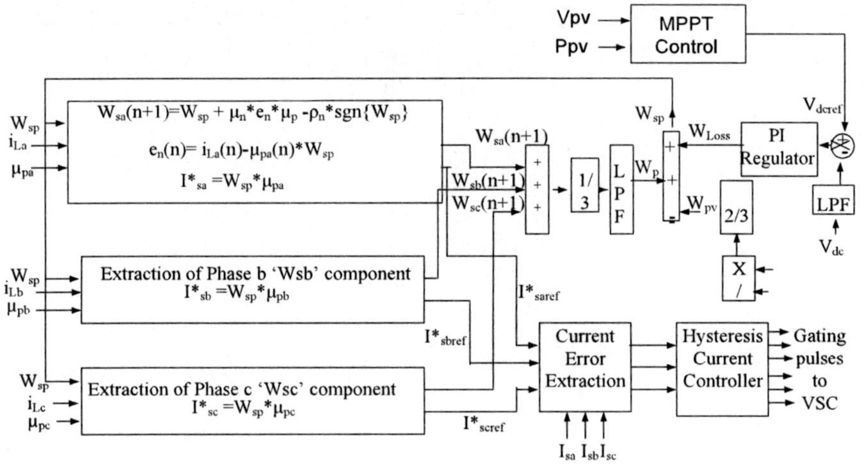

Fig. 2 Proposed control algorithm

controller for generating switching pulses for the converter. RC filters and interfacing inductances are utilized to mitigate voltage and current ripples, respectively.

3 Control Algorithm

Since it is a double-stage system, two control algorithms, i.e., for boost converter and for VSC are used to integrate the PV array to the grid [2].

3.1 VSC Control

From block diagram, the equations governing the control algorithm are elaborated by the following equations. The control algorithm is based on adaptive control algorithm in a closed loop manner. Feed forward term included in the system to take care of environmental changes like insolation and temperature. The control algorithm shown is for all three phases in a similar manner. First weight component of all three phases is calculated and processed as per equations written from 1 to 9. Sigmoid function here is considered for fast convergence. It reaches steady state very fast.

$$e_n = i_{\mathrm{La}}(n) - \mu_{\mathrm{pa}} * W_{\mathrm{sp}} \tag{1}$$

$$\boldsymbol{\mu}_{\mathbf{pa}} = \frac{v_{\mathrm{sa}}}{V_t}, \boldsymbol{\mu}_{\mathbf{pb}} = \frac{v_{\mathrm{sb}}}{V_t}, \boldsymbol{\mu}_{\mathbf{pc}} = \frac{v_{\mathrm{sc}}}{V_t} \tag{2}$$

Feed forward term can be calculated as follows:

$$\mathbf{W}_{\mathbf{pv}}(\mathbf{n}) = \frac{2P_{\mathrm{pv}}(n)}{3V_t} \tag{3}$$

$$\mathbf{V_t} = \sqrt{2/3\left(v_{\mathrm{sa}}^2 + v_{\mathrm{sb}}^2 + v_{\mathrm{sc}}^2\right)} \tag{4}$$

$$\mathbf{W}_{\mathbf{pa}}(\mathbf{n}+1) = \mathrm{W}_{\mathrm{pa}}(n) + \mu_n\mu_{\mathrm{pa}}(n) * e_{\mathrm{pa}} - \rho_n * \mathrm{sgn}\{W_{\mathrm{pa}}\} \tag{5}$$

$$\mathbf{W}_{\mathbf{pb}}(\mathbf{n}+1) = W_{\mathrm{pb}}(n) + \mu_n\mu_{\mathrm{pb}}(n) * e_{\mathrm{pb}}(n) - \rho_n * \mathrm{sgn}\{W_{\mathrm{pb}}\} \tag{6}$$

$$\mathbf{W}_{\mathbf{pc}}(\mathbf{n}+1) = W_{\mathrm{pc}}(n) + \mu_n\mu_{\mathrm{pc}}(n) * e_{\mathrm{pc}}(n) - \rho_n * \mathrm{sgn}\{W_{\mathrm{pc}}\} \tag{7}$$

$$\mathbf{W}_{\mathbf{g}} = 1/3\left(W_{\mathrm{pa}} + W_{\mathrm{pb}} + W_{\mathrm{pc}}\right) \tag{8}$$

$$\mathbf{W}_{\mathbf{sp}} = W_{\mathrm{g}} + W_{\mathrm{c}} - W_{\mathrm{pv}} \tag{9}$$

Generation of switching pulses for proposed adaptive control can be achieved by comparing reference grid current and sensed grid currents and then passed through hysteresis controller.

$$\mathbf{i}_{\mathbf{pa}}^* = W_{\mathrm{sp}} * \boldsymbol{\mu}_{\mathrm{pa}}, \mathbf{i}_{\mathbf{pb}}^* = W_{\mathrm{sp}} * \boldsymbol{\mu}_{\mathrm{pb}}, \mathbf{i}_{\mathbf{pc}}^* = W_{\mathrm{sp}} * \boldsymbol{\mu}_{\mathrm{pc}} \tag{10}$$

4 System Performance

System performance is analyzed for the following conditions.

4.1 Steady-State Performance

Insolation level at 1000 W/m^2 at 25 °C under steady-state operation is considered. The system variables are shown in Figs. 3 and 4a, b.

4.2 Varying Insolation

The insolation changed from 1000 to 500 W/m^2 at 0.35–0.4 s. The solar power output of the solar system is reduced considerably for reduced solar insolation. The load requirement is fulfilled in conjunction with the grid. The performance is found satisfactory as per IEEE 519 standards.

4.3 Unbalanced Nonlinear Load

During dynamic behavior of load unbalancing, the presented topology remains under unbalanced load condition from 0.35 to 0.4 by disconnecting phase ‘a’ from the grid through breaker. System parameters are shown in Fig. 4e, f, while weight parameters are shown in Fig. 4g.

5 Comparative Study

Figure 5 and Table 1 show the performance comparison between proposed and conventional control algorithm. From the fundamental weight comparison as shown in Fig. 5, it is found that the proposed control algorithm having fast transient response takes less settling time. Table 1 shows the values of MSE and settling time comparison with the conventional control algorithm. It is found that proposed control is superior in terms of various parameters as mentioned in the table over conventional control algorithm.

clearpage

6 Conclusion

The proposed control algorithm has been implemented for three-phase three-wire grid-connected system for solar PV and rectifier-based nonlinear load. The VP-ZA-LMS control algorithm system performance has been analyzed under two different

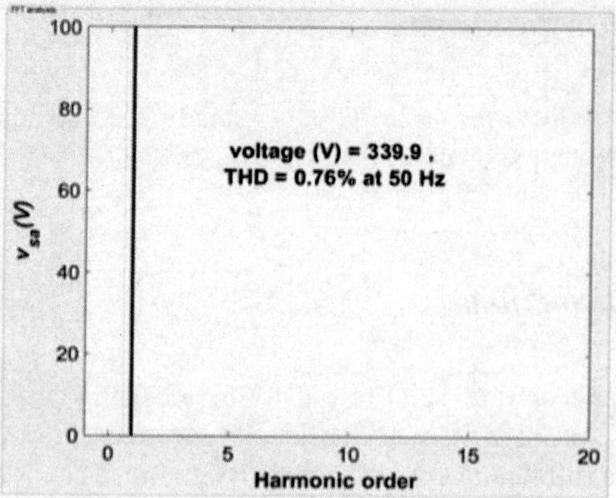

(a)

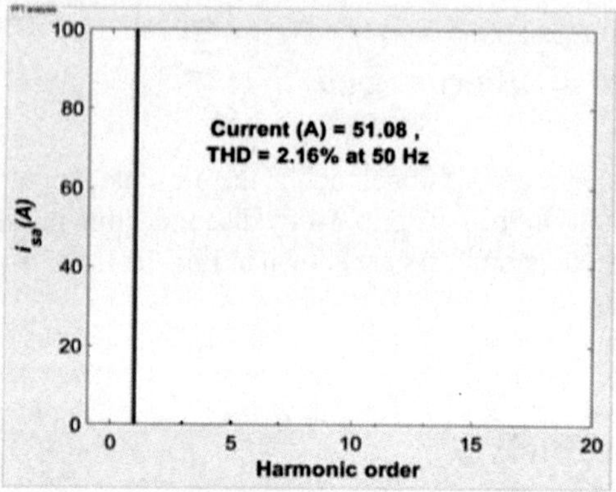

(b)

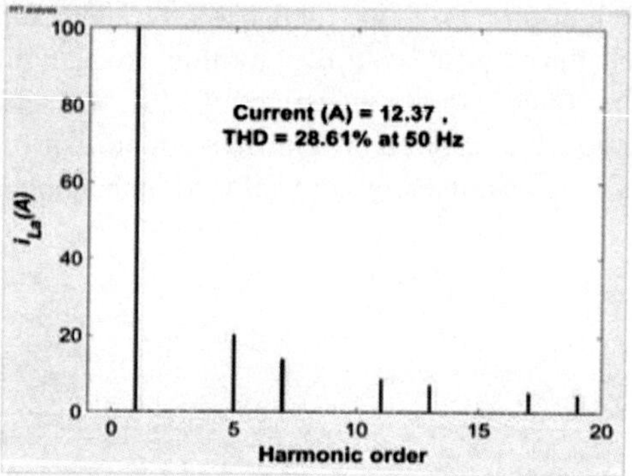

(c)

Fig. 3 **a–c** THD of V_{sa}(0.76%), I_{sa} (2.16) and I_{La}(28.61%) respectively

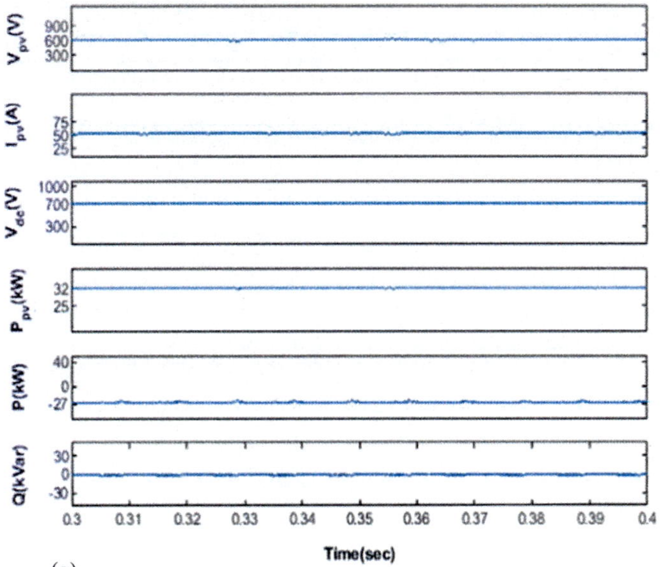

(a)

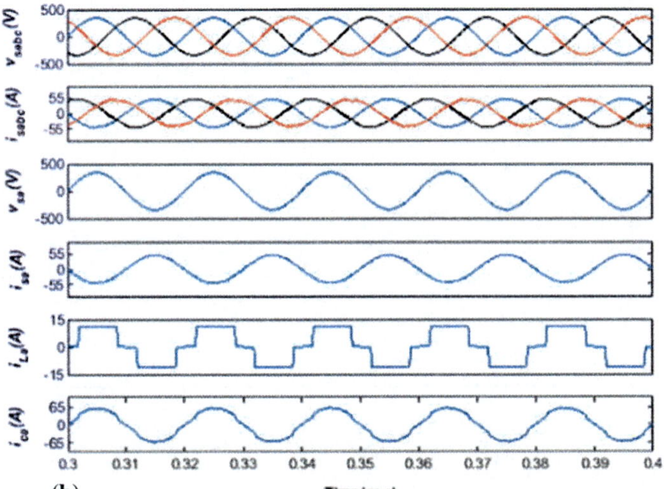

(b)

Fig. 4 **a** Steady-state performance of V_{pv}, I_{pv}, V_{dc} P_{pv}, P and Q, **b** steady-state performance of V_{sabc}, I_{sabc}, V_{sa}, I_{sa}, I_{La} and I_{ca}, **c** dynamic performance with changing insolation of V_{pv}, I_{pv}, V_{dc} P_{pv}, P and Q, **d** dynamic performance of V_{sabc}, I_{sabc}, V_{sa}, I_{sa}, I_{La} and I_{ca}, **e** dynamic performance under unbalanced loading of V_{pv}, I_{pv}, V_{dc} P_{pv}, P and Q, **f** steady-state performance of V_{sabc}, I_{sabc}, V_{sa}, I_{sa}, I_{La} and I_{ca}, **g** dynamic performance under load unbalancing of W_g, W_{Pv}, W_{sp} and W_{pa}

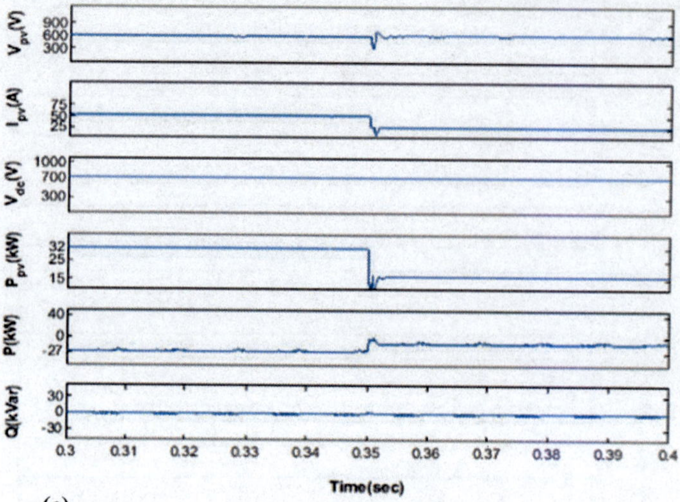

(c)

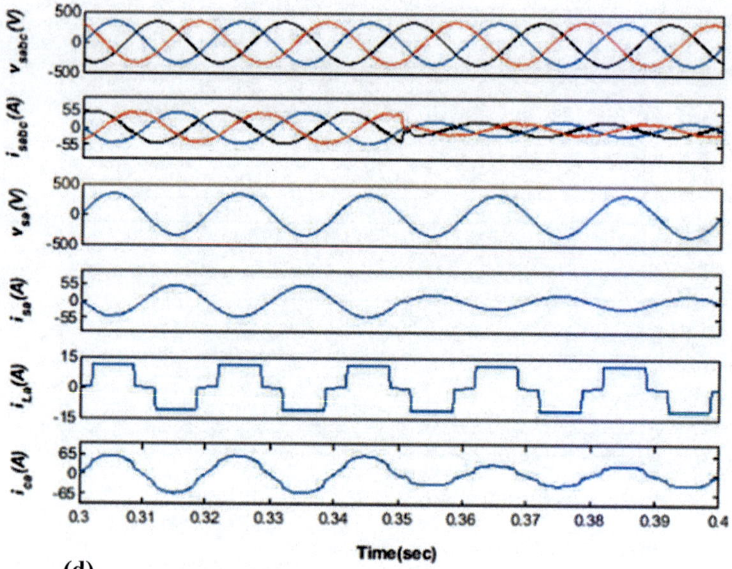

(d)

Fig. 4 (continued)

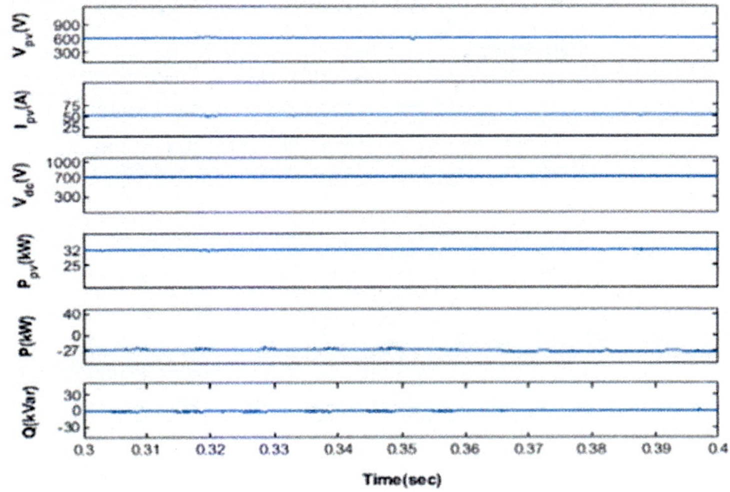

(e)

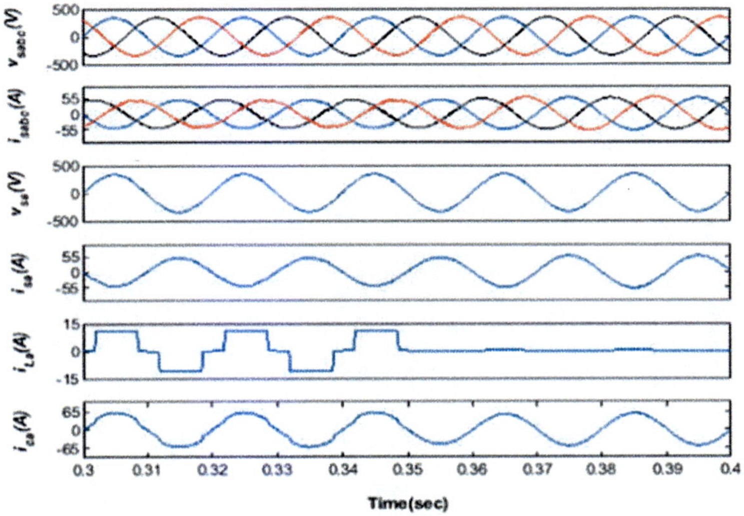

(f)

Fig. 4 (continued)

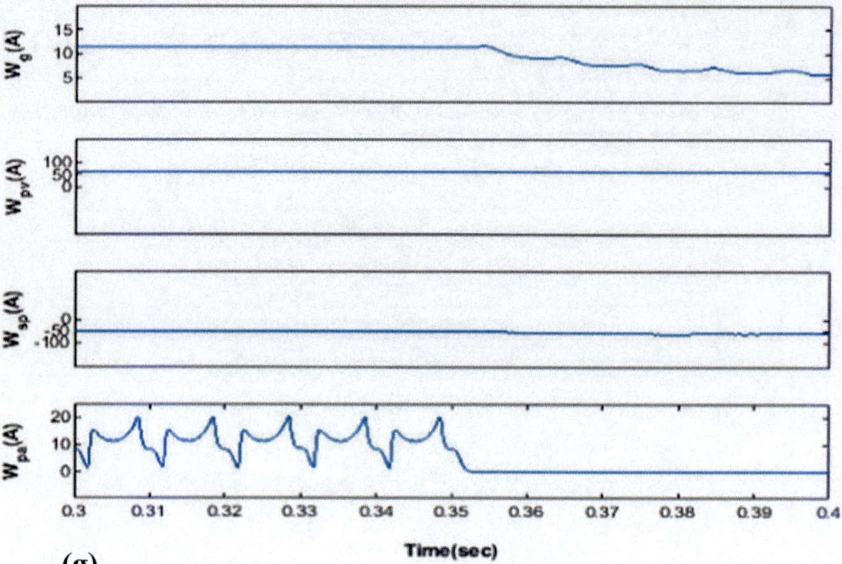

(g)

Fig. 4 (continued)

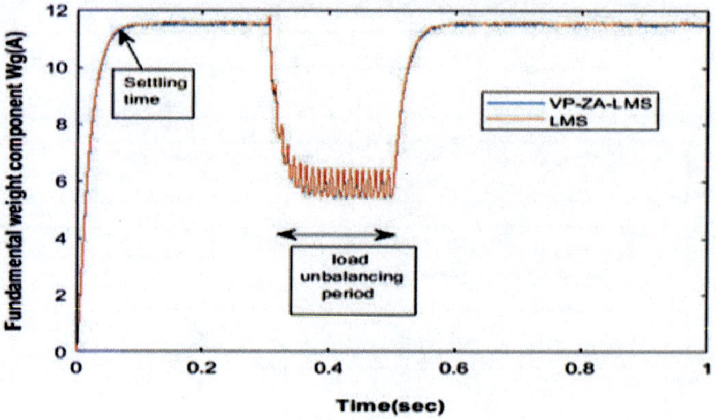

Fig. 5 Fundamental weight comparison with conventional control algorithm

conditions, i.e., steady-state and dynamic-state conditions. In dynamic-state variable, insolation and load unbalancing conditions are considered. For UPF operation, power is transferred to the grid from SPV system. The THD for the proposed control algorithm is analyzed and is within the acceptable limit of IEEE-1519 standards. The system performance for varying solar insolation and load unbalancing is also found satisfactory, meanwhile the exchange of reactive power between PV and grid system is zero.

Table 1 Comparative study

Parameter	LMS control	Proposed VP-ZA-LMS control
Filter type	Adaptive	Adaptive
Structure	More complex	Less complex
Degree	2nd	2nd
Accuracy	good	Better
Static error	High	Low
MSE	15.97	3.34
Sampling time	1 μs	1 μs
Settling time (s)	0.1	0.07

Acknowledgements I express thankfulness to the Integral University, Lucknow, for providing MCN number IU/R&D/2020-MCN000807.

References

1. Villalva MG, Gazoli JR, Filho ER (2009) Modeling and circuit-based simulation of photovoltaic arrays, pp 1244–1254
2. Ibrahim M, Mallick MA, Chankaya M, Iqbal A, Hussain I (2019) Variable parameter resized zero attracting least mean fourth control for grid-tied pv system. Int J Innov Technol Explor Eng 8(11):2489–2493
3. Ibrahim M, Khair A, Ansari S (2011) A review of hybrid renewable/alternative energy systems for electric power generation. IEEE Trans Sustain Energy 2(4):392–403
4. Ibrahim ZM, Tayyab Mohammad M (2015) PV-solar/wind hybrid energy system for GSM/CDMA type mobile telephony base. Int J Eng Res 5(5):74–79
5. Keshavani K, Joshi J, Trivedi V, Bhavsar M (2014) Modelling and simulation of photovoltaic array using matlab/simulink 2(4):3742–3751
6. Dhar S, Scholar PG, Sridhar R, Avasthy ASGV, Scholar PG Modeling and simulation of photovoltaic arrays 4:1–5
7. Mudaliyar S, Mishra S (2017) Coordinated voltage control of a grid connected ring DC microgrid with energy hub. IEEE Trans Smart Grid
8. Debela T, Ensermu G, Bhattacharya A Design, control and simulation of grid connected DC/AC microgrid for residential applications
9. Yongli ZHU, Jianguo Y, Di W Comparative study of two stages and single stage topologies for grid-tie photovoltaic generation by PSCAD/EMTDC
10. Wu T, Member S, Chang C, Lin L, Kuo C (2011) Power loss comparison of single- and two-stage grid-connected photovoltaic systems 26(2):707–715
11. Iqbal A et al (eds) (2020) Soft computing in condition monitoring and diagnostics of electrical and mechanical systems, vol 1096. In: Advances in intelligent systems and computing. Springer, Singapore. https://doi.org/10.1007/978-981-15-1532-3
12. Iqbal A et al (eds) (2020) Meta heuristic and evolutionary computation: algorithms and applications, vol 1096. In: Studies in computational intelligence. Springer, Singapore. https://www.springer.com/gp/book/9789811575709

Fuzzy Discrete Event System (FDES): A Survey

Naveed Jeelani Khan, Gulfam Ahamad, Mohd Naseem, and Qamar Rayees Khan

Abstract The discrete event system (DES) has been deliberated upon for quite a long time. With the advent of fuzzy systems, the research community moved in the direction of fuzzification of the relevant activities. At the outset of the twenty-first century, one such attempt was made by Feng Lin and Hao Ying who brought forward the concept of fuzzy discrete event systems (FDES), which in its simple terms is the fuzzification of DES. The distinction of FDES lies in its ability to model the systems where states are fuzzy and transition from one state to other is not inconsequential, e.g., medical treatment planning, fuzzy control systems, etc. Since the field is of recent origin, the ideas regarding FDES seem to be dispersed. This paper tries to introduce, assemble and organise the basic concepts regarding FDES. We illustrate the development of FDES from a crisp DES and highlight the decision-making process in FDES. Two broad areas of associated use are also discussed as applications.

Keywords Control and automation · Decision-making · Discrete event system · Fuzzy discrete event system

N. J. Khan · G. Ahamad (✉) · M. Naseem · Q. R. Khan
Department of Computer Sciences, BGSB University Rajouri, Rajouri, Jammu and Kashmir 185234, India
e-mail: gulfamahamad.jmi@gmail.com

A. Iqbal et al. (eds.), *Renewable Power for Sustainable Growth*, Lecture Notes in Electrical Engineering 723, https://doi.org/10.1007/978-981-33-4080-0_51

1 Introduction

A system is defined as "a combination of components that act together to perform a function which cannot be achieved by any of the individual parts" [1]. A system is also defined as a subset of the universe chosen for our discourse [2]. Different definitions prevail in literature for different perspectives. However, the objective study of a system reflects that it consists mainly of two components, viz. interacting components and a function that the system is supposed to perform [3]. In automata, these components are represented by respective symbols. An automatic system is shown below in Fig. 1, where arrows denote the events/transitions and the circles indicate the states (Fig. 1).

A system can be classified as time-driven or event-driven [3]. The time-driven systems change states with the variation of time. The continuous state systems, such as an analogue clock, change states with time and hence fall under former category (Fig. 2a). The event-driven systems change states with the occurrence of specific transitions at particular instances of time. The discrete event system changes the states due to the occurrence of events, for example, ON/OFF switch, so it is of the later type (Fig. 2b). There are two broad categories of event-driven systems, one that produces the output on behalf of the present state and another one that provides the output based on both current state and input values [4].

In a system, there are different kinds of events. Some are related to each other while some are merely independent of each other. For example, in a simple system of e-commerce, the placement of order and payment are two dependent events. In contrast, cancellation of order and generation of salary may be two independent events. An event may be any stimuli that generate a response in the system. Each event in a system leads to some state. The state is capable of projecting the behaviour of a system at the particular time interval associated with it [3]. A discrete event system (DES) is a "discrete-state event-driven system" [3]. Its states change with respect to the discrete events that occur at different instances of time. Since the states in a DES are crisp, it is used to model a variety of complex systems, for example, CPU task scheduling, traffic systems, database system,s etc. [5].

The main limitation of DES is that it cannot address the problem where the system possesses vague states. For example, a patient's health state cannot be crisp in nature, because after some prescribed medication from a doctor he/she cannot abruptly turn to well from an unwell state just as a switch does from on to off state. Instead, the patient takes time to transit from poor to stable and thereafter to well state. At some instant, we can only say that the patient is doing well to some degree and

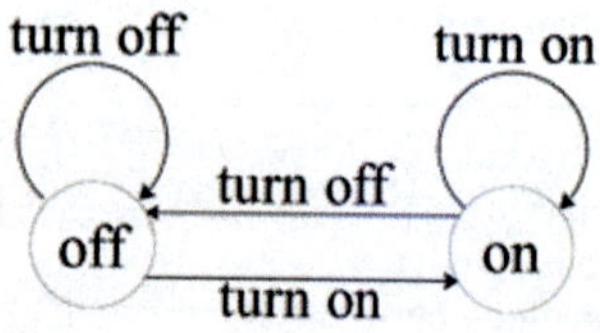

Fig. 1 Simple automaton for ON/OFF switching system

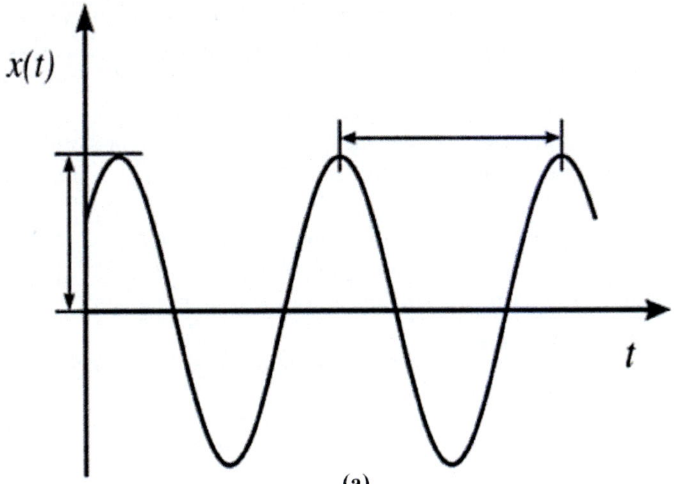

(a)

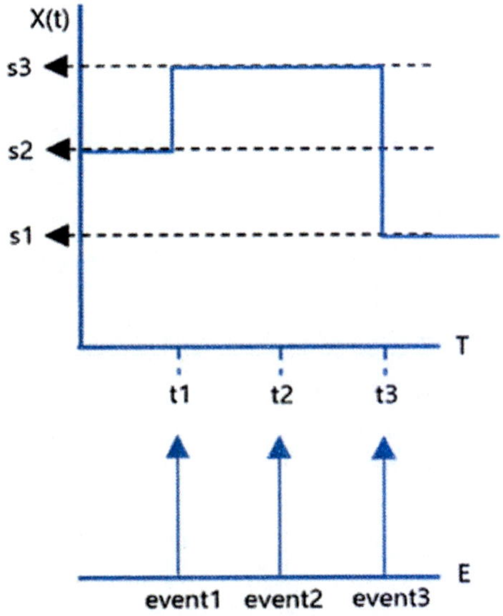

(b)

Fig. 2 **a** A time-driven system showing continuity. **b** An event-driven system showing discreteness

cannot claim with certainty that the patient is in a good state [6]. There exists a certain level of fuzziness in this scenario. Instead of DES, we use the same approach with an introduced degree of fuzziness, i.e., fuzzy discrete event system (FDES) to model such situations/ systems. FDES is a DES which deals with the fuzzy data (vague states). In contrast to DES, which follows the classical crisp set approach,

FDES supports the fuzzy set theory, proposed by Zadeh [7]. The clinical treatment planning is one of the main applications of FDES, identified so far [8]. The theory of state determination and predictability in FDES also helps us in diagnosing a disease [9, 10].

Although fuzzy states and events were known to the research community since the 1960s, no exertive research was performed since 1979 [11]. The concept was reintroduced in 2001 by Lin and Ying [12] with a certain level of clearness. The authors have been continuously exploring the margins of this field as of now [13]. Presently, the formal theory of FDES is in its development phase [14–17]. This approach has also been adopted to model the hybrid systems containing both discrete and continuous components [3, 18].

In order to brush off the vagueness about the topic under discussion, this paper attempts to do a brief and useful survey on FDES. The basic concepts regarding FDES are organised. We discuss the development of an FDES from DES and the techniques associated with it in Sect. 2. Section 3 introduces the decision-making in FDES along with an example to illustrate the concept. Section 4 reveals applications of FDES, described with comprehensive descriptions from the air-conditioning system and disease (such as AIDS) medication selection. Finally, Sect. 5 concludes the paper.

2 Fuzzy Discrete Event System (FDES) Modelling

To analyse a system, we need procedures to design, control and quantify system performance, and that is just what modelling does. A model replicates the behaviour of a system [3]. The behaviour of the system is described by some mathematical equation, as shown in Fig. 3. Crisp DES is modelled using the finite automata. A formal automaton is denoted by $G = (Q, \Sigma, \delta, q^o)$ [19]. The components of finite automata are a crisp set of states (Q), an event set (Σ) that trigger the transition from one state to other and a conversion function (δ) describing the occurrence of the event on some state and the resultant output state. q_o and q_f denote the initial and final states, respectively.

For example, an automaton for a simple on/off switching system, as shown in Fig. 1, may be expressed as:

Fig. 3 Modelling process [3]

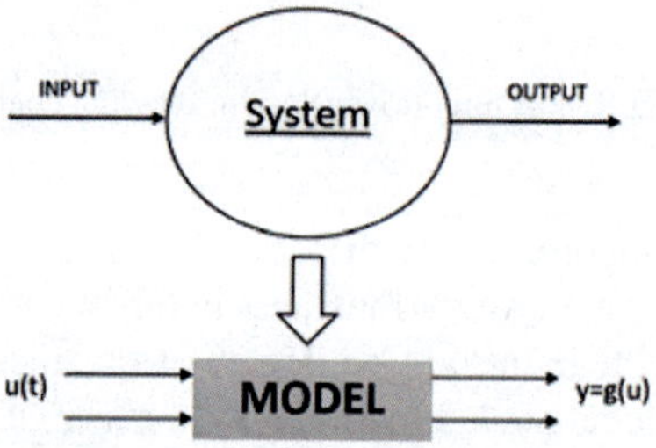

$$Q = \{\text{on, off}\}$$
$$\Sigma = \{\text{turn on, turn off}\}$$
$$\delta = \begin{cases} \text{on}, \forall\Sigma = \text{turn on} \\ \text{off}, \forall\Sigma = \text{turn off} \end{cases}$$
$$q_o = \text{off or on}$$
$$q_f = \text{on or off}$$

As stated earlier, vague states cannot be modelled using DES. Rather, we use FDES for the representation of such systems. Since the states in FDES are not discrete in nature, the above notation proves to be insufficient to represent FDES, and the number of states in FDES is not finite; hence, we use various means to deal with this issue [15, 20]. For FDES modelling, we use a separate notion, as shown in [12, 14 16]. Consider an FDES $G' = (Q', \Sigma', \delta', q^{o\prime})$, where Q' is the set of FDES states, Σ' is the set of FDES events, δ' is the FDES transition function, and $q^{o\prime}$ is the start state [21]. Since the notation with symbols like arrows and circles is not relevant in FDES, the notation as per [12, 14, 16] used is as follows:

The current state, q' is denoted by a vector of 0's, and 1. 0 denotes that the current state value is false for this state indicating the system is not in this state, and '1' denotes that the system is presently in this state. In terminology, this vector is known as a state vector. For example, if we have a system of seven states named as 1, 2, 3, 4, 5, 6, 7 and the current state vector is $q' = [0001000]$, it symbolises that the system is in state 4. The current state from q' is the Jth state where value is 1. The length of this vector is obviously equal to the number of states that our system has. In order to incorporate the fuzzy logic and fuzzy sets [22], we can use the fuzzy state vector as follows:

$q' = [0000.30.700]$ means that the system is simultaneously in two states 4 and 5 with membership of 0.3 for state 4 and 0.7 for state 5, respectively. Intuitively, we can say our system is currently in 30% of state 4 and 70% of state 7. For a patient in the above-mentioned example, we may say that currently, he/she is 30% unwell and 70% well.

The FDES transitions δ' are represented by matrices of events (no of matrices = no of events). For example, if β is any event, then the transition is expressed as:

$$\beta = [\beta_{ij}]_{n\times n} = \begin{bmatrix} \beta_{11} \dots \beta_{1n} \\ \vdots \\ \beta_{n1} \dots \beta_{nn} \end{bmatrix}$$

where $\beta_{ij} = 1$ if the state changes from $\boldsymbol{i}$ to $\boldsymbol{j}$ with the event β and $\beta_{ij} = 0$ if no transition happens with the event β. For example

$\beta = \begin{bmatrix} \beta_{11} & \beta_{12} \\ \beta_{21} & \beta_{22} \end{bmatrix}_{2\times 2} = \begin{bmatrix} 0 & 0 \\ 1 & 0 \end{bmatrix}$ means that the event β transits the system from state 2 to state 1 (as $\beta_{21} = 1$).

In the same manner, as fuzziness is applied in subsection (a), it can be used here as well. It can also be exemplified as follows:

$\beta = \begin{bmatrix} \beta_{11} & \beta_{12} \\ \beta_{21} & \beta_{22} \end{bmatrix}_{2\times 2} = \begin{bmatrix} 0 & 0.1 \\ 0 & 0.4 \end{bmatrix}$ means that it is likely that the system may change its state from state 1 to state 2 with membership (possibility) of 0.1 (as $\beta_{12} = 0.1$) and may remain unchanged in state 2 with a membership of 0.4 (as $\beta_{22} = 0.4$).

The next state is given by the product of q' and β (i.e., $q'.\beta$). For example, let us assume the current state number to be 2 (out of three states named as state one, state two and state three), then

$q' = [010]$ and let the event $\beta = \begin{bmatrix} 0 & 0 & 0 \\ 1 & 0 & 0 \\ 0 & 0 & 0 \end{bmatrix}$, then the next state will be given by their product as

$$q' \cdot \beta = [010] \begin{bmatrix} 0 & 0 & 0 \\ 1 & 0 & 0 \\ 0 & 0 & 0 \end{bmatrix} = [100]$$

which means from state two, our system will next transit to state one with the effect of event β.

When fuzziness is introduced, the next state is given by the max-product operation [7, 12, 23] or composition of fuzzy relations, i.e.,

$$q' o \beta = \max\left[\mu_{q'(i,j)}, \mu_{\beta(i,j)}\right] \tag{1}$$

where μ denotes the product between the subscripts.

Example: if $q' = [0.60.3]$ and $\beta = \begin{bmatrix} 1 & 0.5 \\ 0.8 & 0.4 \end{bmatrix}$, then according to (1)

$$\begin{aligned} q'_o\beta &= [\text{MAX}\{(0.6)(1), (0.3)(0.8)\}, \text{MAX}\{(0.6)(0.5), (0.3)(0.4)\}] \\ &= [\text{MAX}(0.6, 0.24), \text{MAX}\ (0.3, 0.12)] = [0.60.3] \end{aligned}$$

means that our system will remain in the same state after the β event as the resultant state is same as our q'.

In order to diversify the views obtained from the specialists of a particular domain, a separate notion was proposed as an extended FDES theory [16]. It offers to use pure fuzzy numbers instead of the discrete numbers. In this notion, the state is represented by:

$$_k q_i' = \left[{}_k q_1' {}_k q_2' {}_k q' \cdots {}_k q_{n1}'\right] \tag{2}$$

where k represents the k th state, i denotes the fuzzy state and n_1 represents the entire number of fuzzy states of state k. The members are represented by fuzzy representations, for example, if the system is currently in the first state and possesses three fuzzy states, i.e., $k = 1$ and $n_1 = 3$, then according to the extended notion

$$_1 q_i' = \left[\frac{1}{1}\frac{1}{0}\frac{1}{0}\right], \text{ where } {}_1 q_1' = \frac{1}{1}, {}_1 q_2' = \frac{1}{0} \text{ and } {}_1 q_3' = \frac{1}{0}.$$

This representation can handle the aforementioned notation as a special case [16]. It may be noted that in this kind of representation, the numerators stand for the degree of membership while denominators denote the item under discussion. A fuzzy transition matrix represents the event transition accordingly. Online learning algorithms have been developed to make the system learn the event transition matrices [13].

3 Decision-Making in FDES

Mostly fuzzification is used in decision-making where information is imprecise [24]. We try to make the system decide based on vagueness, just like humans do. Given the statement 'it is hot outside', based on the discrete 0 or 1 logic, machines cannot decide how much hot actually the weather is. They cannot determine which clothes to wear and which not because the term 'hot' is a vague term. Once FDES is modelled, the system should be able to make such decisions. A detailed architecture for decision-making in FDES, along with an illustration, exists in [25], Fig. 4. Here, we will only discuss the core procedures briefly.

First of all, we need to identify the necessary factors that affect the process of decision-making. For example, if we are planning a prescription schedule for a patient, then the factors that are crucial for the decision-making of regimen selection may include potency, adverse events, adherence and future drug possibilities as revealed in [8].

Once the essential states are known, the next step is to determine the FDES decision model. This decision model comprises of the following four components: fuzzy automata, corresponding states, events in each automaton and the transition matrices of events [25]. The representation of each component is the same as discussed earlier.

Afterwards, a forward-looking tree, as shown in Fig. 5, is constructed. Each node of this tree represents a fuzzy state (denoted by q'). The root node is the initial state. Next state is calculated as $q' o \beta$. The process goes on until the full tree is ready. L is the number of steps in the tree. The number of nodes depends upon the steps [26].

Since the forward-looking tree contains a number of combinations, we need the configuration (path) with optimal cost. In order to achieve this, performance index

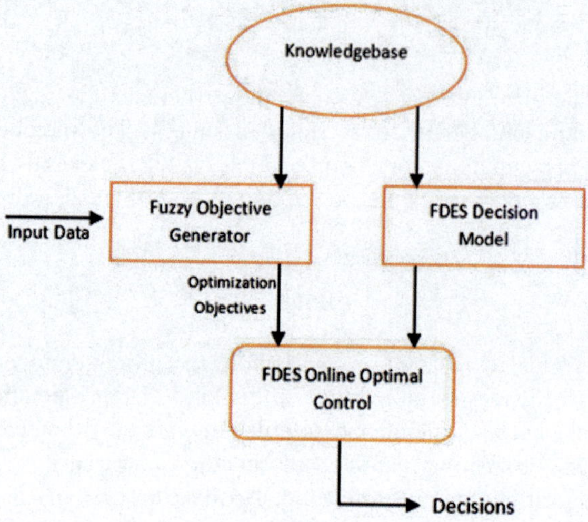

Fig. 4 FDES decision-making and control architecture [25]

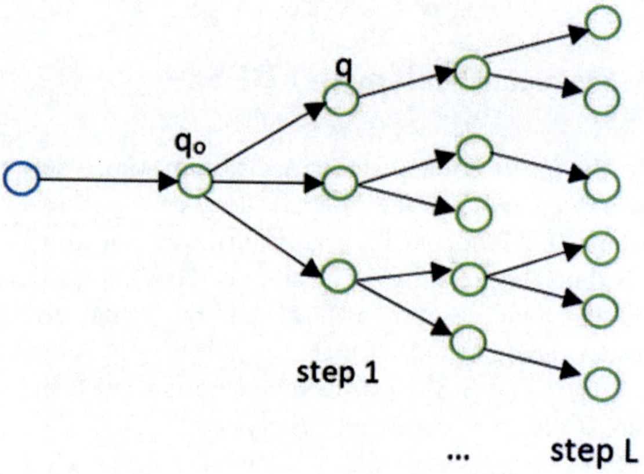

Fig. 5 An *L*-step forward-looking tree [27]

$J(q')$ is calculated against all nodes with the help of any effective function (chosen as per suitability). Effective function outputs the weight-based performance value for each node. The weights denote the grade of impact on the corresponding state. Since weights depend on a number of factors, they can be determined by using the facts from the knowledge base [28]. The path (set of nodes) with maximum $J(q')$ is selected as an optimal path, and the final decision is made. Performance index $J(q')$ is given by:

$$J(q') = w_1^T q_1' + w_2^T q_2' + \cdots + w_n^T q_n' \tag{3}$$

which may be rewritten as:

$$J(q') = \sum_{i=1}^{n} w_i^T q_i' \tag{4}$$

where $w_1, w_2, w_3, \ldots, w_n$ are the weight vectors and $q_1', q_2', q_3', \ldots, q_n'$ are the corresponding state vectors.

For instance, let our calculated J values for forward-looking tree nodes be $x_1, x_2, x_3, \ldots x_n$ where $n = |Q'|$, i.e., number of states. We compare these values or simply arrange them in decreasing order, for instance: $x_3 > x_1 > x_2 > x_4 > \cdots > x_{(n-6)}$.

The above statement means that J value for node 3 is the largest, and hence, we choose this node and then check the nodes under node 3 with the highest J value and select it. Going this way, we get a configuration of nodes and translating the events in the sequence of this configuration is our optimal decision. We take a real example from [25] with calculated J values and highlight the optimal decision path (with events α_1 and β_1) based on greater J value as shown in Fig. 6.

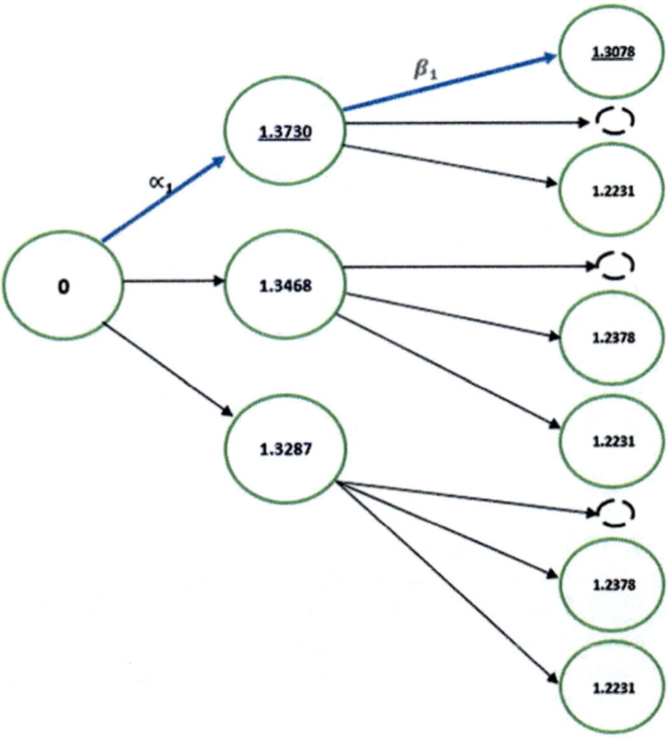

Fig. 6 Optimal decision path based on J value [25]

4 FDES Applications

FDES is applicable in imprecise situations. It finds its application in treatment modelling, economic modelling, control system modelling, etc. [25]. In this section, we discuss two examples of air-conditioning system and HIV/AIDS treatment regimen selection.

4.1 Air Conditioning System

A study of air-conditioning system prevails in [28]. A.C system is a typical example of FDES as it automatically adjusts the temperature of a space according to the requirement and hence possesses fuzzy states instead of discrete ones.

Modelling of Air-Conditioning FDES

D. Li et al. identified the five factors affecting the decision-making in an A.C. system as end supplied air volume, chilling water volume, cooling water volume, the temperature setting of cooling tower fan and water chiller [28]. The respective events corresponding to these factors may be denoted by $\alpha_1, \alpha_2, \alpha_3, \alpha_4$ and α_5.

The three states identified are: raise (R), maintain (M) and lower (L). So, our automaton for such system may look like $G' = (Q', \Sigma', \delta', q_{0'})$, where symbols have their usual meanings. The five factors mentioned may form the five automata for A.C FDES system, i.e., G_1, G_2, G_3, G_4 and G_5, respectively.

G_1: consists of four states, viz. initial (I1), lower state (L1), maintain state (M1) and raise state (R1).

Similarly, G_2 includes I2, L2, M2 and R2.

G_3 includes I3, L3, M3 and R3.

G_4 includes I4, L4, M4 and R4.

G_5 includes I5, L5, M5 and R5.

As an illustration, $q_1 = [000.50]$ means that automaton G_1 is currently in the maintain state (M1) with a membership of 0.5.

Decision-making in A.C FDES

The factors responsible for decision-making are already identified. A.C FDES is also modelled in (a). Now in order to make any decision, we need to construct a forward-looking tree. Every branch of the forward-looking tree is a decision in itself. From so many choices, we need the optimal decision, as stated earlier.

As illustrated in Fig. 7, the states will go on decreasing, e.g., when α_3 occurred, only four events remained ($\alpha_1, \alpha_2, \alpha_4$ and α_5) to happen. Similarly, at the second stage, only three events are left to happen. In this way, only one event remains at last that can occur.

After generating the state combinations, an optimal configuration is needed from all possible amalgamations. This is attained by the effectiveness function $J(q')$, stated in Sect. 3.

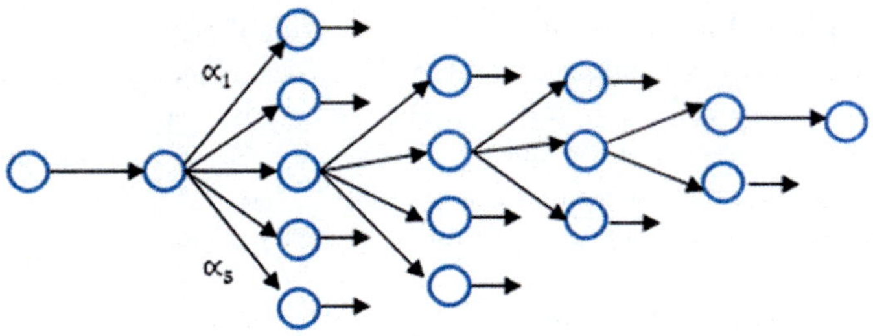

Fig. 7 State reduction [27]

The A.C system will choose the decision path with maximum J value [27]. This configuration will change with the change in inputs, and hence, the temperature will be regulated automatically.

4.2 Treatment Planning

Another important example of the application of FDES is of treatment planning as the treatment process also consists of fuzzy states. Diabetes treatment is illustrated in [29]. Another such example of HIV/AIDS treatment regimen selection process is studied in [26, 30]. This application is vital as HIV/AIDS treatment selection is regarded as a complex medical process [31, 32]. Hao Ying et al. identified the four factors affecting this selection process as potency, adherence, adverse events and future drug possibilities [26, 33–36]. Four famous and significant first-round HIV/AIDS regimens are studied, and after consultation with the field experts, characteristics of regimens (Table 1) are calculated against each factor [30].

Table 1 Data of clinical factors against four popular first-round HIV/AIDS regimens [30]

Regimen name	Potency (%)	Adherence (%)	Adverse events (%)	Future drug options (%)
1 (nelfinavir + zidovudine /lamivudine)	85	55	30	80
2 (efavirenz + zidovudine /lamivudine)	90	80	20	60
3 (nevirapine + zidovudine /lamivudine)	85	85	20	65
4 (abacavir/zidovudine /lamivudine)	80	90	10	85

Modelling of HIV/AIDS treatment regimen selection FDES
The four medical factors mentioned are taken as four state vectors of the fuzzy system denoted by q_1, q_2, q_3 and q_4, respectively. The q_1 comprises of two states, i.e., high and medium, q_2 consists of three states, i.e., easy, moderate and challenging, q_3 has three states, i.e., medium, low and very low, and q_4 has two states, i.e., high and medium. Employing the regimens is identified as a crisp event.

Decisioning in Regimen Selection
The state transition matrices are calculated from the data given in Table 1. The four medical factors are fuzzified in the same way as done in fuzzy controller input variables [37]. The state transition interpretation is the same as shown already. Since q_1 has 2 states, q_2 has 3 states, q_3 has 3 states and q_4 has 2 states, the system has to decide from as many as 36 $(2 \cdot 3 \cdot 3 \cdot 2)$ regimen amalgamations (decisions) and pick the best-suited decision. Out of 36, 4 combinations (viz. medium q_1 and moderate q_4 when q_2 is moderate or challenging and q_3 is moderate or small), are considered to be medically pointless. Out of remaining 32 combinations, effective decision is picked using an effective function. We get four values from the function for each combination corresponding to the four treatment regimens. The value having maximum value is taken as the first regimen, and the one with minimum value is the fourth regimen. Moreover, in order to select the optimal weights, genetic algorithm [38] is employed, and the overall effectiveness measure is increased. The model is validated by comparing the system results with the independent physician results. The valid agreement between the two is evident [30].

5 Conclusion and Future Scope

Fuzzy discrete event system is an extension of discrete event system (DES), that is used to study and model the vague systems. In this paper, we present a brief survey on FDES. The techniques to model the FDES are highlighted. We showed the decision-making process of FDES, along with the real-world application in A.C system and HIV/AIDS treatment regimen selection.

The limited availability of literature in this field was realised. The FDES theory is currently in its development phase, but several efforts and increasing research interest in this field are providing opportunities for the same. The identification of relevant fuzzy areas for implementation of FDES needs to be done effectively. Moreover, the states of FDES being indefinite, effective online optimization is the primary concern.

References

1. Radatz J, ICSS Committee (1997) The IEEE standard dictionary of electrical and electronics terms. IEEE

2. System [online]. Available https://www.britannica.com/science/system-physics
3. Cassandras CG, Lafortune S (2008) Introduction to discrete event systems. Springer, US, Boston, MA
4. Mishra KLP, Chandrasekaran N (2007) Theory of computer science: automata, languages and computation, 3rd edn. Prentice Hall, India
5. Ghabri M-K, Ladet P (1994) An approach for modeling and control of a discrete event system. In: Proceedings of 1994 33rd IEEE conference on decision and control, vol 3, pp 2648–2653. https://doi.org/10.1109/CDC.1994.411547
6. Ying H, Lin F, Sherwin R (2019) Fuzzy discrete event systems with gradient-based online learning. In: 2019 IEEE international conference on fuzzy systems (FUZZ-IEEE), pp 1–6. https://doi.org/10.1109/FUZZ-IEEE.2019.8859012
7. Zadeh LA (1965) Fuzzy sets. Inf Control. https://doi.org/10.1016/S0019-9958(65)90241-X
8. Ying H et al (2004) A fuzzy discrete event system for HIV/AIDS treatment planning. In: 2004 IEEE international conference on fuzzy systems (IEEE Cat. No. 04CH37542), vol 1, pp 197–202. https://doi.org/10.1109/FUZZY.2004.1375717
9. Mekki AO, Lin F, Ying H, Simoff MJ (2017) Fuzzy detectabilities for fuzzy discrete event systems. In: 2017 IEEE international conference on fuzzy systems (FUZZ-IEEE), pp 1–6. https://doi.org/10.1109/FUZZ-IEEE.2017.8015431
10. Benmessahel B, Touahria M, Nouioua F (2017) Predictability of fuzzy discrete event systems. Discret Event Dyn Syst 27(4):641–673. https://doi.org/10.1007/s10626-017-0256-7
11. Manes EG (1981) Fuzzy switching and automata: theory and applications (Abraham Kandel and Samuel C. Lee). SIAM Rev 23(2):271–273. https://doi.org/10.1137/1023058
12. Lin F, Ying H (2001) Fuzzy discrete event systems and their observability. In: Proceedings joint 9th IFSA world congress and 20th NAFIPS international conference (Cat. No. 01TH8569), vol 3, pp 1271–1276. https://doi.org/10.1109/NAFIPS.2001.943730
13. Ying H, Lin F (2019) Online self-learning fuzzy discrete event systems. IEEE Trans Fuzzy Syst:1. https://doi.org/10.1109/TFUZZ.2019.2931254
14. Lin F, Ying H (2002) Modeling and control of fuzzy discrete event systems. IEEE Trans Syst Man Cybern Part B Cybern.https://doi.org/10.1109/TSMCB.2002.1018761
15. Lin F et al Theory for a control architecture of fuzzy discrete event systems for decision making. In: Proceedings of the 44th IEEE conference on decision and control, pp 2769–2774. https://doi.org/10.1109/CDC.2005.1582582
16. Du X, Ying H, Lin F (2007) An extension to the theory of fuzzy discrete event systems. In: NAFIPS 2007—2007 annual meeting of the North American fuzzy information processing society, pp 289–294. https://doi.org/10.1109/NAFIPS.2007.383853
17. Nie M, Tan WW (2012) Theory of generalized fuzzy discrete event system. In: 2012 IEEE international conference on fuzzy systems, pp 1–7.https://doi.org/10.1109/FUZZ-IEEE.2012.6251227
18. Du X, Ying H, Lin F (2010) Fuzzy hybrid systems modeling. In: Annual meeting of the North American fuzzy information processing society, pp 1–6.https://doi.org/10.1109/NAFIPS.2010.5548415
19. Hopcroft JE, Motwani R, Ullman JD (2001) Introduction to automata theory, languages, and computation, 2nd edition. ACM SIGACT News 32(1):60. https://doi.org/10.1145/568438.568455
20. Bisgambiglia P-A, Innocenti E, Bisgambiglia P (2018) Fuzz-iDEVS: an approach to model imprecisions in discrete event simulation. J Intell Fuzzy Syst 34(4):2143–2157. https://doi.org/10.3233/JIFS-171020
21. Cao Y, Ying M, Chen G (2007) State-based control of fuzzy discrete-event systems. IEEE Trans Syst Man Cybern Part B 37(2):410–424. https://doi.org/10.1109/TSMCB.2006.883429
22. Yen J, Langari R (1999) Fuzzy logic: intelligence, control, and information. Prentice Hall
23. Klir GJ, Yuan B (1995) Fuzzy sets and fuzzy logic: theory and applications. Prentice Hall, Upper Saddle River, NJ, 1995. 574 pp. ISBN 0-13-101171-5; J Chem Inf Comput Sci 36(3):619–619. https://doi.org/10.1021/ci950144a
24. Fuzzy Decision-Making. In: Fuzzy logic theory and applications. World Scientific, pp 353–400

25. Lin F, Ying H, MacArthur RD, Cohn JA, Barth-Jones D, Crane LR (2007) Decision making in fuzzy discrete event systems. Inf Sci (Ny) 177(18):3749–3763. https://doi.org/10.1016/j.ins.2007.03.011
26. Lin F et al (2004) Control of fuzzy discrete event systems and its applications to clinical treatment planning. In: 2004 43rd IEEE conference on decision and control (CDC) (IEEE Cat. No. 04CH37601), vol 1, pp 519–524. https://doi.org/10.1109/CDC.2004.1428683
27. Danmei L, Weichun L, Hui Z, Shihuang S (2008) A fuzzy discrete event system control and decision making in air conditioning system. In: 2008 Chinese control and decision conference, pp 3147–3151.https://doi.org/10.1109/CCDC.2008.4597906
28. Li D, Lan W, Zhou H, Shao S (2009) Control of fuzzy discrete event systems and its application to air conditioning system. Int J Model Identif Control 8(2):122. https://doi.org/10.1504/IJMIC.2009.029024
29. Lin F, Ying H (2010) State-feedback control of fuzzy discrete-event systems. IEEE Trans Syst Man Cybern Part B 40(3):951–956. https://doi.org/10.1109/TSMCB.2009.2030785
30. Ying H et al (2007) A self-learning fuzzy discrete event system for HIV/AIDS treatment regimen selection. IEEE Trans Syst Man Cybern Part B 37(4):966–979. https://doi.org/10.1109/TSMCB.2007.895360
31. Boegl K, Adlassnig KP, Hayashi Y, Rothenfluh TE, Leitich H (2004) Knowledge acquisition in the fuzzy knowledge representation framework of a medical consultation system. Artif Intell Med. https://doi.org/10.1016/S0933-3657(02)00073-8
32. P. on the C. P. for T. of H. I. Convened and by the D. of H. and H. S (DHHS) Guidelines for the use of antiretroviral agents in HIV-1 infected adults and adolescents [Online]. Available www.aidsinfo.nih.gov
33. Romanelli F, George K (2002) 20 years of HIV treatment: past, present, and future. J Am Pharm Assoc (Wash)
34. Shafer RW et al (2003) Comparison of four-drug regimens and pairs of sequential three-drug regimens as initial therapy for HIV-1 infection. N Engl J Med 349(24):2304–2315. https://doi.org/10.1056/NEJMoa030265
35. van Leth F et al (2004) Comparison of first-line antiretroviral therapy with regimens including nevirapine, efavirenz, or both drugs, plus stavudine and lamivudine: a randomised open-label trial, the 2NN Study. Lancet 363(9417):1253–1263. https://doi.org/10.1016/S0140-6736(04)15997-7
36. Walmsley S et al (2002) Lopinavir-ritonavir versus nelfinavir for the initial treatment of HIV infection. N Engl J Med. https://doi.org/10.1056/NEJMoa012354
37. Ying H (2000) Fuzzy control and modeling: analytical foundations and applications
38. Melanie M (1996) An introduction to genetic algorithms. MIT Press, Cambridge, MA, p 205. https://doi.org/10.1016/S0898-1221(96)90227-8

Mathematical Modeling of Water Heater and HVAC for Demand Response Analysis

Arshad Mohammad, Md Mustafa Kamal, and Imtiaz Ashraf

Abstract In this paper, demand response-enabled water heater model and heating ventilation air conditioning (HVAC) model is presented. These models are very useful in the analysis of demand responses in home area network. The developed models are based on physical characteristics. For water heater, physical parameters are tank size, temperature set point, and power rating of heating coil. Few environmental factors like ambient temperature, inlet water temperature, and R-value of insulation are also taken care of while modeling the system. House structure, thermal insulation of the house, and other physical characteristics of HVAC such as power consumption, heating/cooling capacity, and heat gain rate have been taken in case of HVAC. These models are demand response enabled, i.e., when demand response (DR) signal is high water heater and HVAC is scheduled, respectively, for their operation. Actual power consumption is collected from the experimental analysis and compared with the model power consumption. Water heater have validated with actual data a deviation of only 3.11% is noticed. A difference of only 11.3% is observed in HVAC which is not much significant in demand response analysis of HVAC system in residential areas.

Keywords Mathematical modeling · Demand response · Water heater · HVAC

1 Introduction

The electricity demand has been increasing sharply nowadays, and the simultaneous growth in electricity generation have various environmental concerns, e.g., increased level of carbon concentration in the atmosphere, etc. [1–3]. Therefore, renewable sources of energy are being promoted which can fulfill the demand without affecting the atmosphere. However, large involvement of sustainable energy resources increases complexity in the network [4].

A. Mohammad (✉) · M. M. Kamal · I. Ashraf
Aligarh Muslim University, Aligarh, India
e-mail: arshad.gb2140@gmail.com

A. Iqbal et al. (eds.), *Renewable Power for Sustainable Growth*, Lecture Notes in Electrical Engineering 723, https://doi.org/10.1007/978-981-33-4080-0_52

Present power grid encountered high-stress condition due to a heavy increase in demand. Transmission line outages and blackouts are the problems which arises in heavy stressed condition. In India, 2012 blackout is recent example of stressed power grid failure in which 9% population of the total world affected [5].

Demand response gives a simple and economic means of providing solution to this problem. Smart grid with demand response program at residential area has capability to deal with these situations. Demand response program intelligently curtails and shifts the load which increases the efficiency and reliability of system.

According to USA department of energy, HVAC and water heater contribute about 35% and 10%, respectively, of total household consumption [6] as shown in Fig. 1. Domestic water heater may be used as an energy storage device during peak load hours. It stores energy during off-peak period and can be properly scheduled to optimize the load profile during peak load hours. Electrical water heater consumes large amount of energy and has a significant effect on the overall system load profile. It can participate actively in demand response program. Power demand for a small house and demand profile of electric water heater are shown in Fig. 2. The load demand profile of a typical house and that of water heater follows the same profile as shown in Fig. 1. The peak hour and off-peak hour occur nearly at the same time. Therefore, power management of residential areas can be maintained by scheduling the ON and OFF periods of electrical water heater [2, 3]. Therefore, HVAC and water heater are very suitable and have very higher potential of demand response in residential areas.

For analysis the potential benefit of water heater and HVAC, mathematical model has developed. For modeling of water heater, the heat transfer model is used. For water heater, in [7], a method based on Monte Carlo rejection has been used to estimate different individual models. The uncertainties in user's behavior have been accounted in state-queuing method [8]. However, in [9], the deterministic parameters of the model has been replaced by normal random variables.

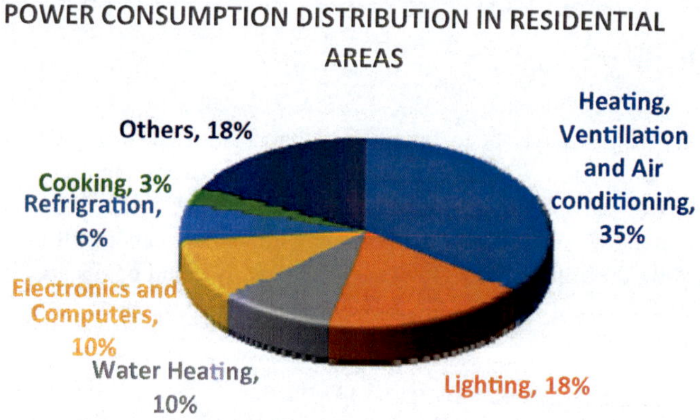

Fig. 1 Energy distribution of different sectors

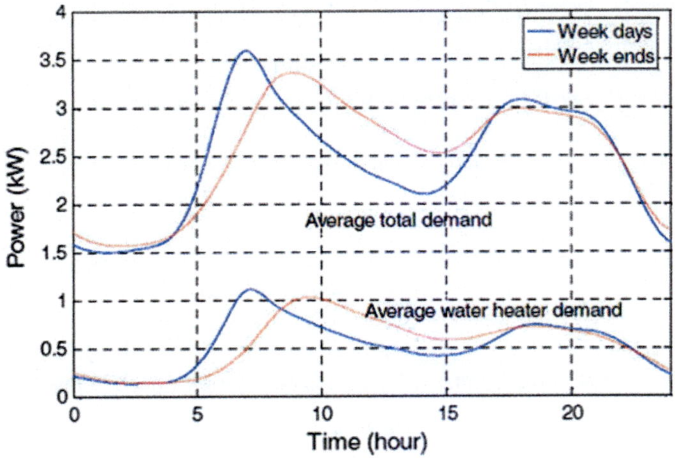

Fig. 2 Load profile of a house

Optimal scheduling of HVAC has higher possibility to decrease peak load demand and reduce electricity bills. Residential buildings have high thermal storage capacity, and using ICT communication network and power electronic devices, it can be achieved. For example, variable speed drives-based heating and cooling units can be used into different DR programs where their energy consumption can be shifted to off-peak from peak hour to reduce electricity bills. The price-based demand response strategies of HVAC units have been widely studied. For example, in [10] cooling/heating rate and input energy, consumption of HVAC units has been taken into consideration to reduce purchased electricity costs with natural gas plant. Moreover in [11], HVAC has been put on its ON/OFF operation for optimizing the electricity bills. Large HVAC model with detailed description have discussed in [12] for achieving demand response.

In this work, a domestic water heater and HVAC are modeled according to their physical and operational characteristics which is similar to real-world situation. These models are interactive, and it can also be interfaced with external DR signal. The developed models are compared with real system data obtained from the practical readings. It shows an excellent result with a negligible deviation from the real system data.

2 Mathematical Models

DR enabled mathematical models are discussed in this section. Models have developed based on their physical characteristics.

2.1 DR Enabled Model of WH

For instant n, the electricity consumption is calculated as:

$$p_{\mathrm{wh}(n)} = S_{\mathrm{wh}(n)}.P_{\mathrm{wh}}.D_{\mathrm{wh}(n)} \tag{1}$$

where

$p_{\mathrm{wh}(n)}$ Electricity consumption in n instant.
P_{wh} Mentioned rated power of water electric heater (kW).
$S_{\mathrm{wh}(n)}$ Power status of WH in instant interval n (0 = OFF; 1 = ON).
$D_{\mathrm{wh}(n)}$ Demand response signal for water heater in n instant (0 = OFF; 1 = ON).

The temperature of water in the electric water heater changes as:

$$\begin{aligned} T_{\mathrm{out}(i+1)} = & \frac{T_{\mathrm{out}(i)}(V_{\mathrm{tank}} - fr_i * \Delta t)}{V_{\mathrm{tank}}} + \frac{T_{\mathrm{in}} * fr_i * \Delta t}{V_{\mathrm{tank}}} \\ & + \frac{1}{8.34} * \left[p_{\mathrm{WH}(i)} * 3412 - \frac{A_{\mathrm{tank}} * \left(T_{\mathrm{out}(i)} - T_r\right)}{R_{\mathrm{tank}}} \right] \\ & * \frac{\Delta t}{60} * \frac{1}{V_{\mathrm{tank}}} \end{aligned} \tag{2}$$

where

T_{out} Temperature of water inside water heater (°F).
T_{in} Temperature of incoming water to water heater (°F).
T_r Room temperature (°F).
fr_i Aquatic movement rate in stint instant i (gpm).
A_{tank} Exterior area of container (ft^2).
V_{tank} Capacity of container (gallons).
Δt Duration of each time slot (minutes).
R_{tank} Thermal confrontation of water boiler (°F ft^{2}h/Btu).

The modeled water heater is fascinated by smart thermostat with temperature tolerance zone, i.e., ΔT. When the temperature of hot water goes below certain level (i.e., $T_{\max} - \Delta T$), water heater coils are ON. Due to heating temperature increases and when temperature of water reaches the maximum specified level, WH coils goes OFF. When hot water temperature is in tolerance zone, the heating coils will remain in their previous state. See (3)

$$S_{\mathrm{wh}(n)} = \begin{cases} 0, & T_{\mathrm{wh}(n)} > T_{\max} \\ 1, & T_{\mathrm{wh}(n)} < T_{\max} - \Delta T \\ S_{\mathrm{wh}(n-1)}, & T_{\max} - \Delta T \le T_{\mathrm{wh}(n)} \le T_{\max} \end{cases} \tag{3}$$

where

T_{max} Maximum warm liquid set value (°F).
ΔT Heat acceptance zone (°F).
$T_{wh(n)}$ Heat of hot liquid in instant n (°F).
$S_{wh(n)}$ Status of water electrical heater in time instant n (0/1:OFF/ON).

The electrical power consumption also depends on demand response signal which is set by the consumer or utility during event DR. This character is attained with the help of home-based energy processor. The DR control signal (D_{wh}) controls the operation of water electrical heater. After $D_{wh} = 0$, it will turn OFF water electric heater, and when $D_{wh} = 1$, it will turn ON the water electric heater.

2.2 DR Enabled HVAC System Model:

DR enabled HVAC model block diagram is shown in Fig. 3. DR event control indicator D_{ac}, alfresco temperature T_{out}, extreme device usual temperature T_{max}, comfort region or allowable temperature range ΔT, and room temperature T_{in} are input to the presented model. HVAC electric power consumption sequence data stands for output of the system. Room temperature time sequence data stands also for output of the model which is feedback to the system for next iteration. Besides these house structure, no of persons and electrical characteristic of HVAC system have also taken into consideration.

The electricity demand of HVAC system for each time instant is calculated as:

$$p_{ac(n)} = P_{ac}.S_{ac(n)}.D_{ac(n)} \tag{4}$$

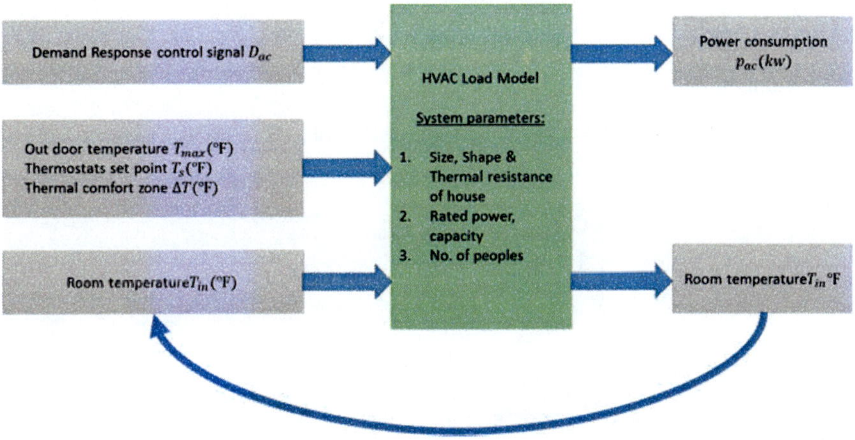

Fig. 3 Block diagram of HVAC

where

$p_{ac(n)}$ Electricity consumption of HVAC in n instant.
P_{ac} Evaluated power consumption of HVAC (kW).
$S_{ac(n)}$ Position of HVAC in time instant n (0 = OFF; 1 = ON).
$D_{ac(n)}$ Demand response signal for HVAC in n instant (0 = OFF; 1 = ON).

For every time n, the volume temperature is observed as:

$$T_{n+1} = T_n + \Delta t \times \frac{G_n}{\Delta c} + \nabla t \times \frac{C_{\mathrm{HVAC}}}{\Delta c} \times S_{\mathrm{ac}} \tag{5}$$

where

T_n Volume temperature (°F) in n time instant.
Δt Duration/period of time space n (hour).
G_n Hotness gain proportion inside the chamber, positive values for space heating and negative values for chilling.
C_{HVAC} Heating/chilling measurements of HVAC.
Δc Energy vital to vary the heat of chamber by 1 °F (Btu/°F).

For every instant *n*, the temperature gain amount G_n is intended as:

$$G_n = \left(\frac{A_{\mathrm{wall}}}{R_{\mathrm{wall}}} + \frac{A_{\mathrm{ceiling}}}{R_{\mathrm{ceiling}}} + \frac{A_{\mathrm{window}}}{R_{\mathrm{window}}} + \frac{11.77\,\mathrm{Btu}}{{}^\circ\mathrm{F} \times \mathrm{ft}^3} \times n_{\mathrm{ac}} \times V_{\mathrm{house}} \right) \times \left(T_{\mathrm{out}(n)} - T_n\right) + H_{\mathrm{p}} \tag{6}$$

where A_{wall}, A_{ceiling}, and A_{window} represent the surface area of side wall, upper – ceiling, and open window, respectively, in ft^2. R_{wall}, R_{ceiling}, and R_{window} are heat resistance of side wall, upper ceiling, and open window, in °F ft^2 h/Btu.

n_{ac} Rate of air variations in respective slot of time.
V_{house} Volume of house in ft^3.
H_{p} Temperature gain from public (Btu/h).

HVAC is also equipped with smart thermostat, and its operation is presented in (7)

$$S_{\mathrm{ac}(n)} = \begin{cases} 0, & T_{\mathrm{ac}(n)} > T_{\max} \\ 1, & T_{\mathrm{ac}(n)} < T_{\max} - \Delta T \\ S_{ac(n-1)}, & T_{\max} - \Delta T \le T_{\mathrm{ac}(n)} \le T_{\max} \end{cases} \tag{7}$$

where

$T_{\max}$ Maximum inside room temperature set point (°F).
ΔT Temperature tolerance zone (°F).
$T_{\mathrm{wh}(n)}$ Temperature of room in interval n (°F).
$S_{\mathrm{wh}(n)}$ Status of HVAC in time interval n (0 = OFF; 1 = ON).

Table 1 Water heater constraints

Parameters	Parameter values
Volume (liter)	35 l
Power rating of heating coil	2.3 kW
Temperature set point	115(°F) – 125 (°F)
Ambient temperature	84 (°F)
R-value	10 ft^2 * °F(Btu/h)
Water consumption	Real water consumption data (ga/m)
Inlet water temperature	Per-minute real temperature captured

3 Model Validation

DR enabled mathematical models are validated against actual power consumption.

3.1 *Water Heater Model*

To validate the mathematical model of water heater, 1-min power consumption of water was captured during our experiment. Water heater data were measured for 16 h from 6 am to 10 pm. Water heater characteristics such as tank size, temperature set point, and power rating of heating coil are given in Table 1. Environmental factors such as ambient temperature, inlet water temperature, and *R*-value of insulation are also given in Table 1.

One-minute captured data and model output of water are shown in Fig. 4. The energy consumption difference is 3.11% which may be tolerable in water heater modeling for domestic purpose.

3.2 *HVAC Model*

To validate, the mathematical model of HVAC real data in 1-min resolution is captured during the experiment. Power consumption was measured for 8 h from 10 PM to 6 AM in the month of October. Ambient temperature was also measured for 1-min resolution. House structure and HVAC specification are given in Table 2 which are considered during experiment.

One-minute measured data of HVAC and model output are shown in Fig. 5. The difference between measured data and model output data is 11.3% which can be tolerated in HVAC modeling for domestic purpose.

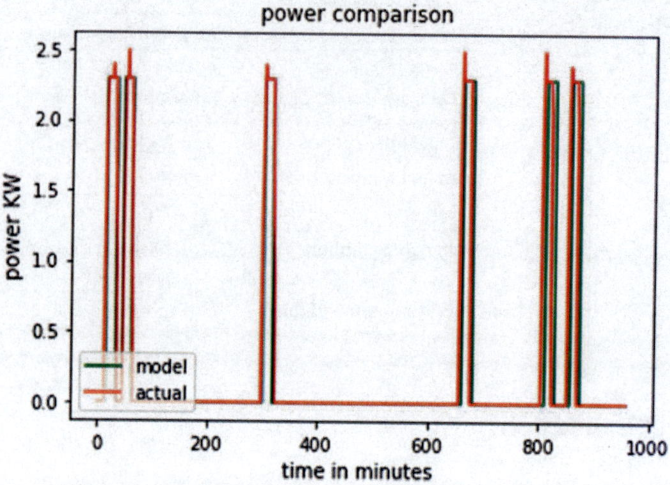

Fig. 4 Actual and model power comparison of water heater

Table 2 HVAC model constraints

Parameters	Parameter values
House area size	2000 + 500 (basement) ft^2
$A_{floor}, A_{ceiling}, A_{wall}, A_{window}$	2000, 2000, 2600, 520 ft^2
$R_{ceiling}, R_{wall}, R_{window}$	49, 13, 2 ft^2 * °F/(Btu/h)
Quantity of individuals	5 people
Measurements of AC element	34,000 Btu
AC heat set value	77 °F
AC power utility	2.3 kW
Room temperature value	84 °F

4 Conclusion

This paper presents DR enabled electrical water heater and HVAC model. The developed mathematical models are created on physical and operative characteristics of electrical water heater load and HVAC, respectively. The models are developed on appliance level which is very useful for analysis of DR potential in residential areas. It is validated against real power consumption data. Modeled power consumption data is similar to captured real system data. Result shows that the proposed load model can be used in distribution network for analyzing demand response potential.

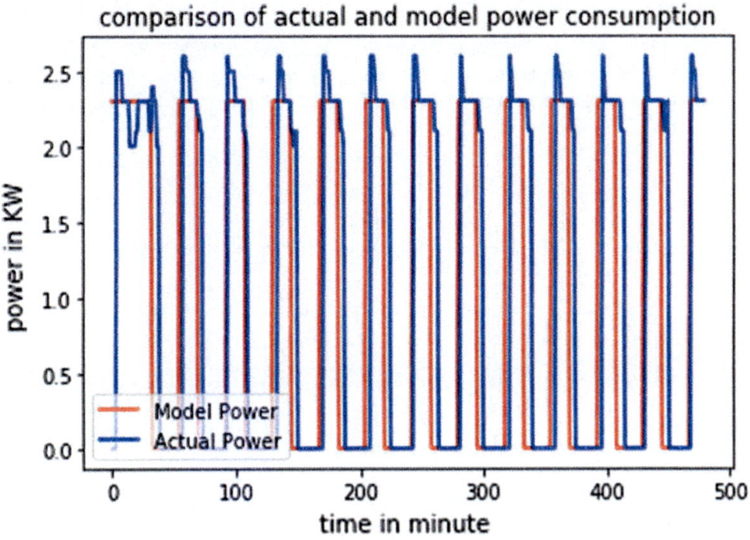

Fig. 5 Actual and model power comparison of HVAC

References

1. Gelazanskas L, Gamage KAA (2014) Demand side management in smart grid: a review and proposals for future direction. Sustain Cities Soc 11:22–30. https://doi.org/10.1016/j.scs.2013.11.001
2. Iqbal A et al (eds) (2020) Soft computing in condition monitoring and diagnostics of electrical and mechanical systems, vol 1096. In: Advances in intelligent systems and computing. Springer, Singapore. https://doi.org/10.1007/978-981-15-1532-3
3. Iqbal A et al (eds) (2020) Meta heuristic and evolutionary computation: algorithms and applications, vol 1096. In: Studies in computational intelligence. Springer, Singapore. https://www.springer.com/gp/book/9789811575709
4. Zhou B, Li W, Chan KW et al (2016) Smart home energy management systems: concept, configurations, and scheduling strategies. Renew Sustain Energy Rev 61:30–40. https://doi.org/10.1016/j.rser.2016.03.047
5. Pina A, Ferrão P, Fournier J et al (2017) Literature review of power system blackouts. Energy Proc 141:428–431. https://doi.org/10.1016/j.egypro.2017.11.055
6. Kim Y (2018) Optimal price based demand response of HVAC systems in multizone office buildings considering thermal preferences of individual occupants buildings. IEEE Trans Ind Inform 14:5060–5073. https://doi.org/10.1109/TII.2018.2790429
7. Dolan PS (1996) Development of a Monte Carlo based aggregate model for residential electric water heater loads 36
8. Member NL, Chassin DP, Widergren SE et al Modeling uncertainties in aggregated thermostatically controlled loads using a state queueing model
9. Nehrir MH, Ieee SM, Jia R et al (2007) Power management of aggregate electric water heater loads by voltage control
10. Guan X, Xu Z, Jia Q (2010) Energy-efficient buildings facilitated by microgrid. IEEE Trans Smart Grid 1:243–252. https://doi.org/10.1109/TSG.2010.2083705

11. Dong Y, Zou B (2015) A research of real-time pricing mechanism and its characteristics. J Power Energy Eng 3:240–249. https://doi.org/10.4236/jpee.2015.34033A
12. De AF, Boaro M, Fuselli D et al (2013) Optimal home energy management under dynamic electrical and thermal constraints. IEEE Trans Ind Inform 9:1518–1527. https://doi.org/10.1109/TII.2012.2230637

Planning and Economical Optimization of Grid-Connected Renewable Energy Resource-Based Microgrid

Md Mustafa Kamal, Imtiaz Ashraf, and Arshad Mohamma

Abstract Microgrid revolutionizes the electric power system for both rural and urban communities. Since microgrid can work in both grid-connected as well as standalone mode, the planning of microgrid must be addressed the cost-effective feasibility as well as the long-term stability of the system. Microgrid planning is a complex process due to its standing alternative goals, available constraints, and uncertainty of renewable energy resources. In the present work, a semi-urban micro-grid is planned with the integration of wind, solar, and battery system. Three types of load (residential, commercial, and industrial) considered for modeling and analysis in four different seasons. Techno-economic and feasibility analysis of proposed microgrid carried out for the semi-urban area in Uttarakhand (India).

Keywords Microgrid · Planning · Cost of energy · Solar · Wind · Optimization · Renewable resource

1 Introduction

Modern societies highly depend on the electric power supply. From the past 45 years, the dependency on the electrical power system significantly increases [1, 12, 13]. The demand of energy is growing rapidly and not possible to entirely fulfill by using conventional energy resources due to its limitation. Renewable energy-based power supply becomes another alternative to supply the demand of energy worldwide. The single energy resource is not able to provide reliable power supply; hence, hybrid energy-based power supply is an option for preserving the feasibility between load and demand [2]. The hybrid energy system integrates more than one energy source to increase the reliability of power supply. The hybrid energy-based microgrid system is the combination of renewable energy resources, traditional energy systems, and storage devices to meet the demand of local areas, both standalone mode and grid-connected mode. Mostly standalone mode of operation of microgrid used in a remote

M. M. Kamal (✉) · I. Ashraf · A. Mohamma
Aligarh Muslim University, Aligarh 202002, India
e-mail: mustafakamalece@gmail.com

A. Iqbal et al. (eds.), *Renewable Power for Sustainable Growth*, Lecture Notes in Electrical Engineering 723, https://doi.org/10.1007/978-981-33-4080-0_53

area with a storage device. During the period in which power supply is not available with renewable energy resource, the storage device can supply the energy [3].

Similarly, the grid-connected mode widely prefers in urban areas during the off period when the electricity price is low power taken from the grid and on the peak period power supplied by renewable and storage device. Hence, the grid-connected mode of operation rise the use of renewable energy resource, increase the reliability in power supply, and decrease the total cost of energy. Renewable energy-based resource with the combination of diesel generators is perceptively used to ensure the electrification of the local community. A combination of the hybrid energy system is more reliable and cost-effective to electrify the rural population. DG structure, which offers electricity for the local community, is diverse from the conventional structure. For the electrification, available energy resources in the locality can combined to form a hybrid energy model, which works on both grid-connected as well as standalone mode [4]. The hybrid model popularly, also known as integrated renewable energy system (IRES) model. The developed model is to provide economical energy supply to the rural population. To integrate renewable energy resources, careful and useful planning is necessary to maximize the benefits and efficiency of customers [5]. The vital factor in the planning of the hybrid energy resource-based system is optimal planning of its component, i.e., solar arrays, number of the wind turbine, generator sets, storage device, and converters, so the formulated objective function maximized/minimized fully satisfy all constraints [6]. IRES models are widely used by the researcher to suggest the electrification of urban and rural communities. Power generation using a modified hybrid system have accomplished several studies worldwide. However, in India, only a few quantities of literature reported on designing of the hybrid energy system. Diverse combination of renewable resources with or without battery proven a cost-effective solution for electrification. Kamal et al. [7–9] modeled the integrated energy model by combing the locally available energy resources and analyses techno-economic and feasibility analysis of the system. Guangquain [10] investigated the scope of electrification using IRES model for the rural communities. They recommended a viable and economical solution of energy for an isolated rural area, where grid connectivity is unreliable. For the fulfillment of energy demand of a specific area, the optimal sizing of the hybrid model is recommended. For the analysis of integrated renewable energy, resource model rural region of Iran is considered for size optimization. Hossain et al. [11] developed a hybrid energy resource-based energy model for the Malaysia. The available energy resource potential estimated and integrated to form the hybrid energy model. The energy model optimizes by using HOMER software for net present cost and per unit cost of energy.

Study Area

State Uttarakhand formed in the year 2000 with Dehradun as its capital in the northern part of India. As the 27th state of India, its boarder touches China in the north, Nepal in the east, Utter Pradesh in the south, and Himachal Pradesh in the west. It has 13 districts, and Tehri Garhwal is selected for the case study of the proposed model. Tehri Garhwal is one of the central provinces in the Himalayan region. The total population of the region is 618,913 (2011 census). Since it is a hilly area, some village has low-density inhabitants.

2 Demand Analysis

A load of study area divided into three categories, i.e., residential, commercial, and industrial. The average use of electricity per customer is 16.85 kWh; hence, daily electricity consumption of the area is 1,341,244.8 kWh. For the commercial use of heavy load such as air conditioner, which used between 10 am to 6 pm, the average load is 2000 MWh. Figure 1 shows the three different loads of study area.

Net load of the system is modeled as

$$\text{Net load} = P_{s,t}^{\text{Wind}} + P_{s,t}^{\text{PV}} - \text{Demand} \tag{1}$$

The amount of power generated by wind and solar systems not able to satisfy the full load requirement of the area. Hence, the diesel generator also placed with the conjunction of renewable energy resources and storage devices. The battery storage system introduced to store the surplus amount of energy in the daytime and used when needed. The microgrid model can be modeled using a four-bus system, which shown in Fig. 2.

The solar photovoltaic system (SPV) consists of several PV modules. Solar radiation data geography located between 28.47 latitude and 78.14 longitudes were taken. The area has yearly average solar insolation which is 5.32 kWh/m^2, and in May, maximum insolation is recorded as 7.420 kWh/m^2. The study area has enormous solar potential; the total annual solar energy potential is recorded as 22,369 kWh/m^2/yr. Figure 3 shows the monthly average insolation and clearness index of the region.

Wind energy generates electricity through the movement of atmospheric air. Turbines change the kinetic energy to the usable power. The speed of wind in the study area is (4–16 m/s), which is suitable for low-speed designed wind turbines. Figure 4

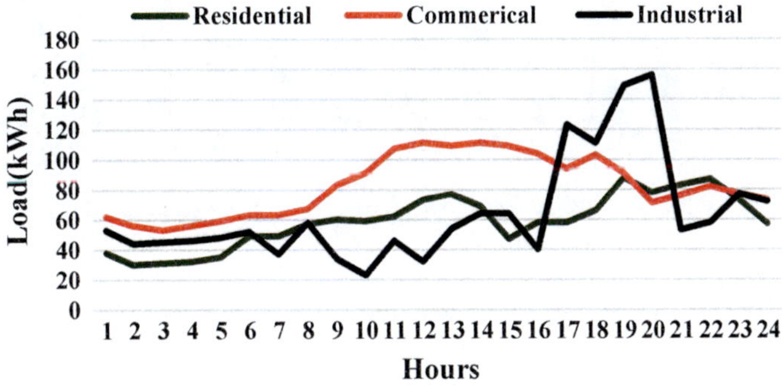

Fig. 1 Load of the area

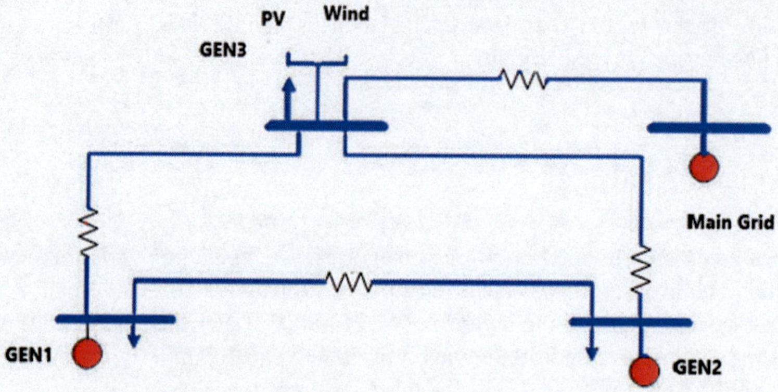

Fig. 2 Grid-connected microgrid system

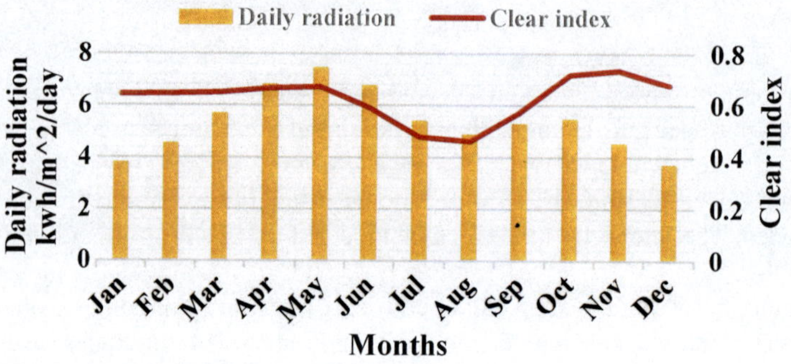

Fig. 3 Monthly average insolation and clear index of the region

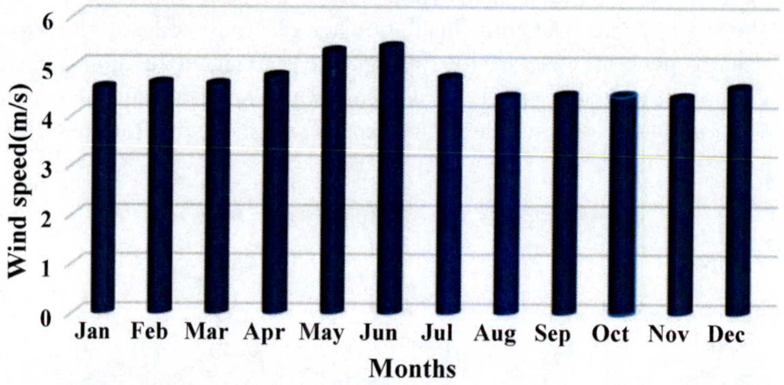

Fig. 4 Monthly average wind speed of region

displays the monthly average wind speed of the study region. Mostly designed wind turbine in India is for 10 m/s.

3 Problem Formulation

To electrify the local community by using solar/wind/diesel/storage and device formed grid-connected microgrid. Three cases have taken for the study of the microgrid system.

Case1: Microgrid Without Renewable Energy Resources
In this mode of operation, the electricity is mainly supplied by generators and the primary grid. The generator capacity of two generators is 100 MW and 150 MW, respectively.

The mathematical formulation of total cost of energy expressed as for peak hours

$$\text{Min } z(x) = \sum_{t=1}^{t=24} \left[(132\text{gen}_{1,t} + 2.64) + \left(99\text{gen}_{3,t} + 2.31\right) \right. \\ \left. + \left(349.80\text{gen}_{4,t} + 2.64\right) + V_{\text{cr}}\left(Z_1 + Z_2 + Z_3\right) \right] \quad (2)$$

For off-peak hours, the total cost of energy is described as

$$\text{Min } z(x) = \sum_{t=1}^{t=24} \left[(132\text{gen}_{1,t} + 2.64) + \left(99\text{gen}_{3,t} + 2.31\right) \right. \\ \left. + \left(95.04\text{gen}_{4,t} + 2.64\right) + V_{\text{cr}}\left(Z_1 + Z_2 + Z_3\right) \right] \quad (3)$$

The constraints are

$$\begin{bmatrix} -S_{12,t} - S_{13,t} + \text{gen}_{1,t} + Z_{1,t} = L_{1,t} \\ S_{12,t} - S_{24,t} - S_{23,t} + Z_{2,t} = L_{2,t} \\ S_{13,t} + S_{23,t} + \text{gen}_{3,t} + Z_{3t} = L_{3,t} \\ S_{24,t} + \text{gen}_{4,t} = L_{4,,t} \\ \text{gen}_{1t} \le 100, \text{gen}_{1t} \le 150, \text{gen}_{1t} \le 320, \\ Z_{1t} \le L_{1,t}, Z_{1t} \le L_{2,t}, Z_{1t} \le L_{3,t} \end{bmatrix} \quad (4)$$

Case 2: Microgrid with Renewable Energy Resources
When the renewable energy resources are integrated in the system, the objective function of the system remains same only few constraint changed

$$\left[-S_{12,t} - S_{13,t} + \text{gen}_{1,t} + Z_{1,t} = L_{1,t} \right.$$

$$S_{12,t} - S_{24,t} - S_{23,t} + Z_{2,t} + P_{\text{wind},t} + P_{\text{pv},t} = L_{2,t}$$
$$S_{13,t} + S_{23,t} + \text{gen}_{3,t} + Z_{3t} = L_{3,t}$$
$$S_{24,t} + \text{gen}_{4,t} = L_{4,,t}] \quad (5)$$

Due to deviation of solar insolation and wind speed that depends on the climatic situation, hence, the total amount of cost is calculated in every season (winter, summer, autumn, and spring).

Case 3: Microgrid with Storage System
To achieve, the reliable supply of energy storage system is introduced in the system. Renewable energy resources cannot supply the demand sufficiently. Hence, renewable energy resource cannot charge battery. Therefore, the purpose of battery system is to store energy in the period of low demand and supply energy at peak hours (Table 1).

$$[-S_{12,t} - S_{13,t} + \text{gen}_{1,t} + Z_{1,t} + P_{\text{dischrging}} + P_{\text{charge}} = L_{1,t}$$
$$S_{12,t} - S_{24,t} - S_{23,t} + Z_{2,t} = L_{2,t}$$
$$S_{13,t} + S_{23,t} + \text{gen}_{3,t} + Z_{3t} = L_{3,t}$$
$$S_{24,t} + \text{gen}_{4,t} = L_{4,,t}] \quad (6)$$

$$\left[E_{t+1,m}^{\text{battery}} = E_{t,m}^{\text{battery}} - \frac{E_{t,m}^{\text{discharge}}}{\Delta_{\text{discharge}}} + P_{t,m}^{\text{charge}} \Delta_{\text{charge}}\right.$$
$$E_t^{\text{battery}}.\text{SOC}_{\text{discharge}} \leq E_t^{\text{battery}}.\text{SOC}_{\text{charge}}$$
$$\left.0 \leq P_{t,m}^{\text{discharge}} \leq P_{t,m}^{\text{charge}}\right] \quad (7)$$

4 Results and Discussion

The proposed work deals with the planning of microgrid with the integration of renewable energy resources and battery systems. For analysis of the system, three cases considered. In the first case, when the energy supplied by the grid and generator sets the total cost of operation for one in peak hour is $224,220.00 and in the off-peak hour $637,659.00. Therefore, the total cost of operation for one year is $234,861,403.50. For case two, when renewable energy resources are available, the microgrid supplies energy to the customer from renewable resources and the primary grid. Renewable energy resources are intermittent; hence, the supply is affected by the different climatic conditions, so based on seasonally available energy, microgrid is analyzed. Figure 5 shows the total cost of energy seasonally. The total cost of electricity per year was $ 222,822,811.20.

Table 1 Parameter used in microgrid modeling

Parameter	Cost ($)
Variable cost coefficient of generator 1	$132/MWh
Fixed cost coefficient of generator1	$2.64/MWh
Variable cost coefficient of generator 2	$99/MWh
Fixed cost coefficient of generator 2	$2.31/MWh
Cost of peak hours from grid	$349.80/MWh
Cost of other hours from grid	$95.04/MWh
Fixed cost coefficient	$2.64/MWh
Value of customer reliability	$52,800/MWh
Investment cost of battery	13,200,000.00
Capacity	100 MWh
Life	5 years
Efficiency	80%

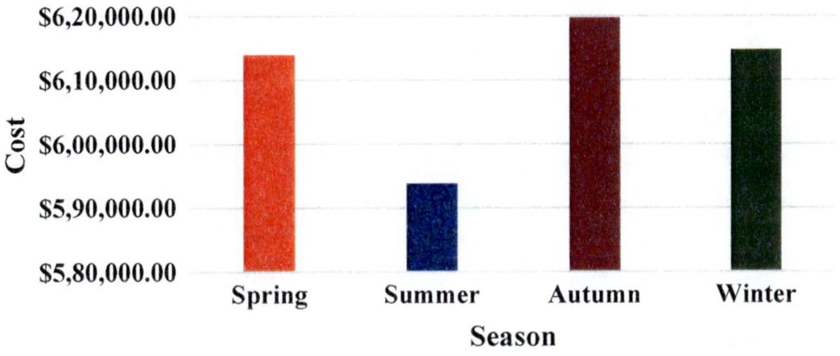

Fig. 5 Seasonal total cost of energy of microgrid with renewable energy resources

For the third case, where the storage system used to provide energy supply when grid supply absent. Figure 6 shows the total seasonal energy cost with the storage device.

Figure 7 shows that the total cost of energy with the storage system was $ 215,071,230.00. A comparison of all three cases suggests that when the battery system introduced in the microgrid system, the total cost reduced with a power system that does not have a battery bank.

For the cost-effective investigation of the whole arrangement, the microgrid system with renewable energy resources and battery is compared to the original model w.r.t equipment and installation cost. The cost of different equipment is shown in Fig. 7.

For the economic evolution of profit of the life cycle of battery for five years with the discount rate of 8%, the cost of microgrid with battery, without battery, and net present value (NPV) is evaluated. Figure 8 shows the five-year projection of different expenses.

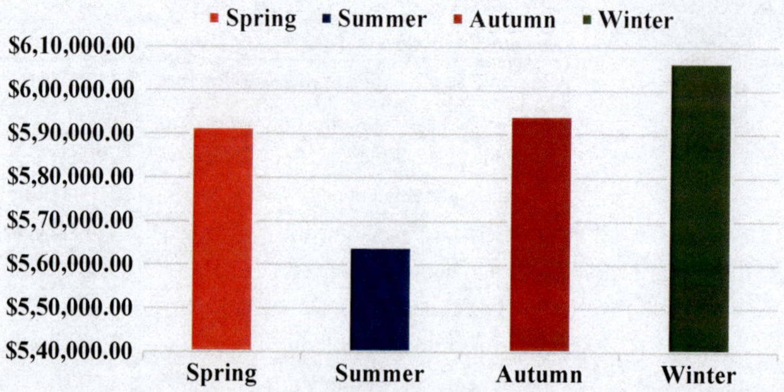

Fig. 6 Seasonal total cost of energy with storage system

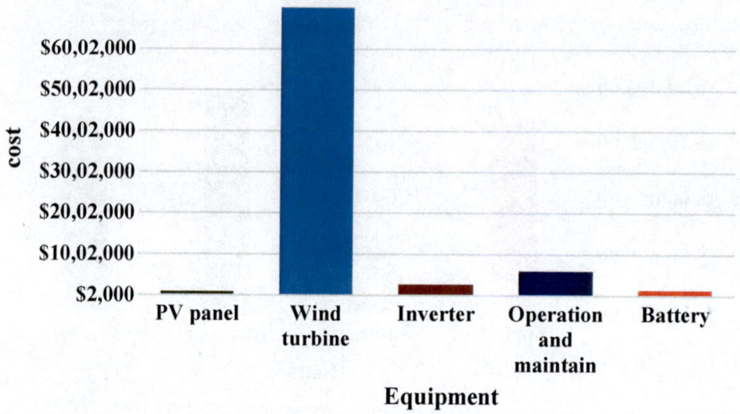

Fig. 7 Total cost of the components

5 Conclusion

In the present work, a hybrid model of renewable energy resources is proposed, which consists of solar/wind/diesel/battery. The suggested microgrid system works with grid-connected mode. The study of the energy system is carried out for the semi-urban area of Uttarakhand state of India. The microgrid system modeled for three different loads, i.e., residential, commercial, and industrial. The proposed framework is analyzed for without renewable, with renewable, and battery, and the model results suggest that the lowest cost of energy for the integrated model with battery system is more cost-effective compared to other systems. Further from the economic analysis with the battery system for a lifetime of five years, it suggests that the model is reliable and beneficial for the local community.

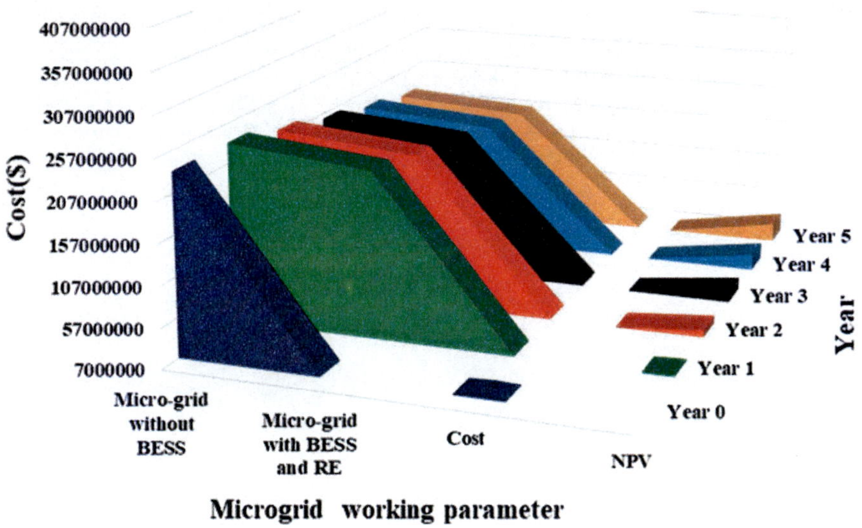

Fig. 8 Total cost of microgrid operation for five years

References

1. Gamarra C, Guerrero JM (2015) Computational optimization techniques applied to microgrids planning: a review. Renew Sustain Energy Rev 48:413–424
2. Junaid M, Kumar A, Mathew L (2017) Techno economic feasibility analysis of different combinations of PV-wind-diesel-battery hybrid system for telecommunication applications in different cities of Punjab India. Renew Sustain Energy Rev 76:577–607
3. Gandini D, de Almeida AT (2017) Direct current microgrids based on solar power systems and storage optimization, as a tool for cost-effective rural electrification. Renew Energy 111:275–283
4. Yadav P, Davies PJ, Palit D (2019) Distributed solar photovoltaics landscape in Uttar Pradesh, India: lessons for transition to decentralised rural electrification. Energy Strateg Rev 26:100392–100401
5. Kala LD, Subbarao PMV (2018) Estimation of pine needle availability in the Central Himalayan state of Uttarakhand, India for use as energy feedstock. Renew Energy 128:9–19
6. Khodayar ME (2017) Rural electrification and expansion planning of off-grid microgrids. Electr J 30:68–74
7. Kamal MM, Ashraf I, Fernandez E (2020) Energy resource allocation for distributed energy resource based power generation in a rural microgrid, pp 571–575
8. Kamal M, Ashraf I, Fernandez E (2019) Energy resource planning for a rural microgrid: comparison of results using different optimization algorithms. In: 2019 2nd international conference on advance and Computing Communication Paradigm, ICACCP 2019, pp 1–6
9. Kamal MM, Ashraf I, Fernandez E, Alam A (2020) Resource allocation, utilization and feasibility study of a rural microgrid, pp 1–6
10. Guangqian D, Bekhrad K, Azarikhah P, Maleki A (2018) A hybrid algorithm based optimization on modeling of grid independent biodiesel-based hybrid solar/wind systems. Renew Energy 122:551–560
11. Hossain M, Mekhilef S, Olatomiwa L (2017) Performance evaluation of a stand-alone PV-wind-diesel-battery hybrid system feasible for a large resort center in South China Sea, Malaysia. Sustain Cities Soc 28:358–366

12. Iqbal A et al (eds) (2020) Soft computing in condition monitoring and diagnostics of electrical and mechanical systems, vol 1096. In: Advances in intelligent systems and computing. Springer, Singapore. https://doi.org/10.1007/978-981-15-1532-3
13. Iqbal A et al (eds) (2020) Meta heuristic and evolutionary computation: algorithms and applications, vol 1096. In: Studies in computational intelligence. Springer, Singapore.https://www.springer.com/gp/book/9789811575709

Smart Streetlight System for Smart Cities Using IoT

Arshad Mohammad, Faiz Ali, M. D. Mustafa Kamal, and Imtiaz Ashraf

Abstract Streetlight is an important utility which consumes large amount of energy, and a big part of energy is wasted due to improper usage. In this paper, a streetlight system has been proposed which controls the illuminance of street lamp according to traffic flow. It will illuminate the street lamp when user is present; otherwise, it will dim the lamp to its 20% illuminance. Smart streetlight system uses IoT platform with motion sensor and light sensor. The system is installed in testbed, and power consumption is observed for single day. The energy-saving for different control strategies are 96.20, 97.34, 97.34% and 95.86 have been observed.

Keywords Smart streetlight · Energy-saving · Power consumption · IoT

1 Introduction

The smart city concept gains popularity among researchers. It provides safety, convenience, comfort and better utilization of resources. Furthermore, it performs task in energy-efficient way. The smart streetlight is one of the most important parts of smart city [1–3]. The lack of proper illumination in streets increases the stealing, robbery and other traffic-associated problems. The current streetlight suffers from high energy consumption, lack of fine control and long maintenance period. Therefore, a smart street lighting system should have the following features:

- Minimized energy consumption: Traditional streetlight will be OFF in daytime, and it will be ON in nighttime. The system should be adaptive, and it will switch ON the light according to traffic. The light will be illuminated when someone is present; otherwise it will dim the luminaire.
- Increased users' satisfaction: The visual comfort should be increased by smart streetlight. It will be increased by switching ON in front luminaire and switching OFF behind luminaire.

A. Mohammad (✉) · F. Ali · M. D. Mustafa Kamal · I. Ashraf
Aligarh Muslim University, Aligarh, India
e-mail: arshad.gb2140@gmail.com

A. Iqbal et al. (eds.), *Renewable Power for Sustainable Growth*, Lecture Notes in Electrical Engineering 723, https://doi.org/10.1007/978-981-33-4080-0_54

- Fine control: In streetlight, each luminaire has its own identity and they can be differentiated from each other. Every streetlight is independent from each other and autonomously controlled by central controller.
- Flexible management: The area of cities increases very sharply. It will increase the demand of streetlight. Street lighting system should have the capability to meet this demand.

In this paper, we propose a smart streetlight based on IoT platform to encourage smart city concept. The proposed streetlight system controls the illumination according to traffic flow. The visual comfort is increased by proper illumination of streetlight. The streetlight is at 20% illumination level when there is no user present to increase the safety.

The traditional way of streetlight control is simple ON/OFF according to daylight. It does not have facility to control the illumination level. Due to technological advancement, software-controlled LED luminaire is available in market. LED streetlight along with motion sensor saves more energy [4–7]. A lighting system with light sensor and motion sensor has capability to save 20–70% energy. Eveliina at el. developed a smart and dynamic street lighting system to control the dimming level according to traffic flow [7]. They used PIR sensor to find the presence and direction of movement. They save 60–70% in different control strategies. Prabu et el. presented a ANN-based smart street lighting system and deployed it in the city of Hosur [8]. This work minimized unwanted utilization and saves 13.5% than static LED system. In [9], GPS-based streetlight has proposed and showed its features and characteristics.

2 Smart Streetlight System

In this paper, we proposed smart street lighting system. It is shown in Fig. 1. It has three sections, and each section has its own functionality. They all work together and form an energy-efficient smart street lighting system. This street lighting system increases visual comfort and minimizes energy consumption. The three sections are as follows:

(1) Intelligent streetlight
The traditional streetlight system simply switches ON/OFF the streetlights according to daylight availability. The proposed streetlight system controls the illuminance of the streetlight according to traffic flow. When public is present on street, intelligent streetlight system switches ON the light to its maximum, and when nobody is present at the street it will dim the light according to different control strategies. Intelligent streetlight system will maintain a minimum illuminance when nobody is present at street to increase the safety.
(2) Efficient and reliable communication network
The network is working in real time. 4G network has been used in this paper. The network should be strong, reliable and secure; therefore, NB-IoT is used

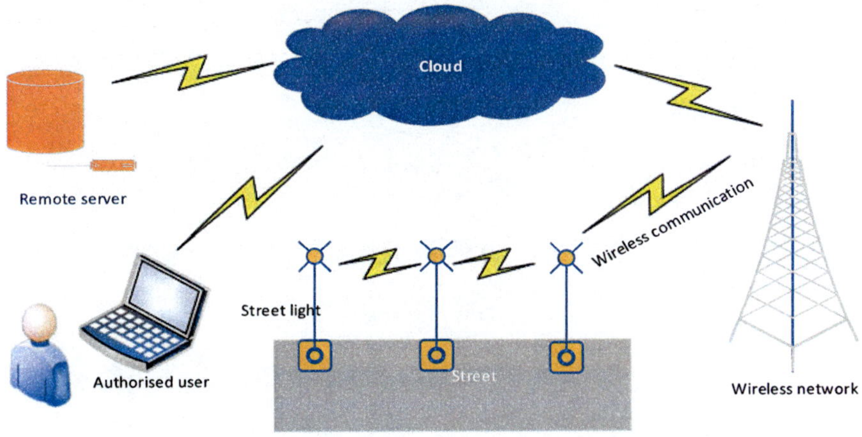

Fig. 1 IoT-based smart streetlight

for communication between server and streetlight. The NB-IoT network has the following features.

- Large number of streetlights are present in the city. The demand increases day by day. Therefore, the communication network has capability to deal with high communication demand.
- The reliability is another important parameter which should meet; otherwise it is useless.
- The security of the system is in such a way that it cannot easily break; otherwise it becomes dangerous.
- The power requirement for communication should be less.

(3) Flexible management: Intelligent streetlight along with efficient network gives the status and information about streetlight. It also provides information about traffic flow. These data are sent to cloud. Cloud provides elastic computing power and large storage at low price. The area of cities is expanding at very fast rate. With the help of cloud computing, expansion of streetlight becomes easy (Fig. 2).

The streetlight is equipped with motion sensor, light sensor and Zigbee networks. The motion sensor provides present information about users. Light sensor quantifies daylight illumination. Zigbee network provides communication to streetlight. The data of different sensors are sent to cloud for analysis. Depending upon user detection, light illumination and weather information, the streetlight is being controlled. If no one is present on street, the streetlight will dim to 20% of its maximum illuminance to provide safety to users. If the person is present, the streetlight is illuminate according to a different control strategy (described in Fig. 3a–d. The direction of movement is also important to controlling the illuminance of streetlight. More illumination is required in front of user than behind. Therefore, streetlight which is in front of user

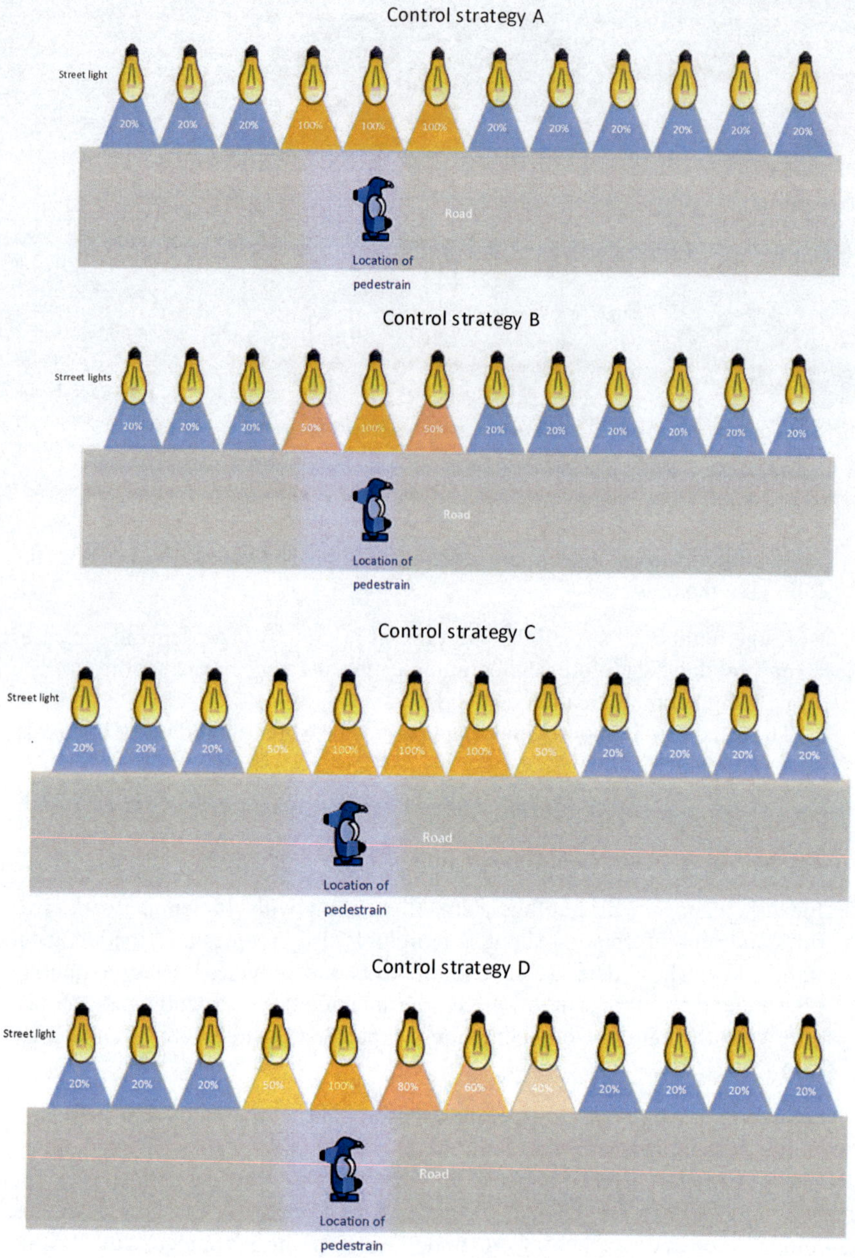

Fig. 2 Control strategies

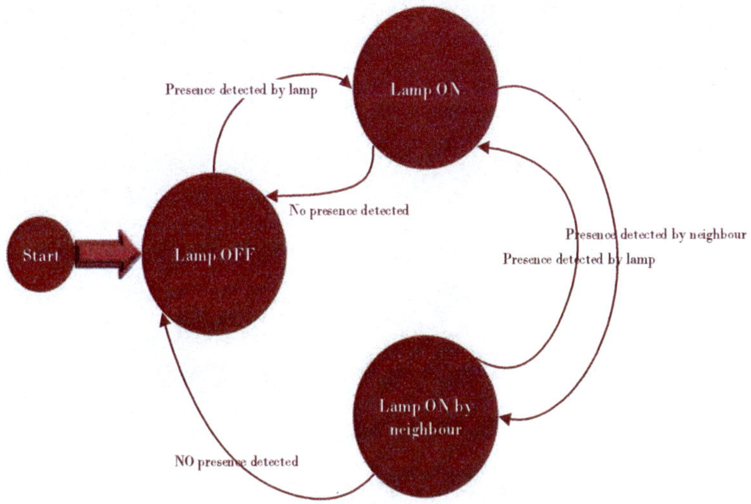

Fig. 3 Sensor operation strategy

will illuminate. A particular streetlight will dim when users passed the streetlight. A particular streetlight is illuminated due to the presence of user and presence of user detected near adjacent streetlight (this phenomenon is shown in Fig. 3). These strategies will increase energy-saving and visual comfort.

3 Testbed and Results

The proposed system is deployed at testbed in university campus AMU, Aligarh. The street which is chosen for testbed is 750 m long. It has 36 number of poles, and each pole had 100-W dimmable LED luminaire. The total power consumption is compared using LED luminaire with and without control strategies. A traffic flow for atypical day is shown in Fig. 4. The energy consumption graph of control strategy *A*, *B*, *C* and *D* for single day is shown in Fig. 5. The energy-saving in different control strategies is 96.20, 97.34, 95.86 and 97.04%.

4 Conclusion

The smart street lighting system has been developed in this proposed work. It minimized the power consumption 96.20%, 97.34, 97.34% and 95.86 by using different control strategies. PIR sensor, daylight sensor and NB-IoT are used in this work. They all worked together and minimized energy consumption and provide user satisfaction. This type of IoT-based smart streetlight is very useful in smart city.

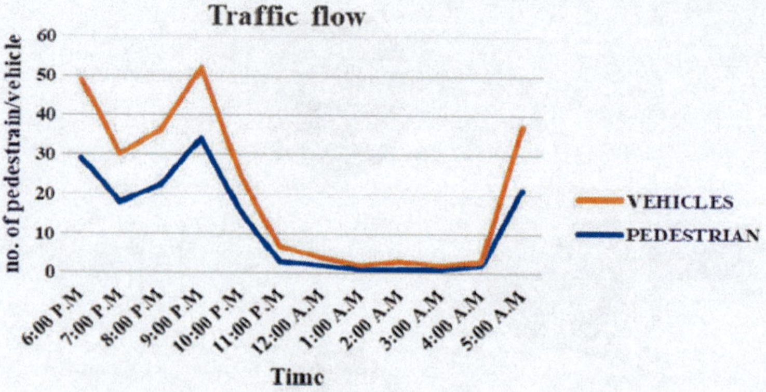

Fig. 4 Traffic flow

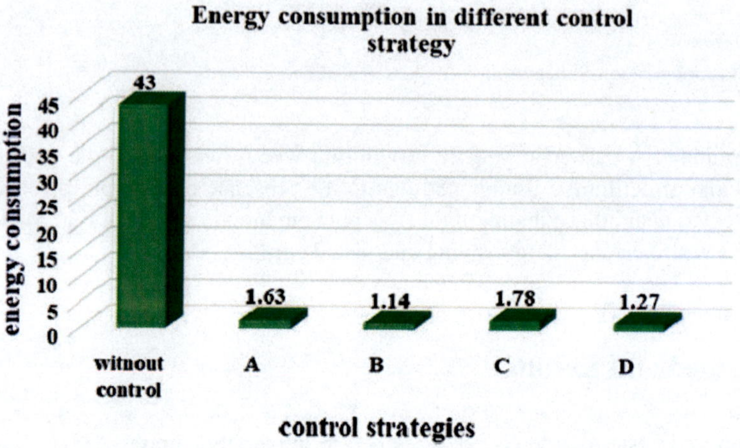

Fig. 5 Energy consumption

References

1. Pasolini G, Toppan P, Zabini F et al (2019) Design, deployment and evolution of heterogeneous smart public lighting systems. Appl Sci 9.https://doi.org/10.3390/app9163281
2. Iqbal A et al (eds) (2020) Soft computing in condition monitoring and diagnostics of electrical and mechanical systems, vol 1096. In: Advances in intelligent systems and computing. Springer, Singapore. https://doi.org/10.1007/978-981-15-1532-3
3. Iqbal A et al (eds) Meta heuristic and evolutionary computation: algorithms and applications, vol 1096. In: Studies in computational intelligence. Springer, Singapore. https://www.springer.com/gp/book/9789811575709
4. Beccali M, Bonomolo M, Lo Brano V et al (2019) Energy saving and user satisfaction for a new advanced public lighting system. Energy Convers Manage 195:943–957. https://doi.org/10.1016/j.enconman.2019.05.070
5. De Paz JF, Bajo J, Rodríguez S et al (2016) Intelligent system for lighting control in smart cities. Inf Sci (Ny) 372:241–255. https://doi.org/10.1016/j.ins.2016.08.045

6. Mahoor M, Salmasi FR, Najafabadi TA (2017) A hierarchical smart street lighting system with brute-force energy optimization. IEEE Sens J 17:2871–2879. https://doi.org/10.1109/JSEN.2017.2684240
7. Juntunen E, Sarjanoja EM, Eskeli J et al (2018) Smart and dynamic route lighting control based on movement tracking. Build Environ 142:472–483. https://doi.org/10.1016/j.buildenv.2018.06.048
8. Mohandas P, Dhanaraj JSA, Gao XZ (2019) Artificial neural network based smart and energy efficient street lighting system: a case study for residential area in Hosur. Sustain Cities Soc 48. https://doi.org/10.1016/j.scs.2019.101499
9. Müllner R, Riener A (2011) An energy efficient pedestrian aware smart street lighting system. Int J Pervasive Comput Commun 7:147–161.https://doi.org/10.1108/17427371111146437

Energy Audit in an Indian University—A Case Study

Hamad Khan, Akif Siddiqui, Aqeel Ahmad Ansari, and Nidal Rafiuddin

Abstract With the 1973 oil crisis and environmental concerns because of burning of fossil fuels and further with the limitations in available energy resources, it has become imperative to harness environmentally friendly non-conventional energy sources and manage energy consumption judiciously. The present work focusses on effective energy management from an audit performed on hall of residences of an Indian university namely Aligarh Muslim University by replacing major inefficient loads by efficient loads and estimates solar energy that can be harnessed. There are about nineteen student residences in Aligarh Muslim University. To ensure maximum electrical energy efficiency and cost saving, while energy auditing on one hall namely Nadeem Tareen was conducted, it was estimated on all other halls of residence. Firstly, all major loads in the hall of residence were identified and their power consumption and quality were measured using appropriate tools. The GridVis 7.2 software was used for this purpose. The major loads are the primary audit concerns as they possess the highest potential for energy saving. Replacing existing lighting and fans with LED bulbs and BLDC fans, respectively, were entertained as conservation opportunities and consequent costs calculated. The rooftop areas of all of the remaining hostels were measured using Google Maps tool, further solar plant capacity was calculated. If the proposed changes are implemented on all hall of residences, an estimated INR 6.3 crore can be saved annually.

Keywords Energy audit · Energy conservation · Energy survey · Solar power plant · BLDC

H. Khan (✉) · A. Siddiqui · A. A. Ansari · N. Rafiuddin
Aligarh Muslim University, Aligarh, UP 202002, India
e-mail: hk20@outlook.com

A. Siddiqui
e-mail: akifsid97@gmail.com

A. A. Ansari
e-mail: aqeelahmad4771@gmail.com

N. Rafiuddin
e-mail: nidal.rafi@gmail.com

A. Iqbal et al. (eds.), *Renewable Power for Sustainable Growth*, Lecture Notes in Electrical Engineering 723, https://doi.org/10.1007/978-981-33-4080-0_55

1 Introduction

This paper presents an energy audit of all buildings of hall of residents in Aligarh Muslim University (AMU) accompanying numerous rooms for residing students. AMU provides residences to around 20,000 students in nineteen halls of residences among which Nadeem Tarin hall is one.

While carrying out this study, major loads are taken into consideration for energy conservation opportunities. In this case, they were fans and lighting sources. The cost of implementation of the recommended opportunities and the calculated payback period and is found to be highly feasible.

Rooftop solar power plant is proposed for this particular case and for other resident building in the entire university and consequent implementation costs and savings are calculated.

2 Case Study

2.1 Building Specifications

Nadeem Tarin (NT) Hall consists of 140 resident rooms, 12 washrooms, office building and a worship place. Each room consists of one fan and two lighting sources which were either tube lights or LED bulbs. Other major loads are air-coolers, geysers and water dispensers. The consumption of these loads on an average is 812.09 units per day. The rooftop area of the same hall is 1750 m^2 and has a potential for a 110-kW solar power plant.

2.2 Lighting System

Firstly, the total number of major lighting loads were identified in NT hall and their consumption was measured. Figure 1 shows the number and proportion of major lighting load in NT.

The kWh difference for one day between a LED and a tube light was found to be about 0.74 units with near same illumination. Hence, replacing all the tube lights in NT by LEDs would save ₹6.5 lacs annually on energy cost. Further, the in-depth power quality analysis was done to compare various parameters of various sources. The payback of replacing tube lights with LEDs is less than a month, and thus the replacement of tube lights by LEDs is proposed to be a very efficient solution.

The harmonic currents are an undesired parameter and were measured to be considerably less in LEDs in comparison with tube lights. Thus, it is suggested to replace all 289 tube lights in the hall with LEDs. The harmonics current of LED is shown in Fig. 2.

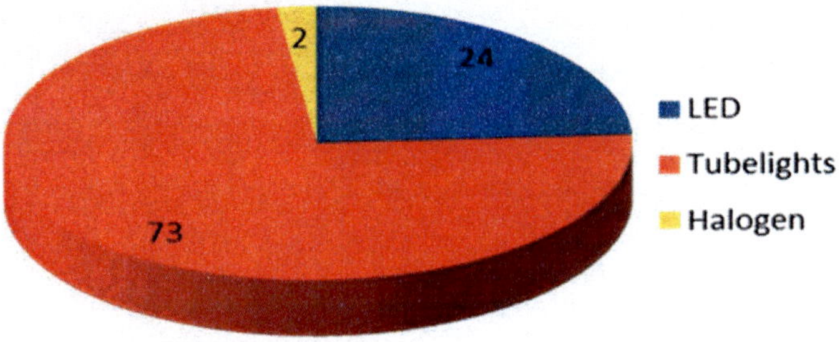

Fig. 1 Various lighting sources in NT

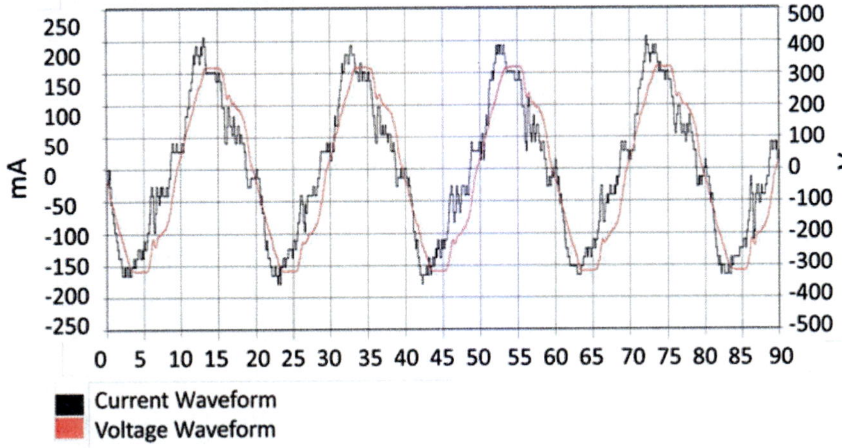

Fig. 2 Voltage and current graphs of LED load

Figure 3 portrays the harmonics generated because of LED bulb. The third harmonic current is less than 10% of the fundamental value.

2.3 *Air Cooling System*

The major air cooling in the building is done by using induction motor fans, where one fan in most of the rooms is allotted while some rooms have two. The consumption of each fan was measured to be 122 W on average. It was noticed that when maintenance was provided to fan by oiling the fan's motor and tightening blade screws, consumption was reduced to 115 W. Hence, the regular maintenance could provide a remarkable figure in energy saving when all the fan's power consumption was taken into consideration. Table 1 enumerates the number of fans installed in NT hall.

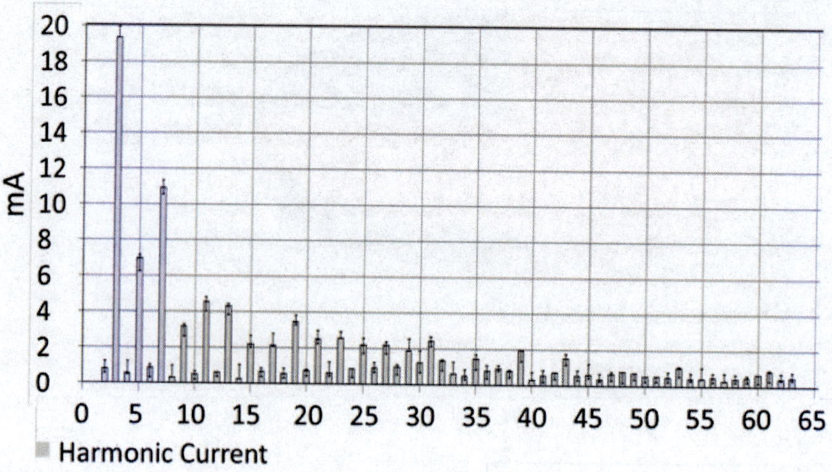

Fig. 3 Harmonic currents in a LED bulb

Table 1 Number of fans in the building

Place	Number of fans
Resident rooms	140
Reading room	11
Dining hall	15
Common room	12
Canteen	2
Worship place	35
Provost office	12
NXC	3
Total	235

Since fans are one of the major loads in the hall, finding an energy-efficient way to produce same or better air cooling results with less power consumption was necessary. Brushless DC fans is one of the alternate efficient possibility to meet the requirements.

To verify this, power quality analysis on both induction motor and BLDC fans was done. This was done by setting up the apparatus as shown in Fig. 4.

The distance and angle forming between fan and the anemometer were measured to keep it constant for both fans.

The fans were switched at different speeds and the power quality results of both the fans were compared. The output was not taken as speed of fan rather the speed or air they deliver at a specific distance underneath. Table 2 shows the results of the fans at their maximum speeds.

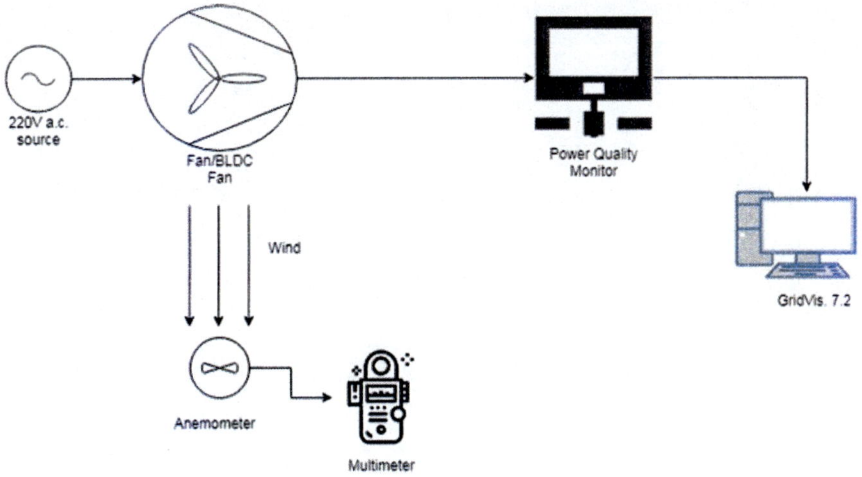

Fig. 4 Experimental setup for power quality measurement

Table 2 Different measurements for both fans

Parameters	Induction motor fan	BLDC fan
Voltage (V)	229	234.26
Current (A)	0.35	0.156
Real power (W)	76.56	34.21
Reactive power (W)	−11.86	−4.59
T.H.D. (%)	5.71	10.398
Power factor	0.98	0.988
Anemometer (m/s)	0.6	1.8

It was observed that at almost half power consumption, BLDC fan gives thrice the amount of wind speed (Fig. 5).

Replacing all 235 fans would save ₹91,613 monthly and ₹1,099,359 annually. The payback period is estimated to be only about 9 months. This energy conservation opportunity is also highly recommended.

3 Solar Power Plant

Nadeem Tarin Hall is only one among nineteen resident halls in the University. The roof area of Nadeem Tarin Hall was measured using an inch tape and was found to be 1750 m^2. Each hall has a good amount of roof area and an on grid solar power plant can be considered for all the halls. However, measuring the roof area of all the halls was not feasible. Thus, a Google Maps tool was used to measure roof area. The

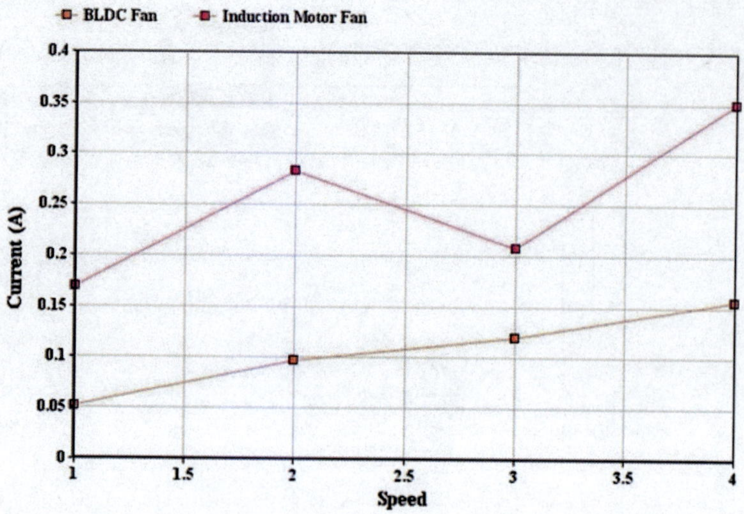

Fig. 5 Current versus wind speed setting for BLDC and induction motor fans

area of Nadeem Tarin Hall using this tool was found 2150 m^2. This error was used to correct the measurements of all other halls. The results of all the halls are shown in Table 3.

Assuming sunshine of 5.5 h in Uttar Pradesh, the average solar irradiation is 1156.39 W/m^2. On an average, a 1KW solar rooftop plant over the year will generate 4.6 kWh of electricity per day.

An online solar power tool by Ministry of New and Renewable Energy (MNRE) [5] gives the following data: 12 m^2 area can provide space for 1 kW plant, feasible capacity for Nadeem Tarin Hall is 110 kW, MNRE current benchmark cost: 53,000 ₹/kW and total financial saving from this would be ₹105,375 monthly and ₹1,264,494 annually. Payback period for this is 4.6 years.

4 Conclusion

With all the measurements, experiments and analysis, we found three conservation opportunities. Two of them can be easily implemented as they are cost efficient. These include replacing all the tube lights with LEDs and replacing all the induction motor fans with brushless DC motor fans. The third conservation opportunity which is comparatively costly is to install a solar power plant on the rooftop of each hall of residence. Implementing these would save considerable assets. Moreover, with around two months of vacations, where the resident students leave for their home, power is hardly consumed in the halls. In this situation, the excess power produced by solar power plant in these residences can be supplied to power grid. Furthermore,

Table 3 Usable area for installing solar plant on rooftop of hall of residences

Hall of residences	Measurements (m^2)	Corrected measurement (m^2)	Usable area (75%) (m^2)
Nadeem Tarin	2150	1750	1312
Sulaiman Hall	6296	4860	3645
M.M. Hall	5035	3887	2915
Sir Syed North	21856	16872	12654
R.M. Hall	3672	2834	2125
Sir Syed South	3917	3023	2267
S.N.	1600	1235	2267
Viqraul Mulk	15912	12284	9213
Sir Ziauddin	2679	2068	1551
Abdullah Hall	2588	1997	1497
Aftab Hall	5520	4260	3195
Allama Iqbal	4661	3597	2697
Habib Hall	3876	2991	2243
Hadi Hassan Hall	3021	2331	1748
Indira Gandhi Hall	3599	2778	2083
Bhim Rao Ambedkar	2977	2298	1723
Begum Sultan Jahan	3727	2876	2157
Bibi Fatima Hall	1715	1323	992
Begum Azizun Nisa	1762	1360	1020

installing low speed wind turbines or wind arrays [6] for harnessing, low wind speed on the rooftops may aid in production of power through wind, which is another aspect of study as the wind speed in Aligarh is low.

Acknowledgements Authors are thankful to Mr. Mohammad Shoeb Alam and Mr. Jamil Khan for their whole hearted support in carrying out practical aspects of this work which would not be possible otherwise.

Reference

1. Turner WC, Doty S (2007) Energy management handbook, 6th edn. The Fairmont Press, Inc., U.S.A, pp23–39
2. Kerimray A, Bektineyev R, Rojas-Solórzano LR (2016) Energy efficiency options for buildings: Insights from buildings energy audit reports in Kazakhstan. In: 4th IET clean energy and technology conference (CEAT 2016), Kuala Lumpur, pp1–6
3. Singh H, Seera M, Idin MAM (2012, Dec) Electrical energy audit in a Malaysian university-a case study. In: 2012 IEEE international conference on power and energy (PECon). IEEE, pp 616–619

4. Wang X, Huang C, Cao W (2010, March) Energy audit of building: a case study of a commercial building in Shanghai. In: 2010 Asia-Pacific power and energy engineering conference. IEEE, pp 1–4
5. Khan BH (2006) Non-conventional energy resources. Tata McGraw-HillEducation
6. Rafiuddin N, Al Saif MY, Ahmad A (2018, Sept) WIND ARRAY: a novel approach to low speed wind harnessing. In: 2018 International conference on computing, power and communication technologies (GUCON). IEEE, pp 578–580

Study of Effectiveness of Autonomous Solar Energy Systems

Prashant Kumar Singh, Rajesh Narayan Deo, Vishal Vajpayee, Vipin Verma, and Vivek Bajpai

Abstract We live in a world where the population is growing day by day, and with the increase in population, our energy demand is also increasing to meet our daily energy consumption (IEEE Standard for Interconnecting Distributed Resources with Electric Power Systems, IEEE Std. 1547, 2003) [1]. This is resulting in shortage of electricity across the world, especially under developed or developing countries. Hence, we need an efficient and effective solar system which can stand alone (IEEE Standard for Interconnecting Distributed Resources with Electric Power Systems, IEEE Std. 1547, 2003 [1]; Obukhov et al. methods of effective use of solar power system 2016 [2]) In this paper, it has been discussed that how effectiveness of autonomous energy systems can be increased and also how an efficient solar system can be utilized for direct cooling and heating with the help of solar chillers without converting it from solar energy into electrical energy.

Keywords Solar energy · Solar power · Solar radiation · Collector · Inverter · Solar heat supply · Reactive power · Solar cells · Solar chillers

1 Introduction

Grid-tied PV system is energy generating system which is connected to grids. Solar photovoltaic power which is created must be operated with the support of inverter connected to grid before putting it for use [1–17]. In this regard, several studies have been done [15–17]. These type of inverters are mounted between the utility grid and solar photovoltaic; [3] it is a combination of some small inverters and every inverter is connected to the single photovoltaic units. According to less cost of energy electronic device and developments in energy renewable technology, then the market of energy renewable has utilize and feel an extra profit and increment in scope in some time. Renewable resource boosts the capacity of energy, according to the decrement of

P. K. Singh (✉) · R. N. Deo · V. Vajpayee · V. Verma · V. Bajpai
Department of Electrical and Electronics Engineering, Galgotias College of Engineering & Technology, Greater Noida 201306, India
e-mail: peek.rajput@gmail.com

A. Iqbal et al. (eds.), *Renewable Power for Sustainable Growth*, Lecture Notes in Electrical Engineering 723, https://doi.org/10.1007/978-981-33-4080-0_56

cost. In many countries, solar power and wind power take a perfect action and give competition to fossil fuels [4–6].

The capacity comprises of solar photovoltaic is more than every form of resources combined. After the use of solar photovoltaic, it is important to study about the barrier in front of generating power and solar photovoltaic [2].

The things to keep in mind while connecting photovoltaic solar energy to grid are as follows [16, 17];

1. Cost.
2. Enhancing reliability.
3. Decreasing the harmonics.
4. Decrementing the number of components.
5. To control the reactive and real power [7].

2 Operation of Solar Inverter

Solar generation which is based on photovoltaic has its disadvantages but it has many advantages to hide it. So, in Gujarat, which is located in India has good potential to increase solar power electrical energy by 20–25% in past 25 years. Solar PV cells are used to directly convert sunlight energy into electrical energy without releasing any pollutant [2, 8, 9]. This generation of electricity is effected by some environmental and physical parameters like temperature and radiation of sunlight on photovoltaic cell. Photovoltaic power which is supplied to utility grid gaining great attention these days, thus, many standards which are mentioned by different authorities of grid monitoring must be followed. Thus, according to the characteristics of photovoltaic panels, total O/P voltage differs due to various temperature, irrigation parameters, and effects of clouds. Therefore, the voltage at input level can differ widely [10] (Fig. 1).

1. Solar panel: Devices used to change light in the form of electricity are known as solar panels because sunlight cannot be exhausted.
2. Comparator: A circuit used for observing two or more electrical pulses.
3. Relay: The electrical circuit that controls one electrical circuit by opening and closing contacts in another circuit.
4. Inverter: The device that is used to convert direct current to alternating current is known as inverter [14].

2.1 *Buck–Boost Converter*

When the value of K is between 0 and 0.5 and converter will work in buck.

When the value of K is between 0.5 and 1, then it will operate in boost mode [9, 12, 15] (Fig. 2).

And when K is equal to 0.5,, then this will be ideal mode. And the relation between output voltage and input voltage is mentioned in Eq. 1.

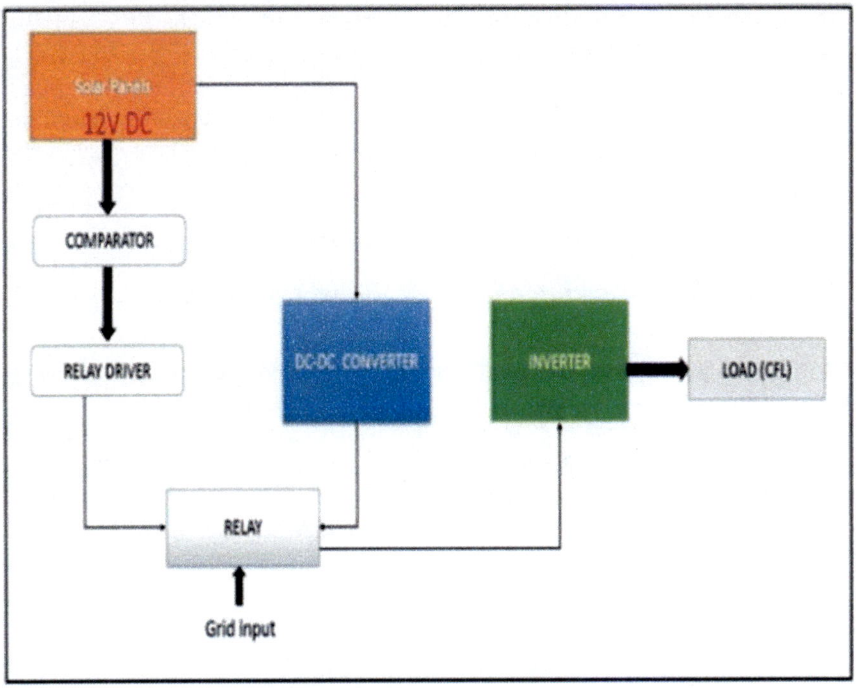

Fig. 1 Diagram of typical solar inverter [8]

Fig. 2 Typical schematic diagram of a basic buck–boost converter [14]

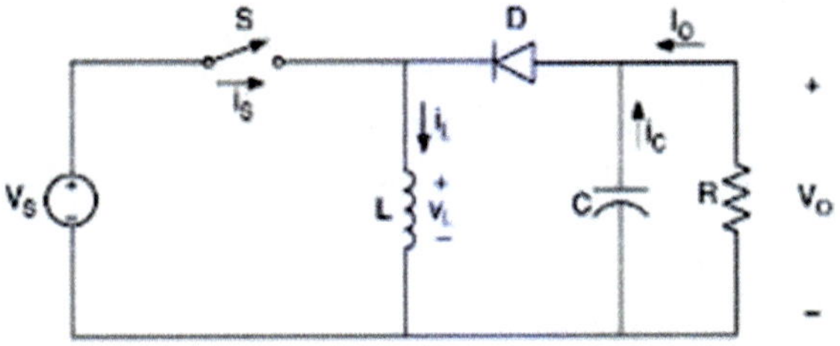

$$V_o = \left(\frac{K}{1 - K} \times V_S \right) \tag{1}$$

2.2 *Functioning Principle of Buck–Boost Converter*

Here, the circuit is made to obtain max efficiency by quality and equipments of current, and it is used for various purposes like as an under voltage [6–8].

Buck converter: The work of a buck converter is to indicate the I/P circuit that resists sudden variations in I/P current. In the case of switch ON, the inductor 'L' stores energy [9].

Boost converter: The inductor 'L' in I/P circuit opposes changes in I/P current. It stores energy when the circuit is open in magnetic form and expenses energy in opposite conditions. [10].

2.3 *Principle of Working of a Buck–Boost Converter*

An inverter changes DC into AC, and hence the devices do not produce any type of energy because energy is generated from a DC power source [1–9] (Fig. 3).

Above figure represents the circuit diagram of single phase half wave bridge inverter 2 MOSFETs $T1$ and $T2$ here are applied to switch devices. Where $T1$ ranges 0–$T/2$, whereas $T2$ ranges $T/2$–T and $T1$–OFF [7].

Therefore, the direction of the current is from B to An in load, the O/P voltage is $-V_s/2$ (Fig. 4).

3 Introduction to Solar Heat Energy (SHE)

The supply of electricity is totally based on demand of the consumer's consumption of the organic fuel and become more possible in the situation that the energy source will cost more. Reducing energy cost can be achieved by fuel economy based on its replacement with renewable source. The use of solar energy nowadays is at all level of power consumption. These are high-power solar electric power stations and individual power installations with no more than 10 kilowatts of power. As

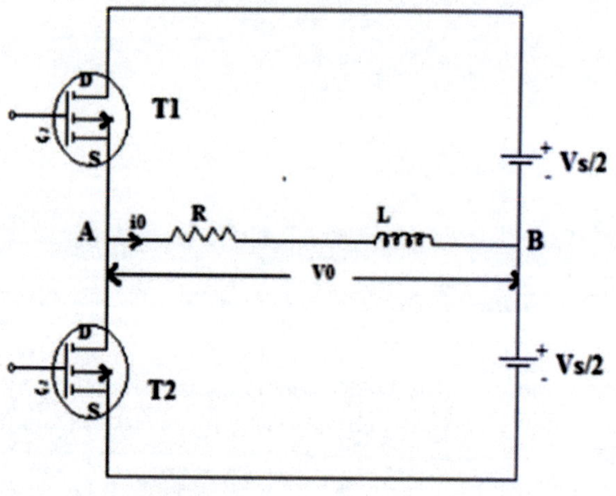

Fig. 3 Single phase half bridge inverter [18]

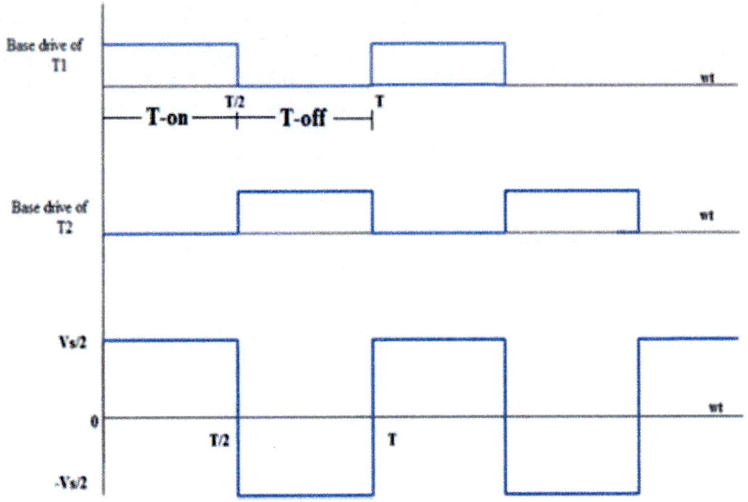

Fig. 4 Single phase half bridge inverter's waveform [13]

an individual low-power installation, solar power systems are used to receive heat energy. Solar heat supply (SHS) systems are a solar energy collection mechanism and the transformation in solar collectors (SC) [11, 12].

1. A large number of different SC constructions nowadays make the best use of solar radiation effectively. At building, SC may be flat or tubular. All SCs can conditionally be diverted into some basic elements: Transparent serves the purpose of shielding absorber from environmental impacts [11, 13].
2. Absorb the object of which soaks and transformation of solar energy in heat energy. Fast absorption to the effect of solar energy while service is the main requirement for the absorber using the convective aspect of SC heat losses.

Solar energy passing with the help of element in the given. It can also be seen that the main component of SC and the element or parameter which define them (Fig. 5).

Solar collector(SC) is connected by the help of pipeline and also with the tank accumulator. Solar heat supply is classified into different schemes of system according to their designation. In the given figure, solar heat supply has one open and one close system [3].

3.1 Formulation and Solution of the Task on Defining the Effectiveness of SHE

To define the method of the effectiveness of SC, it are directly related to incoming energy from area of SC. First choice is the Hottel-Whiller Bliss formula that define the essential heat from the SC area unit as a function in 2 variables—the flux density

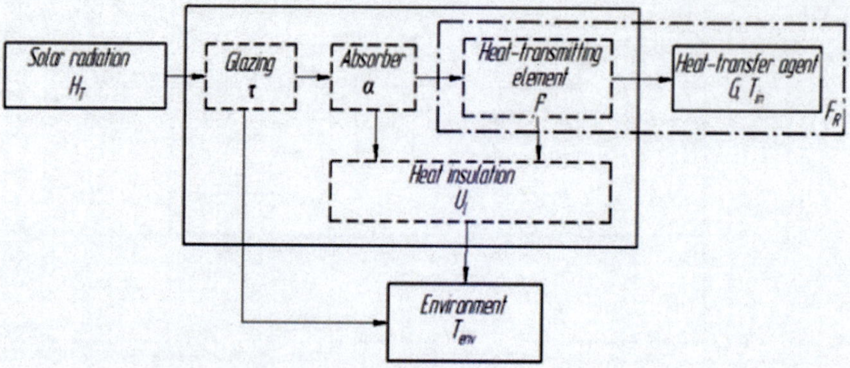

Fig. 5 SC and its basic parameters and elements [2]

of optimized solar waves in the SC HT plane and the difference in heat transfer temperatures at the input of SC H_T solar collector in term of environment T_{env} is given in Eq. 2 [23].

$$Q_{n1} = F_r[H_T \tau_\alpha - U_1(T_{in} - T_{env})] \tag{2}$$

where F_r is heat coefficient of the collector parameter.

U_1 is totally depend the construction of collector. Using conventional glazing materials and selective surfaces for absorbers, this coefficient can be increased to 0, 95 for modern collectors [9].

On formula, the heat extraction coefficient from SC can be defined

$$F_r = \frac{GC_P}{U_1}\left[1 - e^{\frac{-FU_1}{GC_P}}\right] \tag{3}$$

where G is heat transfer consumption with the help of collector.

F is effectiveness of SC work on the interconnection of control and positive parameters is done on the basis of the specified dependencies [14, 15] (Fig. 6).

The max meaning on which the collector's heat extracted coefficient has max. Meaning and it is not worth improving the usage and on the other hand the min. consumption meaning exist depleting which result in a rapid decrease of the SC output.

$$\eta_{SC} = F_r\tau_\alpha - F_r U_1\left[\frac{T_{in} - T_{env}}{H_T}\right] \tag{4}$$

The study of the formula in question allows for the division in two sections. First portion of formula defines technical portion of SC and depends on used materials. Second portion carries out changes from technical portion and explicitly relies on

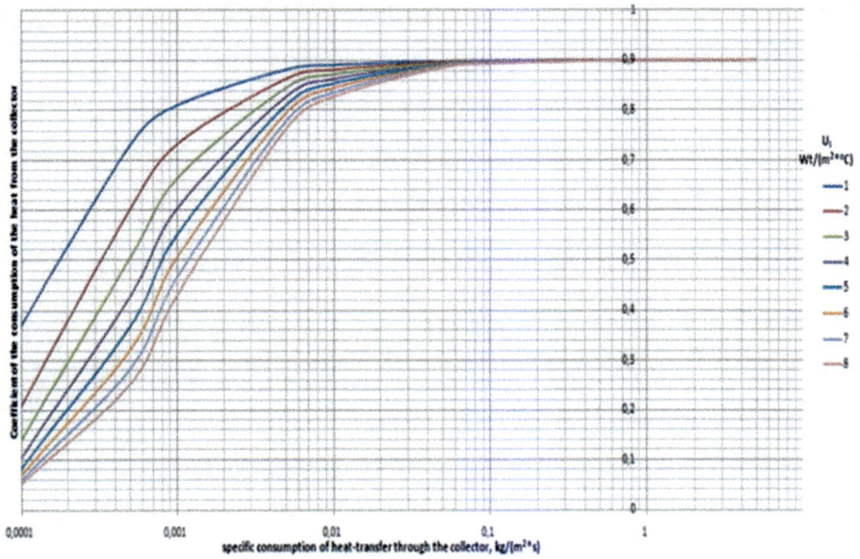

Fig. 6 Control parameters of solar collector and its dependence [11]

control where SC is used and changes are made in the dependency of heat transfer parameters, ambient temperature, and coming solar energy [15].

3.2 Work Study for SC Containing Solar Field

Several SC connecting to each other and joining in the solar fields are used to make the necessary level of heat loading with the aid of solar energy [13].

In case, we have solar field with n quantity of SC, the total useful energy may be presented as and which consist of n amount of collector [10].

For each collector in one solar field, the intensity of the flux of summarized solar radiation in the SC H_T plane will be the same. Then, the flux density of the solar field's condensed solar radiation can be described as the efficiency of solar field which consists of n amount of collectors [15].

$$\eta_{sf} = \frac{\eta_{SC1} + \eta_{SC2} + \eta_{SC3} + \cdots + \eta_{SCn}}{n} \quad (5)$$

This formula shows that the solar field efficiency is the mean quantity and depends on each SC's efficiency.

Increasing the productivity of each SC permits promoting the effective work of the entire solar field [13].

In the solar field, SC can be connected on different schemes with respect to each other. The output energy of the collector may be presented as

$$Q_n = GC_p(T_{\text{out}} - T_{\text{in}}) \tag{6}$$

where T_{out}—temperature of heat transfer agent.

> The output energy of the SC is determined with the help of heat transfer agent. Thereby the parameters of the heat-transfer agent on the output of one collector may be the parameters on the input of the other which in the definite senses of consumption, the temperature of the atmosphere and the isolation that result in a significant decrease in its efficiency and of the whole solar field. [11]

The control of the heat transfer agent's parameters is related to adjusting the specific heat transfer agent consumption through the solar collectors [12].

3.3 *Scheme Development of SC for Solar Field*

To ensure the efficient functioning of the solar heat supply system by adjusting the SC link scheme, the unit is controlled out. Change the SC link scheme between them and also the relay for valve control, photoresistor for relay control in dependence of solar radiation power [13–15] (Fig. 7).

To exclude this scheme's dependency from the power supply grid, the solar batteries are provided as source (SB) electricity. Then, choose pumps that operate on the direct current.

The link scheme of the SHS elements on the SB power supply is given to the control scheme [7] (Fig. 8).

> The provided scheme from four SCs works in this way: the voltage at the output of SB is weak on small power of solar radiation and it is not enough to open the relay KM1 and KM 2 and the valves. All electromagnetic valves are closed which exclude inverse heat transfer agent circulation. [11]

Increasing the strength of solar radiation later causes valves 1, 3, 7, 9 to open and valves 2, 8 to close. The valve switching enables parallel communication of

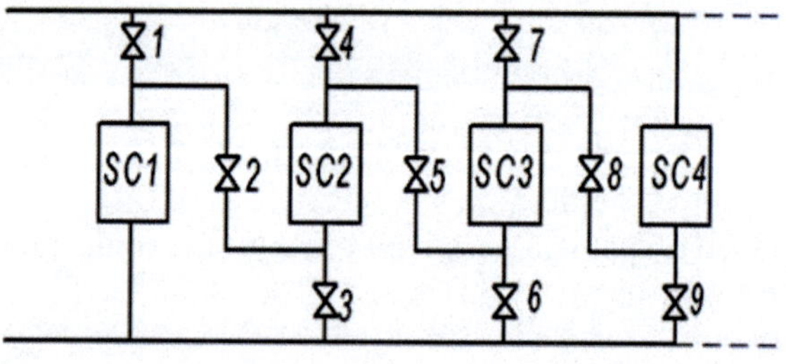

Fig. 7 SC connection in SHS [25]

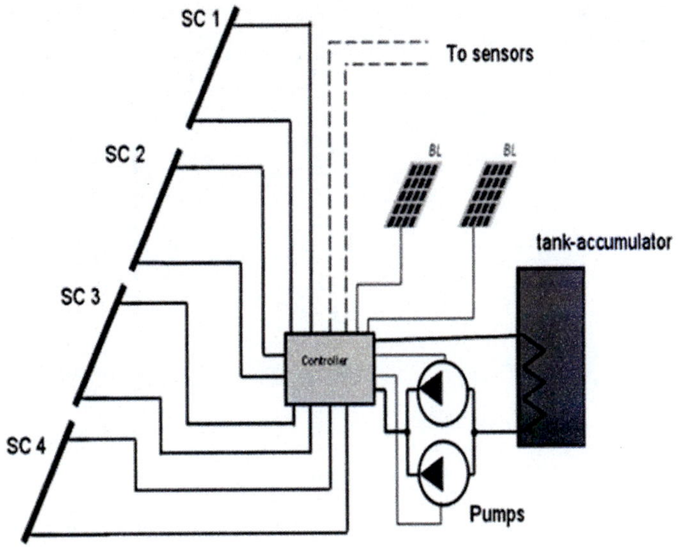

Fig. 8 Basic connection of SHS from power source [22]

SC between the valves. The process of switching the SC link scheme is in reverse sequence upon decrease in solar radiation strength [14] (Fig. 9).

4 Conclusion

The use of high value renewed sources for increasing energy supply expenses. The solar power development makes the substitution of commonly used organic types of fuel, being more insight at that. It is important to give the correct parameter of the heart transfer agent in SC circuit in the dependence of solar insulation and the ambient temperature for working effectively [7]. The worked out SC link schemes and the control algorithm permit offline automatically to give the provided SHS control parameters [9].

A large efficiency buck–boost converter is offered. Initial portion works for buck–boost mode. Therefore, it offers a large range of I/P voltage. The second part of inverter is made up of unfolding circuit which is based on the grid's direction. Thus, from the view of power processing, it is a one step inverter. This happens due to power transfers buck–boost converter [10].

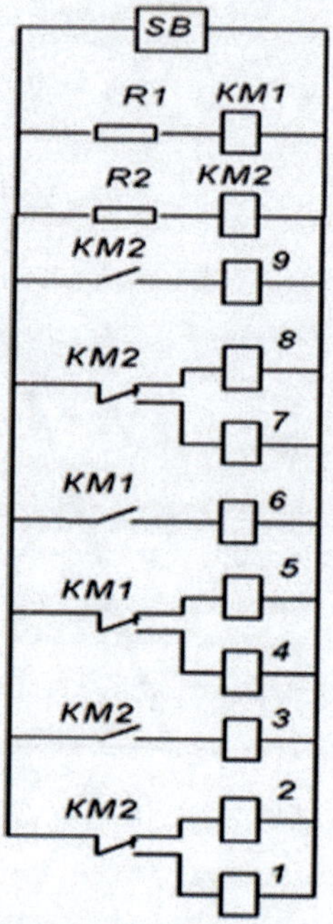

Fig. 9 Electromagnetic valve control mechanism [24]

References

1. IEEE standard for interconnecting distributed resources with electric power systems. IEEE Std. 1547 (2003)
2. Obukhov SG, Plotnikov IA, Sheryazov SK (2016) Methods of effective use of solar power system. In: 2nd international conference on industrial engineering, applications and manufacturing ICIEAM: 2016—Proceedings 2016. https://doi.org/10.1109/ICIEAM.2016.7911015
3. . Watson NR, Miller A (2015) Power quality indices: a review. In: Proceedings of the EEA conference and exhibition, Wellington, New Zealand, 24–26 June 2015
4. Shinjo F, Wada K, Shimizu T (2007) A single-phase grid-connected inverter with a power decoupling function. In: Power electronics Specialists conference. IEEE 17-21, pp 1245–1249
5. Huang Y, Shen M, Peng FZ, Wang J (2006) Z-source inverter for residential photovoltaic systems. IEEE Trans Power Electron 21(6):1776–1782
6. Tsoukalas H (2008) From smart grid to an energy internet: assumption, architecture, and requirements. DRPT 2008, Nanjing, China, 6-9 Apr 2008
7. Eskom's (1998) Renewable energy experience, presented at village power 98, 6–8 Oct 1998
8. Ortiohann E (2009) A general modular design methodology for flexible smart grid inverters. IEEE Explore, 19 Nov 2009
9. Krein PT (1998) Elements of power electronics. Oxford University Press, New York
10. Renewable energy. IEEE J 35(1):275–282
11. Rai GD Non-conventional energy sources. Khanna Publishers, p 591
12. Soonkang F, Park S-J, Eogcho S, Kim J-M (2006) Applied energy 82(3)
13. Lornic J (2009) Feed-in tariffs: Ontario's experience, a blog about energy and the environment, 10 Feb 2009
14. Singh G (2008) National action plan on climate change launched: solar energy to change the face of India, 30 June 2008

15. "Hydro projects" ASAPP medra information group power today, Sep 2009
16. Iqbal A et al (eds) (2020) Soft computing in condition monitoring and diagnostics of electrical and mechanical systems, vol 1096. In: Advances in intelligent systems and computing. Springer, Singapore. https://doi.org/10.1007/978-981-15-1532-3
17. Iqbal A et al (eds) (2020) Meta heuristic and evolutionary computation: algorithms and applications, vol 1096. In: Studies in computational intelligence. Springer, Singapore. https://www.springer.com/gp/book/9789811575709

Analysis of Combined Effect of Temperature and Wind on Solar Power Production

Anurag Sinha, Arun Kumar, Abhishek Tiwari, and Kushal Yadav

Abstract Wind speed (WS) and air temperature are a concern for solar power generation PV industry and policy makers. This causes reduction in direct normal irradiance (DNI) and in turn increases the uncertainty in solar power plant (SPP) output. Therefore, for proper unit commitment and efficiently operating SPP, solar radiation should be properly calibrated. In this work, a detailed study has been performed on combined effect of the wind speed and air temperature on performance of SPP having the different panel types. The monocrystalline (C-si)-type module and multi-crystalline (MC-si)-type module are analysed in this study for Tucson, USA. A 140 W solar panel of area 1.26 m^2 of an SPV panel with 72 cells in series and single string in parallel has been designed and simulated in pvlib-python environment. The aim of this study is to analyse the behaviour of wind speed and air temperature at a location operating SPV system for maximum output. This analysis will help PV industry to know the uncertainty in performance of the PV plant with respect to change in wind speed and air temperature.

Keywords Solar energy · Monocrystalline · Multi-crystalline · Wind speed · Air temperature · Direct normal irradiance

1 Introduction

As per literature available in digital domain [1–21] in this modern era, electricity has become a basic need for human society [1]. There is an exponential growth in world's population [2]. Due to the combined effect of increasing population and modernization, the requirement of electricity has increased many times [3]. With exponential increasing demand, but shortage of technologies due to disunity to fulfil those demand creating shortage of energy [4]. We are continuously tackling the problem of energy shortage and a sustainable energy fulfilment by moving to production of energy

A. Sinha (✉) · A. Kumar · A. Tiwari · K. Yadav
Department of Electrical and Electronics Engineering, Galgotias College of Engineering and Technology, Greater Noida, India
e-mail: Sinha.anurag1999@gmail.com

A. Iqbal et al. (eds.), *Renewable Power for Sustainable Growth*, Lecture Notes in Electrical Engineering 723, https://doi.org/10.1007/978-981-33-4080-0_57

through non-conventional sources of energy. There are several green energy strategies available for the sustainable development [5]. Solar energy production is a very easy approach to harness electrical energy continuously from sun for a long duration of period. PV industry has increased its revenue many times in recent [6]. Although efficiency is low due to which there is less output of a SPP today, continuously work is going on. The unwanted change in different properties with respect to predefined quality can be termed as pollution. Urbanizaton has a very mammoth impact on the ecological system. The direct effect is being seen on change in climatic pattern, wind speeds, rainfall pattern, etc. Due to which there is an increase in uncertainty in SPP production.

This paper is based on analysing the combined effect of temperature and wind speed which can help PV industry experts to predefine their steps for proper unit commitment. This efficient unit commitment will increase the financial viability of the proposed solar power plant.

2 Description of SPV System

There are various types of solar panels available based on their materials by which they are manufactured. In this work, monocrystalline and multi-crystalline-type panels are discussed and used to calculate the feasibility.

2.1 Solar cell's Mathematics

The solar cell output depends on the incident solar radiation and temperature. The relation of different parameter for solar cell's current is shown in Eq. 1:

$$I = I_{\text{ph}} - I_0\left\{e^{\frac{q(V+R_S I)}{\text{AKT}}} - 1\right\} - \frac{V + R_s I}{R_{\text{sh}}} \tag{1}$$

where A is curve fitting factor I_{d}, I_{ph}, is diode current and photocurrent, respectively. And, R_s and R_{sh} shows the series and shunt resistance of the solar cell.

The configuration of the SPV which is used for the simulation is discussed below for monocrystalline and multi-crystalline, shown in Figs. 1, 2 and Tables 1, 2, respectively: (Source: Sandia National Laboratories Module Database).

Monocrystalline SPV system

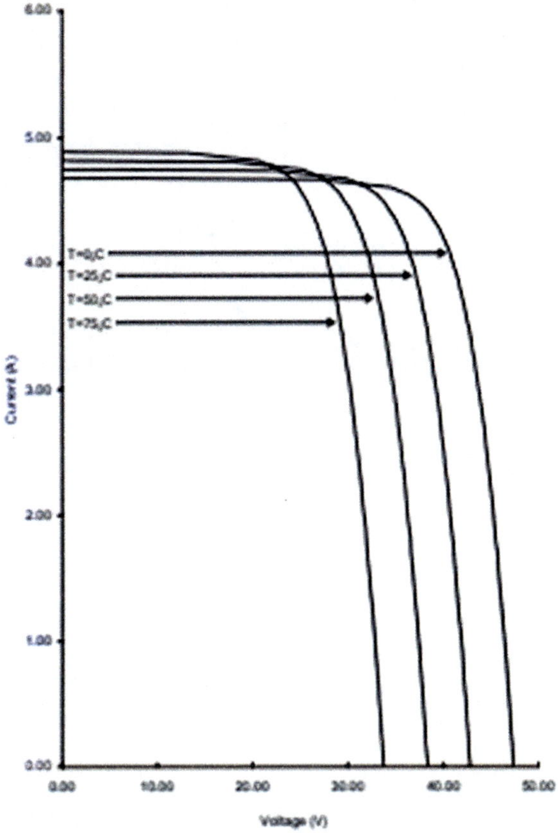

Fig. 1 I-V characteristics of BP solar BP2140S

Module Name: BP Solar BP2140S 2000 (E)

Multi-crystalline PV module is BP Solar SX140 2003 (E)

The inverter for the grid connected power plant is studied in this work which is ABB MICRO 0.25 I-OUTD US 208 V. The inverter efficiency with respect to rated power is shown in Fig. 3, and technical specification is shown in Table 3.

3 Result and Analysis

Here, mono- and multi-crystalline SPV modules have been used for simulation. The location selected is at 'TUCSON' having coordinates (30°, −110°) and altitude of 700 m. Hourly solar radiation frequency has been used in the analysis, and the data for analysis is considered for 2019 and 2020 years (1 year duration).

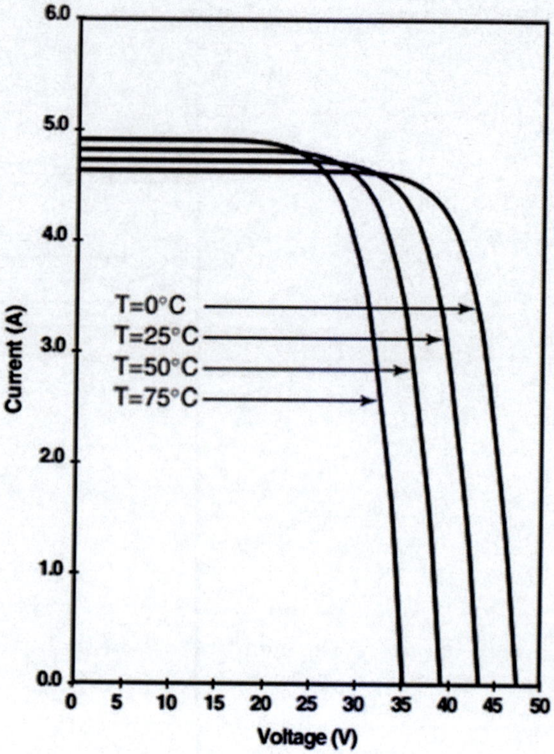

Fig. 2 I-V characteristics of BP solar SX 140

First we are seeing the combined effect of temperature and wind speed at provided location.

Monocrystalline behaviour chart: It shows the power output variation being generated by the BP Solar SX140 with respect to the change in wind speed, shown in Fig. 4.

Polycrystalline behaviour chart: It shows the power output variation being generated by the BP Solar SX140 with respect to the change in wind speed, shown in Fig. 5.

For comparison of monocrystalline and multi-crystalline, performance on change in speed and specific temperature a curve has been plotted of change in power output against each speed (Fig. 6).

4 Conclusion

In this work, the effect of temperature and wind speed on solar panel power production is analysed with pvlib tool. With the increase in temperature of the panel, the output power decreases, whereas on increase in wind speed the power output increases

Table 1 Monocrystalline panel characteristics

Electrical characteristics	
V_{mp} (V)	34.0
I_{mp} (A)	4.16
V_{oc} (V)	42.8
I_{sc} (A)	4.48
P_{mp} (W)	140
Power tolerance	−3.00% ~ 3.00%
V_{max} (V)	600
Fuse rating (A)	15
Panel efficiency	11.1%
Fill factor	73.0%
Temperature coefficients (TC)	
TC of P_{mp}	−0.50%/°C
TC of I_{sc}	0.065%/°C
TC of V_{oc}	−0.38%/°C
Operation conditions	
Operating temperature	−40.0–85.0 °C
NOCT	47.0 °C
Safety rating	Class C

Table 2 PV2 characteristics

Electrical characteristics	
V_{mp} (V)	34.0
I_{mp} (A)	4.11
V_{oc} (V)	42.8
I_{sc} (A)	4.50
P_{mp} (W)	140
Power tolerance	−7.10 to 7.10%
V_{max} (V)	1000
Maximum series fuse rating (A)	15
Efficiency	11.1%
Temperature coefficients (TC)	
TCof P_{mp}	−0.50%/°C
TC of I_{sc}	0.065%/°C
TC of V_{oc}	−0.50%/°C

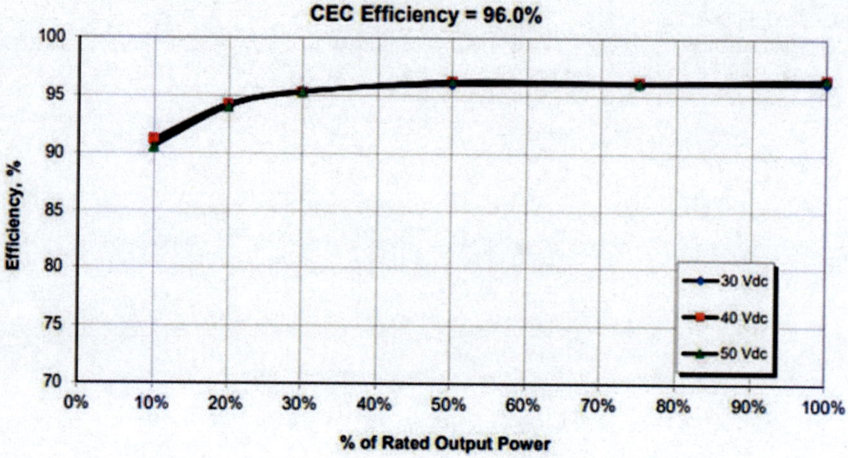

Fig. 3 Efficiency of the inverter over the different rated outputs

Table 3 Technical specification of inverter

Technical data and types		
Type code	MICRO-0-25-I-OUTD	
Nominal output power	250 W	
Rated grid AC voltage	208 V	240 V
Maximum output power	260 W	
Input side (DC)		
Maximum usable DC input power	265[2]Wp	
Maximum PV panel rating (STC)	300 W	
Absolute maximum voltage (Vmax)	65 V	
Start-IP voltage (Vstart)	25 V	
Full power MPPT voltage range	25–60 V	
Operating voltage range	12–60 V[2]	
Maximum usable current (Idcmax)	10.5 A	
Maximum short circuit current limit	12.5 A[2]	
DC connection type	Amphenol H4 PV connector	
Output side (AC)		
Grid connection type	10/2 W	Split-0/3 W
Adjustable voltage range	183–228 V	211–264 V
Nominal grid frequency	60 Hz	
Adjustable grid frequency range	57–60.5 Hz	
Maximum output current	1.20 A	1.04 A
Power factor		
Maximum number of inverters per string	13	15

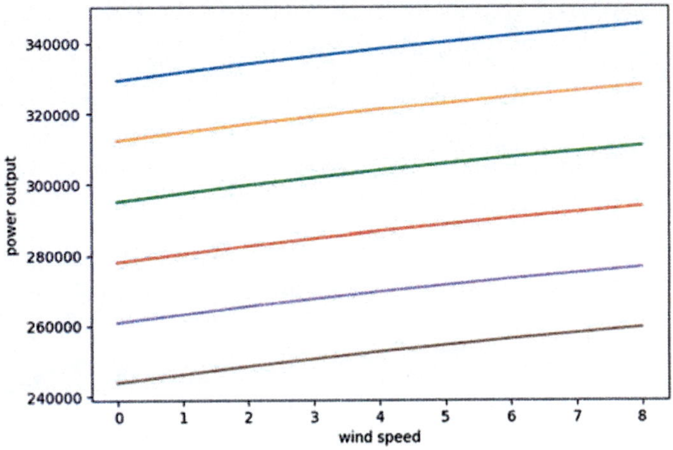

Fig. 4 Wind speed versus solar power for monocrystalline

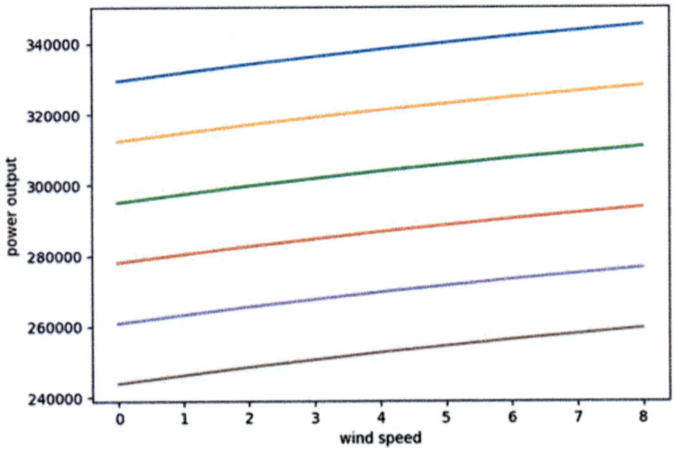

Fig. 5 Wind speed versus solar power for multi-crystalline

in both monocrystalline module and multi-crystalline modules. But the change is nonlinear, monocrystalline shows a degrading behaviour with the change in the output in comparison with the multi-crystalline modules. The table is attached in appendix to show this behaviour on different module.

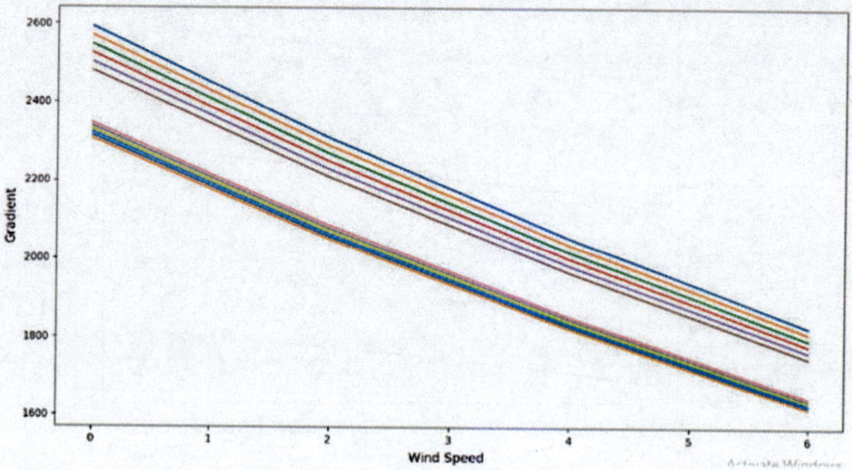

Fig. 6 Comparison of wind speed versus solar power for mono- and multi-crystalline

Appendix

See Tables 4 and 5.

Table 4 Monocrystalline output table

power Output Table	wind speed = 0	wind speed = 2	wind speed = 4	wind speed = 6	wind speed = 8
Temperature = 0	329434	334125	338295	342000	345294
Temperature = 10	312206	316881	321036	324729	328011
Temperature = 20	295037	299696	303837	307517	310788
Temperature = 30	277926	282570	286697	290365	293625
Temperature = 40	260872	265500	269613	273269	276518
Temperature = 50	243877	248489	252588	256232	259469

Table 5 Multi-crystalline output table

		Monocrystalline Power Output Table			
power Output Table	wind speed = 0	wind speed = 2	wind speed = 4	wind speed = 6	wind speed = 8
Temperature = 0	330397	335582	340198	344305	347959
Temperature = 10	311430	316570	321146	325218	328840
Temperature = 20	292631	297726	302262	306298	309889
Temperature = 30	273995	279046	283542	287543	291103
Temperature = 40	255526	260532	264988	268954	272483
Temperature = 50	237222	242183	246600	250531	254028

References

1. Taylor LD (1975) Bell J Econ JSTOR
2. Lutz W, Sanderson WC, Scherbov S (2004) Google book
3. World Energy Outlook (2013) Universal access to modern energy for poor
4. Tanzer M Energy crisis: world struggle for power and wealth
5. Midilli A, Dincer I (2005) Murat Ay-2005-energy policy
6. Sarti D, Einhaus R Solar energy materials and solar cells
7. Sandia National Laboratories Module Database
8. Ineichen P (2016) Validation of models that estimate the clear sky global and beam solar irradiance. Sol Energy 132:332–344
9. Reno M, Hansen C, Stein J (2012) Global horizontal irradiance clear sky models: implementation and analysis. In: Sandia National Laboratories, SAND2012-2389
10. Bird RE, Hulstrom RL (1980) Direct insolation models
11. Stein JS (2012) The photovoltaic performance modeling collaborative (PVPMC). In: Photovoltaic Specialists conference
12. Andrews RW, Stein JS, Hansen C, Riley D (2014) Introduction to the open source pvlib for python photovoltaic system modelling package. In: 40th IEEE Photovoltaic Specialist conference (paper)
13. Holmgren WF, Andrews RW, Lorenzo AT, Stein JS (2015) PVLIB python 2015. In: 42nd Photovoltaic Specialists conference (paper and the notebook to reproduce the figures)
14. Stein JS, Holmgren WF, Forbess J, Hansen CW (2016) PVLIB: open source photovoltaic performance modeling functions for matlab and python. In: 43rd Photovoltaic Specialists conference
15. Holmgren WF, Groenendyk DG (2016) An open source solar power forecasting tool using PVLIB-python. In: 43rd Photovoltaic Specialists conference
16. Pathak PK, Yadav AK (2019) Design of battery charging circuit through intelligent MPPT using SPV system. Solar Energy
17. Boccaletti C, Macilletti M (2011) Electrical storage systems in cogeneration plants based on the solar energy source. In: 2011 international conference on clean electrical power (ICCEP)
18. Moharram KA, Abd-Elhady MS, Kandil HA, El-Sherif H (2013) Influence of cleaning using water and surfactants on the performance of photovoltaic panels. Energy Convers Manage
19. Wang C, Wang T, Li P, Wang Z Recycling of SiCl4 in the manufacture of granular polysilicon in a fluidized bed reactor. Chem Eng J
20. Iqbal A et al (eds) Soft computing in condition monitoring and diagnostics of electrical and mechanical systems, vol 1096. In: Advances in intelligent systems and computing. Springer, Singapore. https://doi.org/10.1007/978-981-15-1532-3
21. Iqbal A et al (eds) Meta heuristic and evolutionary computation: algorithms and applications, vol 1096. In: Studies in computational intelligence. Springer, Singapore. https://www.springer.com/gp/book/9789811575709

High Gain DC-DC Converter for Modular Multilevel Converter Applications

Mohammad Tayyab, Adil Sarwar, and Javed Ahmad

Abstract Solar photovoltaic (PV) is one of the most emerged and widely used alternative source of energy that can support existing conventional sources of energy used in electrical power generation to a larger extent. At present, grid connected solar PV systems are used extensively, but standalone solar PV system plays a very important role in providing electric power to the places where power grid is not available (e.g., some villages located at remote places and remote islands). It can also be used in standalone mode for applications such as toys, calculator, electric vehicle, space crafts, and satellites. This paper presents high gain DC-DC converters which enables the use of low voltage solar PV system for high-voltage applications which can be used as a DC link voltage of modular multilevel converter (MMC). It minimizes the effect of changing weather conditions which continuously changes solar irradiance by varying duty cycle. Simulation of the converter is carried out by using MATLAB/Simulink software under constant load condition and by varying the load at the output.

Keywords High gain converter · Solar photovoltaic · Partial shading · Solar irradiance

1 Introduction

The ever-increasing use of non-renewable sources of energy in the last few decades has increased the environmental pollution to a greater extend [1–10]. Deteriorating environmental conditions and increase in the cost of non-renewable sources of energy have motivated researcher to start working on commonly available renewable sources of energy such as wind, solar photovoltaic (SPV), and fuel cells [1]. Among these commonly available renewable sources, SPV is getting more attention from the

M. Tayyab (✉) · A. Sarwar · J. Ahmad
Department of Electrical Engineering, AMU, Aligarh, UP 202002, India
e-mail: mtayyab1@myamu.ac.in

A. Sarwar
e-mail: adil.sarwar@zhcet.ac.in

A. Iqbal et al. (eds.), *Renewable Power for Sustainable Growth*, Lecture Notes in Electrical Engineering 723, https://doi.org/10.1007/978-981-33-4080-0_58

researchers as consequence of its numerous advantages such as (1) it is environment friendly (2) it is available in abundant in nature (3) it is freely available to everyone and everywhere. However, a SPV module output voltage depends on the environmental conditions which keep on changing with time. Partial shading condition also causes change in output of a solar photovoltaic system which reduces the system output voltage [2, 3]. A high gain boost converter is a more suitable choice for meeting higher voltage demand. It is used for providing high input voltage to electric vehicle, modular multilevel converter [4], DC micro-grid, etc. The conventional DC-DC converter can be used for limited applications due to its unstable operation at higher duty ratio which affects the functionality of the converter. By utilizing reactive component sources, researchers have proposed various high voltage gain converters in recent times. In isolated type high gain converter transformer is used to step-up the input side voltage but at the same time have shortcoming of high ripple current on the input side of the HFT and higher voltage stress on the secondary size of the transformer. Moreover, the additional transformer has certain shortcomings such as (1) it makes the overall converter bulky in size (2) it causes leakage of energy through the magnetic path (3) it requires multi stage power conversion process. In the literature, voltage lift structure (coupled inductor) is used to get higher DC voltage for non-isolated converters. By using the voltage lifting structures, researchers have proposed numerous high gain DC-DC converters [5–10].

In Sect. 2, mathematical analysis is shown. Voltage gain analysis at different duty ratios is carried out in Sect. 3. Section 4 shows simulation results of the proposed converter. Conclusion is provided in Sect. 5.

2 Proposed High Gain DC-DC Converter

A new topology of DC-DC converter with high gain is proposed with single controlled switch for DC link voltage for MMC applications. Figure 1a represents the converter which has two inductors (L_1 and L_2), three capacitors (C_{o1}, $C_{o2,}$ and C_o), and six diodes and a controlled switch. The key features of the proposed converter are (1) it consists of only one controlled switch so the control is simple (2) high voltage gain (3) it is non-isolated type of converter hence compact in size. This converter operates in two modes under continuous current conduction mode by appropriately turning on and off switch (S). Mode-I operation is from 0 to *DT* and mode-II operation is from *DT to T* where *D* is duty ratio.

2.1 *Mode-1 (0 to* DT*)*

During the first mode of operation, switch S is in ON condition and diode *D*1, *D*2, and *D*6 are conducting while *D*3, *D*4, and *D*5 are reversed biased. The voltage across the inductor L_1 during this mode of operation can be obtained by applying KVL as

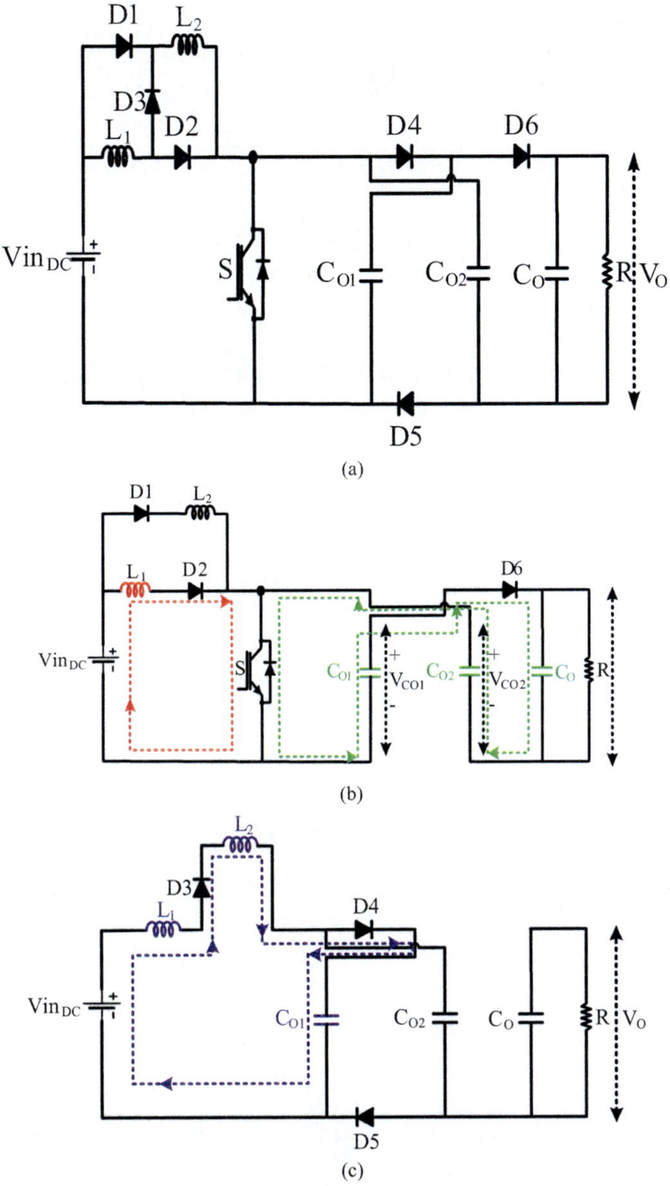

Fig. 1 Circuit diagram of **a** proposed converter, **b first mode of operation, and c** second mode of operation

shown in Fig. 1b

$$VL_1 = V_{in} \tag{1}$$

$$V_o = V_{C01} + V_{C02} \tag{2}$$

2.2 Mode-2 (DT to T)

During this mode, the switch S is in OFF condition $D3$, $D4$, and $D5$ which are conducting and diode $D1$, $D2$, and $D6$ are reverse biased. The voltage equations for the inductor L_1 and L_2 can be obtained by using Fig. 1c as given below.

$$VL_1 + VL_2 = V_{in} - VC_{01} \tag{3}$$

$$VL_1 = \frac{V_{in} - VC_{01}}{2} \tag{4}$$

$$V_{C01} = V_{C02} \tag{5}$$

Now applying Volt-Sec balance across inductor L_1

$$\int_0^T VL_1(t).\mathrm{d}t = 0 \tag{6}$$

$$V_{in} \times DT + \left(\frac{V_{in} - VC_{01}}{2}\right) \times (1 - D)T = 0 \tag{7}$$

$$V_{in} \times D + \left(\frac{V_{in} - VC_{01}}{2}\right) \times (1 - D) = 0 \tag{8}$$

$$V_{in} \times D + \frac{V_{in}}{2} - \frac{VC_{01}}{2} - \frac{V_{in}}{2} \times D + \frac{VC_{01}}{2} \times D = 0 \tag{9}$$

$$\frac{V_{in}}{2} \times D + \frac{V_{in}}{2} - \frac{VC_{01}}{2} + \frac{VC_{01}}{2} \times D = 0 \tag{10}$$

$$VC_{01}(1 - D) = V_{in}(1 + D) \tag{11}$$

$$VC_{01} = \frac{V_{in}(1 + D)}{(1 - D)} \tag{12}$$

$$VC_{01} = \frac{V_o}{2} = \frac{V_{in}(1 + D)}{(1 - D)} \tag{13}$$

$$V_o = \frac{2V_{in}(1 + D)}{(1 - D)} \tag{14}$$

By including non-idealities of inductor, voltage gain can be calculated as follows,

$$\text{Input Power} = \text{Output Power} + \text{Losses}$$

$$V_{in} I_{L1} = V_o I_{D6} + I_{L1}^2 r_{L1} + I_{L2}^2 r_{L2} \tag{15}$$

$$I_{D6} = I_o = \frac{V_o}{R} \tag{16}$$

$$I_{D4} = \frac{I_{L2}}{2} = I_{C01OFF} \tag{17}$$

$$I_{C01ON} * D = I_{D6} \tag{18}$$

$$I_{C01OFF} = \frac{I_{C01ON} * D}{1 - D} \tag{19}$$

$$I_{C01OFF} = \frac{V_o}{R(1 - D)} \tag{20}$$

Now,

$$I_{L1} = I_{L2} = 2 * \frac{V_o}{R * (1 - D)} \tag{21}$$

$$I_{in} = 2 * I_{L1} * D + (1 - D) * I_{L1} \tag{22}$$

$$I_{in} = I_{L1} * D + I_{L1} \tag{23}$$

$$I_{in} = I_{L1} * (1 + D) \tag{24}$$

$$I_{in} = 2 * \frac{V_o}{R * (1 - D)} * (1 + D) \tag{25}$$

Now putting the values of I_{D6}, I_{L1} and I_{L2} in terms of output voltage in Eq. (15).

$$V_{\text{in}}^2 * \frac{V_o}{R*(1-D)} * (1+D) = V_o\frac{V_o}{R} + \left(2*\frac{V_o}{R(1-D)}\right)^2 r_{L1} + \left(2*\frac{V_o}{R(1-D)}\right)^2 r_{L2} \tag{26}$$

$$V_o^2\left(\frac{1}{R} + \left(\frac{2}{R(1-D)}\right)^2 r_{L1} + \left(\frac{2}{R(1-D)}\right)^2 r_{L2}\right) = V_{\text{in}}^2 * \frac{V_o}{R*(1-D)} * (1+D) \tag{27}$$

$$V_o\left(\frac{1}{R} + \left(\frac{2}{R(1-D)}\right)^2 r_{L1} + \left(\frac{2}{R(1-D)}\right)^2 r_{L2}\right) = \frac{2V_{\text{in}}}{R*(1-D)} * (1+D) \tag{28}$$

$$V_o\left(\frac{1}{R} + \frac{4}{R^2(1-D)^2} r_{L1} + \frac{4}{R^2(1-D)^2} r_{L2}\right) = \frac{2V_{\text{in}}}{R*(1-D)} * (1+D) \tag{29}$$

$$V_o\frac{R(1-D)^2 + 4r_{L1} + 4r_{L2}}{R^2(1-D)^2} = \frac{2V_{\text{in}}}{R*(1-D)} * (1+D) \tag{30}$$

$$V_o = \frac{2V_{\text{in}}R(1-D)}{R(1-D)^2 + 4r_{L1} + 4r_{L2}} * (1+D) \tag{31}$$

By assuming $r_{L1} = r_{L2} = r_L$ Eq. (31) can be written as,

$$V_o = \frac{2V_{\text{in}}R(1-D)}{R(1-D)^2 + 8r_L} * (1+D) \tag{32}$$

Equation (14) represents gain of the DC-DC converter under ideal condition, and Eq. (32) represents gain of the converter under practical condition.

3 Voltage Gain Analysis and Comparison

Voltage gain analysis of the converter is discussed in this section. Figure 2 shows the plot of voltage gain (per unit) at different duty ratio under ideal and non-ideal conditions. Under ideal condition all the diodes, switch, inductor, and capacitors are assumed to ideal and for different duty ratio output voltage is been calculated and simulated as shown in Table 1. Output voltage is also calculated and simulated for different duty ratios by considering internal resistances of the inductors which is shown in Table 1.

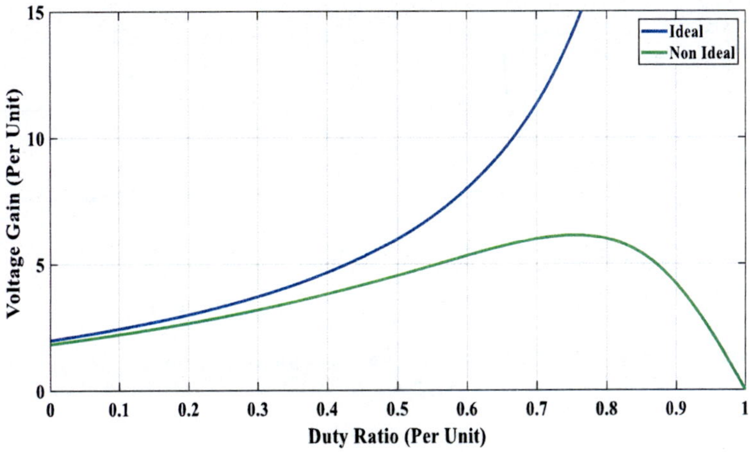

Fig. 2 Voltage gain plot with changing duty ratio

Table 1 Calculated and simulated output voltage under ideal and non-ideal condition

Duty Ratio	Calculated output voltage (ideal) (V)	Simulated output voltage (ideal) (volts)	Calculated output voltage (practical) inductor (volts)	Simulated output voltage (practical) (volts)
0.1	29.33	24.45	26.69	22.26
0.2	36	30.56	32	27.12
0.3	44.5	38.2	38.3	32.6
0.4	56	48.2	45.8	39.3
0.5	72	61.55	54.5	46.7
0.6	80.3	96	64	54.3
0.7	107.4	136	72	59.85
0.8	216	145.1	72	58
0.9	456	165.5	50.66	39.7
0.95	936	120	28.36	22
0.99	4776	28.1	5.96	9.2

4 Simulation Results

The simulation is carried out by considering input voltage (V_{in}) = 12 V, load resistance (R) = 100 Ω, inductor (L_1, L_2) resistances as $r_{L1} = r_{L2} = r_L = 1\ \Omega$ and $D = 0.6$. Figure 3 shows the inductor current i_{L1} and i_{L2} plot in both the modes. Figure 4 shows the plot of input voltage, both capacitor voltage (i.e., V_{c1} and V_{c2}) and output voltage. Figure 5 shows the plot of voltage and current under varying load. It is seen from the obtained results that voltage at the output reaches 80 V by providing 12 V input at 60% duty cycle.

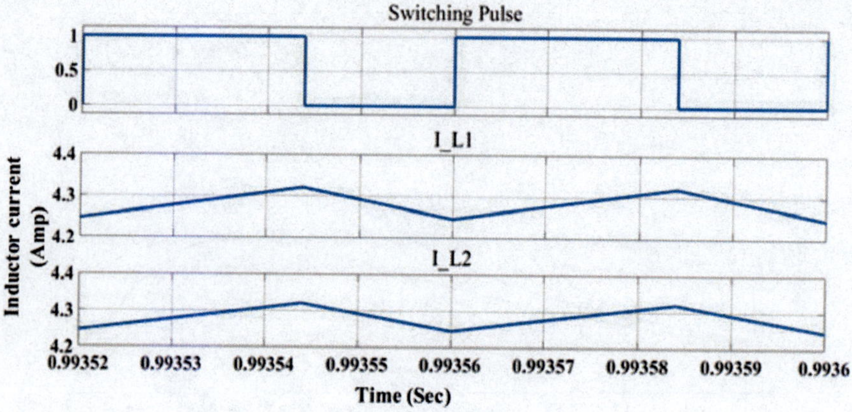

Fig. 3 Inductor current variation plot with time

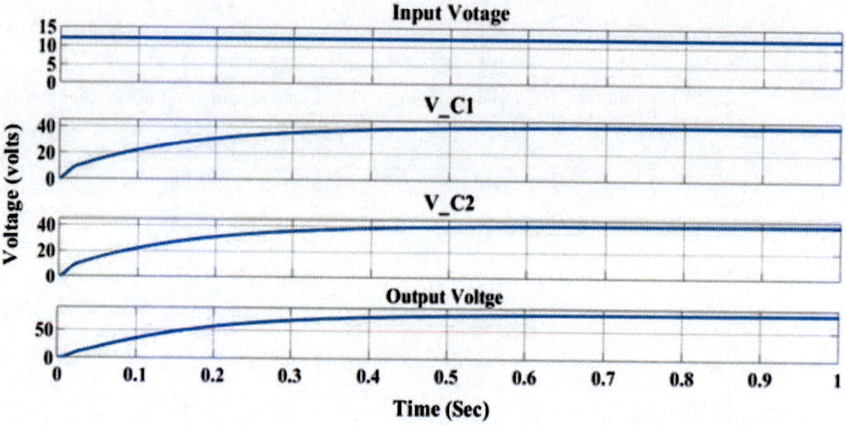

Fig. 4 Input, output, and capacitor voltage plots with time

5 Conclusion

A new circuit topology of high gain converter for providing DC link voltage to MMC is proposed. High voltage gain and less complexity in the control circuitry are the advantages of this converter. The working of the proposed converter in continuous current conduction mode is presented in this paper. Additionally, the mathematical voltage gain expression under ideal condition and by including inductors resistance is derived. Overall comparison of calculated and simulated output voltage under ideal and non-ideal condition is carried out at different duty ratio. As per the obtained results for 60% duty ratio voltage gain is more than 6 which justifies that it can be used for providing higher output voltage to the system.

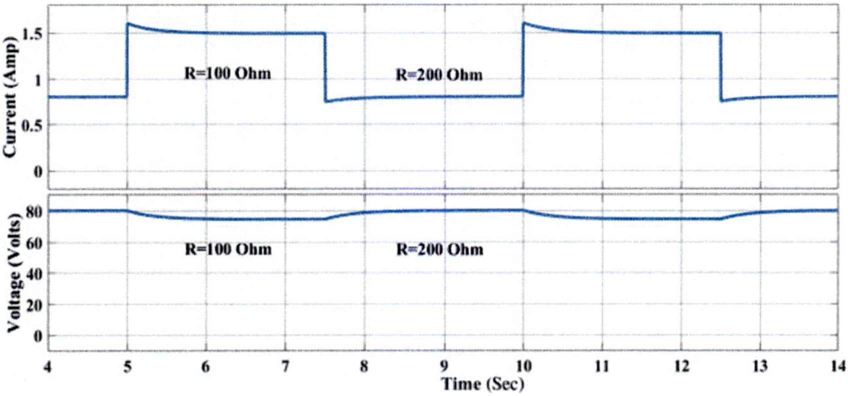

Fig. 5 Voltage and current plot with varying load

Acknowledgements The authors acknowledge the support provided by CSIR under JRF category. CSIR sanction number is 09-112(0569)-2017-EMR-1.

References

1. Maroti PK, Padmanaban S, Holm-Nielsen JB, Sagar Bhaskar M, Meraj M, Iqbal A (2019) A new structure of high voltage gain SEPIC converter for renewable energy applications. IEEE Access 7:89857–89868
2. Pervez I, Sarwar A, Tayyab M, Sarfraz M (2019) Gravitational search algorithm (GSA) based maximum power point tracking in a solar PV based generation system. In: 2019 i-PACT, Vellore, India, pp 1–6
3. Das M, Agarwal V, Novel high-performance stand-alone solar PV system with high-gain high-efficiency DC-DC converter power stages. IEEE Trans Ind Appl
4. Tayyab M, Sarwar A (2019) Submodule capacitor voltage balancing of modular multilevel converter. In: 2019 International conference on electrical, electronics and computer engineering (UPCON), Aligarh, India, pp 1–5
5. Srivastava A, Prabhakar M (2014) High gain hybrid DC-DC converter. In: Proceedings of 2014 IEEE 2nd International Conference on Electrical Energy Systems ICEES 2014, pp 218–222
6. Babalou M, Dezhbord M, Alishah RS, Hossein Hosseini S (2019) A soft-switched ultra high gain DC-DC converter with reduced stress voltage on semiconductors. In: 2019 International power electronics, drive systems and technologies conference PEDSTC 2019, pp 677–682
7. Singh G, Srivastava S, Lalwani R, Tewari N, Prabhakar M (2019) High gain DC-DC converter based on hybrid switched-inductor topology for PV application. In: IEEE international conference on system, computation, automation and networking (ICSCAN), pp 1–6
8. Meinagh FAA, Yuan J, Yang Y (2019) New high voltage gain DC-DC converter based on modified quasi Z-source network. In: Proceedings of 2019 IEEE international conference on compatibility, power electronics and power engineering CPE-POWERENG 2019, pp 1–6

9. Iqbal A et al (eds) (2020) Soft computing in condition monitoring and diagnostics of electrical and mechanical systems, vol 1096. In: Advances in intelligent systems and computing. Springer, Singapore. https://doi.org/10.1007/978-981-15-1532-3
10. Iqbal A et al (eds) (2020) Meta heuristic and evolutionary computation: algorithms and applications, vol 1096. In: Studies in computational intelligence. Springer, Singapore. https://www.springer.com/gp/book/9789811575709

State Flow Realization and Performance Evaluation of Selected MPPT Techniques

Sheikh Tawfique Elahi, Zaheeruddin Ansari, Syed Mohd Saad, and M. Saad Bin Arif

Abstract The efficiency of the PV system is very much dependent on the efficiency of the panels and its operation at the maximum power point, apart from environmental effects. Various high-efficiency panels are now commercially available as well as more and more efficient materials are still developing for it. Another critical issue, to improve the efficiency of the system, is to operate the panel at the maximum power point, which can be achieved using various available maximum power point tracking (MPPT) techniques. Implementation of MPPT in the PV system is a challenging task that can be made simple using state flow block (SFB) in MATLAB/Simulink. State flow makes the interface more accessible, and in many cases, it can also improve the performance of the MPPT tracker. This work presents the state flow model implementation of fractional open circuit voltage (FOCV) and fractional short circuit current (FSCC) MPPT techniques. Conventional method, Perturb and Observe (P&O), has also been implemented in SFB. Moreover, MPPT techniques implemented using SFB are compared with the conventional method in Simulink. It has been found that the SFB is easy to understand, reduces the complexities of algorithms and makes the user interface intimately. Also, less fluctuations and settling time have been observed in SFB than the conventional block program, in P&O, almost 14.2% fluctuations reduced in SFB and for FOCV fluctuation reduced from 13.22% to 11.05%. Results obtained show SFB implementation that outperforms the conventional method on many parameters.

Keywords Solar PV · MPPT · State flow · P&O · FOCV · FSCC

S. T. Elahi · Z. Ansari · S. M. Saad · M. S. B. Arif (✉)
Department of Electrical Engineering, Z. H. College of Engineering and Technology, Aligarh Muslim University, Aligarh 202002, India
e-mail: saadbinarif@gmail.com

A. Iqbal et al. (eds.), *Renewable Power for Sustainable Growth*, Lecture Notes in Electrical Engineering 723, https://doi.org/10.1007/978-981-33-4080-0_59

1 Introduction

Energy is one of the basic and essential requirements of human life; at the present, total annual energy consumption of the world is estimated as 500 Hexajoules. India is the sixth-largest consumer of energy in the world [1–3]. Solar power is the most abundant source of energy. It is readily available as solar insulation and possibly as wind energy. In the form of electromagnetic radiation, Sun sends out energy. The energy which is available as solar energy is 178 billion MW, about 20,000 times the requirement of the market worldwide. Generally, solar panels can convert 15% of solar energy into usable energy. Solar PV's output is of poor quality with reduce efficiency, so the MPPT technique is required to extract the maximum power from the PV module and always operate at MPP. MPPT monitors the operating point of solar PV at which high power output is obtained [1]. The MPPT regulates stresses or current without reliance on power. A variety of algorithms are designed to track this operating condition as mentioned several examples in [2, 3].

In MATLAB, there are different blocks which can be used to implement the algorithms and analyse their performances. But these blocks need very complex calculations and lots of time in the simulation. SFB is blocked also known as a chart in the MATLAB. There have some researches done by different authors doing analysis using SFB. Like authors in [4] focused on power balance with load shedding and output limited by the PV system. They consider system quality and grid insecurity by smart grid communications in that article. While in [5], SFB is designed to accurately model, understand and evaluate electronic systems. A new state flow-based maximum power point monitoring technique is proposed in [5]. Authors in [6] implemented only the P&O technique of MPPT using SFB and gave a comparative study between MATLAB and experimental results. Also, in [7], the authors implemented a developed high-performance method of MPPT using SFB. But here they have applied it for the wind turbine. A variable phase MPPT system, based on an appropriate solar cell model, is proposed in [8] for faster and more precise tracking operation. The latest MPPT method allows a convergence speed of second order, compared to traditional MPPT approaches, which always function as a convergence of one order. A comparative study between P&O and InC implemented in state flow is presented in [9].

In this paper, we have presented different MPPT techniques for maximum solar power extraction from a PV panel using SFB. P&O, FOCV [10] and FSCC [11] are the three techniques that have been implemented in Simulink using SFB. The results obtained were satisfactory, and they were also compared with results available in the literature implementing conventional block of Simulink. The paper is divided into 5 sections. This introduction is in Sect. 1. Mathematical model of the PV module and the buck–boost converter that we used in all the Simulink models is discussed in Sect. 2. Then in Sect. 3, the discussion of SFB and our implemented techniques and their algorithms are presented. And in Sect. 4 and Sect. 5, results and comparative studies have been discussed and then conclusion of the work.

2 System Modelling

The solar PV system consists of different blocks such as PV module, MPPT, power converters and load, as shown in Fig. 1a. In case when the system is connected to the AC load as well as with DC load, then the inverter is also used. That is the output of the panel is given to DC converter integrated with MPPT. And the output of the MPPT is connected to the inverter for getting AC power as well as charge controller for charging battery storage system connected with the system. Variety of high-efficiency PV modules are nowadays available commercially and can be used with same specification in this system as well. As far as DC power converters are concern, efficient buck–boost converter is generally used as it is robust and its control is less complex. In this work, we have selected DC-DC buck–boost converter, different selected MPPT is used, and its output is converted into pulses using PWM and given to converter's duty control which performs the load matching.

2.1 PV Module

A PV cell can be modelled as shown in Fig. 1b in which Ip current is working as a source Id current is flowing through diode parallel to it, and I_{sh} current is flowing through shunt resistance. I_{out} is the output current which is coming at the terminal through equivalent series resistance, and terminal voltage is coming out after the voltage drop of series resistance from shunt resistance voltage. So, a PV cell can be described as an equivalent circuit diagram from there its characteristics can be drawn, and other related mathematical calculations can be performed.

$$I_{\text{out}} = I_p - I_s\left(e^{\frac{q(V_{pv}+I_{\text{out}}R_s)}{nK_t}} - 1\right) - \frac{V_{pv} + IR_s}{R_{\text{sh}}} \tag{1}$$

where

I_p = Current across source, I_{sh} = Current across the R_{sh}, I_d = Current across diode, I_s = Saturated current, I_{out} = Current across the output, t = Temperature in

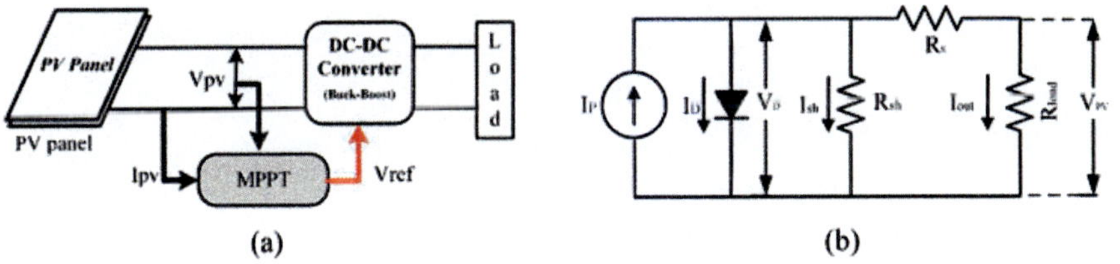

Fig. 1 **a** Schematic sketch of MPPT implemented in standalone PV system and **b** PV cell equivalent circuit

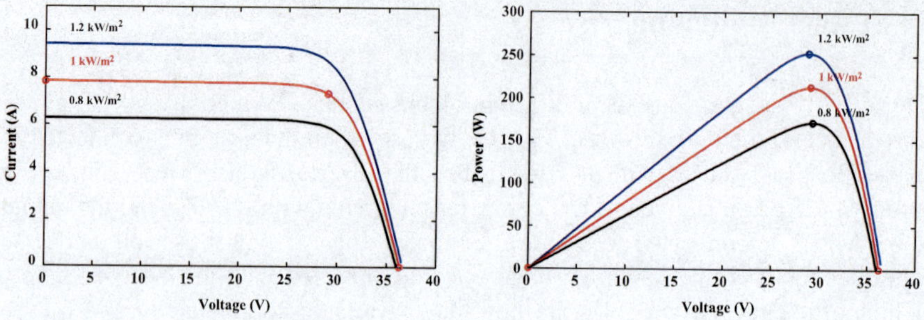

Fig. 2 Characteristics of PV module

kelvin, $q =$ Single electron charge, $V_{pv} =$ Voltage across output and $k =$ Boltzmann constant.

Variation in the output module characteristics with same atmospheric condition and at different solar irradiance are presented in Fig. 2. It shows different power output at different insolation greater the effective photons in solar rays greater will be power generation.

2.2 Buck–Boost Converter

It is a power electronic converter capable of producing variable dc output voltage. Inductor is used as an energy storing element in this converter which is utilised when required using controlled switching of the switch. In MPPT, a DC to DC converter is used to adjust the load resistance to the PhotoVoltaic cell's internal resistance until it transforms maximum power from the source to load. The voltage amount is regulated by duty ratio if D is greater than 0.5, then the voltage output (V_{out}) is greater than the voltage input (V_{in}), D is less than 0.5, the V_{out} is less than the V_{in} and $D = 0.5$ is the same.

Here,

$$f = \frac{1}{T};\, D = \frac{T_{ON}}{T} \; and \; \frac{V_o}{V_{in}} = \frac{-D}{1-D} \tag{2}$$

3 State Flow Implementation

Three MPPT techniques, namely P&O, FOCV and FSCC, have been implemented using state flow in MATLAB, and their performance was evaluated. We have taken

Table 1 Values of the variables used for selected module

S. No.	Parameters	Value
1	P_{max} (W)	213.15
2	V_{mp} (V)	29
3	I_{mp} (A)	7.35
4	V_{oc} (V)	36.3
5	I_{sc} (A)	7.84
6	R_{sh} (Ω)	313.0553
7	R_s (Ω)	0.39381
8	Total cell in a module	60
9	Temperature co-efficient of V_{oc}	−0.361%/°C
10	Temperature co-efficient of I_{sc}	0.102%/°C
11	Diode ideality factor	0.98119

one solar PV module considering following specifications for state flow simulation of techniques mentioned above (Table 1).

3.1 State Flow—Implementation and Analysis

State flow is a handy tool in MATLAB which gives the insight of different states of block. It becomes straightforward for the designer to understand how different blocks are used because it provides the pictorial view of inside machine activity. Sate flow gives the flexibility of doing pictorial programming from Fig. 3a–e of state flow different states have been shown when it runs in Simulink all the links glow according to their condition so it will be easier for programmers to identify the error. In the conventional block system, the program executes every block, and it takes a lot of time, while in the SFB system, the whole calculation is done within a single block. Thus, a lot of simulation time can be saved.

In Fig. 3a, first initialization block glow which shows the calculation of the algorithm starts with the given initial conditions, Fig. 3b calculates the equation first time for given initial values, Fig. 3c after checking status it runs the duty block and decreases duty, Fig. 3d returns block glow where PV voltage is taken according to changed duty. Afterwards, the close loop will be made, and the new loop form calculation block will start executing. Figure 3e shows the increment of duty if $V_{best}< 0$, rest of the cycle will be done as for $V_{best}> 0$. At every step, the condition is checked, and in blocks, corresponding calculation occurs. In our algorithm, we have made voltage and current from solar PV as input to state flow and duty ratio as output. This duty ratio is fed to DC to DC buck–boost converter through PWM. We can use SFB to generate target code for a machine [5].

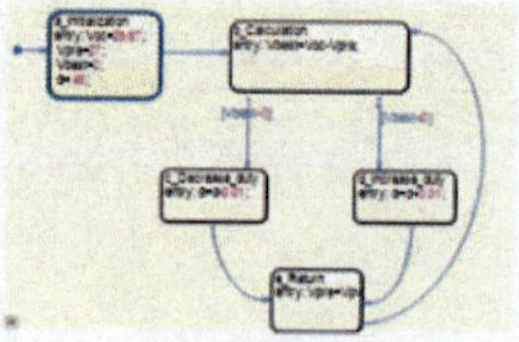

(a) Initialization of state flow

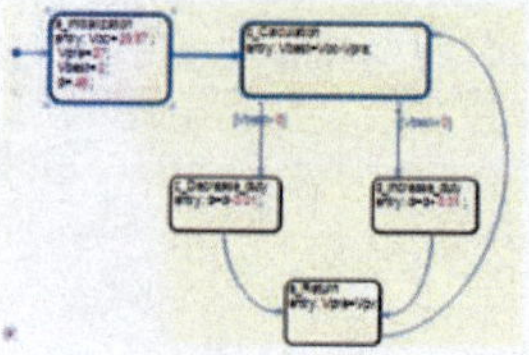

(b) State from Initialization to Calculation

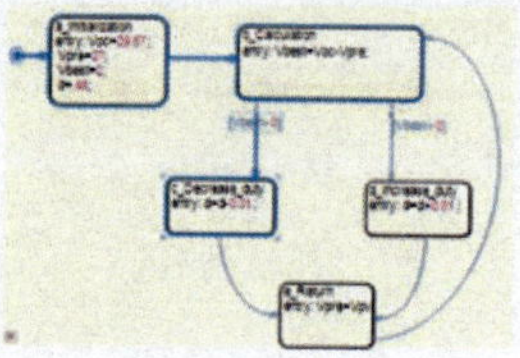

(c) State to decrease the duty cycle (assumed V_{best}>0)

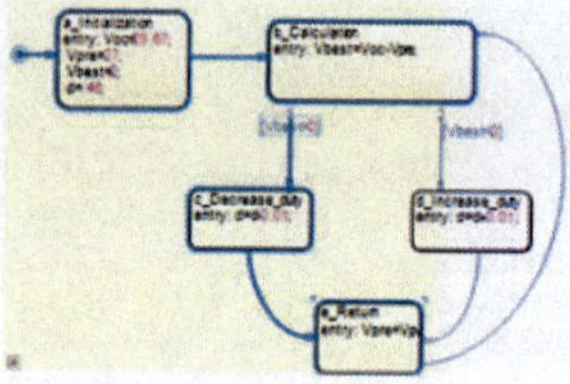

(d) Returning state of the flow

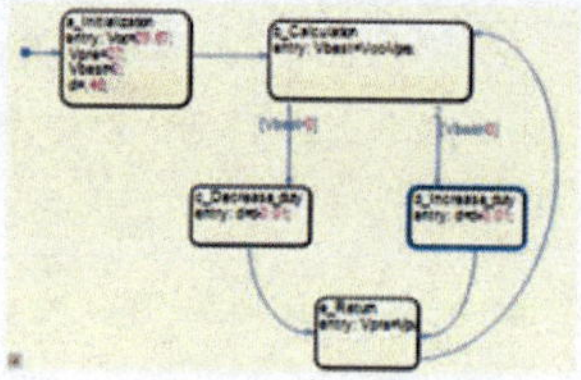

(e) Previous steps will be followed as same in increasing duty

Fig. 3 Different states of the SFB while executing the program

3.2 *Perturb and Observe*

It is among the methods commonly employed and hybridized. The scientific meaning of the word perturb is to change the natural or usual condition of some unit. This approach is about studying specific changes and seeking MPP. The voltage and current are determined in this method, and the resulting power (P_1) is estimated (Fig. 4) and assuming a small variation of the DC to DC converter voltage (ΔV) or duty cycle (Δd).

3.3 *Fractional Open Circuit Voltage*

With changing environmental conditions (irradiance change), there is a single point at which solar PV gives maximum power. Somehow anyone of current at MPP (I_{mpp}) or voltage at MPP (V_{mpp}) is tracked and with the help of that we can draw maximum power from the module. There is a linear relationship between VOC and V_{mpp}. The relationships are as follows, FOCV, $V_{\text{mpp}} = \text{KocVoc}$; Koc varies between 0.78 and 0.92 [10], so with the help of this relationship, we can easily trace the maximum powerpoint (Fig. 5).

3.4 *Fractional Short Circuit Current*

The PV module characteristic of a given constant irradiance state PV is set, and there is a linear relationship between I_{mpp} (present at MPP) and I_{sc}. The slope can range from 0.85 to 0.92, where slope varies between 0.85 and 0.92 [11]. Thus, the following expression will be obtained, FSCC, $I_{mpp} = K_{sc} I_{sc}$; K_{sc} varies between 0.85 and 0.92 (Fig. 6).

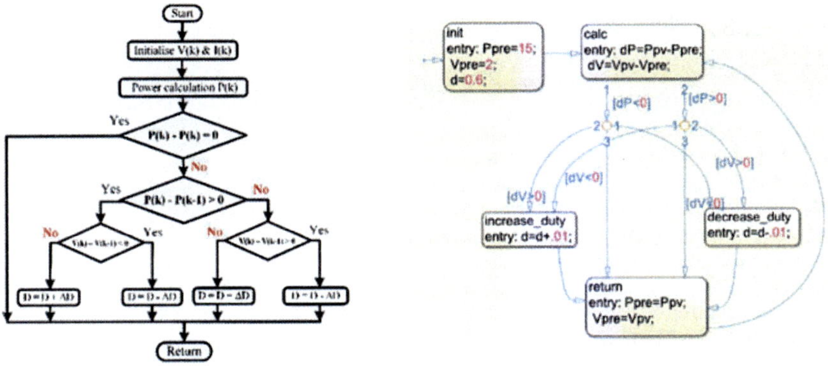

(a) Flowchart of Conventional P&O (b) P&O Implementation in State Flow

Fig. 4 P&O algorithm implemented using **a** Conventional and **b** state flow method

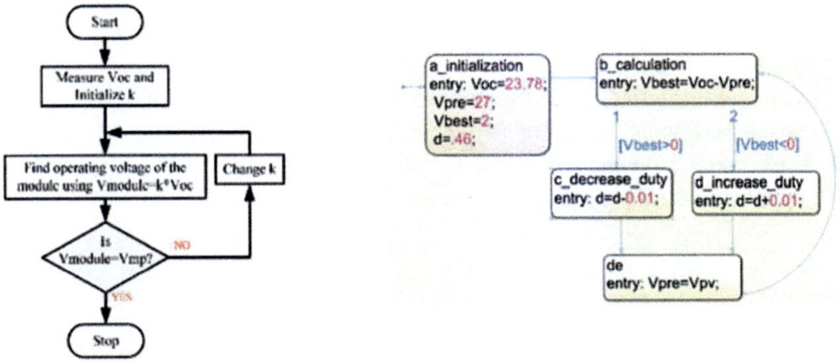

(a) Flowchart of conventional FOCV (b) FOCV Implementation in State Flow

Fig. 5 FOCV algorithm implemented using **a** Conventional and **b** state flow method

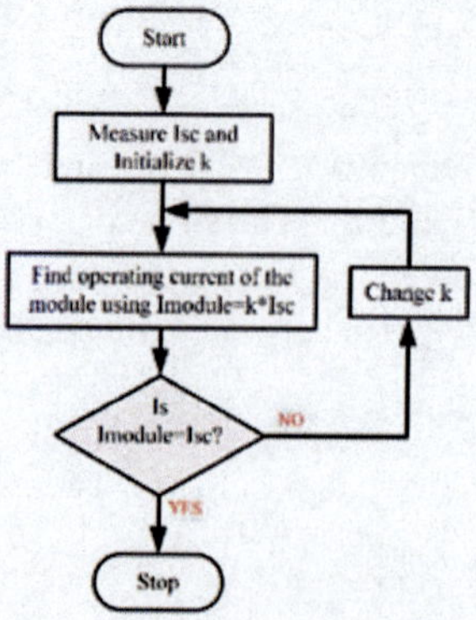

(a) Flowchart of conventional FSCC

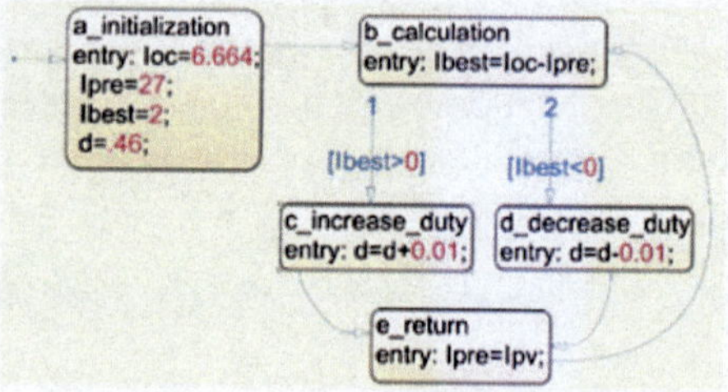

(b) FSCC Implementation in State Flow

Fig. 6 FSCC algorithm implemented using **a** Conventional and **b** state flow method

4 Results and Comparative Analysis

Results of implemented techniques are obtained from SFB-based models, and inter-comparisons are also made. Results of the above methods executed through the conventional process are obtained under the same conditions and comparisons are made from with the results obtained using state flow (Figs. 7, 8 and 9).

For comparing the result of obtained from different techniques, a solar panel of 213 w at 1000 w/m^2 irradiance and 25 °C temperature is taken. The P&O tracking and settling time implementation with SFB are 0.039 and 0.045 s, respectively, for FOCV tracking and settling time 0.024 s and 0.025 s, respectively, and for FSCC tracking and settling time 0.023 s and 0.024 s, respectively, it means FSCC has fastest-tracking and has fast settling time. Further results have been showcased in Table 2.

In terms of efficiency, P&O has higher efficiency, FSCC has less efficiency and FOCV has more efficiency than FSCC but less than P&O. FSCC and FOCV have less efficiency because it is based on approximation.

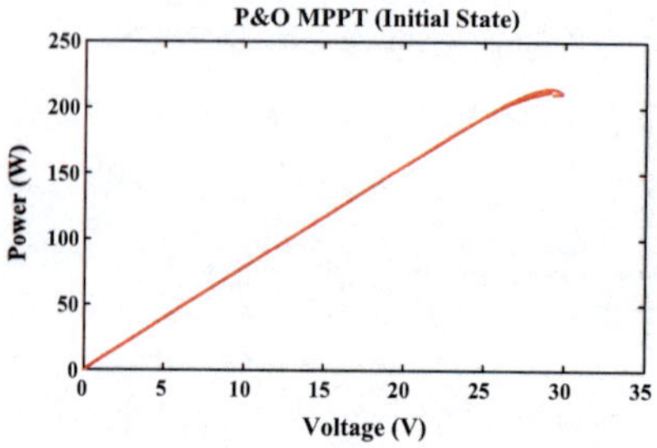

(a)

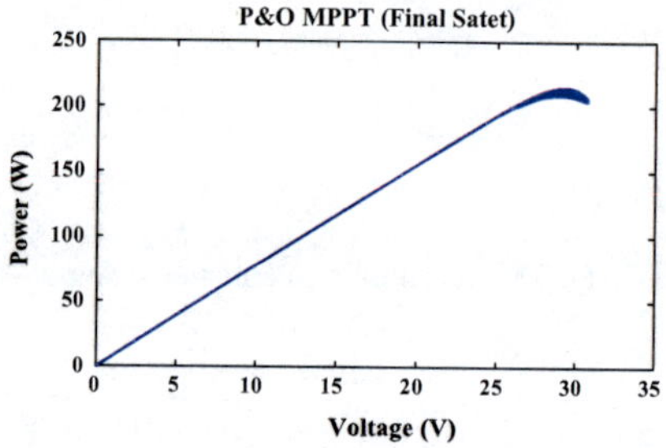

(b)

Fig. 7 P-V curves implementing P&O method. **a** Initial state and **b** final state

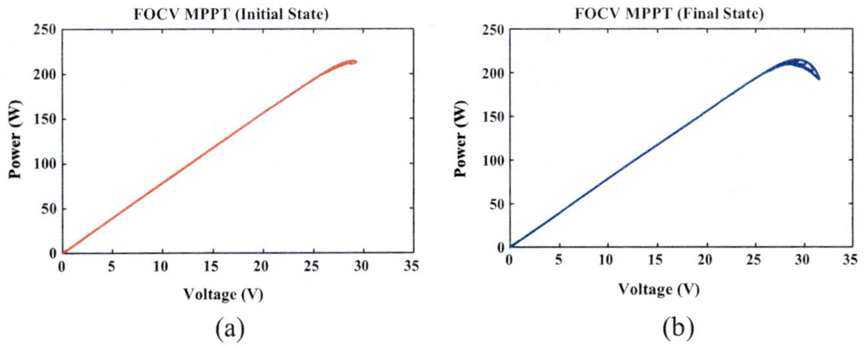

Fig. 8 P-V curves implementing FOCV method. **a** Initial state and **b** final state

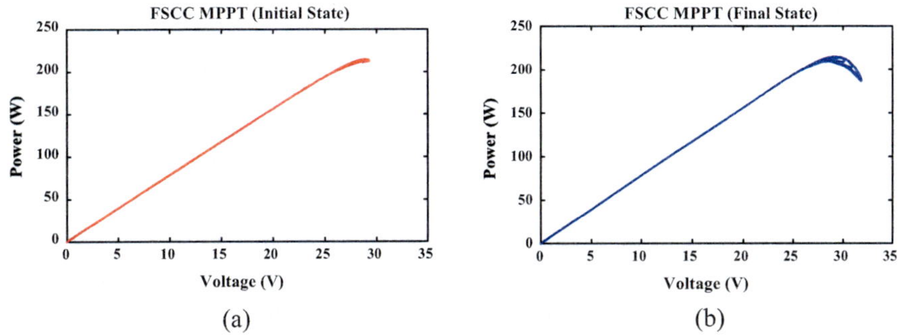

Fig. 9 P-V curves implementing FSCC method. **a** Initial state and **b** final state

Table 2 Results obtained implementing different techniques

Technique	Tracking Time	Settling time (s)	Input parameters	Controlling parameter	Output voltage (V)	Output current (A)	Maximum power (W)	Efficiency (%)
P&O	0.039	0.045	V_{PV}, P_{PV}	Duty	28.6	7.391	211.4	99.24
FOCV	0.024	0.025	V_{OC}	Voltage	29.88	6.963	208.1	97.69
FSCC	0.023	0.024	I_{SC}	Current	30.58	6.594	201.7	94.70

4.1 Comparative Analysis Between SFB and Conventional Block

Comparison between P&O, FOCV and FSCC techniques implemented using SFB and conventional block method has been carried out. Results are shown in Figs. 10 and 11 as well as presented in detail in Table 3. Conventional P&O has fluctuation of 15.7% with settling time of 0.06 s and SF realization of P&O improves fluctuation to 1.5% and settling time reduced to 0.045 s. Improvement in the performance is also

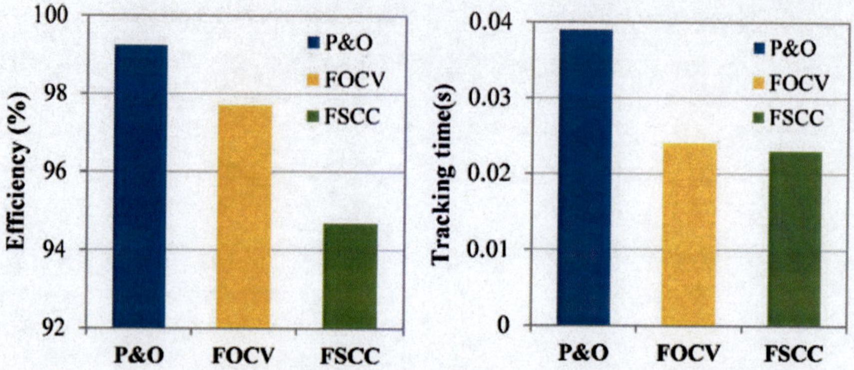

Fig. 10 Efficiency and tracking time comparison of the implemented techniques in SFB

Fig. 11 Comparison based on settling time

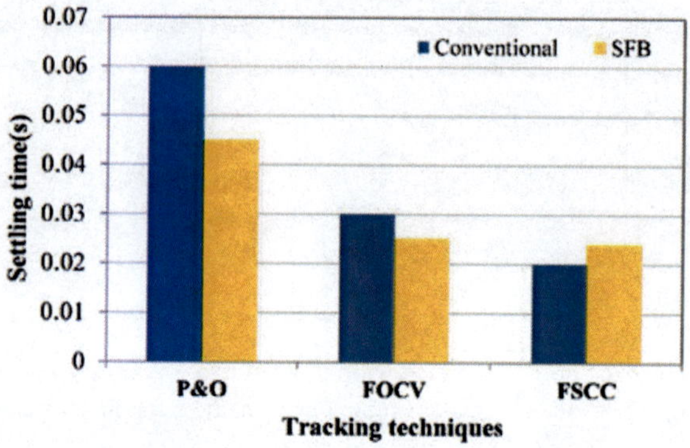

Table 3 Comparison of results obtained from SFB and conventional

		Settling time (s)	Simulation response	Fluctuations	Implementation feasibility
P&O	Conventional	0.06	Slow	15.7%	Complex
	SFB	0.045	Fast	1.5%	Simple
FOCV	Conventional	0.030	Slow	13.22%	Moderate
	SFB	0.025	Fast	11.05%	Simple
FSCC	Conventional	0.02	Fast	–	Moderate
	SFB	0.024	Slow	18.5%	Simple

observed in FOCV technique. It is observed that the simulation time is faster in case of SFB due to its simple single block programming. Also, reduced fluctuations have been observed in SFB than the conventional block program, in P&O, almost 14.2% fluctuations reduced in SFB and FOCV fluctuation reduced from 13.22 to 11.05% (fluctuations as calculated by taking difference of upper and lower peak divided by average value in steady state). In the case of FOCV, settling time is lesser in SFB than in conventional block model.

5 Conclusion

In this work, selected MPPT techniques, i.e. P&O, FOCV and FSCC, have been implemented using state flow model. And their performance is evaluated based on different parameters such as efficiency, tracking, settling time and max power achieved. It has been found that, among all the techniques implemented in state flow, FSCC is the fastest technique with tracking time of 0.023 s. Also, the input parameter in case of FSCC and FOCV is one, whereas P&O requires two parameters to be sensed for tracking. Results obtained are also compared with that obtained using same techniques implemented using the conventional methods of implementation in Simulink. It has been observed that using state flow model of implementation has many advantages over the conventional methods for selected techniques. Such as, settling time in case of P&O is improved, implementation complexity is very much reduced by using SFB-based models. It also minimized steady-state power fluctuations, improved transient efficiency and faster start-up operation.

References

1. Tajuddin M, Arif M, Ayob S, Salam Z (2015) Erratum to the 'Perturbative methods for maximum power point tracking (MPPT) of photovoltaic (PV) systems: a review. Int J Energy Res 39:1153–1178; Int J Energy Res 39(12):1720–1720
2. Iqbal A et al (eds) (2020) Soft computing in condition monitoring and diagnostics of electrical and mechanical systems. In: Advances in intelligent systems and computing, vol 1096. Springer, Singapore. https://doi.org/10.1007/978-981-15-1532-3
3. Iqbal A et al (eds) (2020) Meta heuristic and evolutionary computation: algorithms and applications. In: Studies in computational intelligence, vol 1096. Springer, Singapore. https://www.springer.com/gp/book/9789811575709
4. Husain MA, Tariq A, Hameed S, Arif MSB, Jain A (2017) Comparative assessment of maximum power point tracking procedures for photovoltaic systems. Green Energy Environ 2(1):5–17
5. Ali US, Veeraraghavulu D, Niveditha M, Priyadharshini N, Sandhiya P (2016) Stateflow based incremental conductance MPPT of a photovoltaic system using Z-source DC-DC converter. In: 2016 Biennial international conference on power and energy systems: towards sustainable energy (PESTSE). IEEE, pp 1–6
6. Shourov EC, Sajid A, Sabzehgar R, Roshan YM (2018) A novel implementation of the state flow approach for maximum power point tracking of photovoltaic systems. In: 2018 9th IEEE international symposium on power electronics for distributed generation systems (PEDG). IEEE, pp 1–6
7. Maher RA, Abdelsalam AK, Dessouky YG, Nouman A, High performance state-flow based MPPT technique for micro WECS
8. Piegari L, Rizzo R (2010) Adaptive perturb and observe algorithm for photovoltaic maximum power point tracking. IET Renew Power Gener 4(4):317–328

9. Subudhi B, Pradhan R (2012) A comparative study on maximum power point tracking techniques for photovoltaic power systems. IEEE Trans Sustain Energy 4(1):89–98
10. El Mentaly L, Amghar A, Sahsah H (2017) Comparison between HC, FOCV and TG MPPT algorithms for PV solar systems using buck converter. In: 2017 International conference on wireless technologies, embedded and intelligent systems (WITS). IEEE, pp 1–5
11. Kollimalla SK, Mishra MK (2013) A new adaptive P&O MPPT algorithm based on FSCC method for photovoltaic system. In: 2013 International conference on circuits, power and computing technologies (ICCPCT). IEEE, pp 406–411

Comparative Analysis of TPS and EPS of IBDC for Power Management

Mohammad Aslam Alam, Kulsoom Fatima, and Ahmad Faiz Minai

Abstract In this paper, extended phase shift and traditional phase shift performance is analyzed for controlling isolated bidirectional DC-DC converter; along with this, study of current stress and backflow power is done. Continuous utilization of IBDC is increasing day by day for different applications involving interfacing standalone renewable energy sources to UPS, hybrid electrical vehicles (EHVs) and to other utility grids. When the extended phase shift control topology implemented on IBDC, it enhances overall performance of the system: It amplifies the power transmission range, flexibility of the system as well as it reduces the current stress and enhances net efficiency of the system. This topology is helpful in foreseeing converter performance inclusive of losses that take place in active as well as passive component. Operation of extended phase shift control topology and its operating modes is analyzed using MATLAB/Simulink-14 in this paper.

Keywords IBDC · TPS · EPS · Power flow · Current stress · Phase shift control

1 Introduction

Nowadays, microgrids are new reality for the interconnected loads, and various kinds of distributed energy resources and storage system are joined together and further integrated into the grid. In current scenario, flexible electricity system is required to minimize environmental impact which led to establishment of the microgrids [1–3].

Microgrids are driven in two modes (a) grid-tied mode and (b) standalone mode. Microgrid is connected to central grid in grid-tied mode, where it operates in

M. A. Alam (✉) · K. Fatima · A. F. Minai
Department of Electrical Engineering, Integral University, Lucknow, India
e-mail: aslamintegral@gmail.com

K. Fatima
e-mail: kulsoomfatima666@gmail.com

A. F. Minai
e-mail: fzminai@gmail.com

A. Iqbal et al. (eds.), *Renewable Power for Sustainable Growth*, Lecture Notes in Electrical Engineering 723, https://doi.org/10.1007/978-981-33-4080-0_60

parallel with other distributed energy sources; power sharing in this interconnection is achieved at common exchange point of microgrid. The major advantages of microgrids is that they can be disconnected from the central grid and can be used as standalone system; various distributed energy sources work as standalone grid for generation of electricity in standalone system. For both the systems, i.e., grid-tied mode and standalone mode, demand and supply must be balanced which is transmitted from central grid and distributed energy sources [4, 5].

Due to various aspects such as social, political, economical and environmental interest, many types of hybrid system are used. Due to uncertainty in power generation, storage devices are used for the uninterrupted power supply of the local loads [6, 7].

For power balancing in distributed system and storage device in the microgrid, the BDC are employed. These converters are interface between distributed energy resource like solar PV array, wind, fuel cell, etc., and storage system like super capacitor or battery bank [6]. IBDC can be implemented in other forms by interfacing high voltage bus and low voltage bus. Basically, BDC is categorized as (a) non-isolated BDC and (b) isolated BDC. In contrast with traditional BDC, this converter is more advantageous: easy soft switching control, increased reliability, bidirectional energy flow and electrical isolation. Galvanic isolation of these converters is required as long as system flexibility, safety standard is taken into consideration (Fig. 1).

For controlling IBDC, there are mainly two methods: (a) traditional PWM control, (b) phase shift control. In traditional PWM control, H-bridge IBDC is depicted in Fig. 2. Cross-connected switches of H-bridge on primary side are operated in such a manner (S_1, S_4) and (S_2, S_3) are turned ON to change high voltage V_1 in inversion mode, whereas on secondary side, (Q_1–Q_4) is turned OFF and current conducts via free-wheeling diode (M_1–M_4) for rectification mode. Due to this operation, power is conveyed to secondary side (V_2) from primary side (V_1); vice versa for reverse power flow, operation of (S_1–S_4) and (Q_1–Q_4) is inter-changed. This traditional control

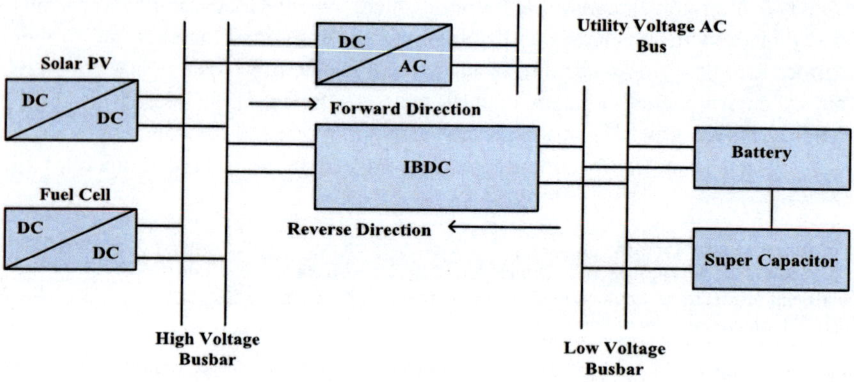

Fig. 1 Implementation of IBDC for power flow management

Fig. 2 IBDC configuration

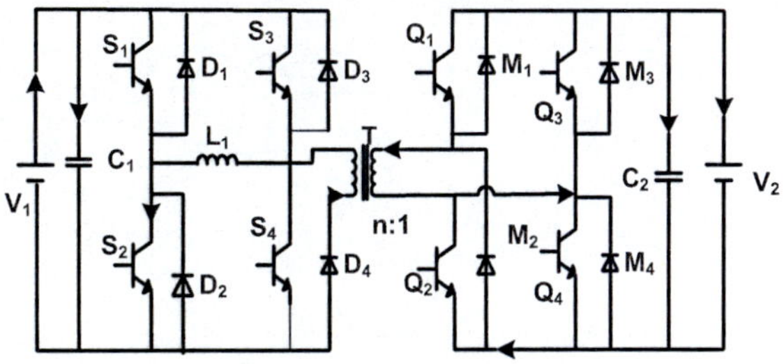

technique is easy to implement and simple, but the drawbacks with this method are poor dynamic system performance, and output voltage range is limited [7–9].

In traditional phase shift control, for generation of square waves, primary and secondary sides of transformer are cross-connected switches of H-bridge (H_1 and H_2) which are turned ON; when phase is shifted, it results in change of voltage for transformer leakage inductor (L) for controlling magnitude and direction of power flow.

Phase shift control is advantageous such as small inertia, improved performance and easy soft switching control. Major disadvantages of this method are power flow depends on transformers leakage inductor which results in power flow and current stress. When the ratio of primary voltage (V_1) and product of turns ratio of transformer (n) and secondary voltage (V_2) i.e., V_1/nV_2 changes from unity; the overall efficiency is reduced, power loss and magnetic loss in devices are increased [10].

2 Traditional Phase Shift (TPS) Control Topology

For phase shift control, T-type equivalent circuit takes place of transformer; it is pre-assumed that open circuit and leakage inductance of transformer is substantially smaller than the magnetizing inductance. Analogous circuit of IBDC for phase shift control represented by two square wave voltage sources tied over inductor (L) is depicted in Fig. 3. Inductor (L) represents aggregate for inductor L_1 (auxiliary) and transformer leakage inductance.

Output of H_1 and H_2 on V_1 and V_2 side is represented by V_{H1} and V_{H2}, respectively, where i_L is inductor current and V_L is inductor voltage. Adjusting the phase shift between V_{H1} and V_{H2}, magnitude and direction of power flow are regulated. Here, forward power flow is taken from V_1 to V_2[11]. Output waveform of IBDC by the TPS control is depicted in Fig. 4; here, D is primary: secondary voltage of isolation transformer which lies in between 0 and 1. This is half of switching period and P_{in} is transit of transmission power. For the analysis, it is assumed that $V_1 \geq nV_2$. Output waveform of IBDC by EPS control is depicted in Fig. 5, and there is a phase difference allying primary current and voltage, because V_{H1} and V_{H2} are output voltages of primary and secondary side of bridge which is associated with the inductor. Inductor current (i_L) is inversed with primary voltage (V_{H1}) for the time interval of $(T_0 - T_0')$

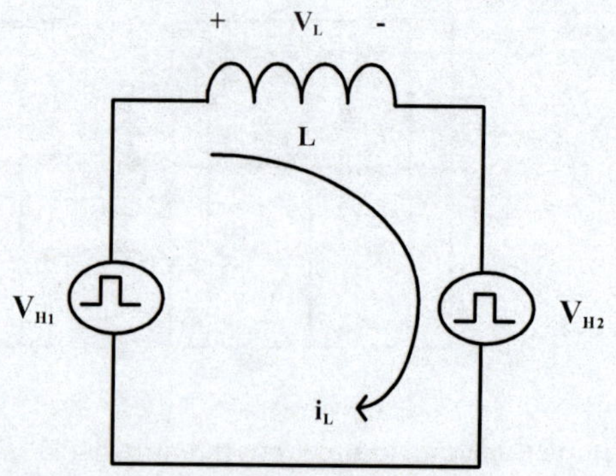

Fig. 3 Phase shift controller

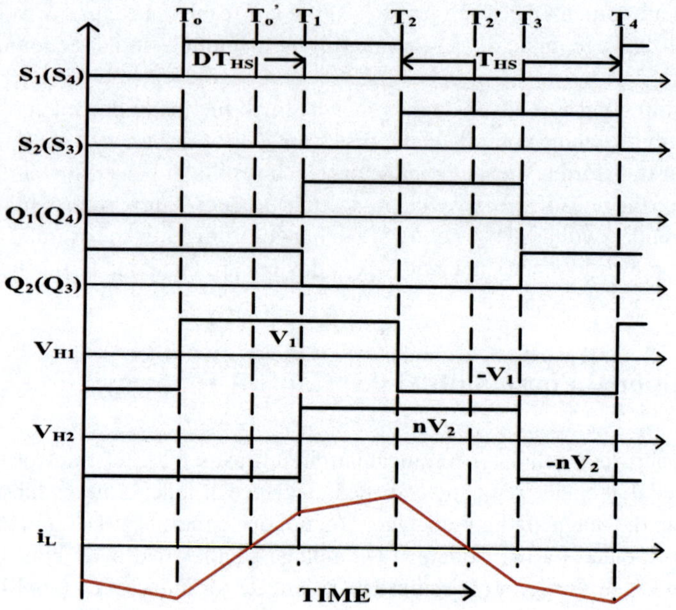

Fig. 4 Waveform under TPS control

and $(T_2 - T_2')$ as depicted in Fig. 4. In this duration power is send to secondary side (V_2), whereas leftover power is send back to primary side (V_1). Hence, power and current stress increases, which lead to disadvantages in circuit such as less magnetic components and power in devices, and hence, the overall efficiency of the system decreases [12].

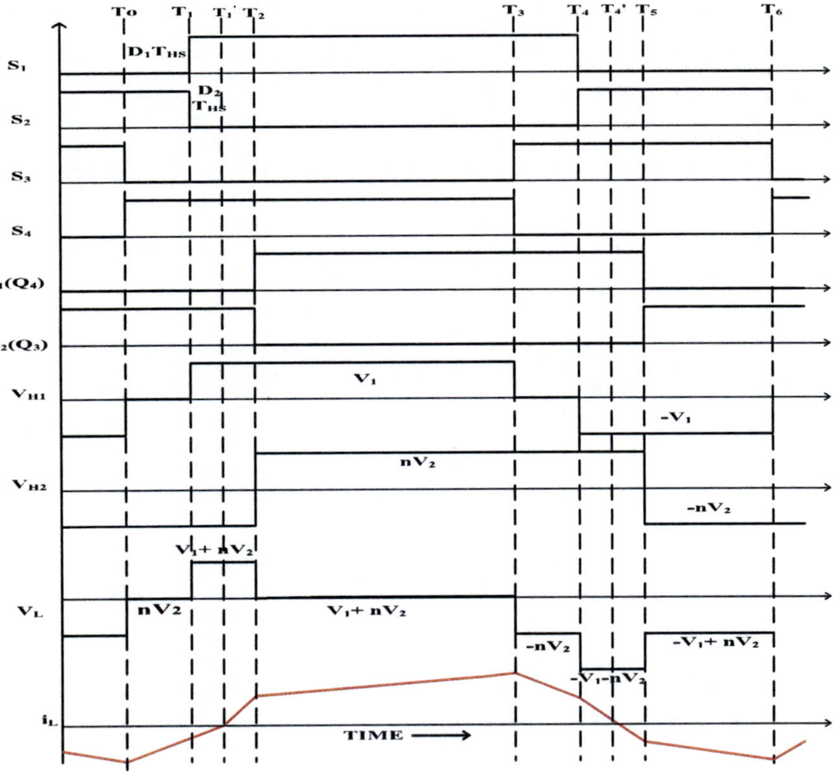

Fig. 5 Waveform under EPS control

3 Extended Phase Shift (EPS) Control Topology

In IBDC, deduction of backflow power is achieved when output for primary side of converter is not confined to square waveform with the 50% fraction of primary and secondary voltage of isolation transformer (i.e., D). Assuming S_1 and S_4 are not having similar triggering pulse rather has phase shift ratio, D [13, 14].

Figure 5 depicts that primary voltage will adopt three levels rather than traditional two level due to this inductor current (i_L) also changes. From Fig. 2, ratio of switch S_1 and S_4 in first leg or S_2 and S_3 in second leg of the H-bridge is named as D_1 where D_1 lies in between or equal to 0 and 1, whereas in D_2, the outer phase shift ratio (primary: secondary voltage for isolation transformer) lies in between or equal to 0 and 1. Total range of D_1 and D_2 is $0 \leq D_1 + D_2 \leq 1$ [15, 16]. Advantage of EPS control includes reduced current stress, increases the total power transmission range, and flexibility of the system is also enhanced.

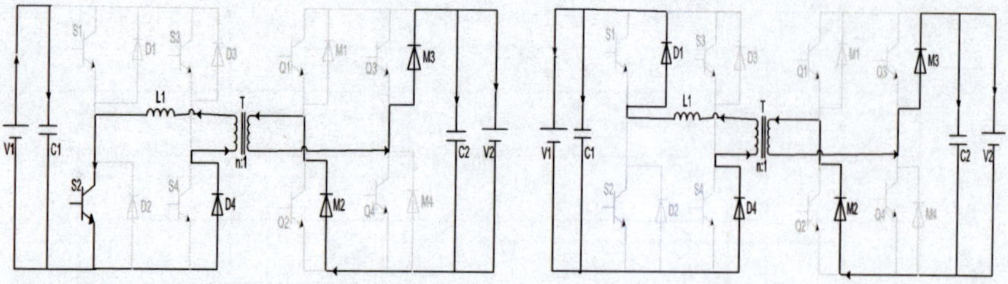

(a) (b)

Fig. 6 (a). (b)

3.1 Mode of Operation in IBDC for Extended Phase Shift

For the analysis, converter is at steady-state operation [17]. Switching cycle is analyzed for eight distinct operating modes is depicted in Fig. 6a–j.

Operating mode 1 (T_0–T_1)

Figure 6a is analogous circuit for operating mode 1 in which before T_0, the inductor (i_L) is in negative direction because S_2, S_3 are conducting; but at T_0, S_3 is turned OFF and S_4 starts conducting; but at zero current, D_4 starts conducting, but on the secondary side, M_2 and M_3 are conducting so that current is carried from the inductor (L). The secondary output voltage is V_2. Inductor current (i_L) decreases linearly and is specified by

$$i_L(t) = i_L(T_0) + \frac{nV_2}{L}(T - T_0) \tag{1}$$

Operating mode 2 (T_1–T'_1)

Figure 6b is analogous circuit for operating mode 2 in which at T_1, S_2 is turned OFF and S_1 starts conducting at zero current. The inductor current (i_L) remains in negative direction and is transferred from inductor (L) to primary voltage V_1 across the diode D_1 and D_4, but on secondary side, inductor current (i_L) is transferred to V_2 by diode M_2 and M_3. Inductor current (i_L) decreases up to zero and specified by

$$i_L(t) = i_L(T_1) + \frac{V_1 + nV_2}{L}(T - T_1) \tag{2}$$

Operating mode 3 (T'_1–T_2)

Figure 6c is analogous circuit for operating mode 3 in which at T'_1, S_1, S_4, Q_2 and Q_3 are conducting, but at T_2, Q_2 and Q_3 are turned OFF. Inductor current (i_L) diverges from negative value to positive value and is specified by Eq. (2).

Operating mode 4 (T_2–T_3)

Figure 6d is analogous circuit for operating mode 4 in which at T_2; at zero current (Q_1, Q_4) is turned OFF. whereas (Q_2, Q_3) is turned ON and diode M_1 and M_4 starts

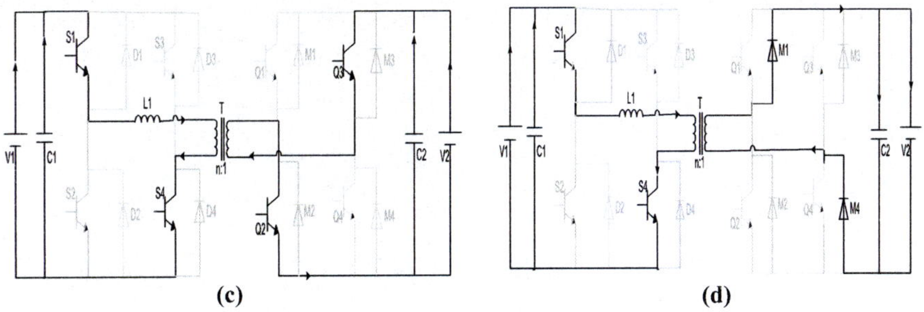

Fig. 6 (c). (d)

conducting to the secondary H-bridge. At T_3, S_4 is turned OFF. Inductor current (i_L) increases linearly and is specified by

$$i_L(t) = i_L(T_2) + \frac{V_1 - nV_2}{L}(T - T_2) \tag{3}$$

Operating mode 5 (T_3–T_4)

Figure 6e is analogous circuit for operating mode 5 in which at T_3, at zero current S_3 and D_3 start conducting, S_4 is turned OFF, whereas on secondary side current is withdrawn from the inductor (L) to secondary output voltage V_2 through the diode M_1 and M_4. At T_4, S_1 is turned OFF. Inductor current (i_L) decreases linearly is and specified by

$$i_L(t) = i_L(T_3) + \frac{-nV_2}{L}(T - T_3) \tag{4}$$

Operating mode 6 (T_4–T_4')

Figure 6f is analogous circuit for operating mode 6 in which S_1 is introduced at zero current whereas S_2 is set back and i_L is carried to primary side via diode D_1 and D_3 whereas on secondary side voltage V_2 is carried out through the diode M_1 and M_2. Inductor current (i_L) decreases up to zero and is specified by

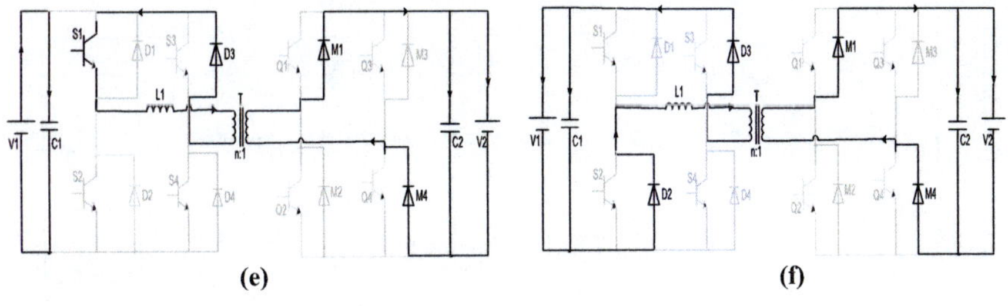

Fig. 6 (e). (f)

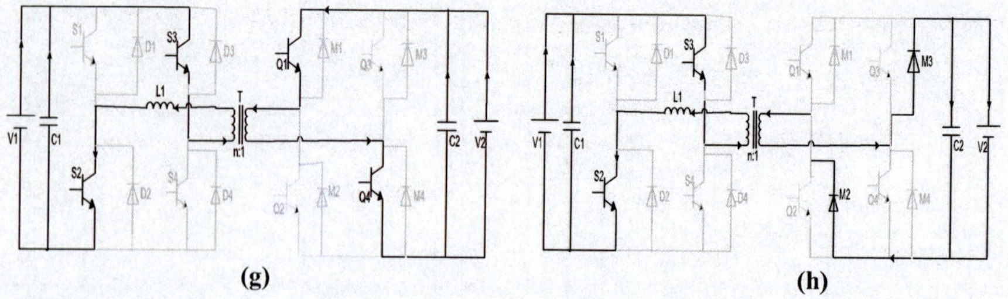

Fig. 6 (g). (h)

$$i_L(t) = i_L(T_4) + \frac{-V_1 - nV_2}{L}(T - T_4) \tag{5}$$

Operating mode 7 (T_4'–T_5)

Figure 6g is analogous circuit for operating mode 7 in which at T_4', S_2, S_3, Q_1 and Q_4 are turned ON. At T_5, Q_1 and Q_4 stop conducting. Inductor current (i_L) increases linearly and is specified by Eq. (5).

Operating mode 8 (T_5–T_6)

Figure 6h is analogous circuit for operating mode 8 in which at T_5, zero current (Q_2, Q_3) is turned ON, whereas (Q_1, Q_4) is turned OFF. Diode M_2 and M_3 starts conducting at T_6, S_3 is turned OFF. Inductor current (i_L) is specified by

$$i_L(t) = i_L(T_5) + \frac{-V_1 + nV_2}{L}(T - T_5) \tag{6}$$

From Fig. 5, transformer primary voltage V_{H1} is 0 and backflow power is also 0 from mode 1 to 5 due to which overall backflow power becomes less for power supply. For T_1 or T_4, inductor current (i_L) is decreased to zero because of this backflow power shall be neglected. As depicted in Fig. 5, for eliminating backflow power, modes 2 and 6 will be replaced by mode 2' and 6'.

Operating mode 2' (T_1–T_1')

Figure 6i is analogous circuit for operating mode 2' in which at T_1', S_4, Q_2, Q_3 and diode D_2 are conducting. At T_1, switch S_2 is turned OFF. Inductor current (i_L) gains

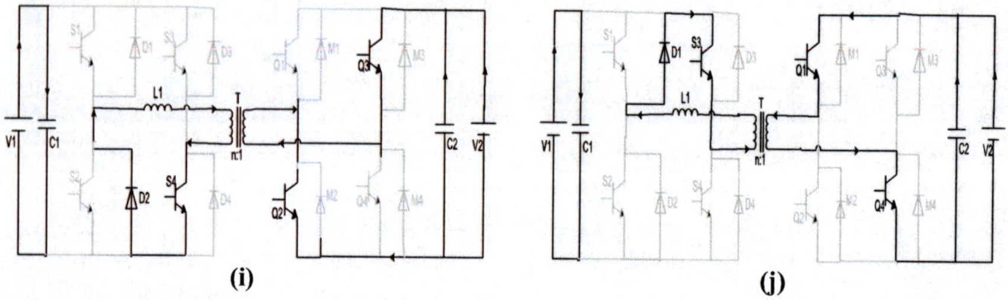

Fig. 6 (i). (j)

the positive value from the negative value and will increase linearly and specified by

$$i_L(t) = i_L(T_0) + \frac{nV_2}{L}(T - T_0) \tag{7}$$

Operating mode 6' (T_4-T'_4)

Figure 6j is analogous circuit for operating mode 6' in which at T'_4, S_3, Q_1, Q_4, and D_1 starts conducting. At T_4, S_1 is turned OFF. Inductor current (i_L) increases linearly and is specified by

$$i_L(t) = i_L\left(T'_4\right) + \frac{nV_2}{L}\left(T - T'_4\right) \tag{8}$$

4 Simulation Result

For validation of EPS control topology, system parameters taken into consideration are specified in Table 1.

Comparative analysis of EPS and TPS is done by implementing both the control topologies on MATLAB/SIMULINK-14. It is clear from output waveforms that phase shift for TPS (Fig. 7a) has limited output range, whereas for EPS (Fig. 7b) output range is extended; load current and load voltage for TPS (Figs. 8a and 9a) have harmonics, whereas for EPS (Figs. 8b and 9b) has reduced harmonics. For improvement of system, EPS control of the IBDC is implemented for power supply

Table 1 System parameters

Parameters	Numerical values
Capacitance	6 μF
Transformer ratio	1:4
Inductance of IBDC	0.0001 H
Input voltage	100 V
Switching frequency	30 kHz
Output voltage	120 V

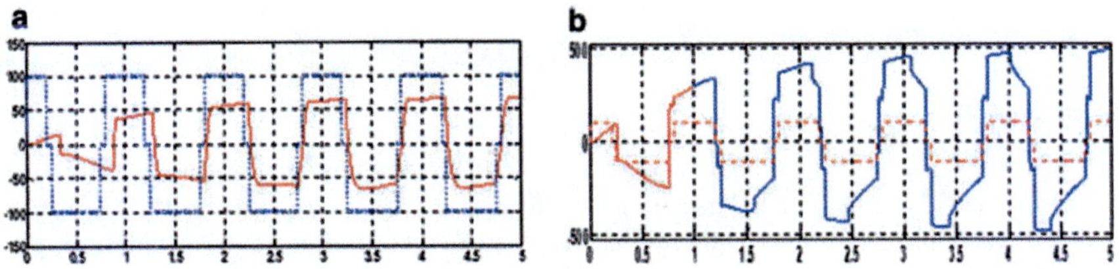

Fig. 7 **a** Phase shift for TPS control, **b** phase shift for EPS control

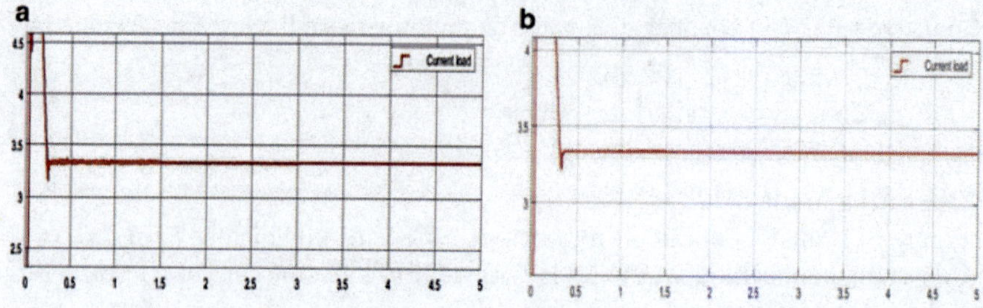

Fig. 8 **a** Load current for TPS, **b** load current for EPS

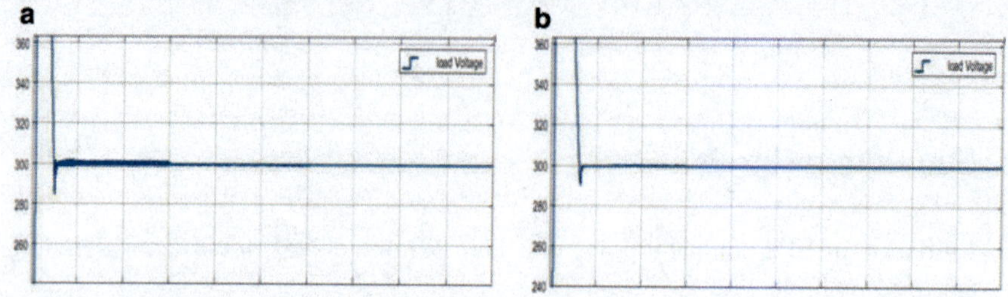

Fig. 9 **a** Load voltage for TPS, **b** load voltage for EPS

to microgrids. Major advantages of EPS control topology are small circulating power, less current stress and expands power transmission range.

5 Conclusion

IBDC is a sustainable component to deal with power management between renewable energy source and storage systems. To encounter inbuilt disadvantages of TPS, a novel EPS topology is implemented. From theoretical and experimental analysis, it is clear that EPS control has numerous features: The overall power transmission is increased, increased service life of devices; it also reduces switching frequency losses, current stress as well as power flow. Its principle of operation is simple and easy to implement which enhanced its overall efficiency.

References

1. Deshpande A, Karnataki K, Darshana K, Deshpande P, Mitavachan H, Shankar G (2015) Smart renewable energy micro grid for Indian scenarios. In: 2015 International conference on advanced computing and communications (ADCOM), Chennai, pp 22–26

2. Iqbal A et al (eds) (2020) Soft computing in condition monitoring and diagnostics of electrical and mechanical systems. In: Advances in intelligent systems and computing, vol 1096. Springer, Singapore. https://doi.org/10.1007/978-981-15-1532-3
3. Iqbal A et al (eds) (2020) Meta heuristic and evolutionary computation: algorithms and applications. In: Studies in computational intelligence, vol 1096. Springer, Singapore. https://www.springer.com/gp/book/9789811575709
4. Nara K, Hasegawa J (1997) A new flexible, reliable, and intelligent electrical energy delivery system. Electr Eng Jpn 121(1):26–34
5. Lasseter RH (2011) Smart distribution: Coupled microgrids. Proc IEEE 99(6):1074–1082
6. Alam MA, Singh N, Singh VP (2018) Impact of PPAM for power modulation in solar & battery bi-directional DC-DC converter. In: 2018 International conference on power energy, environment and intelligent control (PEEIC), Greater Noida, India, pp 29–33
7. Oggier GG, Garcıa GO, Oliva AR (2011) Modulation strategy to operate the dual active bridge dc–dc converter under soft switching in the whole operating range. IEEE Trans Power Electron 26(4):1228–1236
8. Krismer F, Kolar JW (2009) Accurate small-signal model for the digital control of an automotive bidirectional dual active bridge. IEEE Trans Power Electron 24(12):2756–2768
9. Tenti P, Paredes HKM, Mattavelli P (2011) Conservative power theory, a framework to approach control and accountability issues in smart microgrids. IEEE Trans Power Electron 26(3):664–673
10. Zhao B, Yu Q, Sun W (2012) Extended-phase-shift control of isolated bidirectional DC–DC converter for power distribution in microgrid. IEEE Trans Power Electron 27(11):4667–4680
11. Fatima K, Alam MA, Minai AF (2019) Optimization of solar energy using ANN techniques. In: 2019 2nd International conference on power energy, environment and intelligent control (PEEIC), Greater Noida, India, pp 174–179
12. Wang KR, Lee FC, Lai J (2000) Operation principles of bi-directional full-bridge DC-DC converter with unified soft-switching scheme and soft starting capability. In: Proceedings of IEEE application power electronics conference, pp 111–118
13. Alam MA, Singh N (2018) Analysis of switching topology for advanced isolated bi-directional DC-DC converter. In: 2018 2nd International conference on electronics, materials engineering & nano-technology (IEMENTech), Kolkata, 2018, pp 1–8
14. Wu TF, Sun KH, Kuo CL, Chang CH (2010) Predictive current controlled 5-kW single-phase bidirectional inverter with wide inductance variation for dc-microgrid applications. IEEE Trans Power Electron 25(12):3076–3084
15. Hua B, Chris M (2008) Eliminate reactive power and increase system efficiency of isolated bidirectional dual-active-bridge DC–DC converters using novel dual-phase-shift control. IEEE Trans Power Electron 23(6):2905–2914
16. Minai AF, Tariq A (2011) Analysis of cascaded multilevel inverter. In: Proceedings of IICPE10, CD-ROM, 28–30 Jan 2011. ISSN: 978-1-4244-7882-8/11/$26.00 ©2011IEEE at NSIT, New Delhi
17. Fan H, Li H (2010) A novel phase-shift bidirectional DC–DC converter with an extended high-efficiency range for 20kVA solid state transformer. In: Proceedings of IEEE energy conversion congress expo (ECCE), pp 3870–3876

Level Shifted Carrier-Based Pulse Width Modulation for Modular Multilevel Converter

Mohammad Tayyab and Adil Sarwar

Abstract Analysis of level shifted carrier-based PWM technique (LS-PWM) applied to the modular multilevel converter (MMC) is carried out in this paper. It can be classified as phase disposition (PD), phase opposition disposition (POD) and APOD pulse width modulation. The MMC which is used for simulation purpose consists of two half-bridge submodules in each phase arm which can produce five-level voltage waveform. PD-PWM and POD-PWM techniques are applied to the MMC for providing switching signals to the half-bridge switches. The simulation of MMC is done in MATLAB/Simulink software, and the results from these two techniques are discussed.

Keywords Modular multilevel converter · Level shifted pulse width modulation · Half-bridge submodules

1 Introduction

As the voltage and power level increase for numerous applications like HVDC power transmission, medium and high rating drives, high power solar PV plant flexible alternating current transmission system and traction system, etc., the rapid advancement of various multilevel converters has emerged in the past two or three decades [1–5].

The neutral point clamped multilevel converter consists of a single DC source, and it achieves required number of voltage levels by series connected capacitors without using power transformer. Number of capacitor requirement become more, and the problem of capacitor voltage balance arises when voltage level increases. The flying capacitor multilevel converter has capacitors which are within the phase leg. Selection of capacitors of different ratings and initialization of these capacitors

M. Tayyab (✉) · A. Sarwar
Electrical Engineering Department, AMU, Aligarh, UP 202002, India
e-mail: mtayyab1@myamu.ac.in

A. Sarwar
e-mail: adil.sarwar@zhcet.ac.in

A. Iqbal et al. (eds.), *Renewable Power for Sustainable Growth*, Lecture Notes in Electrical Engineering 723, https://doi.org/10.1007/978-981-33-4080-0_61

are major drawbacks of FC topology which limits the use of these converters to high-voltage applications [6, 7].

The advantages of MMC include the output voltage waveform quality, elimination of DC link capacitor and easy submodule capacitor voltage balancing [8]. LS-PWM are used for controlling the switching operation of submodule switches of MMC.

The operating principle of MMC is discussed in Sect. 2. Multicarrier-based PWM techniques are briefly discussed in Sect. 3. Simulation results of MMC with PD-PWM and POD-PWM are presented in Sect. 4, and the conclusion is summarized in Sect. 5.

2 Modular Multilevel Converter

MMC consists of N number of submodules as shown in Fig. 1. Inductor L_o is connected between the arms to reduce the circulating current between the arms. Each submodule is made up of half-bridge configuration which provides two states either 0 or total submodule voltage (V_{sm}) as shown in Table 1. Each submodule capacitors voltage is calculated as V_{dc}/N. Each submodule capacitor voltage is shown in Table 2.

3 Multicarrier Pulse Width Modulation Techniques

These techniques are classified as phase shifted (PS) and LS-PWM. In PS-PWM, carrier waves are shifted by some phase angle at same level while in case of LS-PWM carrier waves are level shifted. The LS-PWM technique consists of carrier waves equal to the number of submodules in MMC, and they are distributed vertically along the modulating wave amplitude range. Figure 2 represents examples of the three LS-PWM carrier dispositions.

4 Simulation Results

A three-phase MMC simulation is carried out by using MATLAB/Simulink software. Simulation parameters are shown in Table 3. LS-PWM technique is used for providing switching signals to the submodules used in MMC. Fig. 3 shows the simulation results by using Phase disposition pulse width modulation scheme. Fig. 4 shows the results obtained by using phase opposition disposition pulse width modulation scheme.

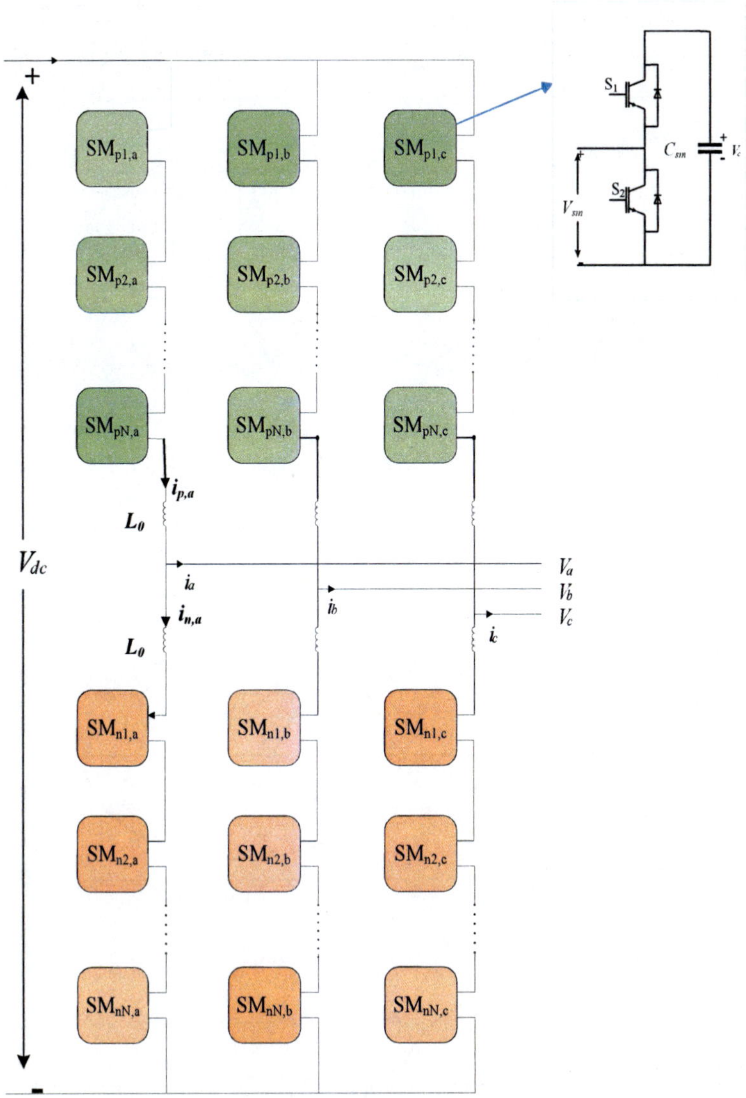

Fig. 1 Three-phase MMC

Table 1 Submodule states

Submodule state	S_1	S_2	V_{sm}
On state	Turned on	Turned on	V_c
Off state	Turned off	Turned off	0

Table 2 Submodule capacitor voltage states

Upper and lower arm current	SM state	Capacitor state	Submodule capacitor voltage (V_c)
Positive	On	Charging state	Will increase
	Off	Bypassed state	Remain unchanged
Negative	On	Discharging state	Will decrease
	Off	Bypassed state	Remain unchanged

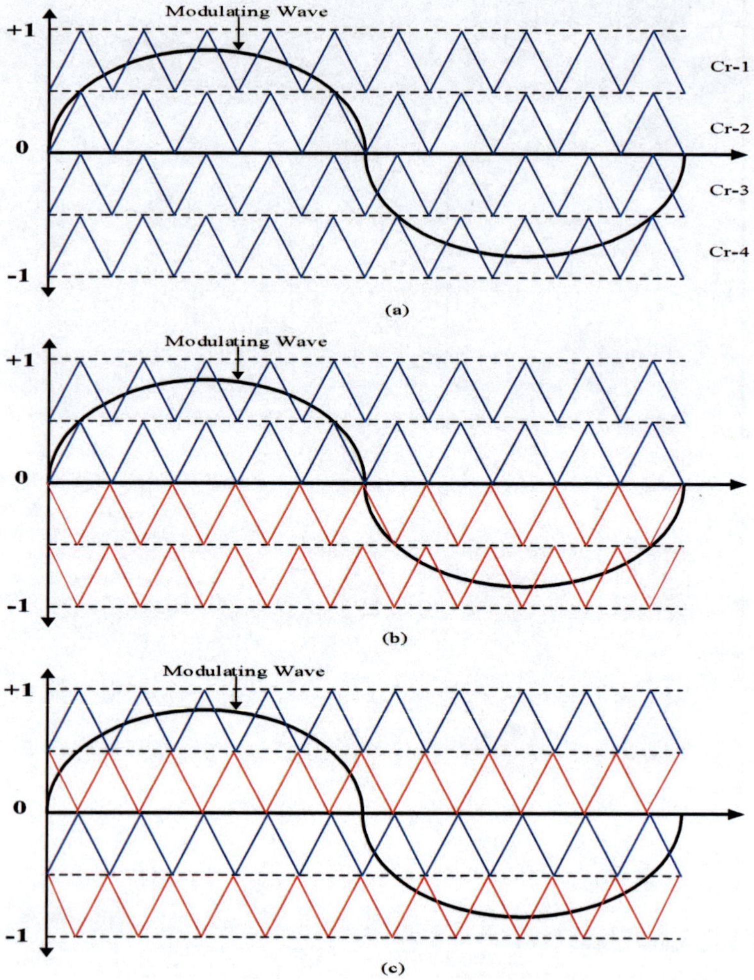

Fig. 2 Level shifted pulse width modulation schemes a PD, **b** POD, **c** APOD

Table 3 Simulation parameters

Number of submodule (N)	2
Type of submodule	Half-bridge
DC link voltage	400 V
AC output voltage (L-L)	230 V
Submodule capacitance	6000 μF
Arm inductance	1.6 mH
Output frequency	50 Hz
Carrier frequency	3000 Hz
Modulation index (m_a)	0.95

5 Conclusion

In this paper, PD-PWM and POD-PWM technique is used. Load specifications are ($R = 20\ \Omega$ and $L = 32$ mH). LS-PWM technique cannot provide balance voltage to each submodule of MMC and hence voltage balancing across the SM capacitors is achieved by gating signal rotation. The output voltage waveform, load current waveform, phase-A submodule capacitor voltage profile and total harmonic distortion (THD) are plotted at 0.95 modulation indices (m_a) for both techniques. Obtained results show that POD-based pulse width modulation technique provides better results in turns of THD and PD-PWM provides better capacitor voltage balancing.

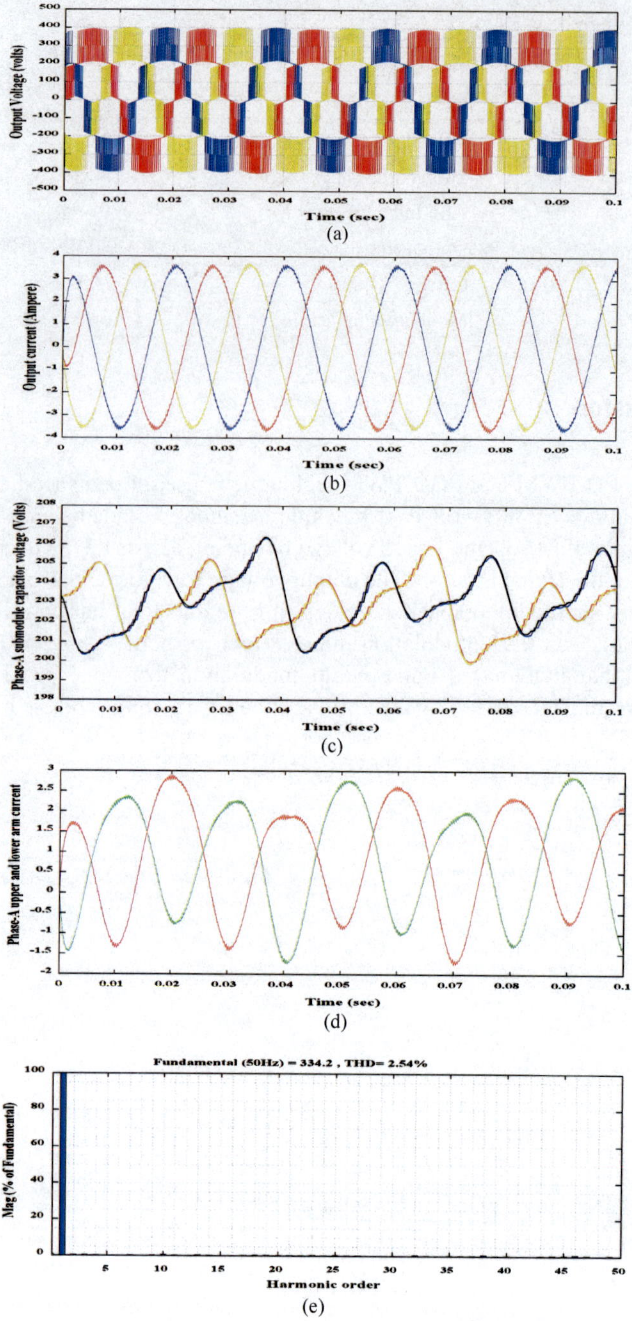

Fig. 3 PD-PWM simulation **a** voltage, **b** current, **c** phase-A submodule capacitor voltage profile, **d** phase-A arm current, **e** THD profile

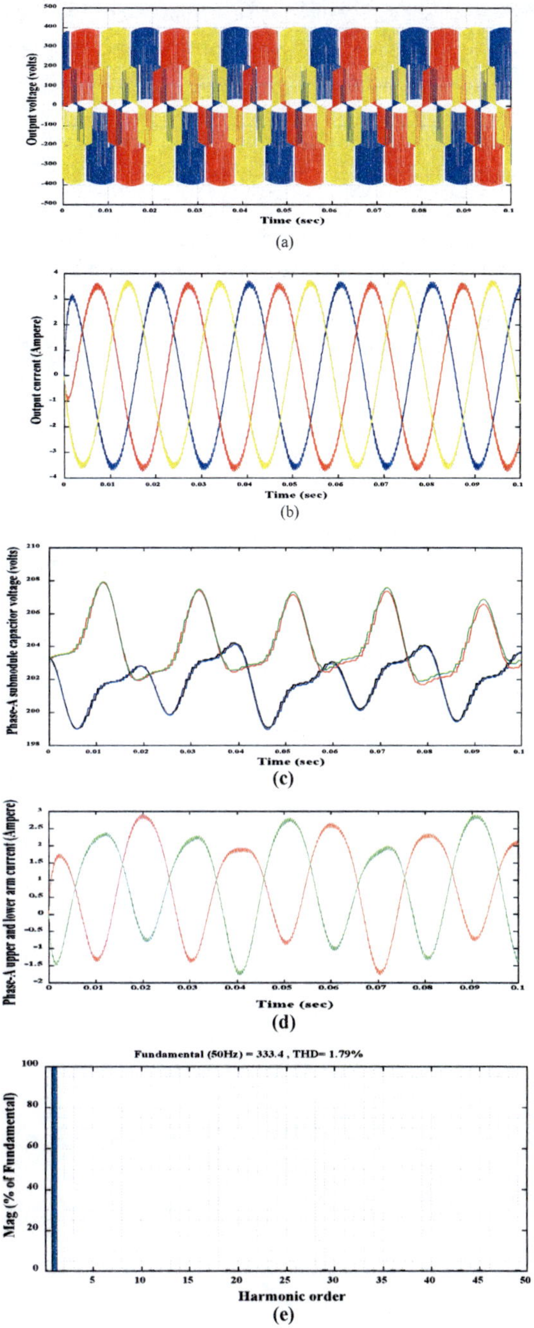

Fig. 4 POD-PWM simulation **a** voltage, **b** current, **c** phase-A submodule capacitor voltage profile, **d** phase-A arm current, **e** THD profile

Acknowledgements Author acknowledges CSIR-New Delhi for Junior Research fellowship (09-112(0569)-2017-EMR-I).

References

1. Siddique MD, Mekhilef S, Shah NM et al (2019) Low switching frequency based asymmetrical multilevel inverter topology with reduced switch count. IEEE Access 7:86374–86383
2. Shu H, Lei S, Tian X (2019) A new topology of modular multilevel converter with voltage self-balancing ability. IEEE Access 7:184786–184796
3. Li G, Xin Y, Jiang S, Wang W, Du J (2019) Design of MMC hardware-in-the-loop platform and controller test scheme. CPSS Trans Power Electron Appl 4(2):143–151
4. Iqbal A et al (eds) (2020) Soft computing in condition monitoring and diagnostics of electrical and mechanical systems. In: Advances in intelligent systems and computing, vol 1096. Springer, Singapore. https://doi.org/10.1007/978-981-15-1532-3
5. Iqbal A et al (eds) (2020) Meta heuristic and evolutionary computation: algorithms and applications. In: Studies in computational intelligence, vol 1096. Springer, Singapore. https://www.springer.com/gp/book/9789811575709
6. Wang S, Liu T, Huang X, Yu Y (2019) Capacitor voltage balancing control with reducing the average switching frequency in MMC. J Eng 2019(16):2375–2380
7. Xue C, Shen K, Ji Y, Jiang B (2014) Capacitor voltage balancing of modular multilevel converter. Dianli Zidonghua Shebei/Electric Power Autom Equip 34(7):27–31
8. Tayyab M, Sarwar A (2019) Submodule capacitor voltage balancing of modular multilevel converter. In: 2019 International conference on electrical, electronics and computer engineering (UPCON), Aligarh, India, pp 1–5

A Mathematical Approach to Speech Enhancement for Speech Recognition and Speaker Identification Systems

Rohun Nisa, Haweez Showkat, and Asifa Baba

Abstract In order to cope with acoustic degradation where clean sample of speech, free of interference and noise, prior to recognition stage, and identification–verification system, an efficient recognition and authentication of a particular speaker are necessary. In this paper, an approach for enhancement of speech is implemented using Fourier transform followed by spectral subtractive principle in upgrading speech signal contaminated due to noise. This methodology is employed in efficient recognition system for speech and identification–verification system of speaker as the degraded signal complicates hearing and understanding of speech signal. A Fourier transform approximates and derives spectrum of corrupted speech, and the spectral subtractive algorithm suppresses the amount of noise from noise spectrum to achieve clean signal.

Keywords Additive white Gaussian noise · Fourier transform · Spectral subtraction · Speaker identification · Speech enhancement · Speech recognition · Temporal convolutional neural network · Machine learning

1 Introduction

As we reside in native surroundings filled with noise and disturbance, there is generally unwanted noise associated with the signals particularly speech that hinders the processing of signals in original form. Noise and unwanted interference affect human–human and human–machine communications among varied fields which include degrading the properties of speech involving intelligibility together with

R. Nisa (✉) · H. Showkat · A. Baba
Department of Electronics and Communication Engineering, Islamic University of Science and Technology, Awantipora, Kashmir 192122, India
e-mail: rohunnisa@islamicuniversity.edu.in

H. Showkat
e-mail: haweezsk@gmail.com

A. Baba
e-mail: asibababa@gmail.com

A. Iqbal et al. (eds.), *Renewable Power for Sustainable Growth*, Lecture Notes in Electrical Engineering 723, https://doi.org/10.1007/978-981-33-4080-0_62

quality, identification corresponding to particular speaker, and recognition of speech [1–3]. Noise is generated everywhere, characteristics of which are either known or unknown. The field involving removal of noise and interference out of disrupted speech by incorporating variants of signal processing methodology constitutes speech processing. The different categories that comprise processing of signals include coding, enhancement, recognition, and synthesis particularly of speech. To frame the voice communication comfortable, natural, and practical, digital signal processing techniques are required [4]. Applications of speech communication requiring the noise reduction algorithms include answering machines, freehand communication, hard-of-hearing aids, localized and remote distance telecommunications, mobile and car phones, multiparty conferencing, noisy manufacturing and cockpits, teleconferencing systems, and voice over Internet Protocol (VoIP).

Normally, the word noise describes the undesirable signal that hinders and disrupts the analysis, processing, transmission, and reception of required informative acoustic signal. In order to achieve desirable representation and suppression of impact of noise, it becomes necessary to classify the concerning terminology of noise into respective four subclasses defined as follows: *additive noise* is the interference that gets associated with the signal due to varied sources when transmitted via communication channel, *interfering signals* that arise when multiple speakers are communicating at a time, *reverberation* is the effect of sound that remains after the sound is produced and is particularly due to multipath propagation, and *echo* is the sound reflection that reaches the listener after delay and arises mainly because of mixed link among microphones and loudspeakers. To take into account the corresponding problems mentioned, numerous speech signal processing techniques are employed including *reduction in noise or enhancement of speech*, *separation of source and speaker*, *de-reverberation of speech*, and *cancelation and suppression of echo* [5].

The signal analyzed through microphone is usually a representation of pure signal of speech with undesirable noise effect, resulting in corrupted signal and main challenge being to deal with background noise that causes degradation of signal of interest. The foremost consideration of suppression algorithms of noise is thus to recover and restore clean speech in original form given the superimposed signal to achieve the following essential goals: enhancing perceptual speech quality corrupted due to noise, improving objective performance criteria including intelligibility and signal-to-noise ratio (S/N or SNR) and enhancing the robustness of remaining applications of speech processing techniques comprising echo suppression and cancelation, coding of speech, recognition and synthesis of speech, particularly to noise [6].

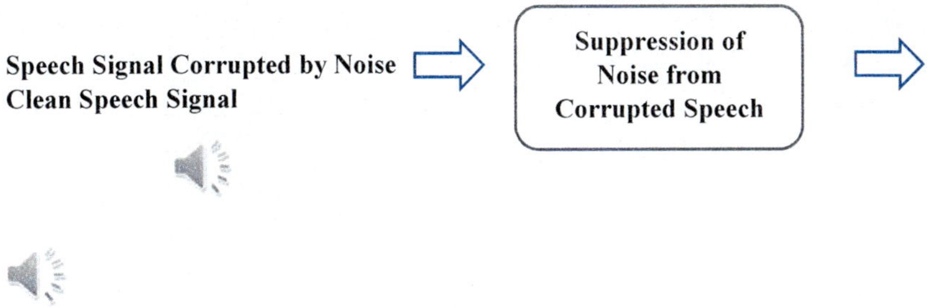

Fig. 1 Speech enhancement system

The presence of unwanted background noise in the acoustic signal severely impacts the functionality and execution involving speaker identification–verification (SIV) process that results in the reduction of recognition rate. Such systems are usually employed and incorporated before any SIV systems for enhancing the working of such systems to achieve better results, as depicted in Fig. 1.

2 Methodology Employed

Various speech enhancement methods are employed for reducing the noise in speech signal, among which spectral subtractive method is popular and commonly used method in real-world applications [1, 5]. Other traditional speech enhancement algorithms comprise statistical model-based methodologies, subspacing procedures, and binary masking principles. The spectral subtraction method including Fourier transform domain relies on eliminating the spectrum related to noise from noise corrupted speech in magnitude form obtained via Fourier transform, giving the enhanced clean speech signal as output [6]. The work on noise reduction techniques started with the novel contribution including two patents by Schroeder [7, 8] who put forward an application of analog method for spectral magnitude subtraction algorithm. After that, Boll [9] through his explanatory work specified the digital domain representation of spectral subtraction algorithm. Lim and Oppenheim [10] in form of their milestone effort represented the noise suppression problem by considering the already existing algorithms and forming a comparison. Their work explained the usefulness of reduction and suppression in noise from noise corrupted signal to upgrade the signal intelligibility and quality.

The noise reduction challenges are numerous in quantity. Pertaining to single channel where signal is recorded by one microphone, and of multichannel where signal is recorded by more than one microphone, there is the necessity to derive an optimal solution for removing as much undesirable noise as possible without degrading the standards including quality of speech signal and its intelligibility for purpose of communication. The proposed work presents a combination of Fourier transform decomposition of noise corrupted signal together with spectral subtraction to enhance speech signal for improvement of speaker identification–verification process.

2.1 Segmentation and Framing of Speech Signal

A speech signal is usually not stationary in real sense, but is typically considered quasi-stationary for short period of time. The main rationale being the glottal system and the features of such system do not change instantly [11]. Particularly for definite units of sound in a language called as Phonemes, the characteristics of speech usually stay unchangeable and are short approx. 5–100 ms time period. As such, application of traditional signal processing techniques becomes practical to be incorporated during short time span. Normally, speech processing is applied by considering very short windows including overlapping followed by analyzing and processing of such windows, referred to as frame. Thus, a speech signal, typically stationary in windows of suppose 20 ms, is partitioned and segmented into frames of 20 ms, corresponding to N samples given as

$$N = t_{fs} f_s \tag{1}$$

where t_{fs} forms the time frame step and f_s comprises frequency of sampling of signal.

Figure 2 depicts the segmentation of speech signal into short window frames. The overlapping of frames is shown with the corresponding first part of frame overlapped with the previous frame and remaining part with the next frame. The time frame step t_{fs} specifies the time duration among the start time of corresponding frame.

The duration from the beginning of new frame up to the end of current frame is referred to as overlap time t_{o}. Following from these considerations, the frame length t_{fl} is represented as

$$t_{\mathrm{fl}} = t_{\mathrm{fs}} + t_{\mathrm{o}} \tag{2}$$

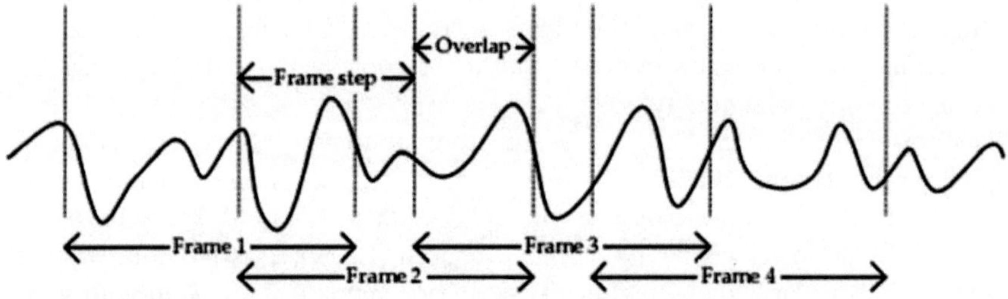

Fig. 2 Segmentation of speech into frames

Thus, the window is of length t_{fl}, which corresponds to $t_{fl} f_s$ samples.

In this method, frames are taken to be about 25 ms long, and audio file is taken to be of 16 kHz. This corresponds to 0.025 s*16,000 samples/s = 400 samples in length. We are using an overlap of 50% that constitute about 200 samples. So, the first frame will start at 0 instant, second frame will start at 200, third frame will start at 400, etc., indicated by frame1, frame2, and frame3 in the figure.

2.2 Decomposition in Fourier Transform Domain

Considering the quasi-stationary feature of speech for processing, the analysis involving speech is done taking short segmented windows referred frames and applying short time domain of Fourier transform (STFT) on respective individual short segment, yielding Fourier spectrum on corresponding frame [12]. Getting the noise corrupted signal as input is the combination of speech in clean form and corrupted due to additive noise. The model is represented as

$$y[\eta] = x[\eta] + s[\eta] \tag{3}$$

where $y[\eta]$, $x[\eta]$, and $s[\eta]$ represent the sampled noise corrupted signal, pure signal, and additive noise, with the assumption of additive noise having average time domain value of zero, not varying together with speech signal, η being the discrete index of time [13].

Now the STFT of the noise corrupted signal y (η) will thus be represented by

$$Y(\eta, \varpi) = \sum_{l=-\infty}^{\infty} y(l)w(\eta - l)e^{-j2\pi\varpi l/N} \tag{4}$$

where ϖ constitutes the discrete frequency index, N as the duration of frame (in samples), l as the frame number, and $w(\eta)$ as speech analysis function referred to window function. While considering the processing of speech signal, the Hamming window is usually used having duration range of typically 20–40 ms [14]. Windowing is required as the analysis of input signal involves processing of samples that are finite, resulting in discontinuation of respective frames. Such discontinuities among the corresponding frames are eliminated by employing windowing, resulting in smooth end of frames and getting connected accurately to the start of upcoming frame [15].

2.3 Reconstruction of the Signal

To construct the improved clean signal, $x(\eta)$, another transform referred as inverse STFT is applied on modified speech spectrum and continuing with the incorporation of least-squares overlap-add synthesis, depicted as

$$x(\eta) = \frac{1}{W_\theta(\eta)} \sum_{l=-\infty}^{\infty} \left[\left(\frac{1}{N} \sum_{\varpi=0}^{N-1} Y(l, \varpi) e^{\frac{j2\pi\eta\varpi}{N}} \right) w_\tau(l-\eta) \right] \quad (5)$$

where $w_\tau(\eta)$ represents the function referred as synthesis window, with $W_\theta(\eta)$ represented as

$$W_\theta(\eta) = \sum_{l=-\infty}^{\infty} w_\tau^2(l-\eta) \quad (6)$$

Usually, the synthesis window employed is Hanning window, depicted as

$$w_\tau(\eta) = \begin{cases} 0.5 - 0.5\cos\left(\frac{2\pi(\eta+0.5)}{N}\right), & 0 \le \eta \le N \\ 0, & \text{otherwise} \end{cases} \quad (7)$$

2.4 Spectral Subtractive Principle

Spectral subtractive principle forms the practical and useful method that is employed for the suppression of ambient noise from signal. The method relies on the regeneration of spectrum involving magnitude of a signal with background noise associated with signal and subtracting the average noise spectrum approximation obtained from Fourier transform from noise corrupted signal spectrum. The scenarios involving processing of signals at receiver with communication channel are contaminated by noise, and the corrupted signal is usually encountered at the receiver end. For such circumstances, local average impact of noise is considered on spectrum of signal [16]. The addition of additive noise on signal thus raises the average value and variance of magnitude spectrum of a signal, as depicted in Fig. 3.

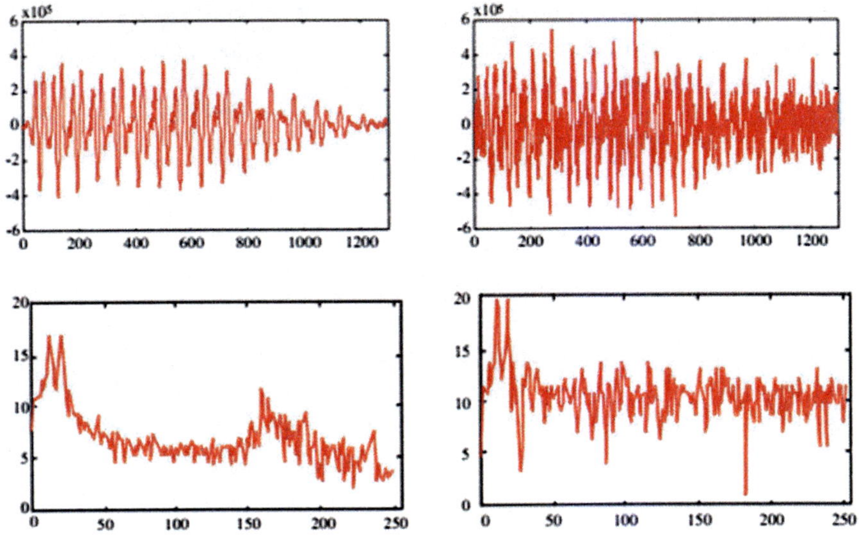

Fig. 3 Impact of noise on signal pertaining to time and frequency domain

Due to variant time characteristics of speech, the signal analysis is achieved and done using frame-by-frame analysis by incorporating short time domain of Fourier transform (STFT) on signal depicted by Eq. 4, illustrated as

$$Y(\eta, \varpi) = X(\eta, \varpi) + S(\eta, \varpi). \tag{8}$$

With the assumption of independent relation among speech signal and background noise, the corresponding magnitude spectrum of corrupted signal $y[\eta]$ is represented without cross terms and depicted as

$$|Y(\varpi)|^2 = |X(\varpi)|^2 + |S(\varpi)|^2 \tag{9}$$

To obtain the spectrum involving improved clean signal, an approximate of corrupted signal spectrum is eliminated out of input signal spectrum, represented as

$$\left|\dot{X}(\varpi)\right|^2 = |Y(\varpi)|^2 - \left|\dot{S}(\varpi)\right|^2 \tag{10}$$

The other application of spectral subtractive principle involves the realization as filter referred as spectral subtractive filter, mathematically represented as product of corrupted spectrum pertaining to speech by noise and the spectral subtractive filter (SSF), depicted as

$$\left|\dot{X}(\varpi)\right|^2 = \left(1 - \frac{\left|\dot{S}(\varpi)\right|^2}{|Y(\varpi)|^2}\right)|Y(\varpi)|^2 \tag{11}$$

$$\left|\dot{X}(\varpi)\right|^2 = \hat{H}^2(\varpi)|Y(\varpi)|^2 \tag{12}$$

where $\hat{H}(\varpi)$ represents the function referred as gain function, related to spectral subtractive filter (SSF) which is considered as filter with zero phase, having the representation of magnitude varying among the range $0 \leq \hat{H}(\varpi) \leq 1$, given as,

$$\hat{H}(\varpi) = \left\{\max\left(0, 1 - \frac{\left|\dot{S}(\varpi)\right|^2}{|Y(\varpi)|^2}\right)\right\}^{1/2} \tag{13}$$

For reconstructing the signal, phase spectrum characteristics of speech are taken into account. The usual method in determining the phase or angle variation of corresponding corrupted speech is relating the angle variation of noise degraded speech to the phase of clean signal obtained after suppression. Thus, the approximation of speech regarding a short segment frame is expressed as

$$\dot{X}(\varpi) = \left|\dot{X}(\varpi)\right|e^{j<Y(\varpi)} \tag{14}$$

$$\dot{X}(\varpi) = \hat{H}(\varpi)Y(\varpi) \tag{15}$$

From this, it follows that an approximate waveform of speech in time domain can be reconstructed using inverse Fourier transform. The sequence followed in performing speech enhancement with Fourier transform and spectral subtraction approach is depicted by flow diagram in Fig. 4.

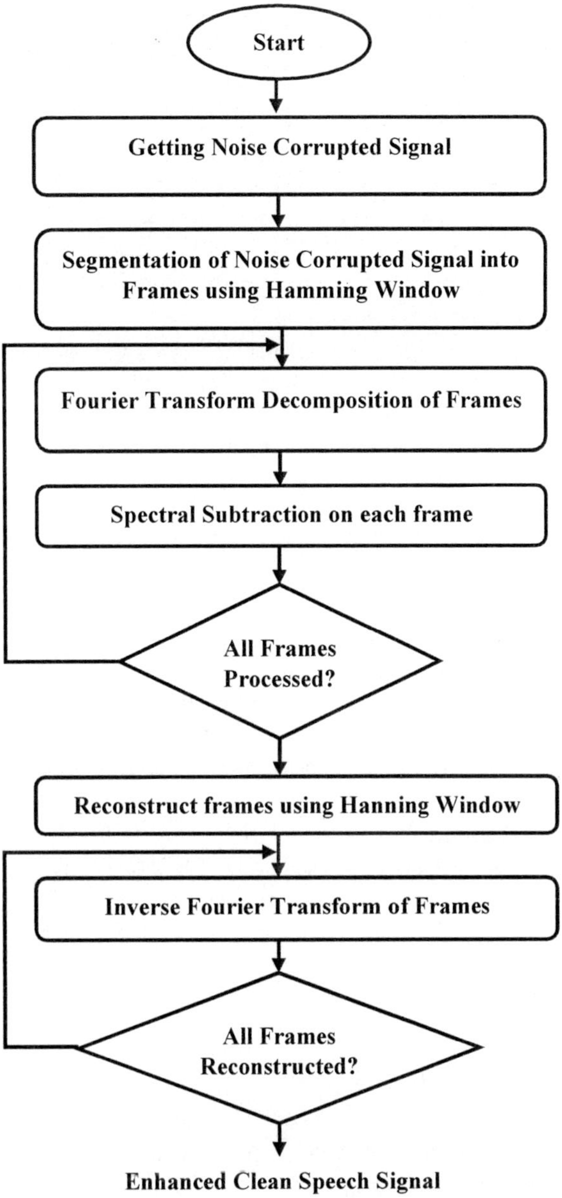

Fig. 4 Enhancement of noise corrupted speech signal approach

3 Program Code

```
clear all
[speech, fs] = wavread ( 'noisy_signal.wav' ); % read the input noisy speech signal
NFFT = 512; % N-point FFT - use of 512 point FFT
len_window = 0.025; % length of frame window (in milliseconds) taken as t_fl
overlap_window = 0.0125; % overlap time (in milliseconds) taken as t_o
frames_n = frame_signal ( speech, len_window * fs, overlap_window * fs,
@hamming ); % hamming window
cplx_spc = fft ( frames_n, NFFT, 2 ); % spectrum in complex form
mag_spc = abs ( cplx_spc ).^2; % spectrum of noisy speech signal in terms of
magnitude
phs_spc = angle ( cplx_spc ); % spectrum of noisy speech signal with phase
% spectral subtractive principle giving modified spectrum in terms of magnitude
noise_est = mean ( mag_spc (1:3, :) ); % noise estimated from first three frames
cln_spc = mag_spc – repmat ( noise_est, size ( mag_spc, 1), 1 ); % subtract
noise_est from mag_spc
cln_spc ( cln_spc < 0 ) = 0; % negative spectrum of magnitude discarded from clean
spectrum
% reconstruction of frames
rcnstrctd_frames_n = ifft ( sqrt (cln_spc ).* exp ( phs_spc), NFFT, 2 );     % inverse
FFT used
rcnstrctd_frames_n = real ( rcnstrctd_frames_n (:, 1:len_window * fs) ); % with
small residuals (complex)
enhanced_signal = deframe_sig_n( rcnstrctd_frames_n, length (speech),
len_window * fs, overlap_window * fs, @hamming);
            plot ( enhanced_signal ); % wave plot of enhanced signal
            sound ( enhanced_signal, fs ); % listen to enhanced signal
```

4 Experimental Results

While enhancing the speech signal, the main rationale is suppressing the noise from corrupted speech to upgrade the signal intelligibility together with quality. Signal quality forms the subjective performance measure that evaluates to what degree the speech sounds fine and thus includes the characteristics as naturalness, roughness of noise, etc., and intelligibility forms an objective performance measure that determines how much the signal is understood.

The experiment is conducted on two speakers, taking one male voice and female voice considering the coded speech database of ITU-T P-series recommendations [17]. This coded speech database comprises the sentences with varying durations

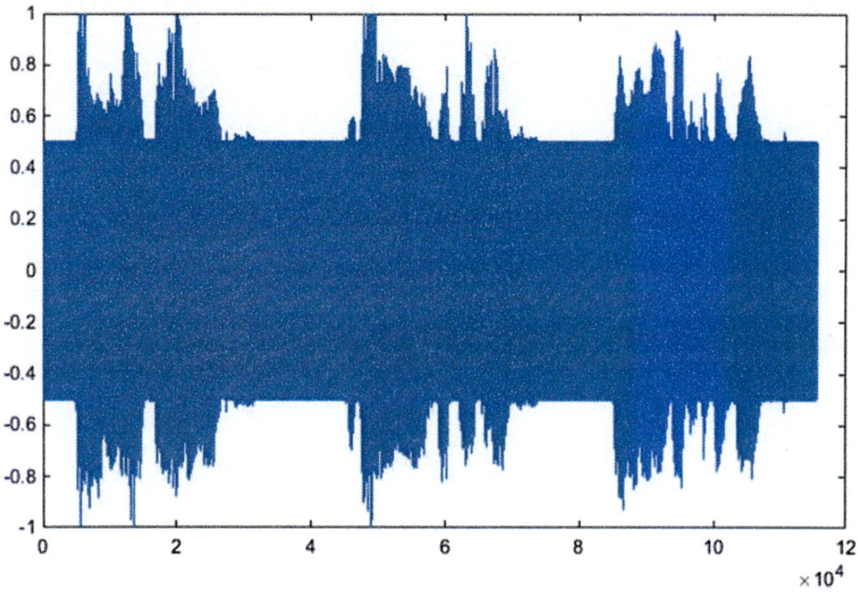

Fig. 5 Audio wave of corrupted female voice

that are uttered in diversified languages and accent. These uttered sentences are corrupted by noise particularly additive noise having contrasting signal-to-noise ratio (S/N or SNR) to authenticate and verify particular speaker and for speech recognition purpose, while incorporating this speech enhancement method as prior treatment to such systems. The experiment results of the methodology involving spectral subtractive principle on female corrupted voice and the enhanced female voice are depicted in Figs. 5, 6, 7, and 8. With the incorporation of algorithm, noise is shown to be removed from the signal, resulting in understandable speech signal.

5 Conclusion

In this paper, the procedure of enhancing the speech of interest incorporating Fourier transform domain and spectral subtractive principle is shown that suppresses the noise associated with speech signal. Further, this method is employed before the recognizer system for speech and speaker identification process to lessen the undesirable impact of noise and interference on speech, resulting in the improvement of speech quality and speech intelligibility. The experimental results show the audio wave of speech signal with and without the effect of noise together with the spectrum of both the signals. The waveform shows the removal of noise from female voice and deriving the clean voice free from ambient noise.

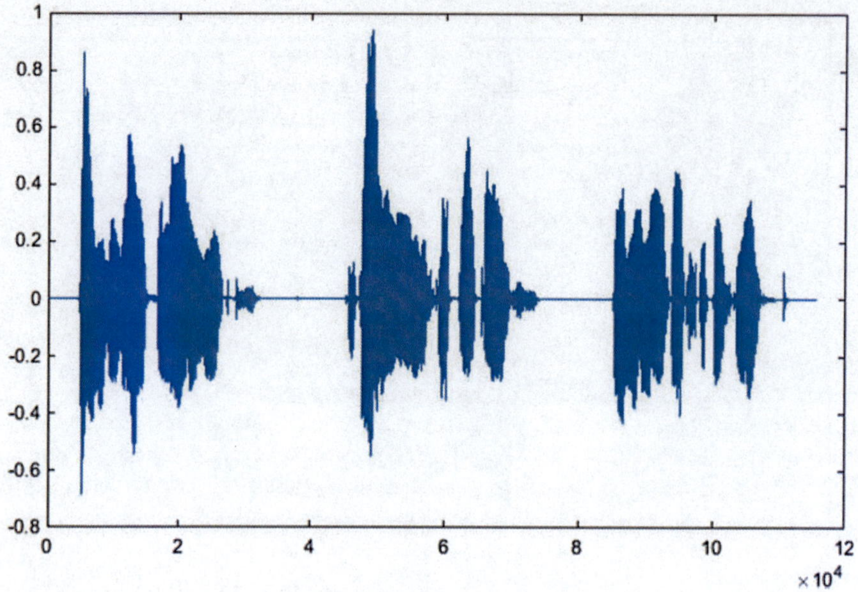

Fig. 6 Audio wave of enhanced female voice

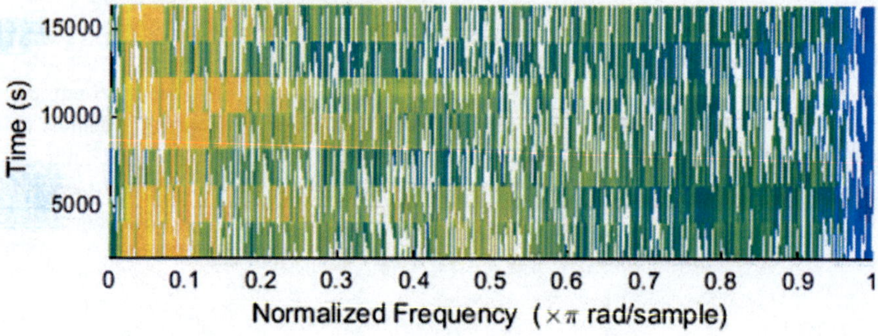

Fig. 7 Spectrum representation of corrupted female voice

6 Future Considerations

In the future work, we will implement speech enhancement in real-time systems involving time domain by employing fully convolutional neural network known as temporal convolutional neural network (TCNN), a hybrid deep learning approach. This method will involve the training of model in a speaker and noise unconstrained procedure and will include few parameters to train the model. This will explore deep neural network architecture pertaining to time domain analysis of speech enhancement. This research will further incorporate the analysis of additional speech

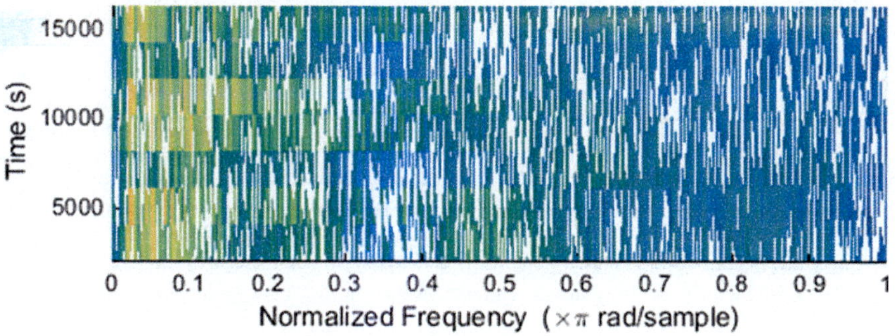

Fig. 8 Spectrum representation of enhanced female voice

processing tasks including de-reverberation of speech, echo suppression and cancelation, source separation, and speaker separation using TCNN model so as to upgrade SNR, quality, and intelligibility of speech signal.

References

1. Benesty J, Makino S, Chen J (2005) speech enhancement. Springer, Berlin, Heidelberg. https://doi.org/10.1007/3-540-27489-8
2. Iqbal A et al (eds) (2020) Soft computing in condition monitoring and diagnostics of electrical and mechanical systems. In: Advances in intelligent systems and computing, vol 1096. Springer, Singapore. https://doi.org/10.1007/978-981-15-1532-3
3. Iqbal A et al (eds) (2020) Meta heuristic and evolutionary computation: algorithms and applications. In: Studies in computational intelligence, vol 1096. Springer, Singapore. https://www.springer.com/gp/book/9789811575709
4. Karam M, Khazaal HF, Aglan H, Cole C (2014) Noise removal in speech processing using spectral subtraction. J Signal Inf Process 5:32–41. https://doi.org/10.4236/jsip.2014.52006
5. Benesty J, Sondhi MM, Huang Y (2008) Springer handbook of speech processing. Springer, Berlin, Heidelberg. https://doi.org/10.1007/978-3-540-49127-9
6. Loizou PC (2007) Speech enhancement theory and practice, 2nd edn. CRC Press Taylor and Francis, London
7. Schroeder MR (1965) U.S. Patent No. 3180936, filed 1 Dec 1960, issued 27 Apr (1965). https://patents.google.com/patent/US3180936
8. Schroeder MR (1968) U.S. Patent No. 3403224, filed 28 May 1965, issued 24 Sept (1968). https://patents.google.com/patent/US3403224
9. Boll SF (1979) Suppression of acoustic noise in speech using spectral subtraction. IEEE Trans Acoust Speech Signal Process 27(2):113–120. https://doi.org/10.1109/TASSP.1979.1163209
10. Lim JS, Oppenheim AV (1979) Enhancement and bandwidth compression of noisy speech. IEEE Proc 67:1586–1604. https://doi.org/10.1109/proc.1979.11540
11. Andersen BB, Dyreby J, Kjærskov FH, Mikkelsen OL, Nielsen PD, Zimmermann NH, Jensen B (2004) Bandwidth extension of narrowband speech using linear prediction. Aalborg University, Denmark, Worksheets
12. Paliwal K, Wójcicki K, Schwerin B (2010) Single-channel speech enhancement using spectral subtraction in the short-time modulation domain. Speech Commun 52:450–475. https://doi.org/10.1016/j.specom.2010.02.004

13. Upadhyay N, Karmakar A (2015) Speech enhancement using spectral subtraction-type algorithms: a comparison and simulation study. In: 11th International multi-conference on information processing 2015, Procedia Comput Sci 54:574–584. https://doi.org/10.1016/j.procs.2015.06.066
14. Paliwal K, Wójcicki K (2008) Effect of analysis window duration on speech intelligibility. IEEE Signal Process Lett 18(15):785–788. https://doi.org/10.1109/LSP.2008.2005755
15. Sen S, Dutta A, Dey N (2019) Audio processing and speech recognition concepts techniques and research overviews. Springer, Singapore. https://doi.org/10.1007/978-981-13-6098-5
16. Vaseghi SV (2000) Advanced digital signal processing and noise reduction, 2nd edn. Wiley & Sons Ltd, New York. https://doi.org/10.1002/0470841621
17. ITU-T Test Signals for Telecommunication Systems. https://www.itu.int/net/itu-t/sigdb/genaudio/Pseries.htm

Cost Analysis of 18 kW Solar Photovoltaic System for Smart Cities Growth in India

Iram Akhtar, Ward Ul Hijaz Paul, Sheeraz Kirmani, and Mohammed Asim

Abstract Smart city is an innovative idea that provides the people to use the advanced technology and growth of the particular city at the same time. Smart cities are expanding much fast, and they provide the new ideas which influence the development and policy making. Power systems for smart cities growth require the need of different renewable-based energy sources like solar energy source and wind energy system. When various factors are considered for large-scale use of these sources, the cost is a key concern of different renewable energy-based system. Therefore, designing aspects and cost analysis of the solar energy system is necessary nowadays. So, this paper presents the cost analysis of 18 kW solar energy system for smart cities growth in India. This work also provides the economic scheduling of grid-connected solar energy system for smart cities development. In comparison with grid-connected solar energy system and off-grid solar energy system that is used for solar energy generation, the proposed grid-connected solar energy system is best suited for smart cities development in India.

Keywords Solar energy system · Utility · Smart cities · Cost

I. Akhtar (✉) · W. U. H. Paul · S. Kirmani
Department of Electrical Engineering, Faculty of Engineering & Technology, Jamia Millia Islamia, New Delhi 110025, India
e-mail: iram1208@gmail.com; akhtariram12@gmail.com; alquranwalhadith@gmail.com

W. U. H. Paul
e-mail: wardulhijazpaul@gmail.com

S. Kirmani
e-mail: sheerazkirmani@gmail.com

M. Asim
Department of Electrical Engineering, Integral University, Lucknow, India
e-mail: masim@iul.ac.in

A. Iqbal et al. (eds.), *Renewable Power for Sustainable Growth*, Lecture Notes in Electrical Engineering 723, https://doi.org/10.1007/978-981-33-4080-0_63

1 Introduction

With the quick depletion of fossil fuels reserves, energy security and ecological factors, there is the need to fulfill the future demand, so renewable energy sources are used, and these sources are quickly developed. There are various schemes are available in the literature for integration of different renewable energy-based sources to the utility grid [1–4]. The cost-effective system is necessary to connect the grid with different renewable energy sources. The control system should be employed because of changing behavior of the wind and solar energy sources. Economic scheduling is also an important factor for smart cities growth, whereas designing aspects of the solar energy system is necessary nowadays so different models are existing in the literature. The stability of solar energy system depends on the solar radiation variation because the solar radiation is constant throughout the day; it varies with time so output is affected. Hence, effective architecture is required to cover the overall system. Further, there is need of control scheme of boost converter and solar modules to extract the maximum from [5–8].

Therefore, control algorithm is needed for this system. The solar energy system is a fresh source of incessant and reliable source of energy to satisfy the present energy demand. The solar energy system should provide the two-way communication for smart cities development so that it can give the huge saving and key factor of any system to decrease the overall cost.

In the present paper, growth of the solar energy system-based smart cities is defined in Sect. 2, whereas in Sect. 3, cost analysis of 18 kW solar energy-based system is presented. Further, economic scheduling of solar energy-based system is discussed in Sect. 4. Results and discussion are discussed in Sect. 5. Lastly, concluding reports are presented in Sect. 6.

2 Growth of the Solar Energy System-Based Smart Cities

The solar power system design is based on the site location, value of solar radiation, type, etc. The main components of solar power system are solar modules, charge controllers, battery and inverter as presented in Fig. 1. In case of grid-connected solar energy-based system, storage system is not required, therefore the cost will be decreased. A 18 kW grid-connected solar energy-based system with the help of inverter is considered. The relationship between the voltage and current is defined by Eq. 1.

$$I = I_{\text{solar}}\left[\exp\left(\frac{V + IR_{\text{series}}}{kT_s}\right) - 1\right] - \frac{V + IR_{\text{series}}}{R_{\text{shunte}}} \quad (1)$$

The 18 kW solar system is considered for smart cities development in small level like commercial building, clinic, etc., and the grid-connected system gives the huge

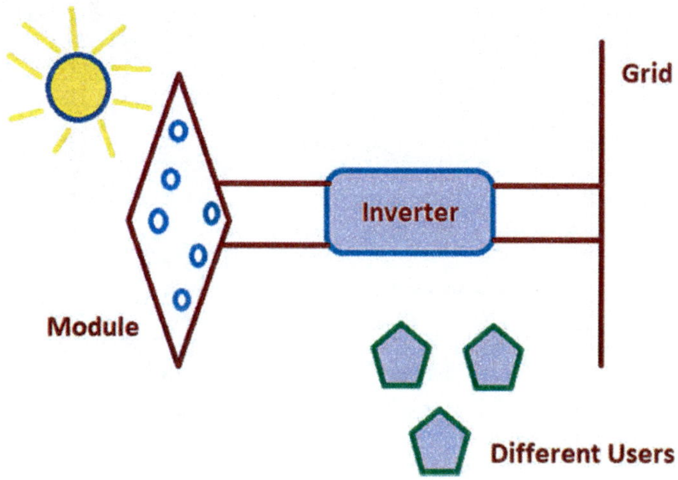

Fig. 1 Grid-connected solar energy system for the smart cities growth

saving and develops the smart cities with the help of two-way communication with net metering scheme. The net solar energy system provides the huge amount of saving in early basis, whereas the solar module requirement can be determined by considering the total power needed and single watt ration of the particular solar module.

The total number of solar modules needed = 18,000/270 = 67.

Therefore, 67 modules of 270 W are required for the proposed system. The specifications of the solar modules are as follows:

$$P\mathrm{max} = 270\,\mathrm{W}$$

$$V_{\mathrm{mp}} = 29\,V_{\mathrm{dc}}$$

$$I_{\mathrm{mp}} = 9.31\,\mathrm{A}$$

$$V_{\mathrm{oc}} = 34\,\mathrm{V}$$

$$I_{\mathrm{sc}} = 9.50\,\mathrm{A}$$

Efficiency per cell area is 18.62%, and this system is reliable and effective to send power to grid also in smart way (Table 1).

Whereas the inverter size depends on the appliances power so for the safety purpose, the inverter size should be considered carefully. So, three inverters are taken with 6 kW capacity. When there is need of electricity, the inverter should provide the constant power without the harmonics; it should be effectively controlled with the

Table 1 Solar modules specifications

Factors	Values
R_{series}	0.28 Ω
R_{shunt}	280 Ω
Series resistance apparent	0.39 Ω
Width	872 mm
Weight	18 kg
Cell area	1.45 m^2
Cell in series	67
Cell in parallel	1

proper control scheme. The geographical location is also the main concern when the proposed system is installed.

3 Cost Analysis of 18 kW Solar Energy System

The overall cost of grid-connected 18 kW system can be known by taking the cost of different components to het the cost-effective approach for smart cities development. The overall cost of proposed system with grid-connected mode and off-grid mode form smart city growth is presented in Tables 2 and 3, respectively.

Table 2 Overall cost of 18 kW solar energy system—grid connected

Item	Cost (INR)
PV modules	945,659
Inverter	235,600
Setting, wiring	123,000
Other misc.	50,000
MPPT regulator	320,000
Total	1,674,259

Table 3 Overall cost of 18 kW solar energy system—off-grid

Item	Cost(INR)
PV modules	945,659
Battery	310,000
Inverter	235,600
Setting, wiring	123,000
Other misc.	50,000
MPPT regulator	320,000
Total	1,984,259

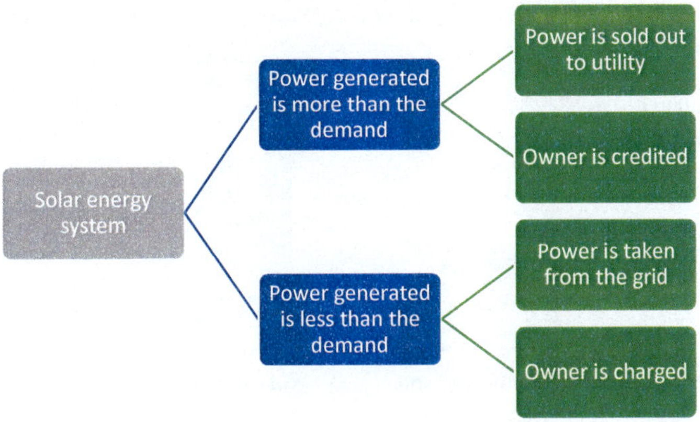

Fig. 2 Economics scheduling scheme for the proposed system

This analysis provides the overall saving of 310,000 INR with the same value of solar energy system in grid-connected mode. Storage system is essential in off-grid mode so increases the overall cost of the system. Therefore, the cost of grid-connected solar energy system is low in comparison with off-grid system. The smart cities can provide higher value of solar energy because of the tall buildings. So, smart energy scheme is required for the development of the smart city to track the generated power, distribution, storage and fulfilling the different demand at a particular time.

4 Economic Scheduling for Solar Energy System in Smart City

The economic scheduling is very significant for effective and reliable operation of system. If the proposed system generates more power as needed by the connected load, then this extra power can be given to the utility grid, and if this system generates less power demanded by the connected load, then needed power can be taken from the grid. For this purpose, net metering scheme base meter is employed, it provides the two-way communication, if power is given to grid, then owner is credited, and if power is taken from the grid, the owner is charged for the unit consumed. Economic scheduling scheme is shown in Fig. 2.

5 Results and Discussions

The solar energy system in grid-connected mode is the most cost-effective way to develop the smart cities. The cost of 18 kW solar energy system in 2020 is around

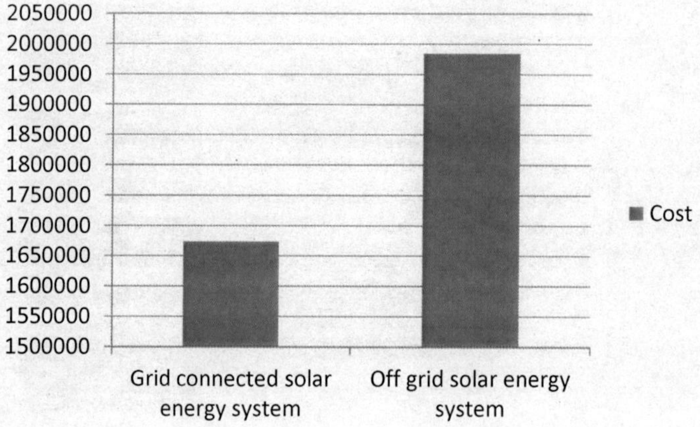

Fig. 3 Saving per year

1,674,259 INR without the storage system and cost of 18 k W solar system with battery storage is around 1,984,259 INR.

The proposed system can generate around 72 units in a day and total 26,280 units in a year. So these units need not to be purchased from the main grid. Per unit cost of electricity generation from the solar energy system is low in comparison with thermal-based power generation system. Figure 3 shows the saving using this system. Once the solar system is implemented in the smart city, then different actions are made for the execution to the implementation of the technique to handle the different challenges of development of the smart city. Once the idea of solar energy system-based smart city and the main challenges are known, then a prepared technique to the act chances of the task has been established. The cost analysis and economic scheduling of the 18 kW solar energy system have been discussed. The initial cost of installing the solar energy is high, but it gives the saving. One unit electricity generated from the grid generates the 0.814 kW carbon emissions. This entire system produces total 26,280 units in a year so 21,391.92 kg carbon emission is decreased by using this system.

6 Conclusion

This paper highlights various factors of smart cities development with solar energy system. This work also presents the cost analysis and economic scheduling of the grid-connected solar energy-based system to electrify the smart city in India. If the generated power from the proposed system is higher than the demand, then extra power is given to the grid and owner is credited for power supply to the grid. Further, if power generated from the proposed system is less than the demand, then extra power is taken from the grid and owner is charged for the units consumed. Therefore, this system provides the two-way communication for smart cities development. This system gives the huge saving and key factor of any system to decrease the overall cost.

References

1. Kirmani S, Jamil M, Akhtar I (2017) Bi-directional power control mechanism for a microgrid hybrid energy system with power quality enhancement capabilities. Int J Renew Energy Res (IJRER) 7(4):1962–1969
2. Kirmani S, Jamil M, Akhtar I (2017) Effective low cost grid-connected solar photovoltaic system to electrify the small scale industry/commercial building. Int J Renew Energy Res 7(2):797–806
3. Iqbal A et al (eds) (2020) Soft computing in condition monitoring and diagnostics of electrical and mechanical systems. In: Advances in intelligent systems and computing, vol 1096. Springer, Singapore. https://doi.org/10.1007/978-981-15-1532-3
4. Iqbal A et al (eds) (2020) Meta heuristic and evolutionary computation: algorithms and applications. In: Studies in computational intelligence, vol 1096. Springer, Singapore. https://www.springer.com/gp/book/9789811575709
5. Besarati SM, Padilla RV, Goswami DY, Stefanakos E (2013) The potential of harnessing solar radiation in Iran: generating solar maps and viability study of PV power plants. Renew Energy 53:193–199
6. EL-Shimy M (2009) Viability analysis of PV power plants in Egypt. Renew Energy 34:2187–2196
7. Asim M, Tariq A, Sarwar A (2009) Simulation and analysis of a directly coupled solar PV based water pumping system. J Electr Eng 2(3):72
8. Asim M, Tariq M, Mallick MA, Ashraf I (2016) An improved constant voltage based MPPT technique for PMDC motor. Int J Power Electron Drive Syst (IJPEDS) 7(4):1330–1336

A Survey of Recent Trends in Two-Stage Object Detection Methods

M. F. Ansari and K. A. Lodi

Abstract Object detection deals with locating an object in an image or a video and identifying its class label. In this regard, integration of object detection techniques with deep learning has revolutionized the area of computer vision. Furthermore, the ability of a deep neural network to directly learn feature representation from images has significantly improved object detection models. The present survey analyzes and systematically provides a comprehensive overview of typical two-stage object detection methods with deep learning and summarizes the most popular benchmark dataset for object detection.

Keywords Object detection · Deep learning · Two-stage object detection

1 Introduction

Deep learning (DL) has empowered computer vision to effectively learn image features. It should be noted that most object detectors use a deep neural network as their backbone architecture to extract a feature from images and a detection network to detect objects in images or videos. An object detection method locates objects of a particular category in images or videos. It has fascinated researchers over the last decade. This technology has been applied to humans and society in the form of self-driving cars, face detection, activity recognition, pedestrian detection medical imaging, robotics, object counting, and crop monitoring. Recent developments in computational power have significantly contributed to the development of object detection techniques [1–8].

M. F. Ansari (✉)
Department of Computer Science, Aligarh Muslim University, Aligarh, India
e-mail: mfansari2395@gmail.com

K. A. Lodi
Department of Electrical Engineering, Aligarh Muslim University, Aligarh, India

A. Iqbal et al. (eds.), *Renewable Power for Sustainable Growth*, Lecture Notes in Electrical Engineering 723, https://doi.org/10.1007/978-981-33-4080-0_64

Several benchmark datasets for instance KITTI, Caltech, MS COCO, PASCAL VOC, and Open Image V5 have played an important role in improving object detection. Organizations maintain a public dataset containing images and videos and information needed how to use them; anyone can download these datasets and conduct experiments. Presently, detection based on deep neural network can be divided into two classes:

- Two-stage detector
- One-stage detector.

R-CNN [1] and its different types are examples of two-stage detectors, whereas YOLO [2], its variants are one-stage detectors. Two-stage detectors are highly accurate in terms of localization and classification, whereas one-stage detectors have greater speed in terms of real-time detection. The stages of a two-stage detector can be specified; for instance, in faster R-CNN [4], first stage is called the region proposal network (RPN), which proposes a bounding box; in the next stage, features are pulled out from these boxes with the aid of the RoI pool (RoI pooling) operation. Architecture of a two-stage detector can be viewed in Fig. 1. One-stage detectors, on the other hand, directly predict bounding boxes from input images and the corresponding class label of each box. Architecture of a two-stage detector can be viewed in Fig. 2.

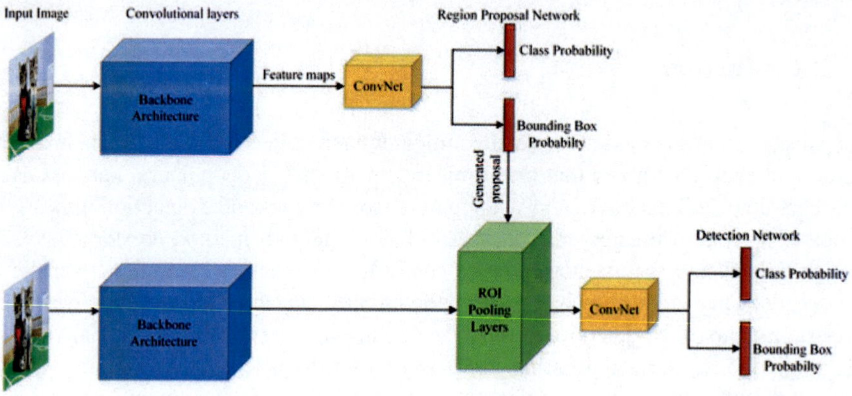

Fig. 1 Two-stage detector

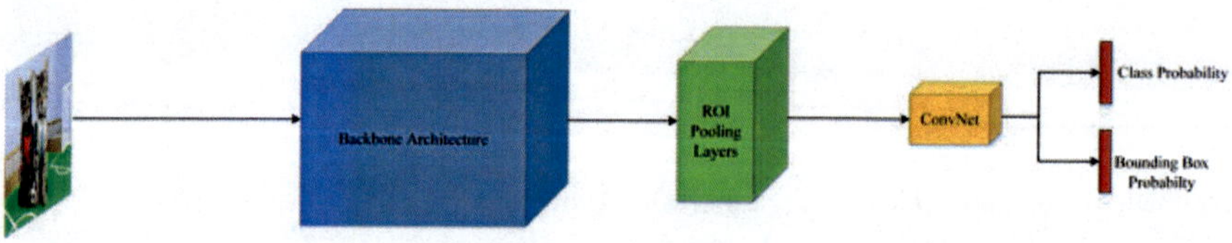

Fig. 2 One-stage detector

The main objective of this survey paper is to provide an all-inclusive understanding of DL-based two-stage object detection. The authors have reviewed numerous papers and their contribution to object detection; although many survey papers have been published on object detection on border way, but the literature lacks paper focusing on recent development and the recent start of art which have achieved great success in two-stage object detection. Furthermore, the authors have provided a summary on convolutional neural network (CNN) architecture, which serves as the backbone network for a feature extractor in the detection task and is described as the most popular two-stage detector. The authors have also summarized an explanation of popular benchmark datasets for object detection and evaluation metrics.

Section 2 presents the problem definition. Section 3 discusses the popular backbone architecture for object detection. In Sect. 4, authors have covered detail description of two-stage detection methods. Section 5 summarizes information about the application of object detection. Finally, conclusion is given in Sect. 6.

2 Problem Definition

Handcrafted features were one of the main limitations in terms of obtaining good accuracy computer vision tasks. However, with the rise of DL methods, the accuracy of solving vision problems has improved significantly. One of the major problems was object classification, which refers to categorizing all objects present in images into their respective classes. Object detection, also described as object category detection, is a more complex task than classification, as it involves predicting the class of a particular object and its precise location from an input image.

3 Popular Backbone Architecture

The primary requirement of good object detection is to learn a good feature representation. If the learned features are good enough, high accuracy in terms of object detection can be achieved. The popular backbone DCNN architecture widely used in object detection is AlexNet, VGGNet, ResNet, InceptionNet, and ResNeXt.

AlexNet was the first network architecture to be proposed by Krizhevesky in 2012 [3]. It possessed the ability to learn good representation from input images, with a minimal number (8) of layers. It has improved accuracy by a huge margin in the ILSVRC classification challenge [4]. VGG-16, with 16 layers, was based on AlexNet. After further increasing the number of the layers to 20 o network witnessed a dip in accuracy. In [5], the concept of a skip connection was introduced and the new ResNet was proposed, which reduced difficulties pertaining to optimization. This network can be extended to 100 layers with only a few parameters, as compared to VGGNet and AlexNet. Later, its various variants were proposed.

4 Detection Scheme Build on Deep Learning

In a two-stage detector, the first stage is used to generate the proposal in which potential objects can be present. During the second stage, predictions are made based on the generated proposal. The current two-stage detector can more accurately predict an object's location based on benchmark datasets.

4.1 R-CNN

R-CNN was the first network to be formed on CNN. After the success of the CNN in classification tasks, Ross Girshick proposed the R-CNN network for object detection. The R-CNN detects objects in three phases:

(i) Region generation phase
(ii) Extraction of feature phase
(iii) Prediction phase for classification and regression.

In the first phase, the R-CNN makes use of selective search algorithm (SRA) to select important regions in every input image; the selected regions are known as proposed regions. The advantage of using selective search is that it searches 2000 regions where objects can be present. In the second stage, the selected regions are cropped, resized, and fed into the CNN. At this phase, the CNN produces a 4096 dimensional feature vector as output. In the final step, classification and bounding box prediction happen. Architecture of the R-CNN can be viewed in Fig. 3.

The R-CNN considerably improved the object detection performance of traditional algorithms by a huge margin. However, it still has a few flaws:

(i) The extraction of features from the 2000 selected regions through a deep CNN requires a long computational time.
(ii) Optimization is difficult, as the network is divided into three stages.
(iii) Computational time is a lot for test images.

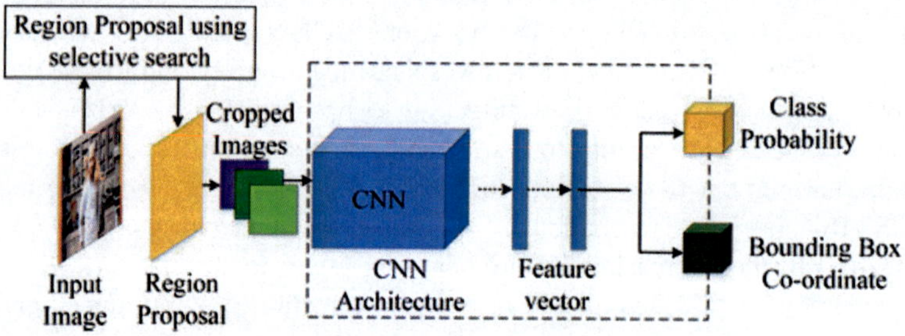

Fig. 3 R-CNN

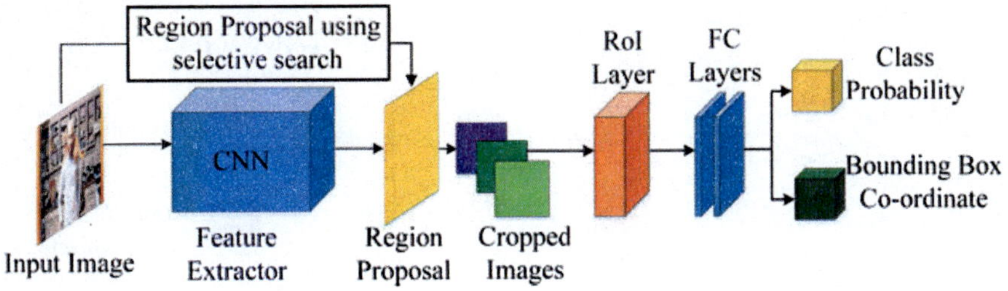

Fig. 4 Fast R-CNN

4.2 *Fast R-CNN*

After the addressing constraint of the SPPNet and R-CNN, Ross Girshick et al. proposed a fast detection algorithm called fast R-CNN. It is same as the R-CNN, except that the generated region is fed to the CNN. It takes the whole image as input and feeds it to the CNN to obtain the convolutional features map. After convolution, the feature map goes from the RoI layer, which generates the reshaped feature with a fixed size. Fixed features are fed to the classification and regression layers to predict class labels and bounding boxes, respectively. Fast R-CNN extracts features from entire images, whereas R-CNN uses 2000 regions to extract features. This saves immense time during training and testing. Architecture of the fast R-CNN can be viewed in Fig. 4.

4.3 *Faster R-CNN*

Both previous networks were based on traditional SRA, which were slow, time-consuming, and capture only low-level features in the features map. Faster R-CNN, developed by [4], uses the RPN to generate regions based on the CNN. The RPN generates regions from input images by feeding them into the CNN. It also increases the generation of region proposals with the aid a common set of CNN layers with detector network. After being generated, the regions are changed using RoI pooling layer. The image is then fed to the classification and regression layer for label classification and offset prediction. Faster R-CNN achieved relatively better results with respect to object detection benchmark datasets for instance MSCOCO, Pascal VOC, and ILSVRC [5]. The stages of the network have been outlined in Fig. 5.

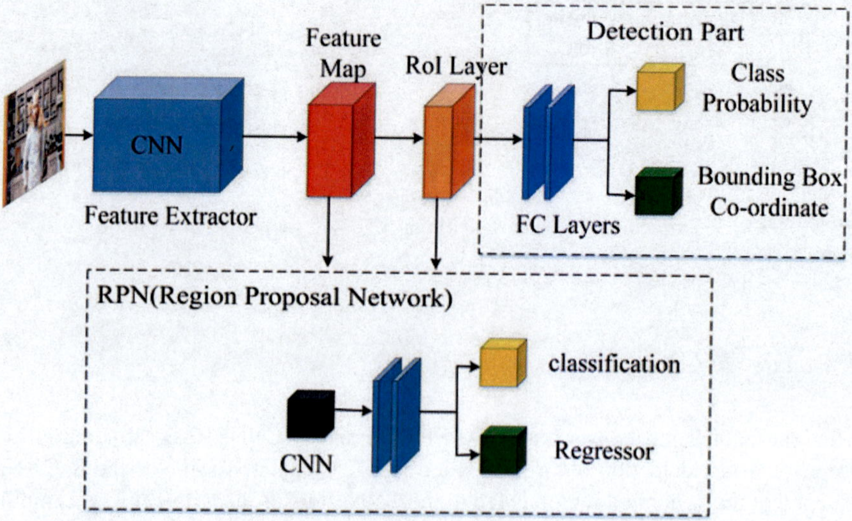

Fig. 5 Faster R-CNN

4.4 *R-FCN*

In R-FCN [5], the fully connected (FC) layers that follow the RoI layers are removed, and all major complex features are assigned before the RoI layers. The R-FCN generates position-sensitive maps, which contain information about position regarding distinct classes. The position-sensitive RoI layer is applied to pull out features from score maps. The R-FCN makes use of simple average voting on extracted features from the RoI layer to generate a class vector. At last, the Softmax function is performed on this vector to predict the class score. Architecture of the R-FCN can be viewed in Fig. 6.

4.5 *Mask R-CNN*

For pixel level detection, He et al. [6] developed the instance segmentation algorithm, the mask R-CNN. This can be viewed as an extension of the faster R-CNN. The Mask R-CNN uses a two-phase strategy. In the first phase, it uses the RPN to generate regions where objects might be present. In the second phase, it foresees the binary mask based on the feature map. A mask-generating branch based on CNN is used to better capture the relevant areas. The mask R-CNN uses RoI align layer, in place of the RoI layer with backbone architecture. Mask R-CNN is simple to accomplish and achieves better accuracy in terms of the instance segmentation task. Figure 7 shows the architecture of the mask R-CNN.

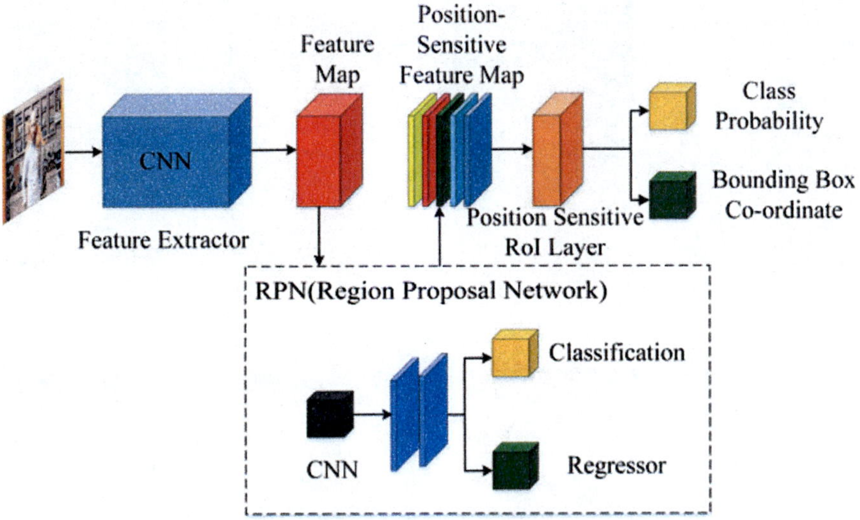

Fig. 6 R-FCN

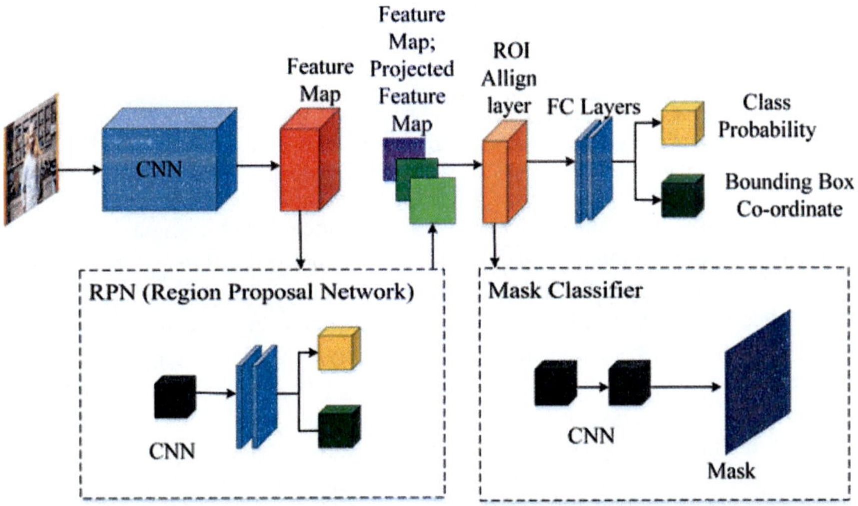

Fig. 7 Mask R-CNN

5 Object Detection Application

There are wide range object detection application in real-world scenarios, spanning from social to personal levels (Fig. 8).

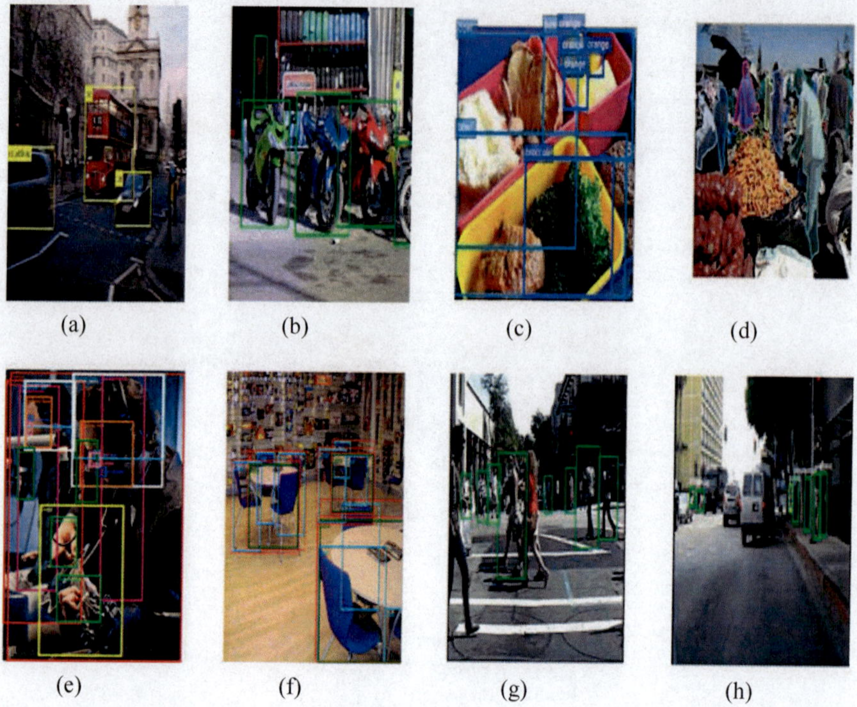

Fig. 8 Benchmark datasets, example **a**, **b** are from Pascal Voc dataset, example **c**, **d**, are from MS COCO dataset, **e**, **f** are from Open Image v5, example **g**, **h** from Caltech dataset

- **Face detection**: Detection of faces is a very prominent area in computer vision; it involves detecting human faces from images or videos. It has many applications for instance in security, health care, advertisements, and so on.
- **Pedestrian Detection**: It is worth noting that several specific datasets have been published on pedestrian detection. The Euro City Persons dataset, for example, contains information regarding pedestrians, cyclists, and other riders in traffic areas.
- **Text Detection**: Text detection deals with detecting text area in images or videos text detection have many applications for example in identifying vehicles by reading number plates, in assisting visually impaired persons.

6 Conclusion

Over the last few years, with the advancement of DL, object detection tasks have evolved rapidly. In this survey, the authors reviewed the modern literature on object detection, covering all relevant information about two-stage object detection and describing backbone architecture. The authors also covered the popular benchmarks of object detection and evaluation matrix. The authors even attempted to cover all terminologies in a deterministic manner to allow the survey to better compress object detection based on deep learning.

References

1. Girshick R, Donahue J, Darrell T, Malik J (2014) Rich feature hierarchies for accurate object detection and semantic segmentation. In: Proceedings of IEEE computer society conference on computer vision and pattern recognition, pp 580–587. https://doi.org/10.1109/CVPR.2014.81
2. Redmon J, Divvala S, Girshick R, Farhadi A (2016) You only look once: Unified, real-time object detection. In: Proceedings of IEEE computer society conference on computer vision pattern recognition, pp 779–788. https://doi.org/10.1109/CVPR.2016.91
3. Girshick R (2015) Fast R-CNN. In: Proceedings of IEEE international conference on computer vision, ICCV 2015, pp 1440–1448. https://doi.org/10.1109/ICCV.2015.169
4. Ren S, He K, Girshick R, Sun J (2015) Faster r-cnn: towards real-time object detection with region proposal networks. In: NeurIPS
5. Dai J, Li Y, He K, Sun J (2016) R-FCN: Object detection via region-based fully convolutional networks. In: Advances in neural information processing systems, pp. 379–387
6. He K, Gkioxari G, Dollár P, Girshick R (2020) Mask R-CNN. IEEE Trans Pattern Anal Mach Intell 42:386–397. https://doi.org/10.1109/TPAMI.2018.2844175
7. Iqbal A et al (eds) (2020) Soft computing in condition monitoring and diagnostics of electrical and mechanical systems. In: Advances in intelligent systems and computing, vol 1096. Springer, Singapore. https://doi.org/10.1007/978-981-15-1532-3
8. Iqbal A et al (eds) (2020) Meta heuristic and evolutionary computation: algorithms and applications. In: Studies in computational intelligence, vol 1096. Springer, Singapore. https://www.springer.com/gp/book/9789811575709

A Review Paper on Analysis, Planning of Electricity Generation in Turkey, United Arab Emirates and Germany and Comparison with India

Rahat Ullah Khan, Gazali Ashraf, Gaurav Srivastava, Mohammad Hamza Habib, and Israr Ahmad

Abstract This paper focuses on the comparison of various methods and quantity of electrical power generation in the countries as Turkey, United Arab Emirates and Germany including sources of electricity and comparison with India is highlighted. In this paper, there is an outlook of various power generation techniques adopted, and the amount of the power with the help of various means are highlighted and put for the immediate comparison. It is also discussed that a brief outlook of power generation and their comparison show the significance of the geographical, climatic and political situations emphasis on the methods of their power generations, respectively. In different countries, the nature of electrical power generation has shown with graphical form on the account of percentage utilization of energies in recent year. Various power generation methods like thermal, wind, solar, gas, nuclear, hydro electricity, other renewable energy sources, etc., are discussed in details. A detailed comparison of cost of generated power and cost of a single unit per person according to the respective country has been discussed. By knowing this, it may be analyzed that different sources of power generation can be compared with the power generated in India. With the help of this, analysis and comparing to the utilization of different types of sources for the purpose of generation of electricity by different countries can be understood including source for production of electricity as compared to renewable source of energy.

Keywords Electrical power generation · Different countries electrical power comparison · Types of electrical energy generation sources

1 Introduction

In present scenario of electrical energy generation, most of the nations totally relay on the exhaustive sources of energy generations. One day all such sources will get an end, and one day all these countries will turn their energy generation resources

R. U. Khan (✉) · G. Ashraf · G. Srivastava · M. H. Habib · I. Ahmad
KIET Group of Institutions, Ghaziabad, Uttar Pradesh, India
e-mail: rahat.khan@kiet.edu

A. Iqbal et al. (eds.), *Renewable Power for Sustainable Growth*, Lecture Notes in Electrical Engineering 723, https://doi.org/10.1007/978-981-33-4080-0_65

renewable sources for their energy harnessing, hence it is a necessity to endorse the renewable sources in the countries for electricity generation [1–8].

This paper creates the evaluation of various power generation methods for different countries. It is well known that major advantages of renewable sources of generation over non-renewable sources are less air pollution, non-extensible sources of energy, although the generation of electricity in renewable sources is quite less than the generation by non-renewable sources. In addition, the efficiency of non-renewable sources is much better than renewable sources.

1.1 Importance of this Analysis and Study

The importance of this study is with the help of this analysis it could be easy to understand the energy scenario in between different demographically situated nations. One can easily understand the detailed analysis of the generated power quantity and the price of generation of electrical energy and by how many means of sources these countries are generating electricity.

2 Electrical Power Generation Schemes of Turkey

Turkey has various types of electrical power plants. Below are the details along with their generating capacities as shown in Fig. 1 (Table 1).

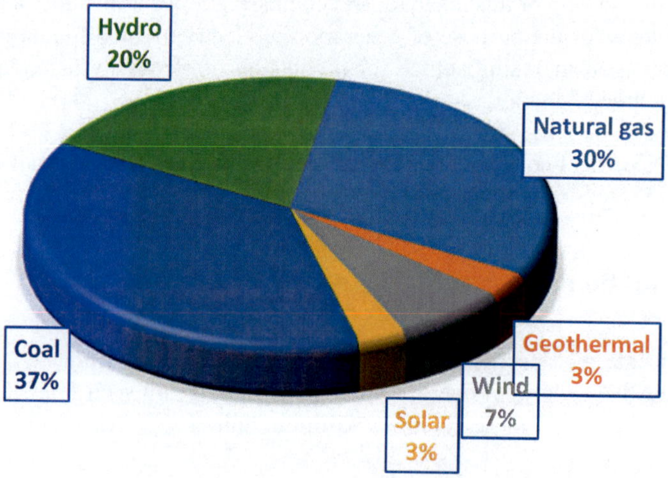

Fig. 1 Different types of power generated in Turkey (generated power shown in %)

Table 1 Different power plants in Turkey

Hydro electric power plant	*Solar power plant*
Total Number of Watt Power Produce from Hydro Plant-35 GW (Total Number of Plants-71)	Total Power Produce from Solar Plant-45 MW (No. of Total Power Plants-02)
Number of wind power plant	*Geothermal power plants*
Total Power Produce from Wind Plant-1023.4 (MW) Total Power Plant-11	Total Power Produce from Geothermal Power Plant-108.2(MW) Total No. of Plant-02
Natural gas power plant	*Coal power plant*
Total Power Produced From Natural Gas Power Plants-1942 (MW) Total Power Plant-10	Total Power Produced from Coal Power Plant-17,181(MW) Total Power Plant-22

2.1 Energy Policy of Turkey

Turkey fulfills its one-fourth of energy demand from National Resources of Renewable and Non-Renewable Energy. Most of the remaining part of its electrical energy is supplied by imports of oil and gas, which is imported from Russia, Iraq and Iran. According to the data of 2018, energy import costs $43 billion. Government of turkey is planning for half of the electricity to be generated from renewable energy resources available in Turkey. By 2023, the capacity targets of 34.0 GW for hydro electricity, 20.0 G-watts for wind, 5.0 G-watts for solar PV, 1.0 GW for biomass and 1.0 G-watts for geothermal resources [1].

2.2 Electricity Price in Turkey

For household = 0.090 US$ per kWh and for industry = 0.107 US$ per kWh.

2.3 Different Power Plant Install and Generation Capacity in Turkey

Turkey's consumption of electricity is over 6 Hex joules of principal energy per year, over 20 (MW/h) per person. The 88% of energy is produced from the fossil fuels. The energy policy of Turkey to reducing fossil fuel imports in future, which are over 20% of import costs [2] (See Fig. 2).

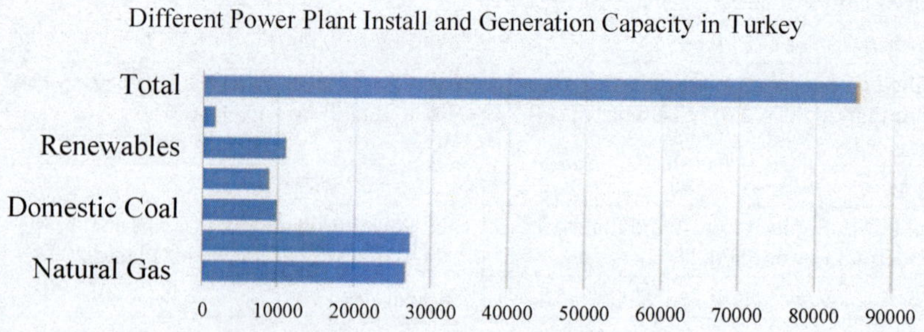

	Natural Gas	Hydraulic	Domestic Coal	Import coal	Renewables	other	Total
installed capacity	26638	27273	9872	8794	11000	1623	85200
Installed Share	31	32	11	10	13	3	100
Generation	108.1	18.3	44	51.1	26.5	7.5	295.5
Generation Share	37	20	15	17	10	1	100

Fig. 2 Different power plant install and generation capacity in Turkey

2.4 Cost of Generation of Different Power Plants

See Table 2.

3 Electrical Power Generation Schemes and Plan of Germany

Any power generation plant is an industrial necessity which is used to generate electric power with the help of one or more than one generators.

The generation of electricity is mainly from the primary resource of energy mainly for any electrical or non-electrical utilities in various industries. In 2018, German

Table 2 Cost of generation of different power plant

S. No.	Power plants	Cost (US$) (per kwh)
1	Hydroelectric power plant	0.073
2	Wind power plant	0.073
3	Geothermal power plant	0.105
4	Solar power plant	0.133
5	Biomass power plant	0.133

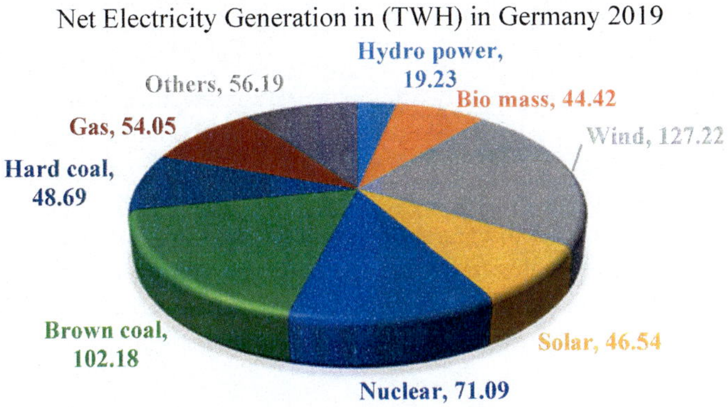

Fig. 3 Net electricity generation in (TWH) in Germany 2019

electrical power generation capacity was 540 T-Wh of electrical energy of which 40% was explored by renewable energy resources, 38.0% from good quality coal and 8.0% energy by natural gas. On the other hand, nuclear power generation has been decreased slightly in the year 2013 to 2014, and electricity generated from hard brown coal and gas-fired power plants decreased drastically by 3 to 9.5 to 13.8%, respectively. Germany will withdraw the electricity generation from nuclear power till maximum by 2022, and as of 2019, coal powered plants are planned to be pulled out by 2038.

3.1 Power Generation Capacity of Germany

Germany was holding the world's biggest photovoltaic installed capacity until 2014, and up to 2016, it was the third with 40 GW solar power generation. Also the world's largest third country whose installed wind power capacity is 50 GW and stands for second for offshore wind, with over 4 GW in the world rank, installed capacity: 198.45 GW (2018), share of renewable energy: 38.2% (2018) share of fossil energy: 61.8% (2018) (Fig. 3).

3.2 Energy Policy of Germany

On the date January 26, 2019, as per the final report presented by the commission, which foresees a path toward the gradual elimination of coal-fired power generation till end of year 2038, if possible by 2035. This will make Germany, the only industrialized country to choose out of both the energy, i.e., nuclear and coal energy.

3.3 *Future Energy Plan of Germany*

It has been cleared that the property of climate changes is obvious to observe, and nations across the world are taking steps to how to reduce orangery gas emission in an effort to bound global temperature rises. In reality, Germany intends to eliminate all coal-fired power generation by 2038, and nuclear power by 2023, considerably changing the energy market in the country. According to Berlin-based policy institute Agora Energiewende, the speed and scope of the Energiewende are brilliant. Energiewende: Germany's quest to achieve a complete renewable energy future [3].

4 Electrical Power Generation Schemes and Plan of United Arab Emirates

Presently, there are more than 27 giga-watts of installed capacity to generate electrical power across the seven emirates by using natural gas, which is the cleanest fossil fuel available in the region. The different sources of electricity production have been shown in Fig. 4.

4.1 *Solar Power Plants in UAE*

The addition to renewable power generation in UAE, it is the primary source of electrical energy which is solar power, plays a great role when it competes with stable sources of normal electrical supply.

Solar energy is presently considered majorly promising renewable energy technology for the UAE as the country is sacred with copious sunshine.

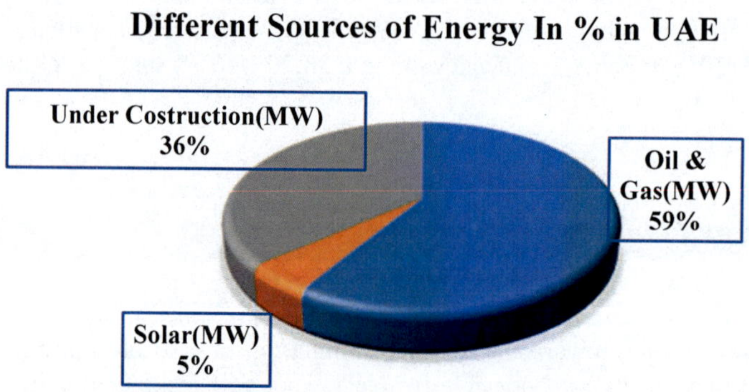

Fig. 4 Different type of source capacity of electrical energy in UAE

Table 3 Names and capacity of the different oil and gas power plants in UAE

Electricity prices per kWh	Date	AED	USD
Households	01/06/2019	0.296	0.081
Business	01/06/2019	0.366	0.1

The government approximates that savings between $01 billion and $3.7 billion might be achieved by achieving its renewable energy generation target and now believes that the savings could be even greater with the changed outlook for vestige fuels and renewable energy generation and supply prices [4, 5].

4.2 Energy Policy of UAE

The total length of transmission line used in UAE is 3305 km out of which 400 kV line is having length of 876 km and that of 132 kV line length is 2429 km. There are total 202 substations out of which 18 are 400 kV substations and 184 are 132 kV substations shown in Table 3.

5 Electrical Power Generation in India

5.1 Energy Policy of India

India is the only country in nearby area where the cost of generation of solar power is approximately 15% less than cost of electricity generated by coal. According to the study, balance cost of solar power generation in India is approximately at around $38.20 MWh (Rs. 2.62/unit), which is the lowest also and followed by Australia with the cost of $52.70/MWh (Rs. 3.62 per unit) including China at $ 61.20/MWh (Rs. 4.20/units). India's levelized cost of onto dry land wind energy generation, which is approximately at $48.90/MWh (Rs. 3.36 per unit), is cheaper in the countries in comparison with Australia with the cost of $74.6 per MWh which translate to Rs. 5.13 per unit.

5.2 Power Generation in India

See Fig. 5.

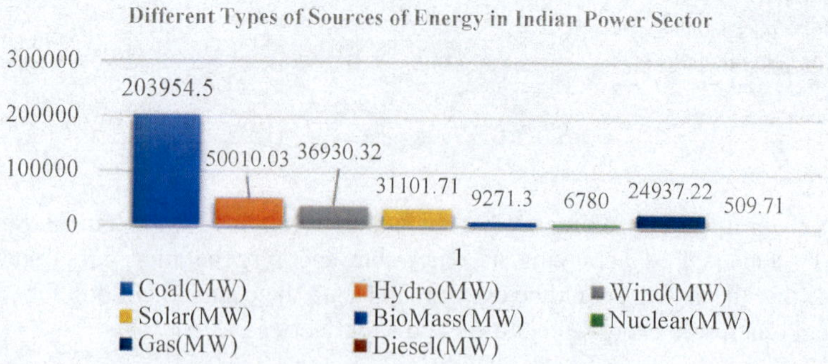

Fig. 5 Different type of sources of energy in India

5.3 *Future Plan for Power Generation of India*

India plans to have 20 GW of installed nuclear capacity by 2020 and 63GW by 2032. The country aims to generate 25% of its total electricity from nuclear energy by 2050. India plans to have 20 GW of installed nuclear capacity by 2020 and 63 GW by 2032. India currently has 37 nuclear reactors planned or under construction. This shows the country's commitment to plans to generate 25% of its electricity from nuclear energy by 2050. However, it also needs to deal with uranium shortages and public opinion, reports Global data. India plans to have 20 GW of installed nuclear capacity by 2020 and 63 GW by 2032 [8]. India is also planning to use the recovered plutonium from spent fuel to trial in the prototype fast breeder reactor (PFBR). It intends to develop a program for using its vast thorium reserves.

6 Electrical Power Generation Schemes in India and Comparison in Between Turkey, Germany and UAE

6.1 *Comparison of Power Sector of India, Turkey, Germany and UAE*

See Table 4.

6.2 *Comparison of Cost of Electricity Per Unit of Turkey, UAE, Germany and India*

See Fig. 6.

Table 4 Comparison of different power sector of Turkey, UAE, Germany and India

Type of energy source/countries	Turkey	UAE	Germany	India
Hydro power (MW)	35,000	N.A.	19,230,000	50,010.03
Wind (MW)	1023.4	N.A.	127,220,000	36,930.32
Solar (MW)	45	1300	46,540,000	31,101.71
Coal (MW)	17,181	2400	150,870,000	203,954.5
Oil & Gas (MW)	1942	15,841	54,050,000	25,446.93
Nuclear (MW)	N.A.	N.A.	71,090,000	6780

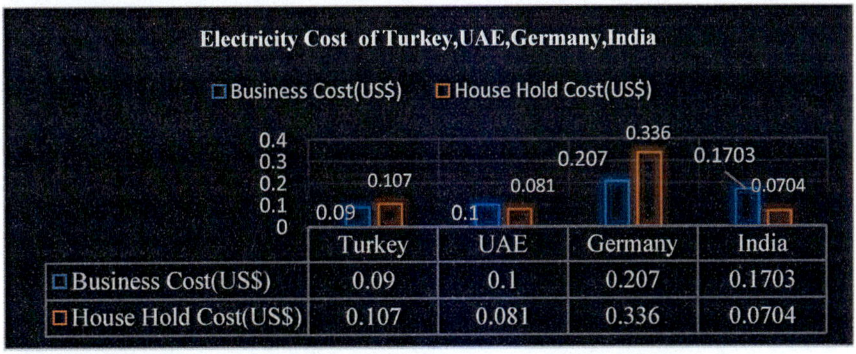

Fig. 6 Different cost of electricity per unit of Turkey, UAE, Germany and India

7 Conclusion

It has reviewed various types of energy for the countries like Turkey, United Arab Emirates, Germany and India in details. During the review of these countries, it has been found that energy trends of these countries, like Germany, have very wide scope of the energy generation, and the per-capita power consumption is more in comparison with other countries. On the basis of the type of fuel used in bulk is the coal for Germany, Turkey and India, and on the other hand oil and gas are the most prominent fuel used in UAE to generate electrical power. Based on nuclear energy generation, Germany is leading in these comparative nations. In comparison with renewable energy resources, wind and solar are the most domination resources of energy harnessing for the Turkey, Germany and India in comparison with UAE, and on the other hand the energy policy of UAE is encouraging in upcoming years in comparison with other countries if we consider the per person demand of the electrical power. The cost of electricity per unit is business/commercial which is the cheapest in comparison with all the countries and the cost of domestic/household electricity in India is leading among these countries.

References

1. Sengül, Ü.; Eren, M.; Shiraz, S.E.; Gezder, V.; ¸Sengül, A.B. Fuzzy TOPSIS method for ranking renewable energy supply systems in Turkey. Renew. Energy 2015, 75, 617–625
2. Uluta¸s, A.; Karaca, C. Selection of Renewable Energy Sources for Sustainable Development and an Economic Model Proposal for Countries. In Emerging Economic Models for Global Sustainability and Social Development; IGI Global: Hershey, PA, USA, 2019; pp. 65–83
3. Mezher T, Dawelbait G, Abbas Z (2012) Renewable energy policy options for Abu Dhabi: drivers and barriers. Energy Policy 42:315–328
4. Harder E, Gibson JM (2011) The costs and benefits of large-scale solar photovoltaic power production in Abu Dhabi. United Arab Emirates. Renew Energy 36:789–796
5. Farghal SA, Abdel Aziz MR (1988) Generation expansion planning including the renewable energy sources. IEEE Trans Power Syst 3(3):816–822
6. Iqbal A, et al. (eds), Soft computing in condition monitoring and diagnostics of electrical and mechanical systems, vol 1096. In: Advances in intelligent systems and computing. Springer, Singapore, 2020. doi:https://doi.org/10.1007/978-981-15-1532-3
7. Iqbal A, et al. (eds), Meta Heuristic and Evolutionary Computation: Algorithms and Applications, vol 1096. In: Studies in Computational Intelligence. Springer, Singapore, 2020. doi:https://www.springer.com/gp/book/9789811575709
8. Central Electricity Authority (2015) Annual Report 2014–2015; Government of India, Ministry of Power. New Delhi, India, Central Electricity Authority

A Transformer-Less Ultra-Gain Switched Inductor Boost Converter for DC Microgrid Applications

Md. Samiullah, Imtiaz Ashraf, and Atif Iqbal

Abstract In a DC microgrid, the integration of renewable energy sources requires a DC-to-DC converter for a better communication and energy transfer. The conventional boost converter endures many snags owing to its operation at an extremely large duty ratio to obtain a substantial suitable high voltage at its output terminals. This paper introduces a novel converter which is capable of attaining a significantly high gain at a nominal duty ratio. The converter structure is free from transformer, coupled inductors, or any cascading complex networks. The proposed topology uses switched inductors along with a capacitor and diode to act as a boosting element. Apart from a detailed topological description, the converter is analyzed in steady-state modes of CCM and DCM. The analytical findings of gain factor other relevant parameters have been well verified by the simulation results at the end.

Keywords DC microgrid · DC–DC converters · High gain · Switched inductor

1 Introduction

The demand of energy is continuously growing owed to many obvious reasons behind. The renewable energy sources have therefore gained a deep attention and being promoted incessantly [1–13]. Solar photovoltaics (PVs) are increasingly being integrated into power distribution in order to trim down the burden on the conventional power system [12, 13]. Power electronic converters act as a mainstay for the appropriate utilization of these solar PVs owing to their intermittency and uncertainty behavior [2–4]. Distributed generation and microgrids are providing a source for interfacing a solar PV into the traditional structure of the power system. A DC-to-DC converter is therefore a prime requirement for properly linking the solar photovoltaic and other sources to DC microgrids [5, 6]. The output voltage at the terminals

Md. Samiullah (✉) · I. Ashraf
Aligarh Muslim University, Aligarh, India
e-mail: md_samiullah@zhcet.ac.in

A. Iqbal
Qatar University, Doha, Qatar

A. Iqbal et al. (eds.), *Renewable Power for Sustainable Growth*, Lecture Notes in Electrical Engineering 723, https://doi.org/10.1007/978-981-33-4080-0_66

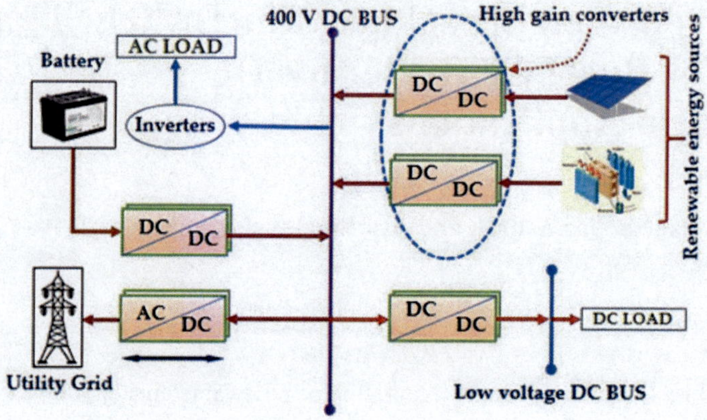

Fig. 1 A typical DC microgrid

of these generating stations are very low (12–40 V) which needs either the operation of the conventional DC–DC boost converter at an extremely large duty ratio or some high gain boost converters [7]. The traditional single switch boost converter when operated at some large duty ratio causes many detrimental effects on the circuit and may result in reduced efficiency. The most common effects while a boost converter is working with higher duty ratio include voltage spikes across switches and diodes, more losses in MOSFETs, impaired transient response, reverse recovery issues in diodes, and more EMI [3, 8, 9]. Therefore, coming up with some high gain converters which must be capable of providing high voltage output at low or moderate duty ratio is the only option to explore (Fig. 1).

In the available literature, there exists many converters which are intended to give higher gain for different applications. Some of them are suitable for medical devices, while some others are designed for traction purposes [4]. Moreover, for microgrid applications too, there are many isolated and non-isolated types of DC–DC converters which could attain significantly high gain. A boost converter realized after using a transformer or coupled inductors can easily attain sufficiently high gain by varying the turn ratio [10]. However, these converters are bulky and heavy in size along with some other constraints such as leakage inductance and core saturation [8, 11]. On the other hand, non-isolated converters do not use any transformers and therefore comparatively compact in size. They mostly use any boosting network such as voltage lift, switched inductor, and switched capacitor, interleaving and cascading of one another. Each of the boosting circuit has its own pros and cons different from others. Nevertheless, there does not exist a single converter conveying all the desired benefits and fits at all the applications [1].

This paper introduces a novel topology for non-isolated type of DC-to-DC converter which is free from any transformer and coupled inductor. The proposed topology incorporates many features in it such as highest gain factor with optimal devise counts, low voltage stress across switches, and simple control. Moreover, low voltage R_{DS}(ON) MOSFET can be used which reduces the conduction loss and increases the converter efficiency. Switched inductor network in the circuit allows parallel charging and series discharging of the inductors which helps the converter to easily attain a significant high voltage. Furthermore, the switches being controlled simultaneously reduces the control complexity. Therefore, the proposed converter is a compact package of transformer-less circuit involving switched inductor facilitated with switched capacitor as well to yield high voltage at the output with an improved efficiency.

2 The Proposed Topology

The circuit layout for the proposed topology is depicted in Fig. 2. The converter design includes four inductors L_{11}, L_{12}, L_{21} and L_{22}, seven diodes D_{11}, D_{12}, D_{21}, D_{22}, D_1, D_2 and D_o along with three capacitors C_1, C_2, C_o and two switches S_1 and S_2. The duo switches of the converter circuit S_1 and S_2 are triggered by the same gate pulses and does not involve a complex control strategy. Meanwhile, the identical condition is assumed for the devices being used in the circuit to evade the complexity and difficulty in analysis such as the switches are assumed to be ideal along with no voltage drop across diodes in forward conduction mode and no ESR in inductances and capacitances.

2.1 *Analysis of the Converter During CCM*

The continuous conduction mode analysis has been carried out with the ideal conditions of the semiconductor devices so as to evaluate the steady-state voltage gain of the converter. The characteristics waveform during CCM has been plotted and shown in Fig. 3. The plot shows the voltages across and through the elements of the converter circuit. The analysis during CCM is done in the following two modes.

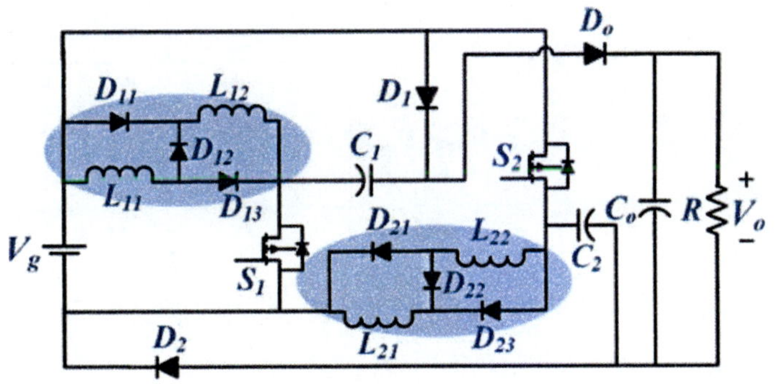

Fig. 2 Proposed converter

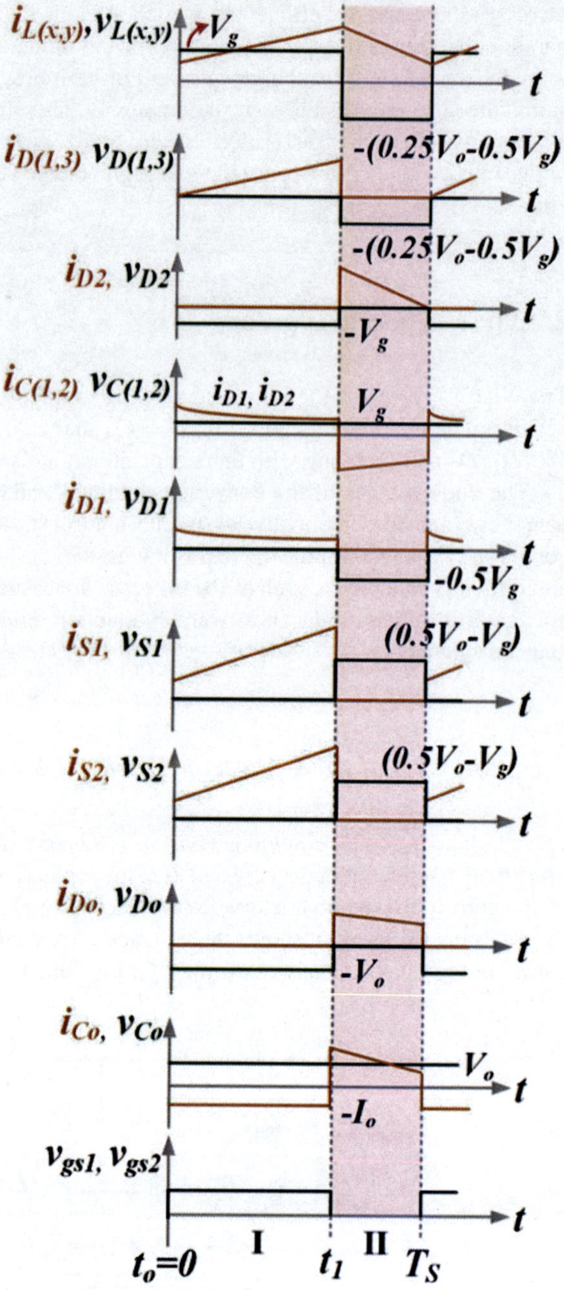

Fig. 3 Characteristics plot during different modes of conduction of the converter

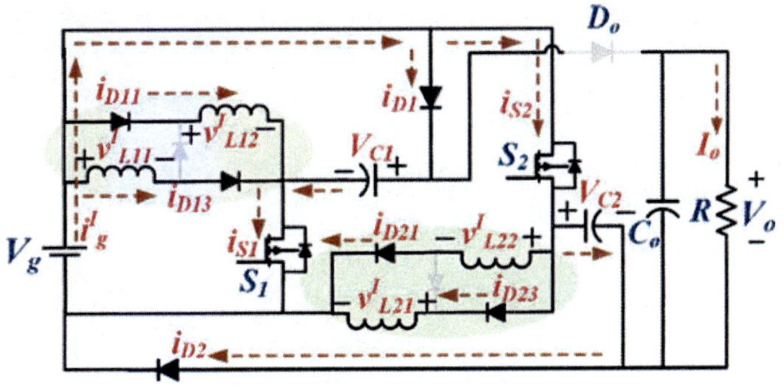

Fig. 4 Converter structure during Mode I of CCM and DCM

Mode I: (Duty period) from 0 to t_1: In this interval, the duo switches are turned to their ON state and the converter follows the topology structure as presented in Fig. 4. All the four inductors are magnetized in parallel through the switches S_1, and S_2. Diodes D_1 and D_2 are forward biased such that capacitors C_1 and C_2 also get charged up to the input voltage. The output voltage is maintained by the capacitor C_o. The voltages across the inductors during this mode are given as follows (Fig. 5):

$$\begin{aligned} v_{L11} &= v_{L12} = v_{L21} = v_{L22} = V_g \\ v_{C1} &= v_2 = V_g \end{aligned} \tag{1}$$

Mode II: (OFF-duty period) from t_1 to T_S: This is the period when both switches are turned OFF and the circuit current adopts the path as it is depicted in Fig. 4. In the equivalent circuit, it can be seen that the inductors L_1 and L_2 are discharging in series through diode D_1. The diode D_o being forward biased provides a conduction path for transferring energy from the input to the load and the capacitor C_o. The voltage across inductors can be evaluated as:

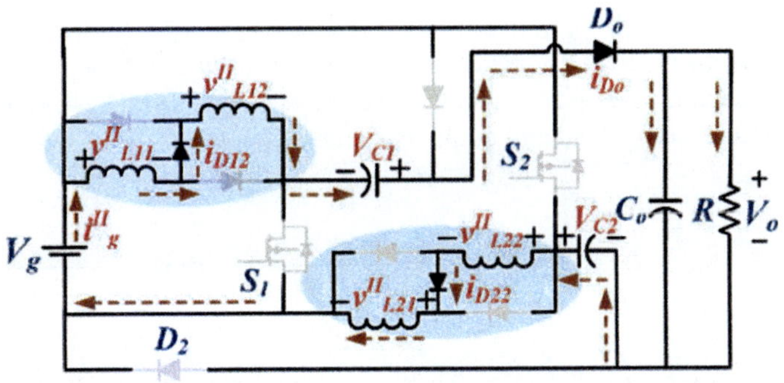

Fig. 5 Converter structure during Mode II of CCM and DCM

$$v_{L11} = v_{L12} = v_{L21} = v_{L22} = \frac{3V_g - V_o}{4}$$
$$v_{C1} = v_2 = V_g \tag{2}$$

Using principle of volt-second balance on inductor L_1, we come about with the following equations:

$$\int_0^{dT_s} V_g \mathrm{d}t + \int_{\mathrm{d}T_S}^{T_s} \frac{3V_g - V_o}{4} \mathrm{d}t = 0 \tag{3}$$

where, T_s represents the time period of the gate pulses for switches and d is the duty ratio of the circuit.

After solving (3), the following relation for voltage gain during CCM is obtained:

$$G_{\mathrm{CCM}} = \frac{V_O}{V_g} = \frac{3 + D}{1 - D} \tag{4}$$

2.2 Analysis During DCM

A brief analysis of the converter in DCM is done in order to come up with the boundary conditions such as the critical values for inductors and other parameters can be analyzed. The DCM has three modes of operation discussed below:

Mode I (from 0 to t_1): This mode of operation is same as that of mode I of the CCM converter. In this mode, the peak currents $I_{\mathrm{L11max}} = I_{\mathrm{L12max}} = I_{\mathrm{L21max}} = I_{\mathrm{L22max}}$ are same as that of change in inductor currents. Therefore,

$$\Delta i_L = I_{L11\mathrm{max}} = I_{L12\mathrm{max}} = I_{L21\mathrm{max}} = I_{L22\mathrm{max}} = \frac{V_g}{L} \mathrm{d}T_S \tag{5}$$

Mode II (from t_1 to t_2): Switches are turned OFF during this mode. The input voltage V_g, inductors L_{11}, L_{12}, L_{21}, and L_{22}, capacitors C_1 and C_2 are series connected to supply energies to the load and output capacitor. The values of inductor currents are given as

$$I_{L11\mathrm{max}} = I_{L12\mathrm{max}} = I_{L21\mathrm{max}} = I_{L22\mathrm{max}} = \frac{V_o - V_g - 2v_C}{4L} d_2 T_S = \frac{V_o - 3V_g}{4L} d_2 T_S \tag{6}$$

Mode III (from t_2 to T_S): The duo switches of the converter are still turned OFF. The circuit adopts the current path as it is presented in Fig. 6. The stored energy in the inductors has fallen down to zero. Only the output capacitor C_o is providing

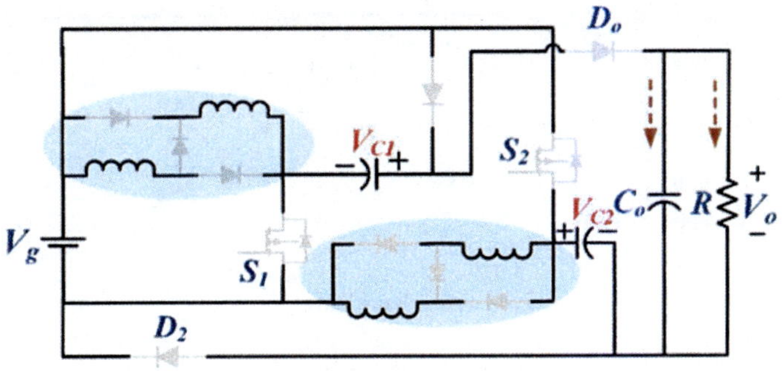

Fig. 6 Converter structure during Mode III of DCM

power to the output terminal at the load R. From (5) and (6), d_2 is obtained as

$$d_2 = \frac{4dV_g}{V_O - 3V_g} \tag{7}$$

The average current through the capacitor during each switching period can be calculated as follows

$$I_{co} = \frac{\frac{1}{2}d_2 T_S I_{L11\max} - I_o T_s}{T_s}$$

or,

$$I_{co} = \frac{1}{2}d_2 I_{L11\max} - I_o \tag{8}$$

Under steady-state condition, current I_{co} is zero. So, (8) can be written as follows with the values from Eqs. (6) and (7).

$$\frac{2d^2 V_g^2 T_s}{L(V_o - 3V_g)} = \frac{V_o}{R} \tag{9}$$

After solving Eq. (9), the following equations for voltage gain during DCM are obtained;

$$G_{\mathrm{DCM}} = \frac{V_o}{V_g} = \frac{3}{2} + \sqrt{\frac{9}{4} + \frac{2d^2}{\tau_L}} \tag{10}$$

where

$$Normalized\ inductor\ time\ constant,\ \tau_L = \frac{Lf_s}{R}$$

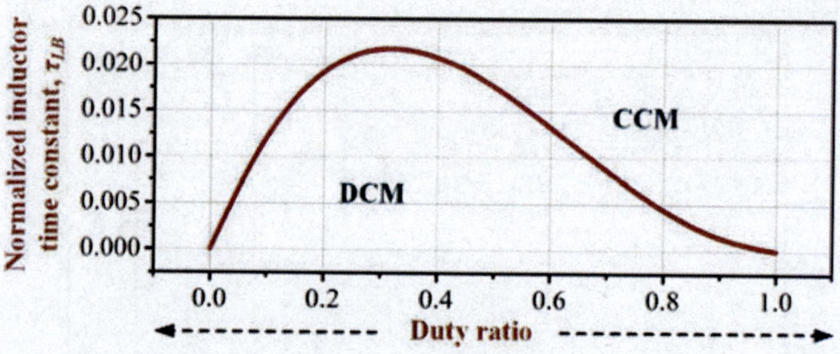

Fig. 7 Curve showing the boundary characteristics between CCM and DCM

2.3 *Operating Condition at the Boundary of CCM and DCM*

A basic conceptual background states that the voltage gain of the converter right some around the boundary of CCM and DCM is equal. The following relation defines the boundary condition and Fig. 7 presents the same,

$$\tau_{\text{LB}} = \frac{d(1-d)^2}{2(3+d)} \tag{11}$$

3 Results and Discussion

The proposed converter is studied and its behavior analysis is done for a DC input of 100 V through simulation in MATLAB. The typical values chosen for the simulation are presented in Table 1. The analytical studies helped in the selection of equipment. For the converter to operate in CCM, more than critical value inductances have chosen and the results are impressive as desired. The converter operation is carried out at a duty ratio of 60% at a frequency of 10 kHz. Figure 8 shows the input and output voltages in a single plot where a high voltage of 890 V at the output can be clearly seen for an input of 100 V. This validates the high voltage conversion with the voltage conversion ratio of the calculated one in Eq. (4). Figure 9 presents the current

Table 1 Simulation parameters

Parameters	Values
Input voltage	100 V
Duty ratio	60%
Switching frequency	10 kHz
Inductances	$L_{11} = L_{12} = L_{21} = L_{22} = 50$ mH
Capacitances	$C_1 = C_2 = 220$ µF, $C_o = 400$ µF

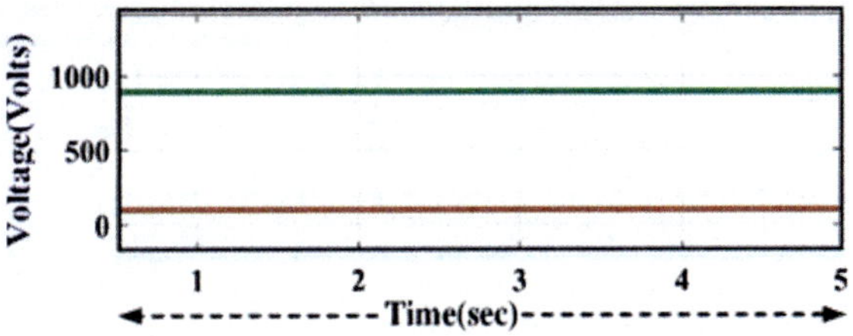

Fig. 8 Input and high voltage output

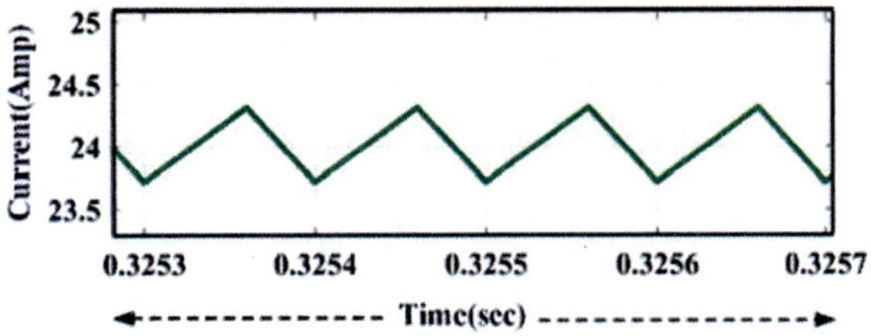

Fig. 9 Inductor current through L_{11}

waveforms through the inductors where charging and discharging of inductors are taking place in duty and OFF-duty periods, respectively. Moreover, the low ripples in the output voltage also low justifies the proper selection of the capacitance. Figures 10 and 11 show the voltages across switches S_1 and S_2, respectively. It can be observed that voltage stress across these switches is not much high even for very high voltage outputs and low rating MOSFETs can be used.

Figures 12 and 13 depict the voltages across the capacitors C_1 and C_2, respectively, where it is clearly seen that the voltages across these capacitors are constant and these voltages are attained by the capacitor once the transient period is over. The ripples can further be minimized by increasing the value of the capacitance. The voltages

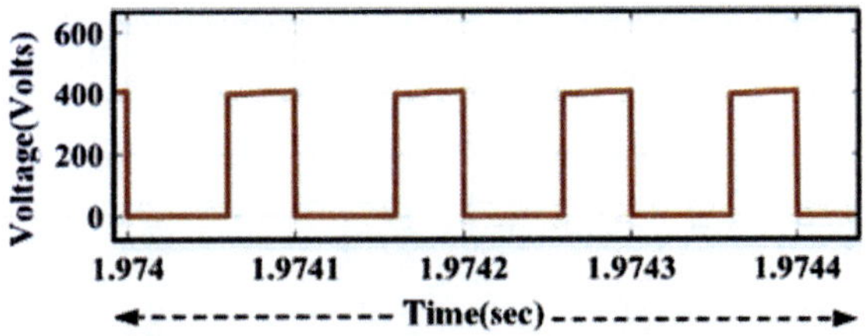

Fig. 10 Voltage across switch S_1

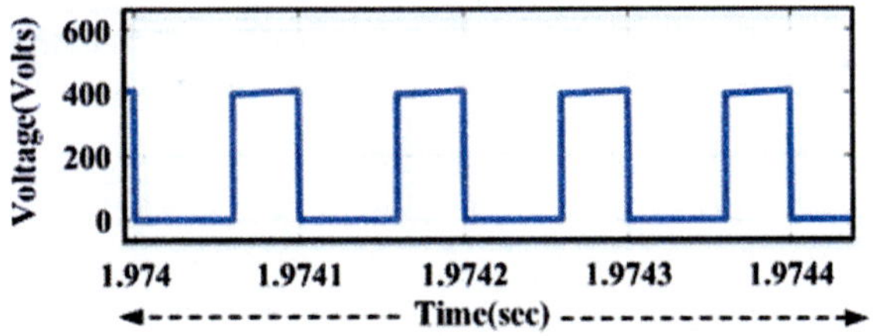

Fig. 11 Voltage across switch S_2

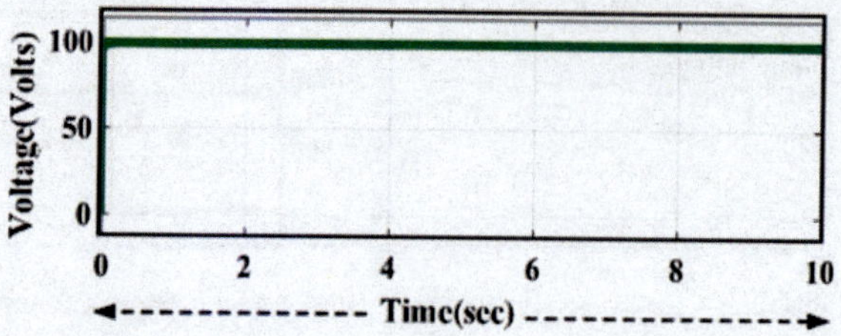

Fig. 12 Voltage across the capacitor C_1

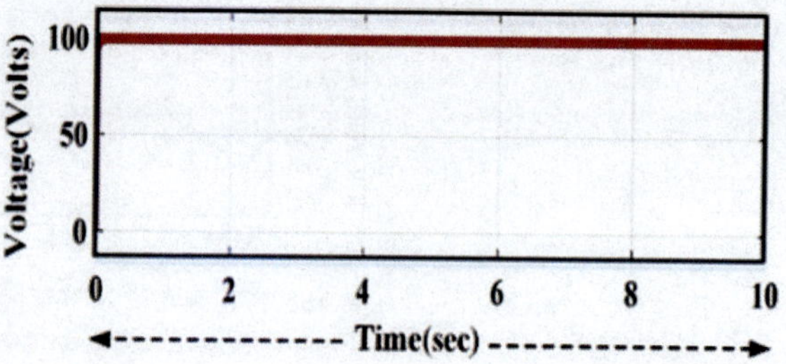

Fig. 13 Voltage across the capacitor C_2

across diodes D_{13} and D_o are shown in Figs. 16 and 17, respectively. Although the converter is designed for very high voltage applications, the diodes do not realize an extreme large potential across them and, therefore, low PIV diodes can be used (Figs. 14 and 15).

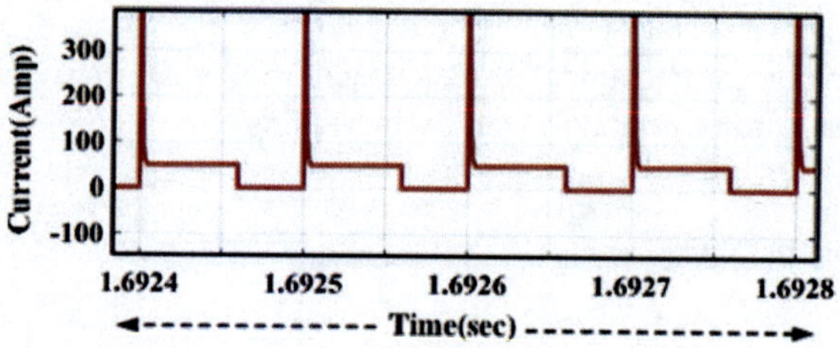

Fig. 14 Waveform of switch S_1 current

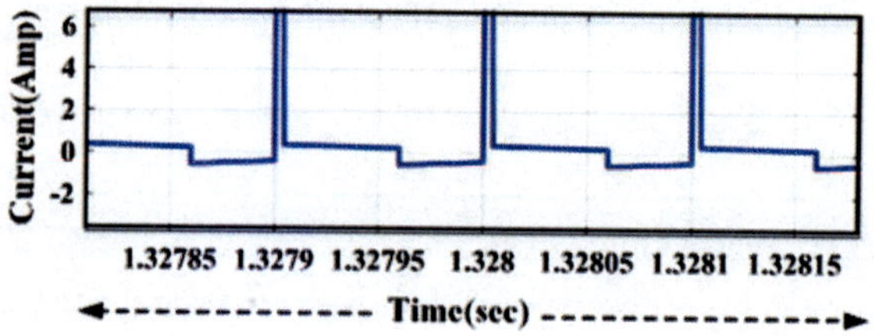

Fig. 15 Waveform of switch S_2 current

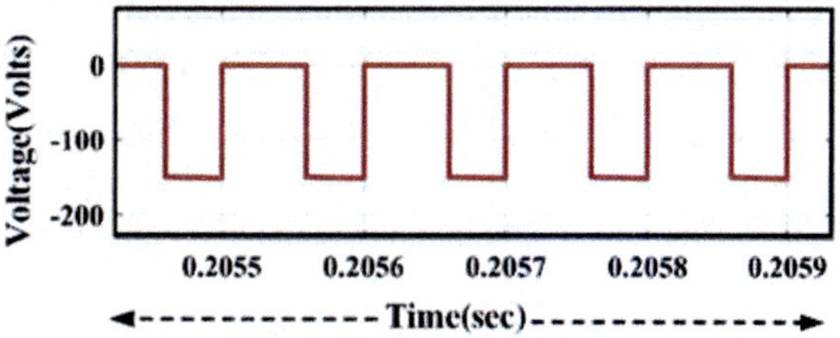

Fig. 16 Voltage waveform across the diode D_{13}

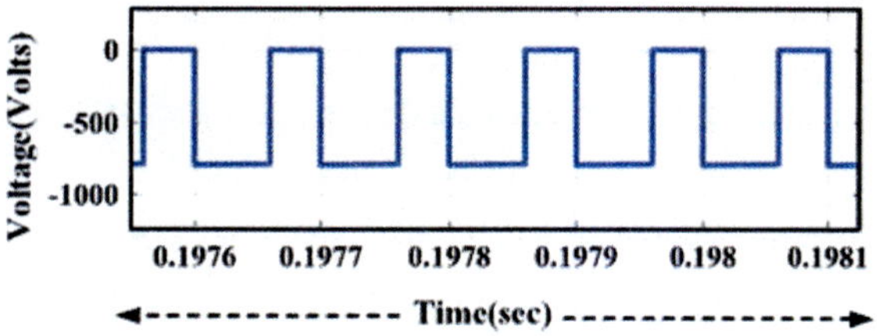

Fig. 17 Voltage waveform across the diode D_o

4 Conclusion

A non-isolated ultra-gain DC-to-DC boost converter has been presented in the paper. The converter is analyzed during its steady state to check its suitability for the applications it is designed for. The proposed converter has been well verified for DC micro-grid applications owed to its high gain behavior at a very low duty ratio with reduced stress across semiconductor devices. The structure of the converter is simple having two switches working simultaneously with the same gate pulses. The converter avails the advantages of switched inductor and capacitors for stepping up the voltage by magnetizing them in parallel and demagnetizing in series. The analytical behavior of the converter is read in CCM and DCM along with the boundary conditions and later the performances are verified through simulation results carried out at a switching frequency of 10 kHz.

References

1. Sri Revathi B, Prabhakar M (2016) Non isolated high gain DC-DC converter topologies for PV applications—a comprehensive review. Renew Sustain Energy Rev 66:920–933. https://doi.org/10.1016/j.rser.2016.08.057
2. Walker GR, Sernia PC (2004) Cascaded DC-DC converter connection of photovoltaic modules. IEEE Trans Power Electron 19:1130–1139. https://doi.org/10.1109/TPEL.2004.830090
3. Liu H, Li F (2016) A novel high step-up converter with a quasi-active switched-inductor structure for renewable energy systems. IEEE Trans Power Electron 31:5030–5039. https://doi.org/10.1109/TPEL.2015.2480115
4. Forouzesh M, Siwakoti YP, Gorji SA et al (2017) Step-Up DC-DC converters: a comprehensive review of voltage-boosting techniques, topologies, and applications. IEEE Trans Power Electron 32:9143–9178. https://doi.org/10.1109/TPEL.2017.2652318

5. Gu Y, Chen Y, Zhang B et al (2019) High step-up DC-DC converter with active switched LC-network for photovoltaic systems. IEEE Trans Energy Convers 34:321–329. https://doi.org/10.1109/TEC.2018.2876725
6. Babaei E, Jalilzadeh T, Sabahi M et al (2019) High step-up DC-DC converter with reduced voltage stress on devices. Int Trans Electr Energy Syst 29:1–24. https://doi.org/10.1002/etep.2789
7. Tang Y, Fu D, Wang T, Xu Z (2015) Hybrid switched-inductor converters for high step-up conversion. IEEE Trans Ind Electron 62:1480–1490. https://doi.org/10.1109/TIE.2014.2364797
8. Kim P, Lee S, Park J, Choi S (2011) High step-up interleaved boost converters using voltage multiplier cells. In: 8th International conference on power electronics—ECCE Asia "Green World with Power Electron ICPE 2011-ECCE Asia, pp 2844–2851. https://doi.org/10.1109/ICPE.2011.5944782
9. Jalilzadeh T, Rostami N, Babaei E, Maalandish M (2019) Ultra-step-up DC-DC converter with lowvoltage stress on devices. IET Power Electron 12:345–357. https://doi.org/10.1049/iet-pel.2018.5356
10. Starzyk JA, Jan YW, Qiu F (2001) A DC-DC charge pump design based on voltage doublers. IEEE Trans Circuits Syst I Fundam Theor Appl 48:350–359. https://doi.org/10.1109/81.915390
11. Shahir FM, Babaei E, Farsadi M (2018) Analysis and design of voltage-lift technique based non-isolated boost dc-dc converter. IET Power Electron 11:1083–1091. https://doi.org/10.1049/iet-pel.2017.0259
12. Iqbal A et al. (eds) (2020) Soft computing in condition monitoring and diagnostics of electrical and mechanical systems. In: Advances in intelligent systems and computing, vol 1096. Springer, Singapore. https://doi.org/10.1007/978-981-15-1532-3
13. Iqbal A et al (eds) (2020) Meta heuristic and evolutionary computation: algorithms and applications. In: Studies in computational intelligence, vol 1096. Springer, Singapore. https://www.springer.com/gp/book/9789811575709

A Comparative Analysis of Small- and Medium-Scale Industrial Development in Amhara Region, Ethiopia

Syed Abid Hussain, Ziaul Hassan Bakhshi, and Ahsanullah Mohsen

Abstract North Wollo is one of ten zones of the Amhara Area of northern Ethiopia with resources and promises of a state that might be extremely industrialized. This research measures the status of industrial development in the country, associating it with other states in the country. Preceding estimate and findings were also evaluated, as certain instances of existing small- and medium-scale enterprises (SMEs) within the state capital, Akure, were explored with findings presented to determine the peculiar difficulties countering against their success and level of profitability. The problems were found to be lack of availability of credit facility, sufficient infrastructures, and poor or weak management. Population size of 500 Small- and Medium-Scale Industrial Development in North Wollo Amhara Region, Ethiopia. In which researcher study the sample of 100 respondents, for 14 questions by each respondent. Likert's scale has been used for better analysis. The alpha value for the 14 items is 0.714, signifying that the elements have moderately strong internal consistency.

Keywords Social · Cultural · Entrepreneurship · Environment · Economic growth · Industry small and medium

S. A. Hussain (✉)
Department of Management, Faculty of Economics, Bakhtar University, Kabul, Afghanistan
e-mail: syedabidh38@gmail.com; dr.abid@bakhtar.edu.af

Z. H. Bakhshi
School of Business Studies and Technology, Indian Institute of Management and Technology, University, Meerut, UP, India
e-mail: bakhshistat@gmail.com

A. Mohsen
Management and Business Administration, Kabul University, Kabul, Afghanistan
e-mail: Ahsanullah.mohsen@ku.edu.af

A. Iqbal et al. (eds.), *Renewable Power for Sustainable Growth*, Lecture Notes in Electrical Engineering 723, https://doi.org/10.1007/978-981-33-4080-0_67

1 Introduction

In previous years, it has become noticeable that, small- and medium-scale enterprises (SMEs) contribute to job creation and economic development [22]. It is always believed that entrepreneurship causes the wealth creation and economic in various societies [18]. Entrepreneurship is considered to process where individuals find, develop, make, and explore different entrepreneurship activities [20]. According to Schumpeter [17], there is a strong association among economic development of nations with the participation of highly motivated people who are engaged in entrepreneurial activities.

Haggen [18], believes suggests that technological change is the reason for economic development and this process is achieved by creativity of industrialists. Another research claims that the existence of small-scale industries results in indorsing pure entrepreneurship in several countries. This in turn aids the invested capital to utilize larger opportunities and local technology to develop at a faster rate [15].

Hence, most of the economic development behaviors are depended on the entrepreneurial activities of the residents of these countries. Many developing countries, including Ethiopia perceives the small-scale firms and the hub for their economic development.

According to Hoseltiz [10], several factors influences the entrepreneurship activities, ranging from coordination and availability productive resources, uncertainty direction, introducing new innovations in the subject matter, and skills. Additionally, various elements are responsible for motivating or demotivating the business persons like, entrepreneurs' need for independence, better financial and monetary position, self-fulfillment, being one's own boss, and so on. Personal characteristics of individuals namely their age, gender, background, qualification, and work experience have statistically significant influence on them for entrepreneurship decisions [11].

Christopher [19] research shows that the main reason for commencing small businesses is the economic benefit that business owners' gains. Furthermore, high demand for output was among the promising factors. The justification for the establishment of small businesses is greater employment opportunities, equal distribution of income, which finally will result to improved living standard of people. It is also necessary to found institution which can guide the business persons in technology and counseling in order to aid the economy.

Max Weber offers the theory stating socio-culture setting affects entrepreneurship. He claims, society has a significant role in emerging entrepreneurs. Due to the fact that, people acquire their values form the society's values, morals, and institutional structure [5].

Social system and culture are both components of the sociocultural environment. Socio culture environment refers to intangible factors affecting the behavior of individuals, their living style, perception, and relationships. Equally, social culture environment shows a group of elements which develops the personality of a person and this in turn affects attitudes, beliefs, norms, values, and habits. Mentioned factors originate from educational background, social conditioning of the people, culture, and religion [3].

Gree and Thurnik [7] challenged that economic growth cannot be continued without the growth of existing SMEs and the continual formation of new SMEs. The management of SME's is significant element in commercial growth and the setting in which industries operate [4]. Study on the causes of industry failure is a wide-ranging research area and numerous cautious difficulties have been recognized by different scholars [16]. The targeted area chosen in the region, which is located North Wollo is one of 10 zones of the Amhara Region of northern Ethiopia. This study region was selected due to the developmental attractiveness of the area and its diversity of the economy which include both a superior industrial sector and a usual agricultural sector.

The SMEs do not have modern indication and are regularly uninformed concerning the up-to-date technologies existent in the global marketplace. Occasionally, they lack administrative skills, entrepreneurial comprehension, and technology demanding education to run a capable SME. This kind of tricky usually succeeds in the rural-based zones. Information Communication and Technology (ICT) can perform a bigger role for SMEs for they face firmer rivalry from the nations (Indonesia, China, Philippines, and Thailand). Therefore, the demand of the hour is upgradation of equipment. For competitive SME's, they need to use the benchmarking and best practices on international level and use the innovation of technology on regular basis.

The other substantial motives behind adoption of innovative technology are requisite to keep on to regulatory standards related to (i) waste discharge norms, (ii) product quality certifications, and others. A condition for SMEs to remain providing to global markets with severe governing policies, policy maintenance and numerous arrangements applied by administrative institutions in supporting technology upgradation, need for upgradation of technology and removal of continuation support for current technology may power SMEs to adopt new technology, and ultimately, lack of accessibility of essential level of skills among the workforce even for ordinary occupations an expanded cause overdue SMEs for fast technology reception.

According to Ayyagari et al. [2], mainly in developing countries, SMEs face plentiful of institutional barriers to a restricted access to finance, a deficient juridical system and legal implementation as well as weak property rights security. The World Bank [21] in assistance with the International Finance Corporation annually links business regulations for local companies and ranks the common of countries (189 in 2013) in the world according to their comfort of doing business for small businesses and how their regulations established over the last years. The World Bank [21], therefore, uses ten general areas of importance for businesses: set up a business, treatment of construction permits, getting electricity, recording property obtaining access to credit, shielding investors' rights, paying taxes, exchange internationally, imposing contracts, winding up failure proceedings, and retaining staff.

2 Statement of the Problem

Culture forms social and economic establishments and a foundation of character development. Due to globalization and worldwide integration, the conventional identification is modified which results in the contribution of intra-national culture for developing the entrepreneurial behavior [12]. In order to understand the business intention, it is essential to study culture. Beliefs and social values of individuals regarding the entrepreneurship influence the motivation for entrepreneurship. In the meantime, the skills are also crucial element for entrepreneurial intention. It is more likely to become entrepreneurs who have better entrepreneurship skills compared to those who do not have [12].

This paper assesses the role of socio cultural factors' role in initiating entrepreneurial intentions. Factors like, family support, friends and family expectations, educational background have an impact on evolving entrepreneurship. There is a direct and positive relationship between the formation of fresh businesses and economic growth and prosperity of the countries. Many research results constructed on psychological and economic approaches claim a significant impact of socio cultural factors on the process of entrepreneurial development of the countries.

Therefore, the main objective of this paper is combined all elements of theoretical viewpoint, the sociocultural factors, and business activities. In this study, a suitable structure to grow potential research discovering the sociocultural issues that impact the business growth. It is an essential to know the behavior of women entrepreneurs looking for finances since internal and external factors related with such entrepreneurs can help justifying problems by enclosing suitable policies and awareness plans. Cressy [6] continues that female business holders use limited sources of finance than the males. Further, it has been established that women use more internal sources of finance than male complements [13]. Hisrich and Brush [9] originate that women trust exclusively on personal savings mainly for startup and employed wealth.

3 Objectives of the Study

The aim of this research is to value the impact of sociocultural aspects on entrepreneurship development of the SMEs. The main goal of the analysis is to find the significant impacts of sociocultural elements on the entrepreneurship development of the SMEs and to form the constructive prospects of progressive SMEs.

The key research questions are:

1. What are the foremost factors that impact the entrepreneurial growth?
2. Is sociocultural factors are a foundation for developing entrepreneurial progression?
3. Does family business exposure assist in forming and emergent entrepreneurial endeavors?

4 Research Design and Sample Size Determination

The study is descriptive and exploratory research. It comprises of qualitative and quantitative research. Qualitative research was applied to explain the nature of relationship with phenomena and to attain a methodical insights of evidence to get into the complexity of study problems. However, quantitative research was applied to provide numerical measurement and analysis of the magnitude and extent of the problem. Regarding sources of data, primary and secondary sources was considered for producing valued and pertinent information. Primary data was gathered throughout field work study, interview, questionnaire, and focus group discussions. Likert scaling gives respondents a set of statements that they asked, i.e., 1 = Strongly Disagree to 5 = strongly agree. Secondary data will be gathered from formally available data.

5 Sampling Procedure

In Wollo region has been classified into two areas North Wollo and South Wollo. For the better results of research, the researcher was targeted as are mainly industrially developed towns. The target population for this study was all owner-manager (SME) manufacturing sector and service sector(Steel, Textile, oil, Agricultural, Brewery, Mobile assembly, Mineral spring water, soft drinks, construction, transport and wood and bakery) in the region, and concerned government offices across wollo region of selected areas like Kombalcha, Dessie, Hike, Woldia, Mersa, Kobo, Siringa.

To establish the sample frame, lists of manufacturing sectors will be considered with observation and local information. To get homogeneity of sample units, stratified random sampling technique was applied for collecting the data. Population size of 500 Small and Medium-Scale Industrial Development in North Wollo Amhara Region, Ethiopia. In which researcher study the sample of 100 respondents, for 14 questions by each respondent. Likert's scale has been used for better measurment. The alpha value for the 14 items is 0.714, signifying that the elements have moderately strong internal consistency (Table 1).

6 Data Analysis and Techniques

To evaluate the reliability of the measurement instrument used in this research is Croanbach Alpha Test which is most widely used coefficient was evaluated. Also

Table 1 Reliability Statistics

Cronbach's Alpha	Number of items
0.714	14

inter-item correlation, mean, mode standard deviation method were used to analyzes the data.

Analysis was carried out in two approaches, i.e., qualitative and quantitative. The qualitative was done using in-depth interview, focus group discussions, and document analysis. The quantitative was done in two stages. The first was at initial phase and second at multivariate level. At beginning phase, descriptive statistics and correlation matrix was built. Descriptive statistics was applied to define, associate, and contrast different matters linked to entrepreneurial growth and sociocultural factors considering the required features. In multivariate analysis, we run multiple regressions using cross sectional. Collected data was analyzed by using SPSS software. Overall, this research tries to explore the current condition of the variables and association among them in a real-world application of the Wollo region, both qualitative and quantitative research methods and processes have been used to explore relevant information for this analysis.

7 Conclusions

With the suitable condition of capital, intelligence, education facility and own creativity, active youth can be turned into a decent entrepreneur. These variables are supported by sociocultural and economic aspects of a nation.

Poverty extinction has been the main objective of small boldness development in most developing nations. Small- and medium-scale businesses signify 80% of industrial base of most of the developed countries [14].

Global Competitiveness Index (2014–2015), the top five difficult issues for doing business in Ethiopia are: foreign currency regulations, incompetent government bureaucracy, access to funding, corruption, and insufficient supply of infrastructure. This is maintained by results of a 2014 public-private dialogue for the National Business Agenda, 28 where firms identified the top five important constraints as: tax, access to funding, and so on. The sector contributing to economy is immense but extremely vulnerable to socio-economic changes, some of them able to survive more than 5–6 years [1].

The results show that important obstacles has to be solved. Problems such as development of adequate infrastructure, access to credit, supply of raw materials, and education of employees to be employed by the SMEs also do exist. Thus, all these problems have to be solved concurrently for the general prosperity of SMEs in the international market. Finally, the research will guide the MSMEs which are still to arrive the global market as to what are the cautious problems, how they can be solve and how can the state aid them to have access to international markets in far future.

References

1. Audet J, St-Jean E (2007) Factors affecting the use of public support services by SME owners: evidence from a periphery region of Canada. J Dev Entrepreneurship 12(02):165–180
2. Ayyagari M, Demirguc-Kunt A, Maksimovic V (2011) Small vs. young firms across the world: contribution to employment, job creation, and growth. The World Bank
3. Bennett Peter D, Kassarjian Harold H (1972) Motivation and personality. Consumer
4. Churchill NC, Lewis VL (1983) The five stages of small business growth. Harv Bus Rev 61(3):30–50
5. Christopher KJ (1974) Socio-psychological factors influencing the adoption of the innovation of starting a small industry unit—a research study. SIET Institute, Hyderabad, p 109
6. Cressy R (1992) The theory of the opportunistic entrepreneur. Small Bus Econ 4(4):267–271
7. Gree A, Thurnik C (2003) Firm selection and industry evolution: the post country performance of new firm. J Evol Econ 4(4):243–264
8. Haggen EE (1961) How economic growth begins: a study on the theory of social change. Cambridge Mass Institute of Technology, pp 11–20, 31–35
9. Hisrich RD, Brush CG (1985) The woman entrepreneur: characteristics and prescriptions for success. Lexington, MA: Lexington Books
10. Hoselitz BF (1952) Entrepreneurship and economic growth. Am J Econ Sociol 12(1):97–110
11. Kristiansen S, Furuholt B, Wahid F (2003) Internet cafe entrepreneurs: pioneers in information dissemination in Indonesia. Int J Entrepreneurship Innov 4(4):251–263
12. Linan F (2008) Skill and value perceptions: how do they affect entrepreneurial intentions? Int Entrepreneurship Manage J 4:257–272
13. Loscocco KA, Robinson J, Hall RH, Allen JK (1991) Gender and small business success: an inquiry into women's relative disadvantage. Soc Forces 70(1):65–85
14. Mathew PM (1999) Small enterprises and regional development challenges and choices, 20
15. Morch von der Fehr NH (1995) The African entrepreneur: evidence of entrepreneurial activity and firm formation in Zambia (No. 15095, p 1). The World Bank
16. Olawale F, Garwe D (2010) Obstacles to the growth of new SMEs in South Africa: a principal component analysis approach. Afr J Bus Manage 4(5):729–738
17. Schumpeter JA (1967) The theory of economic development. Oxford University Press, New York, p 621
18. Tilley F, Young W (2009) Sustainability entrepreneurs: could they be the true wealth generators of the future? Green Manage Int 55:79–92
19. Van de Ven H (1993) The development of an infrastructure for entrepreneurship. J Bus Ventur 8(3):211–230
20. Venkataraman S (1997) The distinctive domain of entrepreneurship research. In: Katz JA, Brockhaus RH (eds) Advances in entrepreneurship, firm emergence and growth. JAI Press, Greenwich, CT, pp 119–138
21. World Bank Group (2013) Global financial development report 2014: financial inclusion (vol 2). World Bank Publications
22. Wube MC (2010) Factors affecting the performance of women entrepreneurs in micro and small enterprises

An Improved 15-Level Asymmetrical Multilevel Inverter with Reduced Switch Count

Zeeshan Sarwer, Marif Daula Siddique, Adil Sarwar, and Saad Mekhilef

Abstract This paper discusses a new and modified circuit topology of a 15-level multilevel inverter having the advantage of lower switch count. The presented topology is having asymmetrical configuration. The superiority of the proposed topology is proved by the fact that it requires lower switches for generating same output levels and at the same time, it is also having lower value of reverse blocking voltage of switches. The detailed comparison with other topologies is explained and shown in tabulated form. The simulation results for different R and RL loading conditions and also for different modulation index are shown in paper. MATLAB/SIMULINK is used for the simulation. Hardware results are also discussed in the paper to validate the simulation results. These results are taken with an experimental prototype. A brief summary is also presented in the last section of the presented work.

Keywords Multilevel inverter · Nearest level control · Total standing voltage

Z. Sarwer (✉) · A. Sarwar
Electrical Engineering Department, Z.H.C.E.T., Aligarh Muslim University (AMU), Aligarh, India
e-mail: zeeshan.sarwer@zhcet.ac.in

A. Sarwar
e-mail: adil.sarwar@zhcet.ac.in

M. D. Siddique · S. Mekhilef
Power Electronics and Renewable Energy Laboratory (PEARL), Department of Electrical Engineering, University of Malaya, Kuala Lumpur, Malaysia
e-mail: marifdaula1@gmail.com

S. Mekhilef
e-mail: saad@um.edu.my

A. Iqbal et al. (eds.), *Renewable Power for Sustainable Growth*, Lecture Notes in Electrical Engineering 723, https://doi.org/10.1007/978-981-33-4080-0_68

1 Introduction

There has been a significant increase in the interests on the non-conventional sources of energy due to the exhaustion of reserved fossil fuels. From many such types of possible sources, the two important sources are the solar energy and the energy extracted from wind. Solar energy, in particular, is gaining quite a significant attraction among the researchers. In solar PV system, the incident solar energy on the solar panels is converted to the electrical form of energy. Since the output from the solar is in the form of DC, there is a need for DC-AC conversion. An inverter is used to perform this conversion process.

Nowadays, multilevel inverters (MLI) are gaining popularity. The levels are significantly higher in a multilevel inverter as against its two-level counterpart. Better quality of power and reduced harmonic distortion is the merits of these inverters over its two-level counterpart. Moreover, the stepped output voltage waveform in multilevel inverters resembles closely to a sinusoidal waveform. MLIs are used in industrial drives, renewable energy, distributed generation, HVDC, etc. [1]. Over the last few decades, there has been a significant amount of work done in area of multilevel inverters. The three conventional structures of MLI are (1) diode-clamped MLI, (2) Flying capacitor MLI, and (3) Cascaded H-bridge (CHB) MLI. The flying capacitor suffers from the drawback of necessity of greater number of capacitors. Moreover, capacitor voltage balancing is also an issue which needs attention in this topology. The problem with the diode-clamped structure is the balancing of DC link voltages and higher number of diode requirement. The CHB-type inverter is an improvement over the two discussed topologies in a way that it requires fewer number of components [2]. Further, the CHB MLI can be subdivided as symmetric topology and asymmetric topology. The symmetrical CHB multilevel inverter contains same type of DC sources whereas in asymmetric MLI, the DC voltage sources are of unequal magnitude. It is also possible to achieve a multilevel output using a switched-capacitor structure. Such type of inverter is known as switched-capacitor multilevel inverters (SCMLI) [3].

Researchers are trying to eliminate the issues in different MLI configurations and make them viable to use in different applications. One area in which there has been a lot of work in recent times is to propose new topologies with lower total number of power electronic components. One such topology consisting of nine switches and three DC sources is discussed in [4]. Both symmetrical and asymmetrical configurations are explained. In the symmetric MLI, seven levels are produced whereas by using it in asymmetric configuration, 15 levels are obtained. The authors in [5] have proposed a new asymmetrical MLI which is able to generate 17 levels. The proposed structure, named as square T-type module, contains 12 switches and diodes along with three DC sources. The extension of the structure to produce more voltage levels is also discussed. Similarly, the work presented in [6–9] discusses new structures of MLI for producing higher levels using lower number of switches.

This paper introduces an improved asymmetrical multilevel inverter topology for producing 15 levels. It requires ten switches and three different DC voltage sources to attain the desired levels. The paper is structured as: Sect. 1 contains the introduction in which a brief literature review is explained. Section 2 elucidates the proposed topology which contains the description of topology along with reverse blocking voltage calculations. This section also shows the various conducting modes of the proposed MLI and the technique used for modulation which is required for the output voltage regulation. The detailed comparison is discussed and explained in Sect. 3. The comparison is on based on the number of switches, blocking voltage of switches and the number of DC sources, etc. The different simulation results and its experimental validation are elaborated and explained in Sect. 4 of the paper. These results are shown for different R and RL loads. This section is followed by a brief conclusion in Sect. 5.

2 Proposed Multilevel Inverter Structure

2.1 Proposed Circuit for the Multilevel Inverter

The proposed MLI topology is illustrated in Fig. 1. It consists of ten power semiconductor switches and three different DC voltage sources. With this circuit configuration, it is possible to achieve 15-level output from the inverter. The switches are numbered from S_1 to S_{10}. The three voltage sources are labelled as V_1, V_2 and V_3. The values of DC voltage sources are chosen as, $V_3 = V_{dc}$, $V_1 = 3V_{dc}$ and $V_2 = 4V_{dc}$. Figure 2 shows the nature of voltage waveform which will be obtained at the

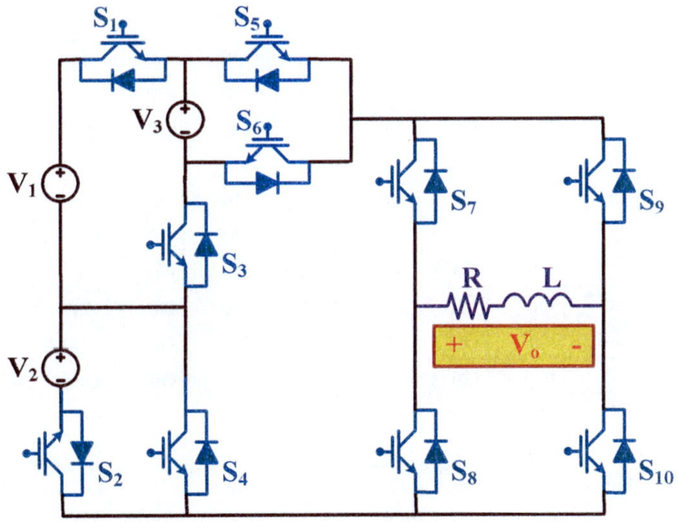

Fig. 1 Proposed circuit topology

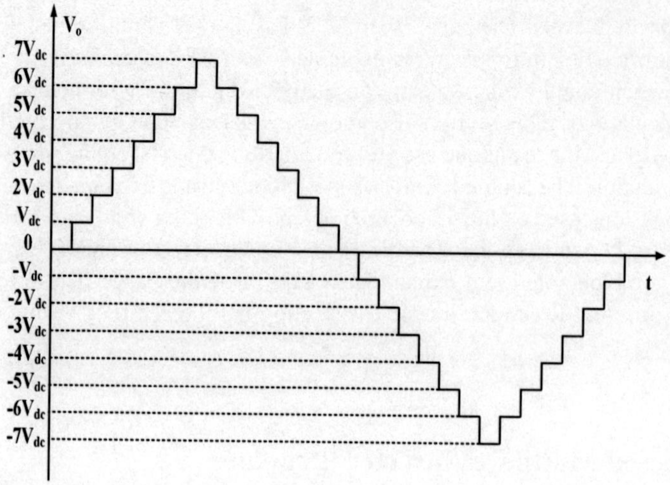

Fig. 2 Load voltage waveform

output stage of the converter. The switching pattern for the proposed structure is described in Table 1. It can be seen that seven positive and seven negative levels are achieved in addition with a zero voltage level making it a total of 15 levels. To achieve a desired level, a specific set of switches have to be switched On while the

Table 1 Switching table

Switching states of proposed MLI										Output
S_1	S_2	S_3	S_4	S_5	S_6	S_7	S_8	S_9	S_{10}	
Off	Off	On	On	On	Off	On	Off	Off	On	V_{dc}
On	Off	Off	On	Off	On	On	Off	Off	On	$2V_{dc}$
On	Off	Off	On	On	Off	On	Off	Off	On	$3V_{dc}$
Off	On	On	Off	Off	On	On	Off	Off	On	$4V_{dc}$
Off	On	On	Off	On	Off	On	Off	Off	On	$5V_{dc}$
On	On	Off	Off	Off	On	On	Off	Off	On	$6V_{dc}$
On	On	Off	Off	On	Off	On	Off	Off	On	$7V_{dc}$
Off	Off	Off	Off	Off	Off	Off	On	Off	On	0
Off	Off	On	On	On	Off	Off	On	On	Off	$-V_{dc}$
On	Off	Off	On	Off	On	Off	On	On	Off	$-2V_{dc}$
On	Off	Off	On	On	Off	Off	On	On	Off	$-3V_{dc}$
Off	On	On	Off	Off	On	Off	On	On	Off	$-4V_{dc}$
Off	On	On	Off	On	Off	Off	On	On	Off	$-5V_{dc}$
On	On	Off	Off	Off	On	Off	On	On	Off	$-6V_{dc}$
On	On	Off	Off	On	Off	Off	On	On	Off	$-7V_{dc}$

remaining ones should be Off. However, for obtaining zero voltage level, i.e., $V_o = 0$, two different possibilities are there. One is to make the switches S_8 and S_{10} ON and the other one is to bring S_7 and S_9 switches in On state.

Figure 3 explains the different conducting states of the structure for obtaining various positive voltage levels. The path shown in red color is the conducting path. For obtaining the first voltage level, i.e., $V_o = V_{dc}$, the switches S_3, S_4, S_5, S_7 and S_{10} should be On. Similarly for obtaining a voltage level of $2V_{dc,}$, the switches which should remain On are S_1, S_4, S_6, S_7 and S_{10}. The negative levels can also be drawn using the switching states shown in Table 1.

2.2 *Total Standing Voltage*

One of the important parameter taken for consideration in designing an inverter is the rating of switches as it ultimately affects the cost, size, and efficiency of the circuit topology. For determining the switch rating, the voltage which is to be blocked by the switch should be accurately known. The sum of all the switches' reverse blocking voltage is termed as total standing voltage (TSV). Thus, lower value of TSV for any topology means that lower rating can be switches can be used in the circuit which will reduce the cost and increase the efficiency.

If V_{s_i} represent the blocking voltage of ith switch, then the TSV for 'n' number of switches can be written as

$$\text{TSV} = \sum_{i=1}^{n} V_{s_i} \tag{1}$$

In the present case, the individual blocking voltages of the switches are as follows:

$$V_{s_1} = V_{s_3} = 2V_{\text{dc}},\ V_{s_2} = V_{s_4} = 4V_{\text{dc}},$$
$$V_{s_5} = V_{s_6} = V_{\text{dc}},\ \text{and}\ V_{s_7} = V_{s_8} = V_{s_9} = V_{s_{10}} = 7V_{\text{dc}}$$

Therefore, the total standing voltage can be written as

$$\text{Total Standing Voltage} = 2 \times (2V_{\text{dc}} + 4V_{\text{dc}} + V_{\text{dc}}) + (4 \times 7V_{\text{dc}}) = 42V_{\text{dc}} \tag{2}$$

Moreover,

$$\text{TSV}_{\text{p.u.}} = \frac{\text{TSV}}{V_{\text{o,max}}} = 6 \tag{3}$$

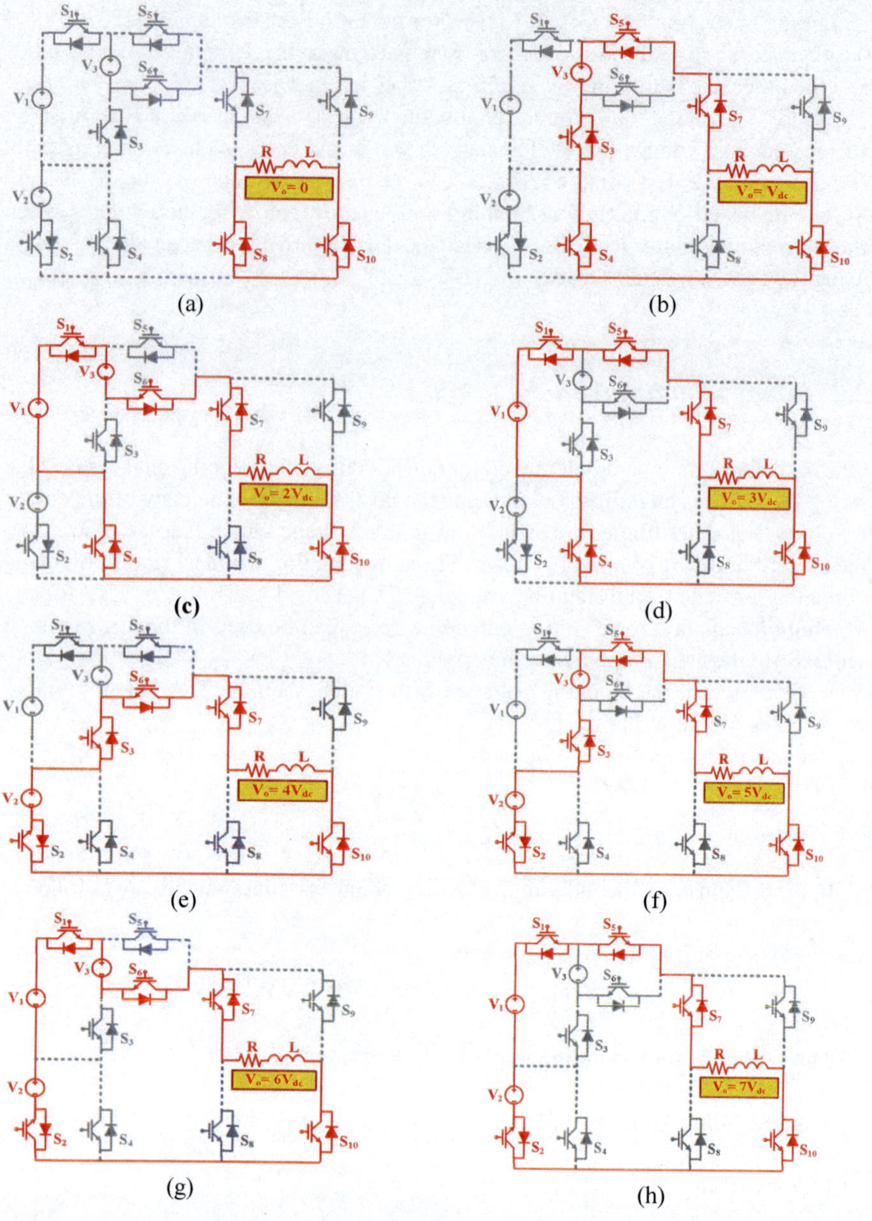

Fig. 3 Positive voltage states of the proposed topology

2.3 Nearest Level Control Modulation Technique

The control of output voltage is required when a converter is designed. For this purpose, it is imperative to use different modulation techniques. In this work, nearest level control method is used. The reference sinusoidal signal is compared against various level shifted carrier signals and a resultant stepped waveform is obtained. It operates at fundamental frequency of 50 or 60 Hz.

For 'N' output levels, the switching angles for this modulation technique can be evaluated by using the following relation:

$$\beta_p = \sin^{-1} * \left(\frac{p - 0.5}{n}\right), \text{ in which} n = (N-1)/2 \tag{4}$$

where $p = 1, 2, 3, \ldots, n$.

3 Comparative Analysis

For having a clear understanding about the merits and improvement in the proposed structure, a comparative study has been done in which the presented topology is compared with other similar kind of multilevel inverter. The different parameters included in comparative analysis are number of levels (N_L), number of switches (N_{sw}), number of bidirectional switches (N_{bd}), number of DC voltage sources (N_{dc}) and TSV. Since the number of levels are different in the considered topologies, therefore, two more parameters 'N_L/N_{sw}' and 'TSV/N_L' are included. Table 2 provides the details of the topologies taken in the comparison.

Ten switches are used for producing 13 levels in [6] whereas in the proposed topology, 15 levels are produced using ten switches. All the topologies utilize four DC sources except the proposed topology which requires three DC sources. Also, no bidirectional switch is required for the proposed structure. The ratio N_L/N_{sw} for the proposed topology is higher than [5–8]. Also, the factor TSV/N_L is lower than [5, 7].

Table 2 Comparative study of different topologies

Topology	N_L	N_{sw}	N_L/N_{sw}	N_{bd}	N_{dc}	TSV	TSV/N_L
[5]	17	12	1.42	3	4	32	2.82
[6]	13	10	1.3	2	4	32	2.46
[7]	17	12	1.42	2	4	48	2.82
[8]	17	12	1.42	2	4	40	2.35
[Proposed]	15	10	1.5	0	3	42	2.8

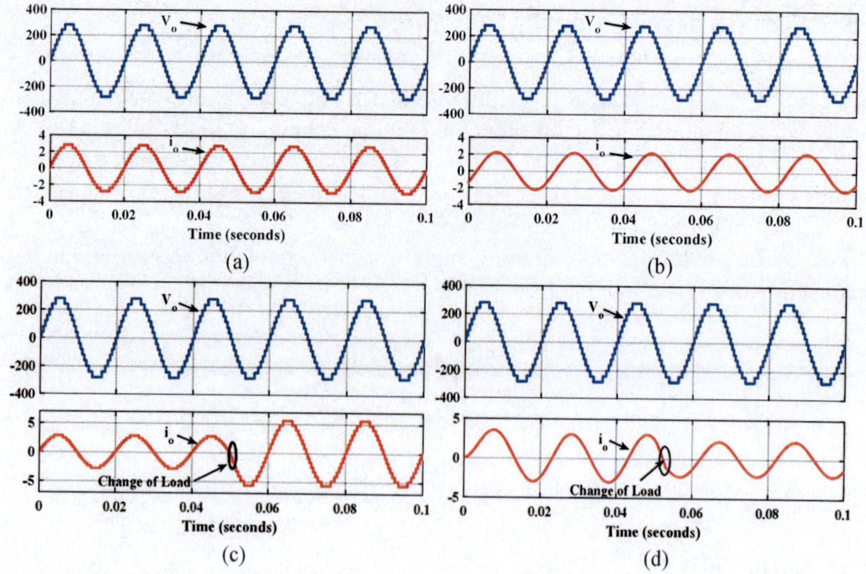

Fig. 4 Topology simulation outcomes **a** Z = 100 Ω, **b** Z = 100 Ω + 250 mH, **c** load variation from Z = 100 to 50 Ω, **d** load variation: Initial—Z = (50 Ω +250 mH), Final—Z = (100 Ω + 250 mH)

4 Results and Discussion

4.1 Simulation Results

Figure 4 shows the simulation results of the proposed MLI. Different types of R and RL loads are taken for evaluating the performance of proposed structure. Output voltage and output current are obtained for each case and shown below. These results are obtained in MATLAB/Simulink. Figure 4a, b represents the waveforms for fixed loads having $Z = 100\ \Omega$ and ($100\ \Omega + 250$ mH), respectively. In Fig. 4c, d, the condition of load change is depicted. Figure 4c shows the results for change in load from $Z = 100$ to $50\ \Omega$ whereas in Fig. 4d, the load is changed from $Z = (50\ \Omega +$ 250 mH) to ($100\ \Omega + 250$ mH). All the results clearly indicate the generation of 15-level stepped waveforms of output voltage and current. However, at the instant of load change, the current waveform will change accordingly as is visible from the obtained results.

4.2 Experimental Results

With the help of an experimental prototype, the hardware results of the proposed MLI are obtained and are depicted below in Fig. 5. The load voltage is reflected in

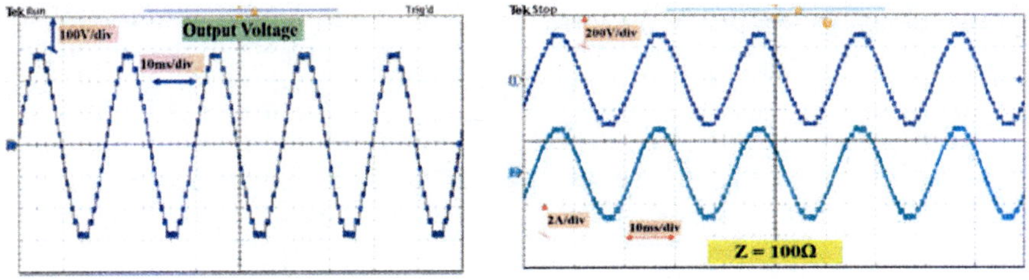

(a) (b)

Fig. 5 Experimental findings **a** load voltage waveform, **b** $Z = 100\ \Omega$

Fig. 5a. It can be clearly seen that a 15-level waveform is obtained which validates the result obtained in simulation. Figure 5b represents the waveforms for load voltage and current for $Z = 100\ \Omega$. However, in all the cases, the experimental results which are obtained are validating the simulation results.

5 Conclusion

This work proposes a modified and improved asymmetrical 15-level multilevel inverter. The improvement of the proposed is the requirement of lower number of power electronic components. A comparative analysis, comprising of different parameters, of the proposed inverter in this work with other topologies is also discussed, which proves the supremacy of the proposed structure of MLI. The effectiveness of circuit topology is also seen by taking various simulation and experimental results. The results are taken for varying load settings. The proposed topology successfully generates 15-level output voltage. Further, the simulation results are also verified by taking results from an experimental prototype.

References

1. Akagi H (2017) Multilevel converters: fundamental circuits and systems. Proc IEEE 11(105):2048–2065
2. Vijeh M, Rezanejad M, Samadaei E, Bertilsson K (2019) A general review of multilevel inverters based on main submodules: structural point of view. IEEE Trans Power Electron 10(34):9479–9502
3. Ahmad A, Anas M, Sarwer A, Zaid M, Alam A, Sarwer Z (2019) Realization of an asymmetric switched-capacitor multilevel inverter using nearest level control. In: 2nd IEEE international conference on power energy, environment & intelligent control (PEEIC2019), pp 522–527, Greater Noida, India
4. Siddique MD, Mekhilef S, Shah NM, Sarwar A, Iqbal A, Memon MA (2019) A new multilevel inverter topology with reduce switch count. IEEE Access 7:58584–58594

5. Samadaei E, Sheikholeslami A, Gholamian SA, Adabi J (2018) A square T-Type (ST-Type) module for asymmetrical multilevel inverters. IEEE Trans Power Electron 2(33):987–996
6. Samadaei E, Gholamian SA, Sheikholeslami A, Adabi J (2016) An envelope type (E-Type) module: asymmetric multilevel inverters with reduced components. IEEE Trans Ind Electron. 11(63):7148–7156
7. Alishah RS, Hosseini SH, Babaei E, Sabahi M (2016) A new general multilevel converter topology based on cascaded connection of submultilevel units with reduced switching components, DC sources, and blocked voltage by switches. IEEE Trans Ind Electron 11(63):7157–7164
8. Shalchi Alishah R, Hosseini SH, Babaei E, Sabahi M (2017) Optimization assessment of a new extended multilevel converter topology. IEEE Trans Ind Electron. 6(64):4530–4538
9. Hosseinzadeh MA, Sarbanzadeh M, Yazdani L, Sarbanzadeh E, Rivera M (2018) New submodule inverter for cascaded multilevel inverter with reduced number of switch counts. In: Proceedings—2017 IEEE Southern power electronics conference, pp 1–6, Puerto Varas, Chile

THD Analysis of 5-Level, 7-Level and 9-Level CHB—Multilevel Inverters Using SPWM Switching Approach

Isarar Ahamad, A. J. Ansari, and Atif Iqbal

Abstract The advantages of multilevel inverter over traditional inverter mainly for high power applications are very high. Due to increase in voltage level, the multilevel technology is very attractive for high power application and also for photovoltaic application due to the need of several sources on DC side. The output voltage waveforms of multilevel inverters are nearly sinusoidal, due to this, inverter efficiency improves, and total harmonic distortion (THD) minimizes. This paper basically presents the THD analysis, waveform pattern, harmonic spectrum and output voltage of 5-level, 7-level and 9-level inverters using a approach of sinusoidal pulse width modulation (SPWM). This has been analyzed as the inverter level increases the total harmonic distortion decreases and output voltages improve. The multilevel inverter models are based on MATLAB/Simulink.

Keywords Multilevel inverter · Sinusoidal pulse width modulation · THD · Simulink models

1 Introduction

Numerous scientists demonstrate that the multilevel inverters can now one day assimilate into many industries of low and high-level voltage applications such as flexible AC transmission system devices, HVDC transmission system, motor drives and renewable power systems [1–3]. Multilevel inverter is a method to transform the power for the phase voltage in output side, and it is identical to a sine wave with less

I. Ahamad (✉) · A. J. Ansari
Department of Electrical and Electronics Engineering, Integral University, Lucknow, India
e-mail: israr@iul.ac.in

A. J. Ansari
e-mail: ajansari@iul.ac.in

A. Iqbal
Department of Electrical and Electronics Engineering, Qatar University, Doha, Qatar
e-mail: atif2004@gmail.com

A. Iqbal et al. (eds.), *Renewable Power for Sustainable Growth*, Lecture Notes in Electrical Engineering 723, https://doi.org/10.1007/978-981-33-4080-0_69

THD value [4]. Multilevel inverter advantages are basically related to the conventional two-level inverter, which generates stair case output voltage, high power efficiency, low value of harmonic, improved electromagnetic compatibility and losses minimize. Multilevel inverters have generally been categorized into three manners: Diode-clamp, flying capacitor and cascaded H-bridge multilevel. The problem of harmonic is a major issue with multilevel inverter [5]. Thus, many methods are introduced for harmonic removal in multilevel inverter, such as sinusoidal pulse width modulation, multi-carrier and selective harmonic pulse width modulation [6]. Equate the above all methodology, sinusoidal pulse width modulation is a better and prominent control technique method generally used in power electronic converter circuit. It has major advantages such as minimum losses, less harmonic performance and easy to implement process. In this analysis, only one modulating and one carrier signal is used to construct a generalized gating signal system. The signal produced by the pulse steering circuit established at the current work is directed into various switches. The one of main disadvantages of inverter is switches requirement that is higher number, and therefore, the system complexity increases.

2 Multilevel Inverter Cascade Topology

The multilevel inverter consists of a series of single complete H-bridge converters, each with a separate DC bus. Each bridge contains four IGBT switches. Multilevel inverter generates the output voltage near-sinusoidal wave from various DC sources get from solar cells, batteries, etc. [7]. Using distinct combinations of the four switches could obtain AC output. Every level will produce three different power outputs $+V_{dc}$, 0 and $-V_{dc}$ that is shown in Fig. 1 [8]. The output of multi level inverter voltage is addition of all the different outputs of the inverter. Every dynamic device of H-bridge is switched only at fundamental frequency, and by moving its positive and negative phase legs, every H-bridge is able to produce a square wave form. As the number of bridges added in cascaded manner, the level of inverter increases. Every switching device always operates for 180° irrespective of pulse width of square wave so that each active device's current stress is equalized by this switching process [9].

To obtain the wave form with preferable harmonic range, lowering switching frequency and for high voltage, this topology is ideal. The four main benefits are a multilevel inverter over the traditional inverter. The first is significance of voltage on every switch is minimized according to series connection of the switches. So, the rated voltage and therefore whole power of the inverter could be increased. Second is the rate of change of voltage is minimized due to the swing of lower voltage for every switching cycle. Third is the harmonic distortion that is minimized as levels of output increase. The fourth is to achieve reduced ambient noise and electromagnetic interference.

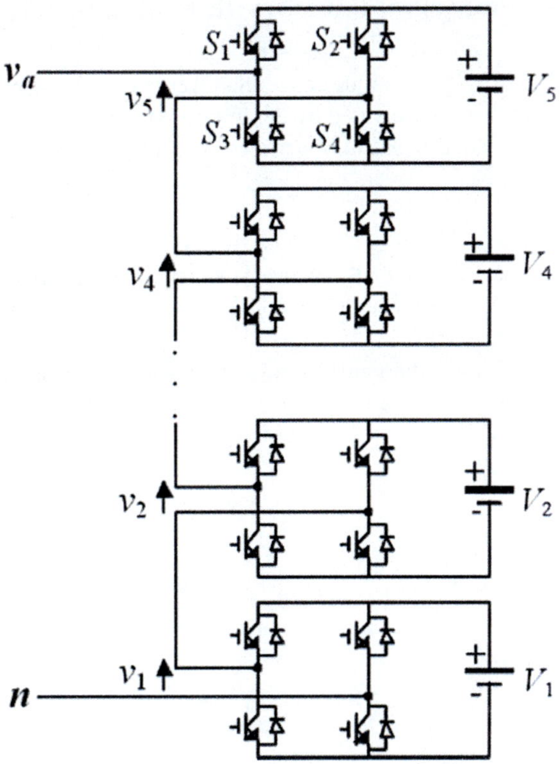

Fig. 1 H-Bridge multilevel inverter

3 Modeling of CHB—Multilevel Inverter

The equation of the inverter output is

$$V_{oi} = V_{\mathrm{dc}}(S_{1i} - S_{2i}) \tag{1}$$

DC current at input is

$$I_{\mathrm{oi}} = I_{\mathrm{dc}}(S_{1i} - S_{2i}) \tag{2}$$

where

(1) $i = 1 \ldots n$ and $n =$ number of full H-bridge inverters
(2) $I_{\mathrm{oi}} =$ output current
(3) S_{1i} and S_{2i} are switches of upper side MLI.

Voltage at output side is given by [10]:

$$V_{\mathrm{on}} = \sum V_{oi} \quad \text{where } i = 1, 2 \ldots n \tag{3}$$

4 Sinusoidal Pulse Width Modulation Technique

The SPWM technique carries harmonics of several orders in the phase voltage wave form that is dominated by the fundamental ones are n and $n \pm 2$ in order, where n is described as the ratio of carrier to modulating frequency.

$$n = f_c/f_m$$

The SPWM method gives the complete desired output voltage with a reference waveform (modulation signal) connected to the triangular carrier wave of high frequency.

In Fig. 2, the generation of the switching pulse is to be generated with the all combination of carrier pulse and reference sinusoidal pulse. Modulation index is the ratio of sinusoid of A_m amplitude, and the triangular carrier's amplitude is A_C that is given by [11].

$$m = A_m/A_c$$

where "m" is represented as the modulation index. The applied output voltage is controlled by regulating the modulation index by using SPWM method displayed in Fig. 3.

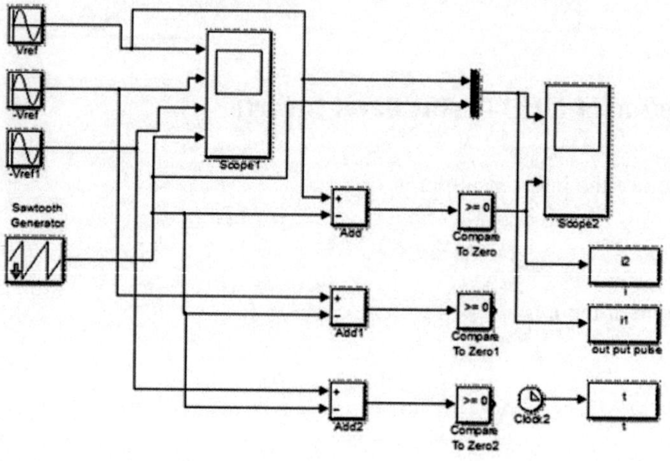

Fig. 2 SPWM switching model

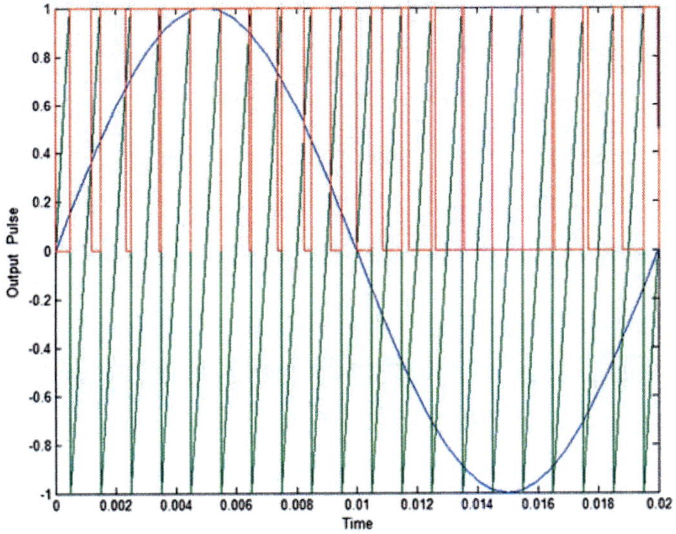

Fig. 3 Output carrier signal wave form. Reference signal > Carrier signal, output will be high (red line indication). Reference signal < Carrier signal, output will be low (missing red line)

5 Working of Multilevel Inverter

To working the multilevel inverter basically the SPWM technique uses for generating switching pulse. In this technique, the reference signals are compared to the carrier signal. If this signal is larger than carrier signal, the output is high, and if the signal is less than carrier signal, the output is low [12]. The high output is applied to the inverter legs as switching pulses. This switching pulse is applied to the each inverter legs that contain four IGBT switches as displayed in Fig. 4 to generate the output voltages level [13].

For five levels, it uses the two inverter legs, seven level, three inverter legs and nine-level inverter, four inverter legs are used as shown in Figs. 5, 6, and 7.

The corresponding outputs are given in Figs. 8, 9 and 10 [14]. The corresponding THD results have been evaluated and find low as the level increases. That is shown in Table 1.

6 Simulation of Multilevel Inverters

The simulation has done for these models that are stated below. The harmonic spectrums have been fined on FFT window and fined THD result. Every inverter legs have their own switching technique. The switching angles of reference wave are equally divided according to the number of switches with proper phase shifting of 180°. In

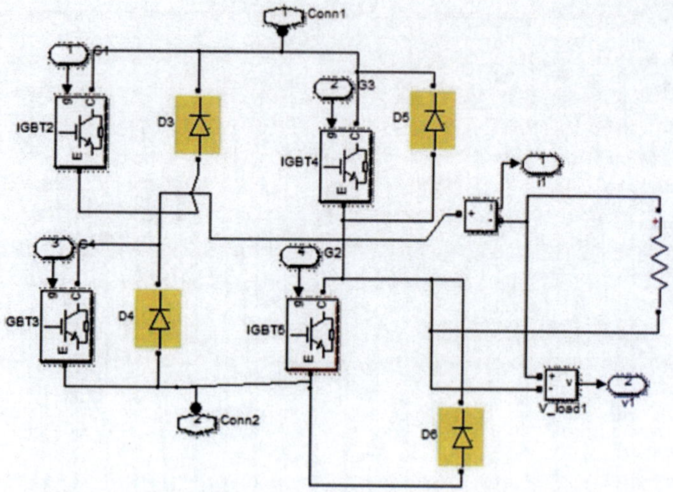

Fig. 4 One leg of H-bridge multilevel inverter

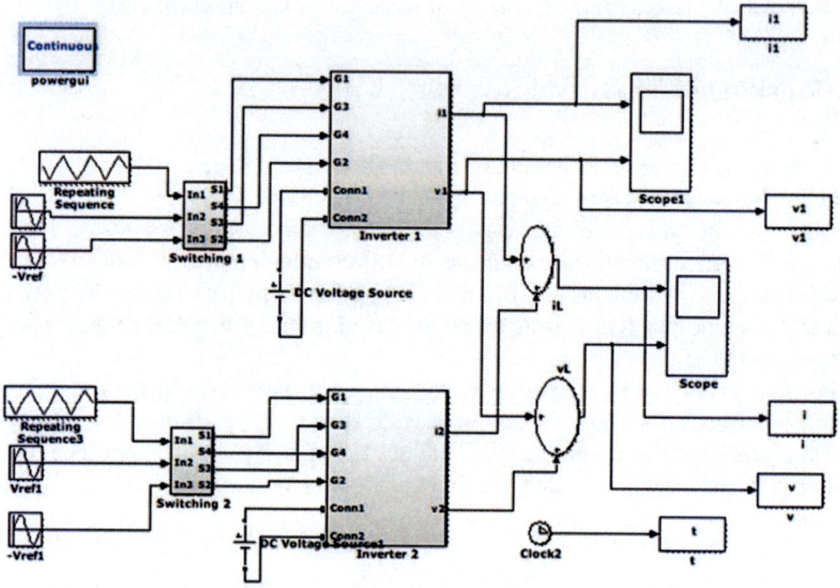

Fig. 5 Simulink model of five-level inverter

this switching technique, carrier signal is compared with reference signal, and then, compared signal is applied to the corresponding IGBT switches to these inverters model as are stated below.

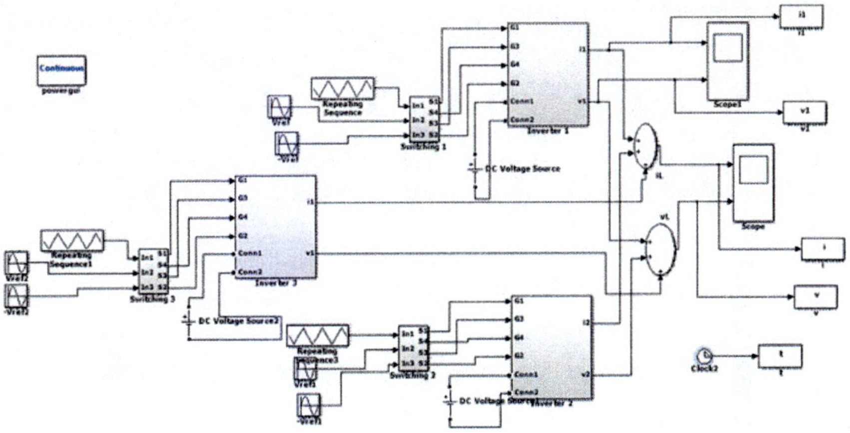

Fig. 6 Simulink model of seven-level inverter

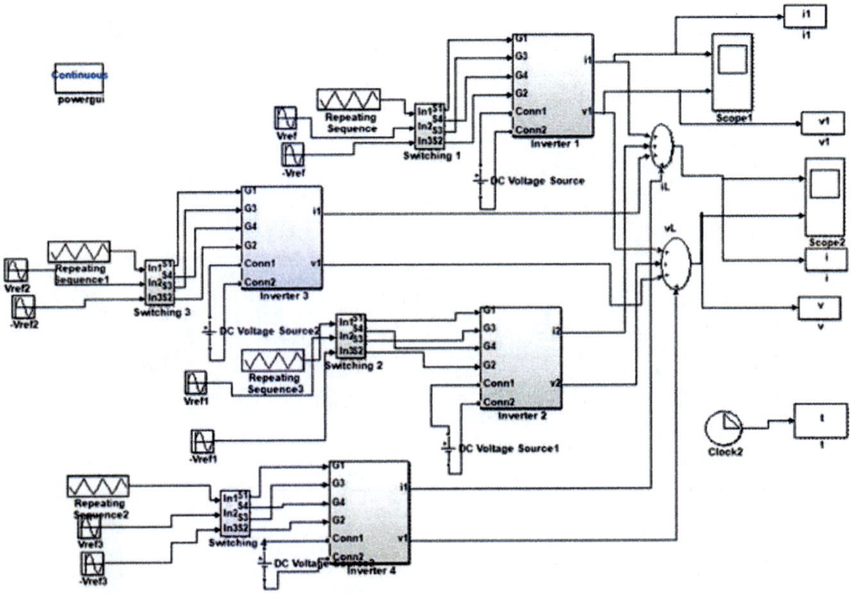

Fig. 7 Simulink model of nine-level inverter

7 Conclusion

This paper discusses the SPWM control design for five-, seven- and nine-level inverters. MATLAB/Simulink environment has been developed for these kinds of inverters, and THD has been analyzed. From the simulation result, it is clear that as the level of inverter increases the output becomes more and more sinusoidal, and total

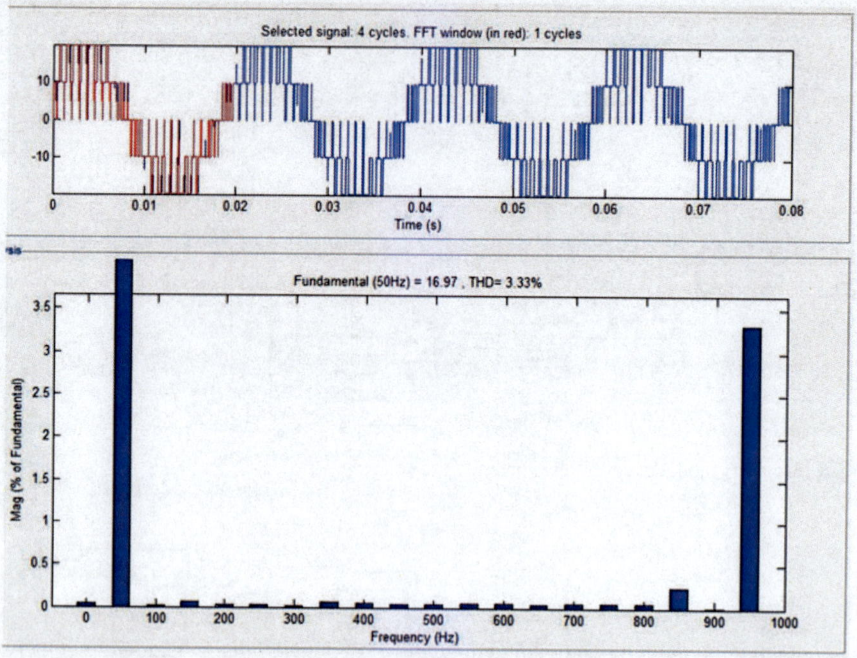

Fig. 8 Five-level output wave form

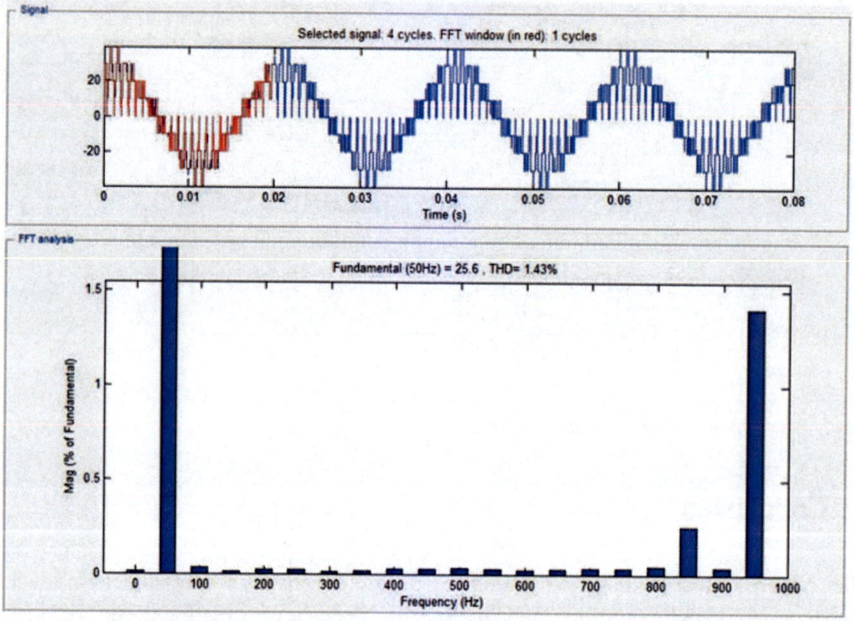

Fig. 9 Seven-level output wave form

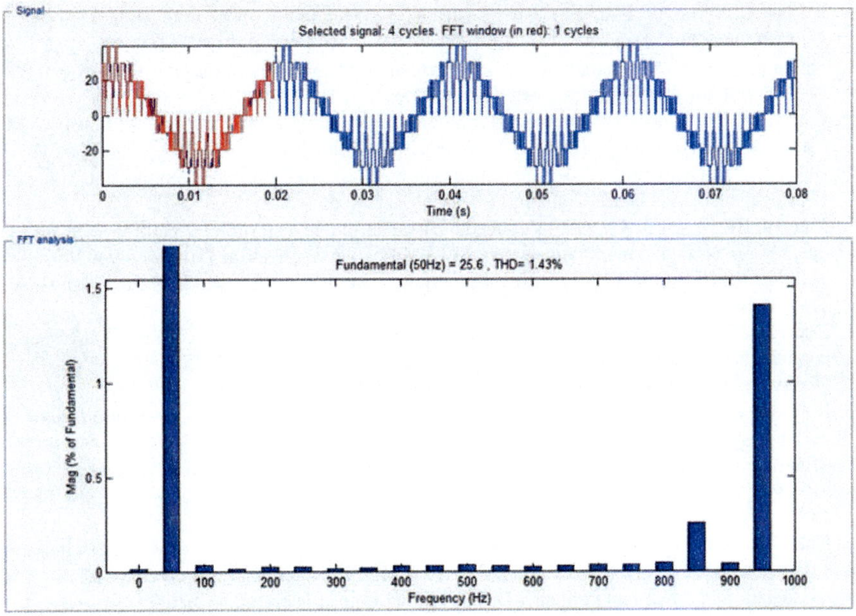

Fig. 10 Nine-level output wave form

Table 1 THD result of multilevel inverters

S. No.	Types of level of inverter	Total harmonic distortion (THD) (%)
1	5 level	3.33
2	7 level	3.26
3	9 level	1.43

harmonics distortion (THD) decreases. This paper is also useful for the PV array and grid system.

Acknowledgements The author would like to acknowledge Integral University for providing the MCN number "IU/R&D/2020 – MCN000840".

References

1. Singh B, Mittal N, Verma KS (2012) Multi-level inverter: a literature survey on topologies and control strategies. Int J Rev Comput 10:1–16
2. Shadab MM, Khan MS, Asim M, Alam Q (2019) Solar PV based grid-tied multilevel inverter topology

3. Raghav R, Raza A, Asim M (2015) Distributed power flow controller-an improvement of Unified power flow controller. Develop 2(5)
4. Sarkar PR, Khan FA, Ahamad I (2015) Performance enhancement of multilevel inverter in PV system with new topology. Int J Adv Eng Res Dev 2:301–306
5. Carrara G, Gardella S, Marchesoni M, Salutari R, Sciutto G (1992) A new multilevel PWM method a theoretical analysis. IEEE Trans Power Electron 7(3):497–505
6. Babaei E, laali S, Alilu S (2014) Cascaded multilevel inverter with series connection of novel H-bridge basic units. IEEE Trans Ind Electron 61(12):6664–6671
7. Mcgrath BP, Holmes DG (2006) Reduced n PWM harmonic distortion for multi level inverters operating over a wide modulation range. IEEE Trans Power Electron 21(4):941–949
8. Tolbert LM, Habetler TG (1999) Novel multilevel inverter carrier based PWM method. IEEE Trans Ind Appl 35(5):1098–1107
9. Rodríguez J, Lai JS, Peng FZ (2002) Multilevel inverters: a survey of topologies, controls, and applications. IEEE Trans Ind Electron 49(4):724–739
10. Hongyang W, Xiangning H (2000) Research on PWM control of cascade multilevel converter. In: Proceedings of the third international conference on power electronics and motion control, pp 1099–1103
11. Minai AF, Usmani T, Mallick MA (2016) Performance analysis of multilevel inverter with SPWM strategy using MATLAB/SIMULINK. J Electr Eng (JEE) 16(4):428–433. ISSN 1582–4594
12. Khan AA, Minai AF, Alam Q, Mallick MA (2014) Performance analysis of various switching scheme in multilevel inverters using MATLAB/ SIMULINK. Int J Curr Eng Technol 4:718–724 (Association of American Publisher U.S.A, Global Impact Factor: 2.552)
13. Park SJ, Kang FS, Lee MH, Kim CU (2003) A new single-phase five level PWM inverter employing a deadbeat control scheme. IEEE Trans Power Electron 18(3):831–843
14. Rahim NA, Chaniago K, Selvaraj J (2011) Single-phase seven-level grid connected inverter for photovoltaic system. IEEE Trans Ind Electron 58(6):2435–2443

Performance Assessment of Variable Speed Induction Motor by Advanced Modulation Techniques

Khadim Moin Siddiqui, Mohd. Khursheed, Rafik Ahmad, and Fazlur Rahman

Abstract Nowadays, induction motor with different modulation techniques is being used in the industries with better performance. The modulation technique makes induction motor to use it in variable speed application efficiently. In the past, SPWM inverter-based variable speed induction motor has been widely accepted in the industries with satisfactory performance. But, still, industries demanding better performance of the variable speed induction motor with less harmonics in the inverter voltage. This research paper discussed two modulation techniques for induction motor performance comparison purpose first is older sinusoidal modulation technique and second is newly space vector modulation technique. The space vector-based advanced power electronics modulation technique with squirrel cage asynchronous motor is providing better results as compare to sinusoidal modulation technique. Henceforth, advanced power electronics modulation technique will be the better choice for the industries to use asynchronous motor with benefitted performance. Herein paper, the direct torque control scheme is applied together with advanced modulation technique in the efficient way for achieving better results.

Keywords Power electronics modulation techniques · Squirrel cage asynchronous motor · Direct torque control · IGBT inverter · Time domain analysis · Performance assessment

K. M. Siddiqui (✉) · R. Ahmad · F. Rahman
Babu Banarsi Das National Institute of Technology and Management (BBDNITM), Lucknow 226030, India
e-mail: siddiquikhadim@bbdnitm.ac.in

R. Ahmad
e-mail: rafik8329@gmail.com

F. Rahman
e-mail: fazlurrahman09@bbdnitm.ac.in

Mohd. Khursheed
Integral University, Lucknow 226026, India
e-mail: khursheed20@gmail.com

A. Iqbal et al. (eds.), *Renewable Power for Sustainable Growth*, Lecture Notes in Electrical Engineering 723, https://doi.org/10.1007/978-981-33-4080-0_70

1 Introduction

At desired output voltage and frequency, a power electronics inverter transforms direct power into alternating power. Though, obtained voltage is highly depends on the switching condition of the power electronics switches [1–9]. But, the inverter introduces harmonics in the induction motor's voltage, currents, etc. [10]. Therefore, for proper operation of the induction motor, it is important to choose suitable PWM control technique with minimum harmonics for inverter-fed induction motor drives.

Generally, SPWM inverter technique is being used with induction motor in many applications and this technique is popularly used in industries. But, the sinusoidal pulse width modulation (SPWM) control technique produces large harmonics in the inverter output voltage. Therefore, the speed variation is not possible largely by SPWM control technique in the inverter-fed induction motor drives [11, 12]. Therefore, this paper presents a solution of above-discussed problem by space vector modulation (SVPWM) control technique.

In the present time, the digital signal processing (DSP) techniques are used with induction motor for better accuracy and performance. The DSP techniques are also used in the health monitoring analysis of induction motor. The realization in the digital circuitry of the SPWM control technique is not easy because it produces distorted voltage including unfiltered harmonics consequently fundamental component in the voltage is decreased [13]. Therefore, to solve this problem, we are introducing SVPWM-based variable speed asynchronous motor model setup in this paper. Same rating squirrel cage asynchronous motor has been used with SPWM and SVPWM control techniques for efficient performance comparison.

For computation of duty cycle of power electronics devices, the SVPWM control technique has become the better solution than SPWM control technique and it has also been verified in this paper. For making digital realization effective and easy, the SVPWM technique is the preferred technique over other techniques because it uses space vectors. The SVPWM technique also provides large modulation index range to achieve better voltage in the output [6, 10, 11].

Since the induction motor is always a best choice in the industries for large number of applications due to its large number of advantages. The induction motor is highly compatible with power electronics converters with efficient control strategies. Therefore, this research paper deals the concert analysis of squirrel cage asynchronous motor with advanced power electronics modulation techniques. The mathematical analysis of SVPWM control technique has been discussed in [3]. The paper presents SPWM and SVPWM control-based squirrel cage asynchronous motor simulation models.

A model-based improved direct torque control method is implemented by with the help of three-phase three-level inverter for reducing the flux and torque ripples of induction motor [10]. A virtual voltage space vector-based DTC induction motor drive has been discussed. For reducing torque ripple, twelve vectors DTC is used and for reducing common model voltage across the windings, authors increased number of vectors [11]. Invariable switching frequency DTC technique is discussed for reducing torque ripple of low speed induction motor [12]. A neutral point clamped 15-switch 3-level inverter is proposed in [13] for improving induction motor drive performance.

In the past, many researchers have worked to improve induction motor performance with the help of direct torque control method. But, all above-mentioned techniques have complexity in implementation with high cost. Therefore, in the present work, adjustable speed induction motor drive implemented precisely with direct torque control method and advanced modulation control technique. The prime objective of this article was the performance analysis of squirrel cage asynchronous motor with direct torque control for transient condition by advanced power electronics modulation technique with minimizing complexity.

2 Result Analysis of Induction Motor by SPWM Control Modulation Technique

SPWM inverter-based variable speed squirrel cage asynchronous motor drive is given in Fig. 1. Through this figure, the transient analysis of the motor has been done and obtained desired results. The simulation model is formed in the shape of block diagram.

Specification of induction motor (IM) is as follows:

- Rotor type: Squirrel cage
- Induction motor: 3 Hp or 2.3 kW
- Speed: 1430 RPM
- Frequency: 50 Hz.

The parameters of the inverter are as follows:

- Rated frequency of sinusoidal wave: 50 Hz
- Frequency of triangular wave: 1650 Hz
- Frequency modulation factor: 33
- Switching frequency corresponds to 33 × 50 = 1650 Hz.

For simulation of smoothing reactor, the leakage inductance of the stator is set twice to its original value and motor's inertia prevents the noise. For transformation of *abc* to *dq* or vice versa, stationary reference frame is used. The simulation speed

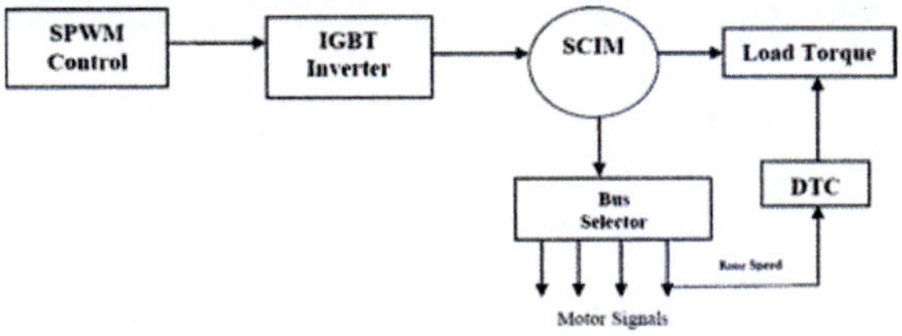

Fig. 1 Setup of SPWM control squirrel cage asynchronous motor

and also accuracy of results will be affected by this reference frame. Therefore, one needs to choose correct reference frame as per necessity of the application. The line to line nominal root mean square (RMS) voltage of the IM is obtained 220 V by the given equation:

$$V_{\text{line - to - line}} = \frac{m}{2} \times \sqrt{1.5} \times V_{\text{DC}} \tag{1}$$

2.1 *Result Analysis by SPWM Control Technique*

The results in symmetrical asynchronous motor state are given in Fig. 2. For the analysis of SPWM control technique, there are four motor signals are used and also shown in Fig. 2. For symmetrical condition, nominal parameters are used in the motor. From results, it has been observed that all four motor signals achieved their stable state after 0.3 s.

The rated speed of the motor is 1430 rpm and it also achieve it stable state after 0.3 s. It is observed in the obtained results that, in the starting, the motor stator current and electromagnetic torque have large amplitude and within few seconds it starts falling down and got the stable condition. Therefore, one can say that the results are satisfactory for SPWM control-based variable speed induction motor. The momentary characteristics of the induction motor signatures may be monitored clearly for further analysis purpose. The obtained results under motor healthy state are given in Fig. 2a–d. The magnitude of motor current is 96 A in peak with 68 A root mean square value. Therefore, the chopped gesticulate gives 311 V at 220 V.

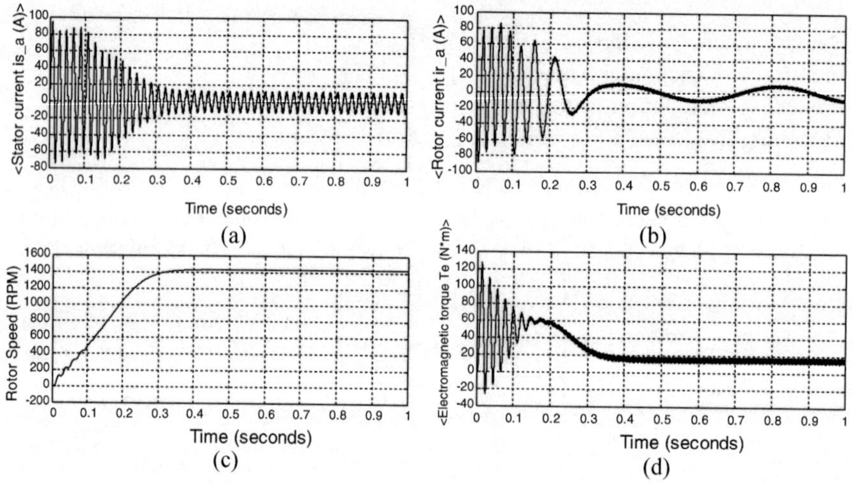

Fig. 2 Motor symmetrical condition with SPWM control

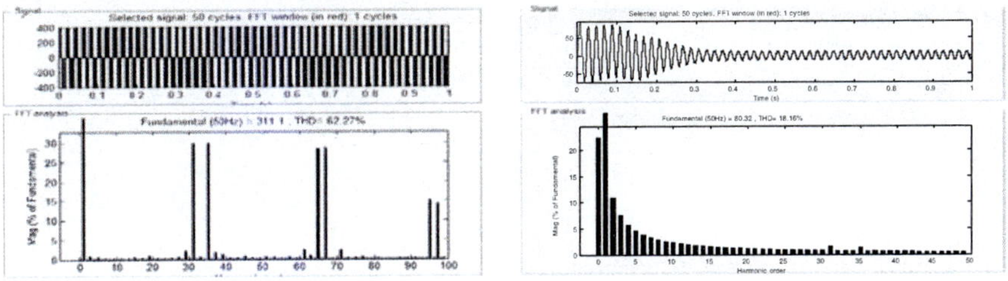

(a) (b)

Fig. 3 Total harmonic distortion analysis

The waveform of electromagnetic torque represents that after 0.3 s, it attains stable condition. In the starting, there is sturdy oscillation in the electromagnetic torque. Therefore, the mean value of rated load torque has been observed and that is 15 N m.

After zooming the stator current waveform, one may observe that, the harmonic components are taken place at switching frequency and it may be removed by the inductance. The THD is shown in Fig. 3 for motor current and inverter voltage. The harmonic investigation of SPWM control asynchronous motor has been done for two signals; first is inverter voltage and second is stator current. The inverter voltage signal along with THD is given in Fig. 3a. The fundamental magnitude of the inverter voltage is obtained 311 V for modulation index $m = 0.9$ as shown in Fig. 3a. This value can also be verified theoretically.

The harmonics are shown in the percent of the fundamental component and which is multiples of carrier frequency and also giving expected values. The harmonic analysis motor current has also been done and shown in Fig. 3b. The THD in current has been observed 18.16% along with fundamental component 80.32 A.

In the next section, the same induction motor will be used for analysis purpose along with SVPWM control technique for further improvement.

3 Investigation of Motor Results by Advanced Modulation Technique

This part gives the result analysis of induction motor controlled by SVPWM. The squirrel cage asynchronous motor fed into SVPWM inverter control has been used for analysis purpose as shown in Fig. 4. The model has been converted into the form of block diagram. The SVPWM 'modulator block' is used to generate firing pulses. The chopping frequency is computed to 1650 Hz and the 'magnitude angle' is input reference vector. The constant V/f block has been used in the controlling of speed of the motor and all specifications of the motor and inverter are same as we used in the previous section.

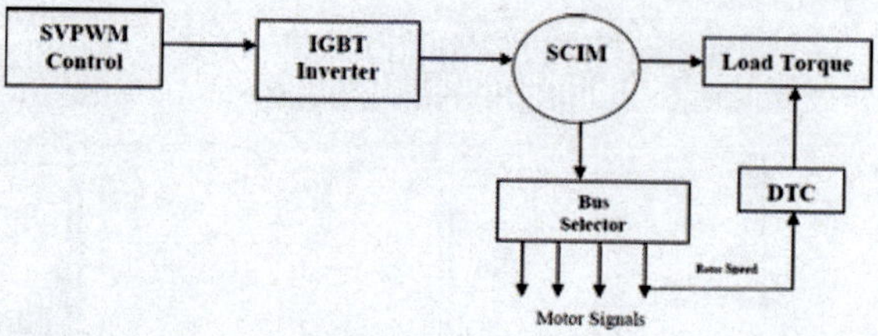

Fig. 4 Setup of SVPWM control squirrel cage asynchronous motor

3.1 Result Analysis by SVPWM Control Technique

The simulation results of the motor in the balanced supply (or healthy mode) motor state are given in Fig. 5. For the analysis of SVPWM control technique, there are four motor signatures are used and also displayed in Fig. 5. In the symmetrical condition, nominal parameters are used in the motor. From results, it has been observed that all four motor signals achieved their stable state after 0.1 s. The rated speed of the motor is 1455 rpm and it also achieve it stable state after 0.1 s. It is observed in the obtained results that in the starting, the motor stator current and electromagnetic torque have large amplitude and within few seconds it starts falling down and got the stable condition. Therefore, one can say that the results are much encouraging for SVPWM control-based variable speed induction motor. The momentary characteristics of the motor signatures are monitored clearly for further analysis purpose. The obtained result under motor healthy or symmetrical state is given in Fig. 5.

The THD investigation by space vector pulse width modulation control with squirrel cage asynchronous motor is done for two signals; first is for inverter voltage and second is for stator current. The THD in inverter voltage is shown in Fig. 6a. The fundamental inverter voltage magnitude is obtained more compared well with

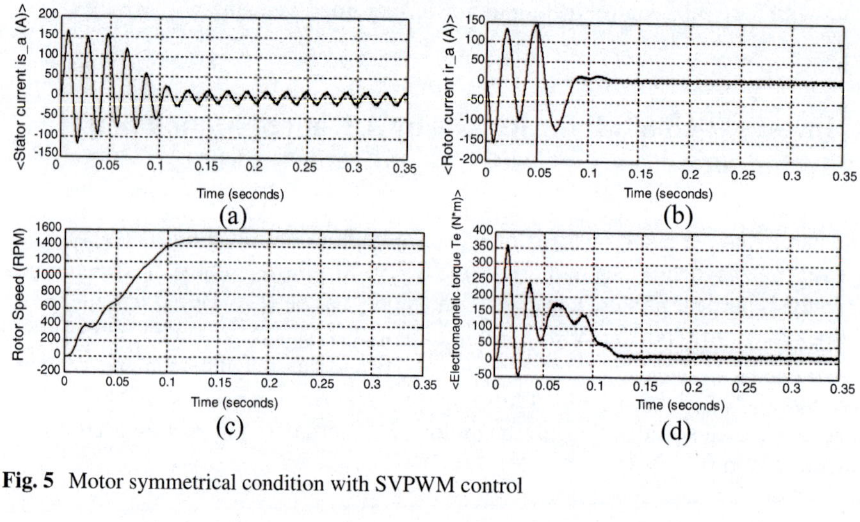

Fig. 5 Motor symmetrical condition with SVPWM control

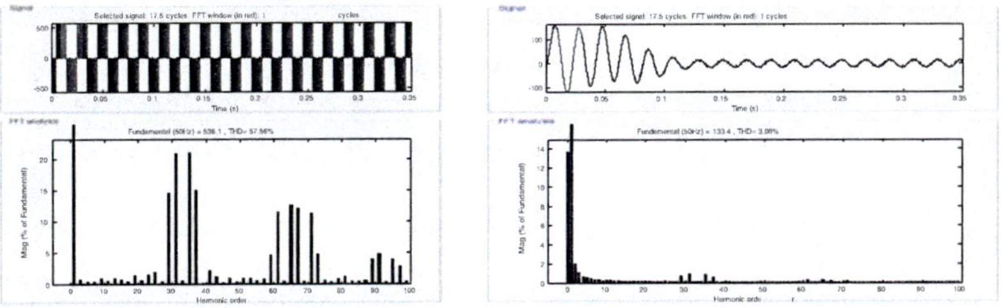

(a) (b)

Fig. 6 THD analysis by SVPWM technique

the theoretical value with 0.9 modulation index. The multiple carrier frequency ($n \times 33 \pm k$) gives expected harmonics around it.

The harmonic analysis of stator current of the motor has also been done and given in Fig. 6 and observed THD 3.08% for motor current along with fundamental component 133.4 A. After observing all waveforms, one may articulate that the SVPWM control technique gives much better results as compared to SPWM control technique.

4 Performance Assessment of SVPWM Inverter Control and SPWM Inverter Control

In this section, the performance comparison of space vector PWM control inverter-driven asynchronous motor drive and SPWM inverter-driven asynchronous motor drive. The SVPWM inverter control technique gives much better results with reduced harmonics as compared to SPWM inverter control method. Following conclusions have been made after observing results of SVPWM inverter control and SPWM inverter control. Same rating asynchronous machine has been used for analysis purpose with SVPWM and SPWM.

1. Inside SPWM inverter control technique, machine signatures attained stable state after time 0.3 s, but in SVPWM control technique, machine parameters attained the stable state after 0.1 s. A significant improvement has been done by SVPWM inverter control technique.
2. In the SPWM inverter control technique, the rotor achieves its stable speed after 0.3 s, i.e., 1430 rpm, but in SVPWM inverter control technique, the rotor speed achieves stable state after 0.1 s, i.e., 1455 rpm for same rating and parameters of the induction machine. Therefore, smooth control of induction machine can be done by SVPWM inverter control technique.
3. The main advantage of space vector PWM control is that it decreases distortion in the line to line voltage. Therefore, by SVPWM technique, wide range of variable speed control is possible.

4. Since SVPWM inverter contains low switching losses; therefore, harmonics in the inverter voltage have been reduced consequently better dc bus utilization.
5. Semiconductor switches contain low switching losses, therefore, digital realization will be much easy with improved performance than SPWM inverter control technique.
6. In the SVPWM inverter control technique, much smooth waveforms are obtained as compared to SPWM inverter control technique.
7. The harmonics in the induction motor stator current have also been reduced and also significant improvement in the fundamental value.
8. Theoretically, the SVPWM technique generates a sinusoidal signal that gives a superior voltage to the induction machine by lesser THD in the output line to line voltage and this thing has also been verified by the obtained results.
9. In the case of SVPWM inverter control, the fundamental value achieved at frequency 50 Hz, i.e., 536.1 with reduced harmonic and in case of SPWM inverter control achieved fundamental value 311 V with same frequency. Therefore, the windings of the used motor load provide proficient use of the DC supply voltage.

5 Conclusions

In present work, adjustable speed induction motor drives have been discussed with different power electronics modulation control technique. The performance analysis of squirrel cage asynchronous machine has been completed by both control techniques. The SVPWM inverter control technique gave effective results as compared to SPWM inverter control technique. Same rating induction motor has been used with both control techniques for clear comparison purpose. It has been clearly observed that the SVPWM control technique produces less harmonics in the line to line voltage. Therefore, a wide range of speed control may be done in the efficient way. The SVPWM control technique used a reference vector for analysis purpose; therefore, the digital implementation is being easy with broad modulation index range for output line to line voltage. Therefore, due to these advantages, one can say that the SVPWM inverter-fed asynchronous motor technique shall be employed in the industries in lieu of SPWM inverter-fed asynchronous motor drives largely in the future. The key strength of the present work is to apply advanced modulation technique with direct torque control method jointly for achieving the best results with reduced THD in inverter voltage and reduced torque ripple.

References

1. Siddiqui KM, Sahay K, Giri VK (2014) Simulation and transient analysis of PWM inverter fed squirrel cage induction motor drive. J Electr Eng 7(3):9–19 (i-manager's)
2. Wang F (2002) Sine-triangle versus space-vector modulation for three-level PWM voltage-source inverters. IEEE Trans Ind Appl 38(2)

3. Siddiqui KM, Sahay K, Giri VK (2015) Performance and analysis of space vector PWM inverter fed squirrel cage induction motor drives. J Electr Eng 8(3):18–30 (i-manager's)
4. 519–2014—IEEE Recommended practice and requirements for harmonic control in electric power systems. https://doi.org/10.1109/IEEESTD.2014.6826459
5. Sarika EP, Raj RSP (2015) Performance comparison of direct torque control of two level and three level neutral point clamped inverter fed three phase induction motor. In: Proceedings of IEEE international conference on advances in green energy
6. Ammar A, Kheldoun A, Metidji B, Talbi B, Ameid T, Azzoug Y (2018) An experimental assessment of direct torque control and model predictive control methods for induction machine drive. In: Proceedings of IEEE international conference on electrical sciences and technologies in Maghreb (CISTEM)
7. Gouri K, Bindu VR (2018) PWM direct torque control of induction motor drive with reduced torque ripple. In: Proceedings of IEEE international conference on conference on control, communication, and computing (IC4), Thiruvananthapuram, India. https://www.ieeexplore.ieee.org/document/8530939
8. Iqbal A et al (eds) (2020) Soft computing in condition monitoring and diagnostics of electrical and mechanical systems. In: Advances in intelligent systems and computing, vol 1096. Springer, Singapore. https://doi.org/10.1007/978-981-15-1532-3
9. Iqbal A et al (eds) (2020) Meta heuristic and evolutionary computation: algorithms and applications. In: Studies in computational intelligence, vol 1096. Springer, Singapore. https://www.springer.com/gp/book/9789811575709
10. Jnayah S, Khedher A (2019) DTC of induction motor drives fed by two and three-level inverter: modeling and simulation. In: Proceedings of 19th IEEE international conference on sciences and techniques of automatic control and computer engineering (STA)
11. Suresh S, Rajeevan PP (2018) Virtual space vector based direct torque control schemes for induction motor drives. In: Proc. 8th IEEE India Int. Conf. on Power Electronics, India. https://doi.org/10.1109/IICPE.2018.8709454
12. Naganathan P, Srinivas S (2020) Direct torque control techniques of three level H-bridge inverter fed induction motor for torque ripple reduction at low speed operations. IEEE Trans Ind Electron 67(10). https://doi.org/10.1109/TIE.2019.2950840
13. Pan L, Zhang JR, Zhang J, Wang K, Pang Y, Wang B (2020) A new three-level neutral-point-clamped fifteen-switch inverter for independent control of two three-phase loads. IEEE J Emerg Sel Top Power Electron 8(3). https://doi.org/10.1109/JESTPE.2019.2903806

Performance Analysis of Grid Connected Solar Energy Conversion System Under Varying Conditions

Arvind Kumar Sharma

Abstract The ever increasing demand of electrical energy and depletion of fossil fuels have necessitated the use of renewable energy sources for electricity generation. Solar energy is one of the most promising renewable energies used for electricity generation. Photovoltaic (PV) effect of solar energy is most widely used for solar energy conversion. In this work, a PV-based solar energy conversion system (SECS) is simulated and connected to grid at distribution level of 415 V, 3 phase, 50 Hz. The performance of this system is analyzed under varying conditions. The grid connected SECS presented in this work is a combination of PV array which converts solar irradiance into electrical energy, a DC–DC converter for converting this electrical energy into a constant voltage DC, and this DC is finally converted into 3 phase, 415 V, 50 Hz AC by using a voltage source inverter which is connected to distribution system with the help of a PLL. Complete system is simulated in MATLAB/Simulink. The system is analyzed for active and reactive power delivery and power quality under varying environmental conditions.

Keywords Solar power · PV system · Renewable energy · Solar energy conversion system · Distributed generation

1 Introduction

India is getting sunny sky for approximately 300 days of the year. Average horizontal radiation from sun ranges from 3 to 6.5 kW h/m^2 per day which can reach to 7.5 kW h/m^2 per day during summer in northern part of India [1–3]. The radiation remains at its maximum level during summer season when the consumption of electrical energy also remains at its maximum level. Due to these facts, solar energy conversion systems are the best suited renewable energy sources in India [2, 3].

Step by step modeling of an improved PV array in MATLAB/Simulink has been presented by conference paper [4]. Design of a three-phase grid tie SPWM inverter

A. K. Sharma (✉)
KIET Group of Institutions, Delhi-NCR, Ghaziabad, India
e-mail: arvind.sharma.en@kiet.edu

A. Iqbal et al. (eds.), *Renewable Power for Sustainable Growth*, Lecture Notes in Electrical Engineering 723, https://doi.org/10.1007/978-981-33-4080-0_71

has been presented in seminar [5]. Literature presents design of enhanced PLL for synchronizing the inverter to the grid [6]. Design of various models of grid connected solar power plants has been presented in the literature [7]. Performance analysis of solar power plant (SPP) has been presented in both grid connected and OFF-grid modes for fulfilling the load demand [8]. Effects of different control strategies on performance of grid tie SPP have been done by Tochi et al. [9]. Dharval et al. have presented the effects of feed forward and current control method for real-power injection to the grid from inverter. The PV source has been linearized by using maximum power point tracking (MPPT) technique. Some researchers analyzed SECS for partial shading conditions and proposed reconfiguration methods to enhance the power output of the system [10]. A strategy presented in [11] is found effective in keeping the total harmonic distortion (THD) well below the level of 5%.

This article presents a detailed design and simulation of PV-based solar energy conversion system (SECS) along with its performance analysis under varying irradiance. The system has been simulated by connecting different sub-systems like PV module, DC–DC boost converter, MPPT, inverter, and PLL.

2 Simulation of Sub-systems for PV-Based SECS

This section presents the detailed design and simulation of all the sub-systems of grid connected SECS. Finally, all the sub-systems are connected to form the complete system.

2.1 *Modeling and Simulation of PV Array, DC–DC Boost Converter with MPPT*

The array is designed to deliver 18 kW power at STC by connecting three strings in parallel. Each string contains 40 PV modules connected in series. Each module PM-150 is developed by connecting 36 polycrystalline-Si cells in series.

In this work, 700 V DC is used by the inverter for conversion into three-phase, 415 V, 50 Hz AC output. The voltage level is maintained constant by DC–DC boost converter. Also, it is ensured to harness maximum power under varying irradiance with the help of an intelligent MPPT. Duty cycle of the converter for maintaining constant 700 V with input of 620 V under varying environment is evaluated as

$$D = 1 - \frac{V_{in}}{V_{out}} \tag{1}$$

This becomes,

$$D = 1 - \frac{V_{in}}{V_{out}} = 0.114 \quad (2)$$

Inductor and capacitance used are of 1 mH and 150 micro-farad, respectively. The converter is associated with an intelligent MPPT which alters the duty cycle of the boost converter switch according to the requirement.

2.2 Modeling and Simulation of Inverter and Filter

A 24 kW current controlled voltage source inverter is designed and simulated in this work. The inverter uses six IGBT switches operating at 10 kHz frequency. Bipolar sine wave and pulse width modulation method are used for converter control. The ripples produced in the output due to switching of IGBT are suppressed by using a LCL filter. A damping resistance is provided in series with the capacitor to reduce the effect of resonance.

2.3 Modeling and Simulation of Inverter Controls

The inverter controller controls the DC link voltage to a constant value by controlling active power delivery to grid. The controller also ensures synchronization of inverter supply to the grid for transferring the acceptable quality of power. Synchronous reference frame control or d-q control has been implemented to control the output of the inverter.

3 Simulation of Grid Connected PV-Based SECS

Complete solar energy conversion system connected to low voltage distribution system of the grid which is simulated by integrating all the components discussed in previous sub-sections as shown in Fig. 1. The system is capable to deliver three-phase power of 18 kW at 415 V, 50 Hz. In this system, the grid is represented by a three-phase source with suitable parameters. A resistive load of 10 kW is connected near the point of common coupling for study purpose.

4 Simulation Results and Discussions

The system designed is made to run for a period of 2 s at different solar irradiance ranging from 400 to 1000 W/m^2. The irradiance is raised in a step of 200 W/m^2

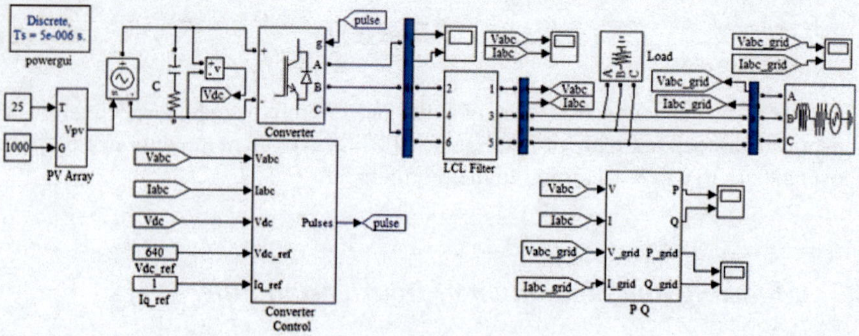

Fig. 1 Simulink model of grid tied PV-based SECS

after running for 0.5 s at present irradiance value. The simulation results obtained by running the system under above-mentioned conditions are presented in this section.

Output voltage profile of the SECS is presented in Fig. 2 which is maintained at 415 V (1 pu) during entire range of irradiance variation. It indicates that the DC–DC boost converter is efficiently maintaining the DC link voltage level. The window of zoomed view in Fig. 2 shows the wave shape of output voltage of the SECS. It is having vary low distortions and hence shaping like perfect sine wave.

The inverter current increases with the rise of solar irradiance level as shown in Fig. 3. It is seen that the current varies from 0.5 pu to 1 pu over entire range of irradiance (from 400 to 1000 W/m^2).

The LCL filter is also seen working perfectly by a very low value (0.53%) of total harmonic distortions (THD) during entire range of irradiance. Figure 4 presents the report of FFT analysis of the output voltage wave.

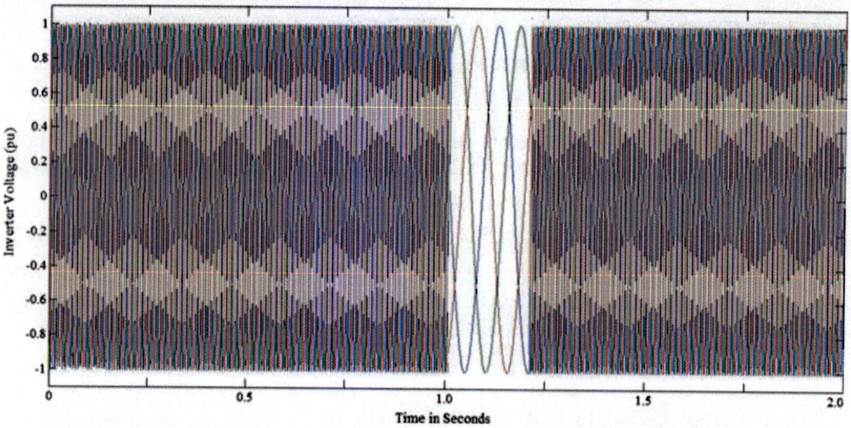

Fig. 2 Output voltage profile of the inverter under varying solar irradiance

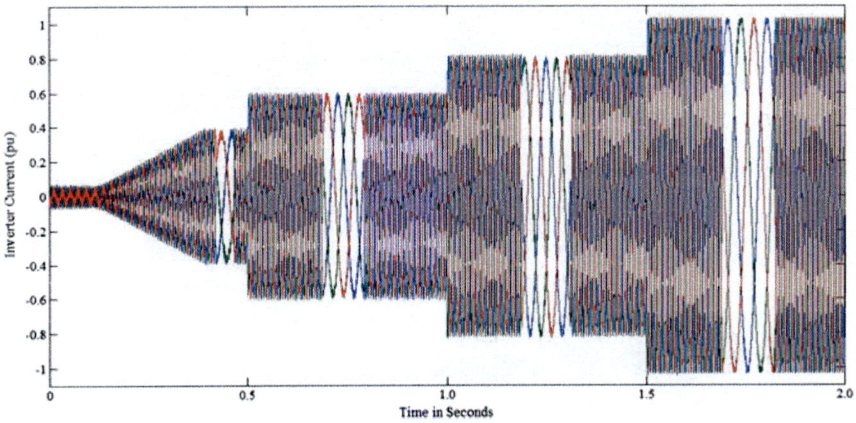

Fig. 3 Current supplied by inverter under varying solar irradiance

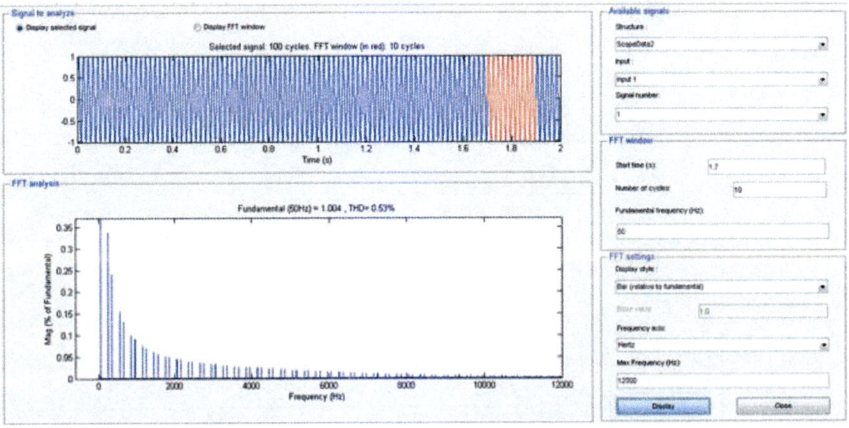

Fig. 4 FFT analysis report of voltage of the inverter under varying solar irradiance

The distortion in current wave is also seen very low as indicated in zoomed windows at all the irradiance level in Fig. 3. The FFT report of current wave indicates 1.73% THD which is well below than acceptable value. The report is reflected in Fig. 5.

It is seen from Fig. 6 that the power supplied by the inverter to the grid rises with the rise in solar irradiance. Inverter delivers a true power of 0.36 pu at 400 W/m^2 which rises to 0.57, 0.77, and 1 pu for 600, 800, and 1000 W/m^2 solar irradiance, respectively. The inverter takes initializing time of approximately 0.35 s in building up the voltage and connection to the grid.

A resistive load of 10 kW (0.5 pu) is connected near the point of common coupling. It is seen from Fig. 7 that the grid delivers the complete load of 0.5 pu till the SECS is connected to the grid at 400 W/m^2 at this instance the power delivered by grid

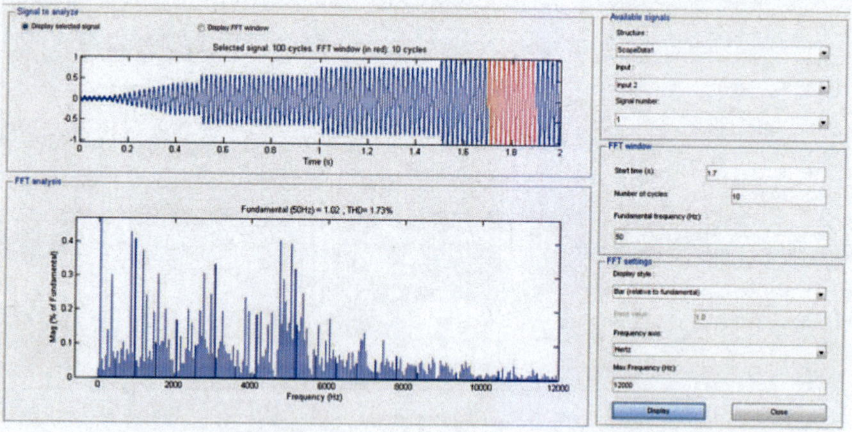

Fig. 5 FFT analysis report of inverter current under varying solar irradiance

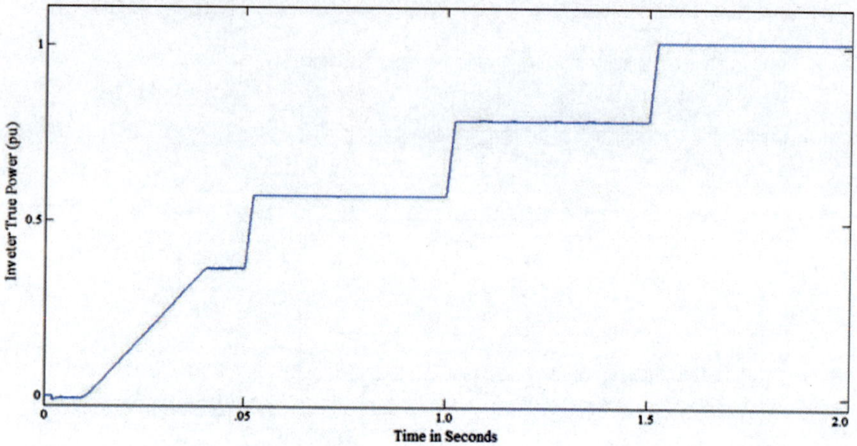

Fig. 6 Power delivered to grid by SECS

drops to 0.2 pu. Remaining power is supplied by the SECS at this time as shown in Fig. 7. Comparison of Figs. 6 and 7 indicates that at higher value of irradiance, when the inverter power generation is more than 10 kW, the inverter feed the excess true power to the grid. Negative grid power after 0.5 s of simulation in Fig. 7 indicates that the grid is receiving true power.

Figure 8 indicates that the inverter controllers are capable to maintain the entire operation at unity power factor by maintaining the reactive power zero for all the values of irradiance.

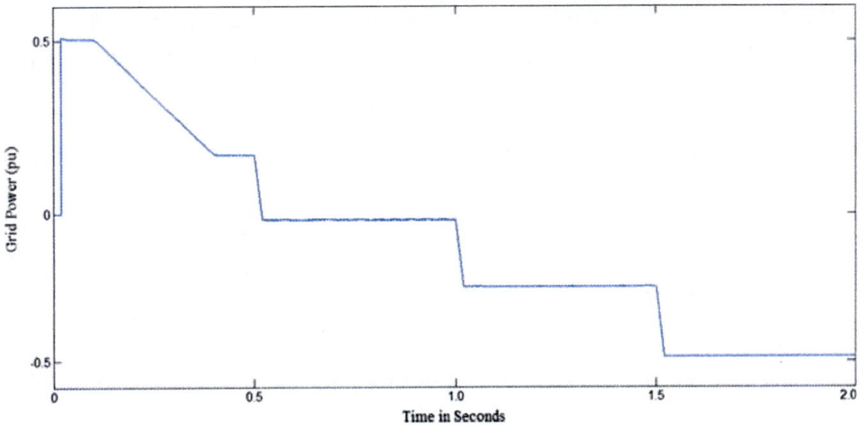

Fig. 7 Grid power

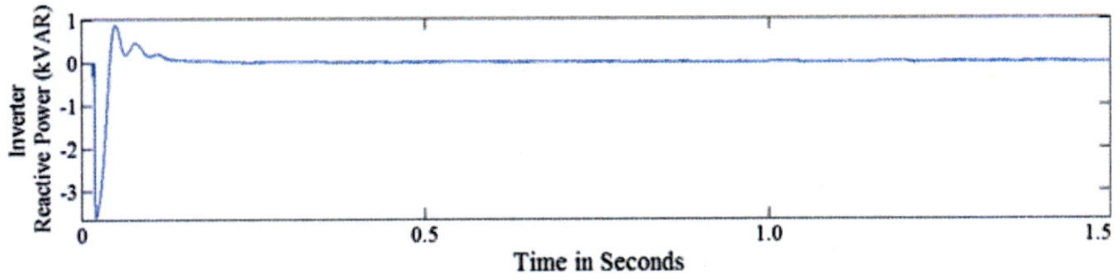

Fig. 8 Reactive power supplied by SECS

5 Conclusions

Design, modeling, simulation, and performance analysis of a grid tied SECS has been presented in this work. The system was simulated in MATLAB/Simulink environment. The system was operated under varying solar irradiance ranging 400–1000 W/m^2.

An exhaustive performance analysis was done on the basis of quantitative comparison of the performance of SECS under different values of solar irradiance. PV module PM-150 performed well under varying environmental conditions which indicates the accuracy of model. It is seen that the DC–DC boost converter is maintaining the DC link voltage level in association with intelligent MPPT efficiently. THD of 0.53% in voltage wave and 1.73% in current wave indicate very effective function of LCL filter. Inverter controllers have performed well above the expectations. This is indicated by proper load sharing, feeding to the grid in case of excess generation, and intact connection with the grid under varying irradiance after synchronization. The reactive power supplied to the grid is maintained at zero throughout the operation thus the power factor of the operation is maintained at unity.

The small size of the system may be useful for the people planning to install rooftop SECS with an intention of supplying surplus power to the grid during day.

References

1. Jamil B, Siddiqui AT, Akhtar N (2016) Estimation of solar radiation and optimum tilt angles for south-facing surfaces in humid subtropical climatic region of India. Eng Sci Technol Int J 19(4):1826–1835
2. Iqbal A et al (eds) (2020) Soft computing in condition monitoring and diagnostics of electrical and mechanical systems. In: Advances in intelligent systems and computing, vol 1096. Springer, Singapore. https://doi.org/10.1007/978-981-15-1532-3
3. Iqbal A et al (eds) Meta heuristic and evolutionary computation: algorithms and applications. In: Studies in computational intelligence, vol 1096. Springer, Singapore. https://www.springer.com/gp/book/9789811575709
4. Jalil MF, Khatoon S, Nasiruddin I (2018) Improved design analysis of photovoltaic modules and arrays under varying conditions. In: 3rd international innovative applications of computational intelligence on power, energy and controls with their impact on humanity, CIPECH-18. IEEE, Ghaziabad, India, pp 68–75
5. Hadianto NR, Mustaghfiri Fifi HS, Joke P, Erik T (2019) Design and implementation of three-phase grid-connected inverter for PV system. In: International seminar on intelligent technology and its applications, ISITIA. IEEE, Surabaya, Indonesia, pp 138–143
6. Priya MAJ, Ashok Kumar B, Senthilrani S (2018) Phase locked loop for controlling inverter interfaced with grid connected solar PV system. In: National power engineering conference, NPEC. IEEE, Madurai, India, pp 1–6
7. Sahay KB, Prabuddha A, Yadav V (2018) Modeling and simulation of grid connected 1 MW solar PV power plant. In: International conference and utility exhibition on green energy for sustainable development, ICUE. IEEE, Phuket, Thailand, pp 1–5
8. Swetalina B, Singh K, Narendra Kumar M, Mishra S (2017) A study on performance analysis of 75kWp grid connected solar power plant at MEMS. In: International conference on computer, electrical & communication engineering, ICCECE. IEEE, Kolkata, India, pp 1–5
9. Tochi N, Barry R (2017) Voltage control grid connection requirements for renewable power plants connected to the electricity transmission system in Nigeria. In: PES power Africa. IEEE, Accra, Ghana, pp 584–588
10. Raval DY, Munjani PJ, Mansoori NR (2017) Design, simulation and performance analysis of grid connected photo-voltaic system with decoupled current control. In: Nirma University international conference on engineering, NUiCONE. IEEE, Ahmedabad, India, pp. 1–7
11. Nasiruddin I, Khatoon S, Jalil MF, Bansal RC (2019) Shade diffusion of partial shaded PV array using odd-even structure. Sol Energy 181:519–529

Comparative Evaluation of Bipolar and Unipolar Control Technique-Based Cascaded H-bridge Multilevel Inverter

Naved Khan, Md Abdullah Ansari, Wasif Dilshad, and Khaliqur Rahman

Abstract This paper is focused on analyzing single-phase five-level cascaded H-bridge inverter followed by bipolar control scheme for full range of linear modulation indices. Sine–triangle-based pulse width modulation (PWM) technique has been employed to get five-level staircase waveform as output. Performance parameters like RMS output voltage, total harmonic distortion (THD), switching stress, power loss and efficiency have been calculated as per the formulae and graphically compared with unipolar switching scheme for full linear modulation range. MATLAB/Simulink software was used for simulation.

Keywords Bipolar · Multilevel · Modulation · Total harmonics distortion

1 Introduction

Rapid rate of increment in conventional energy source depletion is causing more stress on energy generation sector. Hence, newer non-conventional energy sources like sun, wind, water, etc., are exploited by researchers on a large scale [1–3]. Their intermittent availability poses a great hurdle in proper utilization [4]. Power received from these resources is needed to be converted to AC of suitable rating which is performed by power electronic inverters [5]. Conventional two-level inverters were

N. Khan (✉) · Md. A. Ansari · W. Dilshad · K. Rahman
Department of Electrical Engineering, Zakir Hussain College of Engineering and Technology, Aligarh Muslim University, Aligarh, UP 202002, India
e-mail: nk130685@gmail.com

Md. A. Ansari
e-mail: maansari11@myamu.ac.in

W. Dilshad
e-mail: wasif.dilshad138@gmail.com

K. Rahman
e-mail: er.khaliqurrahman@gmail.com

A. Iqbal et al. (eds.), *Renewable Power for Sustainable Growth*, Lecture Notes in Electrical Engineering 723, https://doi.org/10.1007/978-981-33-4080-0_72

inefficient due to low efficiency, higher harmonic content, high voltage stress on each switch, etc. [6, 7]. But with the introduction of multilevel inverters (MLIs), output voltage profile is improved, with requirement of low-size filters [8].

MLIs find their application in numerous areas which include motor drives, electric vehicle drives, back-to-back frequency link systems, etc. [9, 10]. Output voltage obtained is nearly sinusoidal which is attained by switching combination of switches according to switching scheme [7]. Higher the levels in output voltage, better is the efficiency of inverter because of low harmonics [11, 12]. Several classical topologies have been developed which include [2, 3]:

(1) Cascaded H-bridge;
(2) Flying capacitor;
(3) Diode clamped.

Among three topologies, cascaded H-bridge topology which requires isolated DC sources is the most practicable topology for implementation in medium as well high power applications because of its simple construction and modularity [13, 14]. Other topologies like flying capacitor and diode clamp face the problem of voltage unbalance and excess number of capacitors and diodes, respectively [15, 16]. And the number of elements increase quadratically with the increase in voltage levels [17].

This paper is based on modeling, analysis and comparative evaluation of bipolar and unipolar switching scheme of cascaded H-bridge inverter. Section 2 describes about the general as well as proposed topology. Section 3 discusses about performance parameters of the proposed topology. Section 4 compares bipolar and unipolar schemes. Section 5 finally concludes the paper.

2 Description of the Proposed Topology

2.1 Generalized Structure of the Proposed Topology

Figure 1 depicts the generalized circuit design employing single DC source and four power electronic switches per unit for n-cell topology. Each cell is a source of three voltage levels, namely $+E_{DC}$, $-E_{DC}$ and 0. When S_1 and S_2 are turned on, E_1 at the output becomes equal to E_{DC}. Similarly, $-E_{DC}$ and 0 are obtained by closing switches (S_3, S_4) and (S_1, S_3) or (S_4, S_2), respectively.

$$E = E_1 + E_2 + \cdots + E_n \tag{1}$$

where E_1, E_2, ..., E_n represents the output voltage of each cell.

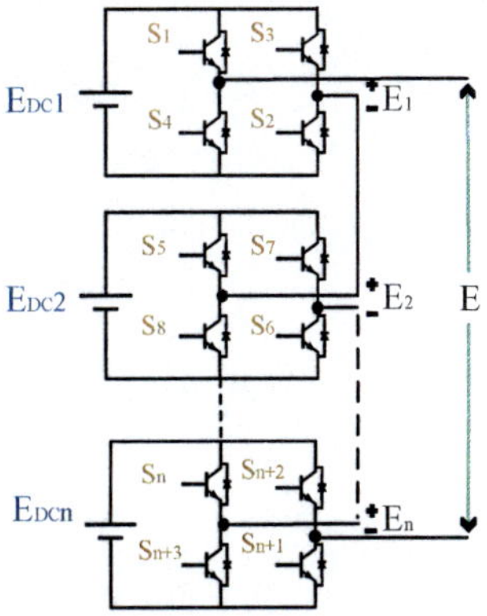

Fig. 1 N-cell single-phase CHB

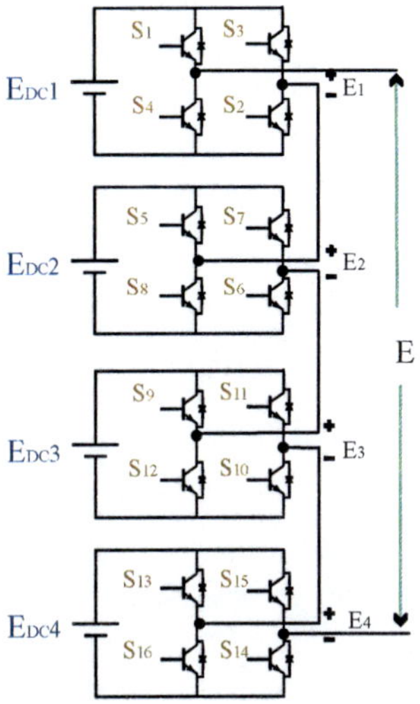

Fig. 2 Proposed topology

2.2 *Proposed Topology*

Figure 2 illustrates four-unit cascaded H-bridge topology supplied by four isolated DC sources. Each power electronic switch is connected in parallel with freewheeling diode to prevent from any voltage spike across the load.

2.3 *Control Scheme*

For obtaining proper n-level staircase output waveform, all the switches are needed to be triggered in specific pattern according to the switching states. All the DC sources are considered equal, i.e.,

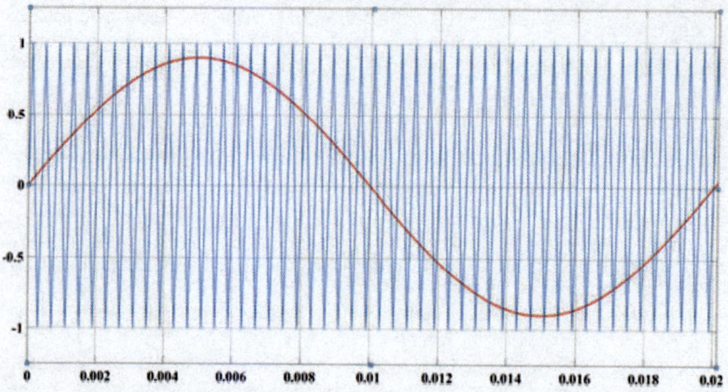

Fig. 3 Sine–triangle PWM technique

Table 1 Switching sequence of the proposed topology

State	$V_o\downarrow$	S_1 S_2	S_3 S_4	S_5 S_6	S_7 S_8	S_9 S_{10}	S_{11} S_{12}	S_{13} S_{14}	S_{15} S_{16}
1	$4E_1$	1	0	1	0	1	0	1	0
2	$2E_1$	1	0	1	0	1	0	0	1
3	0	1	0	1	0	0	1	0	1
4	$-2E_1$	0	1	0	1	0	1	0	1
5	$-4E_1$	0	1	0	1	0	1	0	1

$$E_{\mathrm{DC1}} = E_{\mathrm{DC2}} = E_{\mathrm{DC3}} = E_{\mathrm{DC4}} = E_1 \tag{2}$$

In this paper, sine–triangle PWM technique is studied. In bipolar scheme, two switches are switched together at a time so as to make pole voltage bipolar. From Fig. 3, it can be seen that reference sine wave of fundamental frequency (50 Hz) is compared with triangular carrier waves (5000 Hz). Both signals are passed to a comparator to make gate pulse signal which is given to two switches simultaneously (S_1 S_2). Complement of this signal is given to off-diagonal switches (S_3 S_4). Output voltage is $+E_{\mathrm{DC}}$ and $-E_{\mathrm{DC}}$ only. Other legs are switched by phase shifting carrier waves only by 90°.

Table 1 shows switching states corresponding to each output voltage. Diagonal and off-diagonal switches are switched in complement to each other.

Circuit parameters are as follows:

Component	Specification	Number
Carrier signal	Triangular $F_c = 5000$ Hz	4
Reference signal	Pure sinusoidal $F = 50$ Hz	1
Voltage sources	$E_{\mathrm{DC1}} = 100$ V, $E_{DC2} = 100$ V, $E_{\mathrm{DC3}} = 100$ V, $E_{\mathrm{DC4}} = 100$ V	4
Switches with anti-parallel diodes	IGBT, Ron = 1e−3 Ω	16

(continued)

(continued)

Component	Specification	Number
Load	$R = 10\ \Omega, L = 20$ mH	1

3 Performance Analysis

3.1 *Output Voltage*

Since output voltage being staircased form contains fundamental along with other harmonic components, the useful output is only the fundamental part. And hence, RMS value of fundamental is calculated which is given by:

$$V_{o,\mathrm{rms}} = \sqrt{\frac{\int_0^T V_o^2(t)\mathrm{d}t}{T}} \tag{3}$$

where $V_o(t)$ is the output voltage having periodicity T.

3.2 *Total Harmonic Distortion (THD)*

Coefficients of frequency terms other than fundamental frequency constitute to noise or harmonics in the system. Quantitative term used for determining their contribution is known as THD.

Since,

$$\mathrm{THD} = \sqrt{\frac{V_{\mathrm{rms}}^2 - V_1^2}{V_1}} \tag{4}$$

where V_1 and V_{rms} are, respectively, the fundamental and RMS values of output voltage.

3.3 *Switching Stress*

Switching stress is the overall stress developed in the switches during current flow in the circuit. It depends only on peak value of voltage and current and not on the pulse duration.

$$S = \sum_{j=1}^{N} V_j I_j \tag{5}$$

where V_j and I_j are the peak voltage and current, respectively, across switch j.

3.4 Switching Loss

It is the actual loss that occurs due to switching as well as conduction of switches. Since we are taking ideal switches with ON or OFF delay during switching, the only loss is due to ON state resistance (Ron) only. Average conduction loss is calculated by taking product of instantaneous voltage and current averaged over a complete cycle.

$$P_{\text{SW}} = P_{\text{loss,ON}} + P_{loss,\text{OFF}} \tag{6}$$

3.5 Efficiency (%)

Finally, efficiency of the inverter is calculated by calculating input power and output power.

$$P_{in} = \sum_{k=1}^{2} I_{s,k} V_{s,k}$$

where V_s and I_s are the average input voltage and input current of both source V_1 and V_2, respectively.

$$P_{\text{SW}} = V_o I_o \cos(\varphi)$$

V_o and I_o are the RMS value of **fundamental** output voltage and current, respectively. It is so because only the fundamental power is useful and rest harmonic power is wasted in the form of heat and is not useful.

4 Comparative Evaluation of Bipolar and Unipolar Scheme

Output load waveform obtained corresponding to amplitude modulation index 0.95 is shown in Fig. 4.

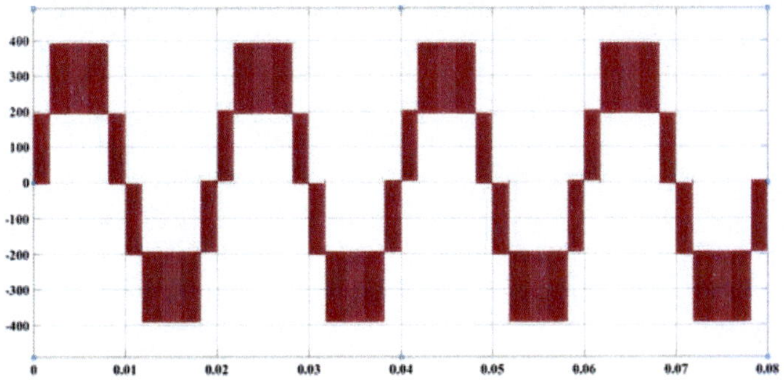

Fig. 4 Output load waveform for ma = 0.95

Table 2 Comparison of unipolar and bipolar switching scheme

	Unipolar	Bipolar
Output levels	**5**	**5**
Switches	8	16
DC sources	2	4
Load	RL (10 Ω, 20 mH)	RL (10 Ω, 20 mH)

Unipolar scheme is applied by switching switches independently such that output voltage changes from either 0 to + or 0 to − not + to −. Bipolar scheme is applied by switching diagonal switches such that output voltages are changed from + to −

For the same circuit parameters, unipolar scheme is compared with the proposed scheme. Table 2 shows circuit parameters in which five-level output is generated using two and four H bridges for unipolar and bipolar schemes, respectively.

4.1 Output Voltage Transferred to Load

For comparing both schemes, RMS output is normalized by dividing with base voltage, i.e., 400 and 200 V, for bipolar and unipolar, respectively. Figure 5 shows the graph.

4.2 Total Harmonic Distortion (THD)

Harmonic comparison is done graphically in Fig. 6. Both schemes are nearly equivalent in terms of THD.

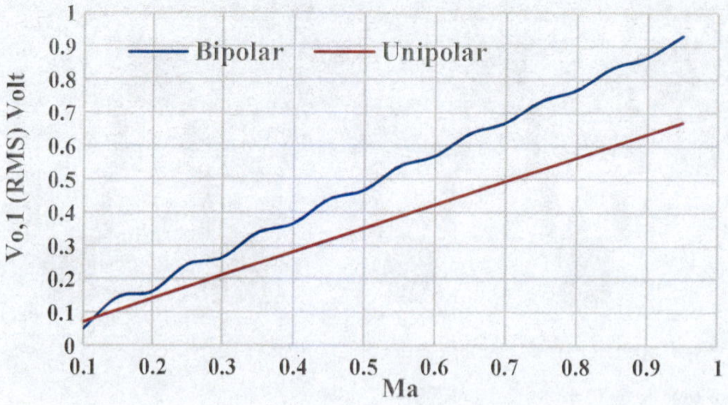

Fig. 5 Output load voltage versus modulation index (Ma)

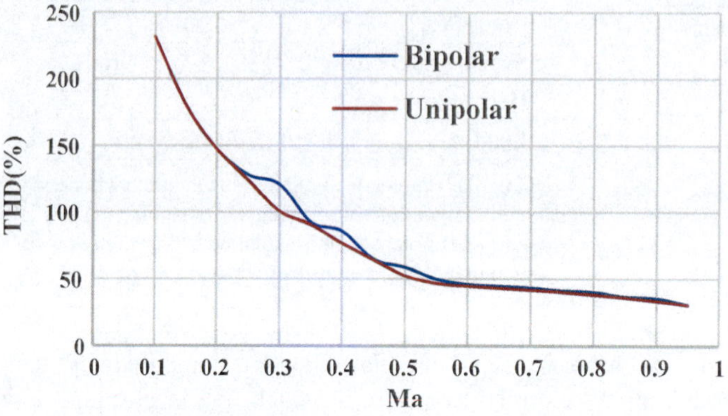

Fig. 6 THD(%) versus modulation index (Ma)

4.3 Switching Stress

Figure 7 depicts the switching stress for both cases. Peak voltage in each case for every value of *ma* is same, i.e., 100 V.

4.4 Switching Loss

Switching losses are directly related with number of switches which is greater in case of bipolar scheme. Figure 8 proves the assumption showing higher switching loss for bipolar than unipolar scheme.

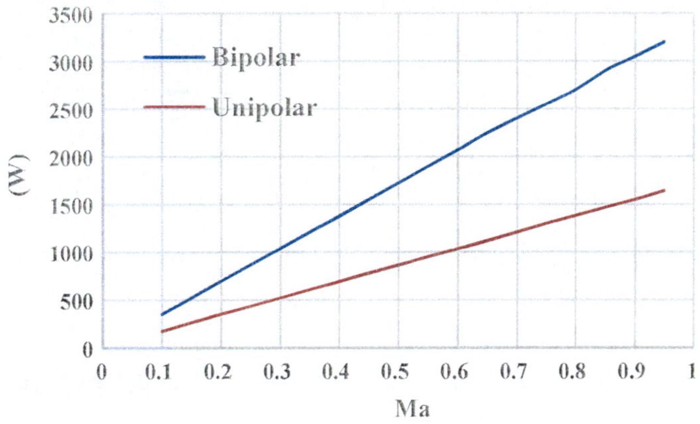

Fig. 7 Switching stress versus modulation index (Ma)

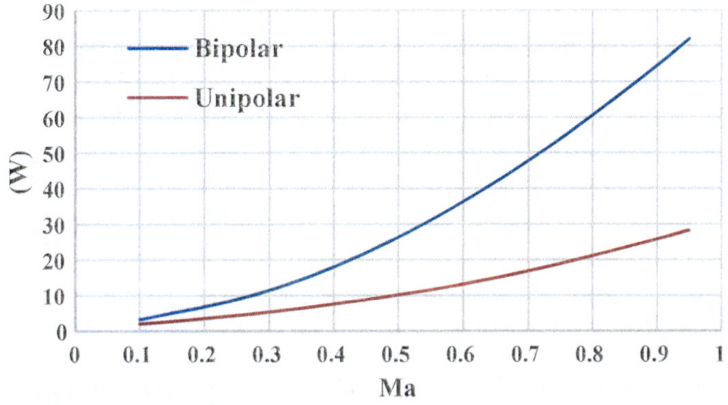

Fig. 8 Switching loss versus modulation index (Ma)

4.5 Efficiency

Figure 9 shows efficiency relationship with amplitude modulation index (ma). Both schemes show saturation with increase in ma. However, unipolar scheme maintains higher efficiency than bipolar scheme for higher values of ma.

5 Conclusion

This paper focused on modeling as well as comparative analysis of both unipolar and bipolar schemes of switching single-phase cascaded H-bridge multilevel inverter

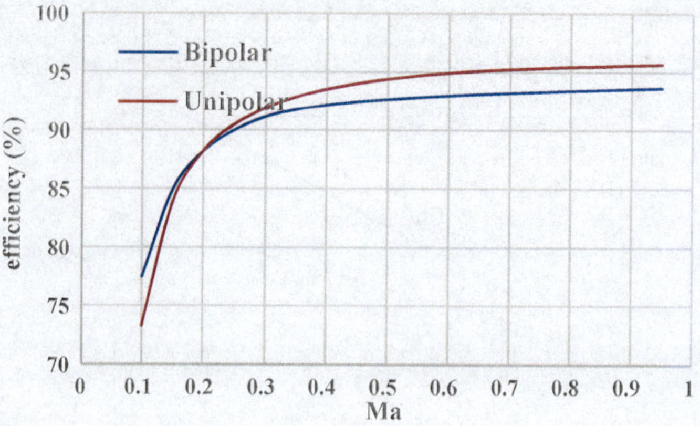

Fig. 9 Efficiency (%) versus modulation index (Ma)

topology. It was found that bipolar scheme is highly inefficient in terms of components. Also, bipolar scheme was found to have lower values of output fundamental component of load voltage and efficiency and higher values of THD, switching stress and switching losses which make it impractical in comparison with unipolar scheme. Hence, it can be concluded that unipolar scheme is better than bipolar scheme in all the above described performance parameters.

References

1. Hani Vahedi MT (2019) Single-DC-source multilevel inverters. Springer, Switzerland
2. Iqbal A et al (eds) (2020) Soft computing in condition monitoring and diagnostics of electrical and mechanical systems. In: Advances in intelligent systems and computing, vol 1096. Springer, Singapore. https://doi.org/10.1007/978-981-15-1532-3
3. Iqbal A et al (eds) (2020) Meta heuristic and evolutionary computation: algorithms and applications. In: Studies in computational intelligence, vol 1096. Springer, Singapore. https://www.springer.com/gp/book/9789811575709
4. Khan BH (2009) Renewable energy resources, 2nd edn. Tata MC Graw Hills
5. Xu H, Toliyat HA, Petersen LJ (2001) Five-phase induction motor drives with DSP-based control system. In: IEMDC 2001—IEEE international electric machines and drives conference, vol 17, issue no. 4, pp 304–309
6. Khamooshi R, Namadmalan A (2016) Converter utilization ratio assessment for THD optimization in cascaded H-bridge multi-level Inverters. IET Power Electron. 9(10):2103–2110
7. Nami A, Zare F, Ghosh A, Blaabjerg F (2011) A hybrid cascade converter topology with series-connected symmetrical and asymmetrical diode-clamped H-bridge cells. IEEE Trans Power Electron 26(1):51–65
8. Vahedi H, Shojaei AA, Dessaint LA, Al-Haddad K (2018) Reduced DC-link voltage active power filter using modified PUC5 converter. IEEE Trans Power Electron 33(2):943–947
9. Siddique MD, Mekhilef S, Shah NM, Memon MA (2019) Optimal design of a new cascaded multilevel inverter topology with reduced switch count. IEEE Access 7:24498–24510
10. Ansari A, Mohammad A, Tariq M, Riyaz A (2020) Residential energy conservation using efficient home appliances. Int J Innov Technol Explor Eng 9(3):3457–3465

11. Tarassodi P, Siadatan A, Keshani M (2019) Single-phase multi-level inverter suitable for symmetrical and asymmetrical photovoltaic (PV) applications. In: 2019 IEEE 28th international symposium on industrial electronics, pp 980–984
12. Ansari MA, Lone JA, Tariq M (2020) Performance analysis and comparative evaluation of two-winding multi-tapped transformer based nine-level inverter. In: International conference on power electronics, control and automation (ICPECA), New Delhi, India, pp 1–5
13. Suresh Y, Panda AK (2013) Investigation on hybrid cascaded multilevel inverter with reduced dc sources. Renew Sustain Energy Rev 26:49–59
14. Ounejjar Y, Al-Haddad K, Grégoire LA (2011) Packed U cells multilevel converter topology: theoretical study and experimental validation. IEEE Trans Ind Electron 58(4):1294–1306
15. Ahmed M, Sheir A, Orabi M (2018) Asymmetric cascaded half-bridge multilevel inverter without polarity changer. Alexandria Eng J 57(4):2415–2426
16. Rahman K, Meraj M, Mahajan SB, Iqbal A (2019) Single phase ZAC-source AC-AC converter with high buck and boost voltage conversion capability. IEEE Trans Ind Electron 1–1
17. Loh PC, Holmes DG, Lipo TA (2005) Implementation and control of distributed PWM cascaded multilevel inverters with minimal harmonic distortion and common-mode voltage. IEEE Trans Power Electron 20(1):90–99

Performance Comparison of Dual Stage Photovoltaics-Based Water Pumping Systems

Anuradha Tomar, Ayush Mittal, and Amruta Pattnaik

Abstract With the advancement in technologies and aiming toward the reduction of carbon footprints, the energy sector is headed toward renewable energy resources, among which solar energy is being utilized using photovoltaic technology. PV generates DC output and it is necessary to vary the level of the DC output in accordance with the respective PV application. In the following paper, a comparison of the performance of DC level converters specifically for the application of standalone PV-based water pumping and irrigation system is presented. This paper includes the performance study of Buck, Boost, SEPIC, and Cuk converters.

Keywords Photovoltaics (PV) · Water pumping system · Irrigation · MPPT · Buck · Boost · Sepic · Cuk

1 Introduction

Energy crises and climate change are among the top priority for every country around the globe. High dependency on non-renewable and conventional energy resources has led to the over-utilization and degradation of resources, environment and has created the necessity to switch for alternate resources of energy [1–3]. Solar being abundant in nature with zero cost for transportation and availability throughout the surface and makes it in the top of the priority list for research among renewable energy resources and utilizing it effectively with maximum efficiency [4]. Lack of

A. Tomar
Eindhoven University of Technology, Eindhoven 5600 MB, The Netherlands
e-mail: eranu28@gmail.com

A. Mittal
HMR Institute of Technology and Management, New Delhi, India
e-mail: am74333@hotmail.com

A. Pattnaik (✉)
Dr. Akhilesh Das Gupta Institute of Technology and Management, Delhi, India
e-mail: amruta.pattnaik@rediffmail.com

A. Iqbal et al. (eds.), *Renewable Power for Sustainable Growth*, Lecture Notes in Electrical Engineering 723, https://doi.org/10.1007/978-981-33-4080-0_73

electrical energy with frequent cuts and discontinuous supply, standalone PV systems are mesmerizing the interest of researchers among which PV-based water pumping (or irrigating) system, especially in rural areas with immense insolation, is gaining popularity over a period of time due to generation at the load end, and moreover, the system can be installed in remote areas with no transmission lines installed [5]. A typical PV-based water pumping system comprises a PV module energizing the electric motor and helping in pumping the underground (or surface) water using a bore (or surface) pump [6] which is utilized for irrigation purposes and other applications. The hybrid system allows storing the unused electrical energy in batteries which can be utilized for other small applications such as—for lighting and backup.

The standalone PV systems are used to fed electrical loads (AC or DC), and therefore to match the nature and requirement of load, the output of PV is passed through multiple electronic circuits—DC–DC converters (D/DC) and inverters—among which D/DC plays a vital role of varying the DC from one voltage (V) level of the PV output to another level with the help of electronic circuits comprising—inductors, capacitors, diodes, etc. Among various types of converters proposed in the literature, some examples are—Buck, Boost, Buck–Boost, Cuk, SEPIC, Zeta, Luo converter, etc. [7].

Undoubtedly, researches had shown that the solution for effective and efficient working of a standalone PV system is only possible if maximum power is extracted from the radiation and conveyed to the respective machine (or system). The conversion of irradiation to electrical energy faces many challenges, among which—low efficiency of conversion and high cost of PV cells are prominent. PV cell V-I characteristics as shown in Fig. 1 is nonlinear in nature and conversion rate is highly affected by the weather conditions, temperature, and irradiance; maximum power point (MPP) shifts over the V-I plot and to overcome the dynamic nature of MPP, and MPP tracking (MPPT) techniques are exerted to cope up the power to optimum level. Figure 2 portrays few of the proposed MPPT techniques from the literature [8].

Fig. 1 PV cell characteristics [9]

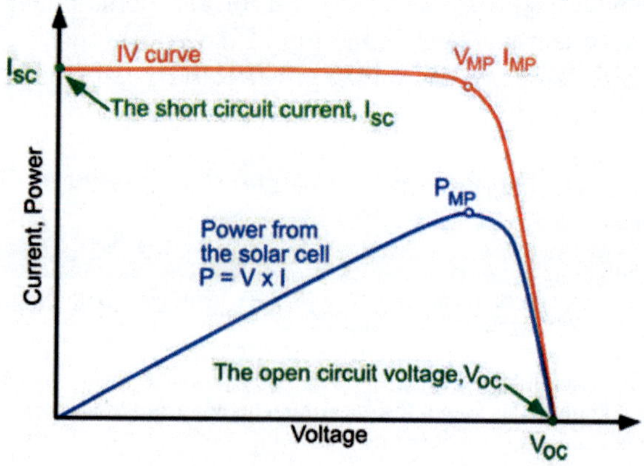

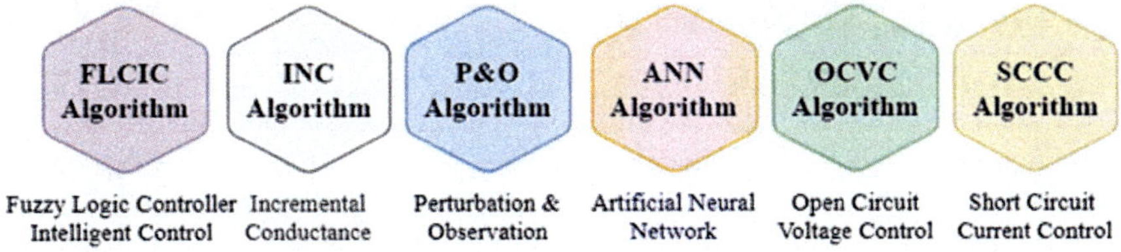

Fig. 2 Some MPPT techniques [8]

Due to immense research work, for the last few years in the field of renewable energy with priority in solar energy, the literature mentions several DC level converters, which creates a difficulty in prioritizing a befitting D/DC for a particular application. In this paper, a performance comparison of Buck, Boost, SEPIC, and Cuk converters for PV water pumping system is presented.

2 System Description

2.1 Configuration of the System

The system that is considered for comparison of variety of D/DC is of traditional and prevailing SPV system, supplying the energy to BLDC motor coupled water extracting (or irrigation) system, which mainly comprises of PV module, D/DC, DC/AC converter, BLDC motor, and its controller. The MPPT controller is positioned at the site for the purpose of tracking the MPP for extracting the PV array's maximum power, and the pulse generator is installed to administer, supervise, and control the DC level converter(s) [10, 11] (Fig. 3).

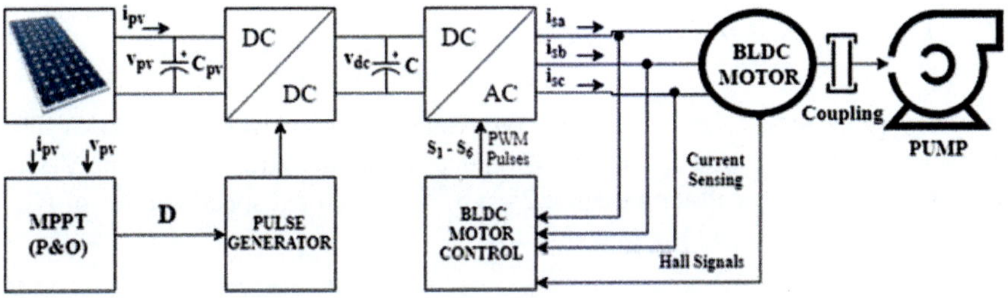

Fig. 3 System configuration [12]

2.2 Working of the System

Assuming the system to be ideal, the losses due to transmission, inversion, and conversion are neglected and all the electric power is constant throughout the process from generation to load end. The demand for electrical power is produced and photons are converted by PV array and is supplied to the BLDC motor and pump-set via various equipment and control unit, among which—D/DC and voltage source inverter (VSI) being primary and important. The PV array's electrical output serves as the input (I/P) power source for the D/DC and due to ideal system, the total power I/P is conveyed and conducted to the output section of the D/DC which acts as an I/P power source for the VSI (DC/AC converter). As per the prompt, the MPPT algorithm P&O, commanded pulses from pulse generator operates multiple IGBTs/MOSFETs used in the DC level converter. Feedback of V and current from the PV serve the basis of P&O algorithm and contribute toward pulling out of maximum power from PV system via generating an optimum value of duty cycle. The DC/AC conversion is performed using VSI and is further supplied to the BLDC motor to steer the mechanically coupled water pump. The VSI is plied and utilized in fundamental frequency switching through an electronic commutation of BLDC motor-assisted by its built-in encoder [12–14].

3 Design of the Nominated System

The system nominated comprises of the multiple factors and parameters along with various components among which design of solar photovoltaic, MPPT controller, and design of the various DC level shifters (Boost converter, Buck converter, SEPIC converter, and Cuk converter) are the essential ones which needs to be addressed.

3.1 Design of Solar Photovoltaic (SPV)

The system proposed uses the PV module type SunPower® to make SPR–305–WHT with a power capacity of $P_{\mathrm{mp}} = 305$ W under standard testing condition (STC: irradiance 1000 W/m^2, cell temperature 25 °C). The electrical specifications of PV module are listed in Fig. 4 [15].

3.2 MPPT Controller

Among various proposed MPPT algorithms in the literature, the simplified and well-established algorithm used in this study are the P&O. In the P&O MPPT algorithm,

Electrical Data		
[illegible]		
Peak Power (+/-5%)	Pmax	305 W
Rated Voltage	Vmp	54.7 V
Rated Current	Imp	5.58 A
Open Circuit Voltage	Voc	64.2 V
Short Circuit Current	Isc	5.96 A
Maximum System Voltage	IEC, UL	1000 V, 600 V
Temperature Coefficients		
	Power	-0.38% / °C
	Voltage (Voc)	-176.6 mV/°C
	Current (Isc)	3.5 mA/°C

Mechanical Data	
Solar Cells	96 SunPower all-back contact monocrystalline
Front Glass	4.0 mm (5/32 in) tempered
Junction Box	IP-65 rated with 3 bypass diodes
Output Cables	900 mm length cables / Multi-Contact connectors
Frame	Clear anodized aluminum alloy type 6063
Weight	24 kg, 53 lbs

Fig. 4 Specifications of SunPower® make SPR–305–WHT

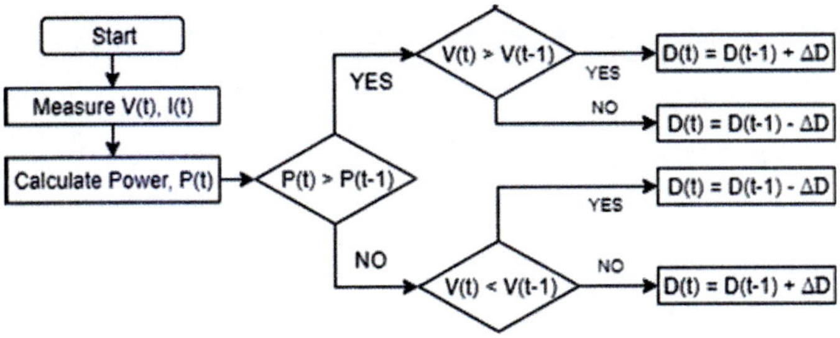

Fig. 5 P&O MPPT controller [16]

the solar output V is first perturbed and made unsettled, and accordingly, the output power is calculated. If power keeps on increasing along with voltage, then the voltage is increased; further, if at any point the power starts decreasing then the voltage is reduced to compel and evaluate the possible maximum power point. The variations in operating voltage are maintained and controlled by a constant value. Also, because of persistent desire for MPP, the P&O algorithm keeps the panel output power oscillating around MPP. Figure 5 depicts the block diagram of the MPPT control unit which is used in this study. MPPT controller's output is supplied to the DC/DC which acts as the power I/P source for DC/AC inverter [16, 17].

3.3 *Design of Various Converters*

To collate the performance and accomplishment of the standalone PV-based water irrigation and pumping system with several D/DC, the following converters are designed using the MATLAB/Simulink as per the design specification and schematics provided in Table 1 [7, 18–20].

Table 1 Various DC/DC converters along with schematic diagram

	Name of the Converter along with the Schematic Diagram	Parameters
BOOS T CONV.	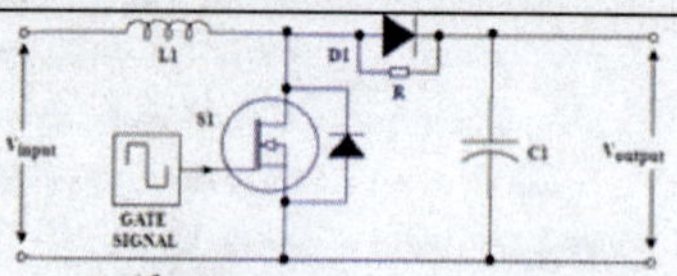	L1 = 0.9mH C1 = 10mF (0.06R) Frequency = 4.6kH
BUCK CONV.	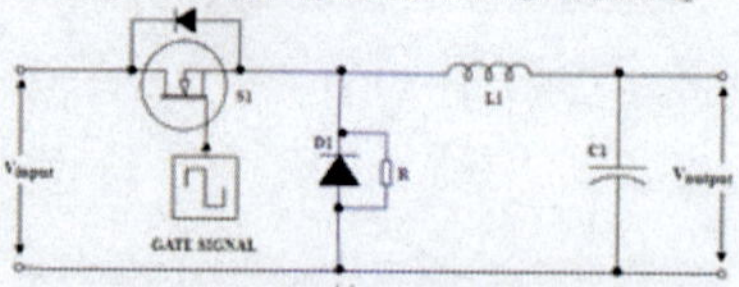	L1 = 400μH C1 = 440μF Frequency = 50kH
SEPIC CONV.	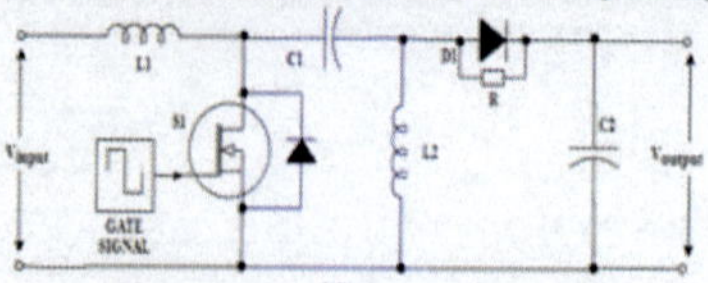	L1 = 100μH L2 = 100μH C1 = 100μF C2 = 100μF Frequency = 10kH
CUK CONV.	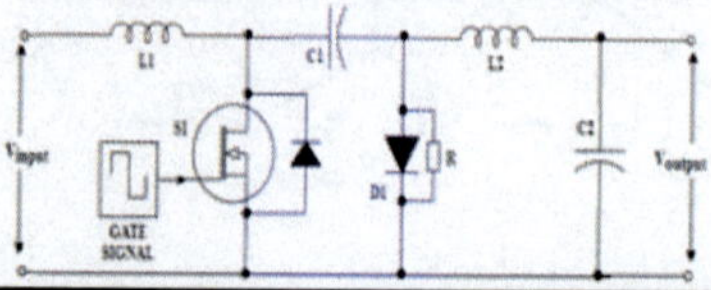	L1 = 100μH L2 = 100μH C1 = 100μF C2 = 100μF Frequency = 10kH

4 Result and Analysis

Figure 3 portrays the schematic version of the MATLAB simulated PV-based water pumping system and Table 1 shows various DC level shifters along with parameters that are used for performance analysis. Each system thus obtained is made to run for 200 s with irradiance level shifting from 800 W/m^2 (0–110 s) to 500 W/m^2 (110–120 s) and further shifting the level and remaining constant at 600 W/m^2 as depicted in Fig. 6.

The variation in the voltage of DC bus in accordance with irradiance level is plotted against time (Fig. 7). Also, the DC bus power shows a considerable amount of change with respect to the time (Fig. 8) which effects the BLDC motor performance

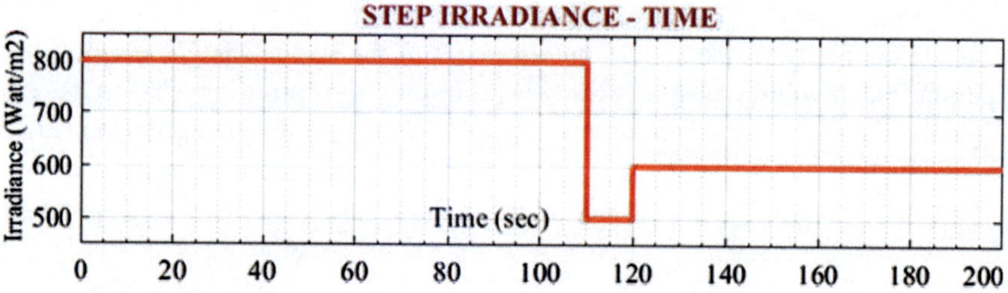

Fig. 6 Level shifting of irradiance level with time

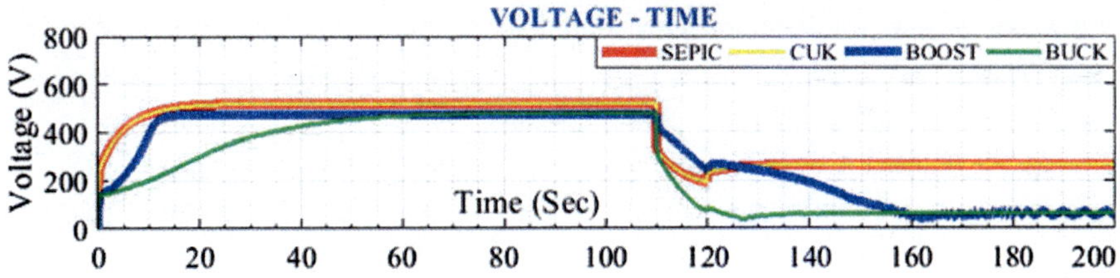

Fig. 7 Voltage versus time plot

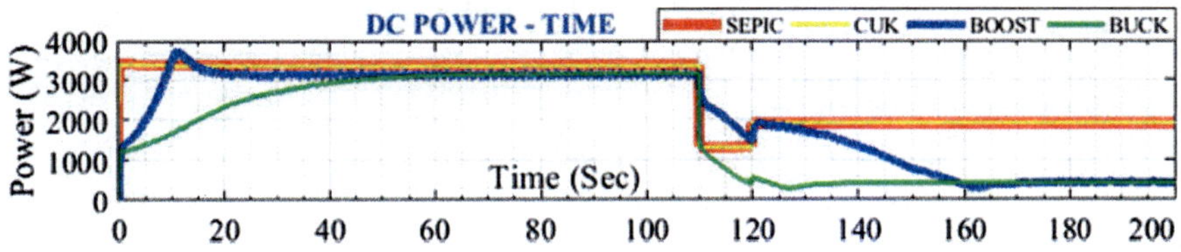

Fig. 8 DC bus power versus time plot

with varying parameters of speed developed (Fig. 9) and mechanical power delivered (Fig. 10). The variation in all four parameters can be studied and analyzed using the data provided in Tables 2 and 3.

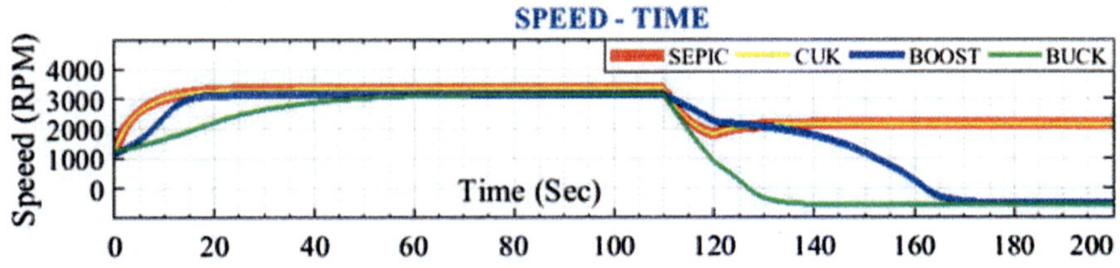

Fig. 9 Variation of motor speed versus time plot

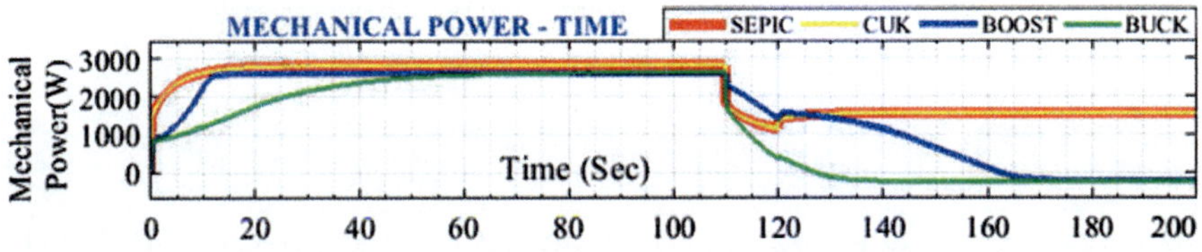

Fig. 10 Varying mechanical power developed by BLDC motor versus time plot

Table 2 Data of voltage and DC bus power with respect to time and irradiance level

Time/irradiance	Voltage (V)				Power DC (W)			
	SEPIC	Cuk	Boost	Buck	SEPIC	Cuk	Boost	Buck
20/800	512	511	475	320	3387	3367	3217	2345
40/800	519	519	476	426	3389	3371	3178	2954
60/800	520	519	476	467	3390	3371	3122	3121
80/800	520	519	456	477	3378	3369	3170	3177
100/800	520	519	456	479	3378	3368	3181	3160
110/500	520	519	456	479	3389	3360	3220	3151
120/600	195	202	249	83	1331	1330	1460	551
140/600	263	263	188	62	1921	1907	1402	411
160/600	264	264	57	63	1921	1909	353	416
180/600	264	264	64	63	1921	1912	428	417
200/600	264	264	64	64	1915	1912	428	417

Table 3 Data of speed and mechanical power with respect to time and irradiance level

Time/irradiance	Speed (RPM)				Mechanical power (W)			
	SEPIC	Cuk	Boost	Buck	SEPIC	Cuk	Boost	Buck
20/800	3264	3259	3110	2206	2784	2782	1729	2597
40/800	3317	3314	3133	2880	2814	2811	2361	2598
60/800	3318	3316	3134	3087	2816	2813	2560	2598
80/800	3318	3316	3134	3138	2816	2813	2608	2598
100/800	3318	3316	3134	3150	2815	2813	2613	2598
110/500	3318	3316	3134	3146	2815	2812	2575	2575
120/600	1798	1797	2240	900	1127	1193	410	1393
140/600	2118	2116	1761	−545	1529	1532	−238	1125
160/600	2123	2121	253	−564	1531	1534	−248	104
180/600	2123	2121	−508	−564	1532	1533	−248	−228
200/600	2123	2121	−508	−564	1532	1533	−248	−228

5 Conclusion

With the aim to collate the performance of the D/DC (Buck, Boost, SEPIC, and Cuk), the designing parameters for the mentioned converters are taken different and it is tried to make the output stable during the initial phase of irradiance level. After studying the various plots and analyzing the tabular data, it is observed that the converters need to be designed and considered specially as per application requirement, otherwise, system performance is adversely impacted. With various sensors and protection schemes installed—such as overvoltage/under-voltage protection and over-current protection— it is practically not possible to achieve speed and mechanical power output with negative values, as motor will stop to safeguard the equipment and preventing failures. The performance of SEPIC and Cuk converter is ascertained to be more apt and relevant for PV-based irrigation systems, as such systems are more exposed to varying/mismatched operating conditions. Therefore, the study revealed that the application and purpose had better be the basis for the selection and design of the DC level shifters.

References

1. Tomar A, Mittal A, Sharma S (2018) PV-Piezo hybrid grid connected system, ICPEICES, Delhi, India, pp 748–753
2. Iqbal A et al (eds) (2020) Soft computing in condition monitoring and diagnostics of electrical and mechanical systems. In: Advances in intelligent systems and computing, vol 1096. Springer, Singapore. https://doi.org/10.1007/978-981-15-1532-3
3. Iqbal A et al (eds) (2020) Meta heuristic and evolutionary computation: algorithms and applications. In: Studies in computational intelligence, vol 1096. Springer, Singapore. https://www.springer.com/gp/book/9789811575709
4. Chandel SS, Naik MN, Chandel R (2015) Review of solar photovoltaic water pumping system technology for irrigation and community drinking water supplies. In: Renewable and sustainable energy reviews. Last Accessed 4 Apr 2020
5. Mergoul K, Laarabi B, Barhdadi A (2018) Solar water pumping applications in Morocco: state of the art. In: IRSEC, Rabat, Morocco, pp 1–6. https://doi.org/10.1109/IRSEC.2018.8703015.
6. Tomar A, Mishra S (2017) PV energy benefit estimation formulation for PV water pumping system. In: UPCON, Mathura, pp 44–48
7. Pachauri RK, Chauhan YK (2016) Modeling and simulation analysis of PV fed Cuk, SEPIC, Zeta and Luo DC-DC converter. In: ICPEICES, Delhi
8. Pakkiraiah B, Sukumar GD, Senthilarasu S (eds) (2016) Research survey on various MPPT performance issues to improve the solar PV system efficiency. J Sol Energy 1–20. https://doi.org/10.1155/2016/8012432
9. https://www.pveducation.org/pvcdrom/solar-cell-operation/iv-curve. Accessed 4 Apr 2020
10. Tomar A, Nguyen PH, Mishra S (2020) SEPIC-MISO converter based PV water pumping system—an improved performance under mismatching conditions. In: 2020 IEEE 9th power india international conference (PIICON) In: SONEPAT, India, pp 1–5. 10.1109/PIICON49524.2020.9112907
11. Kumar R, Singh B (2016, May–June) BLDC motor-driven solar PV array-fed water pumping system employing zeta converter. IEEE Trans Ind App 52(3):2315–2322
12. Kumar R, Singh B (2015) BLDC motor driven water pump fed by solar photovoltaic array using boost converter. In: Annual IEEE India Conference (INDICON), New Delhi, pp 1–6. https://doi.org/10.1109/INDICON.2015.7443676
13. Mishra AK, Singh B (2016) Design of PV powered SR motor driven irrigation pumps utilizing boost converter. In: UPCON, Varanasi, pp 264–268
14. Tomar A, Mishra S (2016) Multi-input single-output DC-DC converter based PV water pumping system. In: ICPEICES, Delhi, pp 1–5
15. https://www.pocosolar.com/wp-content/themes/twentyfifteen/pdfs/SunpowerSolarPanels/sunpower_305wht_spec_sheet.pdf. Last Accessed 4 Apr 2020
16. Gupta AK, Saxena R (2016) Review on widely-used MPPT techniques for PV applications. In: International conference on innovation and challenges in cyber security, pp 270–273. https://doi.org/10.1109/ICICCS.2016.7542321

17. Malla S (2020) Perturb and observe (P&O) algorithm for PV MPPT. https://www.mathworks.com/matlabcentral/fileexchange/39641-perturb-and-observe-p-o-algorithm-for-pv-mppt, MATLAB Central File Exchange. Retrieved 7 July 2020
18. Khan MTA, Ahmed MR, Ahmed SI, Khan SI (2012) Design and performance analysis of water pumping using solar PV. In: ICDRET 2012, Dhaka, pp 1–4
19. Lin R, Pan C, Liu K (2009) Family of single-inductor multi-output DC–DC converters. In: International conference on power electronics and drive systems, Taipei, pp 1216–1221
20. Kumar R, Singh B (2014) Solar PV array fed water pumping system using SEPIC converter based BLDC motor drive. In: Eighteenth national power systems conference (NPSC), Guwahati, pp 1–5. https://doi.org/10.1109/NPSC.2014.7103820

Buck–Boost Converter with no Dead or Overlap-Times

Muhammad Ado, M. Saad Bin Arif, Awang Jusoh, and Abdulhamid Usman Mutawakkil

Abstract A DC–DC synchronous buck–boost converter that is capable of operation with or without dead or overlap-time is proposed. The topology of the converter is derived from the topology of quasi-impedance source converter where two capacitors and two inductors are used for second-order filtering and energy storage. However, its gain is identical to that of the conventional buck–boost converter (BBC), thereby qualifying it as a BBC. The elimination of dead or overlap-time reduces control complexities and permits higher frequency operation which implies the use of smaller capacitors and inductors. Operation of the converter was successfully verified using control signals without dead or overlap-time using a 50 kHz prototype and its output voltage is compared with the ideal theoretical values.

Keywords Buck–boost converter · Shoot-through · Z-source · Quasi-Z-source

1 Introduction

Power electronics converters are electronic devices responsible for converting an electric signal from one type or magnitude to another [1]. AC–AC converters vary the magnitudes of AC signals while DC–DC converters vary the magnitude of DC voltages. AC–DC converters (rectifiers) convert AC to DC signals, whereas DC–AC converters (inverters) convert DC to AC signals. For applications involving DC signals with varying magnitudes, DC–DC converters are employed to stabilize their

M. Ado · M. S. B. Arif · A. Jusoh (✉)
School of Electrical Engineering, Universiti Teknologi Malaysia, Skudai, Malaysia
e-mail: awang@utm.my

M. S. B. Arif
e-mail: saadbinarif@gmail.com

M. Ado · A. U. Mutawakkil
Deptment of Physics Faculty of Physical Science, Bayero University, Kano, Nigeria

M. S. B. Arif
Department of Electronics Engineering, Z. H. College of Engineering and Technology, Aligarh, Muslim University, Aligarh, India

A. Iqbal et al. (eds.), *Renewable Power for Sustainable Growth*, Lecture Notes in Electrical Engineering 723, https://doi.org/10.1007/978-981-33-4080-0_74

magnitude [2]. To reduce losses, they use switches mostly transistors or diodes to vary the magnitude. DC–DC converters are classified as switched-mode or linear-mode base on the operation of the transistors. Switched-mode power converters have higher efficiency but contain harmonics hence they require filters.

To further improve the efficiency of switched-mode converters, their rectifier diodes are replaced with synchronous MOSFETS because the conduction losses of synchronous MOSFETS are less compared to those of diodes [3]. However, the control of synchronous converters is more complex since the switching of the synchronous MOSFET is not natural. Depending on the circuit topology, dead or/and overlap-times are usually provided.

ISCs redefined the concept of shoot-through (ST) and open circuit (OC) in power converters [4–7]. ST could be defined as the scenario that occurs when all the switches of a complementary switching sets are ON while OC could be referred to as when all the switches are OFF. In synchronous converters, overlap and/or dead-times are usually provided to cater for OC and ST, respectively. These overlap and dead-times distort the waveforms and cause restrictions in switching frequency. The restriction is because at higher frequencies, the chances for the occurrence of OC or ST increase [5]. Also, when dead-times are too high, the efficiency of the converter drops because the body diode which has higher conduction losses than the transistor conducts [3]. Conduction by the body diode for 10% of the conduction time could result in 40% increase in conduction losses [3]; thus, the elimination of dead-time could increase efficiency due to non-conduction by body diode [16]. Thus, the elimination of overlap and dead-times permits operation at higher frequencies which results in reduced size of reactive components, weight, and cost [8–10].

Quasi-Z-source converters (q-ZSCs) are a type of ZSCs shown in Fig. 1a proposed by [11] to address some limitations of classical ZSCs such as simplified control and source stress. DC–DC ZSCs and q-ZSCs were later proposed by [12]. Reference [13, 14] proposed additional topologies. These two new topologies have identical gain [15] and form an additional class [1]. However, their operations have only been verified using signals with dead-time [9, 10]. This paper verifies the operation of one of these topologies without using dead or overlap-time for gate control [13–15]. This is possible because the converter is based on impedance source topology which permits the switches of an H-bridge's common leg or similar network to be

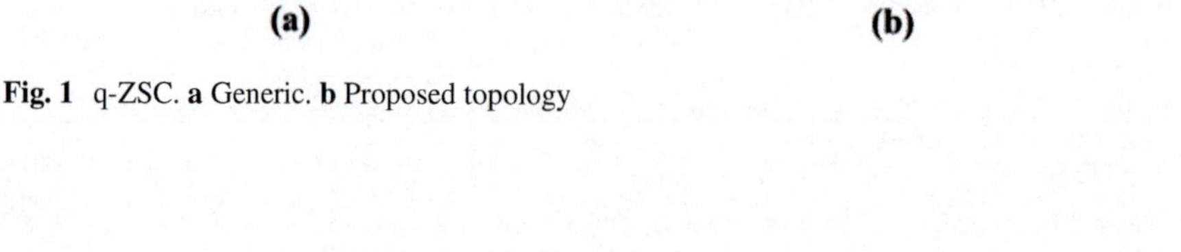

Fig. 1 q-ZSC. **a** Generic. **b** Proposed topology

simultaneously switched. The elimination of overlap and dead-time permits the use of higher switching frequency which edges it over a corresponding classical BBC.

A 50 kHz prototype of the converter was constructed and its operation compared with ideal theoretic gain. An advantage of the elimination of dead and overlap-time is that control is simplified and higher frequency application could be achieved. Higher frequency application permits use of small capacitors and inductors. Use of smaller inductors could imply increase in efficiency because DC resistance of inductors always increases with the inductance value.

2 Derivations and Analysis

This section presents analyses the converter's operations. As discussed earlier, the converter utilizes a synchronous MOSFET. Thus, the converter is operated by complementary switching of the two switches which results in two modes of operation. The modes, mode I and mode II, transit each other without delay or overlap. This implies that delay signals are not required to avoid shoot-through [5]. The duty ratio is with respect to switch S_1.

2.1 Converter Topology

The converter's topology as shown in Fig. 1b is derived from Fig. 1a by interchanging C_1 and V_S and taking the output across C_2 [12, 13].

2.2 Converter Gain

Mode I During this mode with duty ratio D, S_1 is ON and S_2 is OFF. The equivalent circuit is shown in Fig. 2a.

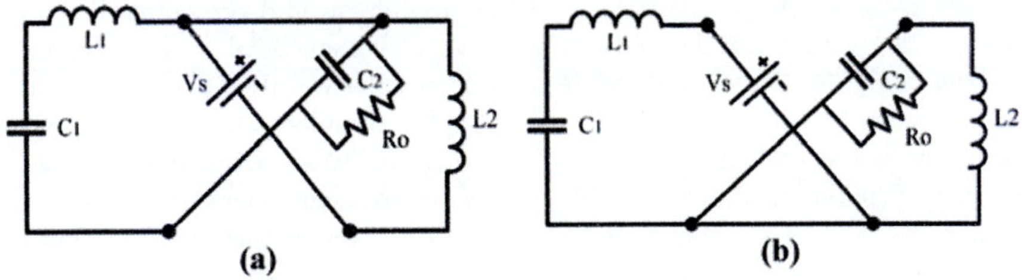

Fig. 2 Operation modes. **a** Mode I. **b** Mode II

$$V_{L1} = V_O - V_{C1} \tag{1}$$

$$V_{L2} = V_S \tag{2}$$

Mode II During this mode with duty ratio $D' = 1 - D$, S_1 is OFF and S_2 is ON. The equivalent circuit is shown in Fig. 2b.

$$V_{L1} = V_S - V_{C1} \tag{3}$$

$$V_{L2} = V_O \tag{4}$$

Applying inductor volts-seconds balance over a switching interval on L_1 and L_2 yields (5) and (6), respectively.

$$\bar{V}_{L1} = DV_O + V_S - V_{C1} - DV_S = 0 \tag{5}$$

$$\bar{V}_{L2} = DV_S - V_O(D - 1) = 0 \tag{6}$$

Solving (6) results in having (7)

$$V_O = -\frac{D}{1 - D} V_S \tag{7}$$

$$A = \frac{V_O}{V_S} = \frac{-D}{1 - D} \tag{8}$$

The expression given in (8) is the proposed converter's ideal gain and is identical to that of the classical BBC.

3 Implementation and Verification

To verify the converter's capability of operation without dead or overlap-time, a prototype was fabricated and tested with a gate signal without dead or overlap-times as shown in Fig. 3a. Atmel ATSAM3X8 microcontroller was used to generate two complementary PWMs with no dead-time delay. The complementary PWM signals were connected to the converter via two gate drivers. Values of parameters used for the verification are shown in Table 1. Asymmetric components were used as suggested by [8, 9]. Two separate operations of the converter were tested for boost mode and buck mode by using duty ratios of 0.65 and 0.35 respectively, as shown in Table 1.

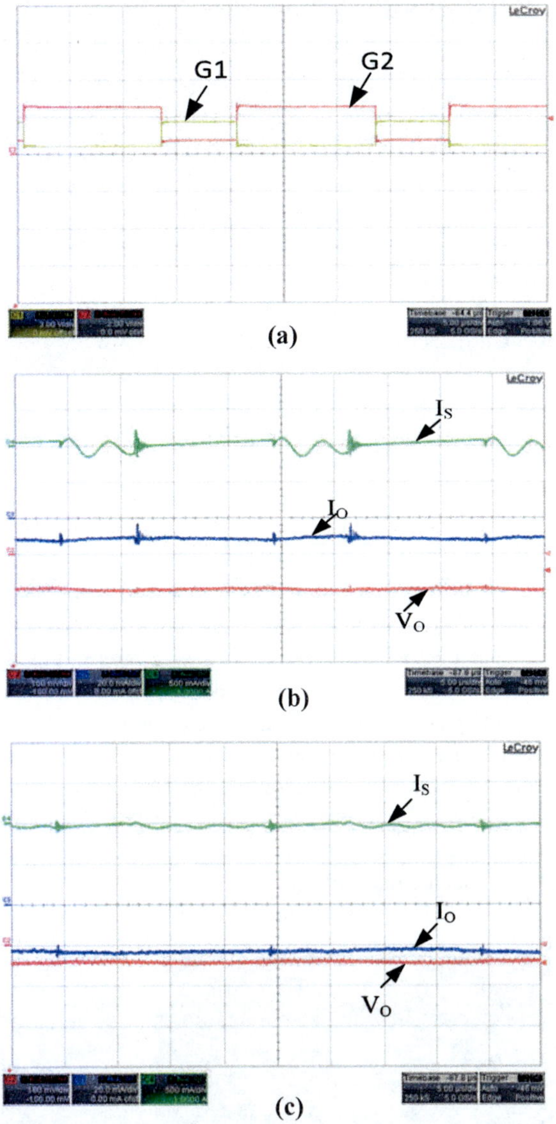

Fig. 3 **a** Gate signal for $D = 0.65$. **b** Steady-state input current (I_S), steady-state output current (I_O), and steady-state output voltage waveforms of the proposed converter at $D = 0.65$. **c** Steady-state input current (I_S), steady-state output current (I_O), and steady-state output voltage waveforms at $D = 0.35$

Table 1 Values of parameters used for the experiment

Parameter	V_g (V)	D_{Buck}	D_{Boost}	f (kHz)	C_1 (μF)	C_2 (μF)	L_1 (μF)	L_2 (μH)	Load (Ω)
Value	3.5	0.35	0.65	50	0.90	0.01	33	470	150

4 Results and Discussion

The gate signals for the boost mode ($D = 0.65$) and buck mode ($D = 0.35$) operation are shown in Fig. 3a with $D = 0.65$ and a frequency of 50 kHz. Results of the experiment are displayed in Fig. 3b, c. Figure 3b, shows the output and input currents and also the output voltage during boost mode, while Fig. 3c shows the output current (I_O) and input current (I_S) and also the output voltage (V_O) during buck mode. The experimental prototype and gate driver are shown in Fig. 4a and the complete setup used for experimenting is shown in Fig. 4b.

A differential voltage probe with a scale of 1:50 was used to display the voltages through a Lecroy WaveSurfer 44MXs-B oscilloscope. The scale implies that the displayed voltages of Fig. 3b, c have to be multiplied by 50 to get the actual voltage value. A multimeter was used to manually measure the average output voltages for both the buck and boost modes. This was to measure the steady-state output voltage (V_O) as against the varying output voltage waveform displayed by the oscilloscope. The voltage of 1.52 V was recorded at $D = 0.35$ and 4.30 V at $D = 0.65$.

The waveforms of Fig. 3b, c confirm the functionality of the proposed buck–boost converter. At D = 0.35, the converter's ideal gain and output voltages are 0.5385 and 1.88 V, respectively. For the boost mode, the ideal gain and output voltages are 1.8571 and 6.5 V, respectively. The experimental output voltages of 4.30 V during boost and 1.52 V during buck mode as shown in Fig. 3b, c are 80.85 and 66.15% of the ideal voltages. These fall in the voltages are due to the parasitic voltage drops by the reactive components [8] and switches.

The input and output current waveforms have ripples due to transients that occur during switching. Another reason for the ripples is because the components were not selected using the converter's design equations. In an upcoming publication, the design equations, ripple expressions current, and voltage rating expressions for the components are derived.

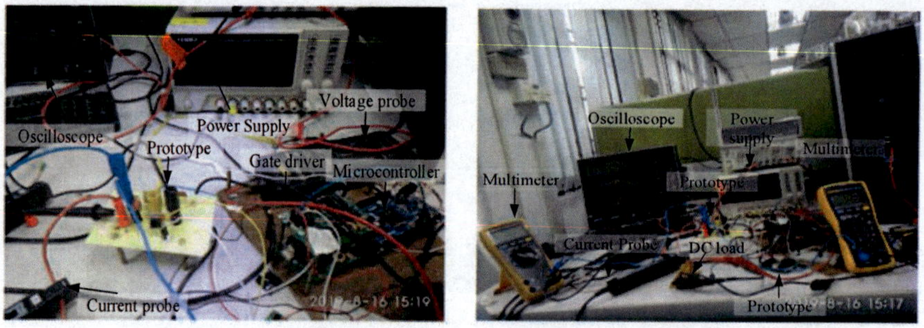

(a) (b)

Fig. 4 Experimental setup. **a** Close view of the prototype. **b** Complete view of the setup

5 Conclusion

The operation of a buck–boost converter without dead or overlap-time is presented. Its boost and buck modes operation were presented with both the experimental and simulation results. The advantage of the converter over traditional buck–boost converter includes the permissibility of higher frequency operation due to the elimination of dead or overlap-time. This implies saving cost and power density because smaller reactive components could be used due to the higher switching frequency. Although the converter used film capacitors rather than the usual heavy and large electrolytic capacitors, ceramic capacitors could also be used to reduce more weight and size.

References

1. Ado M, Jusoh A, Sutikno T, Ado M (2019) Extended family of DC-DC quasi-Z-source converters. Int J Electr Comput Eng 9(6):4540–4555
2. Bin Arif MS, Avob SM, Yahya SM, Mustafa U, Ado M, Khan ZA (2018, October) Effect of Zn-H 2 0 nanofluid back-flow channels on the efficiency and electrical power output of a solar PV panel used in standalone PV system. In: Asia-Pacific power energy engineering conference APPEEC, vol 2018, pp 493–497
3. Acker B, Sullivan CR, Sanders SR (1995) Synchronous rectification with adaptive timing control. In: PESC record—IEEE annual power electronics specialists conference, pp 88–95
4. Yang S, Peng FZ, Lei Q, Inoshita R, Qian Z (2011) Current-fed quasi-z-source inverter with voltage buck-boost and regeneration capability. IEEE Trans Ind Appl 47(2):882–892
5. Peng FZ (2003) Z-source inverter. IEEE Trans Ind Appl 39(2):504–510
6. Ado M, Jusoh A, Sutikno T, Muda MH, Arfeen ZA (2020) Dual output DC-DC quasi impedance source converter. Int J Electr Comput Eng 10(4):3988–3998
7. Ado M, Jusoh A, Nordin NM (2019) Design equations for DC-DC quasi-ZSC. In: 4th IEEE conference on energy conversion (CENCON 2019), pp 194–199
8. Ado M, Jusoh A, Mutawakki AU, Sutikno T (2018) Dynamic model of A DC-DC quasi-Z-source converter. Int J Electr Comput Eng 9(3):1585–1597
9. Ado M, Jusoh A, Muda MH, binti Atan F, Mutawakkil AU (2019) A prototype of DC-DC quasi—Z-source Buck-boost converter. In: 10th IEEE control and system graduate research colloquium, pp 184–188
10. Ado M, Jusoh A, Sutikno T (2020) Asymmetric quasi impedance source Buck-Boost converter. Int J Electr Comput Eng 10(2):2128–2138
11. Anderson J, Peng FZ (2008) Four quasi-Z-source inverters. In: IEEE power electronics specialists conference, pp 2743–2749
12. Cao D, Peng FZ (2009) A family of Z-source and quasi-Z-source DC-DC converters. In: 24th annual IEEE applied power electronics conference and exposition, pp 1097–1101
13. Ado M, Jusoh A, Aziz MJA, Kermadi M, Mutawakkil AU (2018) DC-DC q-ZSC with Buck-Boost converter gain. In: 9th IEEE control and system graduate research colloquium, pp 85–88
14. Ado M, Jusoh A, Ayob SM, Ali MH, Galadanchi GSM (2018) Buck-Boost converter with q-ZSC topology. In: 5th IET international conference on clean energy and techonology
15. Ado M, Jusoh A, Mutawakkil AU, Ayob SM (2018) Two q-ZSCs with efficient Buck-Boost gain. In: 10th IEEE PES Asia-Pacific power and energy engineering conference, pp 252–256
16. Ado M, Arif MSB, Jusoh A, Mutawakkil AU, Danmallam IM (2020) Buck-boost converter with simple gate control for renewable energy applications Energy Sources, Part A: Recovery, Utilization, and Environmental Effects, pp 1–13

The Role of Power Electronics in the Field of Photovoltaic System: A Study

Amruta Pattnaik and Anuradha Tomar

Abstract Renewable energy resources like sun, wind, biomass, geothermal, etc., are abundant in nature, toxic-free, and eco-friendly. Those are sufficient to meet the energy demands which are required by the consumers. Solar photovoltaic energy is one of the clean renewable energies among others. A solar photovoltaic system is gaining popularity as compared to other non-conventional sources because of the sufficient solar irradiation received by the earth. Moreover, the costs of PV modules have been declining. PV system has been included with PV panels, batteries, or any other storage devices as well as power electronic converters. Power electronic converter systems are the only choice to fulfill the requirements in the power system. Therefore, this work aims to discuss the power electronic converter technologies that can be used as interfaces and discuss their merits and demerits. Power electronic converter technologies are used to convert the one form of energy to other energy or vary the energy range as per the load requirement

Keywords Renewable energy · Photovoltaic (PV) system · Power electronics (PE) converters

1 Introduction

Solar power is a non-conventional energy source that might replace fossil fuel-dependent energy sources very soon. In the year 2019, the total power recorded by solar photovoltaics (PV) is 697 GW as compared to other renewable sources according to the International Energy Agency. Non-conventional energy capability is established to grow by 50% from the year 2019–2024, managed by PV energy. Solar PV is one of the non-conventional energies to gain 60% of the estimated development [1, 26, 27]. In *Renewables 2019*s progressed case, non-conventional energy

A. Pattnaik (✉)
Dr. Akhilesh Das Gupta Institute of Technology & Management, Delhi, India
e-mail: amruta.pattnaik@rediffmail.com

A. Tomar
JSS Academy of Technical Education, Noida, India

A. Iqbal et al. (eds.), *Renewable Power for Sustainable Growth*, Lecture Notes in Electrical Engineering 723, https://doi.org/10.1007/978-981-33-4080-0_75

capacity progress could be 26% (1500 GW) greater than in the report's main calculation. Solar systems are a suitable source among other natural energy resources due to the accessibility of plentiful solar light during all seasons. Also, the cost of solar panels becomes less [2]. The number of PV-based energy produce industry is growing faster than other energy industry [3]. The advantages of PV systems are abundant solar energy, zero toxic waste, extensive lifespan, no moving parts. The main shortcomings of renewable energy sources are uncontrollable and unpredictable. So, it becomes quite incapable to fulfil the demand of the consumer side. Moreover, the power quality of the renewable energy sources must be superior. These difficulties can be overcome, by incorporating different conventional and non-conventional energy sources along with a storage system. The improvement of power electronics technologies also creates to have sensitive loads, which are sensitive to power instabilities. This work aims to discuss power electronic converter technologies that can be used as interfaces and discuss their advantages and limitations. The power electronics converter technology is connecting between a solar system and a load which is helped to develop full power from the PV system.

2 Photovoltaic System

PV power varies with the irradiance of sun, temperature, voltage, current, and fill factor. The wiring diagram of a PV cell is as presented in Fig. 1.

The current source represents the photon-generated current. The series resistance (R_s) represents the losses due to the contacts and connections. The leakage currents in the diode are symbolized by shunt resistance (R_{sh}) [4]. Solar intensity is varied according to the weather or atmosphere condition, so the total power required from the PV cell is also varied, it may cause a lack of output power in the PV system. In off-grid PV systems, power fluctuations create unbalance in the PV system, so it needs energy storage. In grid-connected PV systems, power fluctuations are transferred into the grid, but the grid has a sufficient amount of spinning reserve to absorb the fluctuations; otherwise, the grid requires energy storage to avoid fluctuations. Batteries, supercapacitors, flywheels, compressed air, pumped hydro, superconductors, and generation of hydrogen are possible energy storage technologies that can be used in PV systems [5]. The characteristics of batteries and supercapacitors are different from each other. Moreover, photovoltaic systems' characteristics are also different as compared to batteries and supercapacitors. Therefore, interfacing technologies that

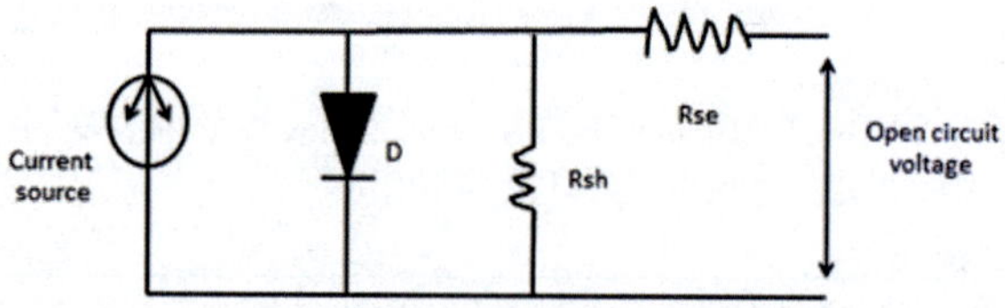

Fig. 1 Wiring diagram of the PV system with series and parallel resistance

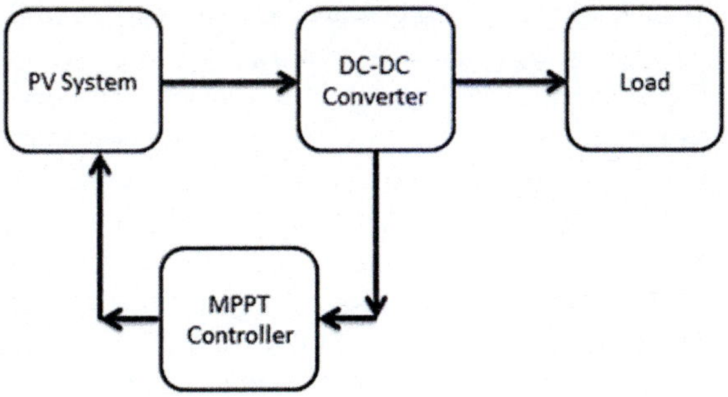

Fig. 2 Block drawing of the PV system

can match those characteristics have become an essential part of modern photovoltaic power systems [6]. Power electronic converter systems are the only choice that can meet all these requirements. Power electronic converter technologies are used either to convert the one form of energy to another form of energy or varied the energy to load. The PV system consists of some number of solar cells that transform the energy of the sunrays straight into electrical energy. Such a system is interconnected in series to get preferred DC voltage and current. It produces electrical energy based on the principle of the photoelectric effect. There are double stages of converter technologies where PE converters are utilized as DC–DC converter in the first stage, and the second stage is DC–AC converter [6]. Due to variation in solar irradiance during the daytime, the yield power is also varied. So, the yield power of solar cells can be maximized at a particular position of the PV system for the maximum intensity of solar irradiance. So, the maximum power point (MPPT) has been used to follow the solar irradiance to get maximum intensity. Therefore, to catch the full power, the PV system should get a suitable solar light intensity for input. With the help of the direct connection of energy storage elements, maximum power can be achieved. It might affect the efficiency of the load which captures the PV output.

Energy storage elements such as batteries and supercapacitors need to be separated from the PV system otherwise productivity of the PV system will be affected. So, the most common way of making this separation is the application of a DC–DC converter, as displayed in Fig. 2 where the PV system is connected to load via a DC–DC converter. The main objective of the DC–DC converter is applied to enhance/reduce the voltage level according to the desired level of voltage at the DC bus to ensure the MPPT operation [6].

3 Power Electronics Device Based DC–DC Converter Topology

This part of the article discusses the different types of converter topologies that are deployed to alter DC into DC. In the above section, the interfacing of the DC converter with solar panels has been discussed. The aim of the DC–DC system here is to change fixed DC voltage into a variable DC voltage. Various types of power electronics devices are used to make a converter. The DC–DC system is operated to increase and decrease the yield energy of the solar PV system [6]. There are different types of converters used to increase and decrease the voltage level. Those are as follows: (a) buck converter, (b) boost converter, (c) buck–boost converter, (d) Cuk converter, (e) single-ended primary inductor (SEPIC) converter. The key point of reference for picking DC–DC converters in solar PV systems is economical consideration, effectiveness, energy flow, and the capability to maintain output unrelatedly of any input deviations. The ripples in the yield voltage of the PV system output must be less [6]. The buck converters are hardly used as DC–DC converters as it reduces the solar PV system output voltage. Figure 3 illustrations the wiring diagram of a buck converter circuit [7, 8].

The usage of the boost converter is presented in Fig. 4 to step up the solar PV output voltage. The power electronic device in particular for a fast process and the switching duty cycle is used to control the voltage output [9, 10].

Veerachary et al. (2001) informed that an interleaved dual boost converter is applied as a connector among the PV systems and the load which is helpful to get full power and eliminated the ripples at both PV side and load side [11]. As a result, the input capacitance value becomes decreased in interleaved dual boost converter PV system The closed-loop boost converter system has advantages like good output

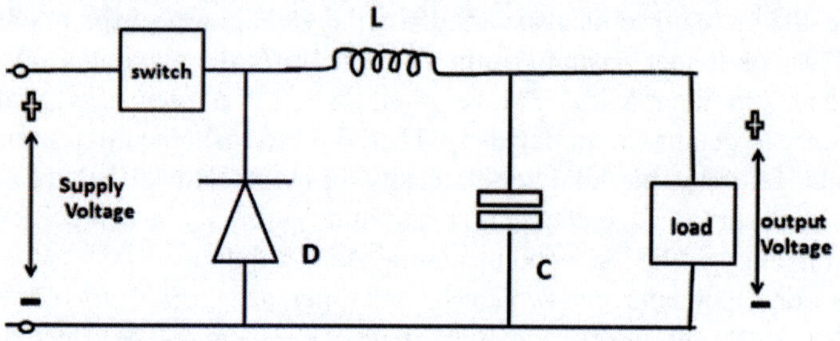

Fig. 3 Circuit diagrams of buck converter

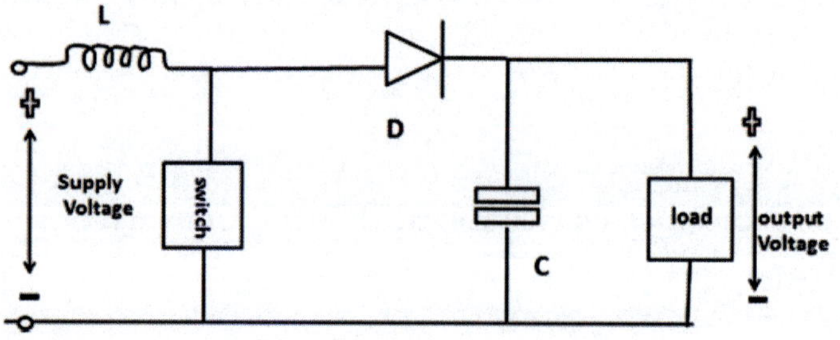

Fig. 4 Circuit diagrams of boost converter

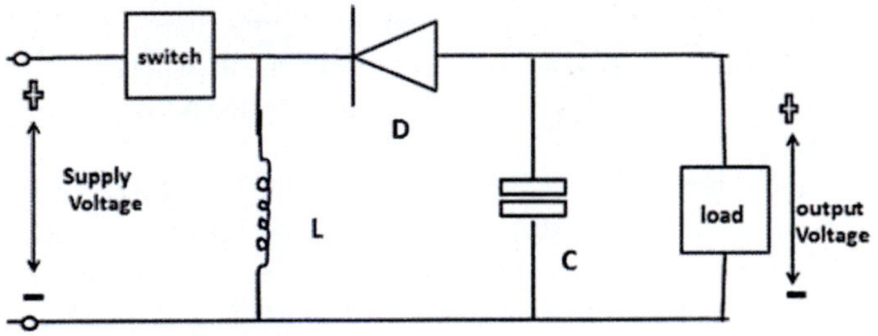

Fig. 5 Circuit diagrams of buck–boost converter

voltage regulation and able to improve the voltage level to the desired level of a stand-alone PV system as reported by Kalirasu et al. (2010) [12]. The boost topology is preferred than a buck because it does not use an extra diode which is in series with the PV system to eliminate reverse current [10].

Buck–boost converter as presented in Fig. 5 is a grouping of a buck and a boost converter. The benefits of the buck–boost converter are better response, better efficiency, and simple controlling method [13]. The shortcoming of this circuit is that it shows increased yield voltage ripple and losses such as conduction, switching losses. However, the losses were reduced by the soft-switched buck–boost converter which was used for PV applications [14, 15].

Buck converters can only lessen the 'V'; boost converters can only raise the voltage. Buck–boost, Cuk, and SEPIC converters can rise or fall the 'V'. Buck–boost converters can give the required output but inverting one [16]. These converters have more component stresses, component sizes, and lesser efficiency. Cuk converter as shown in Fig. 6a solves both of these difficulties using an additional 'C' and 'L'. The drawbacks of both Cuk and buck–boost converters are electrical stress which makes the system overheat [17]. Cuk converter can overcome the limitations of buck–boost converters. SEPIC converter is responsible for a positive controlled output voltage, whereas the buck–boost converter responsible for negative controlled output voltage. The Cuk converter is the twin of the buck–boost converter. The simple SEPIC converter is a reform of the boost and the Cuk model [17]. SEPIC is also basically a boost–buck converter [18, 19].

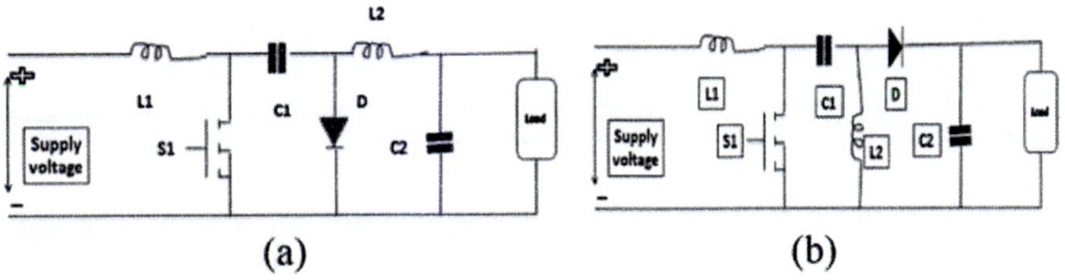

Fig. 6 Circuit diagrams of Cuk (**a**) and SEPIC (**b**) converters

The Fig. 6b represents the circuit diagram of SEPIC. It is a kind of DC–DC converter whose yield is more than, fewer than or equivalent to the PV system contribution. The yield of the SEPIC converter is measured by the duty cycle of the power electronic devices.

Park et al. (2008) reported a new integrated boost-SEPIC (IBS) converter, which provides boost the voltage by the usage of an isolated SEPIC converter [20]. SEPIC converter is developed to maximize the voltage of the PV system. Khateb et al. (2013) reported that the fuzzy logic controller with an MPPT using a SEPIC converter [21]. The novel controller with perturbs and observe methods could improve the performance. It was suitable for high static gain continuing a compact switch voltage. Divya et al. (2014) reported that an enhanced MPPT tracking of the PV system using the deterministic particle swarm optimization technique [22]. Raja et al. (2017) reported that the single-ended primary inductance converter topology operated a buck–boost converter but with no reversal of voltage [23]. It suggested a converter topology that ensured high gain and cost-effective. The advantages of the buck-boost converter model are lesser losses for greater switching frequencies, little ripple with maximum efficiency, and minimum electrical stress on the apparatuses. Natarajan et al. (2017) reported that SEPIC allows a range of DC voltage to adjust and keep a constant output voltage including MPPT technique [24]. Kumar et al. (2017) proposed a model of an integrated Cuk-SEPIC converter topology [25]. Therefore, it had the quality of a simple model and a lesser number of device-based converter. The suggested converter was considered for the combined operation of wind energy and solar energy system. According to Ewerling et al. (2019), the DC–DC SEPIC with a solar power system provided the output voltage without changing polarity [26].

It can be analyzed from the above description that the SEPIC converter can responsible for clarification of the problem related to buck, boost, buck–boost, and Cuk converter. There are so many new studies that are encouraged for the SEPIC converter. Henao-Bravo et al. (2020) reported on the zeta/SEPIC converter with a sliding mode controller (SMC) [28]. Gao et al. (2020) reported DCM soft-switched SEPIC-substructure high-frequency converter whose voltage gain was high and voltage stress was low [27]. It provided a modified SEPIC converter which showed better performance.

4 Conclusion

This paper discussed the PV system with power electronics devices. It mainly focused on different types of interface technologies of a PV system. It was mentioned in the literature that the act of a PV system was influenced by the storage device. So, the DC–DC converter has been used as an interfacing technology between load and PV systems. Moreover, the MPPT system used to maximize the intensity of solar radiation to get maximum power, but it was connected by the DC–DC converter. SEPIC converter performance was noteworthy as reported in the literature. It has enhanced efficiency, performance, and voltage gain. It can eliminate the drawbacks of other converters.

References

1. IEA, WES 2019, IEA, Paris
2. Pattnaik A, Tomar M, Mondal S, Gupta V, Prasad B (2018) Enhancement in PCE of mc-Si solar cell by ZnS NP with PMMA. In: The physics of semiconductor devices: proceedings of IWPSD 2017, pp 215, 399
3. Pattnaik A, Mittal S, Gupta V, Prasad B, Bhoi A (2018) Design, and implementation of 2-level and MI. In: Advances in power systems and energy management. Springer, Berlin, pp 39–46
4. Teo JC et al (2020) Impact of bypass diode forward voltage on maximum power of a photovoltaic system under partial shading conditions. Energy 191:116491
5. Narayanan V, Kewat S, Singh B (2020) Solar PV-BES based microgrid system with multifunctional VSC. IEEE Trans Ind Appl 56(3):2957–2967
6. Liu Y, Liu X, Zhang J, Zhang Y, Zhu Z (2020) A novel maximum power point tracking control strategy for the building integrated photovoltaic system. Energies 13(11):2679
7. Villarreal-Hernandez C, Maldonado JCM, Escobar G, Loranca J, Valdez-Resendiz JE, Rosas-Caro J (2020) Discrete-time modeling and control of double dual boost converters with implicit current-ripple cancellation over a wide operating range. IEEE Trans on Ind Electron
8. Alajmi BN, Marei MI, Abdelsalam I (2020) A Multiport DC–DC converter based on Two-Quadrant inverter topology for PV systems. IEEE Trans on Power Electron 36(1):522–532
9. Dharshan BG, Rajasekar N, Srinivasa Sankarkumar R (2020) Adaptability analysis of particle swarm optimization variants in maximum power tracking for solar PV systems. In: Intelligent computing techniques for smart energy systems. Springer, Singapore, pp 397–409
10. Jaraniya D, Nema RK, Gawre SK (2020) Design and simulation of power electronics interface for modified P & O maximum power point tracking under suddenly varying irradiance. In: 2020 IEEE international students' conference on electrical, electronics, and computer science (SCEECS). IEEE
11. Veerachary M, Senjyu T, Uezato K (2003) Maximum power point tracking of coupled inductor interleaved boost converter supplied PV system. IEE Proc Electr Power Appl 150(1):71–80
12. Kalirasu A, Dash SS (2010) Simulation of closed-loop controlled boost converter for solar installation. Serb J Electr Eng 7(1):121–130
13. Da Luz CMA, Vicente EM, Tofoli FL (2020) Experimental evaluation of global maximum power point techniques under partial shading conditions. Sol Energy 196:49–73
14. Rafi FHM et al (2020) An overview of unbalance compensation techniques using power electronic converters for active distribution systems with renewable generation. Renew Sustain Energy Rev 125:109812
15. Haque AM, Nagal D, Sharma S (2015) Power electronics interface for PV systems. Int J Res Sci Innov (IJRSI)4(I):268–274
16. Zheng H et al (2014) Shading and bypass diode impacts to energy extraction of PV arrays under different converter configurations. Renew Energy 68:58–66
17. Singh AK, Badoni M, Tatte YN (2020) A Multifunctional Solar PV and grid-based on-board converter for electric vehicles. IEEE Trans Veh Technol 69(4):3717–3727
18. Pirog S, Stala R (2009) Power electronic converter for photovoltaic systems with the use of FPGA-based real-time modeling of single-phase grid-connected systems. Bull Pol Acad Sci Tech Sci 57(4):345–354
19. Teke Ahmet, Saribulut Lütfü, Tumay Mehmet (2011) A novel reference signal generation method for power-quality improvement of unified power-quality conditioner. IEEE Trans Power Delivery 26(4):2205–2214

20. Park K-B, Seong H-W, Kim H-S, Moon G-W, Youn M-J (2008) Integrated boost-SEPIC converter for high step-up applications. In: IEEE power electronics specialists conference, Rhodes, pp 944–950
21. El Khateb AH, Abd Rahim N, Selvaraj J (2013) Fuzzy logic control approach of a maximum power point employing SEPIC converter for standalone photovoltaic system. Procedia Environ Sci 17:529–536
22. Divya K, Sugumaran G (2014) DPSO based SEPIC converter in PV system under Partial Shading Condition. World Acad Sci Eng Technol Int J Electr Comput Eng 8(2):1
23. Raja TAS, Senthil R, Nandhakumar A, Santhoshkumar K (2017) Modelling and analysis of SEPIC converter based photovoltaic system. Int Res J Eng Technol (IRJET) 4(10)
24. Natarajan K, Supriya M, Abinaya Y, Nandhakumar P, Samrajesh G (2017) Design of a SEPIC converter for solar PV system. In: Conference on international conference on scientific research of engineering and technology
25. Kumar K, Ramesh Babu N, Prabhu KR (2017) Design and analysis of an integrated Cuk-SEPIC converter with MPPT for standalone wind/PV hybrid system. Int J Renew Energy Res (IJRER) 7(1):96–106
26. Ewerling MVM, Lazzarin TB, Font CHI (2019, December) Proposal of an isolated Two-Switch DC-DC SEPIC converter. In: 2019 IEEE 15th Brazilian Power Electronics Conference and 5th IEEE Southern Power Electronics Conference (COBEP/SPEC), pp 1–6
27. Gao S, Wang Y, Liu Y, Guan Y, Xu DG (2020) A novel DCM soft-switched SEPIC-based high-frequency converter with high step up capacity. IEEE Trans on Power Electron
28. Henao-Bravo EE, Saavedra-Montes AJ, Ramos-Paja CA, Bastidas-Rodriguez JD, Montoya DG (2020) Charging/discharging system based on zeta/sepic converter and a sliding mode controller for dc bus voltage regulation. IET Power Electronics 13(8):1514–1527

A Review Paper on Comparison, Analysis, and Planning of Electricity Generation in Australia, Argentina, New Zealand, Mexico with India

Rahat Ullah Khan, Saksham Yadav, Rajat Srivastava, Shubhendra Dubey, and Shobhit Srivastava

Abstract In this paper, a comparison of the various methods and quantity of electrical power generation of countries like Australia, Argentina, New Zealand, Mexico, and India is highlighted. It is required to get the outlook of the various power generation techniques, and the amount of the power the other countries are generating. It has been also shown the power generation methods of different countries in graphical form and on the account of percentage utilization of energies in recent year. Various power generation methods like thermal, wind, solar, gas, nuclear, and hydro are been taken into consideration. With the help of this, we may analyze different sources of power generation and compare of the above four countries power generated and compared with the power generated in India. With the help of this study and comparison, it can be observed also that the utilization of different types of sources for the purpose of generation of electricity by different countries and how such countries are major dependent on thermal power source for the production of electricity as compared to renewable source of energy.

Keywords Electrical power generation · Different countries electrical power comparison · Types of electrical energy generation sources

1 Introduction

Today's electrical energy scenario, most of the countries depend on the non-renewable sources for the purpose of power generation. Since such sources will be exhausted one day, and the dependency of all the countries for the generation of power will be on renewable sources of energy, there is a need to promote the renewable sources in the countries for the generation of electricity. This paper constitutes the comparison of various power generation methods [1–8] for different countries. The major advantages of renewable sources of generation over non-renewable sources are less air pollution, non-extensible sources of energy, although the generation of

R. U. Khan (✉) · S. Yadav · R. Srivastava · S. Dubey · S. Srivastava
KIET Group of Institutions, Ghaziabad, Uttar Pradesh, India
e-mail: rahat.khan@kiet.edu

A. Iqbal et al. (eds.), *Renewable Power for Sustainable Growth*, Lecture Notes in Electrical Engineering 723, https://doi.org/10.1007/978-981-33-4080-0_76

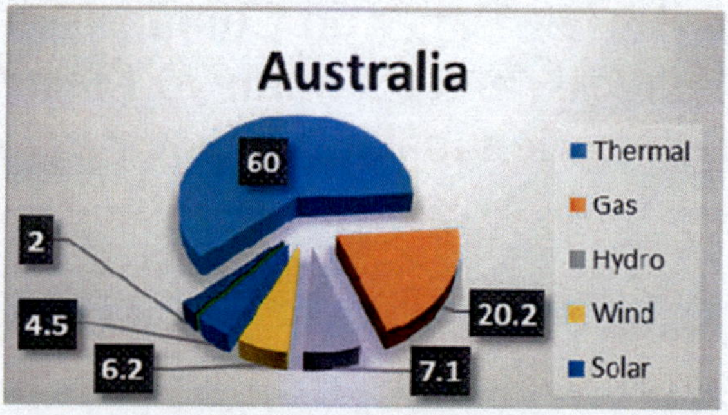

Fig. 1 Power generation in Australia

electricity in renewable sources is quite less than the generation by non-renewable sources. In addition, the efficiency of non-renewable sources is much better than renewable sources.

1.1 Importance

The importance of this study is to with the help of this study a quick reference can be taken out and one can get the immediate information with the help of this study and data [1–4].

1.2 Power Generation of Australia

As shown in the chart, main source of electricity generation in Australia is based on thermal and gas power generation techniques, which constitutes 60 and 20.2% of energy generation, respectively. Various other sources of power generation are hydro, wind, and solar, which have considerable amount of contribution in power generation in Australia (Fig. 1).

1.3 Power Generation in New Zealand

The power sector in New Zealand uses mainly renewable energy sources such as hydropower, geothermal power, and increasingly wind energy. 82% of energy for electricity generation is from renewable sources, making New Zealand one of the lowest carbon dioxide emitting countries in terms of electricity generation. As shown in the chart, the main source of electrical power generation in New Zealand is based on hydro, geothermal, and thermal power generation techniques that constitutes 60.34%, 17.37%, and 15.98% of energy generation, respectively. Various other

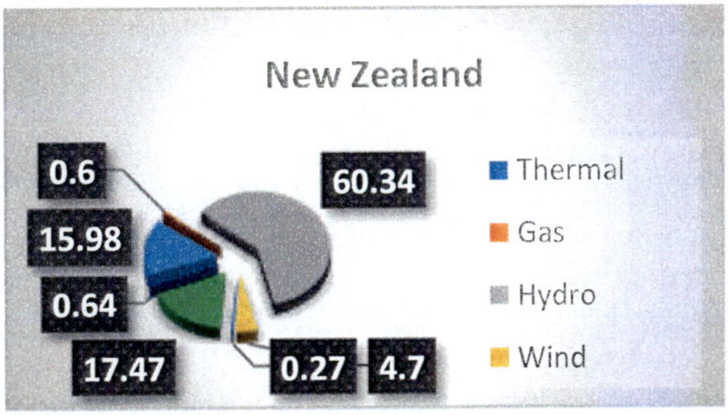

Fig. 2 Power generation in New Zealand

sources of power generation are gas, wind, and solar, which have considerable amount of contribution in power generation in New Zealand [3] (Fig. 2).

1.4 Power Generation in Mexico

Mexico is a country comprising of different varieties of resources for power generation. These varieties consist of power from resources like thermal, hydro, wind, solar, and nuclear. The different compositions of the power plants are as thermal covers over 81% of total power generation and others constitute as 10% of hydro power plants, 3% wind, 3% solar as well as 3% of nuclear power generation as is depicted in charts (Fig. 3).

In comparison with other power plants, thermal is the major one in Mexico due to the presence of excess resources in this sector. In Mexico, after thermal, power is produced mostly by hydro power plants. There are several dams made across the rivers which fulfill the proper requirement of hydro sector.

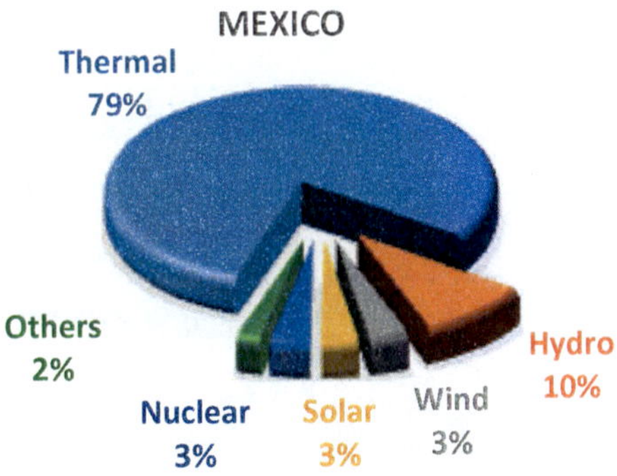

Fig. 3 Power generation in Mexico

1.5 Power Generation in Argentina

The power generation of Argentina mainly depends upon two major sources of energy, namely hydro and thermal with 29.2% and 63.5% of energy generation, respectively. The other sources of power generation include solar, wind, nuclear, and gas which also contributes in power generation of Argentina. Argentina is a country located mostly in the southern half of South America with the population of 44.6 million. The surface area of Argentina is 2,780,400 km^2 [6, 8].

The total expenditure of export of Argentina is 58,427 M$ and that of import is 66,899 M$.

The currency of Argentina is peso with one dollar being equal to 37.1 pesos (Fig. 4).

Argentina is a prominent regional power in the Southern Cone and Latin America and retains its historic status as a middle power in international affairs. Argentina has the second largest economy in South America, the third largest in Latin America (Fig. 5).

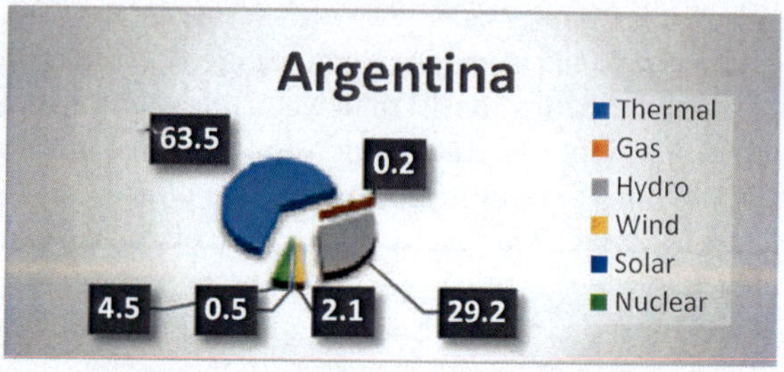

Fig. 4 Power generation in Argentina

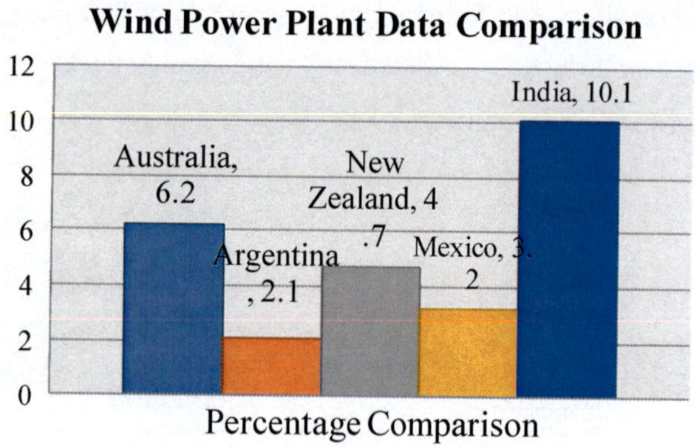

Fig. 5 Power generation in New Zealand

2 Comparison of Different Power Generation Techniques

2.1 Wind Power Plant

Traditionally wind energy is the used to provide the mechanical power through wind turbines to turn electric generators and some traditional work, like milling or pumping. Wind power is a clean and renewable energy and has a much lower impact on the environment in comparison with burning fossil fuels. As we are comparing five different countries (Australia, Argentina, New Zealand, Mexico, and India) altogether on the basis of their respective percentage generation of wind power, we can see a slight difference in power generation through this power plant in different stated countries. However, there is a much less consumption of power through this power plant in Argentina, i.e., 2.1%. India is being as the highest producer of energy by wind power generation techniques among four countries. The potential for wind power generation for grid interaction has been estimated at about 102,788 MW taking sites having wind power density greater than 200 W/m^2 at 80 m hub-height with 2% land availability in potential areas for setting up wind farms @ 9 MW/km^2. In today's scenario, most focus is over developing a world with sustainable power, so encouragement in this sector is mandatory. Renewable energy source is the only way for the development of healthy world [10–12].

2.2 Thermal Power Plant

Thermal power is the main source in most of the countries around the world. This is due to the easy availability of coal and other fossil fuels which are required for generation of thermal power. As we are comparing five different countries (Australia, Argentina, New Zealand, Mexico, and India) altogether on the basis of their respective percentage generation of thermal power, we can see a slight difference in power generation through this power plant in different stated countries. But there is a much less consumption of power through this power plant in New Zealand, i.e., 15.98%. In New Zealand, most dependency of power generation is on renewable resources, so it is less in thermal power generation.

Maximum power generation through thermal is in Mexico, i.e., 78.92%. In Mexico, coal mines are available easily due to its geographical design (Fig. 6).

Mexico is having lesser supplies in different other power plants, so thermal is a reliable power plant here. If we have a look over remaining three countries, they have a similar production capacity in thermal, i.e., around 60%.

In today's scenario, most focus is over developing a world with sustainable power, so decrement in this sector is mandatory. Thermal power plants also produce a high amount of smoke which causes air pollution, so they must be reduced in number and power plants of renewable energies should be planted.

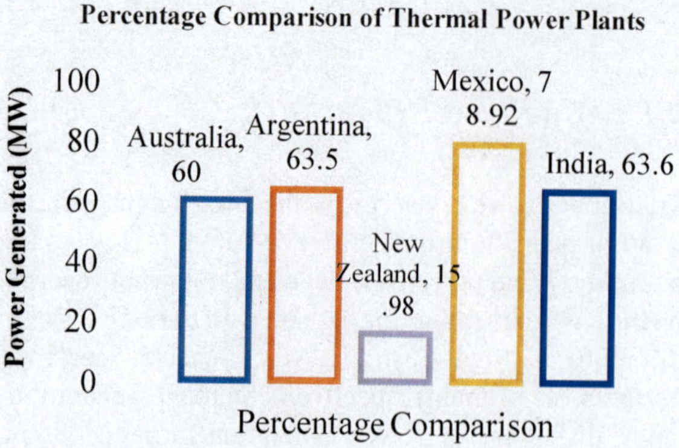

Fig. 6 Different power generation comparison in Australia, Argentina, New Zealand, Mexico with India

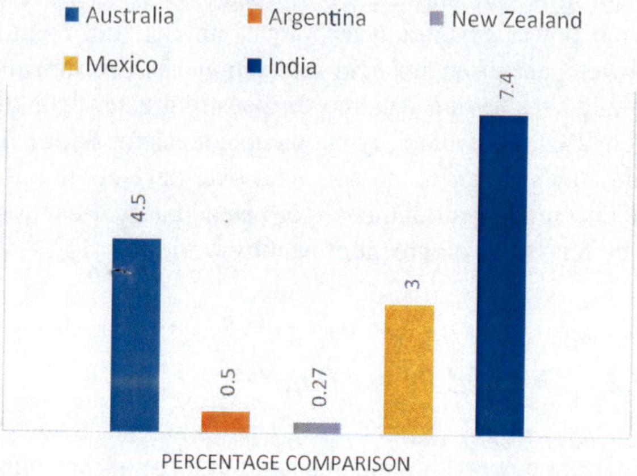

Fig. 7 Percentage comparison of solar power generated in different countries

2.3 *Solar Power Plant*

The solar power generation of Argentina is 0.5% only of the total power generation, whereas the countries like Australia, New Zealand, Mexico, and India generate solar energy of 4.5%, 0.27%, 3%, and 7.4%, respectively. India is being the highest power generation of solar energy among the five countries with 7.4% of its total power generation (Fig. 7).

2.4 *Nuclear Power Plant*

The nuclear power generation of Argentina is 4.5% of the total power generation, whereas the countries like Australia, New Zealand, Mexico, and India generate solar energy of 0%, 0%, 3.03%, and 1.9%, respectively. Argentina is being the highest power generation of nuclear energy among the five countries with 4.5% of its total power generation.

2.5 *Hydro Power Plant*

As we know that the future needs of power generation is based on renewable source of energy so every country has started focusing on this method of power generation. Power generation from hydro power plant has vast acceptance among countries so the data shown in the figure clearly states the current development in this area of power generation. New Zealand is frontrunner among the four countries which include Australia, Argentina, Mexico, and India. New Zealand has 60.34% power generation from hydro power plant. While other countries have very less percent of production of energy from this method, Australia, Argentina, Mexico have 7.1%, 29.6%, and 9.6%, respectively, of power generation from hydro power plant. India is also focusing on power generation on the basis of hydro power plant which have share of 12.9% of total power generation in India.

3 Conclusion

It has been reviewed various types of energy generations for the countries like Australia, Malaysia, Argentina, New Zealand, Mexico with India in detail. During the review of these countries, it has been found that energy trends of these countries Australia has its maximum power generated from coal to run thermal power stations, including New Zealand and Mexico. The thermal power is generated maximum in Mexico in comparison with Argentina, India, and Australia. On the other hands, the second maximum power generations are with the help of gas power stations and hydro power generations. In comparison with the renewable power, energy from generated from wind is that India is leading in comparison with other countries. If solar power is considered, India is producing maximum power with the help of solar and thermal.

References

1. Berrie T (1967) The economics of system planning in bulk electricity supply. Electricity Rev 425–468
2. Nelson T, Orton F (2016) Climate and electricity policy integration: is the South Australian electricity market the canary in the coalmine. Electricity J 29(4):1–7
3. Chakrabarti B, Qureshi W, Nair NKC (2012) Renewable generation and its integration in New Zealand power system. IEEE Power Energy Soc Gen Meet 1–7. https://doi.org/10.1109/PESGM.2012.6345183

4. Winkler H (2005) Renewable energy policy in South Africa: policy options for renewable electricity. Energy Pol 33(1):27–38
5. Wiser RH (2000) The role of public policy in emerging green power markets: an analysis of marketer preferences. Renew Sustain Energy Rev 4(2):177–212
6. Wustenhagen R, Wolsink M, Burer MJ (2007) Social acceptance of renewable energy innovation: an introduction to the concept. Energy Pol 35(5):2683–2691
7. Iqbal A et al (eds) (2020) Soft computing in condition monitoring and diagnostics of electrical and mechanical systems. In: Advances in intelligent systems and computing, vol 1096. Springer, Singapore. https://doi.org/10.1007/978-981-15-1532-3
8. Guzowski C, Recalde M (2008) Renewable energy in Argentina: energy policy analysis and perspectives. Int J Hydrogen Energy 33:3592–3595
9. Iqbal A et al (eds) (2020) Meta heuristic and evolutionary computation: algorithms and applications. In: Studies in computational intelligence, vol 1096. Springer, Singapore. https://www.springer.com/gp/book/9789811575709
10. Central Electricity Authority (2015) Annual report 2014–2015. Government of India, Ministry of Power, Central Electricity Authority, New Delhi, India
11. Gao M, Ding YH, Song SJ, Lu X, Chen XY, McElroy MB (2018) Secular decrease of wind power potential in India associated with warming in the Indian Ocean. Sci Adv 4:1–8
12. Thapar S, Sharma S, Verma A (2018) Key determinants of wind energy growth in India: analysis of policy and non-policy factors. Energy Pol 122:622–638

A Review for Energy Generation Analysis and Comparison in China, Indonesia, and Ireland with India

Rahat Ullah Khan, Biplab Bhattacharyya, Raghav Gupta, Shilpy Tyagi, Shivam Joshi, and Sona Singh

Abstract The energy sector is one of the most important parts of any country, as the major economy of the country depends on the amount of energy generated by that country. Along with the generation of energy, the source of production shares equivalent weight. For a country, it is very important to keep a check on the energy production index to facilitate the development of some alternate sources of energy production. Energy consumption per capita is one of the major indicators, which shows the economic development of the country. The level at which energy is produced in any country reflects the country's level on national and international grounds. It is a common fact that the energy utilization of every human is expanding daily, but the resources utilized for the production process are in a limited quantity. Also, the distribution of the resources is not even throughout all the geographical locations; hence, it becomes another subject of comparison between countries that what are the volume of raw materials consumed for the electricity generation. This study provides a detailed comparison of China, Indonesia, Ireland, and Iceland, including it, an overall comparison of all the countries with India is done. This also helps in understanding that weather our country is leading or lagging in energy production in comparison with other countries.

Keywords Energy generation comparison of different countries · Energy comparison of Indonesia, Ireland, and India · Hydro-electricity

1 Introduction

This research paper is the extensive work done in the field of energy sectors of different countries. We know that every country is different in various aspects and

R. U. Khan (✉) · B. Bhattacharyya
Indian Institute of Technology (ISM) Dhanbad, Dhanbad, Jharkhand, India
e-mail: Rahat.khan@kiet.edu

R. U. Khan · R. Gupta · S. Tyagi · S. Joshi · S. Singh
Department of Electrical and Electronics, KIET Group of Institutions Ghaziabad, Ghaziabad, India

A. Iqbal et al. (eds.), *Renewable Power for Sustainable Growth*, Lecture Notes in Electrical Engineering 723, https://doi.org/10.1007/978-981-33-4080-0_77

the same goes for the energy sector of the countries. There is more than one factor that affects the energy generation and consumption of different countries. These factors vary from geographical, demographic, political, and even environmental. These factors are what make countries different from each other. In this research paper, we are discussing these factors and their effects on the energy sectors of countries like Iceland, Ireland, China, Indonesia, and India [1–4, 10, 11]. We have extensively discussed and compared the energy generation, policies, distribution infrastructure along with different sources used to meet the demand. Finally, we have compared these points with the Indian energy scenario. This paper will provide a deep understanding of the pros and cons of the current technology used in these countries and will showcase the probable are of development which can be achieved by mutual understanding and exchange of information.

2 Power Generation in Different Countries

2.1 Power Generation in China

The Electrical power Sector of China is the largest power sector in the world since early year from 1990. Coal has the largest producer of electricity, which accounts for 66% of the electricity generation. There are the certain expectations that the electrical power generation capacity with the help of coal is to be increased from 960.00 to 1300.00 GW by the end of year 2020 in comparison with year 2016. It has been observed that the common problem faced in China is the voltage drop due to long distance power transmission lines in the entire country.

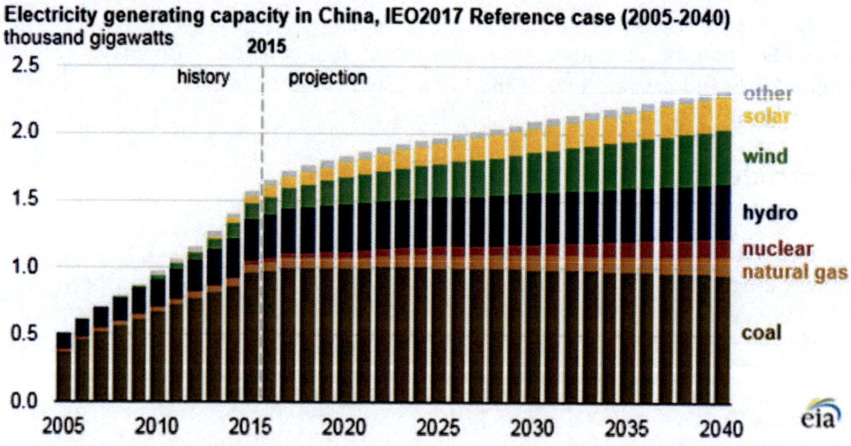

Fig. 1 Energy generation capacity and future projected data for energy production in China

As per Fig. 1, the larger distance inter-district transmission system is implemented after implication of ultra high voltage—UHV system of the voltage rating of 800 kV. On the other hand, hydro-electricity is the second largest source of electrical power generation, which is a renewable, in comparison with coal-based energy generation system. China has exceptional wind resources because of its landmass and a long coastline.

In Fig. 1, as per the China power generation plan up to year 2020, to build 30 nuclear energy power plants with the installed capacity of 80.00 GW. In addition to it, China is the world top 5 market player for solar thermal energy generation and PV energy generation and manufacturing.

2.2 *Power Generation in Indonesia*

Indonesia is producing oil, coal, natural gas, and palm oil and also uses as an energy raw material. In Indonesia, renewable energy potential is high. Solar, wind, hydro, and geothermal energy are the main sources. As per the data analysis, Indonesia is the tenth topmost producer of natural gas with 76 million cubic, i.e., 2.5% of the world population, out of which 36 BCM was exported. Indonesia was the topmost fifth coal manufacturer with the capacity of 263 million tons of solid coal and 38 million tons of brown coal. The capacity of coal in terms of medium and low thermal quality is adequately available. In Indonesia, renewable generation sources are of 5–6% in terms of electricity generation. Biomass primarily holds of about 128 million people are relying upon traditional biomass mainly wood for cooking purpose (Fig. 2).

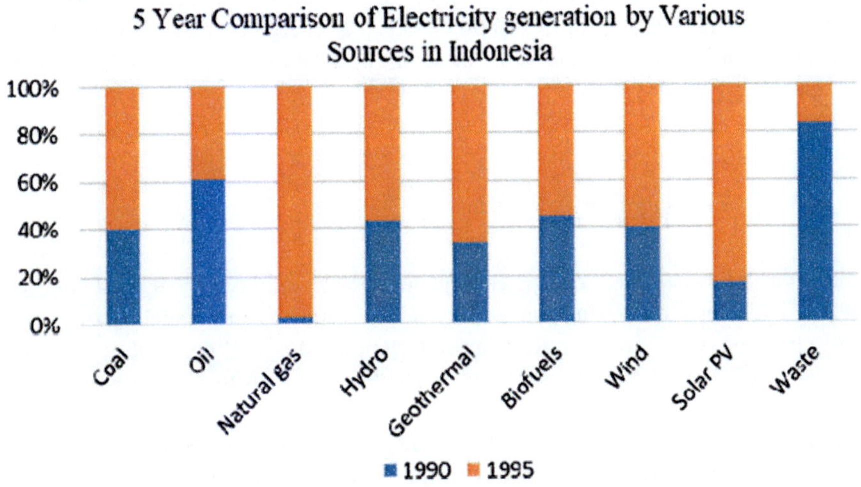

Fig. 2 Comparison of electricity generation by various sources in Indonesia

Indonesia has a set of target of 2 GW installed capacity in hydro-electricity. Indonesia has the third largest install generating capability in the world with of 1.3 GW of installed capacity. In Indonesia, the available wind speeds are not standard to generate electrical energy at large scale [5–8]. Indonesian PV solar sector is quite underdeveloped, although it has major potential of power generation from PV solar energy, but this sector is dedicated specifically set aside from decentralized rural electrification.

2.3 Power Generation in Ireland

Ireland's main source of energy is natural gas and peat, which is used to generate electricity. Fossil fuels account for over 90% of the energy used. About 45% of electricity was generated from natural gas, 96% of which was imported from Scotland. Coal is mostly used solid fuel used in houses for heating purpose. Non-renewable currently account for over 90% of energy consumption in Ireland. The electricity production with nuclear fission is not allowed and illegal according to the Irish Electricity Regulation Acts. [9].

Per capita energy production in Iceland is largest in the world, approximately 55 MWh per person in a year. Iceland industries are fueled by the renewable energy sources; it is the first country in the world who created economy engendered through renewable energy, including of it there is a significant quantity of energy potential still unexploited hydro-electric energy in Iceland. In Iceland, energy generated from geothermal resources is 27% and remaining 73% with hydro-power.

As per government data approximately 20 Terra Watt Hour per annum of estimated geothermal power sources available. The location of Iceland is not favorable in terms of solar insolation, because of high latitude, so solar energy potential is low. There are many agencies checking the scope of solar energy harnessing at desirable level of output (Fig. 3).

2.4 Power Generation in India

India's fourth largest source of energy is nuclear power followed by thermal energy sources, hydro-electricity, and renewable energy resources to generation of electricity. In India, there are 22 nuclear power stations in operating conditions with capacity of 6780 MW and generating total 30,292.91 giga watt hour of electricity [9] (Fig. 4).

In India, coal powered power plants are the largest sources of power. More than 62% of India's electricity demand is meeting through the country's gigantic coal reserves. The total installed capacity of natural gas-based power plants in India is almost 24,508.63 MW. The total installed capacity of major grid-connected diesel-based power plants in India is 993.53 MW.

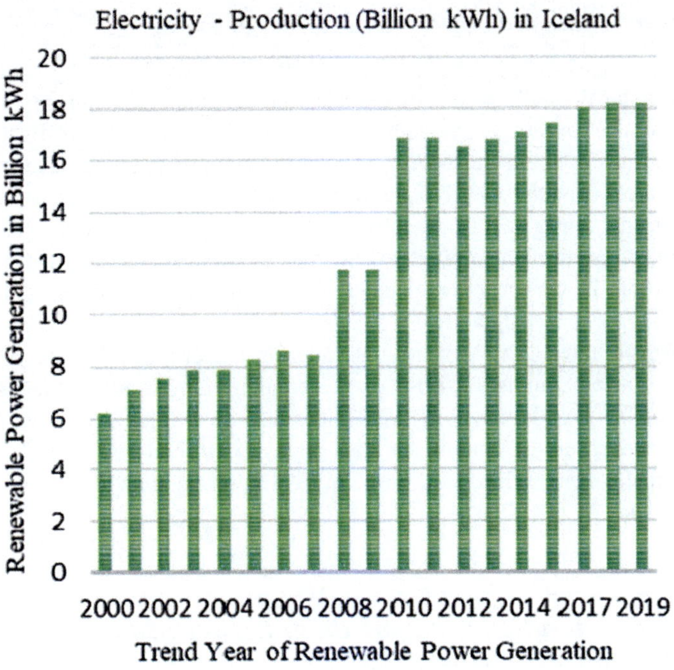

Fig. 3 Trend year of renewable power generation in billion kWh

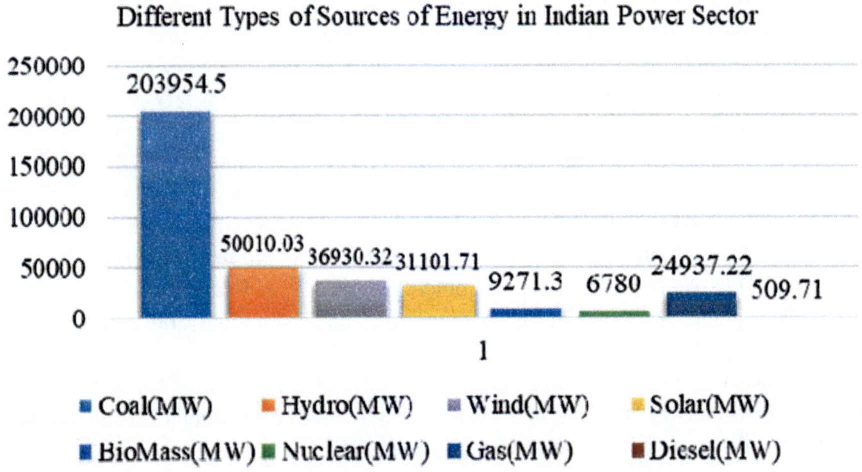

Fig. 4 Different types of sources of energy in Indian power sector

2.5 *Future Plan for Power Generation of India*

India plans to have 20 GW of installed nuclear power generation capacity up to year 2020, and 63 GW by 2032. The country target is to generate 25% of its total electricity from nuclear energy resources by year 2050. India plans to have 20 GW of installed nuclear capacity by 2020 and 63 GW by 2032.

India currently has 37 nuclear reactors planned or under construction. This shows the country's commitment to plans to generate 25% of its electricity from nuclear energy by 2050. However, it also needs to deal with uranium shortages and public opinion and reports global data.

India plans to have 20 GW of installed nuclear capacity by 2020 and 63 GW by 2032. India is also planning to use the recovered plutonium from spent fuel to trial in the prototype fast breeder reactor (PFBR). It intends to develop a program for using its vast thorium reserves. The nuclear deal signed with the USA in 2008 will help develop its nuclear capabilities and further provide a momentum to its future nuclear expansion plans.

2.6 *Power Generation Comparison in Between India and China*

India and China are the two largest promising economies, and their economic developments have been fueled mainly by coal. China burns more of coal to burn than any other country, while India ranking third. The electricity production of India is 1.386 trillion kWh while that of China has 5.883 trillion kWh. The electricity consumption of India is 1.137 trillion kWh while China has 5.564 trillion kWh. India has about 32 giga watts (GW) of wind energy and 12 GW of solar energy generation with combine total generation of 320 GW. In comparison, China has 149 GW of wind and 77 GW of solar power. Exports of electricity in India is nearly 5.15 billion kWh while that of India is 18.91 billion kWh. Electricity from vestige fuels in India is about 71% of total installed capacity while from hydroelectric plants is 12% and from electricity with nuclear plants is 2% of total install capability and from renewable resources is 16% (Fig. 5).

Electricity from fossil fuels in China is about 62% of total installed capacity while hydroelectric plants is 18% and electrical energy from nuclear plants is 2% of total install capability and from renewable resources is 18%.

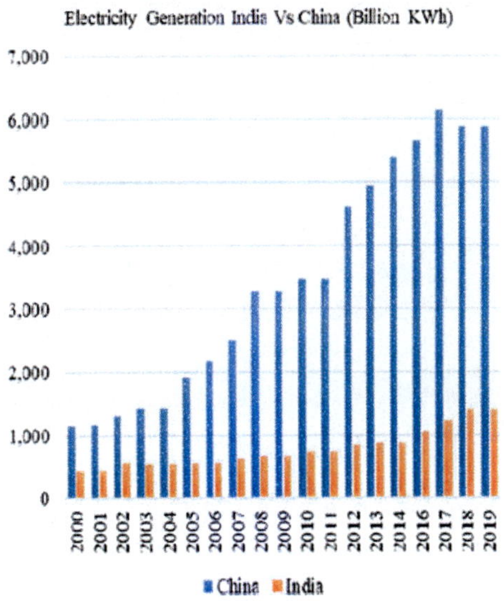

Fig. 5 Year-wise comparison of energy generation in India and China

2.7 Power Generation Comparison in Between India and Indonesia

When we compare India and Indonesia, we obviously see a very large difference in the quantity of generation and consumption India exceeds Indonesia in both energy generation and consumption. India also exceeds Indonesia on the number of resources used for energy production. In this research paper, we see that both India and Indonesia are vastly dependent on the use of fossil fuels for the production of energy. But the major difference between the two countries is the amount of oil production and export. Indonesia leads India in both oil production and oil export this obviously depends upon the large demographic difference. Indonesia also leads India in natural gas reserves, production consumption, and exports. India is one of the developing countries, which has recently tried to find its feet in the nuclear use of energy generation, whereas Indonesia is yet to acquire the use of this technology. India also is ahead of Indonesia in terms of the use of renewable sources for energy production (Fig. 6).

2.8 Power Generation Comparison in Between India and Ireland

When we compare India and Ireland, we do see a very large difference in the quantity of generation and consumption India exceeds Ireland in both energy generation and consumption. India also exceeds Ireland on the number of resources used for energy

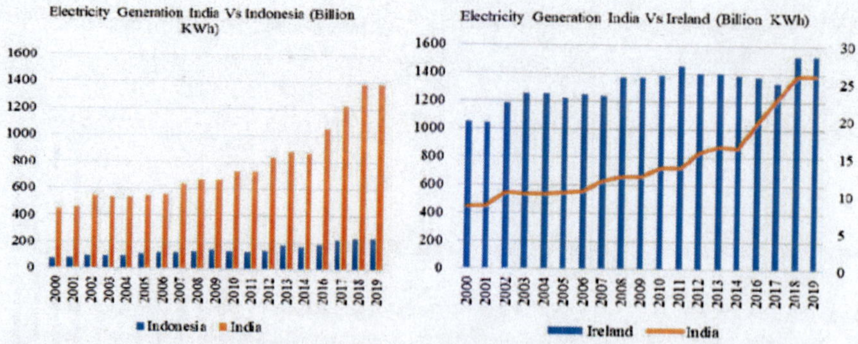

Fig. 6 Year-wise comparison of energy generation in India versus Indonesia and India versus Ireland

production. In this research paper, we see that both India and Ireland are predominantly dependent on the use of fossil fuels for the production of energy. However, the major difference between the two countries is the amount of energy generated from the renewable sources. Ireland leads India for energy generated by the means of renewable resources. In addition, the Ireland is able to achieve the 100% electrification of its country; on the other hand, India has been able to avail electricity to about 84.5% of its urban population (Fig. 7).

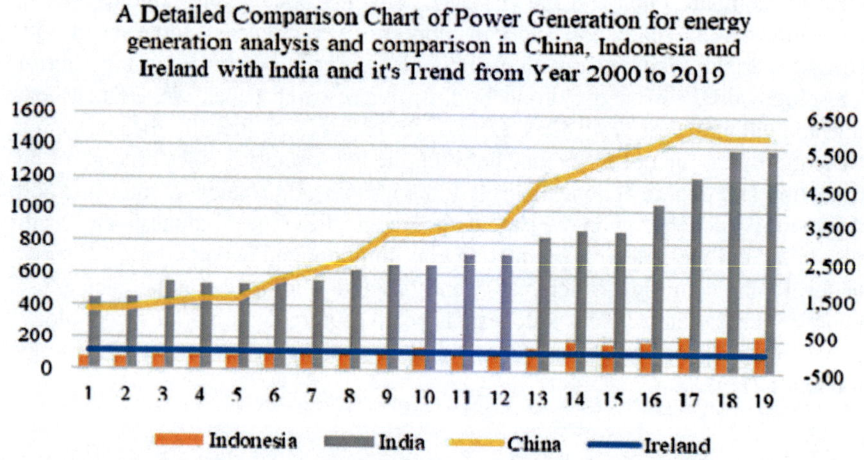

Fig. 7 Detailed comparison chart of power generation for energy generation analysis an comparison in China, Indonesia, and Ireland with India and it is trend from year 2000 to 2019

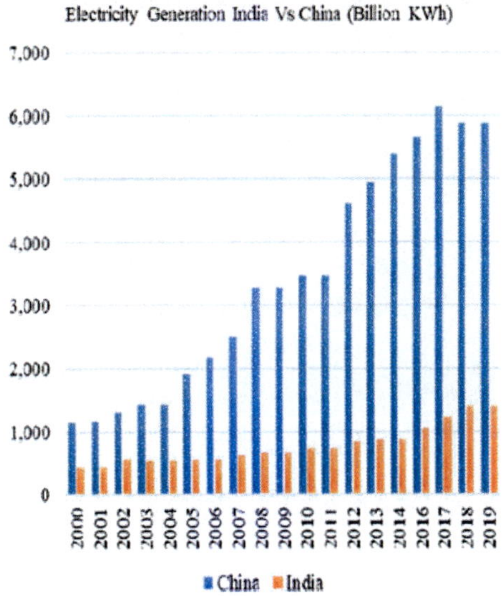

Fig. 5 Year-wise comparison of energy generation in India and China

2.7 *Power Generation Comparison in Between India and Indonesia*

When we compare India and Indonesia, we obviously see a very large difference in the quantity of generation and consumption India exceeds Indonesia in both energy generation and consumption. India also exceeds Indonesia on the number of resources used for energy production. In this research paper, we see that both India and Indonesia are vastly dependent on the use of fossil fuels for the production of energy. But the major difference between the two countries is the amount of oil production and export. Indonesia leads India in both oil production and oil export this obviously depends upon the large demographic difference. Indonesia also leads India in natural gas reserves, production consumption, and exports. India is one of the developing countries, which has recently tried to find its feet in the nuclear use of energy generation, whereas Indonesia is yet to acquire the use of this technology. India also is ahead of Indonesia in terms of the use of renewable sources for energy production (Fig. 6).

2.8 *Power Generation Comparison in Between India and Ireland*

When we compare India and Ireland, we do see a very large difference in the quantity of generation and consumption India exceeds Ireland in both energy generation and consumption. India also exceeds Ireland on the number of resources used for energy

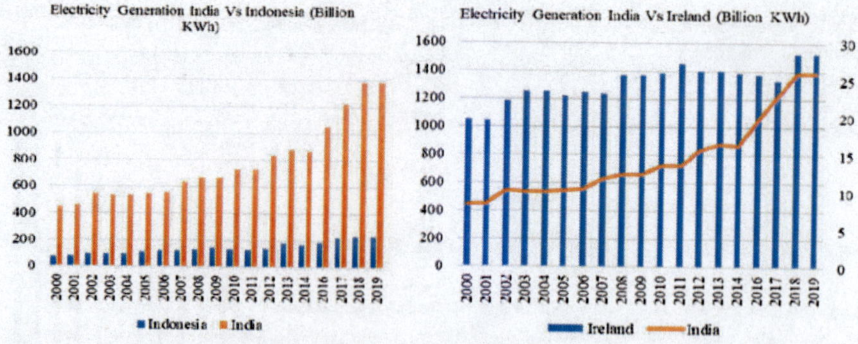

Fig. 6 Year-wise comparison of energy generation in India versus Indonesia and India versus Ireland

production. In this research paper, we see that both India and Ireland are predominantly dependent on the use of fossil fuels for the production of energy. However, the major difference between the two countries is the amount of energy generated from the renewable sources. Ireland leads India for energy generated by the means of renewable resources. In addition, the Ireland is able to achieve the 100% electrification of its country; on the other hand, India has been able to avail electricity to about 84.5% of its urban population (Fig. 7).

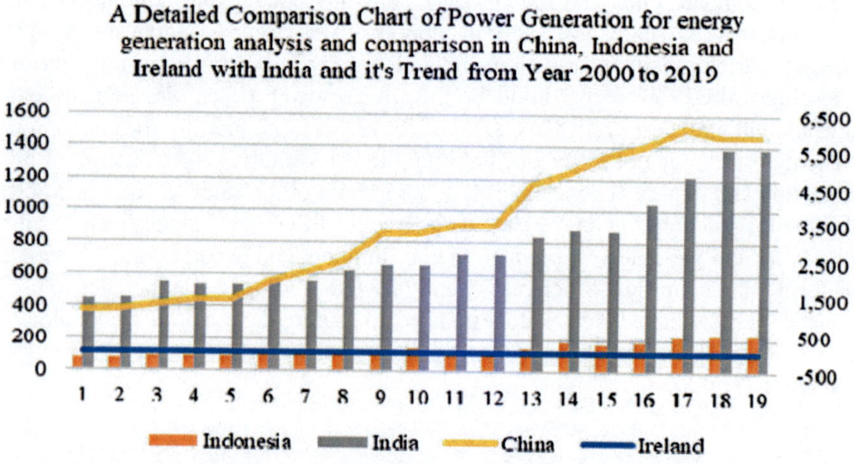

Fig. 7 Detailed comparison chart of power generation for energy generation analysis an comparison in China, Indonesia, and Ireland with India and it is trend from year 2000 to 2019

3 Conclusion

In this paper, review for energy generation analysis and comparison in China, Indonesia, and Ireland with India has been discussed. It has been reviewed various types of energy generations methods to analysis the trend and scope of various power generation trends in these countries. It has been found that energy trends of energy generation are with coal, which is the largest producer of electricity, which accounts for 66% of the electricity generation. Coal powered generating capacity of China would be expected to grow to 1300 GW by year 2020. On the other hand, Indonesia is generating maximum of its power with the waste recycle management system, which is the maximum in the countries included in the review. Also, electricity from fossil fuels in India is about 71% of total install capability while from hydroelectric plants it is 12% and electricity from nuclear plants is 2% of total installed capability and from renewable energy resources it is 16%.

References

1. Lu Z, Zhang Q, Streets DG (2011) Sulfur dioxide and primary carbonaceous aerosol emissions in China and India, 1996–2010. Atmos Chem Phys 11:9839–9864
2. Srikanth R (2018) India's sustainable development goals—Glide path for India's power sector. Energy Pol 123:325–336
3. Lawrenz L et. al (2018) Exploring energy pathways for the low-carbon transformation in India—a model-based approach. Energies 11:3001
4. Energy and Mineral Resource Ministry of Indonesia (ESDM) (2016) Energy Outlook 2016
5. Energy and Mineral Resource Ministry of Indonesia (ESDM) (2017) Peluang investasi dan potensi pengembangan Energi Baru Terbarukan Indonesia, Berdasarkan Peraturan Menteri ESDM Nomor 12 Tahun 2017, Jakarta
6. Kalirajan K, Syed A (2018) The Indonesian energy technology assessment (Ieta), November 2, 2018, pp 1–165
7. Pfenninger S, Staffell I (2016) Long-term patterns of European PV output using 30 years of validated hourly reanalysis and satellite data. Energy 114:1251–1265. https://doi.org/10.1016/j.energy.2016.08.060
8. Strachan P, Lal D, Toke D (2009) Wind power and power politics: international perspectives. Routledge. Sustainable Energy Authority of Ireland. Energy Balance. IWEA, Ireland
9. Kapoor K, Pandey KK, Jain AK, Nandan A (2014) Evolution of solar energy in India: a review. Renew Sustain Energy Rev 40:475–487
10. Iqbal A et al (eds) (2020) Soft computing in condition monitoring and diagnostics of electrical and mechanical systems. In: Advances in intelligent systems and computing, vol 1096. Springer, Singapore. https://doi.org/10.1007/978-981-15-1532-3
11. Iqbal A et al (eds) (2020) Meta heuristic and evolutionary computation: algorithms and applications. In: Studies in computational intelligence, vol 1096. Springer, Singapore. https://www.springer.com/gp/book/9789811575709

3 Conclusion

In this paper, review for energy generation analysis and comparison in China, Indonesia, and Ireland with India has been discussed. It has been reviewed various types of energy generations methods to analysis the trend and scope of various power generation trends in these countries. It has been found that energy trends of energy generation are with coal, which is the largest producer of electricity, which accounts for 66% of the electricity generation. Coal powered generating capacity of China would be expected to grow to 1300 GW by year 2020. On the other hand, Indonesia is generating maximum of its power with the waste recycle management system, which is the maximum in the countries included in the review. Also, electricity from fossil fuels in India is about 71% of total install capability while from hydroelectric plants it is 12% and electricity from nuclear plants is 2% of total installed capability and from renewable energy resources it is 16%.

References

1. Lu Z, Zhang Q, Streets DG (2011) Sulfur dioxide and primary carbonaceous aerosol emissions in China and India, 1996–2010. Atmos Chem Phys 11:9839–9864
2. Srikanth R (2018) India's sustainable development goals—Glide path for India's power sector. Energy Pol 123:325–336
3. Lawrenz L et. al (2018) Exploring energy pathways for the low-carbon transformation in India—a model-based approach. Energies 11:3001
4. Energy and Mineral Resource Ministry of Indonesia (ESDM) (2016) Energy Outlook 2016
5. Energy and Mineral Resource Ministry of Indonesia (ESDM) (2017) Peluang investasi dan potensi pengembangan Energi Baru Terbarukan Indonesia, Berdasarkan Peraturan Menteri ESDM Nomor 12 Tahun 2017, Jakarta
6. Kalirajan K, Syed A (2018) The Indonesian energy technology assessment (Ieta), November 2, 2018, pp 1–165
7. Pfenninger S, Staffell I (2016) Long-term patterns of European PV output using 30 years of validated hourly reanalysis and satellite data. Energy 114:1251–1265. https://doi.org/10.1016/j.energy.2016.08.060
8. Strachan P, Lal D, Toke D (2009) Wind power and power politics: international perspectives. Routledge. Sustainable Energy Authority of Ireland. Energy Balance. IWEA, Ireland
9. Kapoor K, Pandey KK, Jain AK, Nandan A (2014) Evolution of solar energy in India: a review. Renew Sustain Energy Rev 40:475–487
10. Iqbal A et al (eds) (2020) Soft computing in condition monitoring and diagnostics of electrical and mechanical systems. In: Advances in intelligent systems and computing, vol 1096. Springer, Singapore. https://doi.org/10.1007/978-981-15-1532-3
11. Iqbal A et al (eds) (2020) Meta heuristic and evolutionary computation: algorithms and applications. In: Studies in computational intelligence, vol 1096. Springer, Singapore. https://www.springer.com/gp/book/9789811575709

3 Conclusion

In this paper, review for energy generation analysis and comparison in China, Indonesia, and Ireland with India has been discussed. It has been reviewed various types of energy generations methods to analysis the trend and scope of various power generation trends in these countries. It has been found that energy trends of energy generation are with coal, which is the largest producer of electricity, which accounts for 66% of the electricity generation. Coal powered generating capacity of China would be expected to grow to 1300 GW by year 2020. On the other hand, Indonesia is generating maximum of its power with the waste recycle management system, which is the maximum in the countries included in the review. Also, electricity from fossil fuels in India is about 71% of total install capability while from hydroelectric plants it is 12% and electricity from nuclear plants is 2% of total installed capability and from renewable energy resources it is 16%.

References

1. Lu Z, Zhang Q, Streets DG (2011) Sulfur dioxide and primary carbonaceous aerosol emissions in China and India, 1996–2010. Atmos Chem Phys 11:9839–9864
2. Srikanth R (2018) India's sustainable development goals—Glide path for India's power sector. Energy Pol 123:325–336
3. Lawrenz L et. al (2018) Exploring energy pathways for the low-carbon transformation in India—a model-based approach. Energies 11:3001
4. Energy and Mineral Resource Ministry of Indonesia (ESDM) (2016) Energy Outlook 2016
5. Energy and Mineral Resource Ministry of Indonesia (ESDM) (2017) Peluang investasi dan potensi pengembangan Energi Baru Terbarukan Indonesia, Berdasarkan Peraturan Menteri ESDM Nomor 12 Tahun 2017, Jakarta
6. Kalirajan K, Syed A (2018) The Indonesian energy technology assessment (Ieta), November 2, 2018, pp 1–165
7. Pfenninger S, Staffell I (2016) Long-term patterns of European PV output using 30 years of validated hourly reanalysis and satellite data. Energy 114:1251–1265. https://doi.org/10.1016/j.energy.2016.08.060
8. Strachan P, Lal D, Toke D (2009) Wind power and power politics: international perspectives. Routledge. Sustainable Energy Authority of Ireland. Energy Balance. IWEA, Ireland
9. Kapoor K, Pandey KK, Jain AK, Nandan A (2014) Evolution of solar energy in India: a review. Renew Sustain Energy Rev 40:475–487
10. Iqbal A et al (eds) (2020) Soft computing in condition monitoring and diagnostics of electrical and mechanical systems. In: Advances in intelligent systems and computing, vol 1096. Springer, Singapore. https://doi.org/10.1007/978-981-15-1532-3
11. Iqbal A et al (eds) (2020) Meta heuristic and evolutionary computation: algorithms and applications. In: Studies in computational intelligence, vol 1096. Springer, Singapore. https://www.springer.com/gp/book/9789811575709

Author Index

A. Iqbal et al. (eds.), *Renewable Power for Sustainable Growth*, Lecture Notes in Electrical Engineering 723, https://doi.org/10.1007/978-981-33-4080-0